T0260064

Small Carnivores

Small Carnivores

Evolution, Ecology, Behaviour, and Conservation

Edited by

Emmanuel Do Linh San
University of Fort Hare
Alice, South Africa

Jun J. Sato
Fukuyama University
Fukuyama, Japan

Jerrold L. Belant
Michigan State University
East Lansing, MI, USA

Michael J. Somers
University of Pretoria
Pretoria, South Africa

Registered Offices
John Wiley & Sons, Inc., 111 River Street, Hoboken, NJ 07030, USA
John Wiley & Sons Ltd, The Atrium, Southern Gate, Chichester, West Sussex, PO19 8SQ, UK

Editorial Office
9600 Garsington Road, Oxford, OX4 2DQ, UK

For details of our global editorial offices, customer services, and more information about Wiley products visit us at www.wiley.com.

Wiley also publishes its books in a variety of electronic formats and by print-on-demand. Some content that appears in standard print versions of this book may not be available in other formats.

Library of Congress Cataloging-in-Publication Data

Names: Do Linh San, Emmanuel, editor.
Title: Small carnivores : evolution, ecology, behaviour, and conservation /
    edited by Emmanuel Do Linh San, University of Fort Hare, Alice, South Africa,
    Jun J. Sato, Fukayama University, Fukayama, Japan, Jerrold L. Belant,
    State University of New York, Syracuse, USA, Michael J. Somers,
    University of Pretoria, Pretoria, South Africa.
Description: First edition. | Hoboken, NJ : Wiley-Blackwell, 2022. |
    Includes bibliographical references and index.
Identifiers: LCCN 2021041751 (print) | LCCN 2021041752 (ebook) | ISBN
    9781118943281 (cloth) | ISBN 9781118943250 (adobe pdf) | ISBN
    9781118943267 (epub)
Subjects: LCSH: Carnivora.
Classification: LCC QL737.C2 S534 2022 (print) | LCC QL737.C2 (ebook) |
    DDC 599.7–dc23
LC record available at https://lccn.loc.gov/2021041751
LC ebook record available at https://lccn.loc.gov/2021041752

Cover Design: Wiley
Cover Image: Courtesy of Emmanuel Do Linh San

Set in 9.5/12.5pt StixTwoText by Straive, Pondicherry, India
Printed and bound by CPI Group (UK) Ltd, Croydon, CR0 4YY

C9781118943281_010722

This book is dedicated to the memory of Professor Anne Rasa (1940–2020), ethologist and pioneer in the study of social small carnivores.

Source: Photo © South African Tourism.

Contents

List of Contributors

Etienne François Akomo-Okoue
Institut de Recherche en Ecologie Tropicale,
CENAREST, Libreville, Gabon

Chieko Ando
Graduate School of Science, Kyoto University, Kyoto,
Japan

Vonjy Andrianjakarivelo
Wildlife Conservation Society Madagascar Program,
WCS Makira Antseranamborondolo, Maroantsetra,
Madagascar

Alessandro Balestrieri
Department of Earth and Environmental Sciences,
University of Pavia, Pavia, Italy
Department of Environmental Sciences and Policy,
University of Milan, Milan, Italy

Colleen M. Begg
Niassa Carnivore Project, The Ratel Trust,
Rondebosch, South Africa

Keith S. Begg
Niassa Carnivore Project, The Ratel Trust,
Rondebosch, South Africa

Jerrold L. Belant
Department of Fisheries and Wildlife,
Michigan State University, East Lansing, MI, USA

Jennifer A. Bohrman
Department of Biology and Natural Resources,
Frostburg State University, Frostburg, MD, USA

Emily A. Bricker
Department of Biology and Natural Resources,
Frostburg State University, Frostburg, MD, USA

Christina D. Buesching
Department of Biology, The University of British
Columbia, Okanagan Campus, Kelowna, Canada

Laëtitia Buisson
CNRS, UMR 5245 EcoLab (Laboratoire Ecologie
Fonctionnelle et Environnement), Toulouse, France
INP, UPS, EcoLab, Université de Toulouse, Toulouse,
France

Andrew W. Byrne
Centre for Veterinary Epidemiology and Risk
Analysis, School of Veterinary Medicine, University
College Dublin, Dublin, Ireland
Veterinary Sciences Division, Bacteriology Branch,
Agri-Food and Biosciences Institute, Belfast, UK
Department of Agriculture, Food and the Marine,
Dublin, Ireland

Diego F. Castillo
Departamento de Biología, Bioquímica y Farmacia,
Universidad Nacional del Sur, Bahía Blanca, Argentina
GECM, Instituto de Ciencias Biológicas y Biomédicas
del Sur (INBIOSUR), Universidad Nacional del Sur
(UNS) – CONICET, Bahía Blanca, Argentina

Juan-Carlos Ceña
Ingeniero Técnico Forestal, Logroño, Spain

Michael V. Cove
Department of Applied Ecology and NC Cooperative
Fish and Wildlife Research Unit, North Carolina State
University, Raleigh, NC, USA

Jennifer J. Crees
Institute of Zoology, Zoological Society of London, London, UK

Thierry Diop Bineni
Institut de Recherche en Ecologie Tropicale, CENAREST, Libreville, Gabon

Emmanuel Do Linh San
Department of Zoology and Entomology, University of Fort Hare, Alice, South Africa

Robert C. Dowler
Department of Biology, Angelo State University, San Angelo, TX, USA

Johan T. du Toit
Department of Wildland Resources, Utah State University, Logan, UT, USA
Department of Zoology and Entomology, Mammal Research Institute, University of Pretoria, Pretoria, South Africa

Zach J. Farris
Department of Health & Exercise Science, Appalachian State University, Boone, NC, USA

Adam W. Ferguson
Department of Biological Sciences, Texas Tech University, Lubbock, TX, USA
Gantz Family Collection Center, Field Museum of Natural History, Chicago, IL, USA

M. Carmen Ferreras
Departamento de Sanidad Animal (Anatomía Patológica), Universidad de León, León, Spain

Pascal Fournier
GREGE, Villandraut, France

Christine Fournier-Chambrillon
GREGE, Villandraut, France

Mariela G. Gantchoff
Global Wildlife Conservation Center, State University of New York College of Environmental Science and Forestry, Syracuse, NY, USA

Brian D. Gerber
Department of Natural Resources Science, University of Rhode Island, Kingston, RI, USA

Gloria Giralda-Carrera
Servicio de Conservación de la Biodiversidad, Gobierno de Navarra, Pamplona, Spain

Lon I. Grassman Jr
Feline Research Center, Caesar Kleberg Wildlife Research Institute, Texas A&M University–Kingsville, Kingsville, TX, USA

Aaron M. Haines
Department of Biology, Applied Conservation Lab, Millersville University, Millersville, PA, USA

Zoe L. Hanley
Northwest Regional Office, Defenders of Wildlife, White Salmon, WA, USA

Jan Herr
Arrondissement Sud, Administration de la Nature et des Forêts, Leudelange, Luxembourg

Tim L. Hiller
Wildlife Ecology Institute, Helena, MT, USA

Eiji Inoue
Graduate School of Science, Kyoto University, Kyoto, Japan

Lisa Isaacs
Centre for Wildlife Management, University of Pretoria, Pretoria, South Africa

Yuji Iwata
Graduate School of Science, Kyoto University, Kyoto, Japan

Andrew P. Jennings
Small Carnivores – Research and Conservation, Portland, ME, USA

Neil R. Jordan
Centre for Ecosystem Science, School of Biological, Earth and Environmental Sciences, University of New South Wales, Sydney, Australia
Botswana Predator Conservation Trust, Maun, Botswana
Taronga Conservation Society Australia, Taronga Western Plains Zoo, Wildlife Reproduction Centre, Dubbo, Australia

Sarah Karpanty
Department of Fish & Wildlife Conservation, Virginia Tech, Blacksburg, VA, USA

Marcella J. Kelly
Department of Fish & Wildlife Conservation, Virginia Tech, Blacksburg, VA, USA

Thijs Kuiken
Viroscience Lab, Erasmus MC, Rotterdam, The Netherlands

Nathan S. Libal
Carnivore Ecology Laboratory, Mississippi State University, Mississippi State, MS, USA

Mauro Lucherini
Departamento de Biología, Bioquímica y Farmacia, Universidad Nacional del Sur, Bahía Blanca, Argentina
GECM, Instituto de Ciencias Biológicas y Biomédicas del Sur (INBIOSUR), Universidad Nacional del Sur (UNS) – CONICET, Bahía Blanca, Argentina

S. Wayne Martin
Department of Population Medicine, Ontario Veterinary College, University of Guelph, Guelph, Canada

Pierre Philippe Mbehang Nguema
Institut de Recherche en Ecologie Tropicale, CENAREST, Libreville, Gabon

Carlo Meloro
Research Centre in Evolutionary Anthropology and Palaeoecology, School of Biological and Environmental Sciences, Liverpool John Moores University, Liverpool, UK

Michael G.L. Mills
School of Biology and Environmental Sciences, Faculty of Agriculture and Natural Sciences, University of Mpumalanga, Nelspruit, South Africa

Yoshihiro Nakashima
College of Bioresource Science, Nihon University, Fujisawa City, Japan

Ludovic Ngok Banak
Institut de Recherche en Ecologie Tropicale, CENAREST, Libreville, Gabon

Alfred Ngomanda
Institut de Recherche en Ecologie Tropicale, CENAREST, Libreville, Gabon

Chimene Nze-Nkogue
Institut de Recherche en Ecologie Tropicale, CENAREST, Libreville, Gabon

Allan F. O'Connell
USGS Patuxent Wildlife Research Center, Laurel, MD, USA

James O'Keeffe
Centre for Veterinary Epidemiology and Risk Analysis, School of Veterinary Medicine, University College Dublin, Dublin, Ireland
Department of Agriculture,
Food and the Marine, Dublin, Ireland

Francisco Palomares
Departamento de Biología de la Conservación, Estación Biológica de Doñana,
CSIC, Sevilla, Spain

Marie-Lilith Patou
Biotope, Recherche & Développement,
Mèze, France

Kelly J. Pearce
Marine-Estuarine-Environmental Sciences Graduate Program, University of Maryland,
College Park, MD, USA

Felix Ratelolahy
Wildlife Conservation Society Madagascar Program, WCS Makira Antseranamborondolo,
Maroantsetra, Madagascar

Luigi Remonti
Department of Earth and Environmental Sciences, University of Pavia, Pavia, Italy
Institut Agricole Régional,
Aosta, Italy

Timothy J. Roper
Department of Biology and Environmental Science, University of Sussex, Brighton, UK

Aritz Ruiz-González
Department of Zoology and Animal Cell Biology, University of the Basque Country (UPV-EHU), Vitoria-Gasteiz, Spain
Systematics, Biogeography and Population Dynamics Research Group, University of the Basque Country (UPV/EHU),
Vitoria-Gasteiz, Spain
Conservation Genetics Laboratory,
National Institute for Environmental

Protection and Research (ISPRA),
Ozzano dell'Emilia, Italy

Jun J. Sato
Department of Biotechnology, Faculty of Life Science and Technology, Fukuyama University, Fukuyama, Japan

Thomas L. Serfass
Department of Biology and Natural Resources, Frostburg State University, Frostburg, MD, USA

Michael J. Somers
Eugène Marais Chair of Wildlife Management, Department of Zoology and Entomology, Mammal Research Institute, University of Pretoria, Pretoria, South Africa
Centre for Invasion Biology, University of Pretoria, Pretoria, South Africa

Robert S. Sommer
Department of Landscape Sciences and Geomatics, Neubrandenburg University of Applied Sciences, Neubrandenburg, Germany

Sadie S. Stevens
Department of Biology and Natural Resources, Frostburg State University, Frostburg, MD, USA

Richard E. Strauss
Department of Biological Sciences, Texas Tech University, Lubbock, TX, USA

Lourens H. Swanepoel
Centre for Wildlife Management, University of Pretoria, Pretoria, South Africa
Department of Zoology, University of Venda, Thohoyandou, South Africa

Yuji Takenoshita
Department of Children, Faculty of Child Studies, Chubu-Gakuin University, Sika, Japan

Michael E. Tewes
Feline Research Center, Caesar Kleberg Wildlife Research Institute, Texas A&M University–Kingsville, Kingsville, TX, USA

Fermin Urra-Maya
Área de Biodiversidad, Gestión Ambiental de
Navarra, Pamplona, Spain

Marco van de Bildt
Viroscience Lab, Erasmus MC, Rotterdam,
The Netherlands

Stephen M. Vantassel
Wildlife Control Consultant, LLC, Lewistown, PA,
USA

Géraldine Veron
Institut de Systématique, Evolution, Biodiversité
(ISYEB), Muséum National d'Histoire Naturelle,
CNRS, Sorbonne Université, EPHE,
Université des Antilles,
Paris, France

Mieczysław Wolsan
Museum and Institute of Zoology, Polish Academy of
Sciences, Warszawa, Poland

Juichi Yamagiwa
Graduate School of Science, Kyoto University, Kyoto,
Japan

Foreword by Professor Anne Rasa

Small Carnivore Research in Retrospect

This is a book that has been needed for decades. It is the first compendium of recent research on a group of mammals that received almost no attention prior to the early 1970s and has not received enough since – the small carnivores. Although the larger members of the Carnivora were studied earlier than this date, notably the grey wolf, *Canis lupus*, in Europe and the 'Big Cats' in Africa, especially the lion, *Panthera leo*, and have been the objects of numerous studies since, the smaller members of the carnivore group were mostly ignored. Nearly all the scientists involved in the earlier carnivore studies were Europeans, with the notable exception of George Schaller from the USA with his benchmark study of the African lion in the late 1960s.

This tendency for Europeans to be at the forefront of small carnivore field research probably had historical reasons. The early 1970s were the time of the Nature/Nurture Controversy and American zoologists were primarily Animal Behaviourists. They concentrated on learning paradigms using rodents in the laboratory rather than studying natural populations of a variety of animals, as the Europeans were doing. The few ecological studies on small carnivores at that time were run by the US Fish and Wildlife Service and dealt with their control for economic reasons. The European field scientists, however, notably the Dutch, focused mainly on the behaviour and ecology of fish and invertebrate species rather than mammals. Probably, owing to the paucity of European small carnivore species and their relative rarity, due to their perception as 'vermin' and centuries of attempted eradication, very little was being done to expand our basic knowledge of these mammals. An exception to this general trend was Nikolaas Tinbergen's Oxford group in Britain who were the first to study small carnivore populations in the wild. Hans Kruuk's study of the European badger,

Meles meles, and David Macdonald's on the red fox, *Vulpes vulpes*, set a trend in the early 1970s that was to continue for the next decades and expand to other species and other continents.

Apart from the rarity of small carnivores as objects for research throughout most of Europe, another factor detrimental to the study of this fascinating group was that the majority of its members were found in Africa and Southeast Asia, with a much lower number in the Americas. Funds to study these were practically non-existent and almost nothing was known about any of them. Many species were represented only by skins and/or skeletons in museums and just a handful were kept as exotic curiosities in zoos. A quick glance through the first edition of Walker *et al.*'s *Mammals of the World*, published in 1964, shows that less than half of the species listed had actually been photographed (mostly in zoos), many were represented just as museum mounts and some were artist's impressions of what the animal looked like from its preserved skin and skull. The appended descriptions of habits and habitat were little more than descriptions of general geographical regions and biotopes in which the specimens had been collected. Even the taxonomy was confusing. Although that of the canids, felids, and most mustelids was fairly straightforward, the taxonomy of the viverrids, which at that time included the present-day Viverridae, Nandiniidae, **Prionodontidae,** Eupleridae, and Herpestidae, was erratic and questionable, being based primarily on skeletal elements, dentition and pelage. Some genera were extremely broadly based (e.g. *Herpestes*) while others (e.g. *Helogale*) were split into numerous species and subspecies based on geographical range and pelage colour. The latter distinctions were, after subsequent genetic analysis, found to be spurious.

The fact that so little was known about small carnivores at this time was also due to their habits. Most species are solitary, secretive, and nocturnal, all these factors making it difficult to study them and collect data on their habits or even on their presence. The majority of species also inhabit difficult terrain: woodlands, tropical forest, waterways, mangrove swamps and tall grasslands, hardly ideal conditions for observation and data collection. The little data available on reproductive cycles and territorial behaviour of small non-European species originated mostly from incidents of human–small carnivore conflict. Reproductive cycles of the suricate, *Suricata suricatta*, and yellow mongoose, *Cynictis penicillata*, were obtained from post-mortem analysis of animals following mass poisoning in a rabies outbreak area in South Africa. Data on territory size and reproduction of the small Indian mongoose, *Herpestes auropunctatus* (now *Urva auropunctata*), were collected with regard to its negative impact on island ecosystems after its introduction as a biological control agent for snakes and rats. It, therefore, comes as no surprise that the first non-European species to be studied in depth were group-living, diurnal mongooses, such as the dwarf mongoose, *Helogale parvula*, and, later, the banded mongoose, *Mungos mungo*, and suricate which, because they inhabited comparatively open areas, were amenable to direct observation.

As can be seen from many of the contributing chapters of this book, over the past 40 years, this situation has changed remarkably with regard to research on the evolution, taxonomy, general ecology, and behaviour of many previously unknown species. Probably, the most valuable tool developed for small carnivore studies was the miniature radio-telemetry transmitter. For the first time, it was possible to follow individuals without inhibiting their natural behaviour. This was especially useful for nocturnal species or those inhabiting biotopes where direct observation was almost impossible. Prior to this, the only way of collecting data was by the capture–recapture method with marked animals or the time-consuming process of their habituation to human presence or relying on their tracks and signs. The radio-collar meant that individual animals could be located, their home ranges plotted, activity periods recorded and denning sites mapped with accuracy and relatively little trouble. This yielded a wealth of new information on their habits and made inter-species and also inter-habitat comparisons within a species possible. The use of motion-detecting cameras also allowed records to be made of the presence, habitat use, and activity patterns of cryptic species without disturbing their normal behaviour. As a result, the database on small carnivore ecology and behaviour expanded almost exponentially over the next decades. Probably, the most remarkable thing about these general findings, however, was that, apart from the social species, almost all the others followed the same schema in their spatiotemporal patterning, independent of geographical distribution, diel activity pattern, or ecological habitat. The general rule seemed to be that males have larger territories than females and these overlap with those of one or more of the latter. This would be expected when the prey spectrum of most small carnivores, which consists primarily of small prey items such as invertebrates, reptiles, rodents, and birds, is taken into account. Whether this is an example of convergent evolution or a case of retention of evolutionary traits has yet to be analyzed.

Many of the chapters in this book deal with the relationship between humans and small carnivores, which has always been a strained one based primarily on economics. From the start of agriculture and the keeping of small domestic animals, the human attitude towards small carnivores has always been negative, with continuous attempts at their eradication from settled areas. Agricultural practices destroyed – and still destroy – habitats and directly or indirectly result in the exclusion of many species from their historical range, even to the point of practical extinction, e.g. the black-footed ferret, *Mustela nigripes*. Competition with fisheries wiped out whole populations of otters, including sea otters, *Enhydra lutris,* and mink species. The depredations of mustelids, small canids, and felids on domestic livestock, together with their role as disease carriers for rabies, in particular, resulted and still result in their persecution, even in modern times. The only positive attitude towards small carnivores could be found in the fur trade and this attitude could be considered even worse. Their exploitation through ignorance and greed resulted in the population decimation of mainly mustelids, and even the extinction of certain species, especially in North America. Those chapters dealing with human–small carnivore interactions are a valuable contribution to the history of this ongoing conflict. Most importantly, now that our knowledge of their essential role in ecosystem balance has been recognized, the importance of their conservation has come to the fore. Today, positive steps are being taken to repair and prevent the damage done to small carnivore populations in previous centuries, despite the fact that some species are still considered as 'vermin'.

With the major advances in macro-molecular techniques within the last decades, many of the evolutionary and taxonomic puzzles presented by the small carnivores as a group 30 years ago have been elucidated. The introductory chapter gives a brief overview of the phylogenetic relationships among the families currently recognized within the order Carnivora and discusses some taxonomic issues pertaining to this group. It further sets the context by evaluating the research efforts on small carnivores globally through bibliometric analyses. Lastly, the brief review on the distribution, species richness, ecological roles, conservation status, and threats to small carnivores worldwide is of special interest and basic to our understanding of the group as a whole.

In general, this book covers such a wide range of subdisciplines and techniques that it should be considered a solid baseline for further necessary research on this little-known group of highly interesting mammals. As our knowledge regarding how ecosystems function increases, the valuable role of small carnivores and the necessity for their conservation should be regarded as of paramount importance. The topics covered in this book should therefore appeal not only to academics and wildlife researchers, but to the interested layman as well.

Reviewing Process and Acknowledgements

The idea for this book emanated in 2012 when the editors organized a symposium titled 'Small Carnivores in Space and Time' which was held in August 2013, in Belfast, in the framework of the 11th International Mammalogical Congress. We would like to thank all the colleagues who presented their work at, and/or attended the small carnivore symposium, as well as the funding bodies who paid for travel and other related expenses. We are particularly grateful to Ian Montgomery, chair of the congress, for his precious logistic and financial assistance to host our symposium, as well as to all the congress staff whose precious help greatly contributed to the success of our symposium.

The present book, however, does not constitute the proceedings of the small carnivore symposium. It is mostly made of invited contributions, with only about one-third of the chapters corresponding to talks given in Northern Ireland. This approach ensured the production of a more 'balanced' book that covers the main disciplines targeted initially, namely evolution, ecology, behaviour, and conservation.

Each manuscript considered for inclusion in this book has been taken in charge by one or two of the book editors, reviewed by between two and four experts in the related field of study, and, ultimately, accepted or rejected following the same stringent procedure and criteria as for a scientific peer-reviewed journal article. All revised and accepted chapters have then been (re-)reviewed and formatted by EDLS to ensure inter-chapter consistency and highlight or address some previously undetected issues. When needed, chapters were returned to authors for further revision. MJS then (re-)read all chapters for some final language editing and to double-check references,

while JJS and JLB took charge of several other fine-tuning tasks. Due to numerous delays with the book production, several chapters have been accordingly updated along the way to ensure that the information provided at the time of publication is as pertinent as possible.

We sincerely thank the reviewers of the initial book proposal and the whole production team at Wiley for all their help and support, particularly Mahalakshmi Pitchai, Rajalakshmy Devanathan, Mandy Collison and Andrew Harrison. Will Duckworth gave expert advice regarding nomenclatural use and taxonomical aspects, while Keith Aubry kindly shared some precious editing tips. Géraldine Veron, Adam Ferguson, Andrew Kitchener and Alexei Abramov expertly gave input on Appendix B. Alexei Abramov also generously shared his encyclopedic knowledge on Russian small carnivores (among others) on numerous occasions. We are also grateful to the 51 reviewers who generously offered their time and shared their knowledge and scientific expertise of some of the world's small carnivores while reviewing one or two chapters of this book:

Peter Apps, Botswana Predator Conservation Trust, Botswana; **Masakazu Asahara**, Aichi Gakuin University, Japan; **Keith B. Aubry**, USDA Forest Service, Pacific Northwest Research Station, USA; **Bob Bluett**, Illinois Department of Natural Resources, USA; **Luigi Boitani**, Sapienza University of Rome, Italy; **Tim Caro**, University of California, Davis, USA; **Christopher Dickman**, The University of Sydney, Australia; **Robert C. Dowler**, Angelo State University, USA; **Nicole Duplaix**, Oregon State University, USA; **David Eads**, US Geological Survey, Fort Collins Science Center, USA; **Adam W. Ferguson**, Field

Museum of Natural History, Chicago, USA; **Michael H. Ferkin**, The University of Memphis, USA; the late **Ian Gaigher**, Lajuma Research Centre, South Africa; **Daniel Gallant**, Parks Canada, Gatineau, Canada; **Philippe Gaubert**, Université Toulouse III - Paul Sabatier, France; **Al S. Glen**, Manaaki Whenua - Landcare Research, New Zealand; **Juan Pedro González-Varo**, Universidad de Cádiz, Spain; **Lon I. Grassman, Jr**, Texas A&M University–Kingsville, USA; the late **Colin P. Groves**, Australian National University, Australia; **Philipp Henschel**, PANTHERA, USA; **Tim Hiller**, Wildlife Ecology Institute, USA; **Andrew P. Jennings**, SMALL CARNIVORES – Research and Conservation, USA; **Paul G. Jensen**, NYS Department of Environmental Conservation, USA; **Marcella J. Kelly**, Virginia Polytechnic Institute and State University, USA; **Andreas Kranz**, alka-kranz eU, Austria; **Aliza le Roux**, University of Free State, Qwaqwa Campus, South Africa; **Estela M. Luengos Vidal**, Universidad Nacional del Sur, Argentina; **Adrian Marciszak**, University of Wrocław, Poland; **Trevor McIntyre**, University of South Africa, South Africa; **Alessio Mortelliti**, University of Maine, USA; **Yoshihiro Nakashima**, Nihon University, Japan; **Francesco Palomares**, Estación Biológica de Doñana (EBD-CSIC), Spain; **Cino Pertoldi**, Aalborg University, Denmark; **Gilbert Proulx**, Alpha Wildlife Research & Management Ltd, Canada; **L. Miguel Rosalino**, Universidade de Lisboa, Portugal; **Sugoto Roy**, IUCN, Switzerland; **Martin Sabol**, Comenius University, Slovakia; **Theodore Stankowich**, California State University, Long Beach, USA; **David C. Stoner**, Utah State University, USA; **Lourens H. Swanepoel**, University of Venda, South Africa; **Mathias W. Tobler**, San Diego Zoo Wildlife Alliance, USA; **Iain Trewby**, Fauna & Flora International, UK; **Géraldine Veron**, Muséum National d'Histoire Naturelle, France; **Lars Werdelin**, Swedish Museum of Natural History, Sweden; **Bryant White**, Association of Fish & Wildlife Agencies, USA; **Paula A. White**, University of California, Los Angeles, USA; **Gary Witmer**, United States Department of Agriculture, National Wildlife Research Center, USA; **Lesley Wright**, IUCN/SSC Otter Specialist Group, UK; **Takahiro Yonezawa**, Tokyo University of Agriculture, Japan; **Jabi Zabala Albizua**, University of the Basque Country, Spain; **Iñigo Zuberogoitia**, Estudios Medioambientales Icarus SL, Spain.

It is needless to say that we are indebted to all the contributing authors for their hard work and endless patience. We sincerely hope that they will find the final product worth the excruciatingly long wait.

Last but not least, we are grateful to the late Professor Anne Rasa for writing the foreword to this book and providing some insider's view on what research on small carnivores looked like during the second half of the twentieth century. While we are devastated that she passed away on 15 November 2020 and could not see the printed version of this book, we are joyful to be able to follow in her footsteps and hope that this book will continue her legacy.

Part I

Introduction

1

The World's Small Carnivores: Definitions, Richness, Distribution, Conservation Status, Ecological Roles, and Research Efforts

Emmanuel Do Linh San[1,], Jun J. Sato[2], Jerrold L. Belant[3], and Michael J. Somers[4,5]*

[1] Department of Zoology and Entomology, University of Fort Hare, Alice, South Africa
[2] Department of Biotechnology, Faculty of Life Science and Technology, Fukuyama University, Fukuyama, Japan
[3] Department of Fisheries and Wildlife, Michigan State University, East Lansing, MI, USA
[4] Eugène Marais Chair of Wildlife Management, Department of Zoology and Entomology, Mammal Research Institute, University of Pretoria, Pretoria, South Africa
[5] Centre for Invasion Biology, University of Pretoria, Pretoria, South Africa

SUMMARY

Small carnivores – here defined as members of the mammalian order Carnivora with a body mass < 21.5 kg – occur worldwide, including in Oceania, following introductions. They are represented by 210 to 282 species, which correspond to about 90% of terrestrial carnivores globally. Some species are endemic to 1 or 2 countries (sometimes only islands), while others, like the red fox, *Vulpes vulpes*, are present in nearly 90 countries over 5 continents. Small carnivores inhabit virtually all of the Earth's ecosystems, adopting terrestrial, semi-fossorial, (semi-)arboreal or (semi-)aquatic lifestyles. They occupy multiple trophic levels, being primary consumers when feeding on fruits, seeds, and other plant matter, secondary consumers when preying on frugivorous, granivorous, and herbivorous animals, or tertiary consumers when killing and devouring meat-eating animals. Therefore, they play important roles in the regulation of ecosystems, e.g. natural pest control, seed dispersal and nutrient cycling. In areas where humans have extirpated large carnivores, small carnivores may become the dominant predators, which may increase their abundance ('mesopredator release') to the point that they can sometimes destabilize communities, drive local extirpations, and reduce overall biodiversity. On the other hand, one-third of the world's small carnivores are Threatened or Near Threatened with extinction (*sensu* IUCN). This results from regionally burgeoning human populations' industrial and agricultural activities, causing habitat reduction, destruction, fragmentation, and pollution. Overexploitation, persecution, and the impacts of introduced predators, competitors, and pathogens have also negatively affected many small carnivore species. Although small carnivores have been intensively studied over the past decades, bibliometric studies showed that they have not received the same attention given to large carnivores. Furthermore, there is a huge disparity in how research efforts on small carnivores have been distributed, with some species intensively studied, and others superficially or not at all. Regionally, North American and European small carnivores have been the focus of numerous studies, and more research is being progressively conducted in Asia. However, there is a need to increase the research effort in Africa and Central and South America. Encouragingly, the recognition of the importance of the mesopredator release effect and the exponential deployment of camera-traps have started to boost the research effort and scientific knowledge on small carnivores around the world. This book aims at filling a gap in the scientific literature by elucidating the important roles of, and documenting the latest knowledge on, the world's small carnivores. It is divided into four main sections: (i) Evolution, Systematics, and Distribution; (ii) Ecology, Behaviour, and Diseases; (iii) Interspecific Interactions and Community Ecology; and (iv) Interactions with People and Conservation. We hope that the book will appeal to a wide audience and, considering that the field of small carnivore science remains wide open, stimulate much-needed research globally.

Keywords

Bibliometric studies – Carnivora – geographic range – phylogeny – taxonomy – threats – systematics

---

* Corresponding author.

Small Carnivores: Evolution, Ecology, Behaviour, and Conservation, First Edition. Edited by Emmanuel Do Linh San, Jun J. Sato, Jerrold L. Belant, and Michael J. Somers.

What is a Small Carnivore?

This edited book focuses on small carnivores. This naturally calls for a definition of 'small carnivores', particularly because these members of the class Mammalia do not form a distinct – or monophyletic – taxonomic unit. As a start, the term 'carnivore' (from the Latin *carne*, meat, and *vorare*, to eat) is used here as a popular synonym of carnivorans, i.e. mammal species belonging to the order Carnivora. Readers should however bear in mind that the latter term, which is based on a phylogenetic classification (Goswami, 2010), is technically more correct and less confusing, as some carnivorans rarely include meat in their diet. For example, the red panda, *Ailurus fulgens*, almost exclusively eats bamboo, and the kinkajou, *Potos flavus*, primarily feasts on fruits. The ancestors of modern-day carnivores were all meat-eaters and had in common the possession of a set of four carnassial teeth – the two fourth upper premolars and the two first lower molars – that would shear through flesh efficiently (Macdonald, 1992). Carnivores have since evolved and colonized a wide range of habitats, with some species progressively changing their diet to mostly feed on plant matter or insects with corresponding morphophysiological adaptations, including modifications of the carnassial teeth (Ewer, 1973). Similarly, not all meat-eating animals are carnivorans. In fact, the main confusion comes from the fact that the term 'carnivore' can both be used as a synonym of 'carnivoran' (phylogenetic concept) and as an ecological concept (Allaby, 2009), with carnivore then corresponding to the substantive form of the adjective 'carnivorous'. This book, however, does not include other small predator species such as highly carnivorous marsupials, notably several members of the families Didelphidae (endemic to North and South America) and Dasyuridae (found in Australia, Tasmania, Papua New Guinea, and Indonesia). Nonetheless, it is worth mentioning that marsupial carnivores likely play similar ecological roles to those of small carnivorans, and several species of carnivorous marsupials are equally understudied (e.g. Glen & Dickman, 2014).

There is no unanimously accepted definition of what a small carnivore is. Among carnivore biologists involved in conservation and familiar with the specialist groups of the International Union for the Conservation of Nature's Species Survival Commission (hereafter IUCN SSC), small carnivores are implicitly defined as all the terrestrial carnivores that do not belong to the most charismatic carnivore families, namely the cats (Felidae), dogs (Canidae), bears (Ursidae) and hyenas (Hyaenidae). The IUCN SSC possesses specialist groups dedicated to the conservation of each of the above families, while the remaining families have been progressively incorporated into the original Mustelid & Viverrid Specialist Group to form what is currently known as the Small Carnivore Specialist Group. The ensuing corporate definition of small carnivores is obviously arbitrary, especially considering the plethora of small-sized felids and canids that populate the world's ecosystems.

From a purely biological standpoint, there are morphological, ecofunctional, and ecophysiological definitions, and a species considered as a small carnivore by some may be regarded as a medium-sized or even large carnivore by others. The classification may also differ from one ecosystem to another, and vary locally over time, depending on changes in the composition of the carnivore taxocenosis (i.e. a group of sympatric species sharing a common phylogenetic clade). Using body mass as the primary criterion to categorize carnivores, Buskirk (1999) defined mesocarnivores (i.e. medium-sized carnivores) as mammalian predators weighing between 1 and 15 kg. Based on this categorization, only species weighing less than 1 kg would be considered small carnivores, while large carnivores would be those species weighing 15 kg and above (see e.g. Wolf & Ripple, 2018). Other authors regard mesocarnivores (or mesopredators) as small- and mid-sized species weighing less than 15 kg (Roemer *et al.*, 2009), which itself brings confusion to the debate, as this definition disregards the very meaning of the prefix *meso-* which refers to middle or intermediate body size or mass. In contrast, the latter definition is interpreted literally when a species such as the cheetah, *Acinonyx jubatus* (with an adult body mass of 20–65 kg) is defined by some authors as a 'mesopredator' in comparison to the much larger African lion, *Panthera leo* (110–270 kg) (Gigliotti *et al.*, 2020). This raises the question as to which upper body mass threshold should ideally be selected to define mesocarnivores, seeing that other studies rather regard the cheetah as belonging to the African large predator guild (Rafiq *et al.*, 2020) and prefer to focus on the dominant versus subordinate roles played by these predators in interspecific interactions (Marneweck *et al.*, 2019).

Ecologically, carnivores can occupy a broad range of trophic levels (see Fleming *et al.*, 2017 for a review focusing on canids). Although most of them are secondary consumers feeding mainly on herbivorous, frugivorous, and granivorous animals, some are essentially primary consumers feeding on plant matter, fruit, and seeds (Hunter & Barrett, 2018). Further, it is not uncommon for others to act as tertiary or even quaternary consumers (e.g. grass→rodent→ first-order carnivore→second-order carnivore→third-order carnivore). For example, Yang *et al.* (2018) found that up to six mid-sized carnivore species were predated by Amur tigers, *Panthera tigris altaica*, and Amur leopards, *Panthera pardus orientalis*. Therefore, because carnivores also feed on each other (intra-taxocenosis or even intraguild predation), the available behavioural and dietary data suggest that the three following ecofunctional categories could be considered: (i) species who never kill and consume other carnivore species, but who fall prey to larger-sized carnivores; (ii) species who kill and consume other carnivore species and are themselves predated by other carnivores; and (iii) species who kill and occasionally consume other carnivore species, but whose adult individuals (contrarily to young) are not usually killed by other carnivore species. Due to subduing imperatives, these categories would imply that the corresponding members of these trophic levels are of small, medium, and large sizes, respectively. However, body size and mass are not the only factors that determine the occurrence and outcome of such predatory interactions, and the range of carnivore body sizes and masses will vary depending on the local taxocenosis. As a result, body mass thresholds between these three ecological groups of carnivores are not discrete, and, therefore, difficult to determine precisely.

Ecophysiological considerations can provide scientific grounding for categorizing carnivores into body mass groups. Carbone *et al.* (1999) noted a dichotomy in terrestrial carnivore diets, with smaller species feeding on invertebrates and small vertebrates generally weighing less than half their body mass; and larger species essentially preying on large vertebrates that are near their mass. These authors suggested that although intake rates of invertebrate feeders are low, small carnivores could subsist on such a diet because invertebrates constitute a superabundant resource in most ecosystems, and small carnivores have low absolute energy requirements. For larger carnivores, however, invertebrate feeding appears to be unsustainable. Using a simple energetic model and known invertebrate intake rates, Carbone *et al.* (1999) predicted a maximum mass of 21.5 kg above which feeding on small prey is unsustainable. In a follow-up study, Carbone *et al.* (2007) showed that the transition from small to large prey can be predicted by the maximization of net energy gain. While their improved model showed that small prey can sustain carnivores weighing up to 18–45 kg, carnivores weighing above 14.5 kg will achieve a higher net gain by feeding on larger prey. The shift from small to large prey is, therefore, expected and indeed observed in a body mass range of roughly 15–21 kg, depending on the species, season, and location. A recent study that investigated the relationship between stomach capacity and pack-corrected prey mass (i.e. the amount of food available for each member of the predator 'hunting group') confirmed the existence of two main carnivore functional groups, namely small-prey feeders and large-prey feeders (De Cuyper *et al.*, 2019). The majority of large-prey feeders are above, and of small-prey feeders below, a body mass of 10–20 kg. However, both functional groups occur across the whole body mass spectrum, suggesting that the dichotomy might not only be determined by physiology, but also by ecological factors related to body size.

In order to select the species to consider for this book as a whole, we, therefore, decided to define small carnivores as members of the order Carnivora whose average body mass is below 21.5 kg (see species list and average body masses in Appendix A). We, however, acknowledge that because the shift from small to large prey can take place over a large range of body masses, some species weighing between 15 and 21 kg may warrant being regarded as large carnivores. Similarly, the giant otter, *Pteronura brasiliensis*, and the sea otter, *Enhydra lutris*, are considered here as large carnivores based on their average body mass (28 and 30 kg, respectively; Hunter & Barrett, 2018), even though they feed on comparatively small prey (10–40 cm long fish and marine invertebrates, respectively). Living in aquatic and marine environments could possibly complicate access to larger prey – in terms of availability, catching, and subduing – as opposed to what is the case in terrestrial systems. Giant otters are, however, capable of taking large turtles, > 1 m long catfish, caimans to 1.5 m and anacondas to 3 m (Hunter & Barrett, 2018). The case of some bears feeding almost exclusively on plants (e.g. giant panda, *Ailuropoda melanoleuca*) or termites, ants, and fruits (e.g. sloth bear, *Melursus ursinus*) is more puzzling, and this is clearly food for thought for evolutionary biologists and nutritional ecologists (Nie *et al.*, 2019; Jiangzuo & Flynn, 2020). Hence, due to the above-mentioned hurdles and the lack of a current consensus on what a small carnivore is, we have not imposed this definition upon the contributors. As a result, the individual chapters may follow any of the approaches mentioned above, or be based on different body mass thresholds.

This book does not include contributions on any of the 36 species of marine carnivores (Wozencraft, 2005) belonging to the Phocidae (seals), Otariidae (sea lions), and Odobenidae (walrus, *Odobenus rosmarus*) families whose representatives weigh between 90 and 3600 kg and feed on a different range of primarily aquatic prey (Reeves *et al.*, 2002). Due to their shared morphological and ecological characteristics, these aquatic species were previously considered to form an order on their own, the Pinnipedia; however, elevating pinnipeds to the order rank would make the Carnivora an incomplete systematic unit, a so-called paraphyletic taxon (Flynn *et al.*, 2005; Sato *et al.*, 2006).

Although the term 'terrestrial' is used here to qualify all members of the order Carnivora outside the Pinnipedia (previously called the Fissipedia), some terrestrial species are largely aquatic (e.g. otters; otter civet, *Cynogale bennettii*) or may forage in shallow waters (e.g. water mongoose, *Atilax paludinosus*, crab-eating mongoose, *Urva urva*, aquatic genet, *Genetta piscivora*, crab-eating raccoon, *Procyon cancrivorus*, fishing cat, *Prionailurus viverrinus*, and brown bear, *Ursus arctos*) (Garshelis, 2009; Jennings & Veron, 2009; Kays, 2009; Larivière & Jennings, 2009; Gilchrist *et al.*, 2009; Sunquist & Sunquist, 2009). Other species are partly (e.g. martens, *Martes* spp.; genets, *Genetta* spp.; fosa, *Cryptoprocta ferox*; several bear species) to chiefly or exclusively arboreal (e.g. palm civets; oyans

and linsangs; binturong, *Arctictis binturong*; olingos, *Bassaricyon* spp.; sun bear, *Helarctos malayanus*), while Eurasian badgers *sensu lato*, *Meles* spp., and American badger, *Taxidea taxus*, can be regarded as semi-fossorial, as they dig extensive burrows where they spend a significant proportion of their life (Garshelis, 2009; Gaubert, 2009a,b; Goodman, 2009; Jennings & Veron, 2009; Kays, 2009; Larivière & Jennings, 2009; Proulx *et al.*, 2016).

Phylogeny and Number of Families

The scientific revolution created by the advent of molecular biology has enabled taxonomists, systematists, and phylogeographers to shed light on the genetic relationships between extant taxa. These methods have assisted in fine-tuning the phylogenetic tree of the order Carnivora (Figure 1.1). It now regroups 13 terrestrial families; Ursidae does not contain any small carnivores, while several others are comprised entirely of small carnivores (Figure 1.1).

Recent progress in molecular phylogenetics has since revealed the well-resolved carnivoran phylogeny in both major clades of Caniformia ('dog-like' species) and Feliformia ('cat-like' species) to the extent that little remains to be enlightened on the interfamilial relationships. During the last decade or so, the supermatrix approach with nuclear gene sequences under the probabilistic phylogenetic criteria has contributed greatly to the clarification of some enigmatic phylogenetic relationships; e.g. Eupleridae: Yoder *et al.* (2003); Prionodontidae: Gaubert & Veron (2003); Pinnipedia and Ursidae: Flynn *et al.* (2005) and Sato *et al.* (2006); Ailuridae and Mephitidae: Sato *et al.* (2009). In addition to the phylogenetic topological issues, the divergence times among families have also been largely consistent among studies which adopted different fossil calibrations, especially those using mainly nuclear gene sequences (Koepfli *et al.*, 2006; Sato *et al.*, 2009; Eizirik *et al.*, 2010; Fulton & Strobeck, 2010; Meredith *et al.*, 2011). Figure 1.1 shows the currently accepted phylogenetic relationships among families within the order Carnivora, where the divergence times among them were calculated from arithmetically averaging the estimates in the above-mentioned studies. These are based on the taxon-by-characters supermatrix

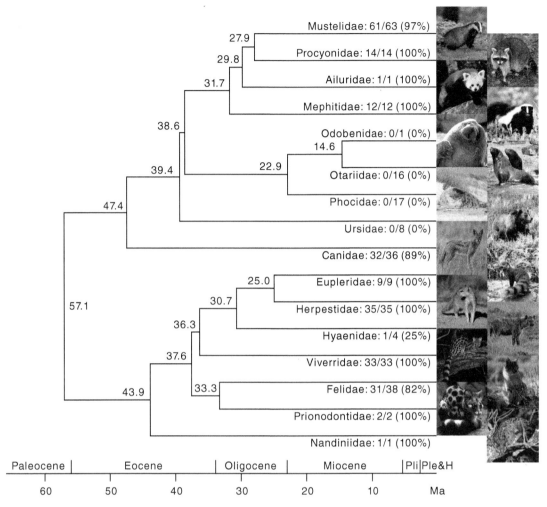

Figure 1.1 Phylogenetic relationships among the 16 families within the order Carnivora summarized on the basis of the estimates in recent molecular phylogenetic studies using the taxon-by-characters supermatrix mainly consisting of more than 5000 bp nucleotide sequences from multiple nuclear genes. Divergence times were arithmetically averaged based on estimates provided in the literature (see references in the text). The percentage proportions of the small carnivores (< 21.5 kg; *n* = 232) in each family are indicated after the family names. Pli = Pliocene, Ple = Pleistocene, H = Holocene. *Source:* Photos © Pascal Gérold (European badger, *Meles meles*), Steven Jansen (northern raccoon, *Procyon lotor*; grey seal, *Halychoerus grypus*), Jacob Dingel (striped skunk, *Mephitis mephitis*), Alex Sliwa (red panda, *Ailurus fulgens*; walrus, *Odobenus rosmarus*; Steller sea lions, *Eumetopias jubatus*; brown/grizzly bear, *Ursus arctos*), Emmanuel Do Linh San (black-backed jackal, *Canis [= Lupullela] mesomelas*; yellow mongoose, *Cynictis penicillata*; spotted hyaena, *Crocuta crocuta*; African wild cat, *Felis lybica*), Nick Garbutt @ www.nickgarbutt.com (ring-tailed vontsira, *Galidia elegans*), Len de Beer (rusty-spotted genet, *Genetta maculata*), Johannes Pfleiderer (Spotted linsang, *Prionodon pardicolor*), and David Mills (African palm civet, *Nandinia binotata*).

consisting of more than 5000 bp nucleotide sequences from multiple nuclear genes. Most of the divergence among families occurred from the middle Eocene to the late Oligocene. Only the marine pinniped families started the diversification in the Miocene. The number of subfamilies within the order Carnivora is still debated and likely to change with future studies.

Number of Species, 'New' Species, and Taxonomic Uncertainties

The number of small carnivore species worldwide is dependent upon both the definition used as well as the handling of several cases of taxonomic uncertainty present within this mammalian order. An in-depth look at species lists in recent compendia on carnivores suggests that the species rank attributed to 233 taxa of terrestrial carnivores is not in dispute. Recent sources list between 245 and 251 carnivore species, and species composition slightly varies among them (Nowak, 2005; Wozencraft, 2005; Wilson & Mittermeier, 2009; Hunter & Barrett, 2011, 2018; Table 1.1). During its 2015/2016 reassessment of the conservation status of carnivores worldwide, the IUCN listed 255 terrestrial carnivore species. However, as more work is completed, additional species are likely to be recognized. For example, the African wolf, *Canis lupaster*, was recently recognized and assessed following a series of research works filled with twists and turns (IUCN,

2021; see other references in Appendix B). Two additional cat species were recognized – but not yet assessed – in 2017 after a thorough and much-needed revision of the taxonomy of the Felidae (Kitchener *et al.*, 2017). A detailed review of the literature indicates that at least an additional 51 taxa, for a total of 309, were claimed, suggested or convincingly demonstrated by some authors to warrant species status (see contentious cases for small carnivores in Appendix B). Domestic and feral carnivores constitute a separate and complex case, which will not be debated here (see Wyrwoll, 2003; Gentry *et al.*, 2004). Considering that there are 23 to 27 large terrestrial carnivore species globally, the number of small carnivore species could,

Table 1.1 Number of extant terrestrial carnivore species recognized by five recent standard references, and number of species in each family.

Suborders and families[a]	Nowak (2005)	Wozencraft (2005)	Wilson & Mittermeier (2009)	Hunter & Barrett (2011)	Hunter & Barrett (2018)	IUCN (2021)	% IUCN (2021)
Caniformia							
Mustelidae	67	58	57	56	60–62	63	**24.6**
Procyonidae	19[b]	14	12	13	13	14	**5.5**
Ailuridae	—[c]	1	1	1	1	1	**0.4**
Mephitidae	—[d]	12	12	12	11	12	**4.7**
Ursidae	8	8	8	8	8	8	**3.1**
Canidae	35	34	35	35	35	35	**13.7**
Feliformia							
Eupleridae	–[e]	8	8	9	7	9	**3.5**
Herpestidae	39	33	34	34	34	35	**13.7**
Hyaenidae	4	4	4	4	4	4	**1.6**
Viverridae	35	35	34	33	33	33	**12.9**
Felidae	38	40	37	37	40	39	**15.2**
Prionodontidae	—[f]	—[f]	2	2	2	2	**0.8**
Nandiniidae	—[g]	1	1	1	1	1	**0.4**
Total	**246[a]**	**248**	**245**	**245**	**249–251**	**256**	**100.0**

The percentage of species in each family as per IUCN (2021) is provided in the last column. See Figure 1.1 for the numbers and percentages of small carnivore species in each family.
[a] For comparative purposes, families are listed in the same order as that provided in Figure 1.1.
[b] This count included the extinct Barbados raccoon, *Procyon gloveralleni*, as well as several other insular populations of the introduced northern raccoon, *P. lotor*; all of those are currently regarded as insular forms or subspecies at best (see Appendix B).
[c] The red panda, *Ailurus fulgens*, was then classified with the Procyonidae (see Sato & Wolsan, Chapter 2, this volume).
[d] The skunks were previously considered as a separate subfamily (Mephitinae) within the Mustelidae.
[e] The Malagasy carnivores were then regarded as direct relatives of mongooses (Herpestidae) and civets (Viverridae) (see Veron *et al.*, Chapter 3, this volume).
[f] The Asian linsangs were formerly classified with the Viverridae. The African linsangs (oyans) are still included in the Viverridae.
[g] The African palm civet, *Nandinia binotata*, was then classified with the Viverridae.

therefore, range from 210 (90.1% of 233 species) to 282 species (91.3% of 309 species). At present, the IUCN (2021) recognizes 232 small carnivore species (90.6% of 256 carnivore species; see full species list in Appendix A), an approach that we have adopted here in view of showing the general trends in the conservation status of small carnivores worldwide. The recognized species are dominated by 5 families: Mustelidae (61 species; 26.3%), Herpestidae (35; 15.1%), Viverridae (33; 14.2%), Canidae (32; 13.8%), and Felidae (31; 13.4%) (Figure 1.1). The remaining 7 families comprise only 40 species, a mere 17.2% of all small carnivore species worldwide.

At least three species of small carnivores have gone extinct over the past centuries – the Falkland Island wolf, *Dusicyon australis*, his continental sister taxon, *D. avus*, and the sea mink, *Mustela macrodon* (Nowak, 2005; IUCN, 2021). However, very few of the recently named species have, in fact, been newly discovered; the validity of others can be questioned. For example, Durbin *et al.* (2010) reported on the discovery of Durrell's vontsira, *Salanoia durrelli*, from the marshes of Lac Alaotra, in Madagascar. This taxon diverged from brown-tailed vontsira, *Salanoia concolor*, in terms of craniodental morphometrics, but exhibited low genetic differentiation at cytochrome *b*. Recent studies by Veron *et al.* (2017) based on mitochondrial and nuclear markers and additional phenotypic characteristics strongly support the existence of a single species within the genus *Salanoia*. Similar cases of rescinded discoveries are reported in Appendix B. A few years ago, Helgen *et al.* (2013) undertook a comprehensive taxonomic revision of the Neotropical olingos. In the process, these authors came to describe a new species, the olinguito, *Bassaricyon neblina*, endemic to the Andes of Colombia and Ecuador. Although this species was newly named, several specimens had previously been collected and misidentified as northern olingo, *Bassaricyon gabbii*. The abovementioned study has now shown that *B. gabbii* occurs only in Central America.

Perhaps the only recent field discovery of a small carnivore species is the Vietnam ferret badger, *Melogale cucphuongensis*, for which two individuals were found in 2005 and 2006 in Cuc Phuong National Park. Both morphological and cytochrome *b* data pointed to a clear distinction from the two other sympatric ferret badger species (Nadler *et al.*, 2011).

However, it is not impossible that this taxon has previously been described under another name and considered synonymous to that of one of the two continental ferret badger species. Clearly, more samples of this newly described taxon would need to be collected and a thorough revision of this genus undertaken. Similarly, the 'discovery' of Lowe's otter civet, *Cynogale lowei*, was based on a single specimen collected in the winter of 1926–1927 in northern Vietnam. Ninety years later, Roberton *et al.* (2017) using microscopic hair and DNA analyses, showed that the type specimen was, in fact, a juvenile Eurasian otter, *Lutra lutra*, thereby refuting the existence of Lowe's otter civet!

The apparent increase in the number of small carnivore species over the past 15 years is primarily due to taxonomic splitting of species. For example, the three to four species of Eurasian badgers *sensu lato* currently recognized result from a progressive split of *Meles meles* (European badger); similarly, the three species of Asian hog badgers, *Arctonyx* spp., were previously considered a single species (see review in Sato, 2016 and references in Appendix B). More generally, a combination of morphological and molecular studies based on a broad range of genetic markers, sometimes coupled with an analysis of biogeographic data, have provided strong evidence for the upgrading or subsuming of some taxa (Appendix B). Several cases are still disputed and require further investigations (Appendix B). It is likely that genome-wide analyses will help clarify the taxonomic rank of several taxa in the near future, as already demonstrated in recent studies (e.g. Koepfli *et al.*, 2015; Gopalakrishnan *et al.*, 2018; Hu *et al.*, 2020).

Geographic Distribution

Small carnivores are present worldwide, with the exception of Antarctica. They occur in Oceania following successful introductions: the red fox, *Vulpes vulpes*, in Australia (including Tasmania) and New Zealand; the stoat, *Mustela erminea*, the least weasel, *M. nivalis*, and the European polecat, *Mustela putorius*, and/or domestic ferret, *M. furo*, in New Zealand; the small Indian mongoose, *Urva auropunctata*, in Hawaii and Fiji; and the Indian brown mongoose, *Urva fusca*, in Fiji. Gantchoff *et al.* (Chapter 20, this volume) report on some of those successful introductions, as well as on the introduction efforts that failed.

Asia (42%) and Africa (37%) both host large proportions of the world's 232 small carnivore species, followed by Central and South America (22%). Species richness is much lower in North America (16%) and Europe (10%). Most small carnivore species ($n = 185$ or 80%) occur on one continent only and, therefore, analyses which take into account the remaining 20% of species present on two or more continents yield similar proportions (Figure 1.2). The red fox and the small

Indian mongoose are present on five continents (islands included), while the stoat, least weasel, European polecat, American mink, *Neovison vison*, and northern raccoon, *Procyon lotor*, occur on four continents, all partially due to introductions.

Although countries differ drastically in surface area, a simple plot of the number of species that occur in a specific number of countries show that pairs of data fall on or very close to the indicated power regression curve (Figure 1.3). No less than 35 species (15%) are endemic to a single country, and among those, 21 (9%)

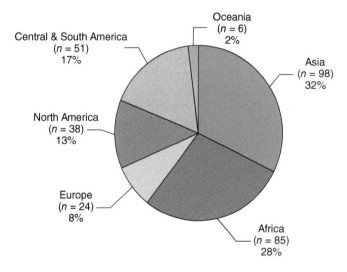

Figure 1.2 Proportional geographic distribution of the 232 small carnivores (< 21.5 kg) recognized by the IUCN (2021). Note that some species occur on two or more continents and are, therefore, counted multiple times ($n = 302$ occurrences). For the purpose of our analysis, Africa includes Madagascar and the nearby Indian Ocean and Atlantic Ocean islands; Central America includes the Caribbean Islands; Oceania comprises Australasia, New Zealand, Polynesia, Melanesia, and Micronesia.

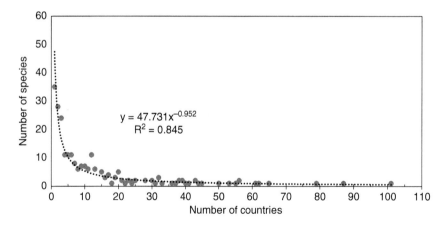

Figure 1.3 The relation between the number of small carnivore species (< 21.5 kg) and the number of countries in which they occur worldwide. The black dotted line represents a fitted power function regression curve. Note that the dot with 101 countries corresponds to *Felis silvestris*. This species has now been split into two, with the European wild cat, *F. silvestris sensu stricto*, present in 34 countries, and the African wild cat, *F. lybica*, in 67 countries occurring from South Africa to China and southern Mongolia.

are island endemics – albeit sometimes present on several islands. Overall, about half of the small carnivore species are present in five or fewer countries, and nearly two-thirds occur in ≤ 10 countries. The remaining one-third consists of species whose distribution spans from 11 to 65 countries. The red fox and the Eurasian otter top the list with 87 and 79 countries, respectively (see caption of Figure 1.3 for the special case represented by the wild cat, *Felis silvestris*).

Conservation Status in an Era of 'Species Inflation'

The latest update of the assessment of the conservation status of small carnivores by IUCN (2021) indicates that about 64% of the 232 species assessed are not currently a source of immediate concern (Figure 1.4; see the full list in Appendix A). Six species (2.6%) could not be assessed due to a deficiency of relevant data; these are three African (Pousargues's mongoose, *Dologale dybowskii*; Ethiopian genet, *Genetta abyssinica*; king genet, *G. poensis*) and three Asian species (Vietnam ferret badger; Sichuan weasel, *Mustela russelliana*; Tonkin weasel, *M. tonkinensis*). Some data have been recently collected on two of these species (D'Haen, 2017; Ferguson *et al.*, 2019), however, while a step in the right direction, the information available is still insufficient to allow a proper evaluation of their

conservation status. The remaining one-third of species are Near Threatened (~11%) or threatened *sensu lato* (~23%). Among the latter, 23 species (~10%) are Endangered or Critically Endangered (Table 1.2). Irrespective of their conservation status, only 27% of the world's small carnivore species have stable populations, whereas a mere 3.4% are increasing. In contrast, 50% of the small carnivore species are currently experiencing population declines, and the trends for the remaining 20% are unknown.

While the Earth is currently facing what some call a 'sixth mass extinction' due to human activities (Ceballos *et al.*, 2010, 2015, 2017), the threats to small carnivores result from hunting and fishing (including deadly by-catches), poaching (often with snares), and persecution which directly reduce small carnivore population sizes (Gray *et al.*, 2018; IUCN, 2021). In addition, industrialization, urbanization, and the expansion of agricultural activities cause the reduction, destruction, fragmentation, and pollution of habitats (see review in Marneweck *et al.*, 2021). In this context, deforestation remains one of the most severe threats to forest-dependent species (Püttker *et al.*, 2020; Rocha *et al.*, 2020). Scenarios based on future land-use change predict that several carnivore species, including small carnivores, will be negatively affected (di Minin *et al.*, 2016). Road-traffic mortality may also severely impact some species or populations. Viruses and organisms (from bacteria to mammals)

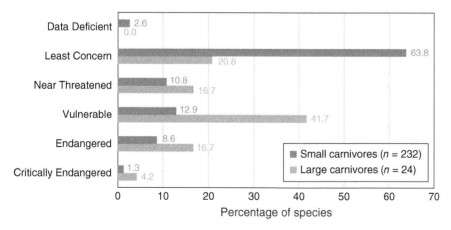

Figure 1.4 *IUCN Red List* categorization of the world's small carnivores (< 21.5 kg) in comparison with that of large carnivores, as assessed majoritarily in 2015 and 2016 (IUCN, 2021).

Table 1.2 A list of the 23 small carnivore species (< 21.5 kg) which are currently listed as Endangered (EN) or Critically Endangered (CR) by the IUCN based on assessments mostly carried out in 2015 and 2016 (IUCN, 2021).

Family and scientific name[a]	Common name	Distribution	IUCN Red List categorization	Main threats/concerns (Asssessors/reference)
Ailuridae				
Ailurus fulgens	Red panda	Bhutan, China, India, Myanmar, Nepal	EN	Relatively high forest loss rate and poor survival in fragmented areas – as the species is dependent on bamboos which are vulnerable in deforested/degraded habitats. High susceptibility to canine distemper. Increase in poaching and live-capture for trade (body parts, pelts, pets). Growing human population and encroachment in red panda habitat (herding, bamboo used for fodder, collection of firewood) (Glatston *et al.*, 2015).
Canidae				
Canis simensis	Ethiopian wolf	Ethiopia	EN	Small population size (~400 adults in seven isolated mountain enclaves). Susceptibility to rabies and canine distemper. Afroalpine range shrinking and increasingly fragmented (Marino & Sillero-Zubiri, 2011).
Cuon alpinus	Dhole	Asia (20 countries)	EN	Depletion of prey base, persecution due to livestock predation, habitat loss, disease transmission from domestic dogs, and possibly interspecific competition. Between 4500 and 10 500 individuals distributed in small and isolated subpopulations, often exhibiting severe fluctuations in numbers (Kamler *et al.*, 2015).
Lycalopex fulvipes	Darwin's fox	Chile	EN	Forest loss and risk of contracting canine distemper from domestic dogs. Population size likely does not exceed 2500 individuals (Silva-Rodríguez *et al.*, 2016).
Eupleridae				
Eupleres major[b]	Western falanouc	Madagascar*	EN	Numerous synergistic threats: widespread hunting, persecution, ongoing habitat conversion and fragmentation, and predation by feral cats and domestic dogs (Hawkins, 2016).
Galidictis grandidieri[c]	Grandidier's vontsira	Madagascar*	EN	Habitat loss and fragmentation, and predation by non-native predators. Population estimate is 3000–5000 individuals. Extent of occurrence comprises a single location and is well below 5000 km^2 (may be as low as 1500 km^2) (Hawkins, 2015a).
Mungotictis decemlineata	Bokiboky	Madagascar*	EN	Widespread and increasing habitat loss and degradation, hunting, persecution, and negative impacts of introduced carnivores (Hawkins, 2015b).

Table 1.2 (Continued)

Family and scientific name[a]	Common name	Distribution	IUCN Red List categorization	Main threats/concerns (Asssessors/reference)
Felidae				
Catopuma badia	Borneo bay cat	Borneo (Indonesia, Malaysia)*	EN	Hunting believed to potentially have a considerable impact. Species prone to untargeted snaring and evidence of capture and export of animals for the pet trade. Area of occupancy estimated to have decreased by 30% between 2000 and 2010 as a result of ongoing forest loss and conversion to oil palm plantations. The smaller protected areas are fragmented and isolated, hence ineffective in conserving such a low-density species (Hearn *et al.*, 2016).
Leopardus jacobita	Andean cat	Argentina, Bolivia, Chile, Peru	EN	Occurs at low densities in patchy rocky habitats associated to wetlands/shrublands. Population estimate is ~2800 individuals. Locally killed by herders (in retaliation for predation) or hunted. Increasing habitat loss and degradation due to expansion of agricultural activities, inadequate livestock management and water extraction. The growing mining and petroleum/gas industry is also of concern (Villalba *et al.*, 2016).
Lynx pardinus	Iberian lynx	Spain, Portugal	EN	Only 156 mature individuals in two subpopulations in 2012, with a total area of occupancy of 1040 km^2. Improved status from CR to EN due to intensive, ongoing conservation actions. At least 1111 individuals in several subpopulations in 2020, but affected by car hits (34 in 2019) and illegal hunting. Future range expansion and population increase highly dependent upon continued reintroductions and maintenance of prey base (mostly rabbits) under epizootic and climate change scenarios (Rodríguez & Calzada, 2015).
Prionailurus planiceps	Flat-headed cat	Brunei, Indonesia, Malaysia, Thailand	EN	Restricted and patchy distribution around wetlands. Wetland destruction/degradation (draining for agriculture; pollution; excessive fishing, hunting, and wood-cutting; clearance of coastal mangroves) and the very high deforestation rate are, therefore, the primary threats. Population size is plausibly < 2500 mature individuals (Wilting *et al.*, 2015a).
Mustelidae				
Lontra felina	Marine otter	Argentina, Chile, Peru	EN	Original range decreased substantially following excessive hunting. Distribution along the Pacific coast is now restricted and patchy, with fragmentation caused by poaching, pollution, and increased human occupation along the seashores. Accidental kills in crab pots reported. Inferred future population decline due to habitat loss and exploitation (Valqui & Rheingantz, 2015).

(Continued)

Table 1.2 (Continued)

Family and scientific name[a]	Common name	Distribution	IUCN Red List categorization	Main threats/concerns (Asssessors/reference)
Lontra provocax	Southern river otter	Argentina, Chile	EN	Original range decreased drastically due to habitat destruction (including removal of vegetation), river and stream canalization, and extensive dredging. Poaching and predation by domestic dogs occasionally reported. Projected population decline of ~50% over the 30 years following the assessment due to accelerated habitat destruction and degradation (for otters using freshwater habitats) and the impacts of intensive fishery activities (for marine otters) (Sepúlveda *et al.*, 2015).
Lutra sumatrana	Hairy-nosed otter	Cambodia, Malaysia, Thailand, Vietnam	EN	Loss of > 50% of the population during the 30 years preceding the assessment caused by illegal trade-driven hunting (meat, medical use, pets), by-catch, pollution and prey depletion due to overfishing. Current decline rates suspected to continue into the future due to burgeoning human population and growing pressure on natural resources (Aadrean *et al.*, 2015).
Melogale everetti	Bornean ferret badger	Borneo (Malaysia)*	EN	Small extent of occurrence (4200 km^2) and area of occupancy (1100 km^2) in two to three locations. Species is rare even in its core range, and possibly threatened by the ongoing land-cover transformations. Climate change on Borneo is projected to particularly threaten highland species such a *M. everetti* because potential upslope range shifts would be impossible (Wilting *et al.*, 2015b).
Mustela lutreola	European mink	Estonia, France, Romania, Russian Federation, Spain, Ukraine	CR	Loss of > 50% of the population in the 10 years preceding the assessment. Predicted to intensify to reach a decline rate > 80% in the following 10 years due to habitat degradation/loss and the effects of introduced species, notably the American mink, *Neovison vison*. Western populations (France and Spain) have a very low genetic variability. Genetic introgression (0.9%) following hybridization (3%) with European polecat, *M. putorius*, occur at low levels (Maran *et al.*, 2016).
Mustela nigripes	Black-footed ferret	USA, Mexico	EN	Extirpated from most of its former range (from northern Mexico to southern Canada) mainly as a result of prairie-dog, *Cynomys* spp., control programmes and sylvatic plague (exotic disease introduced to the wild population). Today occurs in the wild as 17–22 reintroduced, small, and restricted populations which are intensively managed. Four of those are self-sustaining (all in USA) and have a combined range > 500 km^2. Only ~300 wild-born mature individuals in 2015, with two-thirds in the viable populations (Belant *et al.*, 2015).

Table 1.2 (Continued)

Family and scientific name[a]	Common name	Distribution	IUCN Red List categorization	Main threats/concerns (Asssessors/reference)
Procyonidae				
Nasuella meridensis	Eastern mountain coati	Venezuela (Andes)	EN	Very small known extent of occurrence ($770 \, km^2$) with only five confirmed and spatially aggregated locations. Slight but ongoing loss of cloud forests and Paramo habitats leading to population decline (González-Maya & Arias-Alzate, 2016).
Procyon pygmaeus[d]	Pygmy raccoon	Mexico (Cozumel Island)*	CR	Small population (~190–570 mature individuals) confined to Cozumel Island ($488 \, km^2$). Although population size is likely to fluctuate, the overall trend is a rapid decline caused by ongoing human encroachment (urban growth, development, tourism) into the remaining habitat fragments, invasive predators and pathogens, road network expansion and the increasing magnitude/severity of hurricanes (Cuarón *et al.*, 2016).
Viverridae				
Chrotogale owstoni	Owston's civet	China, Lao, Vietnam	EN	Dramatic population decline (> 50%) due to overexploitation, with habitat fragmentation exacerbating the speed at which this civet is extirpated. Exposed to high levels of snaring and other forms of ground-level trapping. Wider occurrence in montane forest and karstic areas, into which industrial snaring has not yet widely spread (Timmins *et al.*, 2016a).
Cynogale bennettii	Otter civet	Brunei, Indonesia, Malaysia, Thailand	EN	Primary habitat (forested, lowland wetlands) and population have declined by ≥ 20% in the 10 years preceding the assessment. Remaining habitat is discontinuous and habitually degraded. Water sources are often polluted. Snaring is also a threat. Population size is plausibly < 2500 mature individuals (Ross *et al.*, 2015).
Viverra civettina[e]	Malabar civet	India (Western Ghats)	CR	This species is possibly extinct because there are no recent reliable records since 1989 despite surveys (including with camera-trapping) in potentially suitable habitats. In any case, population size does not exceed 250 mature individuals (Mudappa *et al.*, 2016).

(Continued)

Table 1.2 (Continued)

Family and scientific name[a]	Common name	Distribution	IUCN Red List categorization	Main threats/concerns (Asssessors/reference)
Viverra megaspila	Large-spotted civet	Cambodia, China (?)[f], Lao, Malaysia, Myanmar, Thailand, Vietnam (?)[f]	EN	Past and projected dramatic population decline (> 50%) over the previous/next 15 years due to the combined effects of trade-driven overexploitation and habitat destruction/fragmentation. Species living almost exclusively in level lowland which is sought for conversion to agriculture, infrastructure, and settlement (Timmins *et al.*, 2016b).

Population trends are deemed decreasing for all species, except for the population of the Iberian lynx, which is increasing. Island endemics are indicated with an asterisk.

[a] When relevant, alternative taxonomic treatments are provided in Appendix A.

[b] Molecular studies of Veron & Goodman (2018) have suggested that *Eupleres major* does not differ from *E. goudotii* (eastern falanouc), but additional studies are needed to confirm this.

[c] Veron *et al.* (2017) have now shown that Grandidier's vontsira should be considered a subspecies of broad-striped vontsira, namely *Galidictis fasciata grandidieri*, and that therefore the conservation status of the latter species should be reassessed.

[d] Treated as a separate species from *Procyon lotor* based on morphometric and genetic differences (McFadden *et al.*, 2008; McFadden & Meiri, 2013). However, a reanalysis of the genetic data currently available suggests that the pygmy raccoon should be treated as a synonym of *Procyon lotor hernandezii* (Wagler, 1831), and, therefore, its conservation status may have to be revised (Louppe *et al.*, 2020b).

[e] Nandini & Mudappa (2010) argued that *Viverra civettina* is possibly the same species as *V. megaspila* following transport of individuals to the Western Ghats (southern India).

[f] This species is regarded as Possibly Extinct in China and Vietnam (Timmins *et al.*, 2016).

transported and introduced purposefully or involuntarily by humans have also brought about a series of challenges for indigenous small carnivores, leading to population size reductions attributable to the negative effects of alien predators, competitors, and pathogens, and this particularly on islands (Cuarón *et al.*, 2004; Farris *et al.*, 2017; Rasambainarivo *et al.*, 2017). The burgeoning human population and wildlife trade in Asia – for meat, medicinal use of body parts and the pet industry – is of particular concern (Siriwat & Nijman, 2018; Nijman *et al.*, 2019; Willcox, 2020). The potential effects of human-driven climate change on small carnivores are difficult to predict with accuracy. Some species are likely to be affected more severely than others, notably those living on islands (Struebig *et al.*, 2015), but climate change may also lead to an increase in the distribution range of several species (Pandey & Papes, 2018). For example, projections based on ecological niche modelling suggest that introduced and invasive species such as northern raccoon and small Indian mongoose will expand their geographic range, and that by 2050, they will have an

increasing influence on ecosystems and biodiversity, particularly in Europe (Louppe *et al.*, 2019, 2020a). More generally, small carnivores may not be directly affected by climate change *per se*, but could theoretically be impacted by a change (reduction or increase) in the abundance of food resources.

The anthropogenic threats listed above are not exclusive to Endangered and Critically Endangered species as they affect and negatively impact Vulnerable, Near Threatened, and Least Concern species as well; hence, the need for at least a periodical reassessment of the conservation status of each species. Figure 1.4 shows that small carnivores generally fare better than large carnivores when it comes to conservation status, with a lower percentage of species being categorized as Near Threatened or threatened *sensu lato*. However, because small carnivore species richness is nine times higher than that of large carnivores, a much larger number of small carnivore species are threatened worldwide. The potentially grave ecosystem-wide implications of the extirpation of large carnivores across the globe are raised later in this chapter.

As discussed above, the apparent increase in the number of small carnivore species worldwide is primarily related to recent species splitting and proposals to elevate some subspecies to species level. While some (suggested) changes were grounded on a solid combination of morphological, molecular, and biogeographic data, other cases relied singly on phenotypic variations or single molecular-marker (often mitochondrial DNA) analyses. The latter, in combination with the recent use of the phylogenetic species concept based on diagnosability, has led to what some call an unwarranted 'species inflation', with some serious conservation implications (see e.g. Zachos *et al.*, 2013). While the term 'species' is generally regarded by governments and most funding bodies as the appropriate taxonomic unit when it comes to conservation, both taxonomic inflation and inertia have benefits and drawbacks (Zachos, 2013; Gippoliti *et al.*, 2018). It is likely that structured and concerted team efforts, such as that demonstrated by the IUCN SSC Cat Specialist Group to revise the taxonomy of the Felidae (Kitchener *et al.*, 2017), would ensure an accurate evaluation of the contentious cases. Periodical literature reviews or reassessments of the taxonomy and systematics/phylogeny of carnivore subgroups are also particularly encouraged (see e.g. Veron, 2010; Helgen *et al.*, 2013; Sato, 2016; Zrzavý *et al.*, 2018; Alvares *et al.*, 2019; Sato & Wolsan, Chapter 2, this volume; Veron *et al.*, Chapter 3, this volume), especially considering the fast pace at which the field of molecular biology is developing. Irrespective of their taxonomic rank, genetic lineages of high conservation priority (either already or yet to be identified) should receive full attention from governments and conservation organizations (Thakur *et al.*, 2018). For the sake of conservation, informed scientists should make it a responsibility to educate the designated authorities about the logic underlying their recommendations. In view of the evolving nature of biological lineages, 'species' or 'subspecies' are somewhat arbitrary terms used to describe unique biological entities (i.e. taxa) that share distinct, transmissible biological characteristics. Considering that recent reference works recognize the existence of between 990 (Wilson & Mittermeier, 2009) and 1233 (Wozencraft, 2005) subspecies of terrestrial carnivores, prioritizing may be called for irrespective of whether a focal taxon is called a 'species' or 'subspecies'. For example, Wilting *et al.* (2016) have convincingly demonstrated the conservation significance of the Javan leopard, *Panthera pardus melas*, subspecies for which only a few hundred individuals still live in the wild. When relevant, zoologists and conservationists should work hand in hand to define Evolutionary Significant Units (Ryder, 1986) as well as Management Units (Moritz, 1994), as notably proposed for the Andean cat, *Leopardus jacobita* (Cossíos *et al.*, 2012), and the Pampas cat, *Leopardus colocola* (da Silva Santos *et al.*, 2018). Conservation actions would then have to be prioritized based on the threat level pertaining to those 'operational units' under the species level.

Ecological Roles of Carnivores

Large carnivores such as lions, tigers, brown bears or grey wolves, *Canis lupus*, or brown bears, have drawn considerable attention from researchers. This is firstly due to their charisma, as indeed humans have long been fascinated by these magnificent creatures. Secondly, emphasis on research effort is a result of the potential of these species to be involved in human–wildlife conflicts (e.g. Rajaratnam *et al.*, 2016; Penteriani *et al.*, 2017; Moreira-Arce *et al.*, 2018; Kelly *et al.*, 2019; Ugarte *et al.*, 2019; Reyna-Saenz *et al.*, 2020) and the ensuing persecution and threats to several species and/or local populations (e.g. Bauer *et al.*, 2015; Jacobson *et al.*, 2016; Trinkel & Angelici, 2016; Durant *et al.*, 2017). Recovery of large carnivore populations and co-existence with humans, however, are not impossible, as exemplified by the current situation in Europe with grey wolves and brown bears (Chapron *et al.*, 2014). Reintroductions and rewilding may also assist in this context (Hayward & Somers, 2009; Wolf & Ripple, 2018; Linnel & Jackson, 2019; but see Alston *et al.*, 2019). Certainly, large carnivores and the 'landscape of fear' they create are vital for the regulation of land and aquatic ecosystems through cascading interactions across trophic levels. This has been dealt with in great depth in a large body of empirical and theoretical literature spanning two decades (e.g. Wright *et al.*, 1994; Palomares *et al.*, 1995; Terborgh *et al.*, 1999; Ripple &

Beschta, 2004; Ray *et al.*, 2005; Johnson *et al.*, 2007; Ripple *et al.*, 2014; Suraci *et al.*, 2016; Leempoel *et al.*, 2019; Hoeks *et al.*, 2020). Recent studies have shown that the presence of top carnivores constrains the distribution and/or abundance of mesocarnivores (Newsome *et al.*, 2017; Davis *et al.*, 2019; Jiménez *et al.*, 2019; Prugh & Sivy, 2020). On the other hand, apex carnivores may facilitate resource partitioning through the provision of carrion to smaller carnivores (Sivy *et al.*, 2018; Ruprecht *et al.*, 2021; cf. Prugh & Sivy, 2020) and facilitate overall species coexistence in both natural and altered ecosystems (Wallach *et al.*, 2015b). Equally, there is strong mathematical support showing that top predators *sensu lato* may induce the evolutionary diversification of intermediate-size predator species (Zu *et al.*, 2015).

Small carnivores, although represented worldwide by many more species, are often mistakenly thought to exert a weaker ecological influence at the ecosystem level. Although they do not impact on the same spectrum of prey as large carnivores, they are similarly important ecosystem regulators through structuring invertebrate and/or small mammal communities, including in rural agro-ecosystems (e.g. Williams *et al.*, 2018), which, in turn, may affect both lower and higher trophic levels (Roemer *et al.*, 2009). They may also be important in seed dispersal (e.g. Koike *et al.*, 2008; López-Bao & González-Varo, 2011; see review in Nakashima & Do Linh San, Chapter 18, this volume), which may both enhance forest regeneration and plant gene flow (Jordano *et al.*, 2007), thus reducing inbreeding depression. The roles of small carnivores in shaping ecosystems have also been shown through accidental introductions, which have been numerous for this group of predators (Boitani, 2001; Gantchoff *et al.*, Chapter 20, this volume). Finally, where large carnivores have been extirpated by humans – directly by persecution or through habitat modifications – small carnivores have or may become dominant predators in these ecosystems. This may alter their abundance, ecological roles, and importance in the corresponding food webs, and sometimes destabilize communities, drive local extinctions, and reduce overall biodiversity (e.g. Trewby *et al.*, 2008; Prugh *et al.*, 2009; Colman *et al.*, 2014; Wallach *et al.*, 2015a; Alston *et al.*, 2019; Cove & O'Connell, Chapter 21, this volume).

Research Efforts on Carnivores

Although small carnivores have been the focus of numerous research projects over the past several decades, smaller species have not received the same attention as that bestowed upon large carnivores. In addition, there is a huge disparity in the research efforts directed toward small carnivores, with some species intensively studied and others superficially or not at all. Brooke *et al.* (2014) reviewed 16 367 peer-reviewed papers focusing on at least one of the 286 species of the order Carnivora (including pinnipeds; Wozencraft, 2005) published from 1900 to the end of 2010. Unsurprisingly, they found that the most charismatic families of terrestrial carnivores were over-represented in the literature; namely Canidae (3387 publications), Felidae (2968), Ursidae (2002), and Hyaenidae (319), the sum of which corresponds to exactly two-thirds of all papers published on terrestrial carnivores. When the number of species per family was taken into account, the ranking remained relatively unchanged, with the Ursidae topping the list (with an average of 250 papers per species), followed by the Canidae (97), Hyaenidae (80), and Felidae (74). By comparison, the best-studied family of (mostly) small carnivores, the Mustelidae, was the subject of an impressive 2968 papers, but this corresponds to an average of only 49 papers per species. However, the above values are somewhat misleading because the distribution of papers within families is further biased toward the large species. Among the Felidae, for example, 52% of the papers focused on the 7 large felids, while the remaining 48% were dedicated to the 33 species of small felids (Z. Brooke, personal communication). Comparable trends were observed for the Canidae, with the notable exception of the 'small' red fox, which generated an imposing 923 publications. Similarly, among the Mustelidae, 4 species accounted for 46% of all publications for this family (rich of at least an additional 55 species); these are the European badger (517 papers), Eurasian otter (369), European polecat (233), and stoat (228). Research emphasis seems to be attributable to a series of additive factors: large distributional range, initial higher research effort by European zoologists, lack of larger carnivores (notably in the UK and Ireland, where natural history studies have always played a preponderant role), the implication of some species in major zoonoses (rabies, bovine tuberculosis, and sarcoptic mange), and introductions to non-native environments leading to undesirable effects on native wildlife. Other species-rich families such as the Herpestidae and Viverridae averaged only 10 and 6 papers per species, respectively. Similarly, low average numbers of papers were obtained for the Mephitidae (18) and Eupleridae (5). Strikingly, there were no records of publications on 28 small carnivore species from 7 families, including 9 mongoose, 6 mustelid, and 5 viverrid species. Among Herpestidae, group-living species such as meerkats, *Suricata suricatta*, and banded mongooses, *Mungos mungo*, have attracted most of the attention, as they have been used as model species to understand the evolution of sociality. Of course, this literature survey would not have picked up 'grey literature' (e.g. technical reports or unpublished academic theses), papers published in small regional journals, or those written in languages other than English. Overall, however, the results clearly confirmed the suspicion that, until the end of 2010, there was a

massive disparity in the research efforts on large vs. small carnivore species globally.

Considering the relatively recent and in-depth bibliometric studies of Brooke *et al.* (2014), and the broad but general scope of this introduction, we did not aim to undertake a detailed analysis of the evolution of the number of papers published over the last decade on each of the >232 species of the world's small carnivores. Rather, we sought to understand whether the situation had improved. To that end, two literature datasets that we retrieved and analyzed provide some interesting insights.

We first ran a search for all the publications containing a combination of keywords referring to small carnivores (see details in Figure 1.5 caption) as a topic using all databases accessible in Web of Science (Clarivate Analytics). The output list was then imported into EndNote X7.8 (Thomson Reuters) and curated, after which all papers that did not focus on carnivoran, terrestrial predators were discarded. This exercise yielded a total of 618 articles published

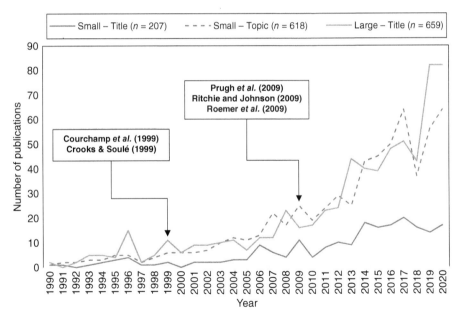

Figure 1.5 Number of scientific publications from 1990 to 2020, as retrieved from the Web of Science database (https://apps.webofknowledge.com). The purple continuous line corresponds to the number (*n*) of publications with the terms 'small carnivores', 'mesocarnivores', 'mesopredators' or 'mid-size(d)/medium-size(d) carnivore(s)/predator(s)' included in the title. The purple dashed line represents all publications with the same keywords as a topic. The green continuous line depicts publications with the terms 'large/top/apex carnivore(s)/predator(s)' in the title. Datasets were curated to only include publications on terrestrial members of the order Carnivora. The years of publication of five seminal papers or reviews on the roles of mesopredators are also indicated in the graph.

between 1990 and 2020. Providing that this subsample is representative of the trends that would have been observed had all publications been searched for (i.e. with the > 232 species scientific and English names used as keywords) and curated, Figure 1.5 suggests that the number of publications on small carnivores has increased in a nearly exponential fashion since 1990, with a significant increase only taking place from 2009 onward. During that year, three seminal review papers were published: one focusing on the ecological role of mesocarnivores (Roemer *et al.*, 2009); another on the rise of mesopredators in ecosystems where apex predators have been extirpated (Prugh *et al.*, 2009) – a phenomenon known as 'mesopredator release' (Soulé *et al.*, 1988; Litvaitis & Villafuerte, 1996; Courchamp *et al.*, 1999; Jachowski *et al.*, 2020); and a third arguing that effective biodiversity conservation requires proper knowledge of predator interactions and understanding of the mechanisms underlying mesopredator release (Ritchie & Johnson, 2009). There is no doubt that these now well-cited publications have highlighted the roles and importance of the smaller carnivores, thereby boosting research effort on them.

In a second step, to infer whether the research efforts were focusing on small carnivores as research models rather than on species *per se*, we restricted the above dataset to publications that included any of the selected keywords in the title. This reduced the total to 207 articles. As apparent in Figure 1.5, the number of publications meeting this new criterion increased over the past decade, but not as sharply as the overall number of publications whose main topic was small carnivores. This difference may indicate that although small carnivores are receiving more attention from the scientific community as model organisms (for example, to test the mesopredator release theory), there might also be an increased focus on the general biology, ecology, and conservation of these fascinating organisms. This discrepancy in numbers seems to be accounted for by the current boom in camera-trapping studies worldwide (Rich *et al.*, 2017), accompanied by an exponential increase in publications (Rovero & Zimmerman, 2016) which have provided insights into the distribution, relative abundance, activity patterns, and interspecific relationships of numerous mammalian species.

As a third and last step in evaluating how small carnivore research fared in comparison with that on the larger species, we searched for all the publications containing a combination of keywords referring to large carnivores in the title of the article (see details in Figure 1.5 caption). This new search yielded a total of 659 curated articles published between 1990 and 2020. The same search with the selected keywords as topic yielded over 83 000 non-curated references. While a large number of those seemed to be unrelated to our target keywords, it is clear that several thousands of papers focusing on large carnivores have been published over the last 10 years. These literature searches strongly suggest that research on carnivores is still disproportionately geared toward the larger species. While it is possible that the observed increase in small carnivore research is partly or even largely due to (and therefore proportional to) an overall increase in the research efforts and scientific publications over the past decade, it is reassuring to see that small carnivores have at least been included in this trend. Whatever the reason(s) behind the nearly exponential increase in the number of publications, it signifies an increase in both our interest and scientific knowledge of small carnivores.

A preliminary examination on the major institutions driving the research and the regions where small carnivore research is taking place reveals that research efforts are not equally distributed. About one-third of the research is carried out in North or Central America, East Africa, and Southeast Asia by teams led by North American scientists. An equal amount of research output is produced by European researchers (UK/Ireland, Spain, France, Germany, Portugal, Italy, and Scandinavia), working in Europe, as well as in North and Equatorial Africa, Arabia, and Southeast Asia. The last third of publications emanate from Australia, South Africa, Brazil, India, China, and the rest of the world, with research generally taking place in the home country of the researchers with the exception of some Japanese researchers working in Equatorial Africa and Borneo.

The general trends depicted above seem to be only partly reflected in the geographic coverage of articles and notes published in *Small Carnivore Conservation* (SCC), the scientific journal of the IUCN SSC Small Carnivore Specialist Group. Indeed, 41% of papers published until March 2017 in SCC comfortably focused on Asian small carnivore species (González-Maya & Ramírez-Chaves, 2017). In contrast, other continents have been largely under-represented, with only 13% and 11% of publications dealing with African and American species, respectively (cf. continental proportions of small carnivore species in Figure 1.2). To remedy this unbalance in coverage, however, SCC has recently encouraged and promoted research in the Americas (and particularly in South and Central America; Schipper *et al.*, 2009) and in Africa (Do Linh San & Somers, 2013) through the production of dedicated Special Issues. Currently, the majority of publications in SCC report on faunal surveys, new range records, and unusual behaviours. Often these reports arise as by-products of research projects focusing on other animal taxa.

For all of the reasons listed above, it now appears that the tide is changing with regards to prioritizing small carnivore studies. We hope to see more small carnivore-dedicated work in the future, particularly where species richness and diversity are high and where knowledge gaps are apparent. In particular, research on threatened and Data Deficient species (*sensu* IUCN, 2021) is of critical importance.

Purpose and Structure of the Book

This book aims at filling a gap in the scientific literature by elucidating the important roles of the world's small carnivores and documenting the latest knowledge acquired on them. Specifically, our plan was to bring together contributions dealing with several aspects of small carnivore evolution, ecology, behaviour, and conservation biology. While several recent books have focused on restricted taxa of small carnivores (Harrison *et al.*, 2004; Santos-Reis *et al.*, 2006; Aubry *et al.*, 2012; Proulx & Do Linh San, 2016; Zalewski *et al.*, 2017), this volume deals with a wide range of species belonging to numerous families. The species of interest predominantly include badgers, martens, otters and allies (Mustelidae), civets, genets and allies (Viverridae), mongooses (Herpestidae), raccoons (Procyonidae), and skunks (Mephitidae). Other taxa, i.e. the smaller representatives of the Canidae and Felidae families, are well represented in

whole-taxocenosis studies, meta-analyses and reviews; several of these species have also been covered comprehensively in books edited by Macdonald & Sillero-Zubiri (2004) and Macdonald & Loveridge (2010), respectively. A recent addition to this series – published during the preparation of this book – deals with several aspects of the biology and conservation of musteloids (Macdonald *et al.*, 2017). This is encouraging and further indicates that small carnivores are progressively enjoying the research attention they deserve.

While a wide range of methods are described and used by the numerous authors of this edited book, no section is specifically dedicated to research techniques. Readers seeking greater details, including more technical information on the methods currently used in carnivore research and more broadly in animal ecology, are encouraged to consult books dedicated to these topics, i.e. edited works by Boitani & Fuller (2000), Long *et al.* (2008), O'Connell *et al.* (2011), Boitani & Powell (2012), Silvy (2012a,b), Meek *et al.* (2014), and Rovero & Zimmerman (2016). The books edited by Gittleman (1989, 1996), Gittleman *et al.* (2001), and Goswami & Friscia (2010) also provide complementary information on several aspects of carnivore behaviour, ecology, evolution, and conservation.

The present volume contains a series of broad reviews (nine chapters) on systematics, biogeography, ecomorphology, ecology, and conservation issues, including some meta-analyses. However, due to the paucity of data available for several disciplines, comprehensive reviews of all topics were not always possible. Instead, some contributors were invited to write chapters summarizing their research work on the autecology, interspecific interactions, and/or conservation of selected small carnivore species (three chapters). Finally, as mentioned above, the field of small carnivore research is just stepping out of its infancy and, thus, the future will continue to be one of fundamental research. Hence, the book also contains 10 original research papers that will bring new insights into a wide range of disciplines, including morphological studies, movement ecology, disease dynamics, and interspecific interactions.

Following the present **introductory chapter (Part I)**, the core of the book is divided into four main parts. Contributions were allocated to one of four sections, even though some chapters deal with a broad range of topics and could have been classified in any of two or even three sections. For example, ecological studies presented here often had important implications for management and conservation, especially those carried out in human-dominated, and hence, modified or degraded landscapes.

Part II consists of five chapters focusing on **evolution, systematics, and distribution**. These topics are in themselves important in terms of general biology and often carry implications for the conservation of biodiversity. Recent molecular phylogenetic approaches have overwritten the traditional taxonomy based on phenotypic characters (e.g. morphology) and have detected many homoplasious convergent evolutions (Springer *et al.*, 2004). Resolving the taxonomic issue and setting the unit are required as a premise in the field of conservation biology (Frankham *et al.*, 2010). Molecular data have also greatly improved our understanding of how the clade genesis is correlated with the Earth's environmental changes (Condamine *et al.*, 2013) or how indices for conservation prioritization could be obtained (e.g. EDGE; Isaac *et al.*, 2007).

The first two chapters of this section deal with taxonomic, evolutionary, and conservation issues by focusing on molecular phylogenetics. In Chapter 2, Jun J. Sato and Mieczyslaw Wolsan review the molecular phylogenetic studies on the long-standing evolutionary puzzle presented by the taxonomic position of the red panda and describe the promising approaches that have reliably elucidated caniform carnivoran phylogeny. They further provide an up-to-date subfamily classification of the family Mustelidae (weasels, martens, otters, badgers, and allies; Caniformia), and show a correlation between lineage diversification and global climatic change. Their review also suggests that the priority for species conservation should be inferred from the supermatrix-based phylogenetic tree instead of the supertree that has commonly been used. In Chapter 3, Géraldine Veron, Marie-Lilith Patou, and Andrew P. Jennings review the molecular systematics of the family Herpestidae (mongooses and allies; Feliformia) as well as the Malagasy 'mongooses' currently classified into Eupleridae, and provide a novel taxonomic framework especially for the Asian mongooses (the genus *Urva*) based on a reliable molecular phylogeny. Using this phylogeny, they also discuss the evolution of life-history traits such as sociality within

this family and conclude that the evolution of sociality occurred once during the Late Miocene to Pliocene in Africa, with more open habitats likely favouring group living.

The following chapters adopt rapidly developing statistical approaches, species distribution modelling (SDM; Svenning *et al.*, 2011), ecomorphology (Werdelin & Wesley-Hunt, 2010), and/or ecometrics (Polly *et al.*, 2011) analyses, combined with environmental variables available in the WorldClim database (Hijmans *et al.*, 2005). These methodologies are useful to assess the evolutionary responses (e.g. adaptation) to the biotic and/or abiotic environmental changes. In particular, adaptation should be considered in defining how to conserve populations or species (Funk *et al.*, 2012). In Chapter 4, Robert S. Sommer and Jennifer J. Crees review the subfossil evidence of the past distributional history of small carnivores in Europe during the Late Pleistocene, and discuss how climatic changes and human activity influenced their distribution. They propose species-specific refugia and post-glacial recolonization patterns for small carnivores in Europe. The inference of the past distribution based on SDM suggests that, in contrast to the case of large carnivores, small carnivores were resilient to human impacts during the Holocene. In Chapter 5, Carlo Meloro examines the ecomorphology of small carnivore assemblages in six independent terrestrial ecosystems and shows that tropical guilds in Gunun Lensang, La Amistad, and Yasuni assemblages possess a higher diversity in mandible shape, while guilds in higher latitudes (Krokonose, Kruger, and Yellowstone) exhibit a shortage of some parts in the mandible morphospace. His ecometrics analysis also shows that the amount of lagomorphs and precipitation in the focal study sites are correlated with a higher morphological disparity, hence highlighting the likely importance of biotic and abiotic environmental changes on the historical community assembly of small carnivore guilds. In Chapter 6, Adam W. Ferguson, Richard E. Strauss and Robert C. Dowler investigate coat-colour variations in the North American hog-nosed skunk, *Conepatus leuconotus*. They present a novel statistical technique to assess quantitative morphological variations and clarify that the amount of the dorsal whiteness in this skunk species is larger in individuals living in open and dry habitats in northwestern parts of the distribution range (Arizona, Colorado, New Mexico, and northern Mexico), while individuals with less dorsal whiteness (or more blackness) are restricted to lower latitude areas, namely the Gulf of Mexico and Central America. Their statistical ecometrics analyses suggest that canopy cover, ground surface moisture, and/or ambient temperature might explain the observed coat-colour variations. Their results are consistent with the intermediate aposematism hypothesis, which posits that there is a selective balance between conspicuousness and crypsis in aposematic animals.

Part III comprises six chapters focusing on **ecology**, **behaviour, and – peripherally – diseases** through contributions that are largely based on or derived from data collected in the field. As alluded previously, basic natural history data are still missing for a majority of small carnivore species, thus, field studies are paramount. In addition to providing baseline biological and ecological information, field studies contribute empirical data to laboratory and computer (modelling) analyses.

A large number of mammals (see review table in Dröscher & Kappeler, 2014), including carnivores, use latrines which are believed to play an important role in olfactory communication. In Chapter 7, Christina D. Buesching and Neil R. Jordan review the available literature and show that detailed knowledge of specific information exchange is still lacking for most carnivores. They present an innovative and well-thought-out research framework to test hypotheses about the function of latrines in carnivores and other species. Relying on their extensive field experience, they then review current knowledge on latrine use in three small carnivore species and test some of the hypotheses put forward to explain the function(s) of latrines. The authors highlight that a combined approach (namely spatial, temporal, individual-based, and signal-content-related) is needed to clarify the function(s) of latrines in different species.

In the UK and Ireland, the European badger makes extensive use of latrines. This species is also a known reservoir and vector of bovine tuberculosis (Gallagher & Clifton-Hadley, 2000). Bovine tuberculosis is an infectious disease that not only causes animal health issues but also economic and logistical hardship to livestock farmers (McCulloch & Reiss, 2017). Extensive knowledge of badger ecology and behaviour is

therefore vital in this context. In Chapter 8, Andrew W. Byrne, James O'Keeffe, and S. Wayne Martin study the movement patterns of European badgers in Ireland based on an extensive mark–recapture dataset collected at a large spatial scale. Their analytical models suggest that density-dependent mechanisms affect badger movement patterns within subpopulations. Movements are generally shorter and less frequent in higher-density than lower-density areas. However, there was no net tendency for badgers to move into higher- or lower-density subpopulations. The authors discuss the implications of these findings for understanding the dynamics of bovine tuberculosis across badger populations in rural areas.

Human activities taking place in agricultural landscapes globally are expected to have some direct and indirect effects on a majority of organisms. In Chapter 9, Diego F. Castillo and Mauro Lucherini investigate the behavioural adaptations of Molina's hog-nosed skunk, *Conepatus chinga*, to human-modified landscapes in Argentina. These authors found considerable behavioural plasticity by skunks that facilitated the species' persistence in their agricultural study site, but conclude that the loss of grasslands through conversion to agricultural land is likely decreasing skunk abundance across the Pampas.

In extreme cases, some highly adaptive small carnivore species have been able to successfully colonize urban areas thanks in part to their eclectic diet (for a recent review, see Gehrt *et al.*, 2010). Due to the high abundance of certain food resources in urban areas, some species may reach higher densities than in their natural habitats, hence becoming synurbic, *sensu* Francis & Chadwick (2012). In Chapter 10, Jan Herr and Timothy J. Roper used radio-telemetry to investigate the activity rhythms of the stone marten, *Martes foina*, in urban areas, and question whether these may be affected by anthropogenic factors. Interestingly, their results strongly suggest that behavioural adaptation in urbanized habitats occurs primarily by temporal avoidance, rather than by tolerance, of human activity.

Other small carnivore species such as the European mink, *Mustela lutreola*, which is listed as Critically Endangered by IUCN (2021), may face more challenging scenarios especially when infectious diseases are among their mortality factors. In Chapter 11, Christine

Fournier-Chambrillon and several co-workers monitored a mink population over nine years in Navarre (Spain) and recorded a population crash concordant with a canine distemper virus outbreak. The population seems to have recovered slowly, but the authors advocate that strong conservation measures for this and all other relictual nuclei populations of the western European mink are urgently needed.

As noted previously, camera-trapping is now a widespread and effective sampling technique in wildlife research (McCallum, 2012; Wearn & Glover-Kapfer, 2019). However, the use of camera-traps requires proper planning and survey design (Ancrenaz et al., 2012; Wearn & Glover-Kapfer, 2017; Kays et al., 2020). Camera-traps are attractive because they allow researchers to detect or confirm the presence of species in targeted areas and determine species richness, diversity, and community structure (Ahumada et al., 2011; several chapters in Meek et al., 2014). They also allow to obtain abundance and density estimates (Carbone et al., 2001; Nakashima et al., 2017), although some concerns were raised about the accuracy of such estimates when the field study design violates key assumptions (Foster & Harmsen, 2012; Sollmann et al., 2013; Tobler & Powell, 2013; Rogan et al., 2019). The use of camera traps also enable to gather preliminary information on activity patterns of target species (Rowcliffe et al., 2014; Frey et al., 2017; Botts et al., 2020) and evaluate their spatial occupancy, distribution, habitat use and daily movements and ranges (e.g. several chapters in O'Connell et al., 2011; Allen et al., 2018; Dechner et al., 2018; Palencia et al., 2019). Lastly, two- or multi-species occupancy models and temporal overlap analyses allow to infer competition or coexistence (Gerber et al., 2012; Haidir et al., 2018; Li et al., 2019; Santos et al., 2019; Mills et al., 2019), as well as predator–prey and other interactions (Cusack et al., 2016; Pudyatmoko, 2019; Dias et al., 2019; Vilella et al., 2020). This technique is particularly applicable to understudied species that are relatively small, nocturnal, elusive, and/or live in dense cover. For example, little is known of the ecology of the African civet, *Civettictis civetta*, including its abundance and density in different habitat types. Lisa Isaacs, Michael J. Somers, and Lourens H. Swanepoel (Chapter 12) provide the first density estimates of the largest African viverrid across a landscape gradient, with densities

reaching as high as 13 individuals/km^2. The authors highlight the value of using 'by-catch' (camera-trap) data obtained during non-invasive surveys of large charismatic species for understanding aspects of small carnivore ecology. More importantly, they emphasize the need to understand small carnivore abundances and the factors that influence them, in particular, guild-level effects and the role of anthropogenic drivers.

Part IV contains seven chapters that focus on **interspecific interactions and community ecology in relation to management and conservation**. In Chapter 13, Zach J. Farris and his colleagues used camera-trapping and a new spatio-temporal technique to demonstrate that the spotted fanaloka, *Fossa fossana*, a native and Vulnerable (*sensu* IUCN, 2021) Malagasy carnivore, is likely excluded from degraded forests by the introduced small Indian civet, *Viverricula indica*. The authors propose management options and correctly highlight that their novel analytical approach has wide-ranging applications for conservation managers working to address the negative impacts of introduced species on indigenous wildlife. Luigi Remonti, Aritz Ruiz-González, and Alessandro Balestrieri (Chapter 14) similarly explore the effects of interspecific competition and habitat constraints on species' distribution ranges and abundance, here focusing on two indigenous nocturnal small carnivores in northern Italy. Using a wide range of methods, these authors convincingly demonstrate that the pine marten, *Martes martes*, which is normally associated with mature coniferous and mixed forests, has experienced a southward range expansion reaching the intensively cultivated Po-Venetian Plain. Colonization took place by using the best-conserved riparian forests as corridors; further expansion southward is constrained only by the River Po. Genetic analyses revealed that this expansion was concomitant with a range contraction of the similar-sized stone marten, probably due to interspecific food competition, as inferred from dietary analyses.

Competition among small carnivores may also take place in space and time. Using remote-camera surveys, Yoshihiro Nakashima and co-workers (Chapter 15) show that most pairs of small carnivores in Moukalaba–Doudou National Park (Gabon) differ in either habitat use or time of activity, which may promote their coexistence in the rainforest across this region. They propose that the relative proportion of small carnivores over space and time likely reflects the degree of degradation of the forest. Hence, maintenance of habitat heterogeneity may be important for the conservation of these species. A long-term camera-trapping programme should prove a very useful monitoring tool. In Chapter 16, Aaron M. Haines, Lon I. Grassman Jr, and Michael E. Tewes explore niche separation through resource partitioning in a carnivore community in north-central Thailand. Using radio-telemetry data, they investigated spatial overlap between individuals from six sympatric carnivore species and correlate this overlap with variables reflecting differences in morphology, habitat use, and activity patterns. Results show that species with greater differences in carnassial tooth length tend to avoid each other spatially, whereas species that exhibit different activity patterns in closed habitat cover display a higher spatial overlap. The study therefore suggests that smaller-toothed carnivores may try to avoid antagonistic confrontations with larger-toothed carnivores. Interspecific competitive killing and intraguild predation (involving prey consumption) are believed to be important factors shaping the spatio-temporal distribution of species in carnivore communities (Palomares & Caro, 1999; Donadio & Buskirk, 2006). With this context in mind, Colleen M. Begg and colleagues (Chapter 17) examined the interactions between honey badgers, *Mellivora capensis*, and other predators in the Kalahari semi-desert, South Africa. During nearly 6000 hours of monitoring and field observations of radio-tagged animals, the authors recorded antagonistic interactions between honey badgers and 12 other carnivore species. Intraguild predation (or attempted) was common, and honey badgers acted both as predators and prey. In addition, foraging associations between honey badgers and seven other predator species (two mammals, five birds) were recorded, mainly commensalist or producer–scrounger interactions with black-backed jackals, *Canis mesomelas*, and pale chanting goshawks, *Melierax canorus*. These fascinating results call for more observation-based research of interspecific interactions at the community-wide scale. Although labour-intensive and often physically demanding, observational studies constitute a much-needed complement to other field and laboratory techniques, and, in many cases, are the only method capable of obtaining specific information (e.g. detailed interspecific interactions).

As noted earlier, not all small carnivores are strictly carnivorous. Some species feed seasonally or exclusively on fruit rather than on animal matter. In Chapter 18, Yoshihiro Nakashima and Emmanuel Do Linh San review the published literature on frugivory and seed dispersal in small carnivores. The authors show that small carnivores play a crucial role as seed dispersers by transporting seeds beyond the crown of the mother plant and subsequently defecating viable seeds. While faeces (and seeds) are often deposited in areas which may be unfavourable for plant growth, in fragmented landscapes, frugivory coupled with long-distance seed dispersal may promote vegetation recovery and enhance plant-genetic diversity.

The fourth section of the book closes with an extensive account by Andrew P. Jennings and Géraldine Veron (Chapter 19) who synthesize the results of their research work on the ecology and distribution of Southeast Asian civets and mongooses. The authors first report on the spatio-temporal ecology of Malay civets, *Viverra tangalunga*, and short-tailed mongooses, *Urva brachyura*, in Sulawesi and/or Peninsular Malaysia, as determined through radio-telemetry. They then report on the predicted distribution of 10 small carnivore species as estimated through ecological niche modelling. Lastly, they present data from their camera-trapping project on Sumatra, where they detected only three small carnivore species. The Malay civet seems to be the most affected by the extensive oil palm plantations. In their discussion, the authors suggest that forest-dependent civet and mongoose species may be threatened by human activities such as forest exploitation (leading to habitat loss, degradation, and fragmentation), hunting, and wildlife trade.

Exploring this topic, **Part V** consists of four chapters focusing on **interactions between small carnivores and people, as well as on conservation issues for small carnivores, their prey, and possible competitors**. In their review on carnivore reintroductions, Breitenmoser *et al.* (2001) listed 12 small carnivore species, including the endangered black-footed ferret, *Mustela nigripes*, which have been reintroduced in an attempt to re-establish populations within their historical range. In other cases, translocations have been used to supplement threatened populations (see Bricker *et al.*, Chapter 23, this volume). This contrasts with the numerous cases of introductions of non-native carnivore species, either domestic animals that have turned feral, animals that have escaped from captivity, animals which have been accidentally transported, or (most frequently) animals that were deliberately released (Boitani, 2001). In Chapter 20, Mariela G. Gantchoff, Nathan S. Libal, and Jerrold L. Belant summarized information on small carnivore introductions worldwide and assessed, irrespective of the cause of introduction, why some have succeeded, yet others failed. They identified 253 documented introduction events of 24 species from five families (out of the nine dealt with by the IUCN SSC Small Carnivore Specialist Group), and show that introduction success is dependent on the absence of congeners and preadaptation to the climatic conditions that prevail in the introduction area, a larger body size coupled with smaller litter size, and a carnivorous diet. Interestingly, the number of individuals introduced did not determine establishment success. As biological invasions become increasingly widespread (Seebens *et al.*, 2018), understanding the biological and environmental factors affecting introduction success is important for conservation and management. Indeed, the numerous introductions of small carnivores worldwide sometimes resulted in adverse effects on threatened and endangered species. Michael V. Cove and Alan O'Connell (Chapter 21) highlight these issues and summarize the effects which have included population reductions and species extinctions, in some instances, with effects that cascade throughout the entire ecosystems. Improved methodologies to understand the relative effects of small carnivores on threatened species will help conservationists develop management strategies that target the primary drivers of species declines.

The importance of small carnivores to humans is dependent on diverse subsistence, social, cultural, economic, and religious values. Tim L. Hiller and Stephen M. Vantassel (Chapter 22) synthesize extensive and diverse literature to highlight the roles of these species in human societies. From early use as subsistence foods to current global marketing of species in the fur trade and concerns from overexploitation as bushmeat, use of small carnivores by humans is complex and widespread. Effective regulation of legal and illegal uses at national and international levels is necessary to ensure the conservation of many species. As an example, the distribution and abundance of North American river otter, *Lontra canadensis*, has been adversely affected by human persecution and habitat degradation. Through telephone and email surveys, Emily A. Bricker and her collaborators (Chapter 23) documented a remarkable recovery of this species, now occupying at least portions of its range in every US state and Canadian province but one, with stable or increasing populations. This recovery was due in part to expansions of extant populations augmented by numerous reintroductions. However, because of vulnerability to overharvest, the authors recommend that conservation strategies include effective harvest management and field monitoring of the species distribution and status to identify threats that may adversely affect their long-term recovery. In an addendum, a subset of the authors also express concern about the rationale for the extensive expansion in the trapping of river otters that has ensued completion of reintroduction projects.

The book closes with an **Appendix section (Part VI)**. For readers interested in particular species, Appendix A lists the 232 species of small carnivores considered by IUCN (2021) and informs on chapters in which each species is dealt with or mentioned in this book. The Appendix also provides alternative scientific and common names (when relevant), average body mass, global distribution, number of countries where species are native or have been introduced, as well as information on conservation status and population trends. More details on the biology, ecology, and conservation status of each species can be found in Wilson & Mittermeier (2009), Hunter & Barrett (2018), and IUCN (2021). Appendix B provides a list of the main small carnivoran taxa that have been the subject of discussions as to whether they should be attributed species or subspecies level. Although detailed, the list is likely not exhaustive considering the large number of subspecies of small carnivores described and considered to date.

In summary, this compendium presents contributions based on a wide range of techniques, which we suggest realistically reflect the state of our current knowledge on the world's small carnivores. Further, it illustrates the varied forms of scientific outputs (reviews, meta-analyses, project syntheses, and original research papers) that current and future small carnivore researchers may be expected to encounter and produce throughout their careers. Many of the world's leading and upcoming small carnivore biologists contributed a chapter, and we hope that this book will become an important reference for all researchers, students, and wildlife professionals working on evolution (including genetics, phylogeography, taxonomy, and systematics), ecology, behaviour, and conservation of the world's small carnivores. Although this volume suggests that a substantial amount of knowledge has been accumulated on at least a subset of small carnivore species, the field of small carnivore science remains wide open. We therefore hope that this book will stimulate much-needed research globally and lead to exciting findings on these fascinating creatures.

People working with large carnivores and mammals, in general, may also find this resource of value, as most of the widely tested or cutting-edge techniques described here can be successfully applied to other species.

Acknowledgements

We are grateful to Len de Beer, Jacob Dingel, Nick Garbutt, Pascal Gérold, Steven Jansen, David Mills, Johannes Pfleiderer, and Alex Sliwa for graciously supplying animal photographs. Zoe Brooke and Angela Glatston kindly provided complementary information on the research efforts on carnivores and the conservation status of the red panda, respectively. Finally, we are indebted to Adam Ferguson, Gilbert Proulx, and Paula White for their constructive comments on a draft of this introductory chapter.

References

Aadrean, A., Kanchanasaka, B., Heng, S., Reza Lubis, I., de Silva, P. & Olsson, A. (2015) *Lutra sumatrana. The IUCN Red List of Threatened Species* 2015, e.T12421A21936999. https://dx.doi.org/10.2305/IUCN.UK.2015-2.RLTS.T12421A21936999.en. Accessed on 22 March 2019.

Ahumada, J.A., Silva, C.E.F., Gajapersad, K., Hallam, C., Hurtado, J. *et al.* (2011) Community structure and diversity of tropical forest mammals: data from a global camera trap network. *Philosophical Transactions of the Royal Society B: Biological Sciences* 366, 2703–2711.

Allaby, M. (ed.) (2009) *Dictionary of Zoology*. 3rd edition. Oxford University Press, Oxford.

Allen, M.L., Farmer, M.J., Clare, J.D.J., Olson, E.R., Van Stappen, J. & Van Deelen, T.R. (2018) Is there anybody out there? Occupancy of the carnivore guild in a temperate archipelago. *Community Ecology* 19, 272–280.

Alston, J.M, Maitland, B.M., Brito, B.T., Esmaeili, S., Ford, A.T., Hays, B., Jesmer, B.R., Molina, F.J. & Goleen, J.R. (2019) Reciprocity in restoration ecology: when might large carnivore reintroduction restore ecosystems? *Biological Conservation* 234, 82–89.

Alvares, F., Bogdanowicz, W., Campbell, L.A.D., Godinho, R., Hatlauf, J., Jhala, Y.V., Kitchener, A., Koepfli, K., Krofel, M., Senn, H., Sillero-Zubiri, C., Viranta, S. & Werhahn, G. (2019) *Old World* Canis *spp. with Taxonomic Ambiguity: Workshop Conclusions and Recommendations*. CIBIO, Vairão, Portugal, May 2019. http://www.canids.org/Old_world_canis_taxonomy_workshop.pdf. Accessed on 4 April 2020.

Ancrenaz, M., Hearn, A.J., Ross, J., Sollmann, R. & Wilting, A. (2012) *Handbook for Wildlife Monitoring Using Camera-Traps*. BBEC II Secretariat, Kota Kinabalu.

Aubry, K.B., Zielinski, W.J., Raphael, M.G., Proulx, G. & Buskirk, S.W. (2012) *Biology and Conservation of Martens, Sables and Fishers: A New Synthesis*. Cornell University Press, Ithaca.

Bauer, H., Chapron, G., Nowell, K., Henschel, P., Funston, P., Hunter, L.T.B., Macdonald, D.W. & Packer, C. (2015) Lion (*Panthera leo*) populations are declining rapidly across Africa except in intensively managed areas. *Proceedings of the National Academy of Sciences of the United States of America* 112, 14894–14899.

Belant, J., Biggins, D., Garelle, D., Griebel, R.G. & Hughes, J.P. (2015) *Mustela nigripes. The IUCN Red List of Threatened Species* 2015, e.T14020A45200314. https://dx.doi.org/10.2305/IUCN.UK.2015-4.RLTS. T14020A45200314.en. Accessed on 22 March 2019.

Boitani, L. (2001) Carnivore introductions and invasions: their success and management options. In: *Carnivore Conservation* (eds J.L. Gittleman, S.M. Funk, D. Macdonald & R.K. Wayne), pp. 123–144. Cambridge University Press, Cambridge.

Boitani, L. & Fuller, T.K. (2000) *Research Techniques in Animal Ecology: Controversies and Consequences.* Oxford University Press, New York.

Boitani, L. & Powell, R.A. (2012) *Carnivore Ecology and Conservation: A Handbook of Techniques.* Oxford University Press, New York.

Botts, R.T., Eppert, A.A., Wiegman, T.J., Rodriguez, A., Blankenship, S.R., Asselin, E.M., Garley, W.M., Wagner, A.P., Ullrich, S.E., Allen, G.R. & Mooring, M.S. (2020) Circadian activity patterns of mammalian predators and prey in Costa Rica. *Journal of Mammalogy* 101, 1313–1331.

Breitenmoser, U., Breintenmoser-Würsten, C., Carbyn, L.N. & Funk, S.M. (2001) Assessment of carnivore reintroductions. In: *Carnivore Conservation* (eds J.L. Gittleman, S.M. Funk, D. Macdonald & R.K. Wayne), pp. 241–281. Cambridge University Press, Cambridge.

Bricker, E.A., Serfass, T.L., Hanley, Z.L., Stevens, S.S., Pearce, K.J. & Bohrman, J.A. (Chapter 23, this volume) Conservation status of the North American river otter in the United States and Canada: assessing management practices and public perceptions of the species. In: *Small Carnivores: Evolution, Ecology, Behaviour, and Conservation* (eds E. Do Linh San, J.J. Sato, J.L. Belant & M.J. Somers). Wiley–Blackwell, Oxford.

Brooke, Z.M., Bielby, J., Nambiar, K. & Carbone, C. (2014) Correlates of research effort in carnivores: body size, range size and diet matter. *PLoS One* 9, e93195.

Buskirk, S.W. (1999) Mesocarnivores of Yellowstone. In: *Carnivores in Ecosystems: The Yellowstone Experience* (eds T.W. Clark, A.P. Curlee, S.C. Minta & P.M. Kareiva), pp. 167–187. Yale University Press, New Haven.

Carbone, C., Mace, G.M., Roberts, S.C. & Macdonald, D.W. (1999) Energetic constraints on the diet of terrestrial carnivores. *Nature* 402, 286–288.

Carbone, C., Christie, S., Conforti, K., Coulson, T., Ginsberg, J.R. *et al.* (2001) The use of photographic rates to estimate densities of tigers and other cryptic mammals. *Animal Conservation* 4, 75–79.

Carbone, C., Teacher, A. & Rowcliffe, J.M. (2007) The costs of carnivory. *PLoS Biology* 5, e22.

Ceballos, G., García, A. & Ehrlich, P.R. (2010) The sixth extinction crisis: loss of animal population and species. *Journal of Cosmology* 8, 1821–1831.

Ceballos, G., Ehrlich, P.R., Barnosky, A.D., García, A., Pringle, R.M. & Palmer, T.M. (2015) Accelerated modern human-induced species losses: entering the sixth mass extinction. *Science Advances* 1, e1400253.

Ceballos, G., Ehrlich, P.R. & Dirzo, R. (2017) Biological annihilation via the ongoing sixth mass extinction signalled by vertebrate population losses and declines. *Proceedings of the National Academy of Sciences of the United States of America* 114(30), E6089–E6096.

Chapron, G., Kaczensky, P., Linnell, J.D.C., Von Arx, M., Huber, D. *et al.* (2014) Recovery of large carnivores in Europe's modern human-dominated landscapes. *Science* 346, 1517–1519.

Colman, N.J., Gordon, C.E., Crowther, M.S. & Letnic, M. (2014) Lethal control of an apex predator has unintended cascading effects on forest mammal assemblages. *Proceedings of the Royal Society B: Biological Sciences* 281, 20133094.

Condamine, F.L., Rolland, J. & Morlon, H. (2013) Macroevolutionary perspectives to environmental change. *Ecology Letters* 16, 72–85.

Cossíos, E.D., Walker, R.S., Lucherini, M., Ruiz-García, M. & Angers, B. (2012) Population structure and conservation of a high-altitude specialist, the Andean cat *Leopardus jacobita. Endangered Species Research* 16, 283–294.

Courchamp, F., Langlais, M. & Sugihara, J. (1999) Cats protecting birds: modelling the mesopredator release effect. *Journal of Animal Ecology* 68, 282–292.

Cove, M.V. & O'Connell, A.F. (Chapter 21, this volume) Global review of the effects of small carnivores on threatened species. In: *Small Carnivores: Evolution, Ecology, Behaviour, and Conservation* (eds E. Do Linh San, J.J. Sato, J.L. Belant & M.J. Somers). Wiley–Blackwell, Oxford.

Crooks, K.R. & Soulé, M.A. (1999) Mesopredator release and avifaunal extinctions in a fragmented system. *Nature* 400, 563–566.

Cuarón, A.D., Martínez-Morales, M.A., Mcfadden, K.W., Valenzuela, D. & Gompper, M.E. (2004) The status of dwarf carnivores on Cozumel Island, Mexico. *Biodiversity & Conservation* 13, 317–331.

Cuarón, A.D., de Grammont, P.C. & McFadden, K. (2016) *Procyon pygmaeus. The IUCN Red List of Threatened Species* 2016, e.T18267A45201913. https://dx.doi.org/10.2305/IUCN.UK.2016-1.RLTS.T18267A45201913.en. Accessed on 22 March 2019.

Cusack, J.J., Dickman, A.J., Kalyahe, M., Rowcliffe, J.M., Carbone, C., MacDonald, D.W. & Coulson, T. (2016) Revealing kleptoparasitic and predatory tendencies in an African mammal community using camera traps: a comparison of spatiotemporal approaches. *Oikos* 126, 812–822.

da Silva Santos, A., Trigo, T.C., de Oliveira, T.G., Silveira, L. & Eizirik, E. (2018) Phylogeographic analyses of the pampas cat (*Leopardus colocola*; Carnivora, Felidae) reveal a complex demographic history. *Genetics and Molecular Biology* 41, 273–287.

Davis, C.L., Rich, L.N., Farris, Z.J., Kelly, M.J., Di Bitteti, M.S. *et al.* (2019) Ecological correlates of the spatial co-occurrence of sympatric mammalian carnivores worldwide. *Ecology Letters* 21, 1401–1412.

Dechner, A., Flesher, K.M., Lindell, C., Veiga de Oliveira, T. & Maurer, B.A. (2018) Determining carnivore habitat use in a rubber/forest landscape using multispecies occupancy models. *PLoS One* 13, e0195311.

De Cuyper, A., Clauss, M., Carbone, C., Codron, D., Cools, A., Hesta, M. & Janssens, G.P.J. (2019) Predator size and prey size-gut capacity ratios determine kill frequency and carcass production in terrestrial carnivorous mammals. *Oikos* 128, 13–22.

D'Haen, M. (2017) A confirmed sighting of Pousargues's mongoose *Dologale dybowskii* in Garamba National Park, Democratic Republic of the Congo. *Small Carnivore Conservation* 55, 69–72.

Dias, D.M., Massara, R.L., de Campos, C.B. & Rodrigues, F.H.G. (2019) Feline predator–prey relationships in a semi-arid biome in Brazil. *Journal of Zoology* 307, 282–291.

Di Minin, E., Slotow, R., Hunter, L.T.B., Montesino Pouzols, F., Toivonen, T., Verburg, P.H., Leader-Williams, N., Petracca, L. & Moilanen, A. (2016) Global priorities for national carnivore conservation under land use change. *Scientific Reports* 6, 23814.

Do Linh San, E. & Somers, M.J. (2013) Editorial – African small carnivores: the 'forgotten Eden'. *Small Carnivore Conservation* 48, 1–2.

Donadio, E. & Buskirk, S.W. (2006) Diet, morphology, and interspecific killing in carnivora. *American Naturalist* 167, 524–536.

Dröscher, I. & Kappeler, P.M. (2014) Maintenance of familiarity and social bonding via communal latrine use in a solitary primate (*Lepilemur leucopus*). *Behavioral Ecology and Sociobiology* 68, 2043–2058.

Durant, S.M., Mitchell, N., Groom, R., Petorelli, N., Ipavec, A. *et al.* (2017) The global decline of cheetah *Acinony jubatus* and what it means for conservation. *Proceedings of the National Academy of Sciences of the United States of America* 114, 528–533.

Durbin, J., Funk, S.M., Hawkins, F., Hills, D.M., Jenkins, P.D., Moncrieff, C.B. & Ralainasolo, F.B. (2010) Investigations into the status of a new taxon of *Salanoia* (Mammalia: Carnivora: Eupleridae) from the marshes of Lac Alaotra, Madagascar. *Systematics and Biodiversity* 8, 341–355.

Eizirik, E., Murphy, W.J., Koepfli, K.-P., Johnson, W.E., Dragoo, J.W., Wayne, R.K. & O'Brien, S.J. (2010) Pattern and timing of diversification of the mammalian Order Carnivora inferred from multiple nuclear gene sequences. *Molecular Phylogenetics and Evolution* 56, 49–63.

Ewer, R.F. (1973) The Carnivores. Weidenfeld and Nicolson, London.

Farris, Z.J., Kelly, M.J., Karpanty, S., Murphy, A., Ratelolahy, F., Andrianjakarivelo, V. & Holmes, C. (2017) The times they are a changin': multi-year surveys reveal exotics replace native carnivores at a Madagascar rainforest site. *Biological Conservation* 206, 320–328.

Ferguson, A.W., Houssein, R.R. & McDonough, M.M. (2019) Noteworthy record of the Ethiopian genet, *Genetta abyssinica*, (Carnivora, Viverridae) from Djibouti informs its phylogenetic position within *Genetta* (Carnivora: Herpestidae). *Mammalia* 83, 180–189.

Fleming, P.J., Nolan, H., Jackson, S.M., Ballard, G.A., Bengsen, A., Brown, W.Y., Meek, P.D., Mifsud, G., Pal, S.K. & Sparkes, J. (2017) Roles for the Canidae in food webs reviewed: where do they fit? *Food Webs* 12, 14–34.

Flynn, J.J., Finarelli, J.A., Zehr, S., Hsu, J. & Nedbal, M.A. (2005) Molecular phylogeny of the Carnivora (Mammalia): assessing the impact of increased sampling on resolving enigmatic relationships. *Systematic Biology* 54, 317–337.

Foster, R.J. & Harmsen, B.J. (2012) A critique of density estimation from camera-trap data. *The Journal of Wildlife Management* 76, 224–236.

Francis, R.A & Chadwick, M.A. (2012) What makes a species synurbic? *Applied Geography* 32, 514–521.

Frankham, R., Ballou, J.D. & Briscoe, D.A. (2010) *Introduction to Conservation Genetics*. 2nd edition. Cambridge University Press, Cambridge.

Frey, S., Fisher, J.T., Burton, A.C. & Volpe, J.P. (2017) Investigating animal activity patterns and temporal niche partitioning using camera-trap data: challenges and opportunities. *Remote Sensing in Ecology and Conservation* 3, 123–132.

Fulton, T.L. & Strobeck, C. (2010) Multiple fossil calibrations, nuclear loci and mitochondrial genomes provide new insight into biogeography and divergence timing for true seals (Phocidae, Pinnipedia). *Journal of Biogeography* 37, 814–829.

Funk, W.C., McKay, J.K., Hohenlohe, P.A. & Allendorf, F.W. (2012) Harnessing genomics for delineating conservation units. *Trends in Ecology & Evolution* 27, 489–496.

Gallagher, J. & Clifton-Hadley, R.S. (2000) Tuberculosis in badgers; a review of the disease and its significance for other animals. *Research in Veterinary Science* 69, 203–217.

Gantchoff, M., Libal, N.S. & Belant, J.L. (Chapter 20, this volume) Small carnivore introductions: ecological and biological correlates of success. In: *Small Carnivores: Evolution, Ecology, Behaviour, and Conservation* (eds E. Do Linh San, J.J. Sato, J.L. Belant & M.J. Somers). Wiley–Blackwell, Oxford.

Garshelis, D.L. (2009) Family Ursidae (Bears). In: *Handbook of the Mammals of the World. Volume 1. Carnivores* (eds D.E. Wilson & R.A. Mittermeier), pp. 448–497. Lynx, Barcelona.

Gaubert, P. (2009a) Family Nandiniidae (African palm civet). In: *Handbook of the Mammals of the World. Volume 1. Carnivores* (eds D.E. Wilson & R.A. Mittermeier), pp. 50–53. Lynx, Barcelona.

Gaubert, P. (2009b) Family Prionodontidae (Linsangs). In: *Handbook of the Mammals of the World. Volume 1.*

Carnivores (eds D.E. Wilson & R.A. Mittermeier), pp. 171–173. Lynx, Barcelona.

Gaubert, P. & Veron, G. (2003) Exhaustive sample set among Viverridae reveals the sister-group of felids: the linsangs as a case of extreme morphological convergence within Feliformia. *Proceedings of the Royal Society B: Biological Sciences* 270, 2523–2530.

Gehrt, S.D., Riley, S.P.D. & Cypher, B.L. (eds) (2010) *Urban Carnivores: Ecology, Conflict, and Conservation*. The Johns Hopkins University Press, Baltimore.

Gentry, A., Clutton-Brock, J. & Groves, C.P. 2004. The naming of wild animal species and their domestic derivatives. *Journal of Archaeological Science* 31, 645–651.

Gerber, B.D., Karpanty, S.M. & Randrianantenaina, J. (2012) Activity patterns of carnivores in the rain forests of Madagascar: implications for species coexistence. *Journal of Mammalogy* 93, 667–676.

Gigliotti, L.C., Slotow, R., Hunter, L.T.B., Fattebert, J., Sholto-Douglas, C. & Jachowski, D.S. (2020) Habitat complexity and lifetime predation risk influence mesopredator survival in a multi-predator system. *Scientific Reports* 10, 17841.

Gilchrist, J.S., Jennings, A.P., Veron, G. & Cavallini, P. (2009) Family Herpestidae (Mongooses). In: *Handbook of the Mammals of the World. Volume 1. Carnivores* (eds D.E. Wilson & R.A. Mittermeier), pp. 262–328. Lynx, Barcelona.

Gippoliti, S., Cotterill, F.P.D., Zinner, D. & Groves, C.P. (2018) Impacts of taxonomic inertia for the conservation of African ungulate diversity: an overview. *Biological Reviews* 93, 115–130.

Gittleman, J.L. (1989) *Carnivore Behavior, Ecology, and Evolution. Volume 1*. Cornell University Press, Ithaca.

Gittleman, J.L. (1996) *Carnivore Behavior, Ecology, and Evolution. Volume 2*. Cornell University Press, Ithaca.

Gittleman, J.L., Funk, S.M., Macdonald, D. & Wayne, R.K. (eds) (2001) *Carnivore Conservation*. Cambridge University Press, Cambridge.

Glatston, A., Wei, F., Than Zaw & Sherpa, A. (2015) *Ailurus fulgens* (errata version published in 2017). *The IUCN Red List of Threatened Species* 2015, e. T714A110023718. https://dx.doi.org/10.2305/IUCN. UK.2015-4.RLTS.T714A45195924.en. Accessed on 11 March 2019.

Glen, A.S. & Dickman, C.R. (eds) (2014) *Carnivores of Australia: Past, Present and Future*. CSIRO Publishing, Collingwood.

González-Maya, J.F. & Arias-Alzate, A.A.A. (2016) *Nasuella meridensis. The IUCN Red List of Threatened Species* 2016, e.T72261777A72261787. https://dx.doi.org/10.2305/IUCN.UK.2016-1.RLTS.T72261777A72261787.en. Accessed on 22 March 2019.

González-Maya, J.F. & Ramírez-Chaves, H.E. (2017) Editorial – global small carnivore conservation: geographical distribution of small carnivore research. *Small Carnivore Conservation* 54, 1–2.

Goodman, S.M. (2009) Family Eupleridae (Madagascar carnivores). In: *Handbook of the Mammals of the World. Volume 1. Carnivores* (eds D.E. Wilson & R.A. Mittermeier), pp. 330–351. Lynx, Barcelona.

Gopalakrishnan, S., Sinding, M.S., Ramos-Madrigal, J., Niemann, J., Samaniego Castruita, J.A. *et al.* (2018) Interspecific gene flow shaped the evolution of genus *Canis. Current Biology* 28, 3441–3449.

Goswami, A. (2010) Introduction to Carnivora. In: *Carnivoran Evolution: New Views on Phylogeny, Form and Function* (eds A. Goswami & A. Friscia), pp. 1–24. Cambridge University Press, Cambridge.

Goswami, A. & Friscia, A. (2010) *Carnivoran Evolution: New Views on Phylogeny, Form and Function.* Cambridge University Press, Cambridge.

Gray, T.N., Hughes, A.C., Laurance, W.F., Long, B., Lynam, A.J., O'Kelly, H., Ripple, W.J., Seng, T., Scotson, L. & Wilkinson, N.M. (2018) The wildlife snaring crisis: an insidious and pervasive threat to biodiversity in Southeast Asia. *Biodiversity and Conservation* 27, 1031–1037.

Haidir, I.A., Macdonald, D.W. & Linkie, M. (2018) Assessing the spatiotemporal interactions of mesopredators in Sumatra's tropical rainforest. *PLoS One* 13, e0202876.

Harrison, D.J., Fuller, A.K. & Proulx, G. (eds) (2004) *Martes and Fishers (*Martes*) in Human-Altered Environments: An International Perspective.* Springer, New York.

Hawkins, F. (2015a) *Galidictis grandidieri. The IUCN Red List of Threatened Species* 2015, e.T8834A45198057. https://dx.doi.org/10.2305/IUCN.UK.2015-4.RLTS.T8834A45198057.en. Accessed on 15 March 2019.

Hawkins, F. (2015b) *Mungotictis decemlineata. The IUCN Red List of Threatened Species* 2015, e.T13923A45199764. https://dx.doi.org/10.2305/IUCN.UK.2015-4.RLTS.T13923A45199764.en. Accessed on 15 March 2019.

Hawkins, F. (2016) *Eupleres major. The IUCN Red List of Threatened Species* 2016, e.T39547A45204313. https://dx.doi.org/10.2305/IUCN.UK.2016-1.RLTS.T39547A45204313.en. Accessed on 15 March 2019.

Hayward, M.W. & Somers, M.J. (eds) (2009) *Reintroduction of Top-Order Predators.* Wiley–Blackwell, Oxford.

Hearn, A., Brodie, J., Cheyne, S., Loken, B., Ross, J. & Wilting, A. (2016) *Catopuma badia* (errata version published in 2017). *The IUCN Red List of Threatened Species* 2016, e.T4037A112910221. https://dx.doi.org/10.2305/IUCN.UK.2016-1.RLTS.T4037A50650716.en. Accessed on 22 March 2019.

Helgen, K.M., Pinto, C.M., Kays, R., Helgen, L.E., Tsuchiya, M.T.N., Quinn, A., Wilson, D.E. & Maldonado, J.E. (2013) Taxonomic revision of the olingos (*Bassaricyon*), with description of a new species, the Olinguito. *ZooKeys* 324, 1–83.

Hijmans, R.J., Cameron, S.E., Parra, J.L., Jones, P.G. & Jarvis, A. (2005) Very high resolution interpolated climate surfaces for global land areas. *International Journal of Climatology* 25, 1965–1978.

Hoeks, S., Huijbregts, M.A.J., Busana, M., Harfoot, M.B.J., Svenning, J.-C. & Santini, L. (2020). Mechanistic insights into the role of large carnivores for ecosystem structure and functioning. *Ecography* 43, 1752–1763.

Hu, Y., Thapa, A., Fan, H., Ma, T., Wu, Q., Ma, S., Zhang, D., Wang, B., Li, M., Yan, L. & Wei, F. (2020) Genomic evidence for two phylogenetic species and long-term population bottlenecks in red pandas. *Science Advances* 6, eaax5751.

Hunter, L. & Barrett, P. (2011) *A Field Guide to the Carnivores of the World.* 1st edition. New Holland Publishers, London.

Hunter, L. & Barrett, P. (2018) *A Field Guide to the Carnivores of the World.* 2nd edition. Bloomsbury, London.

IUCN [International Union for Conservation of Nature] (2021) *The IUCN Red List of Threatened Species. Version 2021-1.* https://www.iucnredlist.org. Accessed on 24 August 2021.

Isaac, N.J.B., Turvey, S.T., Collen, B., Waterman, C. & Baillie, J.E.M. (2007) Mammals on the EDGE: conservation priorities based on threat and phylogeny. *PLoS One* 2, e296.

Jachowski, D.S., Butler, A., Eng, R.Y.Y., Gigliotti, L., Harris, S. & Williams, A. (2020) Identifying mesopredator release in multi-predator systems: a review of evidence from North America. *Mammal Review* 50, 367–381.

Jacobson, A.P., Gerngross, P., Lemeris, J.R. Jr, Schoonover, R.F., Anco, C. *et al.* (2016) Leopard (*Panthera pardus*) status, distribution, and the research efforts across its range. *PeerJ* 4, e1974.

Jennings, A.P. & Veron, G. (2009) Family Viverridae (Civets, genets and oyans). In: *Handbook of the Mammals of the World. Volume 1. Carnivores* (eds D.E. Wilson & R.A. Mittermeier), pp. 174–232. Lynx, Barcelona.

Jiangzuo, Q. & Flynn, J.J. (2020) The earliest ursine bear demonstrates the origin of plant-dominated omnivory in Carnivora. *iScience* 23, 101235.

Jiménez, J., Nuñez-Arjona, J. C., Mougeot, F., Ferreras, P., González, L.M., García-Domínguez, F., Muñoz-Igualada, J., Jesús Palacios, M., Pla, S., Rueda, C., Villaespesa, F., Nájera, F., Palomares, F. & López-Bao, J.V. (2019) Restoring apex predators can reduce mesopredator abundances. *Biological Conservation* 238, 108234.

Johnson, C.N., Isaac, J.L. & Fisher, D.O. (2007) Rarity of top predator triggers continent-wide collapse of mammal prey: dingoes and marsupials in Australia. *Proceedings of the Royal Society B: Biological Sciences* 274, 341–346.

Jordano, P., Garcia, C., Godoy, J.A. & García-Castaño, J.L. (2007) Differential contribution of frugivores to complex seed dispersal patterns. *Proceedings of the National Academy of Sciences of the United States of America* 104, 3278–3282.

Kamler, J.F., Songsasen, N., Jenks, K., Srivathsa, A., Sheng, L. & Kunkel, K. (2015) *Cuon alpinus. The IUCN Red List of Threatened Species* 2015, e.T5953A72477893. https://dx.doi.org/10.2305/IUCN.UK.2015-4.RLTS.T5953A72477893.en. Accessed on 15 March 2019.

Kays, R. (2009) Family Procyonidae (Raccoons). In: *Handbook of the Mammals of the World. Volume 1. Carnivores* (eds D.E. Wilson & R.A. Mittermeier), pp. 504–530. Lynx, Barcelona.

Kays, R, Arbogast, B.S., Baker-Whatton, M., Beirne, C., Boone, H.M. *et al.* (2020) An empirical evaluation of camera trap study design: how many, how long, and when? *Methods in Ecology and Evolution* 11, 700–713.

Kelly, J.R., Doherty, T.J., Gabel, T. & Disbrow, W. (2019) Large carnivore attacks on humans: the state of knowledge. *Human Ecology Review* 25(2), 15–33.

Kitchener, A.C., Breitenmoser-Würsten, C., Eizirik, E., Gentry, A., Werdelin, L. *et al.* (2017) A revised taxonomy of the Felidae. The final report of the Cat Classification Task Force of the IUCN/SSC Cat Specialist Group. *Cat News Special Issue* 11, 1–80.

Koepfli, K.-P., Jenks, S.M., Eizirik, E., Zahirpour, T., Van Valkenburgh, B. & Wayne, R.K. (2006) Molecular systematics of the Hyaenidae: relationships of a relictual lineage resolved by a molecular supermatrix. *Molecular Phylogenetics and Evolution* 38, 603–620.

Koepfli, K.P., Pollinger, J., Godinho, R., Robinson, J., Lea, A. *et al.* (2015) Genome-wide evidence reveals that African and Eurasian golden jackals are distinct species. *Current Biology* 25, 2158–2165.

Koike, S., Morimoto, H., Goto, Y., Kozakai, C. & Yamazaki, K. (2008) Frugivory of carnivores and seed dispersal of fleshy fruits in cool-temperate deciduous forests. *Journal of Forest Research* 13, 215–222.

Larivière, S. & Jennings, A.P. (2009) Family Mustelidae (Weasels and relatives). In: *Handbook of the Mammals of the World. Volume 1. Carnivores* (eds D.E. Wilson & R.A. Mittermeier), pp. 564–656. Lynx, Barcelona.

Leempoel, K., Meyer, J., Hebert, T., Nova, N. & Hadly, E.A. (2019) Return of an apex predator to a suburban preserve triggers a rapid trophic cascade. *bioRxiv*. https://www.biorxiv.org/content/10.1101/564294v1. Accessed on 12 March 2019.

Li, Z., Wang, T., Smith, J.L.D., Feng, R., Feng, L., Mou, P. & Ge, J. (2019) Coexistence of two sympatric flagship carnivores in the human-dominated forest landscapes of Northeast Asia. *Landscape Ecology* 34, 291–305.

Linnel, J.D. & Jackson, C.R. (2019) Bringing back large carnivores to rewild landscapes. In: *Rewilding* (eds N. Pettorelli, S.M. Durant & J.T. du Toit), pp. 248–279. Cambridge University Press, Cambridge.

Litvaitis, J.A. & Villafuerte, R. (1996) Intraguild predation, mesopredator release, and prey stability. *Conservation Biology* 10, 676–677.

Long, R.A., MacKay, P., Zielinski, W.J. & Ray, J.C. (eds) (2008) *Noninvasive Survey Methods for Carnivores.* Island Press, Washington, DC.

López-Bao, J.V. & González-Varo, J.P. (2011) Frugivory and spatial patterns of seed deposition by carnivorous mammals in anthropogenic landscapes: a multi-scale approach. *PLoS One* 6, e14569.

Louppe, V., Leroy, B., Herrel, A. & Veron, G. (2019) Current and future climatic regions favourable for a globally introduced wild carnivore, the raccoon *Procyon lotor*. *Scientific Reports* 9, 9174.

Louppe, V., Leroy, B., Herrel, A. & Veron, G. (2020a) The globally invasive small Indian mongoose *Urva auropunctata* is likely to spread with climate change. *Scientific Reports* 10, 7461.

Louppe, V., Baron, J., Pons, J.-M. & Veron, G. (2020b) New insights on the geographical origins of the Caribbean raccoons. *Journal of Zoological Systematics and Evolutionary Research* 58, 1303–1322.

Macdonald, D.W. (1992) *The Velvet Claw: A Natural History of the Carnivores*. BBC Books, London.

Macdonald, D.W. & Sillero-Zubiri, C. (eds) (2004) *Biology and Conservation of Wild Canids*. Oxford University Press, New York.

Macdonald, D.W. & Loveridge, A.J. (eds) (2010) *Biology and Conservation of Wild Felids*. Oxford University Press, New York.

Macdonald, D.W., Newman, C. & Harrington, L.A. (eds) (2017) *Biology and Conservation of Musteloids*. Oxford University Press, New York.

Maran, T., Skumatov, D., Gomez, A., Põdra, M., Abramov, A.V. & Dinets, V. (2016) *Mustela lutreola*. *The IUCN Red List of Threatened Species* 2016, e.T14018A45199861. https://dx.doi.org/10.2305/IUCN.UK.2016-1.RLTS.T14018A45199861.en. Accessed on 22 March 2019.

Marino, J. & Sillero-Zubiri, C. (2011) *Canis simensis. The IUCN Red List of Threatened Species* 2011, e.T3748A10051312. https://dx.doi.org/10.2305/IUCN.UK.2011-1.RLTS.T3748A10051312.en. Accessed on 11 March 2019.

Marneweck, C., Marneweck, D.G., van Schalkwyk, O.L., Beverley, G., Davies-Mostert, H.T. & Parker, D.M. (2019) Spatial partitioning by a subordinate carnivore is mediated by conspecific overlap. *Oecologia* 191, 531–540.

Marneweck, C., Butler, A.R., Gigliotti, L.C., Harris, S.N., Jensen, A.J., Muthersbaugh, M., Newman, B.A., Saldo, E.A., Shute, K., Titus, K.L., Yu, S.W. & Jachowski, D.S. (2021) Shining the spotlight on small mammalian carnivores: global status and threats. *Biological Conservation* 255, 109005.

McCallum, J. (2012) Changing use of camera traps in mammalian field research: habitats, taxa and study types. *Mammal Review* 43: 196–206.

McCulloch, S.P. & Reiss, M.J. (2017) Bovine tuberculosis and badger culling in England: a utilitarian analysis of policy options. *Journal of Agricultural Environmental Ethics* 30: 511–533.

McFadden, K.W. & Meiri, S. (2013) Dwarfism in insular carnivores: a case study of the pygmy raccoon. *Journal of Zoology* 289, 213–221.

McFadden, K.W., Gompper, M.E. Valenzuela, D.G. & Morales, J.C. (2008) Evolutionary history of the Critically Endangered Cozumel dwarf carnivores inferred from mitochondrial DNA analyses. *Journal of Zoology* 276, 176–186.

Meek, P., Fleming, P., Ballard, G., Banks, P., Claridge, A. Sanderson, J. & Swann, D. (2014) *Camera Trapping: Wildlife Management and Research*. CSIRO Publishing, Collingwood.

Meredith, R.W., Janečka, J.E., Gatesy, J., Ryder, O.A., Fisher, C.A. *et al.* (2011) Impacts of the Cretaceous Terrestrial Revolution and KPg extinction on mammal diversification. *Science* 334, 521–524.

Mills, D., Do Linh San, E., Robinson, H., Isoke, S., Slotow, R. & Hunter, L. (2019) Competition and specialization in an African forest carnivore community. *Ecology and Evolution* 9, 10092–10108.

Moreira-Arce, D., Ugarte, C.S., Zorondo-Rodríguez, F. & Simonetti, J.A. (2018) Management tools to reduce carnivore–livestock conflicts: current gap and future challenges. *Rangelands Ecology & Management* 71, 389–394.

Moritz, C. (1994) Defining 'Evolutionarily Significant Units' for conservation. *Trends in Ecology & Evolution* 9, 373–375.

Mudappa, D., Helgen, K. & Nandini, R. (2016) *Viverra civettina. The IUCN Red List of Threatened Species* 2016, e.T23036A45202281. https://dx.doi.org/10.2305/IUCN.UK.2016-1.RLTS.T23036A45202281.en. Accessed on 16 March 2019.

Nadler, T., Streicher, U., Stefen, C., Schwierz, E. & Roos, C. (2011) A new species of ferret-badger, genus *Melogale*, from Vietnam. *Der Zoologische Garten N. F.* 80, 271–286.

Nakashima, Y. & Do Linh San, E. (Chapter 18, this volume) Seed dispersal by mesocarnivores: importance and functional uniqueness in a changing world. In: *Small Carnivores: Evolution, Ecology, Behaviour, and Conservation* (eds E. Do Linh San, J.J. Sato, J.L. Belant & M.J. Somers). Wiley−Blackwell, Oxford.

Nakashima, Y., Fukasawa, K. & Samejima, H. (2017) Estimating animal density without individual recognition using information derivable exclusively from camera traps. *Journal of Applied Ecology* 55, 735–744.

Nandini, R. & Mudappa, D. (2010) Mystery or myth: a review of history and conservation status of the Malabar Civet *Viverra civettina* Blyth, 1862. *Small Carnivore Conservation* 43, 47–59.

Newsome, T.M., Greenville, A.C., Ćirović, D., Dickman, C.R., Johnson, C.N., Krofel, M., Letnic, M., Ripple, W.J., Ritchie, E.G., Stoyanov, S. & Wirsing, A.J. (2017) Top predators constrain mesopredator distributions. *Nature Communications* 8, 15469.

Nie, Y., Wei, F., Zhou, W., Hu, Y., Senior, A.M., Wu, Q., Yan, L. & Raubenheimer, D. (2019) Giant pandas are macronutritional carnivores. *Current Biology* 29, 1677–1682.

Nijman, V., Ardiansyah, A., Bergin, D., Birot, H., Brown, E., Langgeng, A., Morcatty, T., Spaan, D., Siriwat, P., Imron, M. & Nekaris, K.A. (2019) Dynamics of illegal wildlife trade in Indonesian markets over two decades, illustrated by trade in Sunda Leopard Cats. *Biodiversity* 20, 27–40.

Nowak, R.M. (2005) *Walker's Carnivores of the World*. The Johns Hopkins University Press, Baltimore.

O'Connell, A.F., Nichols, J.D. & Karanth, K.U. (2011) *Camera Traps in Animal Ecology: Methods and Analyses*. Springer, New York.

Palencia, P., Vicente, J., Barroso, P., Barasona, J.A., Soriguer, R.C. & Acevedo, P. (2019) Estimating day range from camera-trap data: the animals' behaviour as a key parameter. *Journal of Zoology* 309, 182–190.

Palomares, F. & Caro, T.M. (1999) Interspecific killing among mammalian carnivores. *American Naturalist* 153, 492–508.

Palomares, F., Gaona, P., Ferreras, P. & Delibes, M. (1995) Positive effects on game species of top predators by controlling smaller predator populations: an example with lynx, mongooses, and rabbits. *Conservation Biology* 9, 295–305.

Pandey, R. & Papes, M. (2018) Changes in future potential distributions of apex predator and mesopredator mammals in North America. *Regional Environmental Change* 18: 1223–1233.

Penteriani, V., Bombieri, G., Fedriani, J.M., López-Bao, J.V., Garrote, P.J., Russo, L.F. & del Mar Delgado, M. (2017) Humans as prey: coping with large carnivore attacks using a predator–prey interaction perspective. *Human–Wildlife Interactions* 11, 192–207.

Polly, P.D., Eronen, J.T., Fred, M., Dietl, G.P., Mosbrugger, V., Scheidegger, C., Frank, D.C., Damuth, J., Stenseth, N.C. & Fortelius, M. (2011) History matters: ecometrics and integrative climate change biology. *Proceedings of the Royal Society B: Biological Sciences* 278, 1121–1130.

Proulx, G., Abramov, A.V., Adams, I., Jennings, A., Khorozyan, I., Rosalino, L.M., Santos-Reis, M., Veron, G. & Do Linh San, E. (2016) World distribution and status of badgers – a review. In: *Badgers: Systematics, Biology, Conservation and Research Techniques* (eds G. Proulx & E. Do Linh San), pp. 31–116. Alpha Wildlife Publications, Sherwood Park.

Proulx, G. & Do Linh San, E. (eds) (2016) *Badgers: Systematics, Biology, Conservation and Research Techniques*. Alpha Wildlife Publications, Sherwood Park.

Prugh, L.R. & Sivy, K.J. (2020) Enemies with benefits: integrating positive and negative interactions among terrestrial carnivores. *Ecology Letters* 23, 902–918.

Prugh, L.R., Stoner, C.J., Epps, C.W., Bean, W.T., Ripple, W.J., Laliberte, A.S. & Brashares, J.S. (2009) The rise of the mesopredator. *BioScience* 59, 779–791.

Pudyatmoko, S. (2019) Spatiotemporal inter-predator and predator–prey interactions of mammalian species in a tropical savannah and deciduous forest in Indonesia. *Mammal Research* 64, 191–202.

Püttker, T., Crouzeilles, R., Almeida-Gomes, M., Schmoeller, M., Maurenza, D. *et al.* (2020) Indirect effects of habitat loss via habitat fragmentation: a cross-taxa analysis of forest-dependent species. *Biological Conservation* 241, 108368.

Rafiq, K., Hayward, M.W., Wilson, A.M., Meloro, C., Jordan, N.R., Wich, S.A., McNutt, J.W. & Golabek, K.A. (2020) Spatial and temporal overlaps between leopards (*Panthera pardus*) and their competitors in the African large predator guild. *Journal of Zoology* 311, 246–259.

Rajaratnam, R., Vernes, K. & Sanday, T. (2016) A review of livestock predation by large carnivores in the

Himalayan Kingdom of Bhutan. In: *Problematic Wildlife: A Cross-Disciplinary Approach* (ed. F.M. Angelici), pp. 143–174. Springer, New York.

Rasambainarivo, F., Farris, Z.J., Andrianalizah, H. & Parker, P.G. (2017) Interactions between carnivores in Madagascar and the risk of disease transmission. *EcoHealth* 14, 691–703.

Ray, J.C., Redford, K.H., Steneck, R.S. & Berger, J. (eds) (2005) *Large Carnivores and the Conservation of Biodiversity*. Island Press, Washington, DC.

Reeves, R.R., Stewart, B.S., Clapham, P., Powell, J.A. & Folkens, P. (2002) *National Audubon Society Guide to Marine Mammals of the World*. Alfred A. Knopf, New York.

Reyna-Saenz, F., Zarco-Gonzalez, M.M., Monroy-Vilchis, O. & Antonio-Nemiga, X. (2020) Regionalization of environmental and anthropic variables associated to livestock predation by large carnivores in Mexico. *Animal Conservation* 23, 192–202.

Rich, L.N., Davis, C.L., Farris, Z.J., Miller, D.A.W., Tucker, J.M. *et al.* (2017) Assessing global patterns in mammalian carnivore occupancy and richness by integrating local camera trap surveys. *Global Ecology and Biogeography* 26, 918–929.

Ripple, W.J. & Beschta, R.L. (2004) Wolves and the ecology of fear: can predation risk structure ecosystems? *BioScience* 54, 755–766.

Ripple, W.J., Estes, J.A., Beschta, R.L., Wilmers, C.C., Ritchie, E.G., Hebblewhite, M., Berger, J., Elmhagen, B., Letnic, M., Nelson, M.P., Schmitz, O.J., Smith, D.W., Wallach, A.D. & Wirsing, A.J. (2014) Status and ecological effects of the world's largest carnivores. *Science* 343, 1241484.

Ritchie, E.G. & Johnson, C.N. (2009) Predator interactions, mesopredator release and biodiversity conservation. *Ecology Letters* 12, 982–998.

Roberton, S.I., Gilbert, M.T.P., Campos, P.F., Salleh, F.M., Tridico, S. & Hills, D. (2017) Lowe's otter civet *Cynogale lowei* does not exist. *Small Carnivore Conservation* 54, 42–58.

Rocha, D.G., Ferraz, K.M.P.M.B., Gonçalves, L., Tan, C.K.W., Lemos, F.G. *et al.* (2020) Wild dogs at stake: deforestation threatens the only Amazon endemic canid, the short-eared dog (*Atelocynus microtis*). *Royal Society Open Science* 7, 190717.

Rodríguez, A. & Calzada, J. (2015) *Lynx pardinus. The IUCN Red List of Threatened Species* 2015, e.T12520A50655794. https://dx.doi.org/10.2305/

IUCN.UK.2015-2.RLTS.T12520A50655794.en. Accessed on 16 March 2019.

Roemer, G., Gompper, M.E. & Van Valkenburgh, B. (2009) The ecological role of the mammalian mesocarnivore. *BioScience* 59, 165–173.

Rogan, M.S., Balme, G.A., Distiller, G., Pitman, R.T., Broadfield, J., Mann, G.K.H., Whittington-Jones, G.M., Thomas, L.H. & O'Riain, M.J. (2019) The influence of movement on the occupancy–density relationship at small spatial scales. *Ecosphere* 10, e02807.

Ross, J., Wilting, A., Ngoprasert, D., Loken, B., Hedges, L., Duckworth, J.W., Cheyne, S., Brodie, J., Chutipong, W., Hearn, A., Linkie, M., McCarthy, J., Tantipisanuh, N. & Haidir, I.A. (2015) *Cynogale bennettii. The IUCN Red List of Threatened Species* 2015, e.T6082A45197343. https://dx.doi.org/10.2305/IUCN.UK.2015-4.RLTS.T6082A45197343.en. Accessed on 16 March 2019.

Rovero, F. & Zimmerman, F. (2016) *Camera Trapping for Wildlife Research*. Pelagic Publishing, Exeter.

Rowcliffe, J.M., Kays, R., Kranstauber, B., Carbone, C. & Jansen, P.A. (2014) Quantifying levels of animal activity using camera trap data. *Methods in Ecology and Evolution* 5, 1170–1179.

Ruprecht, J., Eriksson, C.E., Forrester, T.D., Spitz, D.B., Clark, D.A., Wisdom, M.J., Bianco, M., Rowland, M.M., Smith, J.B., Johnson, B.K. & Levi, T. (2021) Variable strategies to solve risk–reward tradeoffs in carnivore communities. *Proceedings of the National Academy of Sciences of the United States of America* 118, e2101614118.

Ryder, O.A. (1986) Species conservation and systematics – the dilemma of subspecies. *Trends in Ecology & Evolution* 1, 9–10.

Santos, F., Carbone, C., Wearn, O.R., Rowcliffe, J.M., Espinosa, S. *et al.* (2019) Prey availability and temporal partitioning modulate felid coexistence in Neotropical forests. *PLoS One* 14, e0213671.

Santos-Reis, M., Birks, J.D.S., O'Doherty, E.C. & Proulx, G. (eds) (2006) Martes *in Carnivore Communities*. Alpha Wildlife Publications, Sherwood Park.

Sato, J.J. (2016) The systematics and taxonomy of the world's badger species – a review. In: *Badgers: Systematics, Biology, Conservation and Research Techniques* (eds G. Proulx & E. Do Linh San), pp. 1–30. Alpha Wildlife Publications, Sherwood Park.

Sato, J.J. & Wolsan, M. (Chapter 2, this volume) Molecular systematics of the caniform Carnivora and its implications for conservation. In: *Small Carnivores: Evolution, Ecology, Behaviour, and Conservation* (eds E. Do Linh San, J.J. Sato, J.L. Belant & M.J. Somers). Wiley–Blackwell, Oxford.

Sato, J.J., Wolsan, M., Suzuki, H., Hosoda, T., Yamaguchi, Y., Hiyama, K., Kobayashi, M. & Minami, S. (2006) Evidence from nuclear DNA sequences shed light on the phylogenetic relationships of Pinnipedia: single origin with affinity to Musteloidea. *Zoological Science* 23, 125–146.

Sato, J.J., Wolsan, M., Minami, S., Hosoda, T., Sinaga, M.H., Hiyama, K., Yamaguchi, Y. & Suzuki, H. (2009) Deciphering and dating the red panda's ancestry and early adaptive radiation of Musteloidea. *Molecular Phylogenetics and Evolution* 53, 907–922.

Schipper, J., Helgen K.M., Belant, J.L., González-Maya, J., Eizirik, E. & Tsuchiya-Jerep, M. (2009) Editorial – Small carnivores in the Americas: reflections, future research and conservation priorities. *Small Carnivore Conservation* 41, 1–2.

Seebens, H., Blackburn, T.M., Dyer, E.E., Genovesi, P., Hulme, P.E. *et al.* (2018) Global rise in emerging alien species results from increased accessibility of new source pools. *Proceedings of the National Academy of Sciences of the United States of America* 115, E2264–E2273.

Sepúlveda, M.A., Valenzuela, A.E.J., Pozzi, C., Medina-Vogel, G. & Chehébar, C. (2015) *Lontra provocax. The IUCN Red List of Threatened Species* 2015, e.T12305A21938042. https://dx.doi.org/10.2305/IUCN.UK.2015-2.RLTS.T12305A21938042.en. Accessed on 22 March 2019.

Silva-Rodríguez, E, Farias, A., Moreira-Arce, D., Cabello, J., Hidalgo-Hermoso, E., Lucherini, M. & Jiménez, J. (2016) *Lycalopex fulvipes* (errata version published in 2016). *The IUCN Red List of Threatened Species* 2016, e.T41586A107263066. https://dx.doi.org/10.2305/IUCN.UK.2016-1.RLTS.T41586A85370871.en. Accessed on 15 March 2019.

Silvy, N.J. (ed.) (2012a) *The Wildlife Techniques Manual. Volume 1. Research.* 7th edition. The Johns Hopkins University Press, Baltimore.

Silvy, N.J. (ed.) (2012b) *The Wildlife Techniques Manual. Volume 2. Management.* 7th edition. The Johns Hopkins University Press, Baltimore.

Siriwat, P. & Nijman, V. (2018) Illegal pet trade on social media as an emerging impediment to the conservation of Asian otters species. *Journal of Asia-Pacific Biodiversity* 11, 469–475.

Sivy, K.J., Pozzanghera, C.B., Colson, K.E., Mumma, M.A. & Prugh, L.A. (2018) Apex predators and the facilitation of resource partitioning among mesopredators. *Oikos* 127, 607–621.

Sollmann, R., Mohamed, A., Samejima, H. & Wilting, A. (2013) Risky business or simple solution – relative abundance indices from camera-trapping. *Biological Conservation* 159, 405–412.

Soulé, M.A., Bolger, D.T., Alberts, A.C., Wright, J., Sorice, M. & Hill, S. (1988) Reconstructed dynamics of rapid extinctions of chaparral-requiring birds in urban habitat islands. *Conservation Biology* 2, 75–92.

Springer, M.S., Stanhope, M.J., Madsen, O. & de Jong W.W. (2004) Molecules consolidate the placental mammal tree. *Trends in Ecology & Evolution* 19, 430–438.

Struebig, M.J., Wilting, A., Gaveau, D., Meijaard, E., Smith, R.J., The Borneo Mammal Distribution Consortium, Fischer, M., Metcalf, K. & Kramer-Schadt, S. (2015) Targeted conservation to safeguard a biodiversity hotspot from climate and land-cover change. *Current Biology* 25, 1–7.

Sunquist, M.E. & Sunquist, F.C. (2009) Family Felidae (Cats). In: *Handbook of the Mammals of the World. Volume 1. Carnivores* (eds D.E. Wilson & R.A. Mittermeier), pp. 54–168. Lynx, Barcelona.

Suraci, J.P., Clinchy, M., Dill, L.M., Roberts, D. & Zanette, L.Y. (2016) Fear of large carnivores causes a trophic cascade. *Nature Communications* 7, 10698.

Svenning, J.-C., Fløjgaard, C. Marske, K.A., Nógues-Bravo, D. & Normand, S. (2011) Applications of species distribution modeling to paleobiology. *Quaternary Science Reviews* 30, 2930–2947.

Terborgh, J., Estes, J.A., Paquet, P., Ralls, K., Boyd-Heger, D., Miller, B.J. & Noss, R.F. (1999) The role of top carnivores in regulating terrestrial ecosystems. In: *Continental Conservation: Scientific Foundations of Regional Reserve Networks* (eds M.E. Soulé & J. Terborgh), pp. 39–64. Island Press, Washington, DC.

Thakur, M., Wullschleger Schättin, E. & McShea, W.J. (2018) Globally common, locally rare: revisiting disregarded genetic diversity for conservation planning of widespread species. *Biodiversity and Conservation* 11, 3031–3035.

Timmins, R.J., Coudrat, C.N.Z., Duckworth, J.W., Gray, T.N.E., Robichaud, W., Willcox, D.H.A., Long, B. & Roberton, S. (2016a) *Chrotogale owstoni. The IUCN Red List of Threatened Species* 2016, e.T4806A45196929. https://dx.doi.org/10.2305/IUCN.UK.2016-1.RLTS.T4806A45196929.en. Accessed on 16 March 2019.

Timmins, R., Duckworth, J.W., WWF-Malaysia, Roberton, S., Gray, T.N.E., Willcox, D.H.A., Chutipong, W. & Long, B. (2016b) *Viverra megaspila. The IUCN Red List of Threatened Species* 2016, e.T41707A45220097. https://dx.doi.org/10.2305/IUCN.UK.2016-1.RLTS.T41707A45220097.en. Accessed on 16 March 2019.

Tobler, M.W. & Powell, G.V.N. (2013) Estimating jaguar densities with camera traps: problems with current designs and recommendations for future studies. *Biological Conservation* 159:109–118.

Trewby, L.D., Wilson, G.J., Delahay, R.J., Walker, N., Young, R., Davison, J., Cheeseman, C., Robertson, P.A., Gorman, M.L. & McDonald, R.A. (2008) Experimental evidence of competitive release in sympatric carnivores. *Biology Letters* 4, 170–172.

Trinkel, M. & Angelici, F.M. (2016) The decline in the lion population in Africa and possible mitigation measures. In: *Problematic Wildlife: A Cross-Disciplinary Approach* (ed. F.M. Angelici), pp. 45–68. Springer, New York.

Ugarte, C.S., Moreira-Arce, D. & Simonetti, J.A. (2019) Ecological attributes of carnivore–livestock conflict. *Frontiers in Ecology and Evolution* 7, 433.

Valqui, J. & Rheingantz, M.L. (2015) *Lontra felina* (errata version published in 2017). *The IUCN Red List of Threatened Species* 2015, e.T12303A117058682. https://dx.doi.org/10.2305/IUCN.UK.2015-2.RLTS.T12303A21937779.en. Accessed on 22 March 2019.

Veron, G. (2010) Phylogeny of the Viverridae and "viverrid-like" feliforms. In: *Carnivoran Evolution: New Views on Phylogeny, Form and Function [Cambridge Studies in Morphology and Molecules: New Paradigms in Evolutionary Biology]* (eds A. Goswami & A. Friscia), pp. 64–91. Cambridge University Press, Cambridge.

Veron, G. & Goodman, S.M. (2018) One or two species of the rare Malagasy carnivoran *Eupleres* (Eupleridae)? New insights from molecular data. *Mammalia* 82, 107–112.

Veron, G., Dupré, D., Jennings, A.P., Gardner, C.J., Hassanin, A. & Goodman, S.M. (2017) New insights into the systematics of Malagasy mongoose-like carnivorans (Carnivora, Eupleridae, Galidiinae) based on mitochondrial and nuclear DNA sequences. *Journal of Zoological Systematics and Evolutionary Research* 55, 250–264.

Veron, M., Patou, M.L. & Jennings, A.P. (Chapter 3, this volume) Systematics and evolution of the mongooses (Herpestidae, Carnivora). In: *Small Carnivores: Evolution, Ecology, Behaviour, and Conservation* (eds E. Do Linh San, J.J. Sato, J.L. Belant & M.J. Somers). Wiley–Blackwell, Oxford.

Vilella, M., Ferrandiz-Rovira, M. & Sayol, F. (2020) Coexistence of predators in time: effects of season and prey availability on species activity within a Mediterranean carnivore guild. *Ecology and Evolution* 10, 11408–11422.

Villalba, L., Lucherini, M., Walker, S., Lagos, N., Cossios, D., Bennett, M. & Huaranca, J. (2016) *Leopardus jacobita. The IUCN Red List of Threatened Species* 2016, e.T15452A50657407. https://dx.doi.org/10.2305/IUCN.UK.2016-1.RLTS.T15452A50657407.en. Accessed on 22 March 2019.

Wallach, A.D., Izhaki, I., Toms, J.D., Ripple, W.J. & Shanas, U. (2015a) What is an apex predator? *Oikos* 124, 1453–1461.

Wallach, A.D., Ripple, W.J. & Carroll, S.P. (2015b) Novel trophic cascades: apex predators enable coexistence. *Trends in Ecology & Evolution* 30, 146–153.

Wearn, O.R. & Glover-Kapfer, P. (2017) *Camera-Trapping for Conservation: A Guide to Best-Practices.* WWF Conservation Technology Series 1(1). WWF-UK, Woking.

Wearn, O.R. & Glover-Kapfer, P. (2019) Snap happy: camera traps are an effective sampling tool when compared with alternative methods. *Royal Society Open Science* 6, 181748.

Werdelin, L. & Wesley-Hunt, G.D. (2010) The biogeography of carnivore ecomorphology. In: *Carnivoran Evolution: New Views on Phylogeny, Form, and Function* (eds A. Goswami & A. Friscia), pp. 225–245. Cambridge University Press, Cambridge.

Willcox, D. (2020) Conservation status, ex situ priorities and emerging threats to small carnivores. *International Zoo Yearbook* 54, 19–34.

Williams, S.T., Maree, N., Taylor, P., Belmain, S.R., Keith, M. & Swanepoel, L.H. (2018) Predation by small mammalian carnivores in rural agro-ecosystems: an undervalued ecosystem service? *Ecosystem Services* 30, 362–371.

Wilson, D.E. & Mittermeier, R.A. (eds) (2009) *Handbook of the Mammals of the World. Volume 1. Carnivores.* Lynx, Barcelona.

Wilting, A., Brodie, J., Cheyne, S., Hearn, A., Lynam, A., Mathai, J., McCarthy, J., Meijaard, E., Mohamed, A., Ross, J., Sunarto, S. & Traeholt, C. (2015a) *Prionailurus planiceps. The IUCN Red List of Threatened Species* 2015, e.T18148A50662095. https://dx.doi.org/10.2305/IUCN.UK.2015-2.RLTS. T18148A50662095.en. Accessed on 16 March 2019.

Wilting, A., Duckworth, J.W., Hearn, A. & Ross, J. (2015b) *Melogale everetti. The IUCN Red List of Threatened Species* 2015, e.T13110A45199541. https://dx.doi.org/10.2305/IUCN.UK.2015-4.RLTS. T13110A45199541.en. Accessed on 22 March 2019.

Wilting, A., Patel, R., Pfestorf, H., Kern, C., Sultan, K., Ario, A., Peñaloza, F., Kramer-Schadt, S., Radchuk, V., Foerster, D.W. & Fickel, J. (2016) Evolutionary history and conservation significance of the Javan leopard *Panthera pardus melas. Journal of Zoology* 299, 239–250.

Wolf, C. & Ripple, W.J. (2018) Rewilding the world's large carnivores. *Royal Society Open Science* 5, 172235.

Wozencraft, W.C. (2005) Order Carnivora. In: *Mammal Species of the World: A Taxonomic and Geographic Reference.* 3rd edition (eds D.E. Wilson & D.M. Reeder), pp. 532–628. The Johns Hopkins University Press, Baltimore.

Wright, S.J., Gompper, M.E. & DeLeon, B. (1994) Are large predators keystone species in neotropical forests? The evidence from Barro Colorado Island. *Oikos* 71, 279–294.

Wyrwoll, T.W. (2003) Still desiderata: scientific names for domestic animals and their feral derivatives. In: *The New Panorama of Animal Evolution [Proceedings of the 18th International Congress of Zoology]* (eds A. Legakis, S. Sfenthourakis, R. Polymeni & M. Thessalou-Legaki), pp. 683–697. Pensoft Publishers, Sofia – Moscow.

Yang, H., Dou, H., Baniya, R.K., Han, S.Y., Guan, Y., Xie, B., Zhao, G.J., Wang, T.M., Mou, P., Feng, L.M. & Ge, J.P. (2018) Seasonal food habits and prey selection of Amur tigers and Amur leopards in Northeast China. *Scientific Reports* 8, 6930.

Yoder, A.D., Burns, M.M., Zehr, S., Delefosse, T., Veron, G., Goodman, S.M. & Flynn, J.J. (2003) Single origin of Malagasy Carnivora from an African ancestor. *Nature* 421, 734–737.

Zachos, F.E. (2013) Taxonomy: species splitting puts conservation at risk. *Nature* 494, 35.

Zachos, F.E., Apollonio, M., Bärmann, E.V., Festa-Bianchet, M., Göhlich, U. *et al.* (2013) Species inflation and taxonomic artefacts – a critical comment on recent trends in mammalian classification. *Mammalian Biology* 78, 1–6.

Zalewski, A., Wierzbowska, I., Aubry, K.B., Birks, J.D.S, O'Mahony D.T. & Proulx, G. (2017) *The* Martes *Complex in the 21st Century: Ecology and Conservation.* Mammal Research Institute, Polish Academy of Sciences, Białowieża.

Zrzavý, J., Duda, P., Robovský, J., Okřinová, I. & Pavelková Řičánková, V. (2018) Phylogeny of the Caninae (Carnivora): combining morphology, behaviour, genes and fossils. *Zoologica Scripta* 47, 373–389.

Zu, J., Yuan, B. & Du, J. (2015) Top predators induce evolutionary diversification of intermediate predator species. *Journal of Theoretical Biology* 387, 1–12.

Part II

Evolution, Systematics, and Distribution

2

Molecular Systematics of the Caniform Carnivora and its Implications for Conservation

Jun J. Sato[1], and Mieczysław Wolsan[2]*

[1]*Department of Biotechnology, Faculty of Life Science and Technology, Fukuyama University, Fukuyama, Japan*
[2]*Museum and Institute of Zoology, Polish Academy of Sciences, Warszawa, Poland*

SUMMARY

Recent advances in phylogenetic resolution at higher taxonomic levels within the mammalian order Carnivora have been stimulated by the increasing application of nuclear DNA, which is less homoplastic than mitochondrial DNA, and therefore better suited for studying deep-level (e.g. among genera or older) relationships. Immense progress in sequencing nuclear and mitochondrial DNAs from carnivoran species has resulted in a wealth of data in publicly available DNA databases, allowing an improved understanding of phylogenetic relationships at every taxonomic level using the 'total evidence' supermatrix or supertree method. Here, we review recent molecular systematic studies for one of the most enigmatic species, the red panda, *Ailurus fulgens*, and show that the use of nuclear DNA, Bayesian and maximum likelihood phylogenetic inference, and the supermatrix approach have improved the resolution of the phylogenetic position of this species. Secondly, we show that such methodological improvements have also clarified the evolution of the family Mustelidae (weasels, martens, otters, badgers, and allies). We demonstrate this in light of phylogeny, chronology, and historical biogeography and provide an up-to-date subfamily classification of the Mustelidae. Finally, we discuss the implications of molecular systematics to setting and defining conservation priorities on the basis of the EDGE (Evolutionarily Distinct and Globally Endangered) value, and conclude that the supermatrix-based priority setting is preferable to the supertree-based one.

Keywords

Bayesian inference – Caniformia – EDGE – nuclear DNA – supermatrix – systematics

Introduction

Since the advent of DNA amplification and sequencing technologies in the 1980s, methods to sequence DNA have been dramatically developed and we can now obtain the genome sequence of an organism easily using next-generation sequencing techniques (Glenn, 2011; Mardis, 2013). We are currently in the stage where it is possible to apply genome-partitioning approaches and clarify many evolutionary issues in genomic contexts even for non-model organisms (e.g. Ekblom & Galindo, 2011; Rubin *et al.*, 2012; McCormack *et al.*, 2013; Lemmon & Lemmon, 2013; Blaimer *et al.*, 2015; Bragg *et al.*, 2016; Harvey *et al.*, 2016; Jones & Good, 2016; Sato *et al.*, 2019). During the development from traditional phylogenetic to phylogenomic approaches, molecular phylogenetics has revolutionized carnivoran systematics. However, despite the wealth of molecular data, only recently have the major systematic relationships within the

*Corresponding author.

order Carnivora been resolved. The questions that should be asked here are: What was the limiting factor in the resolution of carnivoran species relationships and what was the revolutionizing factor? By reviewing the historical development of data and methods in systematic studies, we can clarify the course that we should follow to complete carnivoran phylogeny. In this chapter, we first review molecular systematic studies of the red panda, *Ailurus fulgens*, as an example, to extract significant factors in data and/or methods which largely contributed to the clarification of the deep-level phylogenetic relationships of the caniform Carnivora. Second, we use the family Mustelidae to show that such factors also helped understand the evolution and classification of the most diversified carnivoran family. Finally, we argue that the phylogeny and chronology of the superfamily Musteloidea elucidated by taking such revolutionizing factors into account would provide a foundation to adequately estimate the evolutionary distinctiveness that should be considered when setting conservation priorities.

A Review of Molecular Systematic Studies of the Red Panda

The red panda (Figure 2.1) constitutes the monotypic family Ailuridae and its current distribution is restricted to the Himalayan region in southern and southeastern Asia (Wozencraft, 2005; Glatston, 2011).

Figure 2.1 Red panda, *Ailurus fulgens*, photographed in a bamboo montane forest at the Himalayan foothills, Singalila National Park, India. *Source:* Photo © Nick Garbutt (www.nickgarbutt.com).

However, paleontological records suggest that related species in Eurasia and North America existed (Morlo & Peigné, 2010). The absence of closely related extant species makes it difficult to estimate the phylogenetic position of this orphaned species. The red panda is a bamboo-feeder, as is the giant panda, *Ailuropoda melanoleuca* (family Ursidae), and shows many phenotypic adaptations independently specialized to this diet, including modified mandible, teeth, and cranial chewing muscles to masticate the fibrous plant and modified forearm to grasp bamboo (Fisher, 2011). In addition, this species exhibits many primitive traits in its cranium and dentition (Wolsan, 1993; Flynn *et al.*, 2000; Morlo & Peigné, 2010). Possession of both specialized (adaptive) and ancestral characteristics may have caused confusion in determining the relatedness of this species to other carnivorans. Consistent with red panda biology, its phylogenetic affinity is a longstanding conundrum in morphological systematic studies. Four cladistic analyses using morphology provided different hypotheses of red panda affinity: (i) closely related to the Procyonidae (Flynn *et al.*, 1988; Wang, 1997); (ii) closely related to the Ursidae (Wozencraft, 1989); (iii) closely related to the clade including Ursidae and Pinnipedia (Wyss & Flynn, 1993); or (iv) placed in an unresolved polytomy among major lineages of the Musteloidea (Wolsan, 1993). Using molecular phylogenetic approaches, the phylogenetic position of the red panda has recently been determined as the closest relative of the clade of the Mustelidae and Procyonidae to the exclusion of the Mephitidae (Figure 2.2 and Table 2.1; Sato *et al.*, 2009, 2012; Eizirik *et al.*, 2010; Yu *et al.*, 2011a). By examining recent literature, including phylogenetic inferences with molecular data (Table 2.1), we extracted three significant developments that facilitated the clarification of the evolutionary origin of this species, as described below.

Impacts of the Use of Nuclear DNA on Red Panda Relationships

Genetic information used in earlier molecular systematic studies of the red panda from 1993 to 1997 (Table 2.1) was primarily mitochondrial DNA (mtDNA). During this period, the phylogenetic position of the red panda was ambiguous, probably because

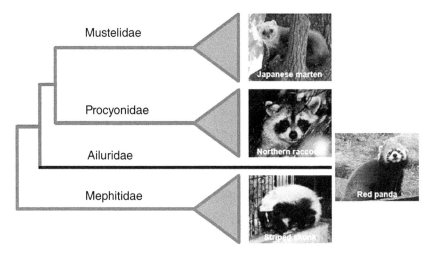

Figure 2.2 Inter-familial relationships within the superfamily Musteloidea strongly supported by recent molecular systematic studies (see text and Table 2.1). *Source:* Photos © Kazutake Hirooka (Japanese marten, *Martes melampus*), Kozue Hiyama (red panda, *Ailurus fulgens*), Tetsuji Hosoda (northern raccoon, *Procyon lotor*), and Jun J. Sato (striped skunk, *Mephitis mephitis*).

of the homoplasious nature of mtDNA ('saturation problem') and poor information in the short sequences used. In murine rodents, phylogeny and divergence time estimations among lineages splitting at 6.0 MYA (million years ago) or older were difficult to correctly estimate using mtDNA (Steppan *et al.*, 2005). In fact, recent studies estimated that the divergence of the red panda lineage occurred ~30 MYA (e.g. Sato *et al.*, 2009, 2012; Eizirik *et al.*, 2010), which is too old for mtDNA phylogenetic inference. Therefore, despite attempts to correct the saturation problem by some researchers (Zhang & Ryder, 1993, 1997; Pecon-Slattery & O'Brien, 1995; Ledje & Arnason, 1996a), red panda affinity was not clarified due to its deep divergence and also because of insufficient phylogenetic information from the short-length nucleotide sequences of examined mtDNA (~2 kb; Table 2.1).

Flynn & Nedbal (1998) were the first to use nuclear DNA (nucDNA) to resolve this issue. nucDNA is more resistant to the saturation problem because of its slow evolutionary rate and may contain more suitable molecular markers for deeper-level phylogenetic inferences (Springer *et al.*, 2001). Also, the time scale estimated from mtDNA (usually an overestimation) has often been improved in other mammalian groups by unsaturated nucDNA variation (e.g. Sato *et al.*, 2016). Use of nucDNA has since increased, but until about 2005, use of mtDNA was more common and estimated

phylogenies may have been subject to the saturation problem (Table 2.1). Consequently, multiple phylogenetic hypotheses were proposed before 2005 (Table 2.1). Nonetheless, between 1998 and 2005, it gradually emerged that the red panda is a major lineage of the Musteloidea, corresponding to independent family status, although the exact relationship remained to be resolved.

As part of a more inclusive study, Sato *et al.* (2006) presented a highly supported phylogenetic hypothesis of the red panda as a second lineage to diverge following the Mephitidae within Musteloidea (Figure 2.2; Table 2.1). This hypothesis was strongly supported by maximum-likelihood (ML) and Bayesian inference (BI) phylogenetic analyses of three nuclear-gene protein-coding exon sequences (Sato *et al.*, 2006; Table 2.1). Later, Fulton & Strobeck (2006) supported this hypothesis using BI of one exon and four intron nucDNA sequences. However, the estimated topology and strength of support for observed relationships in these studies depended on phylogenetic inference, and maximum-parsimony (MP) analyses did not support the hypothesis in both studies. In addition, a topological test did not reject alternative phylogenetic hypotheses (Fulton & Strobeck, 2006), making the red panda phylogenetic position uncertain. Subsequent mtDNA analyses found no statistically significant or consistent phylo-

Table 2.1 Red panda, *Ailurus fulgens*, relationships on the basis of nucleotide sequence data reported since 1993.

Paper	Genetic loci[a]	Characters[b]	Inference[c]	Topology[d]
Zhang & Ryder (1993)	mt (Cyt*b*)	397 bp	MP (ti down-weighted)	(Ai, (Ur, Pr))
	mt (12S rRNA, tRNAs)	438 bp	MP (tv only)	(Ur, (Ai, Pr))
Vrana *et al.* (1994)[e]	mt (Cyt*b*, 12S rRNA) + mor	737 bp + mor (64)	TE MP	(Ca, (Me, (Mu, (Pr, (Ai, (Ur, Pi))))))
Pecon-Slattery & O'Brien (1995)	mt (12S rRNA)	358 bp	NJ	(Ca, (Ur, (Ai, Pr)))
			MP (ti down-weighted)	
			ML	
Ledje & Arnason (1996a)[e]	mt (Cyt*b*)	1140 bp	NJ (ti down-weighted)	((Ca, Pi, (Ur, Ai)), (Me, (Mu, Pr)))
			MP (ti down-weighted)	(Ca, Ur, Pi, Ai, (Me, (Mu, Pr)))
Ledje & Arnason (1996b)[e]	mt (12S rRNA)	890 bp	NJ	(Ca, (Ur + Me), Pi, Ai, (Mu, Pr,))
		890 or 1011 bp	MP	(Ca, Ur, Pi, Me, Ai, (Mu, Pr))
	mt (Cyt*b*, 12S rRNA)	2030 or 2151 bp	MP	(Ca, Ur, Pi, Ai, (Me, (Mu, Pr)))
Zhang & Ryder (1997)	mt (Cyt*b*)	397 bp	MP (ti down-weighted)	(Ai, (Ur, (Pi, (Mu, Pr))))
	mt (12S rRNA, tRNAs)	429 bp	MP (tv only)	(Ca, (Ur, (Pi, (Ai, (Mu, Pr)))))
Dragoo & Honeycutt (1997)	mt (12S and 16S rRNA, tRNA, Cyt*b*)	~1400 bp	MP	((Ca, Ur), (Pi, (Me, (Mu, ((Pr$_1$, Ai), Pr$_2$)))))
	mt (12S and 16S rRNA, tRNA, Cyt*b*) + mor	1647 bp	TE MP	(Ca, (Ur, (Pi, (Me, (Mu, ((Pr$_1$, Ai), Pr$_2$))))))
Flynn & Nedbal (1998)[e]	mt (Cyt*b*) + nuc (*Ttr*) + mor	1991 bp + mor (64)	MP (Cyt*b*: ti down-weighted)	(Ca, (Pi, (Ur, (Ai, (Mu, Pr)))))
	mt (Cyt*b*, 12S rRNA) + nuc (*Ttr*)	2338 bp	ML (Cyt*b*: ti down-weighted)	(Ca, (Ur, (Pi, (Ai, (Mu, Pr)))))
Flynn *et al.* (2000)[e]	mt (Cyt*b*, 12S and 16S rRNA) + nuc (*Ttr*)	3450 bp	MP (Cyt*b*: ti down-weighted, 12S and 16S rRNA: tv only)	(Ca, (Ur, (Pi, ((Ai, Me), (Mu, Pr)))))
			ML	(Ca, (Ur, (Pi, (Ai, (Me, (Mu, Pr))))))
Zehr *et al.* (2001)	nuc (*CanSINE*)	186 bp	MP	(Ca (Pi, (Mu, (Ur (Ai, Pr)))))
			ML	(Ca, (Pi, (Mu, (Pr$_1$, (Pr$_2$, Ai, Ur)))))
Marmi *et al.* (2004)[e]	mt (Cyt*b*) + nuc (*Mel08*)	491 bp	BI	(Pi, (Ai, (Me, (Mu, Pr))))

Table 2.1 (Continued)

Paper	Genetic loci[a]	Characters[b]	Inference[c]	Topology[d]
Yu *et al.* (2004)[e]	nuc (Irbp, *Ttr*)	2 341 bp	MP	(Ca, (Ur, (Pi, (Ai, (Mu, Pr)))))
Flynn *et al.* (2005)[e]	mt (Cyt*b*, 12S rRNA, ND2) + nuc (*Irbp*, *Ttr*, *Tbg*)	6 243 bp	MP	(Ca, (Ur, (Pi, (Ai, Me, (Mu, Pr))))))
			BI*	(Ca, (UR, (Pi, (Ai, (Me, (Mu, Pr))))))
Delisle & Strobeck (2005)[e]	mt (12 protein-coding genes)	10 842 bp	MP	(Pi, ((Mu, Pr), (Ur, (Ai, Me))))
			MP (tv only)	(((Pi, Ur), ((Ai, Me), (Mu, Pr))))
			ML	(Pi, (Ur, ((Ai, Me), (Mu, Pr))))
			BI*	
Domingo-Roura *et al.* (2005)[e]	nuc (*Mel08*)	174–192 bp	ML	(Pi, (Ai, (Me, (Mu, Pr))))
Yu & Zhang (2006)[e]	mt (ND2) + nuc (*Fgb* introns 4 and 7, *Irbp*, *Ttr1*)	4 417 bp	MP (ND2: ti down-weighted)	(Ca, (Ur, (Pi, (Ai, (Mu, Pr)))))
Sato *et al.* (2006)[e]	nuc (*Apob*, *Irbp*, *Rag1*)	3 228 bp	MP	**(Ur, (Pi, (Me, (Ai, (Mu, Pr)))))**
			ML*	
			BI*	
Fulton & Strobeck (2006)[e]	nuc (*Fes*, *Chrna1*, *Ghr*, *Rho*, *Irbp*)	2 974 bp	MP	**(Ca, (Ur, (Pi, (Me, (Ai, (Mu, Pr))))))**
			ML	
			BI*	
			MP-input supertree	
			Bayes-input supertree	
			ML-input supertree	(Ca, (Ur, (Pi, (Me, Ai, (Mu, Pr)))))
Fulton & Strobeck (2007)[e]	nuc (*Chrna1*, *Ghr*, *Irbp*) + mt (CO1, ND2, Cyt*b*)	5 918 bp	MP*	(Ca, (Ur, (Pi, ((Ai, Me), (Mu, Pr)))))
			ML	(Ca, (Ur, (Pi, (Ai, (Me, (Mu, Pr))))))
			BI	
Yonezawa *et al.* (2007)	mt (12 protein-coding genes, 12S rRNA) + nuc (*Ttr*, *Irbp*, *Rag1*, *Apob*, *Chrna1*, *Fes*, *Ghr*, *Rho*)	17 688 bp	ML	**(Me, (Ai, (Mu, Pr)))**
Peng *et al.* (2007)	mt (12 protein-coding genes)	9 975 bp	NJ	(Ca, ((Ur, Pi), ((Me, Pr), (Ai, Mu))))
			ML	(Ca, (Ur, (Pi, ((Ai, Me), (Mu, Pr)))))
Arnason *et al.* (2007)	mt (amino acid sequence from 12 protein-coding genes)	3 601 aa	ML	(Ca, (Ur, (Pi, (Ai, (Me, (Mu, Pr))))))

(Continued)

Table 2.1   (Continued)

Paper	Genetic loci[a]	Characters[b]	Inference[c]	Topology[d]
Yu *et al.* (2008)	mt (ND2) + nuc (*Fgb* introns 4 & 7, *Irbp, Ttr*)	4 272 bp	MP*	(Ca, (Ur, (Pi, ((Ai, Me), (Mu, Pr)))))
			ML	**(Ca, (Ur, (Pi, (Me, (Ai, (Mu, Pr))))))**
			BI	
Sato *et al.* (2009)	nuc (*Apob, Brca1, Irbp, Rag1, vwf*)	5 497 bp	MP*	**(Ca, (Ur, (Pi, (Me, (Ai, (Mu, Pr))))))**
			ML*	
			BI*	
Agnarsson *et al.* (2010)	mt (Cyt*b*)	1 140 bp	BI	((Ai, Ca), ((Ur, Pi), (Pr$_1$, (Mu, (Me, Pr$_2$)))))
Eizirik *et al.* (2010)	nuc (*Adora3, Apob, App, Atp7a, Bdnf, Chrna1, Fbn1, Fes, Ghr, Plp1, Pnoc, Ptprg, Rag2, Rasa2*)	7 765 bp	NJ*	**(Ca, (Ur, (Pi, (Me, (Ai, (Mu, Pr))))))**
			MP*	
			ML*	
			BI*	
Yu *et al.* (2011a)	nuc (*Atp5d-2, Ccng2-2, Ccng2-6, Cidea1, Coro1c-4, Coro1c-5, Fgb* introns 4 & 7, *Guca1b-3, Impal1-6, Ociad1-4, Plod2-13, Plod2-14, Ssr1-5, Tbc1d7-6, Tbk1-8, Tinagl1-1, Tinagl1-3, Ttr1, Wasf1-3, Wasf1-6, Wasf1-7*)	~22 000 bp	ML*	**(Ca, (Ur, (Pi, (Me, (Ai, (Mu, Pr))))))**
			BI*	
Sato *et al.* (2012)	mt (Cyt*b*) + nuc (*Apob, Brca1, Chrna1, Fes, Ghr, Irbp, Rag1, Rho, vwf*)	8 492 bp	MP*	**(Ca, (Ur, (Pi, (Me, (Ai, (Mu, Pr))))))**
			ML*	
			BI*	

[a] mt = mitochondrial DNA; nuc = nuclear DNA; mor = morphology.
[b] bp = base pair; aa = amino acid.
[c] MP = Maximum Parsimony; NJ = Neighbour Joining; ML = Maximum Likelihood; BI = Bayesian Inference; TE = Total Evidence; tv = transversional substitutions; ti = transitional substitutions.
[d] Ai = Ailuridae; Me = Mephitidae; Mu = Mustelidae; Pi = Pinnipedia; Pr = Procyonidae; Ur = Ursidae (including the giant panda, *Ailuropoda melanoleuca*); Ca = Canidae; Boldface = same topology as in Figure 2.1.
[e] Studies listed in the review by Morlo & Peigné (2010).
*Strongly supported (bootstrap value > 90% and/or posterior probability > 0.95).
Source: Adapted from Sato & Wolsan (2012) following the style of Morlo & Peigné (2010) in order to supplement the studies that were not listed in their Table 4.1.

genetic relationships (Table 2.1; Arnason *et al.*, 2007; Fulton & Strobeck, 2007; Peng *et al.*, 2007; Yonezawa *et al.*, 2007). Sato *et al.* (2009) expanded on their previous 2006 study and examined five nuclear gene exon sequences (Table 2.1). They showed a highly supported red panda position as a second lineage to offshoot within Musteloidea (Figure 2.2) using multiple analyses (MP, ML, and BI) and rejected alternative phylogenetic hypotheses using three topological tests: Templeton-test (Templeton, 1983), KH-test (Kishino & Hasegawa, 1989), and AU-test (Shimodaira, 2002). Recent studies using more extensive nucDNA also supported this hypothesis (Eizirik *et al.*, 2010; Yu *et al.*, 2011a; Sato *et al.*, 2012). It is concluded that the less homoplasious nucDNA was important for the clarification of the phylogenetic placement of the red panda, as well as the resolution of the other major caniform carnivoran lineages.

Impact of Bayesian MCMC and Fast-ML Methods on Red Panda Relationships

To obtain robust and reliable phylogenetic hypotheses, we need a pluralistic approach using various phylogenetic strategies based on different optimality criteria. However, the traditional ML method implemented in common phylogenetic programs (e.g. PAUP; Swofford, 2002) became more difficult to implement due to the computational burden imposed by increases in the size of the taxon-by-sequence matrix (Felsenstein, 1978). This was especially the case with the bootstrap analysis (Felsenstein, 1985), which requires repeated searches of the optimal tree. In the ML analysis performed by Sato *et al.* (2009), it took more than three months to complete the 100-repetition bootstrap calculation for the matrix of 51 species by 5497 nucleotide characters using the PAUP software on a Power Macintosh G5 computer (CPU dual 1.8 GHz, memory 512 MB). Flynn *et al.* (2005) did not adopt the traditional ML method in their phylogenetic analysis of a 76 species by 6243 character matrix and, instead, used only MP and BI methods. Delisle & Strobeck (2005) examined 12 protein-coding mtDNA genes (10 842 bp) for 38 species, but did not provide the bootstrap proportion in their ML analyses. The latter two studies may reflect the computational difficulty of the traditional ML method.

Marmi *et al.* (2004) were the first to conduct a BI phylogenetic analysis for caniform carnivoran taxa including the red panda. The BI method is a fast probabilistic phylogenetic method based on posterior probability (PP) as the optimality criterion, using the same evolutionary (substitution) models as in the ML method (Huelsenbeck *et al.*, 2001, 2002). Although the actual posterior probability cannot be calculated, the application of the Markov Chain Monte Carlo (MCMC) approach has enabled approximation within a realistic timeframe (Rannala & Yang, 1996; Mau & Newton, 1997; Larget & Simon, 1999; see also Huelsenbeck *et al.*, 2002). Furthermore, platforms for BI analyses were developed (e.g. MrBayes software; Huelsenbeck & Ronquist, 2001) and accelerated use of this method in systematic studies. After Marmi *et al.* (2004), BI was widely adopted and examined sequence lengths have increased (Table 2.1).

On the other hand, it is known that BI often provides higher support for incorrect phylogenetic relationships (e.g. Douady *et al.*, 2003; Simmon *et al.*, 2004). Therefore, various hypotheses for the phylogenetic position of the red panda have been supported by this method. For instance, Flynn *et al.* (2005), using three mtDNA and three nucDNA sequences, supported the basal position of the red panda in Musteloidea with PP = 1.0 (the highest value; Table 2.1). Delisle & Strobeck (2005) showed close affinity between the red panda and the striped skunk, *Mephitis mephitis*, also with the highest PP value (Table 2.1). Moreover, the topology as given in Figure 2.2 was strongly supported by recent, mostly nucDNA phylogenetic analyses, where the PP values for the relevant clades were almost all equal to 1.0 (Table 2.1; Sato *et al.*, 2009, 2012; Eizirik *et al.*, 2010; Yu *et al.*, 2011a). Thus, more conservative supporting measures, such as the bootstrap proportion, are still needed within the probabilistic framework to increase confidence in the obtained phylogenetic hypothesis. The most recent studies of caniform carnivoran phylogenetics used more than 10 gene sequences with ~7.7–22.0 kb for 16–44 taxa, which would be practically impossible to examine with the traditional ML method (Table 2.1; Eizirik *et al.*, 2010; Yu *et al.*, 2011a; Sato *et al.*, 2012). Therefore, they adopted the recently developed fast-ML search methods implemented in the programs PHYML (Guindon & Gascuel, 2003), GARLI (Zwickl, 2006), and RAxML (Stamatakis *et al.*, 2008). Owing to the development of these fast-ML strategies, pluralistic evaluation of phylogenetic hypotheses by various optimality criteria with different measures of support has been realized with more efficiency than using PAUP (e.g. Sato *et al.*, 2009). Such methodological advance has also contributed to the clarification of the phylogenetic position of the red panda as the second offshoot in Musteloidea, with the skunk lineage being the most basal (Figure 2.2).

Impact of the Supermatrix Approach on Red Panda Relationships

The supermatrix approach combines multiple taxon-by-character matrices into a single 'supermatrix' and simultaneously examines this large combined alignment for phylogenetic inferences (de Queiroz & Gatesy, 2006). It is often compared to the supertree approach and there are many conceptual debates with regard to the clarification of large-scale phylogenetic relationships (Gatesy *et al.*, 2002, 2004; Bininda-Emonds *et al.*, 2003; Gatesy & Springer, 2004;

Bininda-Emonds, 2004a,b). The use of the supermatrix approach has been spurred on by MCMC techniques in the combined BI method, where analyses with independent evolutionary models set for each gene partition can be conducted with less computational burden (Nylander *et al.*, 2004). An update of the MrBayes software (version 3) enabled the application of the independent-model approach (Ronquist & Huelsenbeck, 2003), allowing a precise supermatrix approach using probabilistic methods (de Queiroz & Gatesy, 2006).

One concern in taking advantage of the supermatrix approach is 'missing data' in the data matrix. Missing data correspond to empty cells in the taxon-by-character matrix, which arise as characters may not be determined (e.g. in DNA sequences) or preserved (e.g. in fossils) for some taxa. There is a concern, from morphologists in particular, about the use of sparse data matrices in phylogenetic analysis (see Kearney & Clark, 2003). However, a decade of empirical and theoretical studies has suggested that low resolution in a phylogeny inferred from a supermatrix with abundant missing data is not due to the proportion of missing data, but poor phylogenetic information present in the existing data (e.g. Wiens, 2003, 2006; Philippe *et al.*, 2004; Fulton & Strobeck, 2006; Wolsan & Sato, 2010). Therefore, the collection of as much data as possible, irrespective of the abundance of missing data, has gradually been recognized as a useful strategy for elucidating large-scale phylogenetic relationships (e.g. Wolsan & Sato, 2010).

The supermatrix approach was first used in carnivoran systematics by Flynn *et al.* (2005). Contrary to the recognition that missing data are not problematic, Flynn *et al.* (2005) suggested some adverse effects of missing data on phylogenetic inference. However, their results may have been affected by the large proportion of mtDNA in their data matrix (52.3%; 3266 out of 6243 bp), which is subject to the saturation problem. Nevertheless, since the study of Flynn *et al.* (2005), the use of available data in DNA databases has been considered helpful for clarifying a comprehensive carnivoran phylogeny. Fulton & Strobeck (2006) also examined the effect of missing data on carnivoran phylogenetic inference with a supermatrix comprised of only the less homoplastic

nucDNA. They found that phylogenetic resolution was dependent on the existing data, not missing data, and at the same time, suggested a red panda position that has been supported by recent molecular systematic studies (Figure 2.2). Combined with the other two contributing factors noted above, the recent results inferred from the pluralistic phylogenetic approach based on a supermatrix of multiple nucDNAs have converged towards a consistent phylogenetic hypothesis that the red panda secondly diverged in Musteloidea (Figure 2.2).

We conclude that the use of nucDNA loci, Bayesian and fast-ML inference methods, and the supermatrix approach were revolutionary in clarifying the long-standing phylogenetic conundrum within the caniform Carnivora, namely the phylogenetic position of the red panda. This approach was also effective for the other taxa, including the Mustelidae. Below we show that these methodological advancements also contributed to understanding the evolution and classification of the family Mustelidae.

Evolution of Mustelidae in Space and Time and Subfamily Classification

The Mustelidae is the most speciose family within the order Carnivora, representing ~20% of carnivoran species diversity (59 of 286 species; Wozencraft, 2005). While there is marked variation in body size within this family, from the least weasel, *Mustela nivalis* (less than 50 g; Macdonald, 2009) to the marine-adapted sea otter, *Enhydra lutris* (more than 45 kg; Macdonald, 2009), almost all mustelids can be recognized as small-to-medium-sized animals. The extensive ecomorphological diversity of mustelids allows them to inhabit environments from tropical forest to arctic tundra and from desert to river and coastal sea (Nowak, 1991; Larivière & Jennings, 2009; Macdonald, 2009). Due to their adaptability, mustelids are distributed worldwide. Therefore, understanding the process of mustelid evolution in space and time could provide significant insight into the diversification mechanisms of small carnivores and also offer suggestions for conservation strategies for elusive carnivore species. However, even in the first decade of the twenty-first century during which molecular phylogenetic studies were prevalent, there was no consensus on phylogeny or divergence times and more comprehensive molecular systematic research was needed. Furthermore, since ecomorphological diversity is so intensive, in particular, at the subfamilial level, the subfamily classification of Mustelidae has been highly contentious. For example, Wozencraft (1993) proposed a framework of five subfamilies: Mustelinae, Melinae, Lutrinae, Mellivorinae, and Taxidiinae, while in a more recent classification, Wozencraft (2005) separated this family into only two subfamilies, Mustelinae and Lutrinae. To clarify the spatiotemporal evolutionary reasons for the extensive ecomorphological diversification of this family, and to address the subfamily classification issue, molecular systematic analyses using a pluralistic phylogenetic method with a supermatrix containing many nucDNA sequences have been conducted (e.g.

Koepfli & Wayne, 2003; Sato *et al.*, 2003, 2004, 2006, 2009, 2012; Fulton & Strobeck, 2006; Koepfli *et al.*, 2008; Wolsan & Sato, 2010; Yu *et al.*, 2011b).

Sato *et al.* (2012) used data from 18 genera and 38 species of the family Mustelidae with 8492-bp nucleotide characters from 10 genetic loci, 9 nuclear genes, and 1 mitochondrial gene (Table 2.1). From this data, phylogenetic relationships and divergence times were estimated and eight major mustelid lineages (subfamilies) were identified (Figure 2.3). The chronological analyses based on the Bayesian relaxed molecular clock method (Thorne *et al.*, 1998; Drummond *et al.*, 2006) indicated that the major mustelid lineages (subfamilies and genera) diversified in two radiation events in the Middle and Late Miocene (Figure 2.3). Sato *et al.* (2012) also inferred the ancestral distribution of the mustelid lineages by parsimony (Sankoff & Rousseau, 1975), likelihood (Ree & Smith, 2008), and Bayesian (Pagel &

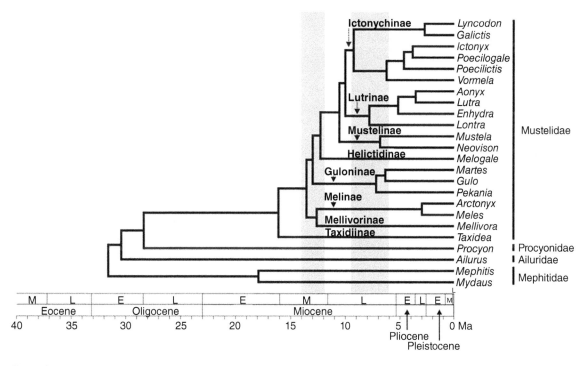

Figure 2.3 A chronogram showing the phylogenetic relationships and divergence times among genera of Musteloidea examined in Sato *et al.* (2012). Chronological data were based on the average value between Multidivtime and BEAST estimates in Sato *et al.* (2012). The shaded parts designate radiation events within Mustelidae. Each family name in Musteloidea is shown on the rightmost side. The eight subfamily names are provided near the ancestral branches. We follow Sato *et al.* (2009, 2012) and Wolsan & Sato (2010) for subfamily names. The letters 'E', 'M', and 'L' in the time scale mean 'Early', 'Middle', and 'Late', respectively.

Meade, 2006) methods and showed that most of the earliest mustelid ancestors occurred in Asia. It is likely that the first radiations among ecomorphologically different lineages (subfamilies) occurred in Asia, implying that Asia could be a central location for the adaptive radiation (Sato *et al.*, 2012). Following the first radiation, the second radiation occurred in a geologically limited time span in the Late Miocene, about 6–9 MYA (Figure 2.3). Interestingly, the first divergence in each major mustelid lineage (e.g. Guloninae, Ictonychinae, Lutrinae, and Mustelinae) was inferred to be inter-continental differentiation between lineages on the Eurasian and North American continents. Hence, it could be argued that in this period, large-scale dispersal from Eurasia to North America may have occurred simultaneously in each major mustelid lineage. One possible reason for the second radiation may be the environmental change occurring in the Late Miocene. For example, global vegetation may have changed from C3-plant forests to C4-plant grasslands 6–8 MYA (Quade *et al.*, 1989; Cerling *et al.*, 1997) and the formation of the Bering Strait was inferred to have occurred 4.8–5.5 MYA (Marincovich & Gladenkov, 1999), which is also consistent with the Late Miocene mustelid dispersal event. This also supports a hypothesis that worldwide environmental changes may have induced inter-continental dispersal of other organisms.

Wolsan & Sato (2010) adopted a supermatrix approach using 52 species and 54 genetic loci (27 965 bp aligned sequence). Although their analysis was focused on the effect of missing data on phylogenetic resolution in the light of empirical data, they proposed a seven-subfamily framework: Guloninae (*Eira*, *Gulo*, and *Martes*), Helictidinae (*Melogale*), Ictonychinae (*Galictis*, *Ictonyx*, *Poecilogale*, and *Vormela*), Lutrinae (*Aonyx*, *Enhydra*, *Hydrictis*, *Lontra*, *Lutra*, and *Pteronura*), Melinae (*Arctonyx* and *Meles*), Mustelinae (*Mustela* and *Neovison*), and Taxidiinae (*Taxidea*). Koepfli *et al.* (2008) and Sato *et al.* (2012) included the honey badger, *Mellivora capensis*, in their phylogenetic analyses and showed that this species constituted a monotypic lineage corresponding to the subfamily Mellivorinae. In addition, Koepfli *et al.* (2008) confirmed with molecular data that *Lutrogale* was a member of the subfamily Lutrinae. Sato *et al.* (2012) further demonstrated that *Lyncodon*, which was the last extant mustelid genus whose phylogenetic position remained to be resolved, was the closest relative of the genus *Galictis* within the subfamily Ictonychinae. Combining the subfamily classification of the family Mustelidae in Wolsan & Sato (2010) with the recent molecular phylogenetic evidence obtained by Koepfli *et al.* (2008) and Sato *et al.* (2012), an eight-subfamily framework would therefore be the most plausible classification in the Mustelidae at this stage (Figure 2.3).

The evolution of the family Mustelidae in space and time was clarified by the same methodological approaches mentioned in the first section, namely nucDNA, Bayesian and fast-ML inference, and the supermatrix approach. Correct inference of the phylogeny and divergence time by taking these methodological factors into account provides us with reliable estimates of endemism for each lineage. In the next section, we caution that using unreliable estimates of endemism could ultimately lead to errors in conservation priority setting.

Implications of Molecular Systematics to Setting and Defining Conservation Priorities

Resources needed for conservation are limited and largely depend on the human economy. In considering what to protect among a vast variety of organisms, it is inevitable to encounter an 'agony of choice' (Vane-Wright *et al.*, 1991). Therefore, we need a criterion for the prioritization of species for conservation. For almost three decades, Phylogenetic Diversity, a measure of biodiversity considering phylogenetic differences, has been considered important in conservation prioritization (e.g. Faith, 1992). Phylogenetic Diversity can reflect the variety of unique or rare features of a species and can capture not only species richness but also functional diversity (Faith, 1992; Safi *et al.*, 2011; Huang *et al.*, 2012; Jono & Pavoine, 2012; but see Hidasi-Neto *et al.*, 2015 for no relation between evolutionary and ecological distinctiveness). Thus, conserving Phylogenetic Diversity could be useful in maintaining a healthy ecosystem. One index utilizing Phylogenetic Diversity is EDGE (Evolutionarily Distinct and Globally Endangered; Isaac *et al.*, 2007).

To prioritize species, this index considers both Evolutionary Distinctiveness, representing Phylogenetic Diversity, and extinction risk inferred from the IUCN Red List categories. The Evolutionary Distinctiveness value is calculated on the basis of the branch length in the phylogenetic tree and the number of descendant taxa from that branch (Isaac *et al.*, 2007). Basically, an endangered species with few related taxa with long branches in its phylogenetic route would have a higher EDGE value. Currently, such incorporation of Evolutionary Distinctiveness is used in the setting of conservation priority. The mammalian EDGE list can be obtained from the EDGE project website (https://www.edgeofexistence.org/).

However, evaluation of Phylogenetic Diversity is highly dependent on how precise the phylogenetic relationships are. When Faith (1992) first introduced the concept of Phylogenetic Diversity, he also noted some cautions as follows: 'the predictive value of PD (Phylogenetic Diversity) depends on having a cladogram that is a reliable estimate of the phylogenetic relationships among the taxa (pp. 8–9)' and 'cladograms based on a small number of characters, or on characters that exhibit large amounts of homoplasy, are probably less reliable (p. 9)'. Most recent studies examining Phylogenetic Diversity for mammals (e.g. Isaac *et al.*, 2007; Collen *et al.*, 2011; Safi *et al.*, 2011, 2013; Huang *et al.*, 2012; Jono & Pavoine, 2012; Hidasi-Neto *et al.*, 2015) are based on the supertree phylogenetic relationships published in *Nature* by Bininda-Emonds *et al.* (2007). This supertree includes the most comprehensive list of species for Mammalia. However, the supertree relationship is not consistent with relationships in other, smaller-scale studies that were often based on the supermatrix method, particularly for carnivoran systematics (e.g. Flynn *et al.*, 2005; Koepfli *et al.*, 2008; Wolsan & Sato, 2010; Sato *et al.*, 2009, 2012). It is likely that this supertree does not present Phylogenetic Diversity scores that are suitable for conservation prioritization, at least for carnivoran species.

We compared the phylogenies and divergence times between the supermatrix and supertree methods for the superfamily Musteloidea. As described previously, Figure 2.3 is the supermatrix chronogram inferred by Sato *et al.* (2012). Figure 2.4 shows the supertree relationships extracted from Bininda-Emonds *et al.* (2007), where there are many unresolved relationships; in addition, this supertree largely reflects a traditional morphological taxonomy that is not supported in recent molecular phylogenetic studies. For example, the Mephitidae is included within the Mustelidae, which is a strong reflection of traditional taxonomy (Figure 2.4). In general, the supermatrix (Figure 2.3) and supertree (Figure 2.4) phylogenies differ markedly. Moreover, the divergence times estimated by the supertree method might be overestimated, where the Oligocene is the main diversification period for the mustelid lineages (Figure 2.4). As we explained above, almost all radiations of the mustelids were inferred to have occurred in the Middle to Late Miocene in the supermatrix method, not the Oligocene (Figure 2.3). This large difference in time estimates suggests a critical problem in using the supertree method for inferring Evolutionary Distinctiveness. In the supertree framework, the calculation of divergence time, the most important factor in determining Evolutionary Distinctiveness value, has several drawbacks because of its indirect estimation, in which date estimates are obtained by fitting source data (e.g. genes) on a given phylogeny (e.g. Jones *et al.* 2005). Therefore, it should be stressed that the Phylogenetic Diversity scores recently presented on the basis of the Bininda-Emonds *et al.* (2007) supertree may not reflect the true evolutionary history of the carnivoran species. It is necessary to devise a more efficient methodology based on the supermatrix in order to obtain a more reliable picture of the evolutionary relationships for conservation.

Recently, Nyakatura & Bininda-Emonds (2012) revised the supertree analysis of Bininda-Emonds *et al.* (1999) for carnivoran species. They concluded that their supertree as depicted in Figure 2.5 was moderately different from the recently published supermatrix tree (e.g. Figure 2.3) and that differences were often observed for higher taxonomic relationships (e.g. among families) due mainly to the failure of the supermatrix analyses to reconstruct commonly evidenced relationships. However, their conclusions were not supported in this study (see Figures 2.2 and 2.4). First, differences in the topology and the divergence times between the supermatrix and supertree approaches were very large. Second, the differences are not concentrated in relationships among distantly related species but are also observed for lower-level relationships (see interrelationships among Ictonychinae genera; dotted

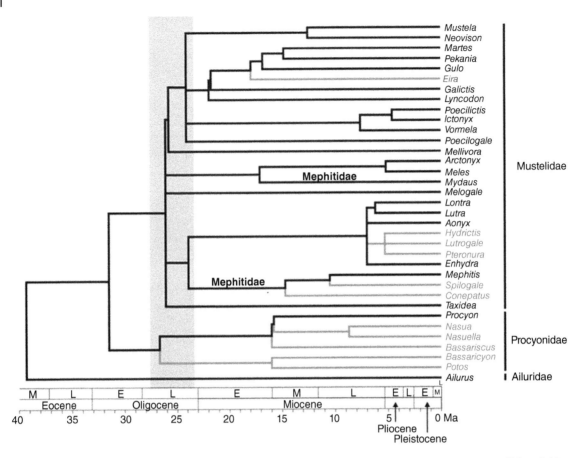

Figure 2.4 A chronogram showing the phylogenetic relationships and divergence times among genera of Musteloidea extracted from tree data (mammalST_bestDates) generated in Bininda-Emonds *et al.* (2007). Each family name for Musteloidea is shown on the rightmost side, except for Mephitidae embedded within Mustelidae. The meaning of the shaded part and letters in the time scale is the same as in Figure 2.2. Branches for the musteloid genera missing in Sato *et al.* (2012) are shown in grey for comparison between both studies.

branches in Figure 2.5). Third, the phylogenetic hypothesis for musteloids has become more consistent in recent supermatrix-based studies (e.g. Sato *et al.*, 2009, 2012; Eizirik *et al.*, 2010; Yu *et al.*, 2011a; Figure 2.2; Table 2.1). Only supertree-based studies have provided inconsistent chronograms (Bininda-Emonds *et al.*, 1999, 2007; Nyakatura & Bininda-Emonds, 2012). It is therefore suggested that the EDGE list presented in Nyakatura & Bininda-Emonds (2012) should be considered with extreme caution.

Agnarsson *et al.* (2010) presented the highest EDGE score for the red panda based on the phylogenetic relationships among almost all species in the order Carnivora (82% of the total carnivoran species) inferred from only mitochondrial cytochrome *b* gene sequences. They showed that the red panda was sister to the dog (Canidae) lineage, suggesting a higher level of Evolutionary Distinctiveness for the red panda lineage due to its long branch. However, the red panda position as a sister to Canidae is highly inconsistent with other recent phylogenetic studies. Their result is probably an artifact caused by the use of only a 1140-bp sequence of a rapidly evolving homoplasious mitochondrial gene for a time scale of ~50 million years. Hence, the saturation problem would have negatively affected the phylogenetic inference. This is a case of

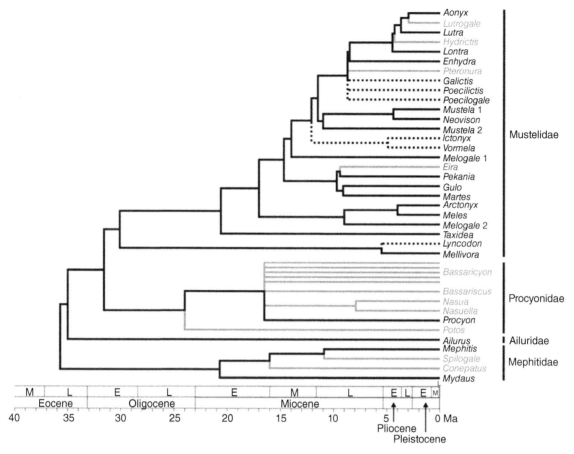

Figure 2.5 A chronogram showing the phylogenetic relationships and divergence times among genera of Musteloidea extracted from tree data (carnivoreST_bestEstimate) generated in Nyakatura & Bininda-Emonds (2012). Each family name for Musteloidea is shown on the rightmost side. The meaning of letters in the time scale and grey branches is the same as in Figures 2.2 and 2.3. Branches for Ictonychinae are shown by dotted lines to highlight the inconsistency with Figure 2.2 topology estimated by Sato *et al.* (2012).

the second caution presented by Faith (1992). The hypothesis proposed by Agnarsson *et al.* (2010) was not supported by recent molecular phylogenetic studies (Table 2.1). Considering the number of extant species and the estimated divergence times for each musteloid family (Figure 2.3), having only one species in the red panda lineage is uncommon compared with the existence of many species in the other families (Mustelidae: 59 species; Procyonidae: 14 species; Mephitidae: 12 species; Wozencraft, 2005). We stress that the red panda has maintained an evolutionarily distinct lineage; however, not as distinct as Agnarsson *et al.* (2010) suggested.

Conclusion

A decade of progress in phylogenetic analysis, such as the adoption of nucDNA, probabilistic Bayesian and likelihood methods, and the supermatrix approach, has produced a reliable and comprehensive evolutionary picture of caniform carnivoran systematics. For example, it helped in clarifying the phylogenetic relationships of enigmatic species such as the red panda as a second offshoot lineage of the superfamily Musteloidea. These methodologies have also contributed to our knowledge of the evolution and classification of the most diversified carnivoran family, the Mustelidae.

Huelsenbeck, J.P. & Ronquist, F. (2001) MRBAYES: Bayesian inference of phylogenetic trees. *Bioinformatics* 17, 754–755.

Huelsenbeck, J.P., Ronquist, F., Nielsen, R. & Bollback, J.P. (2001) Bayesian inference of phylogeny and its impact on evolutionary biology. *Science* 294, 2310–2314.

Huelsenbeck, J.P., Larget, B., Miller, R.E. & Ronquist, F. (2002) Potential applications and pitfalls of Bayesian inference of phylogeny. *Systematic Biology* 51, 673–688.

Isaac, N.J.B., Turvey, S.T., Collen, B., Waterman, C. & Baillie, J.E.M. (2007) Mammals on the EDGE: conservation priorities based on threat and phylogeny. *PLoS One* 2, e296.

Jones, K.E., Bininda-Emonds, O.R. & Gittleman, J.L. (2005) Bats, clocks, and rocks: diversification patterns in Chiroptera. *Evolution* 59, 2243–2255.

Jones, M.R. & Good, J.M. (2016) Targeted capture in evolutionary and ecological genomics. *Molecular Ecology* 25, 185–202.

Jono, C.M. & Pavoine, S. (2012) Threat diversity will erode mammalian phylogenetic diversity in the near future. *PLoS One* 7, e46235.

Kearney, M. & Clark, J.M. (2003) Problems due to data in phylogenetic analyses including fossils: a critical review. *Journal of Vertebrate Paleontology* 23, 263–274

Kishino, H. & Hasegawa, M. (1989) Evaluation of the maximum likelihood estimate of the evolutionary tree topologies from DNA sequence data, and the branching order in Hominoidea. *Journal of Molecular Evolution* 29, 170–179.

Koepfli, K.-P. & Wayne, R.K. (2003) Type I STS markers are more informative than cytochrome *b* in phylogenetic reconstruction of the Mustelidae (Mammalia: Carnivora). *Systematic Biology* 52, 571–593.

Koepfli, K.-P., Deer, K.A., Slater, G.J., Begg, C., Begg, K., Grassman, L., Lucherini, M., Veron, G. & Wayne. R.K. (2008) Multigene phylogeny of the Mustelidae: resolving relationships, tempo and biogeographic history of a mammalian adaptive radiation. *BMC Biology* 6, 10.

Larget, B. & Simon, D.L. (1999) Markov chain Monte Carlo algorithms for the Bayesian analysis of phylogenetic trees. *Molecular Biology and Evolution* 16, 750–759.

Larivière, S. & Jennings, A.P. (2009) Family Mustelidae (weasels and relatives). In: *Handbook of the Mammals of the World.* Volume 1. Carnivores (eds D.E. Wilson & R.A. Mittermeier), pp. 564–656. Lynx Edicions, Barcelona.

Ledje, C. & Arnason, U. (1996a) Phylogenetic analyses of complete cytochrome *b* genes of the order Carnivora with particular emphasis on the Caniformia. *Journal of Molecular Evolution* 42, 135–144.

Ledje, C. & Arnason, U. (1996b) Phylogenetic relationships within caniform carnivores based on analyses of the mitochondrial 12S rRNA gene. *Journal of Molecular Evolution* 43, 641–649.

Lemmon, E.M. & Lemmon, A.R. (2013) High-throughput genomic data in systematics and phylogenetics. *The Annual Review of Ecology, Evolution, and Systematics* 44, 99–121.

Macdonald, D.W. (ed.) (2009) *The Encyclopedia of Mammals.* 2nd edition. Oxford University Press, Oxford.

Mardis, E.R. (2013) Next-generation sequencing platforms. *Annual Reviews of Analytical Chemistry* 6, 287–303.

Marincovich, L. & Gladenkov, A.Y. (1999) Evidence for an early opening of the Bering Strait. *Nature* 397, 149–151.

Marmi, J., López-Giráldez, J.F. & Domingo-Roura, X. (2004) Phylogeny, evolutionary history and taxonomy of the Mustelidae based on sequences of the cytochrome *b* gene and a complex repetitive flanking region. *Zoologica Scripta* 33, 481–499.

Mau, B. & Newton, M.A. (1997) Phylogenetic inference for binary data on dendograms using Markov Chain Monte Carlo. *Journal of Computational and Graphical Statistics* 6, 122–131.

McCormack, J.E., Hird, S.M., Zellmer, A.J., Carstens, B.C. & Brumfield, R.T. (2013) Applications of next-generation sequencing to phylogeographjy and phylogenetics. *Molecular Phylogenetics and Evolution* 66, 526–538.

Morlo, M. & Peigné, S. (2010) Molecular and morphological evidence for Ailuridae and a review of its genera. In: *Carnivoran Evolution: New Views on Phylogeny, Form and Function* (eds A. Goswami & A. Friscia), pp. 92–140. Cambridge University Press, Cambridge.

Nowak, R.M. (1991) *Walker's Mammals of the World. Volume 2.* 5th edition. The Johns Hopkins University Press, Baltimore.

Nyakatura, K. & Bininda-Emonds, O.R.P. (2012) Updating the evolutionary history of Carnivora (Mammalia): a new species-level supertree complete with divergence time estimates. *BMC Biology* 10, 12.

Nylander, J.A.A., Ronquist, F., Huelsenbeck, J.P. & Nieves-Aldrey, J.L. (2004) Bayesian phylogenetic analysis of combined data. *Systematic Biology* 53, 47–67.

Pagel, M. & Meade, A. (2006) Bayesian analysis of correlated evolution of discrete characters by reversible-jump Markov chain Monte Carlo. *American Naturalist* 167, 808–825.

Pecon-Slattery, J. & O'Brien, S.J. (1995) Molecular phylogeny of the red panda (*Ailurus fulgens*). *Journal of Heredity* 86, 413–422.

Peng, R., Zeng, B., Meng, X., Yue, B., Zhang, Z. & Zou, F. (2007) The complete mitochondrial genome and phylogenetic analysis of the giant panda (*Ailuropoda melanoleuca*). *Gene* 397, 76–83.

Philippe, H., Snell, E.A., Bapteste, E., Lopez, P., Holland, P.W.H. & Casane, D. (2004) Phylogenomics of eukaryotes: impacts of missing data on large alignments. *Molecular Biology and Evolution* 21, 1740–1752.

Quade, J., Cerling, T.E. & Bowman, J.R. (1989) Development of Asian monsoon revealed by marked ecological shift during the latest Miocene in northern Pakistan. *Nature* 342, 163–166.

Rannala, B. & Yang, Z. (1996) Probability distribution of molecular evolutionary trees: a new method of phylogenetic inference. *Journal of Molecular Evolution* 43, 304–311.

Ree, R.H. & Smith, S.A. (2008) Maximum likelihood inference of geographic range evolution by dispersal, local extinction, and cladogenesis. *Systematic Biology* 57, 4–14.

Ronquist, F. & Huelsenbeck, J.P. (2003) Mrbayes 3: Bayesian phylogenetic inference under mixed models. *Bioinformatics* 19, 1572–1574.

Rubin, B.E., Ree, R.H. & Moreau, C.S. (2012) Inferring phylogenies from RAD sequence data. *PLoS One* 7, e33394.

Safi, K., Cianciaruso, M.V., Loyola, R.D., Brito, D., Armour-Marshall, K. & Diniz-Filho, J.A.F. (2011) Understanding global patterns of mammalian functional and phylogenetic diversity. *Philosophical Transactions of the Royal Society B: Biological Sciences* 366, 2536–2544.

Safi, K., Armour-Marshall, K., Baillie, J.E.M. & Isaac, N.J.B. (2013) Global patterns of evolutionary distinct and globally endangered amphibians and mammals. *PLoS One* 8, e63582.

Sankoff, D. & Rousseau, P. (1975) Locating the vertices of a Steiner tree in an arbitrary metric space. *Mathematical Programming* 9, 240–246.

Sato, J.J. & Wolsan, M. (2012) [Evolutionary origin of the red panda (*Ailurus fulgens*)]. *Mammalian Science (Honyurui Kagaku)* 52, 23–40. (In Japanese with English abstract).

Sato, J.J., Hosoda, T., Wolsan, M., Tsuchiya, K., Yamamoto, Y. & Suzuki, H. (2003) Phylogenetic relationships and divergence time among mustelids (Mammalia; Carnivora) based on nucleotide sequences of the nuclear interphotoreceptor retinoid binding protein and mitochondrial cytochrome *b* genes. *Zoological Science* 20, 243–264.

Sato, J.J., Hosoda, T., Wolsan, M. & Suzuki, H. (2004) Molecular phylogeny of Arctoids (Mammalia; Carnivora) with emphasis on phylogenetic and taxonomic positions of the ferret-badgers and skunks. *Zoological Science* 21, 111–118.

Sato, J.J., Wolsan, M., Suzuki, H., Hosoda, T., Yamaguchi, Y., Hiyama, K., Kobayashi, M. & Minami, S. (2006) Evidence from nuclear DNA sequences shed light on the phylogenetic relationships of Pinnipedia: single origin with affinity to Musteloidea. *Zoological Science* 23, 125–146.

Sato, J.J., Wolsan, M., Minami, S., Hosoda, T., Sinaga, M.H., Hiyama, K., Yamaguchi, Y. & Suzuki, H. (2009) Deciphering and dating the red panda's ancestry and early adaptive radiation of Musteloidea. *Molecular Phylogenetics and Evolution* 53, 907–922.

Sato, J.J., Wolsan, M., Prevosti, F.J., Delia, G., Begg, C., Begg, K., Hosoda, T., Campbell, K.L. & Suzuki, H. (2012) Evolutionary and biogeographic history of weasel-like carnivorans (Musteloidea). *Molecular Phylogenetics and Evolution* 63: 745–757.

Sato, J.J., Ohdachi, S.D., Echenique-Diaz, L.M., Borroto-Páez, R., Begué-Quiala, G., Delgado-Labañino, J.L., Gámez-Díez, J., Alvarez-Lemus, J., Nguyen, S.T., Yamaguchi, N. & Kita, M. (2016) Molecular phylogenetic analysis of nuclear genes suggests a Cenozoic over-water dispersal origin for the Cuban solenodon. *Scientific Reports* 6, 31173.

Sato, J.J., Bradford, T.M., Armstrong, K.N., Donnellan, S.C., Echenique-Diaz, L.M., Begué-Quiala, G., Gámez-Díez, J., Yamaguchi, N., Nguyen, S.T., Kita, M. & Ohdachi, S.D. (2019) Post K-Pg diversification of the mammalian order Eulipotyphla as suggested by phylogenomic analyses of ultra-conserved elements. *Molecular Phylogenetics and Evolution* 141, 106605.

Shimodaira, H. (2002) An approximately unbiased test of phylogenetic tree selection. *Systematic Biology* 51, 492–508.

Simmon, M.P., Pickett, K.M. & Miya, M. (2004) How meaningful are Bayesian support values? *Molecular Biology and Evolution* 21, 188–199.

Springer, M.S., DeBry, R.W., Douady, C., Amrine, H.M., Madsen, O., de Jong, W.W. & Stanhope, M.J. (2001) Mitochondrial versus nuclear gene sequences in deep-level mammalian phylogeny reconstruction. *Molecular Biology and Evolution* 18,132–143.

Stamatakis, A., Hoover, P. & Rougemont, J. (2008) A rapid bootstrap algorithm for the RAxML web servers. *Systematic Biology* 57, 758–771.

Steppan, S.J., Adkins, R.M., Spinks, P.Q. & Hale, C. (2005) Multigene phylogeny of the Old World mice, Murinae, reveals distinct geographic lineages and the declining utility of mitochondrial genes compared to nuclear genes. *Molecular Phylogenetics and Evolution* 37, 370–388.

Swofford, D.L. (2002) *PAUP*. Phylogenetic Analysis Using Parsimony (* and Other Methods)*. Version 4. Sinauer Associates, Sunderland.

Templeton, A.R. (1983) Phylogenetic inference from restriction endonuclease cleavage site maps with particular reference to the evolution of humans and the apes. *Evolution* 37, 221–244.

Thorne, J.L., Kishino, H. & Painter, I.S. (1998) Estimating the rate of evolution of the rate of molecular evolution. *Molecular Biology and Evolution* 15, 1647–1657.

Vane-Wright, R.I., Humphries, C.J. & Williams, P.H. (1991) What to protect? Systematics and the agony of choice. *Biological Conservation* 55, 235–254.

Vrana, P.B., Milinkovitch, M.C., Powell, J.R. & Wheeler, W.C. (1994) Higher level relationships of the arctoid Carnivora based on sequence data and "Total Evidence". *Molecular Phylogenetics and Evolution* 3, 47–58.

Wang, X. (1997) New cranial material of *Simocyon* from China, and its implications for phylogenetic relationship to the red panda (*Ailurus*). *Journal of Vertebrate Paleontology* 17, 184–198.

Wiens, J.J. (2003) Missing data, incomplete taxa, and phylogenetic accuracy. *Systematic Biology* 52, 528–538.

Wiens, J.J. (2006) Missing data and the design of phylogenetic analyses. *Journal of Biological Informatics* 39, 34–42.

Wolsan, M. (1993) Phylogeny and classification of early European Mustelida (Mammalia: Carnivora). *Acta Theriologica* 38: 345–384.

Wolsan, M. & Sato, J.J. (2010) Effects of data incompleteness on the relative performance of parsimony and Bayesian approaches in a supermatrix phylogenetic reconstruction of Mustelidae and Procyonidae (Carnivora). *Cladistics* 26, 168–194.

Wozencraft, W.C. (1989) The phylogeny of the recent Carnivora. In: *Carnivore Behavior, Ecology, and Evolution. Volume 1* (ed. J. Gittleman), pp. 495–535. Cornell University Press, Ithaca.

Wozencraft, W.C. (1993) Order Carnivora. In: *Mammal Species of the World: A Taxonomic and Geographic Reference. Volume 1*. 2nd edition (eds D.E. Wilson & D.M. Reeder), pp. 279–348. The Johns Hopkins University Press, Baltimore.

Wozencraft, W.C. (2005) Order Carnivora. In: *Mammal Species of the World: A Taxonomic and Geographic Reference. Volume 1*. 3rd edition (eds D.E. Wilson & D.M. Reeder), pp. 532–628. The Johns Hopkins University Press, Baltimore, Maryland.

Wyss, A.R. & Flynn, J.J. (1993) A phylogenetic analysis and definition of the Carnivora. In: *Mammal Phylogeny: Placentals* (eds F. Szalay, M. Novacek & M. McKenna), pp. 32–52. Springer-Verlag, New York.

Yonezawa, T., Nikaido, M., Kohno, N., Fukumono, Y., Okada, N. & Hasegawa, M. (2007) Molecular phylogenetic study on the origin and evolution of Mustelidae. *Gene* 396, 1–12.

Yu, L. & Zhang, Y.-P. (2006) Phylogeny of the caniform carnivora: evidence from multiple genes. *Genetica* 127, 65–79.

Yu, L., Li, Q.-W., Ryder, O.A. & Zhang, Y.-P. (2004) Phylogenetic relationships within mammalian order Carnivora indicated by sequences of two nuclear DNA genes. *Molecular Phylogenetics and Evolution* 33, 694–705.

Yu, L., Liu, J., Luan, P.-T., Lee, H., Lee, M., Min, M.-S., Ryder, O.A., Chemnick, L., Davis, H. & Zhang, Y.-P. (2008) New insight into the evolution of intronic sequences of the β-fibrinogen gene and their application in reconstructing mustelid phylogeny. *Zoological Science* 25, 662–672.

Yu, L., Luan, P.-T., Jin, W., Ryder, O.A., Chemnich, L.G., Davis, H.A. & Zhang, Y.-P. (2011a) Phylogenetic utility of nuclear introns in interfamilial relationships of Caniformia (Order Carnivora). *Systematic Biology* 60, 175–187.

Yu, L., Peng, D., Liu, J., Luan, P., Liang, L., Lee, H., Lee, M., Ryder, O.A. & Zhang, Y.-P. (2011b) On the phylogeny of Mustelidae subfamilies: analysis of seventeen nuclear non-coding loci and mitochondrial complete genomes. *BMC Evolutionary Biology* 11, 92.

Zehr, S.M., Nedbal, M.A. & Flynn, J.J. (2001) Tempo and mode of evolution in an orthologous *Can* SINE. *Mammalian Genome* 12, 38–44.

Zhang, Y.-P. & Ryder, O.A. (1993) Mitochondrial DNA sequence evolution in the Arctoidea. *Proceedings of the National Academy of Sciences of the United States of America* 90, 9557–9561.

Zhang, Y.-P. & Ryder, O.A. (1997) A molecular phylogeny of the Arctoidea. *Chinese Journal of Genetics* 23, 239–246.

Zwickl, D.J. (2006) *Genetic Algorithm Approaches for the Phylogenetic Analysis of Large Biological Sequence Datasets under the Maximum Likelihood Criterion.* PhD thesis, University of Texas at Austin.

3

Systematics and Evolution of the Mongooses (Herpestidae, Carnivora)

Géraldine Veron[1],*, Marie-Lilith Patou[2], and Andrew P. Jennings[3]

[1]*Institut de Systématique, Evolution, Biodiversité (ISYEB), Muséum National d'Histoire Naturelle, CNRS, Sorbonne Université, EPHE, Université des Antilles, Paris, France*
[2]*Biotope, Recherche & Développement, Mèze, France*
[3]*Small Carnivores – Research and Conservation, Portland, ME, USA*

SUMMARY

The Herpestidae is an ecologically and behaviourally diverse family that comprises 25 African and 9 Asian mongoose species. They are slender, small carnivores (ranging from 200 g to 5 kg) that live in Africa, the Middle East, and Asia; one mongoose species is found in Europe and a few species have been introduced in many places in the world. The Herpestidae were initially included in the family Viverridae (civets, genets, and oyans), and previously contained the Malagasy 'mongooses' (subfamily Galidiinae). Molecular systematics and morphological studies have now confirmed that the 'true' mongooses should be placed in a separate family, the Herpestidae, and that the Malagasy 'mongooses' (together with the other Malagasy carnivores) be placed in the family Eupleridae. Recent molecular studies have shown that there are 2 subfamilies within the Herpestidae: the Mungotinae (11 small, social mongooses that occur in Africa) and the Herpestinae (23 larger, non-social mongooses that are found in Asia and Africa). In addition, the genus *Herpestes* has been shown to be paraphyletic; the nine Asian species of *Herpestes* form a monophyletic group and should now be placed in the genus *Urva*. Recent studies have inferred an Early Miocene African origin for the Herpestidae, and a Middle Miocene origin for the Asian mongooses. The evolution of life traits (social organization, activity, and habitat preferences) suggests that mongooses were originally non-social, diurnal, and diversified much more in Africa than in Asia. In this chapter, we review the recent changes in the inter-familial relationships of the Herpestidae and the Malagasy 'mongooses', provide an up-to-date phylogeny of the mongooses, describe the evolution of mongoose life traits, and present the latest systematic classification of the Asian mongooses, based on recent molecular studies.

Keywords

Africa — Asia — biogeography — classification — DNA — *Herpestes* — Feliformia — *Urva*

Introduction

Mongooses are slender, small carnivores, ranging from 200 g (common dwarf mongoose, *Helogale parvula*) to 5 kg (white-tailed mongoose, *Ichneumia albicauda*). They all have a similar morphology and are characterized by their small size, long face and body, short legs, small rounded ears, and long, tapering bushy tails. Most mongooses have a uniform pelage with long coarse hairs that are ringed with different colours in some species, giving a grizzled aspect to the coat. They all share a number of anatomical features, such as the structure of the auditory bullae, specialized ear cartilage, and the presence of an anal pouch. Mongooses are digitigrade and are mainly terrestrial (Gilchrist *et al.*, 2009; Jennings & Veron, 2019).

---

*Corresponding author.

Small Carnivores: Evolution, Ecology, Behaviour, and Conservation, First Edition. Edited by Emmanuel Do Linh San, Jun J. Sato, Jerrold L. Belant, and Michael J. Somers.

Phylogeny of Mongooses and Evolution of Life Traits

Veron *et al.* (2004) showed on the basis of molecular data that the Herpestidae should be split into 2 subfamilies: the Mungotinae (comprising 11 small, social species), and the Herpestinae (containing 23 larger, solitary species) (Figure 3.2). Subsequent molecular studies have confirmed this division (Perez *et al.*, 2006; Patou *et al.*, 2009). Patou *et al.* (2009) inferred an Early Miocene African origin for the Herpestidae, and a Middle Miocene origin for the Asian mongooses.

The yellow mongoose, *Cynictis penicillata* (Figure 3.3a), possesses social traits and was allied to the social mongooses by Wozencraft (1989b), while Gregory & Hellman (1939), Hendey (1974), and Taylor *et al.* (1991) placed it among the solitary mongooses. Based on molecular data, Veron *et al.* (2004) showed that this species should be included in the solitary mongoose group (see Figure 3.2). The morphological features that prompted several authors to consider the yellow mongoose closely related to the social mongooses (Petter, 1969; Wozencraft, 1989b; Veron, 1995) are apparently the result of convergence in ecological and behavioural characteristics (open habitat, insectivorous diet, social family life, diurnal activity, and communal burrows). Although several authors have mentioned that colonies of *C. penicillata* can consist of up to 40–50 individuals (Fitzsimons, 1919; Roberts, 1951; Walker *et al.*, 1964; Dorst & Dandelot, 1972), mean colony sizes of 3.9, 4.1 and 8.0 were observed by Zumpt (1976), Lynch (1980), and Earlé (1981), respectively. According to Earlé (1981), Taylor & Meester (1993), and Wenhold & Rasa (1994), the yellow mongoose hunts alone. Its social behaviour seems only slightly more developed than that of the Egyptian mongoose, *H. ichneumon*, which is a solitary species that occasionally forms social groups (Ben-Yaacov & Yom-Tov, 1983). On the other hand, Balmforth (2004) and Vidya *et al.* (2009) observed large groups of yellow mongooses on farmland, with numerous social interactions and cooperative breeding. However, yellow mongooses are not obligate pack foragers and cooperative breeders as are the true social mongooses (for a review, see Schneider & Kappeler, 2014).

Veron *et al.* (2004), using molecular data, showed that the true social mongooses form a monophyletic group (corresponding to the subfamily Mungotinae) from which the yellow mongoose, *C. penicillata*, is excluded. The social mongooses are characterized by their small size and the presence of long claws on the forefeet. They live in stable groups that are larger than a single-family unit, breed cooperatively, and forage in packs (Gilchrist *et al.*, 2009; Jennings & Veron, 2019). The monophyly of the true social mongooses implies that sociality evolved once in this group. The appearance of extensive areas of grassland during the late Miocene and Pliocene periods favoured the evolution of insectivorous small carnivores that could feed on the abundant insect resources within this habitat. The Resource Dispersion Hypothesis (Macdonald, 1983; Carr & Macdonald, 1986) posits that the quality and dispersion of resources influence the social structure of an animal population in a given habitat, and the abundance of insects in grassland habitats may have facilitated group formation in mongooses (Rood, 1986). Thus, the availability and renewability of their invertebrate food (which decreases the costs of group living), and the higher predation risk in open habitat, may have been the main selective pressures promoting sociality in mongooses (Waser, 1981; Rood, 1986; Palomares & Delibes, 1993). The abundance of shelters (constructed by other animals) that can provide suitable cover also allows communal denning in African social mongooses (Rood, 1986).

Within the Mungotinae, the meerkat, *Suricata suricatta* (Figure 3.4), is the sister taxon of a clade containing the other social mongooses; this is congruent with the recognition of this species as morphologically distinct, which had resulted in it being placed in a separate subfamily, the Suricatinae, by Pocock (1919). The Liberian mongoose, *Liberiictis kuhni*, is the sister taxon of the banded mongooses (*Mungos*), while the dwarf mongooses (*Helogale*) are closely related to the cusimanses (*Crossarchus*) (see Veron *et al.*, 2004). Within the latter genus, four species were described (see Goldman, 1984), *Crossarchus alexandri*, *Crossarchus ansorgei*, *Crossarchus obscurus*, and *Crossarchus platycephalus*, and these were recently reassessed using morphometric and molecular data (see Sonet *et al.*, 2014). Although Sonet *et al.* (2014) confirmed the current taxonomic classification of these four *Crossarchus* species, further studies,

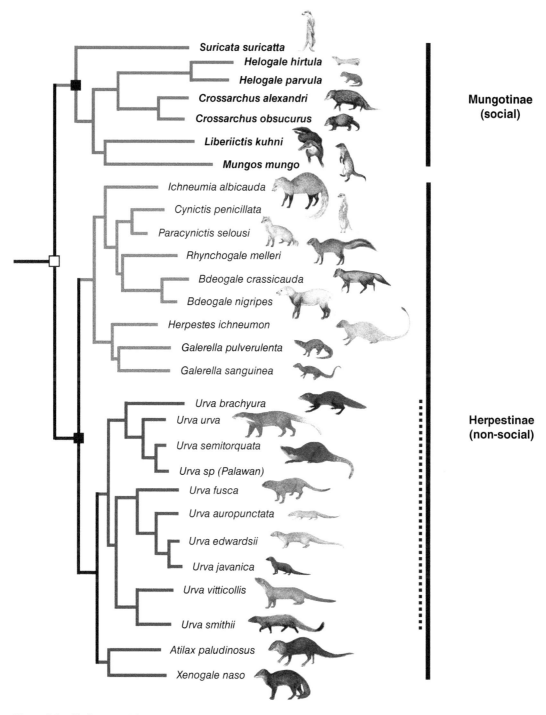

Figure 3.2 Phylogeny of the mongooses, based on Veron *et al.* (2004, 2007, 2015), Perez *et al.* (2006), Gilchrist *et al.* (2009), and Patou *et al.* (2009). The white square indicates the family Herpestidae, the black squares and the black bars on the side indicate the two subfamilies, the Mungotinae and Herpestinae. Species in bold are the true social mongooses (Mungotinae). The dotted line indicates the monophyletic Asian species clade. The branches of the major clades (discussed in the text) are in different colours. *Source:* Animal illustrations from Prater (1971), Payne et al. (1985), and Kingdon (2015). Reproduced by permission of the Bombay Natural History Society, Karen Philipps, and Jonathan Kingdon.

(a)

(b)

Figure 3.3 (a) The yellow mongoose, *Cynictis penicillata*, may share communal burrows with several conspecifics, but it is essentially a solitary forager. Molecular studies indicated that it belongs to the clade of solitary mongooses, the subfamily Herpestinae. (b) Selous's mongoose, *Paracynictis selousi*, was shown to be the sister species of the yellow mongoose. *Source:* Photos © Emmanuel Do Linh San (a) and Peter Apps (b).

Figure 3.4 Meerkats, *Suricata suricatta*, belong to the subfamily Mungotinae. These highly social mongooses stand out as one of the best-studied carnivore species globally. *Source:* Photo © Emmanuel Do Linh San.

using additional genetic markers (e.g. nuclear genes), are needed.

Besides the clade of true social mongooses, Veron *et al.* (2004) also obtained a clade containing the solitary mongooses (including *Atilax*, *Cynictis*, *Galerella*, *Herpestes*, *Ichneumia*, and *Rhynchogale*) that correspond to the subfamily Herpestinae (see Figure 3.2). Mongooses within this subfamily share several morphological characters: first upper premolars present (but variable in *Atilax*), a similar shape of the cheek teeth and of the tympanic bullae (Veron, 1994), and a relatively large size (up to ~5 kg; Gilchrist *et al.*, 2009; Jennings & Veron, 2019). Some mongooses in this group share a particular chromosomal feature: the Y chromosome is translocated onto an autosome, resulting in a different number of chromosomes in the male and female (Wurster & Benirschke, 1968; Fredga, 1972; Pathak & Stock, 1976). This occurs in *Atilax paludinosus* (female: 2N = 36; male: 2N = 35), *Galerella pulverulenta* (40; 39), *Galerella sanguinea* (42; 41), *Urva javanica* (36; 35), *Urva edwardsii* (36; 35), and *Herpestes ichneumon* (44; 43).

Within the subfamily Herpestinae, recent molecular studies have shown that the long-nosed mongoose, *Xenogale naso*, and the marsh mongoose, *Atilax paludinosus*, are sister species (Veron *et al.*, 2004; Perez *et al.*, 2006). These two mongooses have a similar morphology and live in the same habitat types (swamps and riverine forests), but differ in activity, *X. naso* being diurnal and *A. paludinosus* crepuscular (Ray, 1997). Their close relationship had never been proposed before (for review see Bininda-Emonds *et al.*, 1999; Veron *et al.*, 2004). Even though some authors had observed their morphological resemblances, these were believed to be convergences resulting from similar adaptations to their habitat (Orts, 1970; Rosevear, 1974; Ray, 1997). The long-nosed mongoose was generally regarded as the sister taxon of the Egyptian mongoose and, therefore, placed in the genus *Herpestes* (Coetzee, 1977; Kingdon, 1977; Happold, 1987; Corbet & Hill, 1991; Wozencraft, 1993, 2005). Allen (1919) placed it in *Xenogale*, and this was followed by Gregory & Hellman (1939), Rosevear (1974), Ansell (1978), Colyn & Van Rompaey (1994), and Van Rompaey & Colyn (2013). Veron *et al.* (2004) and Perez *et al.* (2006) confirmed that the long-nosed mongoose should be placed in a separate genus, *Xenogale*, based on molecular evidence.

The species of *Galerella* were placed in the genus *Herpestes* by some authors (see Wozencraft, 1989b; Taylor & Goldman, 1993), but their inclusion in a separate genus was supported by allozyme (Taylor *et al.*, 1991) and some morphological data (see Rosevear, 1974), whereas the craniometric study of Taylor & Matheson (1999) did not support the monophyly of this genus. Veron *et al.* (2004) advocated retaining them in *Galerella*, based on their morphological distinctiveness and genetic distance from *Herpestes ichneumon*. Four species are presently recognized in *Galerella* (see Gilchrist *et al.*, 2009; Jennings & Veron, 2019), among which only two have been included in any molecular phylogeny (Cape grey mongoose, *Galerella pulverulenta*, and common slender mongoose, *Galerella sanguinea*). Rapson *et al.* (2012) have advocated that an additional taxonomic entity, *Galerella nigrata*, closely related to *G. sanguinea*, should be recognized as a species. However, this would make *G. sanguinea* paraphyletic, and, thus, further investigations are needed to test the validity of this proposition and to check for any possible hybridization between these taxa.

A monophyletic group of herpestine mongooses comprises *Bdeogale*, *Cynictis*, *Ichneumia*, and *Rhynchogale* (Perez *et al.*, 2006). This group is characterized by a very wide and bushy tail, as well as a large posterior cusp on the third inferior premolar. The cytogenetic data of Fredga (1972) and Wurster & Benirschke (1968) are congruent with this grouping: *Bdeogale*, *Cynictis*, and *Ichneumia* have 2N = 36 chromosomes (i.e. they do not have the chromosome translocation in males that is observed in other species of solitary mongooses) and they have a reduced sexual Y chromosome.

The genus *Bdeogale* is currently separated into three species (Honacki *et al.*, 1982; Wozencraft, 2005): the black-legged mongoose, *B. nigripes*, inhabits the tropical belt from E Nigeria to NE Democratic Republic of the Congo and to N Angola; Jackson's mongoose, *B. jacksoni*, is restricted to SW Kenya; and the bushy-tailed mongoose, *B. crassicauda*, occurs in Kenya, Malawi, Mozambique, Tanzania, Zambia, and Zimbabwe (Wozencraft, 2005). Rosevear (1974) believed that the coat colour characteristics used to separate these species could simply be intra-specific colour variations. However, Perez *et al.* (2006) found 4.7% molecular divergence for the cytochrome *b* gene between *B. crassicauda* and *B. nigripes*, which is similar to that seen between other mongoose species. The third species, *B. jacksoni*, was considered conspecific with *B. nigripes* by Kingdon (1977). In contrast, Kingdon (1997) and Taylor (2013) treated Sokoke bushy-tailed mongoose, *B. omnivora*, from the coastal forests of Northern Tanzania and Eastern Kenya, as a distinct species from *B. crassicauda* due to differences in coat colour and body size. Molecular analyses are needed to help resolve these conflicting cases. According to Thomas (1882), Pocock (1919) and Gregory & Hellman (1939), *Bdeogale* shares many dental features with the white-tailed mongoose, *Ichneumia albicauda*, as well as a similar external morphology. However, *Bdeogale* possesses many foot specializations (e.g. the suppression of the pollex and the hallux, the shortening of the four main digits, and symmetrical toes), as well as expanded molars, which have been said to be related to an insectivorous diet (mainly ants and termites; Kingdon, 1977; Smithers, 1983). Its densely furred coat, muzzle and feet, and short, woolly ears may be a protection against soldier ants or termites (Kingdon, 1977). However, in *B. nigripes*, small vertebrates (mainly rodents and insectivores) have also been reported as an important part of the diet (Ray & Sunquist, 2001).

Meller's mongoose, *Rhynchogale melleri*, was first believed to be closely associated with *Crossarchus* and *Suricata* by Gray (cited by Pocock, 1919) and Thomas (1882). Pocock (1919) suggested instead that it is closest to *Ichneumia* and *Bdeogale*, notably on the basis of dental characters. The body shape of *Rhynchogale* is very similar to that of *Ichneumia* (Kingdon, 1977). However, *Rhynchogale* differs from the other members of the herpestine clade by the absence of the groove on the upper lip (Pocock, 1919), a distinctly snub nose, and the flatness of its molars (Kingdon, 1977). The last feature may have resulted from a dietary adaptation, although some authors have suggested that Meller's mongoose may feed mostly on termites (see Smithers, 1966), whereas others consider that this species has a frugivorous diet (see Pocock, 1919 and Kingdon, 1977); in fact, its biology is almost entirely unknown.

The yellow mongoose, *C. penicillata*, possesses social traits and was allied to the social mongooses by some authors and to the solitary mongooses by others (see above). Pocock (1919) emphasized that the position of *Cynictis* is difficult to establish on the basis of its morphological specializations, but suggested a close relationship with the white-tailed mongoose, *I. albicauda*, on the basis of ear and plantar pad characters. Molecular studies by Veron *et al.* (2004) and Flynn *et al.* (2005) revealed the exclusion of *Cynictis* from the social mongoose group, and Perez *et al.* (2006) showed its inclusion in the bushy-tailed mongoose clade (*Bdeogale* and allies). Selous's mongoose, *Paracynictis selousi* (Figure 3.3b), was considered the sister taxon of *C. penicillata* (Pocock, 1919), and this has been confirmed by molecular results (Flynn *et al.*, 2005).

Veron *et al.* (2004) have shown that social traits appeared at the base of the social mongoose clade (in accordance with the hypothesis of Gorman, 1979). Some male associations (cohesive association of males sharing the same home range) have been observed in non-social mongooses, including slender and Cape grey mongooses (Rood & Waser, 1978; Rood, 1989; Cavallini & Nel, 1990). Waser *et al.* (1994) showed that some male associations in the common slender mongoose can last a very long time (at least seven years). Studies on the Egyptian mongoose have shown some signs of sociality within this species under certain ecological conditions (Ben-Yaacov & Yom-Tov, 1983; Palomares & Delibes, 1993). These findings, in addition to the variable degrees of sociality in yellow mongoose populations, reveal the potential for sociality in this family. Sociality in mongooses evolved only in Africa, and not in Asia, probably due to the opening up of forested habitats in Africa (see above).

Most mongooses are diurnal (Table 3.1; Gilchrist *et al.*, 2009; Jennings & Veron, 2019) and mapping this activity character onto the recent phylogeny of the mongooses allows us to infer the evolution of activity patterns within the Hespestidae. The nocturnal activity appears to have evolved only twice in Africa, within the clade that includes *Bdeogale*, *Cynictis*, *Ichneumia*, *Paracynictis*, and *Rhynchogale*, and in *Atilax*, and once in Asia, in the Indian brown mongoose, *U. fusca*; this might have arisen in order to avoid competition with similar-sized, diurnal mongooses.

Table 3.1   Classification of the Herpestidae showing 2 subfamilies, the Mungotinae (11 small, social species) and the Herpestinae (23 large, solitary species), based on Veron *et al.* (2004, 2007, 2015), Perez *et al.* (2006), Gilchrist *et al.* (2009), and Patou *et al.* (2009).

Subfamily and species	Common English name	Distribution	Social organization	Activity
Mungotinae				
Crossarchus alexandri	Alexander's cusimanse	Africa	Social	Diurnal
Crossarchus ansorgei	Angolan cusimanse	Africa	Social	Diurnal
Crossarchus obscurus	Common cusimanse	Africa	Social	Diurnal
Crossarchus platycephalus	Flat-headed cusimanse	Africa	Social	Diurnal
Dologale dybowskii	Pousargues's mongoose	Africa	Social	Diurnal
Helogale hirtula	Ethiopian dwarf mongoose	Africa	Social	Diurnal
Helogale parvula	Common dwarf mongoose	Africa	Social	Diurnal
Liberiictis kuhni	Liberian mongoose	Africa	Social	Diurnal
Mungos gambianus	Gambian mongoose	Africa	Social	Diurnal
Mungos mungo	Banded mongoose	Africa	Social	Diurnal
Suricata suricatta	Meerkat	Africa	Social	Diurnal
Herpestinae				
Atilax paludinosus	Marsh mongoose	Africa	Solitary	Nocturnal
Bdeogale crassicauda	Bushy-tailed mongoose	Africa	Solitary	Nocturnal
Bdeogale jacksoni	Jackson's mongoose	Africa	Solitary	Nocturnal
Bdeogale nigripes	Black-legged mongoose	Africa	Solitary	Nocturnal
Cynictis penicillata	Yellow mongoose	Africa	Semi-social	Diurnal
Galerella flavescens	Kaokoveld slender mongoose	Africa	Solitary	Diurnal
Galerella ochracea	Somali slender mongoose	Africa	Solitary	Diurnal
Galerella pulverulenta	Cape grey mongoose	Africa	Solitary	Diurnal
Galerella sanguinea	Common slender mongoose	Africa	Solitary	Diurnal
Herpestes ichneumon	Egyptian mongoose	Africa	Solitary	Diurnal
Ichneumia albicauda	White-tailed mongoose	Africa	Solitary	Nocturnal
Paracynictis selousi	Selous's mongoose	Africa	Solitary	Nocturnal
Rhynchogale melleri	Meller's mongoose	Africa	Solitary	Nocturnal
Urva auropunctata	Small Indian mongoose	Asia	Solitary	Diurnal
Urva brachyura	Short-tailed mongoose	Asia	Solitary	Diurnal
Urva edwardsii	Indian grey mongoose	Asia	Solitary	Diurnal
Urva fusca	Indian brown mongoose	Asia	Solitary	Nocturnal
Urva javanica	Javan mongoose	Asia	Solitary	Diurnal
Urva semitorquata	Collared mongoose	Asia	Solitary	Diurnal
Urva smithii	Ruddy mongoose	Asia	Solitary	Diurnal
Urva urva	Crab-eating mongoose	Asia	Solitary	Diurnal
Urva vitticollis	Stripe-necked mongoose	Asia	Solitary	Diurnal
Xenogale naso	Long-nosed mongoose	Africa	Solitary	Diurnal

The general information on distribution, social organization, and activity is taken from Gilchrist *et al.* (2009) and Jennings & Veron (2019). See Appendix A for more attributes of the small carnivore species listed here.

Systematics of Asian Mongooses

The molecular studies of Veron *et al.* (2004) and Perez *et al.* (2006) first indicated that the genus *Herpestes* is not monophyletic, which had not been suggested by previous studies (see Taylor *et al.*, 1991; Taylor & Matheson, 1999; Bininda-Emonds *et al.*, 1999). Patou *et al.* (2009) confirmed that the genus *Herpestes* is paraphyletic, with the two African species that were included in this genus, *H. naso* (now *X. Naso*, see above) and *H. ichneumon*, and the nine Asian *Herpestes* mongooses, belonging to three distinct lineages. Patou *et al.* (2009) showed that the Asian mongooses form a monophyletic clade (see Figure 3.2), and suggested that all Asian mongooses should be placed in a separate genus, which should be *Urva* Hodgson, 1837, Indeed, *H. ichneumon* is the type species of the genus *Herpestes* and, therefore, this name is not available for the Asian species; the earlier genus *Mangusta* Horsfield, 1824, that was utilized for the Javan mongoose, was first used for the Egyptian mongoose, and, therefore, this name is also inappropriate.

Four mongoose species occur in Southeast Asia: the Javan mongoose, *Urva javanica*, the short-tailed mongoose, *U. brachyura*, the collared mongoose, *U. semitorquata*, and the crab-eating mongoose, *U. urva* (Gilchrist *et al.*, 2009; Jennings & Veron, 2019). Another mongoose species, Hose's mongoose, *Herpestes hosei*, was described from Borneo by Jentink (1903) based on an adult female collected from Sarawak in 1893 (the only specimen that has been attributed to this species). This specimen is very similar to *U. brachyura* and Hose's mongoose was not recognized as valid by Corbet & Hill (1992), Wozencraft (2005), or Patou *et al.* (2009), although it was by Payne *et al.* (1985).

The molecular study of Patou *et al.* (2009) found a well-supported sister relationship between the short-tailed mongoose, *U. brachyura*, and crab-eating mongoose, *U. urva*, two species that are often found close to water bodies (G. Veron & A. Jennings, personal observations). This phylogenetic affinity was also suggested by craniometric analyses (Taylor & Matheson, 1999), anatomical characters (such as the posterior development of the ectotympanic bone; Li, 2004; G. Veron, personal observation), and by chromosomal evidence (Fredga, 1972). Their distribution is allopatric, except in Peninsular Malaysia where they both occur (Gilchrist *et al.*, 2009; Jennings & Veron, 2019), and where it is likely that these two species are ecologically separated (see Jennings & Veron, 2011). However, the study by Patou *et al.* (2009) did not include the collared mongoose, *U. semitorquata*. A recent molecular analysis that did incorporate this species (Veron *et al.*, 2015) showed that the collared mongoose is closely related to the crab-eating mongoose, and that the two form the sister group to the short-tailed mongoose. Veron *et al.* (2015) also found that despite Sumatran collared mongooses having a distinctive orange phenotype, they exhibited very little genetic divergence from individuals from Borneo. In contrast, the populations of the short-tailed mongoose from Borneo were strongly divergent from those from Peninsular Malaysia and Sumatra, and they might be separate species (Veron *et al.*, 2015). Within the crab-eating mongoose, a little geographical genetic structure was observed. The findings of Veron *et al.* (2015) suggest that Hose's mongoose is not a valid species, and that mongooses from Palawan Island (in the Philippines), which had been thought to be short-tailed mongooses, did not cluster with the other populations of this species, but were closer to the collared mongoose and, therefore, should be included in the latter species.

The Javan mongoose, *U. javanica*, and the small Indian mongoose, *U. auropunctata*, were considered separate species by some authors (Chasen, 1940; Ellerman & Morrison-Scott, 1951; Hinton & Dunn, 1967; Harrison, 1968; Michaelis, 1972; Ewer, 1973; Medway, 1978; Honacki *et al.*, 1982; Wozencraft, 1989b; Harrison & Bates, 1991; Taylor & Matheson, 1999), with a zone of sympatry on Peninsular Malaysia; or a single species, *U. javanica*, varying in size and colour from west to east (Pocock, 1937, 1941; Bechthold, 1939; Wenzel & Haltenorth, 1972; Lekagul & McNeely, 1977; Macdonald, 1984; Corbet & Hill, 1992; Wozencraft, 1993, 2005; Roberts, 1997; Macdonald, 2001). Veron *et al.* (2007), using mitochondrial DNA, obtained three distinct clades, which corresponded to the Javan mongoose, *U. javanica*, the small Indian mongoose, *U. auropunctata*, and the Indian grey mongoose, *U. edwardsii*, with a mean genetic divergence of 5% between each pair of species. Their analyses supported a sister relationship between *U. javanica* and *U. edwardsii*, but not between *U. javanica* and *U. auropunctata*, as would have been expected (although see Patou *et al.*, 2009). The results of Veron *et al.* (2007) are congruent with the morphometric study of Taylor & Matheson (1999), in which 11 cranial measurements of 15 *Herpestes* species were included in a principal component analysis, and the results showed a clear separation between *U. auropunctata* (specimens from China, India, Kashmir, and Pakistan) and *U. javanica* (Java, Thailand, and Vietnam). Studies of coat colour variation agree with these results. According to Pocock (1941) and Corbet & Hill (1992), specimens from Vietnam and Java (*U. javanica*) are strongly suffused with reddish parts, while specimens from northwest India and Pakistan (*U. auropunctata*) are pale, and those from Assam and Myanmar are darker and greyish. This is in agreement with our observations of museum specimens; those from Thailand, Laos, and Java are darker and reddish (particularly on the head), while those from India, Nepal, and Pakistan are paler.

The distribution of the small Indian mongoose, *U. auropunctata*, stretches from the Arabian Peninsula across the northern Indian subcontinent to Southeast Asia, whereas the Javan mongoose, *U. javanica*, occurs in Southeast Asia. The small Indian mongoose has been introduced to many different parts of the world, mainly on islands (see reviews in Tvrtković & Kryštufek, 1990; Barun *et al.*, 2011; Louppe *et al.*, 2020, 2021; and Gantchoff *et al.*, Chapter 20, this volume). Chinese populations have been assigned to *U. auropunctata* by Ellerman & Morrison-Scott (1951), Michaelis (1972), and Honacki *et al.* (1982), and this was supported by the morphometric studies of Taylor & Matheson (1999); however, the precise collection location of the Chinese specimens that were inspected in their study is unknown. Mongooses from Hainan and southern China have been grouped in the subspecies *H. auropunctatus rubrifrons* by Ellerman & Morrison-Scott (1951). Specimens from Hainan are reddish on the head, and their body colouration is brownish, so their coat colour is closer to *U. javanica* than to *U. auropunctata*. However, their skull measurements are more within the range of that of *U. auropunctata* than that of *U. javanica*. In fact, recent molecular analyses (Veron & Jennings, 2017) suggest that southern China populations belong to *U. javanica*.

The molecular study of Patou *et al.* (2009) showed that the stripe-necked mongoose, *Urva vitticollis*, and the ruddy mongoose, *Urva smithii*, are sister species.

This relationship had never been proposed before, and contradicted the hypothesis of Pocock (1937) that *U. smithii* is very close to *U. edwardsii* and may even be a 'jungle form' of the latter. The stripe-necked and ruddy mongooses live in sympatry in southwestern India and Sri Lanka. They are both forest species and diurnal, but it is suggested that *U. vitticollis* is only found near water bodies, while *U. smithii* occupies a wider variety of habitats (Santiapillai *et al.*, 2000).

The Indian brown mongoose, *U. fusca*, was found by Patou *et al.* (2009) to be the closest relative to a clade containing *U. auropunctata*, *U. edwardsii*, and *U. javanica*. This phylogenetic arrangement had been suggested by Fredga (1972), based on observations of karyotypes, and it invalidates the proposition of Bechthold (1939), who considered *U. fusca* and *U. brachyura* to be conspecific. The Indian brown mongoose is found in India and Sri Lanka. An interesting result of the study by Patou *et al.* (2009) was that a second mongoose species in the Fiji islands (in addition to the introduced small Indian mongoose; as was suggested by Morley *et al.*, 2007), turned out to be the Indian brown mongoose. This means that there had been a recent and undocumented introduction of *U. fusca* since that of a single pair of small Indian mongooses, *U. auropunctata*, from Calcutta in 1883 (Simberloff *et al.*, 2000). Veron *et al.* (2010) highlighted that this is the first known introduction of the Indian brown mongoose to a non-native area, and they suggested that it may have derived from a pair brought from an unknown source to a private zoo in Fiji in the late 1970s. Currently, the Indian brown mongoose co-occurs on Viti Levu Island with the much smaller small Indian mongoose. The Indian brown mongoose appears to be nocturnal (at least in its native range), while the small Indian mongoose is primarily diurnal, which suggests that there may be little interaction between these two species in Fiji.

A New Classification of the Mongooses

A new classification of the mongooses based on Veron *et al.* (2004, 2007, 2015), Perez *et al.* (2006), Gilchrist *et al.* (2009), Patou *et al.* (2009) and Jennings & Veron (2019) is provided in Table 3.1. There are currently

11 species recognized in the Mungotinae, and 23 species in the Herpestinae. The phylogenetic position of Pousargues's mongoose, *Dologale dybowskii*, remains to be investigated with molecular data, but based on morphological evidence (Li, 2004; G. Veron, personal observation), we have placed it within the Mungotinae. There are still some debated taxa (see e.g. Appendix B) and possible cryptic species, and further investigations are needed.

Finally, for most mongoose species, there is very little data on distribution, ecology, population status, and possible threats, rendering the evaluation of their conservation status extremely difficult. On the *IUCN Red List of Threatened Species* (where *Bdeogale omnivora* is recognized valid making a total of 35 species), 29 mongooses were classified as Least Concern (LC), 1 Data Deficient (DD), 3 Near Threatened (NT), and 2 Vulnerable (VU) (IUCN, 2020; see Appendix A). Despite the fact that 31.4% of these species were assessed as having decreasing populations, only 14.3% were placed in a threat category (VU or higher). Mongooses are often considered 'ecologically tolerant' to anthropogenic habitat modifications (see e.g. Zaw *et al.*, 2008) and robust to human threats, such as hunting (IUCN, 2020), but in reality, there is little data available to substantiate these claims or to accurately assess their conservation status. Further research is urgently needed.

Acknowledgements

We are grateful to many researchers, students, and assistants who helped us during our work on the mongooses. In particular, we would like to warmly thank François Catzeflis, Marc Colyn, Regis Debruyne, Dez Fernandez, Jérome Fuchs, Blaise Li, Craig Morley, Mélanie Perez, Daniel Simberloff, Peter Taylor, and Siew Te Wong. Our work was funded by the Programmes Pluriformation/Actions Transversales 'Evolution et Structure des Ecosystèmes' and 'Biodiversité Actuelle et Fossile' (Muséum National d'Histoire Naturelle, French Ministry of Research) and by the UMR 7205 CNRS MNHN. Our molecular work was undertaken at the 'Service de Systématique Moléculaire' (UMS 2700 CNRS MNHN) and we thank Eric Pasquet and Marie-Catherine Boisselier for their support, and the staff of the lab for their help. Our sequencing work was supported by the 'Consortium National de Recherche en Génomique' (agreement no. 2005/67 between Genoscope and Muséum National d'Histoire Naturelle, project 'Macrophylogeny of life' directed by Guillaume Lecointre), and by the network 'Bibliothèque du Vivant', funded by CNRS, Muséum National d'Histoire Naturelle, INRA and CEA (Genoscope). We are indebted to Emmanuel Do Linh San for his invitation to contribute to this book. Drawings in Figure 3.2 were reproduced with the kind permission of the Bombay Natural History Society (from *The Book of Indian Animals* by S.H. Prater, 1971), of Jonathan Kingdon (from *Kingdon Field Guide to African Mammals* by J. Kingdon, 2015) and of Junaidi Payne, Karen Phillipps, and the Sabah Society (from *A Field Guide to the Mammals of Borneo* by Payne *et al.*, 1985). Nick Garbutt (www.nickgarbutt.com), Emmanuel Do Linh San, and Peter Apps kindly provided photographs of selected small carnivore species to illustrate this chapter.

References

Albignac, R. (1973) Monographie des carnivores malgaches. *Bulletin de l'Académie Malgache* 50, 161–163.

Allen, J.A. (1919) Preliminary notes on African Carnivora. *Journal of Mammalogy* 1, 23–31.

Ansell, W.F. (1978) *The Mammals of Zambia*. The National Parks & Wildlife Service, Chilanga.

Balmforth, Z.E. (2004) *The Demographics, Spatial Structure and Behaviour of the Yellow Mongoose, Cynictis penicillata, with Emphasis on Cooperative Breeding*. PhD thesis, University of Sussex.

Barun, A., Hanson, C.C., Campbell, K.J. & Simberloff, D. (2011) A review of small Indian mongoose management and eradications on islands. In: *Island Invasives: Eradication and Management* (eds C.R. Veitch, M.N. Clout & D.R. Towns), pp. 17–25. IUCN, Gland.

Bechthold, G. (1939) Die asiatischen Formen der Gattung *Herpestes*, ihre Systematik, Ökologie,

Verbreitung und ihre Zusammenhänge mit den Afrikanischen Arten. *Zeitschrift für Säugetierkunde* 14, 113–219. (In German).

Ben-Yaacov, R. & Yom-Tov, Y. (1983) On the biology of the Egyptian mongoose, *Herpestes ichneumon*, in Israel. *Zeitschrift für Säugetierkunde* 48, 34–45.

Bininda-Emonds, O.R., Gittleman, J.L. & Purvis, A. (1999) Building large trees by combining phylogenetic information: a complete phylogeny of the extant Carnivora (Mammalia). *Biological Reviews* 74, 143–175.

Bugge, J. (1978) The cephalic arterial system in carnivores with special reference to the systematic classification. *Acta Anatomica* 101, 45–61.

Carr, G.M. & Macdonald, D.W. (1986) The sociality of solitary foragers: a model based on resource dispersion. *Animal Behaviour* 34, 1540–1549.

Cavallini, P. & Nel, J.A. (1990) Ranging behaviour of the Cape grey mongoose *Galerella pulverulenta* in a coastal area. *Journal of Zoology* 222, 353–362.

Chasen, F.N. (1940) A hand list of Malayan mammals. A systematic list of the Malay peninsula, Sumatra, Borneo, and Java, including the adjacent small islands. *Bulletin of the Raffles Museum* 15, 1–209.

Coetzee, C. (1977) Order Carnivora. In: *The Mammals of Africa: An Identification Manual. Part 8* (eds J. Meester & H.W. Setzer), pp. 1–42. Smithsonian Institution Press, Washington.

Colyn, M. & Van Rompaey, H. (1994) Morphometric evidence of the monotypic status of the African long-nosed mongoose *Xenogale naso* (Carnivora, Herpestidae). *Belgian Journal of Zoology* 124, 175–192.

Corbet, G.B. & Hill, J.E. (1991) *A World List of Mammalian Species*. Oxford University Press, Oxford.

Corbet, G.B. & Hill, J.E. (1992) *The Mammals of the Indomalayan Region: A Systematic Review*. Oxford University Press, Oxford.

Dargel, B. (1990) A bibliography on Viverrids. *Mitteilungen aus dem Hamburgischen Zoologischen Museum und Institut* 87, 1–184.

Detry, C., Cardoso, J.L., Heras Mora, J., Bustamante-Alvarez, M., Silva, A. M., Pimenta, J., Fernandes, I. & Fernandes, C. (2018) Did the Romans introduce the Egyptian mongoose (*Herpestes ichneumon*) into the Iberian Peninsula? *The Science of Nature* 105, 63.

Dorst, J. & Dandelot, P. (1972) *Guide des Grands Mammifères d'Afrique*. Delachaux & Niestlé, Paris. (In French).

Earlé, R.A. (1981) Aspects of the social and feeding behaviour of the yellow mongoose *Cynictis penicillata* (G. Cuvier). *Mammalia* 45, 143–152.

Ellerman, J.R. & Morrison-Scott, T.C.S. (1951) *Checklist of Palearctic and Indian Mammals 1758 to 1946*. British Museum, London.

Ewer, R.F. (1973) *The Carnivores*. The World Naturalist. Weidenfeld and Nicolson, London.

Fitzsimons, F.W. (1919) *The Natural History of South Africa*. Longmans & Green, London.

Flower, W.H. (1869) On the value of characters of the base of the cranium in the classification in Order Carnivora and systematic position of *Bassaris* and other disputed forms. *Proceedings of the Zoological Society of London* 1869, 4–37.

Flynn, J.J. & Nedbal, M.A. (1998) Phylogeny of the Carnivora (Mammalia): congruence vs incompatibility among multiple data sets. *Molecular Phylogenetics and Evolution* 9, 414–426.

Flynn, J.J., Neff, N.A. & Tedford, R.H. (1988) Phylogeny of the carnivores. In: *The Phylogeny and Classification of Tetrapods* (ed. M.J. Benton), pp. 73–776. Systematics Association Special Volume. Clarendon Press, Oxford.

Flynn, J., Finarelli, J., Zehr, S., Hsu, J. & Nedbal, M. (2005) Molecular phylogeny of the Carnivora (Mammalia): assessing the impact of increased sampling on resolving enigmatic relationships. *Systematic Biology* 54, 317–337.

Fredga, K. (1972) Comparative chromosome studies in mongooses (Carnivora, Viverridae). 1. Idiograms of 12 species and karyotype evolution in Herpestinae. *Hereditas* 71, 1–74.

Gantchoff, M., Libal, N.S. & Belant, J.L. (Chapter 20, this volume) Small carnivore introductions: ecological and biological correlates of success. In: *Small Carnivores: Evolution, Ecology, Behaviour, and Conservation* (eds E. Do Linh San, J.J. Sato, J.L. Belant & M.J. Somers). Wiley–Blackwell, Oxford.

Gaubert, P. & Veron, G. (2003) Exhaustive sample set among Viverridae reveals the sister-group of felids: the linsangs as a case of extreme morphological convergence within feliformia. *Proceedings of the Royal Society of London B* 270, 2523–2530.

Gaubert, P., Machordom, A., Morales, A., Vicente Lopez-Bao, J., Veron, G., Amin, M., Barros, T., Basuony, M., Djagoun, C.A.M.S., Do Linh San, E.,

Fonseca, C., Geffen, E., Ozkurt, S.O., Cruaud, C., Couloux, A. & Palomares, F. (2011) Comparative phylogeography of two African carnivorans presumably introduced into Europe: disentangling natural versus human-mediated dispersal across the Strait of Gibraltar. *Journal of Biogeography* 38, 341–358.

Gilchrist, J.S., Jennings, A.P., Veron, G. & Cavallini, P. (2009) Family Herpestidae (Mongooses). In: *Handbook of the Mammals of the World. Volume 1. Carnivores* (eds D.E. Wilson & R.A. Mittermeier), pp. 262–328. Lynx, Barcelona.

Goldman, C.A. (1984) Systematic revision of the African mongoose genus *Crossarchus* (Mammalia: Viverridae). *Canadian Journal of Zoology* 62, 1618–1630.

Goodman, S.M. (2009) Family Eupleridae (Madagascar carnivores). In: *Handbook of the mammals of the world. Volume 1. Carnivores* (eds D.E. Wilson & R.A. Mittermeier), pp. 330–351. Lynx, Barcelona.

Goodman, S.M. (2012) *Les Carnivora de Madagascar.* Association Vahatra, Antananarivo.

Gorman, M.L. (1979) Dispersion and foraging of the small Indian mongoose *Herpestes auropunctatus* (Carnivora: Viverridae) relative to the evolution of social viverrids. *Journal of Zoology* 187, 65–73.

Gray, J.E. (1864) A revision of the genera and species of viverrine animals (Viverridae), founded on the Collection in the British Museum. *Proceedings of the Zoological Society of London* 1864, 502–579.

Gregory, W.K. & Hellman, H. (1939) On the evolution and major classification of the civets (Viverridae) and allied fossil and recent Carnivora; a phylogenetic study of the skull and dentition. *Proceedings of the American Philosophical Society* 81, 309–392.

Happold, D.C.D. (1987*) The Mammals of Nigeria.* Clarendon Press, Oxford.

Harrison, D.L. (1968) *The Mammals of Arabia. Volume 2. Carnivora – Artiodactyla – Hyracoidea.* Ernest Benn Limited, London.

Harrison, D.L. & Bates, P.J.J. (1991) *The Mammals of Arabia.* Harrison Zoological Museum Publication, Sevenoaks.

Hendey, Q.B. (1974) The late cenozoic carnivora of the south-western Cape province. *Annals of the South African Museum* 63, 1–369.

Hinton, H.E. & Dunn, A.M.S. (1967). *Mongooses, their Natural History and Behaviour.* Oliver & Boyd, London.

Honacki, J.H., Kinman, K.E. & Koeppl, J.W. (eds) (1982) *Mammals Species of the World.* Allen Press and Association Systematic Collection, Lawrence.

Hunt, R.M. (1987) Evolution of Aeluroidea Carnivora: significance of auditory structure in the Nimravid cat *Dinictis. American Museum Novitates* 2886, 1–74.

Hunt, R.M. & Tedford, R.H. (1993) Phylogenetic relationships within the aeluroid Carnivora and implications of their temporal and geographic distribution. In: *Mammal Phylogeny: Placentals* (eds F.S. Szalay, M.J. Novacek & M.C. McKenna), pp. 53–73. Springer-Verlag, New York.

IUCN [International Union for Conservation of Nature] (2020) *The IUCN Red List of Threatened Species. Version 2020-1.* https://www.iucnredlist.org. Accessed on 25 April 2020.

Jennings, A.P. & Veron, G. (2011) Predicted distributions and ecological niches of 8 civet and mongoose species in Southeast Asia. *Journal of Mammalogy* 92, 316–327.

Jennings, A.P. & Veron, G. (2019) *Mongooses of the World.* Whittles Publishing, Dunbeath.

Jentink, F.A. (1903). A new bornean *Herpestes. Notes from Leyden Museum* 23, 223–228.

Kingdon, J. (1977). *East African Mammals. An Atlas of Evolution in Africa.* Academic Press, London.

Kingdon, J. (1997). *The Kingdon Field Guide to African Mammals.* 1st edition. Academic Press, London.

Kingdon, J. (2015). *The Kingdon Field Guide to African Mammals.* 2nd edition. Bloomsbury, London.

Lekagul, B. & McNeely, J.A. (1977) *Mammals of Thailand.* Kurusapha Ladprao Press, Bangkok.

Li, B. (2004) *Phylogénie des Herpestidae (Mammalia, Carnivora).* Master thesis, Université Pierre et Marie Curie. (In French).

Louppe, V., Leroy, B., Herrel, A. & Veron, G. (2020) The globally invasive small Indian mongoose *Urva auropunctata* is likely to spread with climate change. *Scientific Reports* 10, 7461.

Louppe, V., Lalis, A., Abdelkrim, J., Baron, J., Bed'Hom, B., Becker, A.A.M.J., Catzeflis, F., Lorvelec, O., Ziegler, U. & Veron, G. (2021) Dispersal history of a globally introduced carnivore, the small Indian mongoose *Urva auropunctata*, with an emphasis on the Caribbean region. *Biological Invasions* 23, 2573–2590.

Lynch, C.D. (1980) Ecology of the suricate, *Suricata suricatta* and yellow mongoose, *Cynictis penicillata* with special reference to their reproduction. *Memoirs van die Nasionale Museum Bloemfontein* 14, 1–145.

Macdonald, D.W. (1983) The ecology of carnivore social behaviour. *Nature* 301, 379–384.

Macdonald, D. (ed.) (1984) *The Encyclopedia of Mammals*. Facts on File Publications, New York.

Macdonald, D. (ed.) (2001) *The New Encyclopedia of Mammals*. Oxford University Press, Oxford.

Masters, J.C., Génin, F., Zhang, Y., Pellen, R., Huck, T., Mazza, P.P., Rabineau, M., Doucouré, M. & Aslanian, D. (2021) Biogeographic mechanisms involved in the colonization of Madagascar by African vertebrates: rifting, rafting and runways. *Journal of Biogeography* 48, 492–510.

Medway, L. (1978) *The Wild Mammals of Malaya (Peninsular Malaysia) and Singapore*. Oxford University Press, Kuala Lumpur.

Michaelis, B. von (1972) Die Schleichkatzen (Viverriden) Afrikas. *Zeitschrift für Säugetierkunde* 20, 1–110.

Mivart, S.G. (1882) On the classification and distribution of Aeluroidea. *Proceedings of the Zoological Society of London* 1882, 135–208.

Morley, C.G., McLenachan, P.A. & Lockhart, P.J. (2007) Evidence for the presence of a second species of mongoose in the Fiji Islands. *Pacific Conservation Biology* 13, 29–34.

Neff, N.A. (1983) *The Basicranial Anatomy of the Nimravidae (Mammalia: Carnivora): Character Analyses and Phylogenetic Inferences*. PhD thesis, City University of New York.

Orts, S.G. (1970) Le xenogale de J. A. Allen (Carnivora, Viverridae) au sujet d'une capture effectuée au Kivu. *Revue de Zoologie et de Botanique Africaines* 82, 174–186. (In French).

Palomares, F. & Delibes, M. (1993). Social organization in the Egyptian mongoose: group size, spatial behaviour and inter-individual contacts in adults. *Animal Behaviour* 45, 917–925.

Pathak, S. & Stock, A.D. (1976) Giemsa-Banding and identification of Y-autosome translocation in African marsh mongoose, *Atilax paludinosus* (Carnivora, Viverridae). *Cytogenetics and Cell Genetics* 16, 487–494.

Patou, M.L., McLenachan, P.A., Morley, C.G., Couloux, A., Jennings, A.P. & Veron, G. (2009). Molecular phylogeny of the Herpestidae (Mammalia, Carnivora) with a special emphasis on the Asian *Herpestes*. *Molecular Phylogenetics and Evolution* 53, 69–80.

Payne, J., Francis, C.M. & Phillipps, K. (1985) *A Field Guide to the Mammals of Borneo*. Sabah Society & WWF Malaysia, Sabah.

Perez, M., Li, B., Tillier, A., Cruaud, A. & Veron, G. (2006) Systematic relationships of the bushy-tailed and black-footed mongooses (genus *Bdeogale*, Herpestidae, Carnivora) based on molecular, chromosomal and morphological evidence. *Journal of Zoological Systematics and Evolutionary Research* 44, 251–259.

Petter, G. (1969) Interprétation évolutive des caractères de la denture des Viverridae africains. *Mammalia* 33, 607–625. (In French).

Petter, G. (1974) Rapports phylétiques des Viverridés. *Les formes de Madagascar. Mammalia* 38, 605–636. (In French).

Pocock, R.I. (1915) On some external characters of *Galidia*, *Galidictis* and related genera. *Annales and Magazine of Natural History* 16, 351–356.

Pocock, R.I. (1916) On the external characters of the mongooses (Mungotidae). *Proceedings of the Zoological Society of London* 1, 349–374.

Pocock, R.I. (1919) The classification of the mongooses (Mungotidae). *Annals of Natural History* 23, 515–524.

Pocock, R.I. (1937) The mongooses of British India including Ceylon and Burma, *Journal of the Bombay Natural History Society* 39, 211–245.

Pocock, R.I. (1941) *The Fauna of British India, Including Ceylon and Burma*. Taylor and Francis, London.

Prater, S.H. (1971) *The Book of Indian Animals*. Bombay Natural History Society, Oxford University Press, Oxford.

Radinsky, L. (1975) Viverrid neuroanatomy: phylogenetic and behavioural implications. *Journal of Mammalogy* 56, 130–150.

Rapson, S.A., Goldizen, A.W. & Seddon, J.M. (2012) Species boundaries and possible hybridization between the black mongoose (*Galerella nigrata*) and the slender mongoose (*Galerella sanguinea*). *Molecular Phylogenetics and Evolution* 65, 831–839.

Ray, J.C. (1997) Comparative ecology of two African forest mongooses, *Herpestes naso*, and *Atilax paludinosus*. *African Journal of Ecology* 35, 237–253.

Ray, J.C. & Sunquist, M.E. (2001). Trophic relations in a community of African rainforest carnivores. *Oecologia* 127, 395–408.

Roberts, A. (1951) *The Mammals of South Africa*. Central News Agency, Cape Town.

Roberts, T.J. (1997) *The Mammals of Pakistan*. Revised edition. Oxford University Press, Karachi, Oxford.

Wurster, D.H. (1969) Cytogenetic and phylogenetic studies in Carnivora. In: *Comparative Mammalian Cytogenetics* (ed. K. Benirschke), pp. 310–329. Springer-Verlag, Berlin.

Wurster, D.H. & Benirschke, K. (1968) Comparative cytogenetic studies in the order Carnivora. *Chromosoma* 24, 336–382.

Wyss, A.R. & Flynn, J.J. (1993) A phylogenetic analysis and definition of the carnivora. In: *Mammal phylogeny: Placentals* (eds F.S. Szalay, M.J. Novacek & M.C. MacKenna), pp. 32–52. Springer-Verlag, Berlin.

Yoder, A.D., Burns, M.M., Zehr, S., Delefosse, T., Veron, G., Goodman, S.M. & Flynn, J.J. (2003) Single origin of Malagasy Carnivora from an African ancestor. *Nature* 421, 734–737.

Zaw, T., Htun, S., Po, S., Maung, M., Lynam, A., Latt, K. & Duckworth, J.W. (2008) Status and distribution of small carnivores in Myanmar. *Small Carnivore Conservation* 38, 2–28.

Zumpt, I. (1976) The yellow mongoose (*Cynictis penicillata*) as a latent focus of rabies in South Africa. *Journal of the South African Veterinary Association* 47, 211–213.

4

Late Quaternary Biogeography of Small Carnivores in Europe

Robert S. Sommer[1],* *and Jennifer J. Crees*[2]

[1] Department of Landscape Sciences and Geomatics, Neubrandenburg University of Applied Sciences, Neubrandenburg, Germany
[2] Institute of Zoology, Zoological Society of London, London, UK

SUMMARY

The Late Quaternary distribution history of small carnivores in Europe was strongly influenced by climate and humans. Subfossil records from geological or archaeological excavations can be used to reconstruct spatio-temporal dynamics of species for the Pleistocene and Holocene Epochs. During full glacial conditions, around 60–15 thousand years ago (kya), Central European regions were only permanently colonized by the wolverine, *Gulo gulo*, and both stoat, *Mustela erminea*, and least weasel, *M. nivalis*. From around 14 kya, the European polecat, *M. putorius*, European badger, *Meles meles*, pine marten, *Martes martes*, red fox, *Vulpes vulpes*, and European wild cat, *Felis silvestris*, spread into northern regions due to climate warming (Greenland Interstadial 1) and associated stepwise reforestation of those regions. The Eurasian otter, *Lutra lutra*, colonized Central Europe for the first time during the Early Holocene. In contrast to several other small carnivore species from temperate regions that survived and recolonized Central Europe from refugial regions such as Iberia, the Apennine peninsula, the Carpathians or the Balkans, the otter was restricted to the Apennine peninsula and its low genetic diversity as well as its late arrival may be a consequence of this. The stone marten, *Martes foina*, probably followed Neolithic settlers out of Asia Minor, whereas the origin and colonization pattern of the European mink, *Mustela lutreola*, remains enigmatic. Small carnivores experienced limited distributional shifts during the Holocene in comparison to larger carnivores, probably due to their ecological plasticity and relative resilience to human impacts. The European wild cat disappeared from northern regions after the Holocene thermal optimum and both the pine marten and the European polecat may have experienced some range contraction during the Medieval, but most species broadly maintained their Holocene distributions to the present day.

Keywords

Distribution dynamics – European wild cat – glacial refugia – mustelids – postglacial recolonization

Introduction

The last 60000 years of the Late Quaternary were characterized by numerous rapid climatic oscillations and extreme environmental changes (Dansgaard *et al.*, 1993; Huntley *et al.*, 2003; Wohlfarth *et al.*, 2008). The consequences of these changes are documented by a complex pattern of extinction and colonization for many animal species in Europe (Hewitt, 2000; Sommer & Nadachowski, 2006; Sommer *et al.*, 2007; Stewart, 2008; Sommer & Zachos, 2009; Stuart & Lister, 2012; Crees & Turvey, 2014; Stuart, 2015; Sommer, 2020). Some

---

*Corresponding author.

Small Carnivores: Evolution, Ecology, Behaviour, and Conservation, First Edition. Edited by Emmanuel Do Linh San, Jun J. Sato, Jerrold L. Belant, and Michael J. Somers.
© 2022 John Wiley & Sons Ltd. Published 2022 by John Wiley & Sons Ltd.

characteristic Pleistocene megafaunal species became extinct several thousand years before the ultimate end of the Pleistocene at ~11.7 kya (Stuart & Lister, 2012), whereas many other species that were adapted to steppe or tundra environments, such as the arctic fox, *Vulpes lagopus*, reindeer, *Rangifer tarandus*, or pika, *Ochotona pusilla*, disappeared from Central Europe (CE) only during the end of the Late Glacial or Early Holocene between 12.7 and 11.2 kya (Street & Baales, 1999; Sommer *et al.*, 2014; Sommer, 2020). Other species that were adapted to temperate environments, such as the pond turtle, *Emys orbicularis*, European hedgehog, *Erinaceus europaeus*, or red deer, *Cervus elaphus*, recolonized CE regions from southern glacial refugia (Figure 4.1). From a biogeographical point of view, changes in Europe's mammalian fauna at the end of the Last Glacial were characterized by several different processes that have shaped the recent faunal composition of Europe: (i) extinction of species such as the woolly mammoth, *Mammuthus primigenius*, cave lion, *Panthera spelaea*, cave bear, *Ursus spelaeus*, and woolly rhino, *Coelodonta antiquitatis*; (ii) extirpation of species in CE but maintenance of their distribution in northern tundra and taiga environments, e.g. reindeer and arctic fox; (iii) extirpation of species in CE but maintenance of their distribution in southern steppe and savannah environments, e.g. spotted hyena, *Crocuta crocuta*, European wild ass, *Equus hydruntinus*, and pika; (iv) population turnover in species that were distributed continuously in CE throughout the Pleistocene and the Holocene, e.g. least weasel, *Mustela nivalis*, and grey wolf, *Canis lupus*; (v) recolonization of CE by temperate species from southern glacial refugia, such as the European hedgehog, brown bear, *Ursus arctos*, red fox, *Vulpes vulpes*, and European badger, *Meles meles*; (vi) colonization of Europe by species which appeared during the Holocene

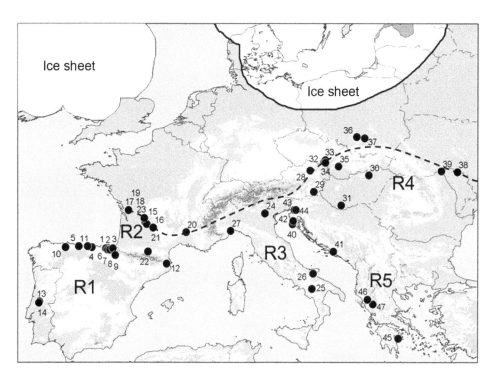

Figure 4.1 Sites with records (black dots) of temperate mammal species during the Last Glacial Maximum (LGM; for a legend, see Sommer & Nadachowski, 2006) which characterize the known main refugial regions: Iberia (R1), Southwestern France (R2), Apennine peninsula (R3), Carpathian surroundings (R4), and the Balkans (R5). A further suggested possible refugial region, the southern Urals, is not displayed here. When discussing the postglacial (or post-LGM) recolonization of Europe in the text, the dashed line is meant to represent the northern limit of the hitherto known refugial regions. The occurrences of forest-bounded species in southern Poland in the sites Nos. 36 and 37 are questionable and discussed as admixtures from other periods.

for the first time, e.g. stone marten, *Martes foina*; and finally (vii) species whose distribution was altered due to climate change and human impacts, e.g. wild horse, *Equus ferus*, and elk, *Alces alces*.

Small carnivores are known to colonize a wide spectrum of biomes. In Europe, these include tundra, taiga (boreal forests), temperate mixed forests, Mediterranean areas, steppe, and desert. The small carnivore fauna of Europe contains species such as the least weasel weasel and the stoat, *Mustela nivalis*, and the stoat, *Mustela erminea*, that are extremely successful due to their ability to adapt to a wide variety of habitats and landscapes. However, certain species such as the Eurasian otter, *Lutra lutra*, have been strongly negatively influenced by habitat loss and landscape fragmentation in western European countries (Honnen *et al.*, 2011). This chapter presents evidence for the Late Quaternary distribution dynamics of small carnivores and a comparison with larger carnivores to inform our understanding of recent patterns of animal biodiversity in Europe.

Late Quaternary Climate History and Consequences for Biome and Faunal Dynamics

The climate and vegetation history of Europe for the Late Quaternary is relatively well documented (e.g. Dansgaard *et al.*, 1993; Björck *et al.*, 1998; Litt *et al.*, 2001, 2003; Barron *et al.*, 2003; Davis *et al.*, 2003; Huntley & Allen, 2003; Hubberten *et al.*, 2004; Renssen *et al.*, 2009). We briefly summarize the environmental history of the last 60 000 years in order to provide context for understanding the colonization history of carnivorous mammals.

From 60 to 27 kya, during the Marine Isotope Stage 3 of the Quaternary, the northern hemisphere was characterized by a generally cold environment punctuated by regular warm intervals, the Greenland Interstadials, which lasted from several hundred years to around 3000 years. The landscape of Europe north of the glacial refuge areas was characterized by open steppe-tundra (so-called mammoth steppe) with an annual mean temperature of −4 to −8 °C (Huijzer & Vandenberghe, 1998). During warmer periods, the mean temperature rose abruptly (within a few years) to around 7–10 °C and led to a

diffusion of animal and plant species from the southern refuge areas (e.g. Balkans or Iberia) to at least 50° latitude, significantly changing regional biotic assemblages. The typical cold-adapted Pleistocene fauna such as the mammoth and the reindeer shifted their ranges in response, although they continued to be distributed in Central Europe alongside temperate species such as the red deer and the European polecat, *Mustela putorius*, for some time (Sommer *et al.*, 2008). The Last Glacial Maximum (LGM), from 26.5 to 19 kya, was characterized by the maximum advance of the ice sheets and all regions of Central Europe were affected by discontinuous permafrost. During this period of cooling, cold-adapted species and open landscapes reached their most southerly extent and temperate species became isolated in southern glacial refugia (Sommer & Nadachowski, 2006; Sommer *et al.*, 2014), although the degree of range isolation (Figure 4.1) differed depending on their ecological tolerance (e.g. adaptation to certain biomes such as mixed deciduous forests, etc.). The onset of the last deglaciation of the northern hemisphere began around 18 kya and until 14 kya, most northern parts of what are now Germany and Poland, as well as the Baltic States, were deglaciated. Mean temperatures rose by about 10 °C from the beginning of the Greenland Interstadial 1, also known as Bølling/Allerød Interstadial (Figure 4.2) and led to environmental change across the whole northern hemisphere. In central Europe, the warming induced the expansion of the birch, *Betula* sp., willow, *Salix* sp., poplar, *Populus* sp., and during the Greenland Interstadial 1c–a (Allerød) pine, *Pinus* sp., also increased its range (Litt *et al.*, 2001, 2003). However, Europe then experienced a brief cool snap, the Younger Dryas, that lasted for 1000 years and caused the forests which had established during the Greenland Interstadial 1c–a to vanish in northern CE (Theuerkauf & Joosten, 2012). The ice core records from northern Europe (which correspond with temperature changes) unequivocally reflect a rapid rise in temperature at the onset of the Holocene (Figure 4.2), which may have happened extremely rapidly, within one to three years (Steffensen *et al.*, 2008; Vinther *et al.*, 2009), and was followed by a slower rise during the Preboreal (PB) and Boreal (Bo) periods (Figure 4.2). This early Holocene warming was associated with a major biome change in CE and the rapid spread of the birch and the pine, later followed by warm-adapted tree taxa such as hazel, *Corylus* sp., oak, *Quercus* sp., and elm, *Ulmus* sp.

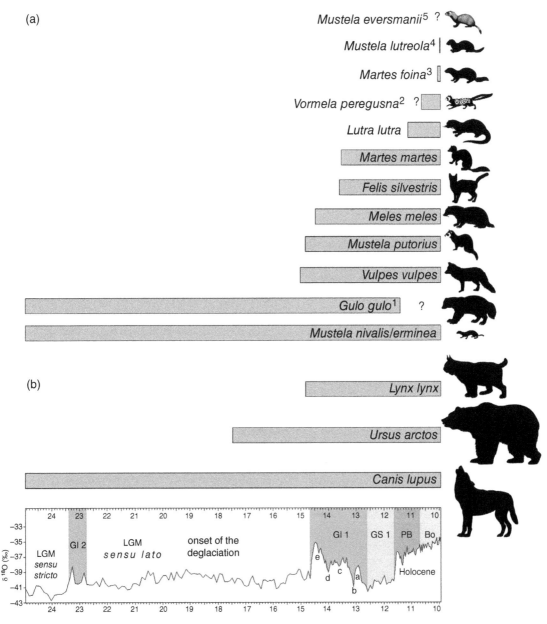

Figure 4.2 Chronological comparison of the colonization history of Central Europe (CE) and areas north of the typical refugial regions (see Figure 4.1) by (a) small carnivores in comparison to (b) larger carnivores after Sommer & Benecke (2004, 2005a,b, 2006). The x-axis of the coloured diagram is the time in thousands of years ago (kya). The black curve indicates the oxygen isotope ratios $\delta^{18}O$ in ‰ from the GISP2 ice core record (using CalPal; Weninger et al., 2008). Ice core event stratigraphy after Björck et al. (1998) and Blockley et al. (2012). LGM: Last Glacial Maximum, GI: Greenland Interstadial (warming epoch), GS: Greenland Stadial (cooling epoch), PB: Preboreal, Bo: Boreal. [1]The hundreds of archaeological sites containing subfossil records of vertebrates give the convincing and general picture that *G. gulo* disappeared in CE with the end of the Younger Dryas Period (GS1). However, the Baltic States were obviously temporarily colonized (probably by single individuals) during migration and dispersal from taiga regions in Scandinavia or north-western Russia. [2]Bulgaria and Danube delta. [3]It is obvious that it colonized Europe during the Neolithic from about 7kya in congruence with the expansion of agriculture and farming economy from SE Europe to CE and Western Europe. [4]*Mustela lutreola* is generally represented extremely rarely (or has been overlooked) in the subfossil record. A few available finds indicate that it may have been present since the Neolithic in the Baltic States and Poland (with probable postglacial origin in north-eastern European refugia) and colonized Europe during the Middle and Late Holocene. [5]During the Weichselian glacial *M. eversmanii* was distributed also in south-western Germany, Switzerland, Austria and, probably, France before the LGM (Krajcarz et al., 2015). With the exception of the LGM record of the Deszcowa Cave (Krajcarz et al., 2015; Figure 4.1, No. 36), there are no further subfossil records outside of the current range since the LGM.

During the Holocene thermal optimum, about 9–5 kya, when annual mean temperatures were up to 2–3 °C higher than today (Renssen *et al.*, 2009), CE was largely forested with oak, elm, lime, *Tilia* sp., and pine dominating in the lowlands and spruce, *Picea* sp., beech, *Fagus* sp., and fir, *Abies* sp., predominating at higher altitudes. In the south, open steppe communities existed. Since 6 kya, vegetation has been increasingly influenced by human activity.

Reconstruction of Past Distributions of Small Carnivores

There are several methods available for reconstructing spatio-temporal dynamics of organisms at the species and/or population level. The phylogeography of species can be reconstructed using recent or ancient DNA. Species distribution modelling (SDM) or climate envelope modelling (CEM) is an increasingly popular technique that does not require direct faunal evidence. Here, the ecological niche (preferred habitat or climatic conditions) is estimated from the environmental parameters at known sites of a species' extant distribution and this information is then used to simulate past potential distribution based on past climatic conditions. When modelling the distribution of a species for past climatic epochs, it is advantageous to ground truth distribution models with data on subfossil records to help evaluate the reliability of results, as demonstrated by Kuemmerle *et al.* (2012) or Prost *et al.* (2013).

Subfossil records from geological or archaeological excavations can also be used to reconstruct spatio-temporal dynamics of species. However, the distribution of subfossil bone records can be strongly influenced by environmental and ecological factors, as well as human influence. For example, subfossil records of the least weasel are rare in archaeological assemblages as it was never hunted by humans and is also not among the main prey of owls, which leave large accumulations of pellets with remains of their prey in cave sediments and which are also a reliable source of information on the past distribution of mice and voles. By contrast, the reindeer was one of the main prey species of both Neanderthals and modern humans and thus the abundance and distribution of its bone remains from archaeological excavations more faithfully reflects the dynamics of its distribution in space and time (Sommer

et al., 2014). In this review, the overwhelming majority of data on subfossil records of small carnivores was recovered from archaeological sites and compiled within the project 'Holocene History of the European Vertebrate fauna' (Benecke, 1999). The age of remains was inferred from the age of an archaeological layer (the majority of cases) or by direct ^{14}C radiocarbon dating of the bone. The age of layers was calculated by relative dating (from the archaeological context of cultural remains) or ^{14}C dating of other bones or charcoal from the same layer. This allowed all information on subfossil assemblages of small carnivore species in Europe to be collected and grouped chronologically. The sites containing fossil records of small carnivores which were evaluated for biogeographic reconstructions are displayed and listed in Sommer & Benecke (2004) and Crees (2013).

The subfossil record consists only of presence data (unlike ecological sampling which can yield presence/absence data). The most common method to assess past range change is therefore extent of occurrence (EOO) or range extent. This is the area measured within a convex hull polygon that encloses all the points (individual subfossil records) with no internal angle measuring more than 180° (IUCN, 2001). However, as EOO measurement can be influenced by sampling bias and can also be skewed by the outermost points, we used bootstrapping to establish null models of range-size expectations. This method 'resamples' the original data, with replacement, to estimate a statistic's sampling distribution and is, therefore, useful when the underlying sampling distribution cannot be assumed normal (Mooney & Duval, 1993). Upper and lower confidence intervals (95%) and mean range obtained from bootstrapping can then be compared to the observed EOO. Only if the observed EOO falls outside these confidence intervals can it be interpreted as representing a genuine, statistically significant deviation from the expected range size for the species (i.e. an increase or decline in range).

Refugial History

The model of glacial refugia as core areas for the survival of thermophilous and/or temperate animal and plant species during unfavourable Pleistocene environmental conditions, and as the sources of

climate-induced forest expansion on influencing post-glacial species ranges is unfortunately not clear. Postglacial climate change was extremely rapid in some areas of Europe, occurring as quickly as within 50 years in Britain for example (Yalden, 1999), but may have taken longer to stabilize in other areas of Europe (Davis *et al.*, 2003). Likewise, palaeoecological evidence indicates that postglacial migrational lags in plant species ranges occurred (Normand *et al.*, 2011) and that forests continued to expand at the beginning of the Holocene during the climatic optimum (Huntley & Birks, 1983; Roberts, 1998; Kleinen *et al.*, 2011). Reconstructed ranges for the European wild cat, pine marten, and European polecat indicate an expanding distribution for the European wild cat and the European polecat from the Mesolithic to the Neolithic (Figure 4.3) suggesting that these species could have continued to increase their postglacial distribution into the first half of the Holocene (Crees *et al.*, 2016). However, as these range expansions occurred within bootstrapped limits, they cannot be considered significant increases. Expanding human populations, for example, could have increased the number and spread of subfossil records from archaeological sites, rather than reflecting a spread in the species themselves. On the other hand, as there are no natural fossil sites containing these species prior to the Neolithic in some regions either, the possibility of continued range expansion cannot be entirely ruled out.

Given the relative climatic stability of the Holocene, the distribution of most mammal species during the Mid-Late Holocene was predominantly influenced by humans. Several large herbivores became extinct or extinct in the wild due to hunting and habitat loss, e.g.

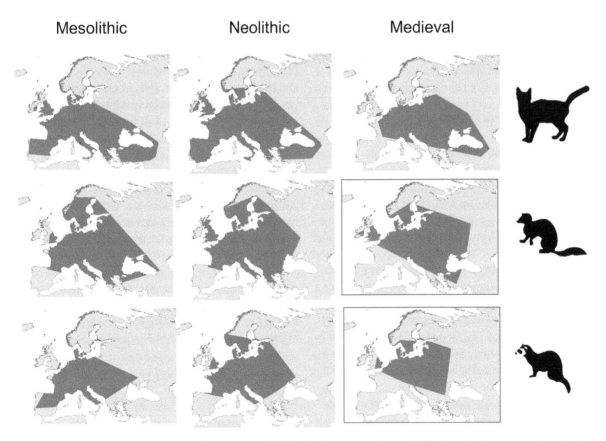

Figure 4.3 Maps showing the extent of occurrence (EOO) for the European wild cat, *Felis silvestris*, the pine marten, *Martes martes*, and the European polecat, *Mustela putorius*, for the Mesolithic (~11.5–7 kya), Neolithic (~7–5 kya), and Medieval (~1.5–0.5 kya). The ranges outlined in red indicate a significant decline.

the aurochs, *Bos primigenius*, and the European bison, *Bison bonasus*. Several large carnivores were extirpated from large parts of their ranges, e.g. the grey wolf, the brown bear, and the Eurasian lynx, *Lynx lynx*, although they survived across much of their respective ranges for longer and declined more recently. By contrast, smaller carnivores were generally more resilient to human impacts and did not suffer widespread declines associated with many larger mammals. Analysis of Holocene range dynamics suggests that the pine marten and the European polecat declined significantly in range only as late as the Medieval period (Figure 4.3). The European wild cat did not significantly decline during any time period during the pre-modern Holocene, although it did disappear from northern peripheral regions of its range, for example, Ireland and Scandinavia, where it only briefly colonized during the climatic optimum (~9–5 kya) before becoming extinct (Lepiksaar, 1986; McCormick, 1999). This was probably due to the unfavourable climate at the end of this period and possibly exacerbated by smaller founder populations and the fact that they were isolated from mainland European populations. It should be noted that all Late Holocene range reconstructions were heavily influenced by Spain, Portugal, and Italy where the subfossil record is poorer. As a consequence, the lack of records in these regions during the Medieval period may have skewed distributions for all three species during this time period where, in fact, all continued to persist (Mitchell-Jones *et al.*, 1999), though perhaps at reduced populations, which may have contributed to their scarcity in the subfossil record.

Conclusion and Future Research

The present study demonstrates that both the subfossil record and molecular data are powerful tools for reconstructing the spatio-temporal dynamics of small carnivore species in Europe. However, it has also revealed gaps in our current understanding and, therefore, we highlight potential future multidisciplinary research directions regarding the Late Quaternary biogeography of small carnivores. For example, the refugial history of the smaller mustelid species in Eurasia remains poorly understood, and the enigmatic Mid- and Late Holocene presence of the wolverine in steppe areas of southeastern Europe also requires further examination (Sommer & Benecke, 2004 and references therein). Another potential area for investigation is the apparent congruency of the stone marten colonization of Europe alongside the expansion of Neolithic farmers from Asia Minor. Finally, we encourage the use of multiple sources of Quaternary data to explore past small carnivore biogeography and the development of analysis techniques to support this research.

Acknowledgements

We are grateful to Martin Theuerkauf (University of Greifswald, Germany) for some comments to an earlier version of the text and to Isabell von Duhn (Lübeck, Germany) for the translation of literature in the French language. Thanks are due to Martin Sabol and an anonymous referee for their comments and suggestions, which which improved the manuscript.

References

Barron, E., van Andel, T.H. & Pollard, D. (2003) Glacial environments II. Reconstructing the climate of Europe in the Last Glaciation. In: *Neanderthals and Modern Humans in the European Landscape during the Last Glaciation: Archaeological Results of the Stage 3 Project* (eds T.H. van Andel & W. Davies), pp. 57–78. McDonald Institute Monographs, Cambridge.

Benecke, N. (1999) The project "The holocene history of the European vertebrate fauna". *Archäologie in Eurasien* 6, 151–161.

Björck, S., Walker, M.J.C., Cwynar, L.C., Johnsen, S., Knudsen, K.-L., Lowe, J.J. & Wohlfarth, B. (1998) An event stratigraphy for the Last Termination in the North Atlantic region based on the Greenland ice core record: a proposal by the INTIMATE group. *Journal of Quaternary Science* 13, 283–292.

Blockley, S.P.E., Lane, C.S., Hardiman, M., Rasmussen, S.O., Seierstad, I.K., Steffensen, J.P., Svensson, A., Lotter, A.F., Turney, C.S.M., Bronk Ramsey, C. & INTIMATE members (2012) Synchronisation of palaeoenvironmental records over the last 60,000 years, and extended INTIMATE event stratigraphy to 48,000 b2k. *Quaternary Science Reviews* 36, 2–10.

Crees, J.J. (2013) *Dynamics of Large Mammal Range Shifts and Extinction: Evidence from the Holocene of Europe*. PhD thesis, Imperial College London and Institute of Zoology, ZSL.

Crees, J.J. & Turvey, S.T. (2014) Holocene extinction dynamics of *Equus hydruntinus*, a late-surviving European megafaunal mammal. *Quaternary Science Reviews* 91, 16–29.

Crees, J.J., Carbone, C., Sommer, R.S., Benecke, N. & Turvey, S.T. (2016) Millennial-scale faunal record reveals differential resilience of European large mammals to human impacts across the Holocene. *Proceedings of the Royal Society B: Biological Sciences* 283, 20152152.

Dansgaard, W., Johnsen, S.J., Clausen, H.B., Dahl-Jensen, D., Gundestrup, N.S., Hammer, C.U., Hvidberg, C.S., Steffensen, J.P., Sveinbjörnsdottir, A.E., Jouzel, J. & Bond, G. (1993) Evidence for general instability of past climate from a 250-kyr ice-core record. *Nature* 364, 218–220.

Davis, B.A.S., Brewer, S., Stevenson, A.C. & Guiot, J. (2003) The temperature of Europe during the Holocene reconstructed from pollen data. *Quaternary Science Reviews* 22, 1701–1716.

Deffontaine, V., Libois, R., Kotlik, P., Sommer, R., Nieberding, C., Searle, J.B. & Michaux, J.R. (2005) Beyond the Mediterranean peninsulas: evidence of central European glacial refugia for a temperate forest mammal species, the bank vole (*Clethrionomys glareolus*). *Molecular Ecology* 14, 1727–1739.

Fortin, M.J. & Dale, M.R.T. (2005) *Spatial Analysis: A Guide for Ecologists*. Cambridge University Press, Cambridge.

Gaston, K. (2009) Geographic range limits of species. *Proceedings of the Royal Society B: Biological Sciences* 276, 1391–1393.

Hewitt, G.M. (1996) Some genetic consequences of ice ages, and their role in divergence and speciation. *Biological Journal of the Linnean Society* 58, 247–276.

Hewitt, G.M. (2000) The genetic legacy of the Quaternary ice ages. *Nature* 405, 907–913.

Honnen, A.-C., Petersen, B., Kaßler, L., Elmeros, M., Roos, A., Sommer, R.S. & Zachos, F.E. (2011) Genetic structure of Eurasian otter (*Lutra lutra*, Carnivora: Mustelidae) populations from the western Baltic sea region and its implications for the recolonisation of northwestern Germany. *Journal of Zoological Systematics and Evolutionary Research* 48, 169–175.

Hubberten, H.W., Andrev, A., Astakhov, V.I., Demikov, I., Dowdeswell, J.A. et al. (2004) The periglacial climate and environment in northern Eurasia during the Last Glaciation. *Quaternary Science Reviews* 23, 1333–1357.

Huijzer, B. & Vandenberghe, J. (1998) Climatic reconstruction of the Weichselian Pleniglacial in northwestern and Central Europe. *Journal of Quaternary Science* 13, 391–417.

Huntley, B. & Allen, J.R.M. (2003) Glacial environments III: palaeo-vegetation patterns in Late Glacial Europe. In: *Neanderthals and Modern Humans in the European Landscape during the Last Glaciation: Archaeological Results of the Stage 3 Project* (eds T.H. van Andel & W. Davies), pp. 79–102. McDonald Institute Monographs, Cambridge.

Huntley, B. & Birks, H.J.B. (1983) *An Atlas of Past and Present Pollen Maps of Europe: 0–13,000 years ago*. Cambridge University Press, Cambridge.

Huntley, B., Alfano, M.J., Allen, J.R.M., Pollard, D., Tzedakis, P.C., de Beaulieu, J.-L., Grüger, E. & Watts, B. (2003) European vegetation during Marine Oxygen Isotope Stage-3. *Quaternary Research* 59, 159–212.

IUCN [International Union for Conservation of Nature] (2001) *IUCN Red List Categories and Criteria, Version 3.1*. IUCN, Gland and Cambridge.

Kleinen, T., Tarasov, P., Brovkin, V., Andreev, A. & Stebich, M. (2011) Comparison of modeled and reconstructed changes in forest cover through the past 8000 years: Eurasian perspective. *The Holocene* 21, 723–734.

Koby, F.E. (1951) Le putois d'Eversmann fossile en Suisse et en France. *Eclogae Geologicae Helvetiae* 44, 394–398.

Kosintsev, P. (2007) Late Pleistocene large mammal faunas from the Urals. *Quaternary International* 160, 112–120.

Kosintsev, P. & Bachura, O.P. (2013) Late Pleistocene and Holocene mammal fauna of the southern Urals. *Quaternary International* 284, 161–170.

Kotlík, P., Deffontaine, V., Mascheretti, S., Zima, J., Michaux, J.R. & Searle, J.B. (2006) A northern glacial refugium for bank voles (*Clethrionomys glareolus*). *Proceedings of the National Academy of Sciences of the United States of America* 103, 14860–14864.

Krajcarz, M.T., Krajcarz, M., Goslar, T. & Nadachowski, A. (2015) The first radiocarbon dated steppe polecat (*Mustela eversmanii*) from the Pleistocene of Poland. *Quaternary International* 357, 237–244.

Kuemmerle, T., Hickler, T., Olofsson, J., Schurgers, G. & Radeloff, V.C. (2012) Refugee species: which historic baseline should inform conservation planning? *Diversity and Distributions* 18, 1258–1261.

Kutschera, V.E., Lecomte, N., Janke, A., Selva, N., Sokolov, A.A., Haun, T., Steyer, K., Nowak, C. & Hailer, F. (2013) A range-wide synthesis and timeline for phylogeographic events in the red fox (*Vulpes vulpes*). *BMC Evolutionary Biology* 13, 1–15.

Lepiksaar, J. (1986) The Holocene history of theriofauna in Fennoscandia and Baltic countries. *Striae* 24, 51–70.

Litt, T., Brauer, A., Goslar, T., Merkt, J., Balaga, K., Müller, H., Ralska-Jasiewiczowa, M., Stebich, M. & Negendank, J.F.W. (2001) Correlation and syncronisation of Lateglacial continental sequences in northern central Europe based on annually laminated lacustrine sediments. *Quaternary Science Reviews* 20, 1233–1249.

Litt, T., Schminke, H.U. & Kromer, B. (2003) Environmental response to climatic and volcanic events in central Eurpoe during the Weichselian Lateglacial. *Quaternary Science Reviews* 22, 7–32.

Loehr, J., Worley, K., Grapputo, A., Carey, J., Veitch, A. & Coltman, D.W. (2006) Evidence for cryptic glacial refugia from North American mountain sheep mitochondrial DNA. *Journal of Evolutionary Biology* 19, 419–430.

McCormick, F. (1999) Early evidence for wild animals in Ireland. In: *The Holocene History of the European Vertebrate Fauna: Modern Aspects of Research* (ed. N. Benecke), pp. 355–372. Verlag Marie Leidorf, Rahden.

McDevitt, A.D., Zub, K., Kawałko, A., Oliver, M.K., Herman, J.S. & Wójcik, J.M. (2012) Climate and refugial origin influence the mitochondrial lineage distribution of weasels (*Mustela nivalis*) in a phylogeographic suture zone. *Biological Journal of the Linnean Society* 106, 57–69.

Mecozzi, B., Coppola, D., Iurino, D.A., Sardella, R. & De Marinis, A.M. (2019) The Late Pleistocene European badger *Meles meles* from Grotta Laceduzza (Brindisi, Apulia, Southern Italy): the analysis of the morphological and biometric variability. *The Science of Nature* 106, 13.

Mitchell-Jones, A.J., Amori, G., Bogdanowicz, W., Kryštufek, B., Reijnder, P.J.H., Spitzenberger, F., Stubbe, M., Thissen, J.B.M., Vohralik, V. & Zima, J. (1999) *The Atlas of European Mammals*. T. & A.D. Poyser, London.

Mooney, C.Z. & Duval, R.D. (1993) Bootstrapping: a nonparametric approach to statistical inference. *Sage University Paper Series on Quantative Applications in the Social Sciences*, 07-095. Sage, Newbury Park.

Mucci, N., Arrendal, J., Ansorge, H., Bailey, M., Bodner, M., *et al.* (2010) Genetic diversity and landscape genetic structure of otter (*Lutra lutra*) populations in Europe. *Conservation Genetics* 11, 583–599.

Münzel, S., Langguth, K., Conard, N. & Uerpmann, H.P. (2001) Höhlenbärenjagd auf der Schwäbischen Alb vor 30.000 Jahren. *Archäologisches Korrespondenzblatt* 31, 317–328.

Normand, S., Ricklefs, R.E., Skov, F., Bladt, J., Tackenberg, O. & Svenning, J-C. (2011) Postglacial migration supplements climate in determining plant species ranges in Europe. *Proceedings of the Royal Society B: Biological Sciences* 278, 3644–3653.

Ponomarev, D., Puzachenko, A., Bachura, O., Kosintsev, P. & von der Plicht, J. (2013) Mammal fauna during the Late Pleistocene and Holocene in the far northeast of Europe. *Boreas* 42, 779–797.

Prost, S., Klietmann, J., van Kolfschoten, T., Vrieling, K., Stiller, M., Nagel, D., Rabeder, G., Guralnick, R., Waltari, E., Hofreiter, M. & Sommer, R.S. (2013) Effects of Late Quaternary climate change on Palearctic shrews. *Global Change Biology* 19, 1865–1874.

Renssen, H., Seppä, H., Heiri, O., Roche, D.M., Goosse, H. & Fichtfet, T. (2009) The spatial and temporal complexity of the Holocene thermal maximum. *Nature Geoscience* 2, 411–414.

Roberts, N. (1998) *The Holocene: An Environmental History*. 2nd edition. Blackwell, Oxford.

Schmitt, T. & Krauss, J. (2004) Reconstruction of the colonization route from glacial refugium to the northern distribution range of the European butterfly *Polyommatus coridon* (Lepidoptera: Lycaenidae). *Diversity and Distributions* 10, 271–274.

Schönswetter, P., Stehlik, I., Holderegger, R. & Triebsch, A. (2005) Molecular evidence for glacial refugia of mountain plants in the European Alps. *Molecular Ecology* 14, 3547–3555.

Sickenberg, O. (1968) Der Steppeniltis (*Mustela [Putorius] eversmanii soergeli* Éhik) in der Niederterrasse der Leine und seine klimageschichtliche Bedeutung. *Eiszeitalter Gegenwart* 19, 147–163. (In German).

Sommer, R.S. (2007) When East met West: the sub-fossil footprints of the west European hedgehog and the northern white-breasted hedgehog during the Late Quaternary in Europe. *Journal of Zoology* 271, 82–89.

Sommer, R.S. (2020) Late Pleistocene and Holocene history of mammals in Europe. In: *Handbook of the Mammals of Europe* (eds K. Hackländer & F.E. Zachos), 16 pp. Springer, Cham.

Sommer, R. & Benecke, N. (2004) Late- and Post-Glacial history of the Mustelidae in Europe (Mustelidae). *Mammal Review* 34, 249–284.

Sommer, R.S. & Benecke, N. (2005a) The recolonisation of Europe by brown bears *Ursus arctos* Linnaeus, 1758 after the Last Glacial Maximum. *Mammal Review* 35, 156–164.

Sommer, R.S. & Benecke, N. (2005b) Late-Pleistocene and early Holocene history of the canid fauna of Europe (Canidae). *Mammalian Biology* 70, 227–241.

Sommer, R.S. & Benecke, N. (2006) Late-Glacial and early Holocene colonisation history of felids in Europe (Felidae): a review. *Journal of Zoology* 269, 7–20.

Sommer, R.S. & Nadachowski, A. (2006) Glacial refugia of mammals in Europe: evidence from fossil records. *Mammal Review* 36, 251–266.

Sommer, R.S. & Zachos, F.E. (2009) Fossil evidence and phylogeography of temperate species: "glacial refugia" and postglacial recolonization. *Journal of Biogeography* 36, 2013–2020.

Sommer, R.S., Persson, A., Wiesecke, N. & Fritz, U. (2007) Holocene recolonization and extinction history of the pond turtle, *Emys orbicularis* (L., 1758) in Europe. *Quaternary Sciences Reviews* 26, 3099–3107.

Sommer, R.S., Zachos, F.E., Street, M., Jöris, O., Skog, A. & Benecke, N. (2008) Late Quaternary distribution dynamics and phylogeography of the red deer (*Cervus elaphus*) in Europe. *Quaternary Science Reviews* 27, 714–733.

Sommer, R.S., Fahlke, J., Schmölcke, U., Benecke, N. & Zachos, F.E. (2009a) Quaternary history of the European roe deer (*Capreolus capreolus*). *Mammal Review* 38, 1–16.

Sommer, R.S., Lindqvist, C., Persson, A., Bringsøe, H., Rhodin, A.G.J., Schneeweiß, N., Široký, P., Bachmann, L. & Fritz, U. (2009b) Unexpected early extinction of the European pond turtle (*Emys orbicularis*) in Sweden and climatic impact on its Holocene range. *Molecular Ecology* 18, 1252–1262.

Sommer, R.S., Kalbe, J., Ekström, J., Benecke, N. & Liljegren, R. (2014) Range dynamics of the reindeer in Europe during the last 25,000 years. *Journal of Biogeography* 41, 298–306.

Statham, M.J., Edwards, C.J., Norén, K., Soulsbury, C.D. & Sacks, B.N. (2018) Genetic analysis of European red foxes reveals multiple distinct peripheral populations and central continental admixture. *Quaternary Science Reviews* 197, 257–266.

Steffensen, J.P., Andersen, K.K., Bigler, M., Clausen, H.B., Dahl-Jensen, D., *et al.* (2008) High-resolution Greenland ice core data show abrupt climate change happens in few years. *Science* 321, 680–684.

Stewart, J.R. (2008) The progressive effect of the individualistic response of species to Quaternary climate change: an analysis of British mammalian faunas. *Quaternary Science Reviews* 27, 2499–2508.

Stewart, J.R., Lister, A.M., Barnes, I. & Dalen, L. (2010) Refugia revisited: individualistic responses of species in space and time. *Proceedings of the Royal Society B: Biological Sciences* 277, 661–671.

Street, M. & Baales, M. (1999) Pleistocene/Holocene changes in the Rhineland fauna in a northwest European context. *Archäologie in Eurasien* 6, 9–38.

Stuart, A.J. (2015) Late Quaternary megafaunal extinctions on the continents: a short review. *Geological Journal* 50, 338–363.

Stuart, A.J. & Lister, A.M. (2012) Extinction chronology of the woolly rhinoceros *Coelodonta antiquitatis* in the context of late Quaternary megafaunal extinctions in northern Eurasia. *Quaternary Science Reviews* 51, 1–17.

Taberlet, P., Fumagalli, L., Wust-Saucy, A.G. & Cosson, J.F. (1998) Comparative phylogeography and postglacial colonization routes in Europe. *Molecular Ecology* 7, 453–464.

Theuerkauf, M. & Joosten, H. (2012) Younger Dryas cold stage vegetation pattern of central Europe-climate, soil and relief controls. *Boreas* 41, 391–407.

Vinther, B.M., Buchardt, S.L., Clausen, H.B., Dahl-Jensen, D., Johnsen, S.J., Fisher, D.A., Koerner, R.M.,

Raynaud, D., Lipenkov, V., Andersen, K.K., Blunier, T., Rasmussen, S.O., Steffensen, J.P. & Svensson, A.M. (2009) Holocene thinning of the Greenland ice sheet. *Nature* 461, 385–388.

von Koenigswald, W., Müller-Beck, H. & Pressmar, E. (1974) *Die Archäologie und Paläontologie in den Weinberghöhlen bei Mauern (Bayern): Grabungen 1937–1967.* Selbstverlag, Institut für Urgeschichte, Tübingen. (In German).

Weninger, B., Jöris, O. & Danzeglocke, U. (2008) *CalPal 2007. Cologne Radiocarbon Calibration & Palaeoclimate Research Package.* http://www.calpal. de. Accessed on 3 May 2014.

Willis, K.J. & Whittaker, R.J. (2000) The refugial debate. *Science* 287, 1406–1407.

Wohlfarth, B., Veres, D., Ampel, L., Lacourse, T., Blaauw, M., *et al.* (2008) Rapid ecosystem response to abrupt climate changes during the last glacial period in Western Europe, 40–16 kyr BP. *Geology* 36, 407–410.

Yalden, D.W. 1999. *The History of British Mammals.* T. & A.D. Poyser, London.

5

Ecomorphological Disparity of Small Carnivore Guilds

Carlo Meloro*

Research Centre in Evolutionary Anthropology and Palaeoecology, School of Biological and Environmental Sciences, Liverpool John Moores University, Liverpool, UK

SUMMARY

Mammalian species composition might change in relation to biotic or abiotic factors depending on the scale of investigation. Ecomorphology is one of the tools that can be employed to understand how species composition changes through space and time. Here, the morphological diversity of small carnivore guilds (defined as a pool of carnivoran species whose body mass is < 7 kg) is explored using 2D geometric morphometrics of mandibles belonging to 61 species. A strong taxonomic signal emerges by looking at mandibular morphospace so that separation of carnivoran families is apparent. Mustelids are the most distinct, being characterized by a short and curved mandibular corpus, while felids exhibit a typical hypercarnivore mandible with no crushing molar area. Overlap occurs between canids, viverrids, and herpestids possibly in relation to their generalized feeding habits and killing behaviours. When species are grouped according to their presence/absence into six carnivoran species-rich ecosystems, an ecogeographical pattern occurs. Guilds from higher latitudes such as Yellowstone (USA) and Krokonose (Europe) together with the Kruger (South Africa) assemblage are highly depleted of mandibular morphotypes. In contrast, guilds from tropical areas (Gunung Lensung, Indonesia; Yasuni, Ecuador; and La Amistad, Panama) exhibit high diversity of mandibular shapes corresponding to higher values of morphological disparity. This latter parameter correlates positively with precipitation variables, supporting a strong influence of climate on the historical community assembly of small carnivore guilds. Clearly, small carnivores can play a key role in ecosystem functioning and more theoretical work is needed to better identify this at multiple spatial and temporal scales.

Keywords

Climate – community assemblage – geometric morphometrics – mandible shape – morphospace – Mustelidae

Introduction

Ecological differences between animal species provide compelling evidences in understanding their distribution through space and time (Rosenzweig, 1995). On the one hand abiotic factors influence species ecology and distribution, on the other hand biotic interactions act as a balancing ecological force that might generate unpredictable patterns. In this regard,

---

*Corresponding author.

Small Carnivores: Evolution, Ecology, Behaviour, and Conservation, First Edition. Edited by Emmanuel Do Linh San, Jun J. Sato, Jerrold L. Belant, and Michael J. Somers.

carnivorans (mammals of the order Carnivora) received considerable attention because they are generally secondary consumers and apex predators in many trophic chains and they include species with a high degree of ecological interactions (Gittleman, 1985; Donadio & Buskirk, 2006; Davies *et al.*, 2007; Davis *et al.*, 2019). Such interactions are significantly documented by direct or indirect competition between taxa due to overlap in a trophic niche or spatial selection (Palomares & Caro, 1999).

On an evolutionary time-scale, the interplay of abiotic and biotic factors is considered the main driving force of carnivoran morphological diversification (Van Valkenburgh, 1999; Wesley-Hunt, 2005; Goswami, 2010), supporting a direct link between carnivoran's ecology and morphology. The term 'ecomorphology' well describes this link (Wainwright, 1994): species are functional units within ecosystems and their function is determined by their anatomy (Polly, 2010; Polly *et al.*, 2011). This innovative ecological concept allows species-specific phenotypes to be re-interpreted into a wider context of community analyses. By looking at functional morphology and how it varies across species, it is possible to predict the potential impact of abiotic and biotic factors on animal communities.

The distribution of carnivoran ecomorphologies can change across the continents (Werdelin & Wesley-Hunt, 2010), although it appears to be unchanged over time when only large taxa are considered (Van Valkenburgh, 1985, 1988, 1995; Meloro, 2011a). In this chapter, I investigate ecomorphologies of small carnivorans (hereafter small carnivores) across different ecosystems in order to identify abiotic or biotic factors responsible for their current assemblage within communities.

The Small Carnivores

Defining small carnivores can be a challenge due to the broad ecological and biological diversity. The suffix 'large' or 'small' relates to how humans perceive animal species. The grey wolf, *Canis lupus*, the tiger, *Panthera tigris*, the lion, *Panthera leo*, or the spotted hyena, *Crocuta crocuta*, are generally associated with large fierce beasts, but smaller forms such as the omnivorous red fox, *Vulpes vulpes*, become difficult to categorize.

The concept of 'guild' (Root, 1967) can help clarify such an issue because it groups all species capable of exploiting the same resource in a similar way. Simberloff & Dayan (1991) provided a broad overview of the use and misuse of guilds, especially in the carnivoran literature, and there is no right or wrong guild definition. Van Valkenburgh (1985, 1988, 1989) pioneered the use of the term 'large' carnivores as a group of Carnivora that includes all species whose average body weight is > 7 kg. On the other hand, Carbone *et al.* (1999) identified an eco-physiological threshold in carnivoran species bigger than 21.5 kg that are generally apex predators with a strong functional role within an ecosystem. Are there any thresholds to define small carnivore guilds? Friscia *et al.* (2006) studied 'small' carnivoran ecomorphologies including all species weighing < 10 kg, while Roemer *et al.* (2009) recently grouped small carnivores as 'mesopredators' whose body mass is < 15 kg.

Defining the 'small' threshold might be problematic, and here I have considered 7 kg as a valid ecomorphological threshold. Due to the main focus on the carnivoran mandible shape, this value is highly appropriate because all taxa above or below this threshold show distinct mandibular morphologies irrespective of their phylogenetic relatedness (Meloro & O'Higgins, 2011). Such a definition is operationally useful as it provides a direct link with previous studies on carnivoran morphological diversity over space and time (Van Valkenburgh, 1985, 1988, 1989; Meloro, 2011a). Consequently, small carnivores are defined here as all members of the order Carnivora whose average body weight is < 7 kg, including taxa from the tiny least weasel, *Mustela nivalis*, that weighs a few hundred grams to the relatively large northern raccoon, *Procyon lotor* (6.4 kg; Gittleman, 1985).

Mandibular Shape in Carnivora

The mandible has a dual function in the mammalian skeleton: (i) it provides support to the developing dentition; (ii) it provides attachment to the main masticatory muscles (temporalis, masseter, and zygomaticomandibularis) (Herring, 1980, 1993). Both these functions are integrated parts of the complex feeding system and can be used to predict feeding adaptations from skeletal morphology only. Early anatomical investigations by Herring & Herring (1974), Greaves (1983, 1985), and Radinsky (1981a,b, 1982) identified a significant association between mandibular morphology and diet in mammals in general and carnivorans in particular. Interestingly, such association did not emerge directly from mandibular metric data that are better descriptors of species' taxonomic affiliation (Crusafont-Pairó & Truyols-Santonja, 1957).

In spite of the significant progresses made in the quantification of complex biological shapes (Adams *et al.*, 2004, 2013; Lawing & Polly, 2009), ecomorphological patterns within carnivorans are still remarkably unchanged: taxonomic differences always emerge when describing mandibular (and skull) morphology, while shape differences between dietary groups are subtle especially after phylogenetic relatedness is taken into account (Meloro *et al.*, 2008, 2011; Figueirido *et al.*, 2010, 2011, 2013; Meloro & O'Higgins, 2011; Prevosti *et al.*, 2012). This is due to the strong interplay between carnivoran feeding adaptations and clade differentiation (Crusafont-Pairó & Truyols-Santonja, 1956, 1957, 1958; Meloro & Raia, 2010): many feeding ecologies can be specific to certain taxonomic groups (e.g. all felids show hypercarnivorous craniodental morphologies related to their strictly meat-eating diet). In particular, the expansion or reduction of molar crushing vs. slicing area (Van Valkenburgh, 1989) drives such patterns of dietary differentiation in carnivorans and it significantly describes differences in mandibular morphologies across species (both small and large, see Popowics, 2003; Friscia *et al.*, 2006; Meloro, 2011b; Asahara, 2013).

Ecomorphological Disparity

Since mandible shape is made up of a complex suite of traits, it requires high dimensional data (e.g. a suite of multiple measurements or functional ratios) to be described in detail. Multivariate techniques such as Principal Component Analysis reduce such data into orthogonal vectors that generally describe what is called a 'morphospace'. Species within the morphospace are dots whose distribution can possibly be influenced by multiple factors.

Foote (1992, 1993) introduced disparity as a way to measure and describe species' distributions in a morphospace. Disparity quantifies the morphospace volume occupied by a specific set of taxa. This metric was generally employed to investigate macroevolutionary patterns such as the relative expansion or contraction of some particular clades relative to others. For Carnivora, Van Valkenburgh (1999) identified a stasis in ecomorphological disparity through time, while Holliday & Steppan (2004) supported a smaller morphospace occupation by hypercarnivorous (strictly meat-eating) species relative to other ecomorphological groups. A recent study by Werdelin & Wesley-Hunt (2010) confirmed such findings although they identified less ecomorphological disparity for canids compared to other clades. Similar disparity values occurred for carnivoran species from different continents. Accordingly, the disparity is computed here as a measure of the small carnivore guild distributions across different continents in the mandibular morphospace. Ecogeographical patterns are expected to occur because previous studies showed that disparity of geographically distinct mammalian assemblages changes with latitude (Shepherd, 1998).

Methods

Mandibles belonging to 61 species of carnivorans were photographed in lateral view and subsequently analyzed using the software tpsDig2 (Rohlf, 2015). This is a subset of data collected by Meloro & O'Higgins (2011) and includes wild-captured adult specimens representative of small ($< 7\,kg$) taxa housed at the Natural History Museum of London. Species selection was drawn from lists of six carnivoran species-rich terrestrial ecosystems (Bio Inventory, source: http://www.ice.ucdavis.edu/bioinventory/bioinventory.html): Krokonose, Czech Republic ($n = 12$); Yellowstone, USA ($n = 8$); Gunung Lensung, Indonesia ($n = 18$); Kruger National Park, South Africa ($n = 12$); Yasuni, Ecuador ($n = 10$); and La Amistad, Panama ($n = 12$) (Table 5.1). Intraspecific variation was not explored here according to other ecomorphological studies that looked at macroevolutionary (i.e. above species level) patterns within the Order Carnivora (e.g. Van Valkenburgh, 1985, 1988, 1989; Christiansen & Adolfssen, 2005; Evans *et al.*, 2007; Polly & MacLeod, 2008; Meloro & O'Higgins, 2011; Meloro, 2011a,c).

Table 5.1 List of small carnivore species guilds geographically partitioned.

Krokonose (n = 10)	Yellowstone (n = 8)	Kruger (n = 12)	Gunung Lensung (n = 18)	Yasuni (n = 10)	La Amistad (n = 12)
Felis silvestris	*Lontra canadensis*	*Atilax paludinosus*	*Aonyx[a] cinereus*	*Eira barbara*	*Bassaricyon gabbii*
Martes foina	*Martes americana*	*Galerella sanguinea*	*Arctogalidia trivirgata*	*Galictis vittata*	*Bassariscus sumichrasti*
Martes martes	*Mephitis mephitis*	*Genetta genetta*	*Catopuma badia*	*Herpailurus yagouaroundi*	*Conepatus semistriatus*
Mustela erminea	*Mustela erminea*	*Genetta maculata*	*Cynogale bennettii*	*Leopardus tigrinus*	*Eira barbara*
Mustela eversmanii	*Mustela frenata*	*Helogale parvula*	*Hemigalus derbyanus*	*Leopardus wiedii*	*Galictis vittata*
Mustela nivalis	*Neovison vison*	*Herpestes ichneumon*	*Lutra sumatrana*	*Mustela africana*	*Herpailurus yagouaroundi*
Mustela putorius	*Pekania pennanti*	*Ichneumia albicauda*	*Martes flavigula*	*Nasua nasua*	*Leopardus wiedii*
Neovison vison	*Procyon lotor*	*Ictonyx striatus*	*Mustela nudipes*	*Potos flavus*	*Mustela frenata*
Nyctereutes procyonoides		*Mungos mungo*	*Paguma larvata*	*Procyon cancrivorus*	*Nasua narica*
Vulpes vulpes		*Otocyon megalotis*	*Paradoxurus hermaphroditus*	*Speothos venaticus*	*Potos flavus*
		Paracynictis selousi	*Pardofelis marmorata*		*Procyon lotor*
		Rhynchogale melleri	*Prionailurus bengalensis*		*Urocyon cinereoargenteus*
			Prionailurus planiceps		
			Prionodon linsang		
			Urva brachyura		
			Urva semitorquata		
			Viverra tangalunga		
			Viverricula indica		

[a] Scientific names are sorted alphabetically. For English names and other attributes of the listed small carnivore species, see Appendix A. *n* = number of species in each guild.

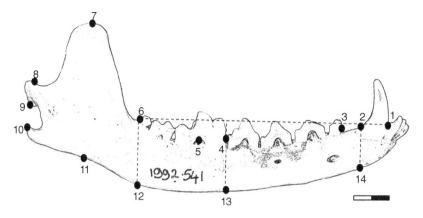

Figure 5.1 The position of landmarks on a mandible outline of red fox, *Vulpes vulpes* (NHM 1992.541). 1–2: anteroposterior diameter of c1; 2–3: diastema length; 3–4: length of the premolar row; 4–6: length of the molar row; 5: projection of the protocone cusp on the m1 baseline; 2–14: the thickness of the mandibular corpus (corpus mandibulae) under the canine; 4–13 and 6–12: the thickness of the mandibular corpus under molar row; 7: the tip of the coronoid process (processus coronoideus); 8–9: maximum depth of the condylar process (processus condylaris); 10: most lateral extreme point of the angular process (processus angularis); 11: the ventral extreme of angular process. The total scale bar equals 1.0 cm.

Two-dimensional (2D) coordinates of 14 landmarks (lnd) were recorded on each mandibular photograph using tpsDig2 (Rohlf, 2015; Figure 5.1). The landmarks functionally describe anatomical features including canine (lnd 1–2), premolar row (lnd 3–4), molar slicing (lnd 4–5) and crushing (lnd 5–6) area, coronoid (lnd 7), condyle (lnd 8–9), and angular process (10–11), as well as mandibular corpus depth (lnd 12–14). Meloro (2011b) and Meloro & O'Higgins (2011) consistently proved the existence of an association between this shape configuration and feeding adaptations in extant and fossil carnivorans.

Generalized Procrustes Analysis (Rohlf & Slice, 1990) was employed to translate, rotate, and scale the 2D landmark coordinates to a unit centroid size (i.e. the square root of the sum of the squared distances of a set of landmarks from the configuration centroid; Bookstein, 1989). The newly registered coordinates (i.e. Procrustes coordinates) were projected into thin-plate spline function, and a weight matrix of affine (Uniform) and non-affine (Partial Warps) components was generated. Relative Warp Analysis (RWA) was subsequently employed using tpsRelw (Rohlf, 2015) to identify orthogonal vectors (Principal Components, here named Relative Warps) that summarize shape variation described by the elements of the weight matrix. Such a procedure detects main shape differences (quantifiable also as Procrustes distances) within the morphospace via thin-plate spline:

deformation grids applied at the onset of each RW extreme score summarize shape deformations from the undeformed score positioned at the origin of each RW axis (the consensus configuration).

Specimens were labelled according to taxonomic affiliation and geographical guild membership (see Meloro, 2011a) to scrutinize patterns of morphospace occupation by small carnivores. Multiple analysis of variance (MANOVA) was employed to test whether taxonomic groups differ significantly in mandible shape using Relative Warp scores as dependent and family as a factor (Meloro *et al.*, 2008). The morphological disparity (Foote, 1992, 1993) was computed to quantify morphospace volume occupied by each geographical guild. In geometric morphometrics, the disparity is obtained as the sum of squared procrustes distances from each species to the grand group mean divided by the number of group members minus 1 (Zelditch *et al.*, 2003, 2004). This is exactly equivalent to the sum of variances obtained from Relative Warp scores for each identifiable group (in this case, the geographic guilds). By using the software IMP (Zelditch *et al.*, 2004), a series of 999 permutations was computed each time to identify 95% confidence intervals around the morphological disparity values. A two-group permutation test was also employed to detect whether differences in disparity values between groups were larger or smaller than expected by chance.

For each geographical guild, factors such as bioclimatic variables and number of species within taxonomic groups potentially predated by small carnivores (Rodentia, Lagomorpha, and Marsupialia; Ewer, 1973) were also quantified using WorldClim database from DIVA GIS (Hijmans *et al.*, 2005) and species lists drawn from the Bio Inventory. Those factors are expected to possibly influence morphospace occupation and volume of small carnivore guilds (see Meloro, 2011a for the case of large carnivores). Due to the small number of guilds analyzed, a Spearman's rank correlation test was employed to explore any possible association between morphological disparity and climatic or biotic factors (i.e. the number of prey species identified in each guild; Meloro, 2011a).

Results

Relative Warp Analysis extracted 24 orthogonal axes with the first nine explaining altogether ~95% of the shape variance. The first two Relative Warps explained 34.39% and 22.49% shape variances, respectively (Figure 5.2). These axes describe clear partitioning of broad taxonomic groups: all mustelids occupy positive RW1 scores and negative RW2 scores, felids show intermediate RW1 scores and highly positive RW2 scores, while canids, herpestids, and viverrids are distinguished for their generally negative RW1 scores; procyonids occupy all areas of the morphospace. MANOVA confirmed such a significant partitioning of RW1/2 morphospace areas by family groups (Wilk's lambda = 0.1326, $F = 18.86$, df = 10, 108, $p < 0.0001$) with mustelids and felids being the most different groups of all the other taxonomic combinations (Table 5.2).

RW1 describes (from negative to positive scores) the relative shortening of the mandibular corpus (corpus mandibulae) due to a smaller premolar row and a curved corpus profile detectable in mustelids. The mandibular ramus (ramus mandibulae) is tall and slender in this group, while it becomes enlarged horizontally and short vertically in small feliform carnivorans such as herpestids and viverrids at the negative RW1 scores. RW2 correlates with changes in the main position of landmark 5 that separates the molar slicing from the crushing area, thus determining the unique condition of hypercarnivorous felids that occupy extreme positive scores on this axis. The

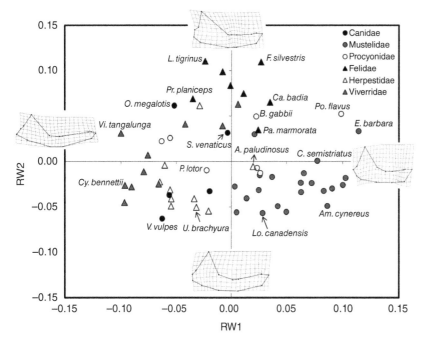

Figure 5.2 Scatter plot of RW1 vs. RW2 for a sample of mandibles belonging to 61 small carnivore species (labelled according to family). Transformation grids visualize shape deformation relative to the mean (regular grid, not shown) at the positive and negative extremes of Relative Warp (RW) axes.

Table 5.2 Probability values for pairwise Hotelling's *t* square comparisons performed using the first two RWs are shown below the diagonal.

	Canidae	Felidae	Herpestidae	Mustelidae	Procyonidae	Viverridae
Canidae	—	**0.04**	1.00	**<0.0001**	1.00	1.00
Felidae	**<0.001**	—	**<0.0001**	**<0.0001**	0.11	**<0.001**
Herpestidae	0.65	**<0.0001**	—	**<0.0001**	0.43	0.36
Mustelidae	**<0.0001**	**<0.0001**	**<0.0001**	—	**<0.001**	**<0.0001**
Procyonidae	0.17	**0.01**	**0.03**	**<0.0001**	—	0.09
Viverridae	0.26	**<0.0001**	**0.02**	**<0.0001**	**0.01**	—

The *p*-values with Bonferroni correction are shown above the diagonal. Significant differences ($p < 0.05$) are indicated in bold.

mandibular ramus is also projected more posteriorly in species at positive RW2, evidencing an almost straight profile for the corpus that is more curved posteriorly in species at the negative end of this axis (Figure 5.2).

When species are labelled according to their guild affiliation, distinct patterns in morphospace occupations occur: the guild of Yellowstone is highly depleted in morphotypes together with that of Kruger (Figure 5.3). Morphologically richer small carnivore guilds are from tropical areas such as La Amistad and Gunung Lensung that exhibit also a higher number of taxa.

Morphological disparity analysis partially confirmed this trend with Kruger and Yellowstone showing the smallest values and La Amistad and Gunung Lensung the highest (Figure 5.4). The 95% confidence intervals are broad and, therefore, no significant differences were detected in disparity values except between Gunung Lensung and Yellowstone, whose disparity difference is higher than expected by chance 95% of the times (Table 5.3). A non-parametric Spearman's rank correlation identified a significantly positive correlation between disparity and climatic precipitation variables, while a strong negative correlation was found with numbers of lagomorph species recorded in each of the analyzed ecosystems (Table 5.4, Figure 5.4).

Discussion

The mandible shape of small carnivores exhibits an evident taxonomic signal and this pattern is no exception in carnivoran datasets. Previous morphometric studies identified a similar degree of morphospace segregation by family both on large extant and fossil carnivorans (Meloro *et al.*, 2008; Figueirido *et al.*, 2010, 2011, 2013; Meloro, 2011a,b, 2012) and all extant carnivorans *sensu lato* (Meloro & O'Higgins, 2011). The most distinct groups of small carnivores are the hypercarnivorous felids and the mustelids (Figure 5.2): small predatory cats show a more reduced molar crushing area than the rest of small carnivore clades while mustelids are distinguished by a more posteriorly curved mandibular corpus. Such main feature of mandibular shape variation is in agreement with earlier investigations on carnivoran skulls (see Radinsky, 1981a,b, 1982) that especially highlighted the unique mustelid condition of masticatory muscles arrangement: the posterior temporalis is generally more developed in this group, thus imposing an almost straight and anteriorly curved configuration in the shape of the ramus mandibulae (Ewer, 1973). This configuration also influences to some extent the glenoid fossa – a structure that provides articulation between the cranium and the mandibular condyle to allow more efficient masticatory loading during the carnassial (lower m1 and upper P4) shear bite. The m1 slicing area is also enlarged as typical of highly carnivorous, predaceous forms (e.g. weasels) but not in such an extreme way as in the felids.

On the opposite area of the mandibular morphospace, small canids, herpestids, and viverrids show considerable overlap. This pattern was already highlighted by Meloro & O'Higgins (2011) and it appears to be the result of more generalized omnivorous feeding adaptations. Small canids, here represented by fox-like

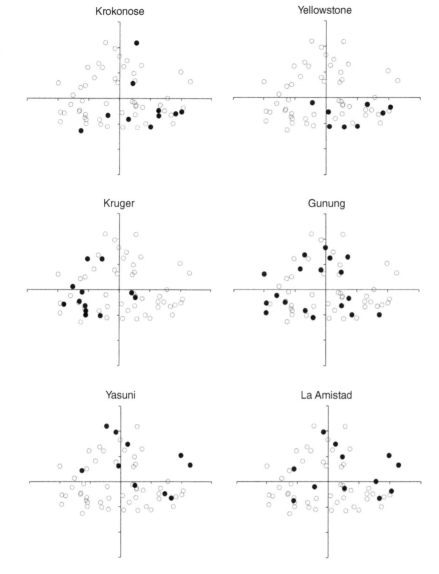

Figure 5.3 Scatter plots of RW1 (*x*-axis, scale −0.15/+0.15) vs. RW2 (*y*-axis, scale −0.15/+0.15) showing each extant small carnivore guild highlighted by closed circles: Krokonose, Czech Republic (*n* = 12); Yellowstone, USA (*n* = 8); Kruger National Park, South Africa (*n* = 12); Gunung Lensung, Indonesia (*n* = 18); Yasuni, Ecuador (*n* = 10); and La Amistad, Panama (*n* = 12).

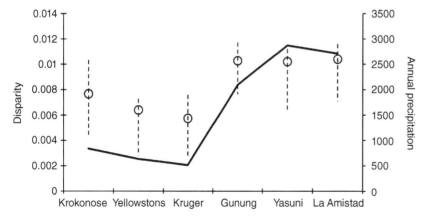

Figure 5.4 Disparity values (circles) computed for morphospace of each extant small carnivore guild superimposed on annual precipitation values (in mm/year). The vertical dotted bars correspond to 95% confidence intervals around morphological disparity values after 999 randomizations. The solid line shows precipitation values from different localities.

Table 5.3 Two-group permutation tests for differences in disparity values between small carnivore guilds.

	Krokonose	Yellowstone	Kruger	Gunung	Yasuni	La Amistad
Krokonose	—	0.95	0.93	0.41	0.93	0.90
Yellowstone	0.0013	—	1.00	**0.05**	0.93	0.82
Kruger	0.0019	0.0007	—	0.24	0.85	0.60
Gunung	0.0026	**0.0039**	0.0045	—	1.00	1.00
Yasuni	0.0025	0.0038	0.0045	0.0001	—	1.00
La Amistad	0.0027	0.0040	0.0047	0.0001	0.0002	—

Below the diagonal are differences in disparity in absolute values. Above the diagonal are p-values after 999 permutations. Significant differences ($p < 0.05$) are indicated in bold. Gunung Lensung is here abbreviated as 'Gunung'.

morphotypes (e.g. *Vulpes vulpes* in Figure 5.2), are mostly distinguished by the negative RW2 scores. On the other hand, the hypercarnivorous bush dog, *Speothos venaticus*, together with the bat-eared fox, *Otocyon megalotis*, occupy more positive RW2 scores towards felids' morphospace (Figure 5.2). The bush dog scores close to the origin of the RW1 axis due to its highly developed molar slicing area (Van Valkenburgh, 1991), while the bat-eared fox occupies a more negative RW1 score due to its longer molar row with undifferentiated m1 as a result of its insectivorous feeding habit (Ewer, 1973).

Viverrids and herpestids also occupy negative RW1 scores and they show a degree of morphospace partitioning. Wesley-Hunt *et al.* (2010) identified an overlap in ecomorphologies of these two families, although this pattern is not detected here, with viverrids being characterized by a much thinner mandibular corpus especially below the molars (extreme negative RW1 scores and slightly negative RW2), while herpestids have a thicker corpus below the enlarged molar crushing area (less negative RW1 and more negative RW2). A thick corpus is correlated with hard food consumption in carnivorans, although this is mostly based on studies about large bone cracker carnivorans (Werdelin, 1989; Raia, 2004; Meloro *et al.*, 2008; Figueirido *et al.*, 2013). The diet of omnivorous and insectivorous mongooses might also include crabs (e.g. the marsh mongoose, *Atilax paludinosus*) or other relatively hard dietary items such as insect exoskeletons (Ray, 1997) that require longer and more rapid masticatory cycles. These cycles impose higher masticatory loadings when compared to more predatory

viverrids that mostly focus their diet on small mammals and other vertebrates (Ewer, 1973).

Another mandibular feature that distinguishes viverrids from herpestids is the much longer angular process in the former group (Figure 5.2). A long angular process suggests more developed superficial masseter fibres whose action includes a forward pulling component: this could be linked to the killing behaviour of genets that use a series of rapid but imprecise bites in contrast to the use of a single precise bite in mongooses (Ewer, 1973).

Procyonids also show an interesting pattern in morphospace occupation with both omnivorous coatis, *Nasua* spp., clustering within viverrids' morphospace (the two white circles that occupy negative RW1 scores and positive RW2 scores; Figure 5.2). The northern raccoon is very close to the consensus configuration (i.e. the origin of RW1/2 axes), while the frugivorous kinkajou, *Potos flavus*, plots near the tayra, *Eira barbara*, a South American mustelid (Figure 5.2). Procyonids exhibit a very high polymorphism in the mandible shape as a possible result of their broad dietary niche differentiation through time. Early members of Procyonidae had a generalized dentition that allowed them to evolve distinct morphologies in relation to more plant-dominant food consumption (Koepfli *et al.*, 2007). The frugivorous kinkajou has always been considered a peculiar form (Figueirido *et al.*, 2010) that occupies here extreme positive scores of RW1 due to its musteloid corpus curvature and expansion of the molar crushing area.

Even if taxonomy and (to some extent) diet are recognized as some of the main factors explaining

Table 5.4 Non-parametric Spearman's rank correlation coefficients (r_s) and p-values between potential prey or bioclimatic variables and morphological disparities of six small carnivore guilds.

Variables	r_s	p
Number of Insectivora	−0.72	0.12
Number of Rodentia	0.77	0.10
Number of Lagomorpha	**−0.88**	**0.05**
Total number of preys	0.66	0.14
Total number of marsupials	0.60	0.18
bio1 = Annual mean temperature	0.14	0.71
bio2 = Mean diurnal range (mean of monthly [max. temp – min. temp])	−0.43	0.36
bio3 = Isothermality (bio2/bio7)(×100)	0.54	0.24
bio4 = Temperature seasonality (standard deviation × 100)	−0.66	0.14
bio5 = Max. temperature of the warmest month	−0.26	0.56
bio6 = Min. temperature of the coldest month	0.55	0.27
bio7 = Temperature annual range (bio5 – bio6)	−0.60	0.18
bio8 = Mean temperature of wettest quarter	−0.03	1.00
bio9 = Mean temperature of driest quarter	0.37	0.42
bio10 = Mean temperature of warmest quarter	−0.03	1.00
bio11 = Mean temperature of coldest quarter	0.43	0.36
bio12 = Annual precipitation	**0.83**	**0.03**
bio13 = Precipitation of wettest month	**0.89**	**0.02**
bio14 = Precipitation of driest month	**0.83**	**0.03**
bio15 = Precipitation seasonality (coefficient of variation)	0.09	0.80
bio16 = Precipitation of wettest quarter	**0.89**	**0.02**
bio17 = Precipitation of driest quarter	**0.83**	**0.03**
bio18 = Precipitation of warmest quarter	**0.94**	**0.01**
bio19 = Precipitation of coldest quarter	0.77	0.10

Significant correlations ($p < 0.05$) are indicated in bold.

species distribution within the small carnivore mandibular morphospace, no geographic patterns become apparent. Both viverrids and herpestids are evidently absent in guilds from Europe and North America (Figure 5.3) that exhibit a higher number of mustelid morphotypes. The Yellowstone guild lacks small felid morphotypes possibly due to the generalized niches of the 'large' bobcat, *Lynx rufus*, and the Canada lynx, *Lynx canadensis*, whose diet can focus mainly on lagomorphs and small rodents. The South African guild shows an opposite trend with a community highly depleted of mustelid morphotypes but enriched with viverrids and herpestids, while Asiatic, Central, and South American morphospaces show a homogenous species distribution in all areas of the morphospace (Figure 5.3). Ewer (1973) and Hunt (1996) already highlighted the mustelid/viverrid–herpestid pattern observed in the Old World and the mandibular morphospace confirms how long-term evolutionary processes generated the species distribution we observe today. This pattern has little influence on morphological disparity which is generally low in European, North American and African guilds (Figure 5.4). The most morphologically diverse communities are detected across the tropics. This observation partially confirms previous findings by Shepherd (1998) who performed a latitudinal survey of morphological disparity in mammalian communities from North America. She concluded that at higher latitudes, species have lower shape diversity than in the tropics. This trend is independent of species number so that no correlation occurs between species richness and morphological disparity (Foote, 1992, 1993).

No association between small carnivore guild disparity and latitude or longitude was detected; however, the inclusion of bioclimatic variables supports a very strong positive influence of precipitation variables. The relative impact of climate on ecomorphologies of carnivoran communities was highlighted by Polly (2010) in a survey on locomotory skeletal traits across North American species. Ecogeographical patterns are also broadly evident in the majority of mammalian groups, including small carnivores (e.g. mustelids; Meiri *et al.*, 2007). In theory, the climate might influence morphological variability of small carnivore species assemblages indirectly via diversification of their potential prey (e.g. rodents). This is clearly not the case: the number of rodent species and other small mammalian prey are not significantly correlated with small carnivore morphological disparity (Table 5.4). Only the number of lagomorph species shows a negative association with disparity. Different explanations can be considered here for such a counter-intuitive pattern:

1) Number of prey species might not be a good predictor of small carnivore morphological disparity simply because it is the wrong metric to consider. Prey biomass is expected to influence more directly small predator populations and eventually their species composition via competitive exclusion (Powell & Zielinski, 1983; Norrdahl & Korpimäki, 1995; St-Pierre *et al.*, 2006).

2) Although lagomorphs are the focal prey species for only a small fraction of small carnivores, their abundance and diversity can strongly impact the feeding behaviour of different species within this guild. For instance, Carvalho & Gomes (2004) studied niche partitioning among four sympatric small carnivores and observed niche convergence between the red fox and the European wild cat, *Felis silvestris*, during periods of an abundance of wild rabbits, *Oryctolagus cuniculus*, thus facilitating their co-existence. This pattern, on a broader and longer evolutionary time scale, might have generated the negative trend we observe today: lagomorph-rich communities might support a higher richness of small carnivores (generally mustelids with higher bite forces; Christiansen & Wroe, 2007) with more similar morphotypes (hence lower disparity values).

The strong negative correlation between the number of lagomorphs and precipitation variables (with bio18, $r_s = -0.99$; with bio13 and bio16, $r_s = -0.94$) indicates variable interaction enforcing the strong impact of climate on small carnivore guilds. High precipitation in tropical areas guarantees food availability in all seasons, facilitating small carnivores to diversify in functional morphotypes (including meat-eaters, frugivores, insectivores). Additionally, small carnivores include a high number of arboreal secondary consumers, whose diversity correlates strongly with tree cover (Louys *et al.*, 2011) and precipitation (Polly, 2010).

The mandibular shape morphospace provides a clear starting point to further explore patterns and processes that influence small carnivore species assemblages. If long-term evolutionary processes characterize their assembly rules, then climatic changes might be a key influence of their morphological diversity. Future studies should combine such an interspecific approach with finer-scale patterns of geographical variation. More ecomorphological approaches are also needed to better identify the degree of interaction and the functional guilds within small carnivores.

Acknowledgements

I am grateful to the staff of the Natural History Museum, London for their invaluable help during the course of my research visits. In particular, P. Jenkins, L. Tomsett, R. Portela-Miguez, A. Salvador, and D. Hills kindly supported me through the years. A big thank you to P. Raia, A. Loy and P. O'Higgins who introduced me to morphometrics and the world of small carnivores. E. Do Linh San kindly invited me to participate in this book, while M. Asahara, J.J. Sato, and L. Werdelin strongly improved the quality and style of this manuscript.

References

Adams, D.C., Rohlf, F.J. & Slice, D.E. (2004) Geometric morphometrics: ten years of progress following the 'revolution'. *Italian Journal of Zoology* 71, 5–16.

Adams, D.C., Rohlf, F.J. & Slice, D.E. (2013) A field comes of age: geometric morphometrics in the 21st century. *Hystrix, the Italian Journal of Mammalogy* 24, 7–14.

Asahara, M. (2013) Unique inhibitory cascade pattern of molars in canids contributing to their potential to evolutionary plasticity of diet. *Ecology and Evolution* 3, 278–285.

Bookstein, F.L. (1989) "Size and shape": a comment on semantics. *Systematic Zoology* 38, 173–180.

Carbone, C., Mace, G.M., Roberts S.C. & Macdonald, D.W. (1999) Energetic constraints on the diet of terrestrial carnivores. *Nature* 402, 286–288.

Carvalho, J.C. & Gomes, P. (2004) Feeding resource partitioning among four sympatric carnivores in the Peneda-Gerês National Park (Portugal). *Journal of Zoology* 263, 275–283.

Christiansen, P. & Adolfssen, J.S. (2005) Bite forces, canine strength and skull allometry in carnivores (Mammmalia, Carnivora). *Journal of Zoology* 266, 133–151.

Christiansen, P. & Wroe, J.S. (2007) Bite forces and evolutionary adaptations to feeding ecology in carnivores. *Ecology* 88, 347–358.

Crusafont-Pairó, M. & Truyols-Santonja, J. (1956) A biometric study of evolution of fissiped carnivores. *Evolution* 10, 314–332.

Crusafont-Pairó, M. & Truyols-Santonja, J. (1957) Estudios masterométricos en la evolución Fisípedos. I. Los módulos angulares á y â. II. Los parámetros lineales P, C, y T. *Boletino Instituto Geologico y Minero España* 68, 1–140. (In Spanish).

Crusafont-Pairó, M. & Truyols–Santonja, J. (1958) A quantitative study of stasigenesis in fissiped carnivores. *Nature* 181, 289–290.

Davies T.J., Meiri, S., Barraclough, T.J. & Gittleman, J.L. (2007) Species co-existence and character divergence across carnivores. *Ecology Letters* 10, 146–152.

Davis, C.L., Rich, L.N., Farris, Z.J., Kelly, M.J., Di Bitteti, M.S. *et al.* (2019) Ecological correlates of the spatial co-occurrence of sympatric mammalian carnivores worldwide. *Ecology Letters* 21, 1401–1412.

Donadio, E. & Buskirk, S.W. (2006) Diet, morphology, and interspecific killing in Carnivora. *The American Naturalist* 167, 524–536.

Evans, A.R., Wilson, G.P., Fortelius, M. & Jernvall, J. (2007) High-level similarity of dentitions in carnivorans and rodents. *Nature* 445, 78–81.

Ewer, R.F. (1973) *The Carnivores*. Cornell University Press, Ithaca.

Figueirido, B., Serrano-Alarcón, F.J., Slater, G.J. & Palmqvist, P. (2010) Shape at the cross-roads: homoplasy and history in the evolution of the carnivoran skull towards herbivory. *Journal of Evolutionary Biology* 23, 2579–2594.

Figueirido, B., MacLeod, N., Krieger, J., De Renzi, M., Pérez-Claros, J.A. & Palmqvist, P. (2011) Constraint and adaptation in the evolution of carnivoran skull shape. *Paleobiology* 37, 490–518.

Figueirido, B., Tseng, Z.J. & Martin-Serra, A. (2013) Skull shape evolution in durophagous carnivorans. *Evolution* 67, 1975–1993.

Foote, M. (1992) Paleozoic record of morphological diversity in blastozoan echinoderms. *Proceedings of the National Academy of Sciences USA* 89, 7325–7329.

Foote, M. (1993) Discordance and concordance between morphological and taxonomic diversity. *Paleobiology* 19, 185–204.

Friscia, A.R., Van Valkenburgh, B. & Biknevicius A.R. (2006) An ecomorphological analysis of extant small carnivorans. *Journal of Zoology* 272, 82–100.

Gittleman, J.L. (1985) Carnivore body size: ecological and taxonomic correlates. *Oecologia* 67, 540–554.

Goswami, A. (2010) Introduction to carnivoran evolution. In: *Carnivoran Evolution: New Views on Phylogeny, Form, and Function* (eds A. Goswami & A. Friscia), pp. 1–24. Cambridge University Press, Cambridge.

Greaves, W.S. (1983) A functional analysis of carnassial biting. *Biological Journal of the Linnean Society* 20, 353–363.

Greaves, W.S. (1985) The generalized carnivore jaw. *Zoological Journal of the Linnean Society* 85, 267–274.

Herring, S.W. (1980) Functional design of cranial muscles: comparative and physiological studies in pigs. *American Zoologist* 20, 283–293.

Herring, S.W. (1993) Functional morphology of mammalian mastication. *American Zoologist* 33, 289–299.

Herring S.W. & Herring S.E. (1974) The superficial masseter and gape in mammals. *The American Naturalist* 108, 561–576.

Hijmans, R.J., Cameron, S.E., Parra, J.L., Jones, P.G. & Jarvis, A. (2005) Very high resolution interpolated climate surfaces for global land areas. *International Journal of Climatology* 25, 1965–1978.

Holliday, J.A. & Steppan, S.J. (2004) Evolution of hypercarnivory: the effect of specialization on morphological and taxonomic diversity. *Paleobiology* 30, 108–128.

Hunt, R.M. (1996) Biogeography of the order Carnivora. In: *Carnivore Behavior, Ecology, and Evolution. Volume 2* (ed. J.L. Gittleman), pp. 485–541. Cornell University Press, Ithaca.

Koepfli, K.-P., Gompper, M.E., Eizirik, E., Ho, C.C., Linden, L., Maldonado, J.E. & Wayne R.K. (2007) Phylogeny of the Procyonidae (Mammalia: Carnivora): molecules, morphology and the Great American Interchange. *Molecular Phylogenetics and Evolution* 43, 1076–1095.

Lawing, A.M. & Polly, P.D. (2009) Geometric morphometrics: recent applications to the study of evolution and development. *Journal of Zoology* 280, 1–7.

Louys, J., Meloro, C., Elton, S., Ditchfield, P. & Bishop, L.C. (2011) Mammal community structure correlates with arboreal heterogeneity in faunally and geographically diverse habitats: implications for community convergence. *Global Ecology and Biogeography* 20, 717–729.

Meiri, S., Dayan, T. & Simberloff, D. (2007) Guild composition and mustelid morphology – character displacement but no character release. *Journal of Biogeography* 34, 2148–2158.

Meloro, C. (2011a) Morphological disparity in Plio-Pleistocene large carnivore guilds from Italian peninsula. *Acta Palaeontologica Polonica* 56, 33–44.

Meloro, C. (2011b) Feeding habits of Plio-Pleistocene large carnivores as revealed by their mandibular geometry. *Journal of Vertebrate Paleontology* 31, 428–446.

Meloro, C. (2011c) Locomotor adaptations in Plio-Pleistocene large carnivores from the Italian peninsula: palaeoecological implications. *Current Zoology* 57, 269–283.

Meloro, C. (2012) Mandibular shape correlates of tooth fracture in extant Carnivora: implications to inferring feeding behaviour of Pleistocene predators. *Biological Journal of the Linnean Society* 106, 70–80.

Meloro, C. & O'Higgins, P. (2011) Ecological adaptations of mandibular form in fissiped Carnivora. *Journal of Mammalian Evolution* 18, 185–200.

Meloro, C. & Raia, P. (2010) Cats and dogs down the tree: the tempo and mode of evolution in the lower carnassial of fossil and living Carnivora. *Evolutionary Biology* 37, 177–186.

Meloro, C., Raia, P., Piras, P., Barbera, C. & O'Higgins, P. (2008) The shape of the mandibular corpus in large fissiped carnivores: allometry, function and phylogeny. *Zoological Journal of the Linnean Society* 154, 832–845.

Meloro, C., Raia, P., Carotenuto, F. & Cobb, S. (2011) Phylogenetic signal, function and integration in the subunits of the carnivoran mandible. *Evolutionary Biology* 38, 465–475.

Norrdahl, K. & Korpimäki E. (1995) Small carnivores and prey population dynamics in summer. *Annales Zoologici Fennici* 32, 163–169.

Palomares, F. & Caro T.M. (1999) Interspecific killing among mammalian carnivores. *The American Naturalist* 153, 492–508.

Polly, P.D. (2010) Tiptoeing through the trophics: geographic variation in carnivoran locomotor ecomorphology in relation to environment. In: *Carnivoran Evolution: New Views on Phylogeny, Form, and Function* (eds A. Goswami & A. Friscia), pp. 374–410. Cambridge University Press, Cambridge.

Polly, P.D. & MacLeod, D.N. (2008) Locomotion in fossil Carnivora: an application of eigensurface analysis for morphometric comparison of 3D surface. *Palaeontologia Electronica* 11, 10A, 1–13.

Polly, P.D., Eronen, J.T., Fred, M., Dietl, G., Mosbrugger, V., Scheidegger, C., Frank, D., Damuth, J., Stenseth, N. & Fortelius, M. (2011) History matters: integrative climate change biology. *Proceedings of the Royal Society B: Biological Sciences* 278, 1131–1140.

Popowics, T.E. (2003) Postcanine dental form in the mustelidae and viverridae (Carnivora: Mammalia). *Journal of Morphology* 256, 322–341.

Powell, R.A. & Zielinski, W.J. (1983) Competition and coexistence in mustelid communities. *Acta Zoologica Fennica* 174, 223–227.

Prevosti, F.J., Turazzini, G.F., Ercoli, M.D. & Hingst-Zaher, E. (2012) Mandible shape in marsupial and placental carnivorous mammals: a morphological comparative study using geometric morphometrics. *Zoological Journal of the Linnean Society* 164, 836–855.

Radinsky, L.B. (1981a) Evolution of skull shape in carnivores, 1: representative modern carnivores. *Biological Journal of the Linnean Society* 15, 369–388.

Radinsky, L.B. (1981b) Evolution of skull shape in carnivores, 2: additional modern carnivores. *Biological Journal of the Linnean Society* 16, 337–355.

Radinsky L.B. (1982) Evolution of skull shape in carnivores. 3. The origin and early radiation of modern carnivores families. *Paleobiology* 8, 177–195.

Raia, P. (2004) Morphological correlates of tough food consumption in carnivores. *Italian Journal of Zoology* 71, 45–50.

Ray, J. (1997) Comparative ecology of two African forest mangooses, *Herpestes naso* and *Atilax paludinosus*. *African Journal of Ecology* 35, 237–253.

Roemer, G.W., Gompper, M.E. & Van Valkenburgh, B. (2009) The ecological role of the mammalian mesocarnivore. *BioScience* 59, 165–173.

Rohlf, F.J. (2015) The tps series of software. *Hystrix, the Italian Journal of Mammalogy* 26, 9–12.

Rohlf, F.J. & Slice, D.E. (1990) Extensions of the Procrustes method for the optimal superimposition of landmarks. *Systematic Zoology* 39, 40–59.

Root, R.B. (1967) The niche exploitation pattern of the blue-gray gnatcatcher. *Ecological Monographs* 37, 317–350.

Rosenzweig, M.J. (1995) *Species Diversity in Space and Time*. Cambridge University Press, Cambridge.

Shepherd, U.L. (1998) A comparison of species diversity and morphological diversity across the North American latitudinal gradient. *Journal of Biogeography* 25, 19–29.

Simberloff, D. & Dayan T. (1991) The guild concept and the structure of ecological communities. *Annual Review in Ecology and Systematics* 22, 115–43.

St-Pierre, C., Ouellet, J.-P. & Crête, M. (2006) Do competitive intraguild interactions affect space and habitat use by small carnivores in a forested landscape? *Ecography* 29, 487–496.

Van Valkenburgh, B. (1985) Locomotor diversity between past and present guilds of large predatory mammals. *Paleobiology* 11, 406–428.

Van Valkenburgh, B. (1988) Trophic diversity in past and present guilds of large predatory mammals. *Paleobiology* 14, 155–173.

Van Valkenburgh, B. (1989) Carnivore dental adaptations and diet: a study of trophic diversity within guilds. In: *Carnivore Behavior, Ecology, and Evolution. Volume 1* (ed. J.L. Gittleman), pp. 410–436. Cornell University Press, Ithaca.

Van Valkenburgh, B. (1991) Iterative evolution of hypercarnivory in canids (Mammalia: Carnivore): evolutionary interactions among sympatric predators. *Paleobiology* 17, 340–362.

Van Valkenburgh, B. (1995) Tracking ecology over geological time: evolution with guilds of vertebrates. *Trends in Ecology & Evolution* 10, 71–76.

Van Valkenburgh, B. (1999) Major patterns in the history of carnivorous mammals. *Annual Review of Earth and Planetary Science* 27, 463–493.

Wainwright, P.C. (1994) Functional morphology as a tool in ecological research. In: *Ecological Morphology: Integrative Organismal Biology* (eds P.C. Wainwright & S.M. Reilly), pp. 42–59. The University of Chicago Press, Chicago.

Werdelin, L. (1989) Constraint and adaptation in the bonecracking canid *Osteoborus* (Mammalia: Canidae). *Paleobiology* 15, 387–401.

Werdelin, L. & Wesley-Hunt, G.D. (2010) The biogeography of carnivore ecomorphology. In: *Carnivoran Evolution: New Views on Phylogeny, Form, and Function* (eds A. Goswami & A. Friscia), pp. 225–245. Cambridge University Press, Cambridge.

Wesley-Hunt, G.D. (2005) The morphological diversification of carnivores in North America. *Paleobiology* 31, 35–55.

Wesley-Hunt, G.D., Dehghani, R. & Werdelin L. (2010) Comparative ecomorphology and biogeography of Herpestidae and Viverridae (Carnivora) in Africa and Asia. In: *Carnivoran Evolution: New Views on Phylogeny, Form, and Function* (eds A. Goswami & A. Friscia), pp. 246–268. Cambridge University Press, Cambridge.

Zelditch, M.L., Sheets H.D. & Fink, W.L. (2003) The ontogenetic dynamics of shape disparity. *Paleobiology* 29, 139–156.

Zelditch, M.L., Swiderski D.L., Sheets H.D. & Fink, W.L. (2004) *Geometric Morphometrics for Biologists. A Primer*. Elsevier Academic Press, London.

6

Beyond Black and White: Addressing Colour Variation in the Context of Local Environmental Conditions for the Aposematic North American Hog-nosed Skunk

Adam W. Ferguson[1,2],*, Richard E. Strauss[1], and Robert C. Dowler[3]

[1] *Department of Biological Sciences, Texas Tech University, Lubbock, TX, USA*
[2] *Gantz Family Collection Center, Field Museum of Natural History, Chicago, IL, USA*
[3] *Department of Biology, Angelo State University, San Angelo, TX, USA*

SUMMARY

Among mammalian carnivores, skunks (family Mephitidae), with their bold aposematic colouration, represent an obvious and interesting group for testing hypotheses associated with colour-pattern evolution. Herein, we introduce and develop a novel technique for quantifying intraspecific variation in colour patterns for the North American hog-nosed skunk, *Conepatus leuconotus*, to test for associations between local environmental conditions and dorsal stripe variation. Using digital photographs of 262 museum study skins in combination with spatially explicit interpolation and modelling techniques, we found that variation in the size and extent of the white dorsal stripe (and, consequently, the extent of black fur along the dorsum) of *C. leuconotus* is non-randomly distributed across the landscape. The extent of dorsal whiteness appears to peak across the southwest desert states of Arizona, New Mexico, and Texas, with reduced whiteness along the Gulf Coast and in Central America, a pattern consistent with Gloger's rule. Generalized dissimilarity modelling revealed that differences in dorsal whiteness were related to differences in canopy cover, ground surface moisture, and temperature variability, whereas random forest analysis found three variables related to minimum temperatures to be the best predictors of variation in dorsal whiteness extent. Such relationships could indicate that skunks with more white along the dorsum (and less black) may benefit (i.e. experience reduced rates of predation) from increased visibility in more arid, open environments, whereas skunks with reduced dorsal whiteness (and increased blackness) may benefit by remaining hidden among the dark understory characteristic of more closed-canopy, aseasonal environments. These results imply that evolutionary trade-offs between conspicuousness and crypsis may be responsible for shaping colour polymorphisms in this aposematic small carnivore.

Keywords

Aposematic colouration − crypsis − ecomorphology − generalized dissimilarity modeling − intraspecific variation − Mephitidae

Introduction

Understanding underlying drivers of colouration in mammals remains an important topic in evolutionary ecology (Caro, 2005a). Mammalian carnivores, whose evolutionary and ecological diversity support a wide variety of colour patterns (Wilson *et al.*, 2009), provide an excellent group in which to investigate the evolution of colour pattern and arrangement. Given this diversity and their charismatic nature, carnivores have long been the focus of research on colour evolution (Pocock, 1908; Stankowich *et al.*, 2011, 2014),

---

*Corresponding author.

Small Carnivores: Evolution, Ecology, Behaviour, and Conservation, First Edition. Edited by Emmanuel Do Linh San, Jun J. Sato, Jerrold L. Belant, and Michael J. Somers.

seems counterintuitive, under certain conditions, such as in forests with patches of light and dark, black and white colouration may actually appear to be cryptic (Caro, 2009). In fact, Gloger's rule, which states that dark colouration is more strongly developed in warm, humid regions compared to cold and dry regions (Gloger, 1833; Searle, 1968), often invokes crypsis as a possible explanation, although the true mechanism behind this pattern remains obscure (Kamilar & Bradley, 2011). In addition, Searle (1968) hypothesized that a darker coat colour could confer some form of selective advantage under conditions of poor illumination, similar to the way in which melanic moths are better concealed than non-melanic forms on smoke-darkened trees. Thus, it seems feasible that colour arrangements in the aposematic North American hog-nosed skunk could be driven by both signalling efficacy and a need to remain cryptic, at least in some capacity. Our study provides one of the first attempts to characterize intraspecific colour polymorphism in skunks using quantitative rather than qualitative approaches and introduces a novel approach that may prove applicable and adaptable to other studies of colour variation in various animal species, including other small carnivores with similar bicoloured pelage patterns.

Methods

Morphological Data

Digital photographs were taken of the dorsum and venter of preserved museum study skins at a native resolution of 3264 × 2448 pixels using a Canon PowerShot SX100IS digital camera mounted on a copy stand. Specimens were photographed from a distance of 75 cm (copy stand centroid to the tip of the camera lens) to minimize impacts of distortion. Specimens were placed flat on the copy stand on clean white paper and photographed once in both dorsal and ventral views, including a specimen identifier and scale bar in each photograph. Specimens were from the following collections: Academy of Natural Sciences of Philadelphia (ANSP), American Museum of Natural History (AMNH), Angelo State Natural History Collection (ASNHC), British Natural History Museum (BNHM), Carnegie Museum (CM), CIIDIR-IPN Unidad Durango (CRD), Denver Museum of Nature

and Science (DMNS), US National Museum of Natural History (NMNH), University of Colorado Museum (UCM), and Texas Tech University Museum (TTU). Geographic coordinates for specimens were obtained by converting textual locality information into latitude and longitude, following the best practices for georeferencing (Chapman & Wieczorek, 2006).

Digital photographs were imported into tpsDIG2 using tpsUtil (http://www.sbmorphometrics.org/) to trace digitally the perimeter of the dorsal stripes from its anterior origin to its termination at the base of the tail. The perimeter of the white portion of the tail was also digitized in this manner. Three landmarks were used to estimate body length and tail length for each specimen: the anterior-most tip of the nose, the base of the tail, and the terminal tip of the tail fur (Figure 6.2, red dots). Although body length measures could be influenced by the subjective placement of landmarks, we were interested in a relative metric of body length to determine whether estimates of dorsal whiteness were influenced by specimen preparation and not a true estimate of body size. Placement of the base of the tail landmark was the most subjective although best efforts were made using the ventral photographs in addition to the position of the hind feet and the maximum tapering point anterior to the tail to locate the true end of the body and beginning of the tail.

Rather than attempting to use a sliding semi-landmark approach (Green, 1996; Bookstein, 1997) to estimate the geometry of the dorsal stripe, a flexible set of closely spaced points for each specimen was used to approximate as faithfully as possible the true geometries of the forms and shapes present in our sample (MacLeod, 1999). Although this method prevented the use of geometric morphometric analyses, it allowed more accurate assessments of dorsal white areas. Ultimately, this method resulted in a two-dimensional image of what is inherently a three-dimensional pattern, a fact that could influence our results by limiting our ability to truly represent the shape and distribution of black and white colour patches, both of which have been shown to influence the efficacy of warning signals in striped skunks, *Mephitis mephitis* (Hunter, 2009).

Each specimen was digitized three times in a non-sequential order to provide replication and error estimates for each digitized polygon. Processing of

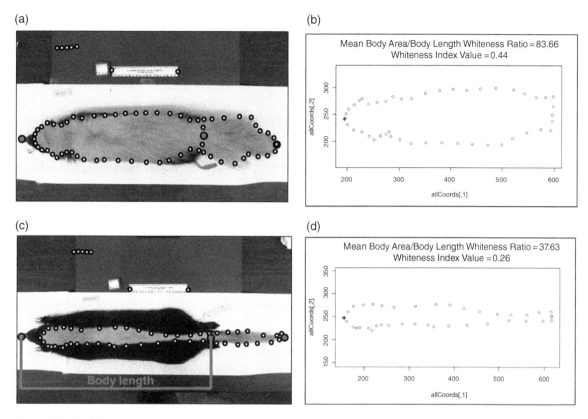

Figure 6.2 (a, c) Depiction of digitization process with landmarks (red dots) and points (yellow dots) used to digitize a polygon of whiteness for the dorsal body and tail of North American hog-nosed skunk, *Conepatus leuconotus sensu lato*. (b, d) Digital outline used to estimate the amount of dorsal body areas containing white pelage, with respective raw whiteness ratios (WR) and whiteness index (WI) values.

the resulting files and estimation of the white area were performed using the R statistical software environment (http://www.r-project.org/). Using the scale bar in each picture, distances in pixels were converted to millimetres to estimate body length (as estimated from the first and second landmarks across replicates) and the area of the white dorsal stripe for each replicate. Body length was used as a standard rather than body area because outlines of flattened preserved specimens were often irregular, depending on how the specimens were prepared. The square root of the white polygon area divided by the body length was used to create a mean, unitless 'whiteness index' (WI), which was used as the response variable in all subsequent analyses. Although whiteness along the dorsum was the only colour directly measured using our technique, a decrease in white coverage should be accompanied by a relative increase in black along the dorsum.

However, direct interpretations of the extent of black could be more systematically misleading than that of the white as the black colour appears more susceptible to distortion due to preparation style (i.e. flat tanned skins or overstuffed animals) and was therefore not directly quantified in this investigation.

Environmental Data

A series of variables derived from remotely sensed satellite and weather-station sources was obtained to examine potential relationships between dorsal colour pattern and environmental conditions. Specifically, we identified variables related to our alternative hypotheses of crypsis and physiological processes (see Table 6.1 for detailed descriptions of environmental variables). For example, metrics of temperature can be tied to physiological processes such as basal metabolic rate

Association (LISA significance maps; Anselin, 1995) were generated using Anselin Local Moran's I values and z-scores estimated with Spatial Analyst's Cluster and Outlier Analysis tool in ArcGIS 10.1. Based on the resulting scores, this tool identifies spatial clusters of features of high or low values while simultaneously identifying spatial outliers. A high positive z-score at an alpha level ≤ 0.05 for a feature indicates that surrounding features have significantly similar values (either high values or low values). A low negative z-score for a feature indicates a statistically significant ($p \leq 0.05$ level) spatial outlier.

Generalized Dissimilarity Modelling

To examine relationships between environmental predictor variables and dorsal-pattern variation, we used Generalized Dissimilarity Modelling (GDM; Ferrier, 2002; Ferrier *et al.*, 2004, 2007). GDM was originally developed for analyzing and predicting spatial patterns of community turnover of beta diversity across large areas (Rosauer *et al.*, 2014), but researchers have successfully adapted this technique to map and predict spatial patterns of morphological and genetic variation in the context of environmental dissimilarity among sampling localities and other similar challenges (Thomassen *et al.*, 2010, 2011; Rosauer *et al.*, 2014 and references therein). One major benefit of GDM lies in its ability to account for nonlinearities typically associated with large-scale ecological data sets (Ferrier *et al.*, 2007). GDM estimates relationships between the dissimilarity of a particular response variable (based on any dissimilarity metric valued between 0 and 1) among sites and the dissimilarity of environmental characteristics among those same sites. GDM analysis was based on pairwise comparisons of each specimen's WI values and the conditions extracted from the 24 environmental variables.

Absolute differences in WI values between individuals were rescaled by dividing each value by the maximum WI value from the series, rescaling the variable from the maximum value to 1. The rescaled difference values in WI between pairs of individuals were then used as the response variable. As no formal method for testing the significance of a GDM model is available, our models were evaluated for significance by permuting the relationship between differences in WI and environmental variables: the response variable was permuted randomly among the 262 individuals, while holding environmental conditions constant. Next, the permuted dataset was used to estimate the amount of deviance explained by this random model using GDM. This process was repeated 1000 times to create a null distribution of deviance-explained values against which to compare the observed deviance-explained value. Three models were tested using GDM: a geographic distance-only model, an environmental conditions-only model, and a full model including both geographic distance and environmental conditions. Geographic distance was estimated using straight-line distances between locality coordinates; although these distances may not be as biologically meaningful, they have been shown to perform as well as more realistic estimates of distance (e.g. least-cost paths; Thomassen *et al.*, 2010). All GDM analyses were carried out using a dedicated package (see most recent version at https://cran.r-project.org/web/packages/gdm/gdm.pdf) in the R statistical software environment following Ferrier *et al.* (2007).

Random Forest Models

Random forest (RF) is a decision-tree technique (De'ath & Fabricius, 2000) that generates a bifurcating tree model based on nodal decisions in which predictor variables maximize homogeneity between two partitions of the response variable (Breiman, 2001a,b; Thomassen *et al.*, 2010). This homogeneity is typically measured by the Gini index (Breiman *et al.*, 1984), and the bifurcation process continues until further partitioning no longer reduces the index value. Model and predictor variable importance is often assessed using a bootstrap technique known as 'bagging' (Breiman, 1996), wherein data are subsampled randomly to create test and training samples for model validation. This whole process is repeated many times, resulting in a final prediction that represents an average of the ensemble tree space (Breiman *et al.*, 1984). The abilities of RF models to allow for nonlinear relationships of predictor variables, and to predict with high accuracy and power, make them well-suited for modelling patterns of variation associated with environmental conditions (Thomassen *et al.*, 2010). The R package ModelMap (Freeman & Frescino 2009) was used to generate an RF-based predictive model of WI variation across the range of *C. leuconotus* and to identify important explanatory variables for variation in WI.

Results

Dorsal-Pattern Variation

A total of 262 (142 males, 90 females, 30 unknown) individual museum skins were photographed and digitized for analysis of dorsal-pattern variation (Figure 6.3a, grey crosses). Years of collection for the specimens ranged from 1885 to 2012, with an average of 1937 and a median of 1915. Plots of replicate values of raw dorsal whiteness ratios versus mean body length indicated that replicate estimates of WIs were relatively precise and that neither specimen preparation (as reflected by the total length of the specimen) nor collection were significantly related to WI estimates (figure not shown). That is to say, WI values were not significantly influenced by the digitization process, specimen preparation, or the source/collection where the specimen originated. Average WI did not differ significantly between males and females ($t = -0.70$, df = 230, $p = 0.48$; Appendices 6.B and 6.C) although they did differ in mean body length ($t = -2.50$, df = 220, $p = 0.01$; Appendices 6.B and 6.C), with males significantly longer than females (Appendices 6.B and 6.C).

Spatial Interpolation – Kriging

Exploratory analyses of the raw data indicated that spatial autocorrelation was present, implying that WI values are more similar between adjacent individuals. The WI were normally distributed although slightly skewed right, with a mean of 0.399 and a standard deviation of 0.052, so no transformation was performed prior to kriging, and a mean of 0.399 was assumed during subsequent analyses. Exploratory trend analysis also indicated the presence of spatial relationships between WI across the YZ (latitude and WI values) and ZX (longitude and WI values) planes, implying decreasing WI values for specimens located at lower latitudes and longitudes. Based on

comparisons with the empirical variogram, a hole-effect model was selected as the best model for the predicted semivariogram for both the simple kriging and standard error surfaces. Surfaces generated with other models yielded qualitatively similar patterns (figures not shown). Optimal parameter values based on the exploratory analysis and other model details can be found in Appendix 6.D. Kriging results indicated that variation in the amount of whiteness on the dorsum of *C. leuconotus* is not distributed randomly across the species' geographic range (Figure 6.3a,b). High WI values were concentrated in the north-western portion of their range (Arizona, Colorado, New Mexico of the United States of America, and northern Mexico; see Figure 6.1a for range demarcation and Figure 6.3a for distribution of WI values), whereas low values were restricted to the Gulf Coast portion of their range (southern Texas, Tamaulipas and Veracruz, Mexico) and Central America (Figure 6.3a). Predicted standard errors were lowest in areas with more specimen records, and highest in areas with few to no specimens (e.g. north-central Mexico; Figure 6.3b).

Spatial Interpolation Cluster Analysis

Cluster analysis yielded similar results to that of kriging (Figure 6.4). Significant clusters of statistically similar high WI values were identified in Arizona, New Mexico, west-central Texas, and northern Mexico. Significant clusters of similar low WI values were identified in southern Texas and north-eastern Mexico, the Transvolcanic Mexican belt, Mexico's central Pacific coast, and in the Central American countries of Honduras and Nicaragua (Figure 6.4). Several statistically significant low-value outliers (in central Texas; Figure 6.4, red dot) were found among high-value individuals in western Texas (Figure 6.4, green dots). Three significantly high-value outliers (Figure 6.4, yellow dots) were recorded close to low-value clusters, one from southern Texas and the other from Mexico's central Pacific coast. A proportion of individuals remained unassigned to either a high or low cluster (Figure 6.4, grey dots), most notably across central Mexico, coastal Oaxaca, and in southern Colorado and Oklahoma. Patterns depicted by the spatial clustering analysis and paralleled by the kriging models indicate that skunks with larger proportions of white along the dorsum are

(a)

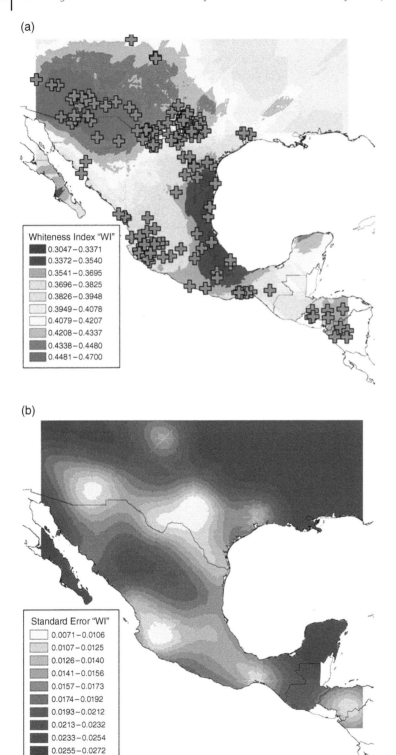

Whiteness Index "WI"
- 0.3047–0.3371
- 0.3372–0.3540
- 0.3541–0.3695
- 0.3696–0.3825
- 0.3826–0.3948
- 0.3949–0.4078
- 0.4079–0.4207
- 0.4208–0.4337
- 0.4338–0.4480
- 0.4481–0.4700

(b)

Standard Error "WI"
- 0.0071–0.0106
- 0.0107–0.0125
- 0.0126–0.0140
- 0.0141–0.0156
- 0.0157–0.0173
- 0.0174–0.0192
- 0.0193–0.0212
- 0.0213–0.0232
- 0.0233–0.0254
- 0.0255–0.0272

Figure 6.3 (a) Simple kriging surface and (b) associated Prediction Standard Error surface based on 262 specimens of *Conepatus leuconotus* (grey crosses) using a hole-effect semivariogram model generated in ArcGIS 10.1 Geostatistical Analyst. Warmer colours indicate higher whiteness indices (WI) (a) and predicted standard errors (b).

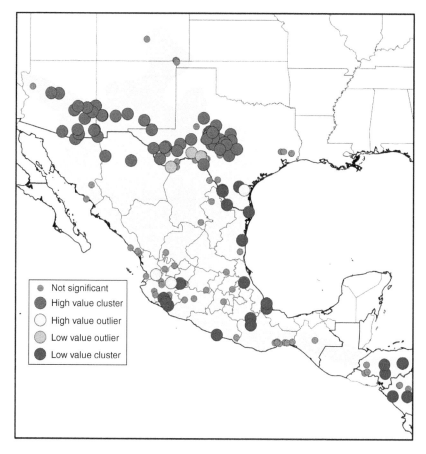

Figure 6.4 LISA significance map (Anselin, 1995) generated using Anselin Local Moran's *I* values and *z*-scores estimated with Spatial Analyst's Cluster and Outlier Analysis tool in ArcGIS 10.1. Grey points indicate individual specimens that could not be significantly assigned to either a high or low cluster based on their relation to spatially proximal individuals.

restricted to northern latitudes and drier conditions across Arizona, New Mexico, and Texas. In contrast, individuals with reduced white dorsal stripe coverage are restricted to tropical and semi-tropical environments, from southern Texas to Central America.

Generalized Dissimilarity Modelling

GDM models for environmental variables only, geographic distance only, and environmental plus geographic distance all accounted for significant amounts of variation compared to their respective random models (Table 6.2). The environmental-only model accounted for 14.2% of the variation in WI values and identified 7 of 24 environmental predictors as significantly contributing to the model: bio1, bio2, bio4, bio7, bio11, QSCAT, and the grand mean of LAI (Figure 6.5; Table 6.2). Of these seven variables, bio2 and GM_LAI had the highest coefficients and largest response curves, indicating their greater contribution to building the model (Figure 6.5c,d). The variable that contributed most to the model, bio2, represents the mean diurnal temperature range [mean of monthly (max temp – min temp)], indicating that regions with greater differences in minimum and maximum temperature values, or seasonality, support skunks with greater differences in WI values. GM_LAI, the grand mean Leaf Area Index over 2005–2009, reflects canopy density as well as seasonality (i.e. leaf loss) and appears to contribute the second most amount of information regarding relationships between differences in WI values and environmental conditions. This relationship

Table 6.2 Deviance explained (variation in the response variable) for the three models generated using Generalized Dissimilarity Modeling (GDM) with contributing model variables.

Model	'True' deviance explained	'Randomized' deviance explained*	Significantly contributing variables (coefficients)
Geographic distance + environmental variables	14.27	0.038 ± 0.018 (0.004–0.113)	bio1 (0.06), bio2 (0.18; 0.03), bio4 (0.007), bio7 (0.004), bio11 (0.06), QSCAT (0.06), GM_LAI (0.12)
Environmental variables only	14.27	0.03 ± 0.018 (0.000–0.108)	bio1 (0.06), bio2 (0.18; 0.03), bio4 (0.007), bio7 (0.004), bio11 (0.06), QSCAT (0.06), GM_LAI (0.12)
Geographic distances only	3.56	0.003 ± 0.005 (0.000–0.044)	Geographic distance (0.16; 0.09)

*Randomized deviance explained = mean of 1000 permuted datasets ± standard deviation (range of values).

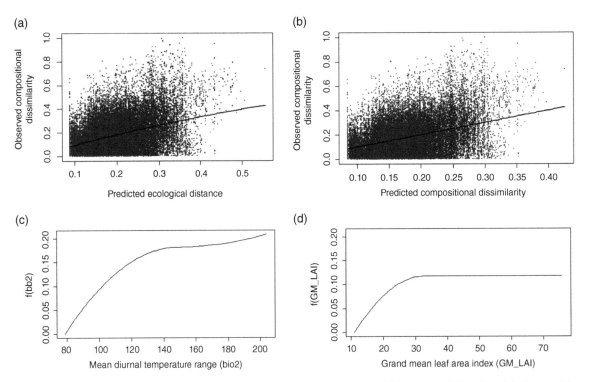

Figure 6.5 (a) Generalized dissimilarity models of the overall fitted model and (b) the overall fitted model after applying the link function for environmental variables only. (c) The two predictors, bio2 (mean diurnal temperature range) and (d) grand mean of Leaf Area Index (GM_LAI) with the highest coefficients and response curves out of the seven variables used to build the model are also presented.

implies that differences in skunk WI values are related to differences in canopy cover, or that greater differences in WI values exist between sites with greater differences in LAI values. The geographic distance-only model accounted for a significant amount of variation in WI values with a deviance-explained value of 3.56 (Table 6.2). When included as the sole predictor, geographic distance appeared to account for less variation than environmental variables alone did (Table 6.2). Combining the environmental and geographic distance

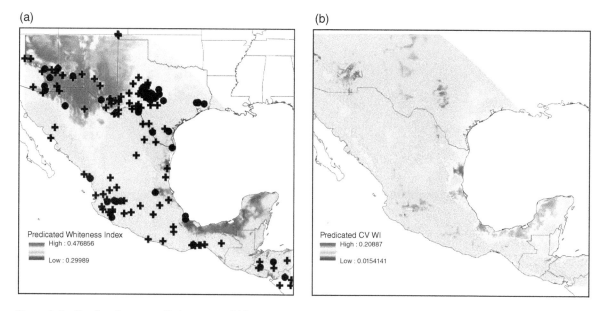

Figure 6.6 Random forest predictive maps of (a) mean and (b) coefficient of variation of whiteness index (WI) values for 262 individual *Conepatus leuconotus* based on 24 environmental predictors using ModelMap package in R. Warmer colours indicate higher WI values. Black crosses represent specimens used to build and train the model; black dots represent specimens used as test points to validate the model.

predictors into a single model changed neither the deviance explained nor the predictor variables selected as significant contributors to the model (Table 6.2).

Random Forest Models

The RF predictive models yielded results similar to both the kriging and cluster analyses, predicting higher WI values in the north-western portion of the range of *C. leuconotus* and lower values along the Gulf Coast (Figure 6.6a). In contrast, the coefficient of variation of the RF predictive models was highest along the Gulf Coast and lowest in the north-western and central portions of the range (Figure 6.6b). Comparing predictor variables between GDM and RF models revealed differences between which variables contributed most to the model. For GDM, bio2 and GM_LAI contributed the most, whereas, for RF, based on the percent increase in mean squared error and nodal purity, bio2 and GM_LAI appeared to contribute little to the model (Figure 6.7), and bio11 was most important, followed by bio1, bio6, bio7, and bio4 (Figure 6.7). Estimates of variable importance between percent increase in mean square error (MSE) and increase in nodal purity did not yield equivalent results, except for

bio11 (mean temperature of the coldest quarter), which was identified as the most important variable under both metrics. Out-of-bag error for the RF levelled off at an MSE of 0.0016 after the generation of 200 trees (figure not shown). Predicted and observed values appeared to be highly correlated (Pearson's product-moment correlation test: $r = 0.74$, df $= 50$, $p = 0.0001$; Figure 6.7b) although predicted values tended to over-predict at low values of WI and under-predict at high WI values (Figure 6.7b).

Discussion

Colour patterns and arrangements play a major role in carnivore natural history (Ewer, 1973). From signalling to camouflage, carnivores rely on the arrangement and patterns of their pelage colour to survive (Ortolani, 1999). Although the importance of colour to carnivores is well documented, a majority of research to date has focused on explaining broad-scale patterns across species or within particular taxonomic groups (Ortolani, 1999; Stankowich *et al.*, 2011, 2014; cf. da Silva *et al.*, 2016). Few studies have investigated the degree of variation within a species; the few that have,

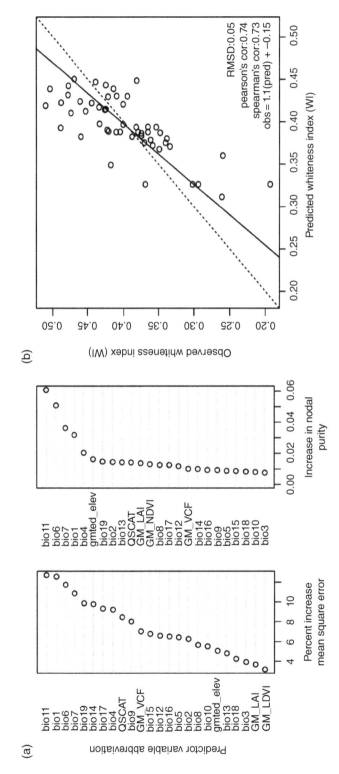

Figure 6.7 (a) Variable importance graph and (b) predicted versus observed plot for random forest models of WI based on 262 individual *Conepatus leuconotus* and 24 environmental predictors (see Table 6.1 for predictor definitions).

have relied on qualitative categorizations to partition variation of what most likely represents a continuous variable (Van Gelder, 1959; Dragoo *et al.*, 2003; Schiaffini *et al.*, 2013; da Silva *et al.*, 2016). Here we used a quantitative approach to describe dorsal stripe pattern variation in the North American hog-nosed skunk to shed light on evolutionary pressures faced by a small, aposematic carnivore distributed over a broad environmental landscape. Our study presents one of the first investigations into intraspecific colour patterns within a small carnivore using quantitative measures and modelling techniques to both map and identify environmental correlates with spatial patterns of morphological variation.

Spatially explicit interpolation and modelling techniques indicate that variation in the size and extent of the white dorsal stripe (and, consequently, the coverage of black fur along the dorsum) of *C. leuconotus* is non-randomly distributed across the landscape (Figures 6.3–6.6). The extent of dorsal whiteness appears to peak in the north-western portion of the species range, with reduced whiteness along the Gulf Coast and Central America. In contrast to patterns associated with white fur along the dorsum, the amount of black fur could be considered greater in more tropical, closed environments of coastal Mexico and Central America and reduced in more arid, open habitats of the desert of southwest and west-central Texas, a pattern in accordance with Gloger's rule (i.e. darker-coloured birds and mammals are typically found in more humid, tropical regions). Adherence to Gloger's rule was similarly hypothesized to support patterns seen in another small carnivore, the jaguarundi, *Herpailurus [= Puma] yagouaroundi*, where grey/dark individuals were found to be significantly associated with moist and dense forests when compared to lighter, reddish forms (da Silva *et al.*, 2016).

GDM and RF analysis were able to identify important explanatory variables for this variation, although the two techniques converged upon different predictors. They differed in their most important variables, with GDM supporting grand mean of Leaf Area Index (GM_LAI) and mean diurnal temperature range (bio2) and RF mean temperature of the coldest quarter (bio11) and minimum temperature of the coldest month (bio6), respectively. These differences do not reflect limitations in predicting whiteness intensity

based on these predictors, but are more reflective of the different approaches of the models to interpreting variation in response variables. GDM attempts to identify which predictor variables account for the *difference* or *dissimilarity* between individuals' WI by comparing those differences to differences found in the environmental conditions associated with the individuals being compared. In our case, individual *C. leuconotus* differed most in their WI between sites that have maximal differences in canopy cover and seasonality, as estimated by satellite-derived measures of LAI and mean annual temperature differences (Figure 6.5). Although only 14.3% of the variation in WI was accounted for by this model, this value was highly significant when compared to randomly permuted samples that broke the association between the differences in WI and predictor values (Table 6.2).

This amount of explained variation parallels results from other studies examining colour variation in terrestrial mammals. Lai *et al.* (2008), examining dorsal colour variation in wild populations of the house mouse, *Mus musculus*, found that 21.6% of the variation in dorsal colour patterns was associated with rainfall. Similarly, Kamilar & Bradley (2011) found that mean evapotranspiration (a measure of the amount of atmospheric water resulting from both evaporation and transpiration) accounted for only 13.7 and 25.2% of the variation in pelage brightness for the dorsal and ventral pelage surfaces of over 200 individuals representing 100 primate species. However, Thomassen *et al.* (2010), using GDM models for a rainforest bird, found that several metrics of environmental conditions, including some used in our own analysis (e.g. QSCAT, LAI) explained upward of 95% of the variation seen in genetic make-up and morphological characteristics among individuals. In addition, Thomassen *et al.* (2011), using similar variables to ours, found that GDMs varied in their performance across both species and traits for seven species of tropical vertebrates, with only 36 of 60 models performing better than random. Although no characters related to colouration were included in their models, the three mammal species (all bats of the genus *Carollia*) examined, consistently displayed the lowest percentage of variation explained, with some significant models accounting for only 12–16% of the total variation (Thomassen *et al.*, 2011). Thus, although our results are consistent with analogous studies on mammal colouration, these values do not reflect those found for multiple characteristics from seven species of tropical vertebrates using the same technique, indicating that other potential and unmeasured factors, e.g. microhabitat characteristics, predator communities, could help explain the amount of variation seen in dorsal stripe patterns of *C. leuconotus*.

Unlike correlative GDM, RF models attempt to identify those variables that best *predict* actual values of the response variable by looking for the amount of variation accounted for by each predictor variable used in the model. In our case, RF models indicated that WI is best predicted using bioclimatic variables alone (plus QSCAT backscatter measurements), with only minimal contributions from Leaf Area Index (GM_LAI). RF models identified only temperature-related variables as important. Considering GDM and RF results together, it appears that variation in the white dorsal stripe of *C. leuconotus* is affected by conditions related to both canopy cover and minimum temperatures, with reductions in whiteness occurring in regions with increased canopy cover and less seasonal climates compared to increased dorsal whiteness which can be found in areas with less annual canopy cover and experiencing cooler temperatures at least during parts of the year.

These results imply that some form of interaction between dorsal whiteness patterns and environmental conditions is occurring among populations of *C. leuconotus*. Of interest is the fact that the RF model predicts lower whiteness indices from southern Veracruz through the Yucatan Peninsula, a geographic area devoid of *C. leuconotus* but occupied by its sister species, the Amazonian hog-nosed skunk, *C. semistriatus*, which shows similar patterns of reduction in white along the posterior portion of its dorsum (Figure 6.8b). Unlike what is observed in *C. leuconotus*, dorsal stripe patterns appear to be relatively conserved in *C. semistriatus* (A. W. Ferguson, personal observation), especially when compared to other hypervariable species such as Molina's hog-nosed skunk, *C. chinga* (Van Gelder & Kipp, 1968; Schiaffini *et al.*, 2013). The dorsal stripe of *C. semistriatus* begins near the top of the head, bifurcates along the dorsum, and terminates prior to reaching the base of the tail, leaving a black posterior region very similar to individuals of *C. leuconotus* from

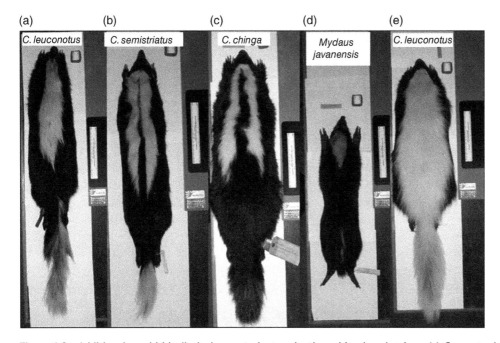

Figure 6.8 Additional mephitids displaying posterior terminating white dorsal stripes: (a) *Conepatus leuconotus* from S Texas; (b) Amazonian hog-nosed skunk, *C. semistriatus* from Veracruz, Mexico; (c) Molina's hog-nosed skunk, *C. chinga* from Peru; and (d) Sunda stink badger, *Mydaus javanensis* from Borneo compared to (e) a solid white dorsum of *C. leuconotus* from Arizona.

the Gulf Coast (Figure 6.8a,b). Other 'tropical forest' species found within the family Mephitidae also seem to maintain a larger portion of white on the anterior versus the posterior portion of their dorsum (Figure 6.8a–d). Taken together with our results for *C. leuconotus*, it appears that reduction of the intensity of white along the posterior of the dorsum and relative increases in black colouration are more common in tropical environments typically characterized by greater canopy cover and more constant temperature regimes.

The fact that variation in dorsal stripe colouration in *C. leuconotus* conforms to Gloger's rule provides a unique opportunity to explore our competing hypotheses of aposematism, crypsis, and physiology. This opportunity stems from the many hypotheses that have been put forward to explain Gloger's rule (Kamilar & Bradley, 2011). Some of the most relevant hypotheses to our system include thermoregulation (Caro, 2005a; Margalida *et al.*, 2008), increased absorption of excessive UV radiation (Caro, 2005a), enhanced water evaporation (Gloger, 1833; Caro, 2005a), and increased background matching for species living in relatively light or dark habitats (Kamilar & Bradley, 2011; Singaravelan *et al.*, 2013). Although other hypotheses, including increased resistance to higher amounts and diversity of microbes in humid regions have been proposed for Gloger's rule, these focus mostly on avian systems and have been less explored in mammalian taxa (Burtt & Ichida, 2004). Considering the balance of whiteness and blackness in the dorsal stripe of *C. leuconotus* in the context of both Gloger's rule and the species' natural history, support is garnered for a balance between aposematic signalling and crypsis via background matching as the most plausible hypotheses.

More precisely, the nearly strictly nocturnal habits of *C. leuconotus* (Dragoo & Sheffield, 2009) indicate that both thermoregulation and reduced exposure to UV radiation seem less feasible as explanatory mechanisms for the observed patterns. Although being nocturnal does not make skunks immune to temperature regimes of their local environment, most species utilize behavioural mechanisms (e.g. switching den sites to match changing climatic conditions; W. Brashear *et al.*, unpublished data) to deal with temperature extremes. In addition, most of the temperature variables selected as important to the models of variation were indicators of minimum temperatures, indicating that the enhanced

cooling experienced by blacker furs in response to warmer and more humid conditions would offer little in terms of explaining the decrease in white and increase in black fur observed for *C. leuconotus* living under these conditions. In addition, no variables related to precipitation were selected for inclusion in the final models, indicating that variables associated with moisture and potentially increased evaporative cooling contribute little toward an explanation of the observed patterns. Finally, their nocturnal nature exposes skunks to little or no UV radiation, limiting the need for increased melanin in response to UV-damaging radiation.

Why then do landscape metrics of minimum temperature and canopy cover appear to correlate with variation in the amount of white and black along the back of *C. leuconotus*? Perhaps the most logical conclusion involves one of the major drivers of natural selection: predation. Although skunks are often thought of as immune to predation due to their well-established defence mechanisms, skunks are preyed upon by a wide array of predators including both aerial and terrestrial species (Dragoo, 2009). Given the fact that white appears highly conspicuous under dark conditions (Searle, 1968), it would seem that whiter individuals would be more obvious to predators. This would be especially true in habitats characterized by greater canopy cover associated with increased vegetation density and less light penetration, where darker conditions would enhance the signalling efficacy of a white animal, making it more obvious to predators. Similarly, in more seasonal habitats characterized by more severe temperature changes and reduced vegetation density at some time of the year, greater light penetration could lead to increasing reflectance from a white surface, again making an individual more conspicuous.

Thus, perhaps there is some advantage conferred upon being inconspicuous in more tropical environments compared to conspicuous in more arid environments, an advantage related to the risk of predation. Being bold and conspicuous may work well in exposed environments characterized by open-pursuit, visual predators such as great-horned owls, *Bubo virginianus*, and coyotes, *Canis latrans*, whereas remaining concealed but without loss of warning abilities (i.e. reduction in whiteness but not a complete loss of white fur) in closed environments characterized by ambush predators (e.g. Neotropical felids) could increase survival through far-field crypsis but coupled with near-field aposematism. For *C. leuconotus* found across the desert southwest, being big and bold may deter predation attempts by size-limited (great-horned owls) or visual (coyotes) predators, whereas for *C. leuconotus* found in the thick Tamaulipan thornscrub, remaining hidden among the shadows of a dense canopy could conceal a skunk's presence to ambush predators such as jaguars, *Panthera onca*.

Although distinguishing causation from correlation is impossible using our approach, our results are consistent with recent findings describing the potential for the evolution of 'intermediate aposematism' in animal species (Tullberg *et al.*, 2005; Ruxton *et al.*, 2009). Although long considered an evolutionary option, especially regarding effects of observation distance on signal efficacy (Endler, 1978), aposematism has traditionally been classified as one of two mutually exclusive anti-predator strategies: conspicuousness or crypsis (Tullberg *et al.*, 2005; Caro *et al.*, 2013). Recent work, however, has provided theoretical (Endler & Mappes, 2004; Mappes *et al.*, 2005; Ruxton *et al.*, 2009) and empirical (Tullberg *et al.*, 2005) examples of how fitness trade-offs between conspicuousness and crypsis could lead to the evolution of aposematism with differing degrees of crypsis. Working hypotheses as to how an intermediate investment in conspicuousness might evolve include aposematism in an environment where some predators are responsive to the signal and others are not (Endler & Mappes, 2004); a distance-dependence function whereby animals appear cryptic from a distance yet aposematic up close (Tullberg *et al.*, 2005; Caro *et al.*, 2013); and a balance between costs of conspicuousness that result in increased predator encounters and benefits of enhanced learned aversion of predators to that signal (Ruxton *et al.*, 2009).

Recently, a distance-dependent function was proposed for the western spotted skunk, *Spilogale gracilis*, which suggested that, from a far distance and under different habitat characteristics, *S. gracilis* appears cryptic to visual predators having similar search images as humans (Caro *et al.*, 2013). This and other studies highlight the potential importance of microhabitat characteristics and behaviour in concealing or enhancing aposematic signals (Ruxton *et al.*, 2009; Caro *et al.*, 2013). Given the association between local environmental conditions and whiteness indices in

C. leuconotus, it is plausible that some sort of intermediate aposematism may be occurring. Perhaps individuals in more canopy-dense and less seasonal environments experience reduced rates of predation through reduction of whiteness along the dorsum, making them less conspicuous than individuals with a solid white back or high whiteness intensity. Alternatively, the concentration of white patches on the periphery of the animal's body (i.e. anterior-mid-dorsal stripes and posterior tail patches) as is seen in other tropical mephitids (Figure 6.8), could also help reduce predation risk through 'disruptive colouration' (Stevens *et al.*, 2006). Of course, other non-adaptive alternative hypotheses could also explain the observed patterns (e.g. genetic drift, structured populations). However, mitochondrial DNA data appear not to support a biogeographic or demographic signature coincident with observed morphological patterns (Ferguson, 2014). Future work including assessments of local predator guilds and non-neutral colour candidate genes (e.g. *Agouti*, *MC1R*, or *KIT*; Kerns *et al.*, 2004; Haase *et al.*, 2007) could shed light on potential selective forces shaping variation in dorsal stripe patterns. Of course, without manipulative experiments under field conditions, separating causation from correlation will remain a problem.

Although our methodology provides an effective approach for quantifying colour variation for aposematic small carnivores, it is not without its limitations. Prepared museum skins can distort the configuration of colour patterns from those exhibited under natural settings, although dorsal-pattern variation was found to be independent of body length (a major form of distortion in museum skins). Also, the years of collection for our specimens do not match the period from which the remotely sensed data were collected, providing a potential source of error if environments have changed significantly. In addition, we were unable to quantify shape dimensionality from our two-dimensional images; body shape has been shown to be important for the recognition and avoidance of skunks by their natural predators (Hunter, 2009). Our images are unable to capture the nuances of behavioural strategies used either to enhance or reduce the signalling of these colour arrangements. In addition, our analyses do not provide a sound assessment of the role of colouration in either inter- or intraspecific communication, although most nocturnal carnivores tend to rely on olfactory signalling for communication purposes (Ewer, 1973). Skunks are known for their behavioural responses to threats, including foot-stomping, lunging, and tail raising or enlargement (Dragoo, 2009). Given the importance of the tail in such displays, quantifying the arrangement and composition of black and white hairs on this part of the body could prove useful, but is difficult due to various ways in which specimens have been prepared (e.g. curved tails, tails bent under the body of the specimen). In fact, research on other carnivores has shown that a whole-body approach to pattern analysis is important for questions of anti-predator defence as markings on different parts of the body may serve different functions (Stankowich *et al.*, 2011). Finally, our interpretation of signals is based on a human-vision system, which may or may not reflect the vision systems of skunk predators, although human and carnivore abilities to detect shapes and patterns are likely similar (Osorio *et al.*, 1999; Gibson *et al.*, 2007).

Despite these limitations, our methodology provides a novel technique for quantifying dorsal-pattern variation for aposematic or contrastingly coloured small carnivores and, therefore, has the potential to be applied to other similarly coloured species such as the honey badger, *Mellivora capensis*, zorilla, *Ictonyx striatus*, or African striped weasel, *Poecilogale albinucha*, to name a few. The application of this technique to quantify variation in what is considered to be a highly polymorphic mephitid, *Mephitis mephitis*, is currently underway and proving to be effective (T. Stankowich, California State University Long Beach, USA, personal communication, 2014). Overall, results indicate that colour patterns in skunks are not simply a black and white issue, but that local environmental conditions could be influencing variation in the arrangement and intensity of these colours within a single species. These findings suggest that further investigations into intraspecific variation in aposematic species such as skunks are warranted. Future tests of the conspicuousness of these variations under different canopy or climatic conditions and examination of differences in other selective factors such as composition of local predator guilds might help elucidate the evolutionary forces shaping colour variation in these and other species.

Acknowledgements

This work would not have been possible without access to the extensive collection of skunks housed in the world's natural history museums: specimens for this paper represent the combined efforts of over 72 collectors across 127 years. Specifically, we would like to thank the following institutions and their associated staff members for assisting us in accessing their collections: American Museum of Natural History – R. Voss; Academy of Natural Sciences Philadelphia – N. Gilmore; Angelo State Natural History Collection – M. Revelez; Instituto Politecnico Nacional's Colección Cientifica de Fauna Silvestre – C. Lopez Gonzalez & D. Garcia; Carnegie Museum of Natural History – S. McLaren; Denver Museum of Natural History – J. Demboski & J. Stephenson; Museum National d'Histoire Naturelle – G. Veron; Natural History Museum – R. Portela Miguez & L. Tomsett; Texas Tech University – R. Baker & H. Garner; University of Colorado Museum – C. McCain; US Geological Survey at the United States National Museum – A. Gardner & S. Peurach. We would also like to thank K. Thrasher, B. Burdett, and M. O'Dell for their assistance in digitizing and geo-referencing specimens and A.T. Peterson, R. Bradley, and R. Baker for comments on earlier drafts of the manuscript. Special thanks to H.A. Thomassen for providing the QUICKSAT imagery. Funding for this project was provided in part through grants to AWF from the American Museum of Natural History's Theodore Roosevelt Memorial Fund, American Society of Mammalogists' Grants-in-Aid, Sigma Xi Grants-In-Aid for Research, The Southwestern Association of Naturalists' Howard McCarley Award, Texas Academy of Science Student Research Award, and Texas Tech University.

References

Anselin, L. (1995) Local indicators of spatial association – LISA. *Geographical Analysis* 27, 93–115.

Bookstein, F.L. (1997) Landmark methods for forms without landmarks: morphometrics of group differences in outline shape. *Medical Image Analysis* 1, 225–243.

Breiman, L. (1996) Bagging predictors. *Machine Learning* 24, 123–140.

Breiman, L. (2001a) Random forests. *Machine Learning* 45, 5–32.

Breiman, L. (2001b) Statistical modeling: the two cultures (with comments and a rejoinder by the author). *Statistical Science* 16, 199–231.

Breiman, L., Friedman, J., Stone, C.J. & Olshen, R.A. (1984) *Classification and Regression Trees*. CRC Press, New York.

Burke, I., Kittel, T., Lauenroth, W., Snook, P., Yonker, C. & Parton, W. (1991) Regional analysis of the central Great Plains. *BioScience* 41, 685–692.

Burtt, E.H. Jr & Ichida, J.M. (2004) Gloger's rule, feather-degrading bacteria, and color variation among song sparrows. *The Condor* 106, 681–686.

Caro, T. (2005a) The adaptive significance of coloration in mammals. *BioScience* 55, 125–136.

Caro, T. (2005b) *Antipredator Defenses in Birds and Mammals*. University of Chicago Press, Chicago.

Caro, T. (2009) Contrasting coloration in terrestrial mammals. *Philosophical Transactions of the Royal Society B: Biological Sciences* 364, 537–548.

Caro, T. (2011) The functions of black-and-white coloration in mammals. In: *Animal Camouflage: Mechanisms and Function* (eds M. Stevens & S. Merilaita), pp. 298–329. Cambridge University Press, Cambridge.

Caro, T., Stankowich, T., Kiffner, C. & Hunter, J. (2013) Are spotted skunks conspicuous or cryptic? *Ethology Ecology & Evolution* 25, 144–160.

Chapman, A. D. & Wieczorek, J. (2006) *Guide to Best Practices for Georeferencing*. Global Biodiversity Information Facility, Copenhagen.

Cuozzo, F.P., Rasoazanabary, E., Godfrey, L.R., Sauther, M.L., Youssouf, I.A. & LaFleur, M.M. (2013) Biological variation in a large sample of mouse lemurs from Amboasary, Madagascar: implications for interpreting variation in primate biology and paleobiology. *Journal of Human Evolution* 64, 1–20.

da Silva, L.G., de Oliveira, T.G., Kasper, C.B., Cherem, J. J., Moraes, E.A., Paviolo, A. & Eizirik, E. (2016) Biogeography of polymorphic phenotypes: mapping and ecological modelling of coat colour variants in an elusive Neotropical cat, the jaguarundi (*Puma yagouaroundi*). *Journal of Zoology* 299, 295–303.

De'ath, G. & Fabricius, K.E. (2000) Classification and regression trees: a powerful yet simple technique for ecological data analysis. *Ecology* 81, 3178–3192.

Dice, L.R. (1940) Ecologic and genetic variability within species of *Peromyscus*. *The American Naturalist* 74, 212–221.

Dragoo, J.W. (2009) Family Mephitidae (Skunks). In: *Handbook of the Mammals of the World. Volume 1. Carnivores* (eds D.E. Wilson & R.A. Mittermeier), pp. 532–563. Lynx Edicions, Barcelona, Spain.

Dragoo, J.W. & Sheffield, S.R. (2009) *Conepatus leuconotus* (Carnivora: Mephitidae). *Mammalian Species* 827, 1–8.

Dragoo, J.W., Honeycutt, R.L. & Schmidly, D.J. (2003) Taxonomic status of white-backed hog-nosed skunks, genus *Conepatus* (Carnivora: Mephitidae). *Journal of Mammalogy* 84, 159–176.

Endler, J. (1978) A predator's view of animal color patterns. *Evolutionary Biology* 11, 319–364.

Endler, J.A. & Mappes, J. (2004) Predator mixes and the conspicuousness of aposematic signals. *American Naturalist* 163, 532–547.

Ewer, R.F. (1973) *The Carnivores*. Cornell University Press, Ithaca.

Ferguson, A.W. (2014) *Evolution of Skunks (Carnivora: Mephitidae) Across the Mexican Transition Zone: Understanding the Influence of Environmental Variation on Morphological and Phylogeographic Patterns*. PhD Thesis, Texas Tech University.

Ferrier, S. (2002) Mapping spatial pattern in biodiversity for regional conservation planning: where to from here? *Systematic Biology* 51, 331–363.

Ferrier, S., Powell, G.V.N., Richardson, K.E., Manion, G., Overton, J.M., Allnutt, T.F., Cameron, S.E., Mantle, K., Burgess, N.D. & Faith, D.P. (2004) Mapping more of terrestrial biodiversity for global conservation assessment. *BioScience* 54, 1101–1109.

Ferrier, S., Manion, G., Elith, J. & Richardson, K. (2007) Using generalized dissimilarity modelling to analyse and predict patterns of beta diversity in regional biodiversity assessment. *Diversity and Distributions* 13, 252–264.

Freeman, E.A. & Frescino, T. (2009) *ModelMap: An R Package for Modeling and Map Production Using Random Forest and Stochastic Gradient Boosting*. USDA Forest Service, Rocky Mountain Research Station, Ogden. http://cran.r-project.org/web/packages/ModelMap/index.html. Accessed on 1 January 2013.

Gibson, B.M., Lazareva, O.F., Gosselin, F., Schyns, P.G. & Wasserman, E.A. (2007) Nonaccidental properties underlie shape recognition in mammalian and nonmammalian vision. *Current Biology* 17, 336–340.

Gloger, C.W.L. (1833) *Das Abändern der Vögel durch Einfluss des Klimas*. A. Schulz, Breslau. (In German).

Green, W. (1996) The thin-plate spline and images with curving features. In: *Proceedings in Image Fusion and Shape Variability Techniques* (eds K. Mardia, C. Gill & I. Dryden), pp. 79–87. Leeds University Press, Leeds.

Greenwood, J.J., Cotton, P.A. & Wilson, D.M. (1989) Frequency-dependent selection on aposematic prey: some experiments. *Biological Journal of the Linnean Society* 36, 213–226.

Groves, C.P., Rajapaksha, C. & Manemandra-Arachchi, K. (2009) The taxonomy of the endemic golden palm civet of Sri Lanka. *Zoological Journal of the Linnean Society* 155, 238–251.

Haase, B., Brooks, S.A., Schlumbaum, A., Azor, P.J., Bailey, E., Alaeddine, F., Mevissen, M., Burger, D., Poncet, P-A., Rieder, S. & Leeb, T. (2007) Allelic heterogeneity at the equine *KIT* locus in dominant white (*W*) horses. *PLoS Genetics* 3, e195.

Hijmans, R.J., Cameron, S.E., Parra, J.L., Jones, P.G. & Jarvis, A. (2005) Very high resolution interpolated climate surfaces for global land areas. *International Journal of Climatology* 25, 1965–1978.

Hunter, J.S. (2009) Familiarity breeds contempt: effects of striped skunk color, shape, and abundance on wild carnivore behavior. *Behavioral Ecology* 20, 1315–1322.

Isaaks, E.H. & Srivastava, R.M. (1989) *An Introduction to Applied Geostatistics*. Oxford University Press, New York.

Kamilar, J.M. & Bradley, B.J. (2011) Interspecific variation in primate coat colour supports Gloger's rule. *Journal of Biogeography* 38, 2270–2277.

Kerns, J.A., Newton, J., Berryere, T.G., Rubin, E.M., Cheng, J-F., Schmutz, S.M. & Barsh, G.S. (2004) Characterization of the dog *Agouti* gene and a *nonagouti* mutation in German shepherd dogs. *Mammalian Genome* 15, 798–808.

Knyazikhin, Y., Martonchik, J., Myneni, R., Diner, D. & Running, S. (1998) Synergistic algorithm for estimating vegetation canopy leaf area index and fraction of absorbed photosynthetically active radiation from MODIS and MISR data. *Journal of Geophysical Research* 103, 32257–32275.

Lai, Y.C., Shiroishi, T., Moriwaki, K., Motokawa, M. & Yu, H.T. (2008) Variation of coat color in house mice throughout Asia. *Journal of Zoology* 274, 270–276.

Larivière, S. & Messier, F. (1996) Aposematic behaviour in the striped skunk, *Mephitis mephitis. Ethology* 102, 986–992.

Long, D.G., Drinkwater, M.R., Holt, B., Saatchi, S. & Bertoia, C. (2001) Global ice and land climate studies using scatterometer image data. *EOS Transactions of the American Geophysical Union* 82, 503–503.

Lovegrove, B.G. (2000) The zoogeography of mammalian basal metabolic rate. *The American Naturalist* 156, 201–219.

MacLeod, N. (1999) Generalizing and extending the eigenshape method of shape space visualization and analysis. *Paleobiology* 25, 107–138.

MacMillen, R.E. & Garland, T. (1989) Adaptive physiology. In: *Advances in the Study of Peromyscus (Rodentia)* (eds J.N. Lane & G.L. Kirkland), pp. 143–168. Texas Tech University Press, Lubbock.

Mappes, J., Marples, N. & Endler, J.A. (2005) The complex business of survival by aposematism. *Trends in Ecology & Evolution* 20, 598–603.

Margalida, A., Negro, J.J. & Galván, I. (2008) Melanin-based color variation in the bearded vulture suggests a thermoregulatory function. *Comparative Biochemistry and Physiology Part A: Molecular & Integrative Physiology* 149, 87–91.

Newman, C., Buesching, C.D. & Wolff, J.O. (2005) The function of facial masks in "midguild" carnivores. *Oikos* 108, 623–633.

Nix, H.A. (1986) A biogeographic analysis of Australian elapid snakes. In: *Snakes: Atlas of Elapid Snakes of Australia* (ed. R. Longmore), pp. 4–16. Australian Government Publication Service, Canberra.

Oliver, M.A. & Webster, R. (1990) Kriging: a method of interpolation for geographical information systems. *International Journal of Geographical Information Systems* 4, 313–332.

Ortolani, A. (1999) Spots, stripes, tail tips and dark eyes: predicting the function of carnivore colour patterns using the comparative method. *Biological Journal of the Linnean Society* 67, 433–476.

Osorio, D., Miklósi, A. & Gonda, Z. (1999) Visual ecology and perception of coloration patterns by domestic chicks. *Evolutionary Ecology* 13, 673–689.

Paruelo, J.M., Oesterheld, M., Di Bella, C.M., Arzadum, M., Lafontaine, J., Cahuepé, M. & Rebella, C.M. (2000) Estimation of primary production of subhumid rangelands from remote sensing data. *Applied Vegetation Science* 3, 189–195.

Pocock, R. (1908) Warning coloration in the musteline Carnivora. *Proceedings of the Zoological Society of London* 78, 944–959.

Prince, S. (1991) Satellite remote sensing of primary production: comparison of results for Sahelian grasslands 1981–1988. *International Journal of Remote Sensing* 12, 1301–1311.

Rosauer, D.F., Ferrier, S., Williams, K.J., Manion, G., Keogh, J.S. & Laffan, S.W. (2014) Phylogenetic generalised dissimilarity modelling: a new approach to analysing and predicting spatial turnover in the phylogenetic composition of communities. *Ecography* 37, 21–32.

Ruxton, G., Speed, M. & Broom, M. (2009) Identifying the ecological conditions that select for intermediate levels of aposematic signalling. *Evolutionary Ecology* 23, 491–501.

Searle, A.G. (1968) *Comparative genetics of coat colour in mammals*. Logos Press, London.

Schiaffini, M.I., Gabrielli, M., Prevosti, F.J., Cardosa, Y.P., Castillo, D., Bo, R., Casanave, E. & Lizarralde, M. (2013) Taxonomic status of southern South American *Conepatus* (Carnivora: Mephitidae). *Zoological Journal of the Linnean Society* 167, 327–344.

Seton, E.T. (1920) Acrobatic skunks. *Journal of Mammalogy* 1, 140.

Singaravelan, N., Raz, S., Tzur, S., Belifante, S., Pavlicek, T., Beiles, A., Ito, S., Wakamatsu, K. & Nevo, E. (2013) Adaptation of pelage color and pigment variations in Israeli subterranean blind mole rats, *Spalax ehrenbergi*. *PLoS One* 8, e69346.

Speed, M.P. (2001) Can receiver psychology explain the evolution of aposematism? *Animal Behaviour* 61, 205–216.

Stankowich, T., Caro, T. & Cox, M. (2011) Bold coloration and the evolution of aposematism in terrestrial carnivores. *Evolution* 65, 3090–3099.

Stankowich, T., Haverkamp, P.J. & Caro, T. (2014) Ecological drivers of antipredator defenses in carnivores. *Evolution* 68, 1415–1425.

Stevens, M., Cuthill, I.C., Windsor, A.M. & Walker, H.J. (2006) Disruptive contrast in animal camouflage. *Proceedings of the Royal Society B: Biological Sciences* 273, 2433–2438.

Thomassen, H.A., Buermann, W., Milá, B., Graham, C.H., Cameron, S.E., Schneider, C.J., Pollinger, J.P., Saatchi, S., Wayne, R.K. & Smith, T.B. (2010) Modeling environmentally associated morphological and genetic variation in a rainforest bird, and its application to conservation prioritization. *Evolutionary Applications* 3, 1–16.

Thomassen, H.A., Fuller, T., Buermann, W., Milá, B., Kieswetter, C.M. *et al.* (2011) Mapping evolutionary process: a multi-taxa approach to conservation prioritization. *Evolutionary Applications* 4, 397–413.

Tullberg, B.S., Merilaita, S. & Wiklund, C. (2005) Aposematism and crypsis combined as a result of distance dependence: functional versatility of the colour pattern in the swallowtail butterfly larva. *Proceedings of the Royal Society B: Biological Sciences* 272, 1315–1321.

Van Gelder, R.G. (1959) A taxonomic revision of the spotted skunks (genus *Spilogale*). *Bulletin of the American Museum of Natural History* 117, 229–392.

Van Gelder, R.G. & Kipp, H. (1968) The genus *Conepatus* (Mammalia, Mustelidae): variation within a population. *American Museum Novitates* 2322, 1–38.

Vignieri, S.N., Larson, J.G. & Hoekstra, H.E. (2010) The selective advantage of crypsis in mice. *Evolution* 64, 2153–2158.

Wallace, A.R. (1867) Mimicry and other protective resemblances among animals. *Westminster Review (London ed.)* 88, 1–43.

Wallace, A.R. (1889) *Darwinism: An Exposition of the Theory of Natural Selection with Some of its Applications*. Macmillan, London.

Willink, B., Brenes-Mora, E., Bolaños, F. & Pröhl, H. (2013) Not everyhing is black and white: color and behavioral variation reveal a continuum between cryptic and aposematic strategies in a polymorphic poison frog. *Evolution* 67, 2783–2794.

Wilson, D.E. & D.A.M. Reeder (2005) *Mammal Species of the World: A Taxonomic and Geographic Reference.* 3rd edition. The Johns Hopkins University Press, Baltimore.

Wilson, D.E., Mittermeier, R.A. & Del Hoyo, J. (2009) *Handbook of the Mammals of the World. Volume 1. Carnivores.* Lynx, Barcelona.

Wood, W. F. (1999) The history of skunk defensive secretion research. *Chemical Educator* 4, 44–50.

Appendix 6.A

Detailed methodology and descriptions for satellite-derived imagery used to estimate environmental predictor variables for subsequent modelling using generalized dissimilarity and random forest procedures.

Satellite remote-sensing data from NASA's Moderate Resolution Imaging Spectroradiometer (MODIS; http://modis.gsfc.nasa.gov) and Quick Scatterometer (QuikSCAT; http://winds.jpl.nasa.gov/missions/quikscat/) was used to infer environmental characteristics that could potentially influence ambient light conditions and therefore signalling efficacy. From the NASA LPDAAC MODIS collection hosted on Earth Explorer (http://earthexplorer.usgs.gov/), we obtained monthly Normalized Difference Vegetation Indices (NDVI; MOD13A3) and a single 8-day interval of the Leaf Area Index/Fraction of Photosynthetically Active Radiation (LAI/FPAR; MOD15A2) for September, February, April, and June, for 2005–2009. NDVI is created from spectral reflectance measurements and is related positively to above ground net primary productivity (aNPP) and biomass (Burke *et al.*, 1991; Prince, 1991; Paruelo *et al.*, 2000). LAI is defined as the one-sided green leaf area per unit ground surface area in broadleaf canopies and is used to characterize plant canopy density (Knyazikhin *et al.*, 1998). The four months chosen represent the seasonality experienced by populations of *C. leuconotus* (i.e. four seasons in the north and wet and dry seasons in the south). Yearly means were estimated by averaging the four monthly values in ArcGIS 10.1 (ESRI, Redlands, CA) over 2005–2009. In addition, we used yearly Vegetation Continuous Field (VCF; MOD44B) values as a measure of the percentage of tree cover. A final grand mean and variance of the five years was calculated and these values were used as final predictors in statistical analyses. Monthly QuikSCAT imagery for 2001 was used to obtain raw backscatter measurements (QSCAT) that reflect attributes related to surface moisture and roughness (Long *et al.*, 2001) and elevation data were obtained from the US Geological Survey and National Geospatial-Intelligence Agency's Global Multi-resolution Terrain Elevation Data 2010 (GMTED2010; http://topotools.cr.usgs.gov/gmted_viewer/). Variables with differing native resolutions (e.g. QSCAT at 2.25 km) were resampled to 1 km resolution using the Reproject function in ArcGIS 10.1 (ESRI, Redlands, CA).

Appendix 6.B

Boxplots of Male and Female Whiteness Index (a) and Mean body length (b) values based on 262 museum skins of the North American hog-nosed skunk, *Conepatus leuconotus*.

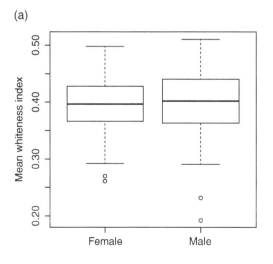

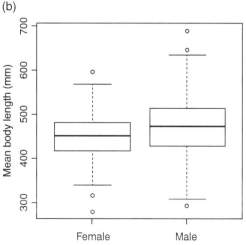

Appendix 6.C

Descriptive statistics for Male and Female whiteness index and Mean body length values based on 262 museum skins of the North American hog-nosed skunk.

Category	Whiteness index (WI)				Body length (mm)			
	Mean	Standard deviation	Variance	Range	Mean	Standard deviation	Variance	Range
Male (*n* = 142)	0.4007	0.0499	0.0025	0.1930–0.5106	473.6	69.1	4779.4	291.7–686.0
Female (*n* = 90)	0.3959	0.0493	0.0024	0.2615–0.4981	452.2	53.9	2909.9	280.6–481.2
Unknown (*n* = 30)	0.4017	0.0657	0.0043	0.2431–0.5133	438.3	57.2	3269.7	291.3–549.7
All (*n* = 262)	0.3992	0.0516	0.0027	0.1930–0.5133	461.6	64.0	4095.2	280.6–686.0

Appendix 6.D

Details of kriging analyses of Whiteness Index values for 262 museum specimens of the North American hog-nosed skunk with semivariogram descriptors and neighbourhood search parameters for both the predicted simple surface and standard error surface.

	Simple kriging surface	Standard error surface
Semivariogram descriptors		
Model	Hole-effect	Hole-effect
Nugget	0.0013	0.0019
Partial sill	0.0015	0.00074

	Simple kriging surface	Standard error surface
Major range	11.85	8.58
Lag size	1.48	1.07
Number of lags	12	12
Neighbourhood search parameters		
Sector type	8 sectors	8 sectors
Copy from variogram	TRUE	TRUE
Neighbourhood type	Standard	Standard
Maximum number of neighbours	5	5
Minimum number of neighbours	2	2
Number of weights	35	35

Part III

Ecology, Behaviour, and Diseases

7

The Function of Carnivore Latrines: Review, Case Studies, and a Research Framework for Hypothesis Testing

Christina D. Buesching[1],* and Neil R. Jordan[2,3,4]

[1]*Department of Biology, The University of British Columbia, Okanagan Campus, Kelowna, Canada*
[2]*Centre for Ecosystem Science, School of Biological, Earth and Environmental Sciences, University of New South Wales, Sydney, Australia*
[3]*Taronga Conservation Society Australia, Taronga Western Plains Zoo, Wildlife Reproduction Centre, Dubbo, Australia*
[4]*Botswana Predator Conservation Trust, Maun, Botswana*

SUMMARY

Latrines are accumulations of two to several hundred faeces resulting from the repeated use of the same defecation sites by the same or several individuals. Many carnivores deposit their faeces in such dedicated latrine sites, which are often shared by several animals either from the same social group or from neighbouring territories. Although latrines are assumed to play an important role in olfactory communication, detailed knowledge of specific information exchange is still lacking. Four different categories of data are important in trying to understand the function of latrines in animal societies: (i) spatial distribution patterns; (ii) temporal usage patterns; (iii) individual visit and contribution patterns; and (iv) information content of the signal. While the spatial distribution of latrines in relation to territory boundaries, landmarks and resources has been studied in a variety of species, only a few studies concentrated on temporal variation in latrine usage. Even fewer studies provide insights into inter-individual differences in visit and contribution patterns or the olfactory information content of latrines. In this review, we outline potential functional hypotheses for latrine use and develop a research framework for the study of latrine function. We then present three model species – European badger, *Meles meles*, meerkat, *Suricata suricatta*, and banded mongoose, *Mungos mungo* – for which we have detailed data for at least three of the four above-mentioned categories, which we will use to test these hypotheses. Throughout the chapter, we review the different techniques used to collect these data in different species, discuss the limitations of using spatial data alone to test functional hypotheses, and highlight the value of a combined approach.

Keywords

Badger – communication – faeces – information – meerkat – mongoose – olfactory – scent-mark

Introduction

Faeces and urine are obligatory metabolic by-products of any heterotrophic diet. As both are intrinsically pungent, readily available and 'free' substances, they are naturally suited for scent-marking (Gosling, 1981, 1985; Macdonald, 1985). In addition, their olfactory profiles can be adapted easily to convey additional information

---

*Corresponding author.

Small Carnivores: Evolution, Ecology, Behaviour, and Conservation, First Edition. Edited by Emmanuel Do Linh San, Jun J. Sato, Jerrold L. Belant, and Michael J. Somers.

about the scent-marking individual (e.g. Kimura, 2001). While urine scent-marks are difficult for humans to study insofar as they are only temporarily visible (Peters & Mech, 1975; Macdonald, 1985), faecal (dropping/scat) surveys can reveal how individuals, or groups, utilize their environment (e.g. Gompper *et al.*, 2006; Zhou *et al.*, 2015a) and can indicate population densities (e.g. Tuyttens *et al.*, 2001; Buesching *et al.*, 2014).

A special case in this context is the use of latrines. Latrines are defined as accumulations of two to several hundred faeces resulting from the repeated use of the same defecation sites by the same or several individuals. Because defecation sites are usually also associated with the deposition of urine and glandular secretions or visually conspicuous marks such as scratch marks (Macdonald, 1980), they are better referred to as 'composite latrines'. The use of such composite latrines or 'midden' sites has been documented throughout the class Mammalia, and includes examples from primates (*Lepilemur* sp. and *Hapalemur* sp.: Irwin *et al.*, 2004; Eppley *et al.*, 2016), ungulates (Leuthold, 1977; Estes, 1991), rodents (e.g. water vole, *Arvicola terrestris*: Woodroffe & Lawton, 1990), and lagomorphs (European wild rabbit, *Oryctolagus cuniculus*: Sneddon, 1991). Latrines have also been documented in some carnivorous marsupials (e.g. spotted-tailed quoll, *Dasyurus maculatus*: Kruuk & Jarman, 1995; Tasmanian devil, *Sarcophilus harrisii*: Owen & Pemberton, 2005) and monotremes (short-beaked echidna, *Tachyglossus aculeatus*: Harris *et al.*, 2019). Despite a conspicuous absence of reports of latrine use in the Ursidae, and relatively few reports in the Felidae, latrine use is probably the most widespread and intensively studied in the Carnivora. Within this order, published evidence of latrine use exists for at least 45 species across seven families (see Table 7.1).

Carnivore latrines are often used by several animals, either from neighbouring territories and/or multiple members of the same social group or pair territory, and even from different species (King *et al.*, 2017; Apps *et al.*, 2019). Nevertheless, despite some knowledge of the distribution of latrines across taxa (Table 7.1), individual patterns of latrine use and behaviour at latrine sites are rarely investigated. As many species produce composite latrines with faeces, urine, and other scent-marks deposited in the same area, the role of latrines in olfactory communication is generally accepted (Brown & Macdonald, 1985). Detailed knowledge of specific information exchange, however, is still lacking, and conclusions drawn about the function of latrines are thus restricted.

In this chapter, we first outline previously suggested hypotheses to explain the function of latrine use. Many of these hypotheses, however, are not mutually exclusive, and there is no published framework for distinguishing between them. Here, we develop such a research framework, identifying four categories of data that we feel researchers should attempt to collect and consider when investigating the function of latrines in any species. We then use this framework to evaluate functional hypotheses for latrine use in three model species – European badger, *Meles meles*, meerkat, *Suricata suricatta*, and banded mongoose, *Mungos mungo* – for which we have detailed data for at least three of the four suggested categories. Throughout the chapter, we will review the different techniques used to collect these data in different species, discuss the limitations of using spatial information alone to test functional hypotheses, and highlight the value of a combined approach. We will present a broad (carnivore-wide) review of latrine use, but where possible, we will refrain from using examples from our three focal species in the general sections, as these will be discussed as in-depth case studies. On occasion, examples from other taxa will be used, but only to discuss hypotheses potentially relevant to explain small carnivore latrine use and where data from the Carnivora are not yet available.

Hypotheses for the Function(s) of Latrines

A variety of hypotheses have been suggested to explain the function(s) of latrine use in different species. The most prominent amongst them relate to *resource acquisition and defence* (here, referred to as *resource ownership*). Others focus on the *information centre hypothesis*, or their role in *orientation*, *parasite avoidance*, or *predator–prey interactions*. In the following, we will discuss each of these hypotheses in turn, illustrating them with examples from a variety of carnivore species before we investigate their validity and limitations in our three focal species.

Table 7.1 Species from the order Carnivora for which latrine use has been documented. Very likely, many more carnivore species use latrines. See Appendix A for selected attributes of species weighing < 21.5 kg.

Family	Scientific name	English name	Selected references
Canidae	*Canis aureus*	Golden jackal	Macdonald (1979)
	Canis latrans	Coyote	Ozaga & Harger (1966), Camenzind (1978), Ralls & Smith (2004)
	Canis simensis	Ethiopian wolf	Sillero-Zubiri & Macdonald (1998)
	Chrysocyon brachyurus	Maned wolf	Kleiman (1972)
	Cuon alpinus	Dhole	Johnsingh (1982)
	Nyctereutes procyonoides	Raccoon dog	Ikeda (1984), Yamamoto (1984), Tsunoda *et al.* (2019)
	Urocyon cinereoargenteus	Grey fox	Trapp (1978)
	Vulpes macrotis	Kit fox	Ralls & Smith (2004)
	Vulpes velox	Swift fox	Darden *et al.* (2008)
Felidae	*Felis catus*	Domestic/feral cat	Molsher (1999)
	Felis silvestris	European wild cat	Piñeiro & Barja (2015)
	Leopardus pardalis	Ocelot	Moreno & Giacalone (2006, 2014), Rodgers *et al.* (2015), King *et al.* (2017)
	Lynx rufus	Bobcat	Bailey (1974)
Herpestidae	*Atilax paludinosus*	Water mongoose	Maddock (1988), Do Linh San *et al.* (2020)
	Crossarchus alexandri	Alexander's cusimanse	Kingdon (1978)
	Cynictis penicillata	Yellow mongoose	le Roux *et al.* (2008), Bizani (2014)
	Galerella pulverulenta	Cape grey mongoose	Mbatyoti (2010)
	Galerella sanguinea	Slender mongoose	Maddock (1988), Zemouche (2018)
	Helogale parvula	Dwarf mongoose	Rasa (1977)
	Herpestes ichneumon	Egyptian mongoose	Maddock (1988), Palomares (1993)
	Mungos mungo	Banded mongoose	Jordan *et al.* (2010)
	Suricata suricatta	Meerkat	Ewer (1963), Jordan *et al.* (2007)
Hyaenidae	*Crocuta crocuta*	Spotted hyena	Kruuk (1972), Vitale *et al.* (2020)
	Hyaena hyaena	Striped hyena	Macdonald (1978)
	Parahyaena brunnea	Brown hyena	Mills *et al.* (1980)
	Proteles cristatus	Aardwolf	Kruuk & Sands (1972), Nel & Bothma (1983), Sliwa (1996)
Mustelidae	*Aonyx capensis*	Cape clawless otter	Jordaan *et al.* (2017)
	Arctonyx collaris	Greater hog badger	Zhou *et al.* (2015b)
	Lontra canadensis	North American river otter	Rostain *et al.* (2004), Green *et al.* (2015), Barocas *et al.* (2016)
	Lontra longicaudis	Neotropical otter	Kasper *et al.* (2008), Santos & dos Reis (2012), Medina-Barrios & Morales-Betancourt (2019)
	Lutra lutra	Eurasian otter	Kruuk & Hewson (1978)
	Martes martes	Pine marten	Lockie (1966), Barja *et al.* (2011)
	Meles anakuma	Japanese badger	Kaneko *et al.* (2009)

(Continued)

Table 7.1 (Continued)

Family	Scientific name	English name	Selected references
	Meles meles	European badger	Kruuk (1978), Buesching *et al.* (2016)
	Mellivora capensis	Honey badger	Begg *et al.* (2003)
	Mustela furo	Domestic ferret	Clapperton (1989)
	Poecilogale albinucha	African striped weasel	Alexander & Ewer (1959)
	Pteronura brasiliensis	Giant otter	Leuchtenberger & Mourão (2009)
Procyonidae	*Bassariscus astutus*	Ringtail	Barja & List (2006)
	Procyon lotor	Northern raccoon	Page *et al.* (1998), Hirsch *et al.* (2014), Kent & Tang-Martinez (2014)
Viverridae	*Civettictis civetta*	African civet	Bearder & Randall (1978), Engel (2000), Bekele Tsegaye *et al.* (2008)
	Genetta genetta	Common/small-spotted genet	Palomares (1993), Barrientos (2006), Espírito-Santo *et al.* (2007)
	Genetta maculata	Rusty-spotted genet	Engel (1998), Blomsterberg (2016), Zemouche (2018)
	Genetta tigrina	Cape genet	Roberts *et al.* (2007), Mrubata (2018), Ziko (2018)
	Paradoxurus hermaphroditus	Common palm civet	Bartels (1964)

Resource Ownership

Latrines are often explained as 'advertisement of ownership' (i.e. the acquisition and defence) of one or several of four main resources. The most frequently defended resource is the territory *per se*; others are the resources that are generally contained within most all-purpose territories: food sources, sleeping and/or breeding sites, and mates. We will now review the evidence for the role of latrines in acquiring or defending each of these resources in turn.

Territories

Territories are relatively stable areas from which residents exclude intruders by some combination of advertisement, threat, and attack (Kaufmann, 1983). The traditional interpretation of latrines and scent-marks is thus that they form a kind of 'scent-fence', representing a 'keep-out' message against intruders (e.g. Hediger, 1949), akin to the suggested role of avian song (e.g. Krebs *et al.*, 1978). To our knowledge, such an effect, however, has not been demonstrated in carnivores and has only been suggested in two rodent species (North American beaver, *Castor canadensis*: Müller-Schwarze & Heckman, 1980; blind mole-rats, *Spalax ehrenbergi*: Zuri *et al.*, 1997), which contrasts markedly with numerous observations of territorial

intrusions by non-resident individuals from many carnivores (e.g. common dwarf mongoose, *Helogale parvula*: Rood, 1983; African lion, *Panthera leo*: McComb *et al.*, 1994).

The failure of the Scent-Fence Hypothesis, however, does not necessarily preclude any role of latrines in territory defence. There are many examples of boundary deposition of faeces in carnivores (e.g. golden jackal, *Canis aureus*: Macdonald, 1979; spotted hyena, *Crocuta crocuta*: Gorman & Mills, 1984), which would be expected intuitively if latrines are indeed a method of demarcating a territory (Johnson, 1973). Macdonald (1980), however, proposed that only group-living species can produce enough faeces to maintain border latrines. To illustrate his point, he gave many examples of social and solitary species that do not scent-mark along their territory borders but do so throughout their home range instead. Indeed, mammals seem to scent-mark throughout their territory where regular patrols and maintenance of border latrines are uneconomical (e.g. Gorman & Mills, 1984; Gorman, 1990). Gorman & Mills (1984) discussed this hypothesis within the Hyaenidae, and suggested that both inter- and intra-specific variations in latrine (and scent-mark) location occur in relation to economic and ecological constraints. For example, all three species of extant hyenas, as well as the closely related aard-wolf, *Proteles cristatus*, use latrine sites, and paste-mark grass stems with a substance excreted from the anal pouch (Sliwa & Richardson, 1998). Generally, those species with relatively large home ranges (e.g. brown hyena, *Parahyaena brunnea*: Mills *et al.*, 1980) scent-mark throughout the territory, whereas those in smaller ranges (e.g. spotted hyena: Kruuk, 1972) scent-mark the border. While border marking gives the potential intruder the earliest warning of transgres-sion, it involves only a single line of defence, which must be relatively continuous and well maintained to ensure detection. This requires the production of a large volume of faeces as well as regular patrols to dis-tribute it, which is not economically feasible where individuals would have to patrol long stretches of the border. Although all hyena species seem to fit this eco-nomically driven pattern of latrine distribution, intra-specific variation in spotted hyena latrine distribution perhaps provides the best illustration. In the Ngorongoro Crater (Tanzania), where an abundant food supply supports large groups in small territories, hyenas position most latrines along territorial borders (Kruuk, 1972). In contrast, small groups of spotted hyenas occupy large home ranges in the Kalahari Desert (South Africa/Botswana) where they adopt a 'hinterland' marking strategy, positioning latrines throughout their territory (Mills & Gorman, 1987).

In reassessing the function of scent-marks in territo-ries, Gosling (1982) proposed the Scent-Matching Hypothesis as an alternative mechanism of how latrines and scent-marks can aid in territorial defence: as territory owners have already made significant investments in the territory, it pays the owner more to defend the territory than it does the intruder to esca-late the conflict in a take-over bid (Maynard-Smith & Parker, 1976; Hammerstein, 1981; Gosling, 1982) as supposed by the Payoff Asymmetry Hypothesis (Dawkins & Krebs, 1978; Krebs, 1982). Thus, when intruders and owners meet, an asymmetry of contest is established. In this context, scent-marks can provide a reliable and honest signal of ownership because only long-term residents will have been able to mark throughout the territory. By comparing scent-marks encountered within the territory with the potential owner's scent, intruders can thus avoid confrontation (e.g. Gosling & McKay, 1990) through scent-matching (Gosling, 1982).

The hypothesis that scent-marks serve to familiarize individuals or provide psychological reassurance to residents, 'making him feel that he belongs in every quarter' (Stoddart, 1980), has been suggested by a vari-ety of authors (e.g. Kleiman, 1966; Seitz, 1969; Mykytowycz, 1970; Ralls, 1971; Ewer, 1973; Walther, 1978; Schilling, 1979). As Gosling (1982) realized, however, most conclusions of this nature result from a lack of supporting evidence for other hypotheses and are not usually based on convincing empirical sup-port. Nevertheless, it could be that as intrusions and encounters are more likely to occur in border regions of the territory in many – if not all – species, the con-centration of scent-marks in this area might serve to provide a home advantage by 'reassuring' the resident. Experimental evidence from European wild rabbits provides some support for this hypothesis, as male rabbits were dominant over others in the presence of their own scent in otherwise neutral arenas (Mykytowycz *et al.*, 1976). This 'resident wins' rule conforms to the predictions of the Scent-Matching Hypothesis (Gosling, 1982).

Nevertheless, animals do not defend territories for space alone but for any resources they may hold. As the motivation for territoriality may differ between the sexes (e.g. African lion: Pusey & Packer, 1997; spotted hyena: Boydston *et al.*, 2001), so may the motivation for latrine use. Falling under the umbrella of defend-able resources are food sources, sleeping and breeding sites, and mates, each of which we will consider in turn below.

Food

In some species, a strong spatial association of latrines with food resources has been recorded. For example, striped hyena, *Hyaena hyaena*, latrines occur close to feeding areas (Macdonald, 1978). Concentrations of faeces around fruiting trees have been described in the grey fox, *Urocyon cinereoargenteus* (Trapp, 1978), and in the greater hog badger, *Arctonyx collaris* (Zhou *et al.*, 2015a), while in some populations spotted hye-nas form temporary latrines close to large kills (Bearder & Randall, 1978). In many species, however, the prey or consumable vegetation is distributed uni-formly or cryptically, thus making it difficult for the human observer to demonstrate a connection. An alternative explanation for the close proximity of latrines to feeding sites is that they signal resource depletion. When individuals den together but forage individually or in small groups (e.g. spotted hyena; Ethiopian wolf, *Canis simensis*), latrine activity may signal resource depletion to the mutual benefit of all group members (*sensu* Eisenberg & Kleiman, 1972). Furthermore, faeces volume and consistency are likely to be honest signals of the type and richness of resources exploited (e.g. Walls *et al.*, 1989), as they vary considerably with diet (e.g. Zhou *et al.*, 2015a). Thus, faeces volume at latrine sites could effectively signal resource exploitation/depletion and maximize foraging efficiency for all group members (Passive Range Exclusion Hypothesis: Stewart *et al.*, 1997) including the marking individual itself (i.e. the Foraging Book-Keeping Hypothesis: Henry, 1977). Latrines of Eurasian otters, *Lutra lutra*, for example, are not associated with territorial boundaries, but instead are thought to function as a spacing mechanism for foraging individuals to maximize feeding efficiency within group territories (Kruuk, 1992). Nevertheless, as we will discuss in more detail later, the spatial association of latrines with particular resources is insufficient evidence of their intended defensive function. Evidence of this is that individuals may remain in the vicinity of abundant food sources over an extended period of time, resulting in the natural accumulation of faeces in these well-used areas without active signal intent.

Although it is possible that latrine use and scent-marking have different or multiple functions even within the same species, some evidence concordant with the resource depletion hypothesis exists from non-latrine scent-marking patterns: African palm civets, *Nandinia binotata*, scent-mark the trees from which they feed (Charles-Dominique, 1978), while red foxes, *Vulpes vulpes,* urine-mark depleted caches (Henry, 1977). In the latter case, it is possible that such marking could reduce the time invested in a subsequent investigation of these sites. However, since the visual effects of prior cache retrieval may be visible from a distance, at least during the day or on moon-lit nights (Macdonald, 1987), it may be unlikely that scent-marking signals (only) resource depletion. Instead, foxes may use the conspicuous sites of former caches to promote urine detection by conspecifics.

In perhaps the first study designed to assess the role of carnivore scats in defending a trophic resource,

Piñeiro & Barja (2015) attempted to determine whether faecal marks deposited by European wild cats, *Felis silvestris*, serve to defend rich food patches. The authors identified scats with a presumed communicative function (i.e. those located on conspicuous substrates, above ground level, at a crossroad or in a latrine) and showed that wild cats deposited faecal scent-marks most often where their main prey (small mammals) are more abundant; a result they interpreted as suggesting that wild cats defend favourable hunting areas. Nevertheless, the possibility remains that wild cats may simply spend more time in these rich food areas, and thus faecal deposition for other purposes, including purely for waste elimination, may therefore occur more frequently in these areas independent of the abundance of food. This, of course, is also a possible explanation for the observation of faeces concentrations close to feeding areas in grey foxes, greater hog badgers, striped hyenas, and spotted hyenas as described above.

Breeding and/or Sleeping Site(s)

Many species defecate predominantly in the vicinity of their sleeping and/or breeding sites: for example, scats of coastal Eurasian otters are deposited more than twice as often within 100 m of holts than elsewhere (Kruuk & Hewson, 1978). As the route into their holts, however, is determined by landing points along the water's edge, otters probably only need to mark these regions, as all other resources are under water. Similarly, latrines at burrow entrances are also reported in yellow mongooses, *Cynictis penicillata* (le Roux *et al.*, 2008).

Mate Acquisition/Defence

Strategies for maximizing reproductive success are sexually dimorphic in most mammals, with mates generally representing a more limiting resource for males than for females (Trivers, 1972; Clutton-Brock, 1988). Various authors have suggested that territoriality acts as a mechanism to deter kleptogamy, that is territorial males attempt to prevent neighbours from gaining reproductive access to resident females, not other physical resources (e.g. Lack, 1966; Wrangham, 1982; Roper *et al.*, 1986). Sillero-Zubiri & Macdonald (1998) suggested a similar hypothesis for the defence of mates in Ethiopian wolves, in which females seek copulations with males from neighbouring packs along territorial borders (Sillero-Zubiri *et al.*, 1996) and may engage in extra-territorial forays (Sillero-Zubiri & Gottelli, 1995). As resident females chase these intruders away but males do not, the authors suggest that these female 'floaters' might use the demographic information contained within scent-marking sites to determine whether a breeding position is available in neighbouring territories, although thus far, no such mechanism has been shown in any species.

Conclusion

Most studies interpreting latrines in the context of territorial defence rely on analyses of the spatial distribution of latrine sites, as they are persistent and often visually conspicuous, making them ideal targets for the study of population densities. The problem is, however, that many previous studies where selective positioning had been 'demonstrated' did not control adequately for the possibility that the study species utilized its home range non-randomly in relation to these features of importance (e.g. grey wolf, *Canis lupus*: Barja *et al.*, 2004). The generation of random control points is seldom sufficient, but rarely – if ever – are shortcomings of this nature acknowledged. The investigation of latrine function thus requires not only correlational analyses of latrine spatial and temporal distribution, but also detailed investigations of individual behaviour at – and responses to – latrine sites and perhaps even the specific deposits within them. In the following, we will review hypotheses formulated on the basis of studies emphasizing other attributes of latrine use, such as temporal patterns, individual-specific behaviour, and olfactory information content.

Information Centre Hypothesis and Reproductive Advertisement

Many species, especially amongst the Carnivora (Brown & Macdonald, 1985), establish composite latrines where animals scent-mark in addition to depositing faeces and urine. The olfactory information available at latrine sites is thus impressive: for example, all carnivores possess paired anal glands, which secrete into the rectum, coating faecal deposits with anal jelly (McColl, 1967), and, in some species, the

secretion has been shown to encode information such as sex (e.g. steppe polecat, *Mustela eversmanni*, and Siberian weasel, *Mustela sibirica*: Zhang *et al.*, 2003; brown bear, *Ursus arctos*: Rosell *et al.*, 2011), group membership (e.g. spotted hyena: Burgener *et al.*, 2008; Theis *et al.*, 2013), and individuality (e.g. giant panda, *Ailuropoda melanoleuca*: Zhang *et al.*, 2008). Scent-presentation experiments have shown that demographic information is discernible at least in some species through olfactory investigation, for example, group discrimination in raccoon dog, *Nyctereutes procyonoides* (Yamamoto, 1984); sex discrimination in domestic dogs, *Canis familiaris* (Dunbar, 1977); individual discrimination in small Indian mongooses, *Urva [= Herpestes] auropunctata* (Gorman, 1976), and spotted hyena (Burgener *et al.*, 2008). In addition, faeces and urine – as metabolic waste products – serve to excrete endocrinological metabolites, such as sex-steroid derivatives, which can be used to determine the animal's reproductive status (e.g. Schwarzenberger *et al.*, 1996). Thus, males of many species can detect female oestrus through investigation of faeces, often employing their vomeronasal organ in the characteristic flehmen response (Kinoshita *et al.*, 2009), such as observed in domestic cats, *Felis catus* (Verberne, 1976; Verberne & DeBoer, 1976). In addition, latrine activity often varies seasonally, generally peaking during the breeding season. For example, common genets, *Genetta genetta*, deposited more faeces at latrines during the peak mating season (Barrientos, 2006), and this pattern was repeated in the marsupial spotted-tailed quoll, though restricted to latrine sites along drainage lines in this species (Ruibal *et al.*, 2011). The mere presence and activity level of latrine sites could thus signal sexual receptivity (Buesching & Macdonald, 2001). Details of which individuals leave these scats are often lacking in such studies, but a radio-collared male Japanese badger, *Meles anakuma*, was recorded to make more frequent visits to neighbouring latrines during the mating season (Kaneko *et al.*, 2009).

Orientation

In many species, latrines are located at conspicuous landmarks, such as ditches, road- or bridge-crossings (e.g. grey wolf: Barja *et al.*, 2004; North American river otter, *Lontra canadensis*: Torgerson, 2014), or particularly big trees (e.g. common genet: Espírito-Santo *et al.*, 2007). Often they are connected with well-travelled paths leading from sleeping sites to foraging grounds or watering holes (e.g. review in Gorman & Trowbridge, 1989); an observation that has caused some authors to liken latrines to signposts along human roads.

Parasite Reduction

In primates (including humans), latrine use has long been suggested to reduce parasite load. Red howler monkeys, *Alouatta seniculus*, for example, use specific sites for defecation, which are characterized by areas free of underlying vegetation, which is interpreted as an adaptation to decrease the likelihood of contaminating potential food sources or arboreal pathways (Gilbert, 1997). Nevertheless, there is currently no evidence that latrine use is also associated with a reduction in parasite load in carnivores. In fact, latrines may have quite the opposite effect, as evidenced by a series of detailed studies on northern raccoons, *Procyon lotor*, with regard to their potential role in the spread of raccoon roundworm, *Baylisascaris procyonis*, which also affects humans (e.g. Page *et al.*, 1998; Logiudice, 2001). Data from proximity-logging collars at 15 latrine sites implicated raccoon latrines as major foci for the infection and spread of *B. procyonis* (Hirsch *et al.*, 2014).

Concealment of Presence to Other Species

Prey animals can avoid predators based on olfactory cues from scent-marks. For example, Mech (1977) demonstrated that prey species may intercept the territorial signs of the grey wolf and keep to the periphery of wolf territories. Similarly, European hedgehogs, *Erinaceus europaeus*, avoid areas scent-marked by European badgers (Ward *et al.*, 1997), and faecal odours of least weasels, *Mustela nivalis*, have been used successfully as olfactory rodent deterrents to protect crops (e.g. Borowski, 1998), whereas Tobin *et al.* (1995) showed that members of three species of wild rats, *Rattus* spp., avoided traps soiled by small Indian mongooses. These findings prompted some authors to explain the faecal covering (i.e. scraping soil over faeces) observed in some carnivores as concealment or anti-prey-detection behaviour (e.g. domestic cat: Feldman, 1994; African wild cat, *Felis lybica*: Estes, 1991; aardwolf: Kruuk & Sands, 1972). The suggestion, however, that aardwolves bury their faeces in middens to avoid detection by their prey (Kruuk & Sands, 1972) is rather unconvincing, as aardwolves feed almost exclusively on *Trinervitermes* termites (Bothma & Nel, 1980). In addition, it has been noted that domestic cats and European wild cats alike bury their faeces only in core areas, and that in all other areas of their range, they leave them in prominent locations (Corbett, 1979; Panaman, 1981; Macdonald, 1985).

Derived Predictions

Each of these four broad hypotheses results in a number of predictions (see Table 7.2), which can then be investigated in the field by collecting an array of data. While many of these hypotheses are not mutually exclusive, we have kept them separate in our effort to assemble a workable framework for the study of the function(s) of latrines generally (Table 7.2 and next section). We have identified the measurable parameters that are important in this context, and have attempted to produce predictions based on these parameters that will allow functional interpretations of latrine use to be drawn more effectively than at present.

Which Parameters Are Important in the Study of Latrines?

At least four categories of data are important in understanding the function(s) of latrines in carnivore societies and formulate an appropriate research framework: spatial distribution patterns, temporal usage patterns, individual visit/contribution patterns, and the information content of the signal (see also Table 7.2). In this section, we will review the available data for each of these categories in turn, with a focus on carnivores. Unfortunately, examples where data for all four categories are available are currently scarce.

Table 7.2 Hypotheses and predictions for the function of latrines.

Category	Hypothesis	Function	1. Spatial distribution		2. Temporal patterns		3. Individual contributions	4. Information content of signals		Predicted behaviour of interloper/receiver
			Broad	Local	Seasonal	Short-term		Chemistry	Discrimination	
1. Resource acquisition and defence	1.1. Territory acquisition/ defence	Defend/ acquire space	Locations to optimize intercepting interlopers (especially those threatening territory integrity). Primarily within territory. Dependent on costs of patrolling and reinforcing signals (e.g. border-marking only possible in small territories).	Locations to optimize intercepting intruders. Substrate and microclimatic site selection to maximize detection and longevity.	Peak when intrusion threat most intense. (Maybe year-round).	Temporally correlated with intruder encounters (especially those threatening territory integrity).	Primarily deposited by territory owners (except where involved in territory acquisition). Sex-dependent (where intrusion or territoriality is sex-biased). Status-dependent (elevated in individuals with more to gain by maintaining territory, e.g. breeding/dominant individuals).	Low volatility ensures long-lived signals (maximize longevity/ minimize distribution costs). Inclusion of anti-microbial components in secretion prolongs 'shelf-life' of scent-marks. Allow association between territory owner(s) and defended areas (in groups, signals should be group-specific; for individual territory owners, signals should be individual-specific).	Solitary: self vs. neighbour/ stranger. Social: own-group vs. neighbour/ stranger. Specific signal degradation pattern encodes information on signal age (time since deposition), to assess the ability of owner(s) to defend area or evaluate the likelihood of being caught intruding. Identification: matching scent of resident to scent of territory (e.g. group, pair, individual discrimination, depending on social system). Possible numerical assessment of territory holders/ group size.	Receivers avoid scent-marked areas/retreat on encountering scents (scent-fence) or modify their interaction with owner(s) when territory scent matches individual scent (scent-match). Neighbour–stranger discrimination in accordance with Dear Enemy/Nasty Neighbour Phenomenon. Over-/ countermarking may occur by intruders in competition.

(Continued)

Table 7.2 (Continued)

Category	Hypothesis	Function	1. Spatial distribution		2. Temporal patterns		3. Individual contributions	4. Information content of signals		Predicted behaviour of interloper/receiver
			Broad	Local	Seasonal	Short-term		Chemistry	Discrimination	
	1.2. Food acquisition/ defence/sharing	Defend/ acquire food	Clustered around large food patches (e.g. grasslands, ponds, rivers).	Clustered around food sources (e.g. fruiting trees or kill sites). In areas of valuable food sources.	Related to seasonal availability/ productivity of food patches. Elevated use when food competition is most intense. Elevated in season of low food availability (increases foraging efficiency).	Temporally associated with encounters with rivals at/ near food resource.	Marking perhaps more common by early arrivals to food (more to lose). Elevated use in presence/proximity of rivals. Can be used by individuals to claim ownership (e.g. food caches)	Individual-specific. Group-specific. Allows owner(s) of resource to be identified.	Self vs. other (group or individual).	Avoid marked areas or approach with heightened alertness. Reduced/absence of food sharing.
		Signal food depletion	More common in (previously) productive hunting/foraging areas. Accumulation as resource is increasingly exploited.	Prominent and on/in vicinity of (depleted) food resources/ caches.	More common in season of low food availability (as it reduces time invested in searching for depleted resources). More common when food is patchily distributed.	Following depletion of food resources at a particular site.	All users of resource expected to mark, especially those who were unsuccessful (i.e. did not feed) at a normally productive food patch. Possibly elevated marking just prior to leaving resource.	Signal age/time since deposition (to assess the likelihood of resource having been replenished).	Signal age.	Reduced visit or foraging/hunting activity in recently marked (i.e. depleted) areas.
		Attract conspecifics to divisible food source(s)	Elevated in areas used for foraging (e.g. edges of productive habitats). Locations maximizing scent-detection distance.	Near divisible resources such as large kill sites, fruiting trees, and other rich patches. Locations maximizing scent-detection distance.	Dependent on seasonal differences in the abundance of divisible food sources.	On arrival at divisible resources.	More likely in social groups with fission–fusion foraging.	Individual-specific: 'friend vs. foe'.	'Friend vs. foe' (familiar vs. unfamiliar individual?).	Recruited to food source. Reduced foraging/ feeding competition at marked food source.

	Avoid detection by potential prey	Located away from hunting areas.	Concealed (e.g. buried) ablutions. Dry sites that may not confer local (microclimatic) protection translating into short life-span of olfactory signal, or conversely thick understorey/hard-to-reach places.	May relate to seasonal differences in the abundance of potential prey.	Cessation/reduction of marking when hunting.	No variation.	Species-specific: generalized predator–prey differences. Low detectability, e.g. involatility.	Prey discriminate predator from non-predator species.	Prey show behaviours reducing risk of predation (e.g. repulsion/increased vigilance) when detecting and identifying predator latrines.
1.3. Breeding/sleeping site defence	Acquire/defend breeding/sleeping site	Clustered around breeding or sleeping sites.	Clustered in immediate vicinity of breeding or sleeping sites (e.g. at burrow entrances, at the basis of sleeping trees, etc.).	Elevated use when competition for sites is most intense (breeding season).		Sleeping sites: all individuals likely to contribute. Breeding sites: predominantly females with dependent young (and potentially the father depending on breeding system).	Individual-specific: self vs. other. Allows owner(s) of resource to be identified.	Self vs. other.	Avoid marked sites completely in solitary breeding/sleeping species; otherwise 'friends' and relatives/mating partner(s) might be attracted; 'foes' repelled.
1.4. Mate acquisition/defence	Acquire mate(s)	Maximize likelihood of intercepting potential mates (when mates are within the same group, may be equally distributed throughout range). Clustered around breeding grounds or at borders.	Maximize likelihood of intercepting potential mates or reproductive rivals (e.g. trails near mates or their foraging areas).	Higher rates in breeding season (or periods when pairs are formed). Over-/counter-marking of potential mate or rival.	Higher rates in presence of potential mate or rival. Higher rates in presence of signals of potential mate or rival. Over-/counter-marking of potential mate or rival.	In presence of potential mate or rival. Over-/countermarking of potential mate/rival.	Sex differences. Sexual status/fitness-related parameters. Individual identity. Relative mark (top vs. lower) position. Kinship/quality. Pair-specific. Relatedness.	Sex. Sexual status, pair-bond. Individual identity. Association of scents of pair with particular pair. Pair-bond strength? Relative (top vs. lower) position of scent-marks. Kinship/quality. Pair-specificity. Related vs. unrelated.	Prefer potential mates who mark more. Sex-specific reaction depending on physiological/reproductive characteristics of receiver. Reaction to scent varies with information content (i.e. individual characteristics of the marker). Respond by mating with individuals that overmark. Prefer potential mates who are successful in overmarking the scents of other potential mates. Prefer unrelated and high-quality mates, but may prefer related helpers.

(*Continued*)

Table 7.2 (Continued)

Category	Hypothesis	Function	1. Spatial distribution		2. Temporal patterns		3. Individual contributions	4. Information content of signals		Predicted behaviour of interloper/receiver
			Broad	Local	Seasonal	Short-term		Chemistry	Discrimination	
		Defend mate(s)	Maximize likelihood of intercepting potential mates or reproductive rivals (when rivals are within the same group, may be equally distributed throughout range).	Maximize likelihood of intercepting potential mates or reproductive rivals (e.g. near breeding grounds or foraging areas). Allo-marking mate. In locations marked by mate or rival.	Higher rates in breeding season (or periods when pairs are threatened).	Higher rates in presence of potential mate or rival. Higher rates in presence of signals of potential mate or rival. Higher rates when mate is receptive.	Overmarking/ competitive countermarking (may be of same sex, for evaluation by female, or of mate for disguising her presence/ advertising 'ownership').	Sex differences. Potential individual differences.	Association of scents of pair with particular pair. Sex. Individual.	Reduced competition for mate in presence of its mate. Attraction to unpaired signals; (relative) repulsion from paired signals.
		Suppress rivals	Maximize likelihood of intercepting potential reproductive rivals (when rivals are within the same group, may be equally distributed throughout range).	Maximize likelihood of intercepting potential reproductive rivals (when rivals are within the same group, may be equally distributed throughout range).	Higher rates in breeding season (or periods when pairs are threatened).	Higher rates in presence of rival or scent-marks of rival.	Over-/ countermarking of potential mate or rival.	Sex differences. Individual identity. Dominance status. Reproductive condition.	Sex. Individual. Dominance status. Reproductive condition.	Undergo reproductive suppression.
2. Other	2.1. Parasite reduction	Reduce parasite load	Avoid areas of high use.	Avoid sleeping and feeding sites. Choose sites that may not confer local (microclimatic) protection.	Related to the life-cycle of parasites.	Elevated at times of high parasite risk.	Used by all. No investigation/ contact with existing scents (especially faeces) at sites.			Contact increases parasite transmission, so avoid contact with latrines.

2.2. Predator avoidance/deterrence	Avoid/deter predator(s)	Risk specific. Greater in areas with elevated predator abundance.	Avoidance: avoid marking areas where predators are present. Potentially conceal ablutions. Strong clustering. Deterrence: elevated in areas where predators are present; latrines very obvious.	Peaks of deterrent marking expected when the threat of predation is highest. Peaks of eavesdropped marking expected when threat of predation is lowest.	Peaks of avoidance or deterrent marking expected when a predator has been detected.	Deterrent scent-marks most common in individuals victim of predation attempts.	Potentially a signal demonstrating that predator has been seen. Potentially a signal suggesting unsuitable or dangerous prey.	Predator interest in prey scent. Predators use scent to locate prey. Reduced predation due to aversion/ overpowering scent.
2.3. Orientation/ familiarization	Navigate	Throughout home range and beyond.	Along trails (or perhaps when away from used trails?).	Elevated in unfamiliar areas.	Elevated in individuals establishing range.	Revisits by depositing individual.	Individual identity. Possibly also signal age.	Individual identity (at least self vs. other). Signal age, potentially travel direction, potentially 'stress level' (i.e. to warn itself of danger?).

Spatial Distribution Patterns of Latrines

Due to the practical difficulties of identifying and observing the behaviour of animals in their natural environment, past field studies most often used remote sampling methods and have thus focused on the spatial distribution of scents in the environment. As scats tend to be more conspicuous than glandular secretions, the majority of studies are therefore biased toward the distribution of faecal samples and latrines (see Macdonald, 1980). Traditionally, the distribution of latrines and their placement in the environment have therefore been mostly investigated in the context of territorial demarcation (Gorman, 1984; e.g. hyenas: Gorman & Mills, 1984; Boydston *et al.*, 2001; Ethiopian wolf: Sillero-Zubiri & Macdonald, 1998). Increasingly, however, studies also relate latrines to other resources (especially food, but also denning sites) and environmental landmarks (Gorman & Trowbridge, 1989). For example, latrines may be used to aid spatial memory in order to optimize foraging efficiency (e.g. Eurasian otter: Kruuk, 1992; Remonti *et al.*, 2011; Almeida *et al.*, 2012); to stake a claim on access to temporally variable resources (e.g. greater hog badger: Zhou *et al.*, 2015a), especially by females to raise young (Gosling,

1986; Mertl-Millhollen, 2006); or to signal the local depletion of resources *sensu* the Foraging Book-Keeping Hypothesis (e.g. red fox: Henry, 1977; greater hog badger: Zhou *et al.*, 2015b).

However, this historical reliance on the spatial distribution of latrines in assessments of their function(s) (and that of scent-marking, more generally) can be problematic. Firstly, the location of scent-marks does not automatically indicate function, as the scent-marking strategy adopted by a species will be affected by economic constraints. Therefore, in actuality, only individuals with small home ranges, or groups with a large number of individuals, may be able to produce enough scent/faeces to demarcate their territorial boundaries effectively (Macdonald, 1980; Gorman & Mills, 1984; Gorman, 1990; Stewart *et al.*, 2001). While the location of scent-marks may provide important clues as to the intended recipients of the signal (Gosling & Roberts, 2001), the spatial location of scents within the environment may be relatively unimportant (Gosling, 1982). Secondly, when making functional interpretations based on the location of scent-marks alone, the signal content is commonly ignored. Figure 7.1 highlights the fallacy of assigning a function based

Figure 7.1 Two signals located on land borders with quite different meanings. While both might be interpreted as territorial signals based on their location alone, information on signal content indicates that only the left one is strictly territorial. The sign on the right, while advertising ownership, actively welcomes individuals. Without knowledge of the signal's information content, the spatial distribution alone does not reliably indicate its function.
Source: Photos © Neil R. Jordan.

entirely on the spatial distribution of signals. Without knowledge of their information content, functional interpretations of latrines on the basis of their location are over-ambitious. Thus, functional investigations of latrines are most likely to succeed where they take into account not only latrine spatial distribution, but also additional sources of data.

Temporal Variation of Latrine Use

Few studies other than those on our focal species (see below) concentrate solely on variations in temporal patterns of latrine use, but there are some that incorporate a time component into spatial distribution surveys. Most of these studies relate to elevated latrine use in the mating season (e.g. common genet: Barrientos, 2006) and are described above. Thus, the mere presence and activity level of latrine sites could signal sexual receptivity (Buesching & Macdonald, 2001). Other studies relate latrine use to seasonality in food resources (e.g. European badger: Pigozzi, 1990; greater hog badger: Zhou *et al.*, 2015a,b), while some report seasonal variation of latrine usage patterns in different habitat types (e.g. Almeida *et al.*, 2012). Many species (e.g. greater hog badger: Zhou *et al.*, 2015b) scent-mark most when resources are either scarce or energetically expensive to acquire (such as earthworms or insect larvae that need to be dug out from the ground), and least when food is most abundant or easily available (such as ripe fruit on the ground). Importantly, temporal patterns of use may also depend on the spatial location of latrines. For example, while spotted-tailed quoll latrines in drainage lines contained more faeces during the mating season, outcrop latrines were most utilized when females were nursing young (Ruibal *et al.*, 2011). This interaction between spatial and temporal factors in latrine use emphasizes the importance of considering data from multiple categories.

The paucity of data on temporal patterns of latrine use relative to spatial distribution data, however, may be explained at least partly by the intractability of many species and populations. Nevertheless, increasingly smaller high-resolution GPS collars will become invaluable in determining temporal patterns of latrine use in small carnivores that are difficult to track with conventional VHF transmitters.

Inter-Individual Differences in Latrine Use

Individual behaviour associated with latrine use is difficult to study and requires not only comprehensive mapping of all latrines within an individual/group territory, but also reliable tracking and observational data for each individual within the territory/group. Given the widespread use of latrines among the Carnivora, relatively little is known about individual-specific behaviour at these sites. In this context, the use of camera-traps opens new avenues to record data on the behaviour of individual animals at latrines. This technology may be particularly applicable for species with individually distinctive pelages, or where individuals can be marked. Additionally, genetic analysis of faeces deposited at latrine sites could be used to determine the sex or identity of individuals using latrines (e.g. Ruibal *et al.*, 2011). However, not all individuals visiting latrines actively deposit faeces at these sites, and genetic analyses alone are therefore likely to underestimate the communicatory importance of latrine sites.

Olfactory Information Content of Latrines

Although latrines serve as olfactory signal stations, the specific information content available to conspecifics at composite latrines has not been decoded completely for any species. Nevertheless, it has been shown that faecal deposits as well as urine can contain information about the reproductive status, age, and dominance status of the marking individual (review in Brown & Macdonald, 1985). In addition to their anal glands described above, some carnivore species possess specialized skin glands (e.g. subcaudal gland of Eurasian badgers *Meles* spp. and hog badgers *Arctonyx* spp.; genel glands of felids) or scratch-mark latrine sites with their claws, probably depositing secretions from their inter-digital glands in the process (e.g. lions and tigers, *Panthera tigris*: Barja & de Miguel, 2010). The olfactory information potentially encoded in these scent types, however, remains largely unresolved, with the exception of the subcaudal gland secretion of European badgers (see below).

As scent-signals degrade over time, scent-marks may also contain information about the approximate time the animal marked/visited the area, and can thus be used as a signal advertising where an individual can be found in space and time. In the context of resource defence, this degradation can be a disadvantage for the signaller as it has to invest time and energy into reinforcing its marks regularly to assert its continued ownership of the resource(s). For the receiver (e.g. the potential intruder into a territory), on the other hand, this time stamp is a definite advantage as it can use this information to judge the risk associated with its 'invasion' (e.g. scent-marks are old, so the territory owner is likely not in this area; or scent-marks are not enforced regularly, so the owner is 'weak') and thus aid in decision-making (e.g. in the foraging context; Zhou *et al.*, 2015a,b).

In the context of reproductive advertisement, on the other hand, information about the age of the scent-mark is beneficial to both males and females. For example, an oestrous female might scent-mark to attract mating partners, but if males keep trying to find and mate with her after her oestrus passed or she has been mated, both parties would expend time and energy unnecessarily from the outdated information. Information on reproductive status (e.g. grey wolf: Raymer *et al.*, 1986) and sex (e.g. steppe polecat and Siberian weasel: Zhang *et al.*, 2003) is available in scents. Many species that use latrines have evolved the ability to determine both reproductive status and sex from scents (e.g. aardwolf: Sliwa & Richardson, 1998).

Unfortunately, for many species, the data available often belong to only one or two of the categories listed above. As a consequence, the interpretation of latrine function in such species is based on limited data and might thus have to be revised if and when data belonging to a different category become available. To test and/or eliminate conclusively any of the specific hypotheses on the function of latrines, however, is possible only if comprehensive data belonging to several of the categories listed in this section are available. In the remainder of this review, we will therefore concentrate on three species as models (European badger, meerkat, and banded mongoose), for which we have detailed data for at least three of the four categories above, to evaluate the hypotheses on the function of latrines outlined in Table 7.2. We will review the different techniques used to collect these data, discuss the limitations of spatial data alone, and highlight the value of a combined approach.

Case Studies on Badgers and Mongooses

The European Badger

The latrine system of European badgers has been studied extensively since the 1970s (Kruuk, 1978). Badgers deposit their faeces in shallow pits (~10–20 cm in diameter and ~5–30 cm in depth), of which several hundred can be aggregated in the same latrine covering up to 400 m^2 (Tuyttens *et al.*, 2001). Latrines are most active in spring, coinciding with a peak in mating activity as well as the cub-rearing season (see Roper, 2010). Each pit can contain one or several faeces (up to several hundred: Stewart *et al.*, 2001) as well as urine and anal gland secretion (Buesching & Macdonald, 2001). In addition, badgers have a specialized skin gland, the subcaudal gland, which they use for squat-marking latrines alongside other objects (Buesching & Macdonald, 2004) and conspecifics (allo-marking: Buesching *et al.*, 2003).

Bait-marking is employed to study latrine use in badgers (Delahay *et al.*, 2000). In this technique, peanuts and treacle are mixed with different coloured indigestible plastic beads (Figure 7.2). If each social group is fed with a different colour, subsequent surveys can thus reveal group- or sett-specific latrine use patterns (see maps in Figure 7.3).

As a variant of this technique, some studies used feeders designed to bait individuals (Stewart *et al.*, 2001; Kilshaw *et al.*, 2009) and thus determine

Figure 7.2 Bait-marking mixture for European badgers, *Meles meles*. *Source:* Photos © Christina D. Buesching.

(a)

(b)

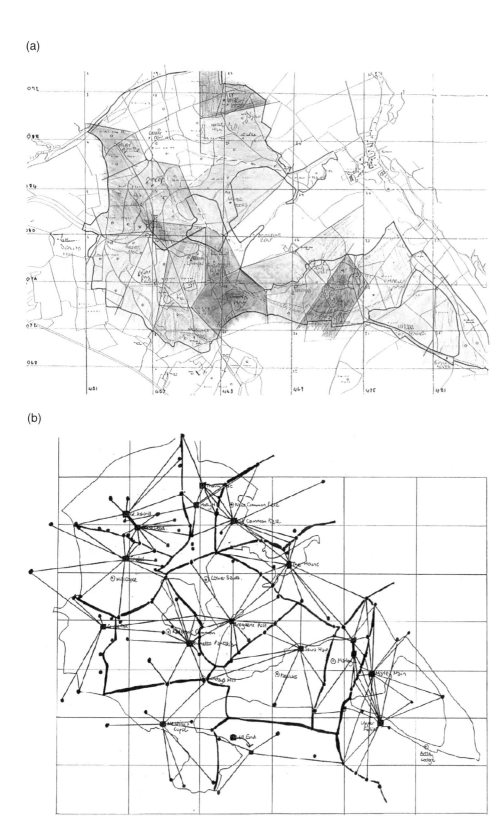

Figure 7.3 Map of Wytham Woods (Oxfordshire, UK) depicting (a) bait returns after bait-marking (i.e. feeding different coloured plastic beads to each social group) and (b) latrines (circles) visited by each social group of European badgers resident at different main setts (squares).

inter-individual variation in latrine use. Bait-marking reveals two types of badger latrines: *hinterland latrines*, which are situated in the interior of the group range and are used exclusively by members of the resident social group, and *border latrines*, which are situated along the perimeter of the group range and frequented by members of all neighbouring groups sharing this border (see review in Roper, 2010). In addition, individual bait-marking evidences that hinterland latrines are used predominantly by females and cubs, whereas males defecate almost exclusively in border latrines (Stewart *et al.*, 2001; Kilshaw *et al.*, 2009; Roper, 2010). Furthermore, individuals appear to frequent the same latrines, and thus mark the same stretch of border, irrespective of their current feeding activity (Kilshaw *et al.*, 2009). Border latrines are situated along well-travelled paths encircling the group range, which are most obvious close to the sett (i.e. the social group's burrow system), but become inconspicuous or disappear altogether in feeding areas (e.g. Loureiro *et al.*, 2007; C.D. Buesching & C. Newman, unpublished data). If plotted on a map, all latrines are spaced regularly, but are situated closer together in the vicinity of the sett (Roper *et al.*, 1986; Buesching *et al.*, 2016), often taking advantage of conspicuous landmarks (e.g. big trees, ditch crossing, road crossing, fence lines: Roper *et al.*, 1986; Stewart *et al.*, 2002).

Video observations show that badgers investigate latrines frequently and intensively by sniffing (Stewart *et al.*, 2002). In scent-playback experiments, faeces (Palphramand & White, 2007) as well as anal gland (Tinnesand *et al.*, 2015) and subcaudal gland secretions (Buesching & Macdonald, 2004; Palphramand & White, 2007) from strangers presented at the sett or at border latrines elicit significantly stronger investigative behaviour and overmarking (where a scent is placed on top of an existing mark) than samples from own-group members or neighbours. Neighbour samples presented at unexpected locations (e.g. at an unshared border latrine) elicit stronger responses than samples presented in an expected context (e.g. at a shared border latrine between the two neighbouring group ranges: Tinnesand *et al.*, 2015; Figure 7.4). Samples from females in oestrus, however, elicit a strong response from all adult males (Tinnesand *et al.*, 2015).

Figure 7.4 European badger with an individually recognizable fur-clip mark defecating and sniffing the ground in response to a transposed anal gland secretion sample from a stranger at a latrine (remote camera set to provide time stamp and picture number on each photo). *Source:* Photos © Helga Veronica Tinnesand.

In chemical analyses using gas chromatography–mass spectrometry (GCMS) analyses, anal- and subcaudal gland secretions have been shown to encode group membership as well as individuality, and to vary with sex, reproductive status, age, and other fitness-related parameters (Buesching *et al.*, 2002a,b,c; Tinnesand *et al.*, 2015; Noonan *et al.*, 2019). The behaviour observed in scent presentation experiments confirms that the differences in the chemical composition of anal- and subcaudal gland profiles are indeed biologically relevant and can be decoded by badgers. In addition, subcaudal gland secretions also decay according to a specific pattern, thus encoding a time component about the age of the scent-mark (Buesching *et al.*, 2002c).

The above-mentioned studies pertaining to aspects of latrine location, usage patterns, and information content have led different authors to interpret their function in different ways, which we will review and discuss here.

Territoriality

Traditionally, badgers have been considered to be mostly territorial (see review in Roper, 2010) and thus many authors interpret their latrine marking system in this context. Border latrines would form a 'first line of defence' with hinterland latrines marking the interior of the territory (Roper *et al.*, 1986). The regular distribution pattern of both latrine types would serve to increase chances that intruders happen across one or several of these scent-marks to enable them to rec-ognize the territory owner(s) through scent-matching (*sensu* Gosling, 1982). In support of this hypothesis (Macdonald, 1980), the use of border latrines in badg-ers is evident mostly in high-density populations, such as described for parts of England (e.g. Gloucestershire: Delahay *et al.*, 2000; Oxfordshire: Macdonald *et al.*, 2008), whereas in mainland Europe, badgers occur at much lower densities, do not form large social groups (Rosalino *et al.*, 2004; Do Linh San *et al.*, 2007a,b), and their latrines are mainly located in the territory hinter-land (Pigozzi, 1990; for a review, see Roper, 2010). In addition, the behavioural reactions of resident badgers to translocated faecal, anal- and subcaudal gland sam-ples observed in scent-playback experiments indicate that badgers show a degree of socio-spatial awareness, conforming to the Dear Enemy Phenomenon (DEP; Fisher, 1954), which predicts that territorial species should get used to the scent of their neighbours, but react more strongly to the scent of unfamiliar (i.e. potentially dispersing) individuals. As badgers also react more strongly to neighbour scent provided at unshared compared to shared borders (Palphramand & White, 2007; Tinnesand *et al.*, 2015), they appear to moderate their response according to the perceived level of threat of the supposed marker (Threat-Level Hypothesis: Temeles, 1994).

Food

Territoriality in badgers is traditionally explained on the basis of the patchy distribution of their main food sources (Resource Dispersion Hypothesis; for a review, see Macdonald & Johnson, 2015). The observation that badgers tend to deposit faeces in latrines immediately prior, during, and after feeding bouts (e.g. Pigozzi, 1992), Kruuk (1992; see also Stewart *et al.*, 2001) sug-gests that faeces volume at latrine sites effectively sig-nal resource depletion, and maximize foraging efficiency for all group members including the marker (*sensu* the Foraging Book-Keeping Hypothesis: Henry, 1977). Stewart *et al.* (1997) suggested that latrines might be situated along the food isopleth (i.e. the line of highest food abundance) between badger setts, after which foraging becomes less profitable. However, as badgers are likely to be important seed dispersers of their food plants (Pigozzi, 1992), the existence of food isopleths along latrine-marked borders could also be a

direct result of their scent-marking habits due to the increased growth of food plants in faecal deposits. Furthermore, faeces volume and consistency are likely to be honest signals of the type and richness of resources exploited (Buesching & Macdonald, 2001) as they vary considerably in appearance and consistency according to diet (Kruuk, 1989).

Contradicting this hypothesis, however, is the fact that latrines are scarce or absent in feeding areas (Roper *et al.*, 1986). Their function in food defence is, therefore, unlikely. Furthermore, the sometimes severe bite-wounds observed in high-density popu-lations, which are explained by some authors as a result of active territorial disputes, do not coincide with peaks in food competition (Delahay *et al.*, 2006). There is also no evidence of dominance hier-archy at feeding sites, either within or between groups, and actual fights are generally avoided (Macdonald *et al.*, 2002).

Dens

Good setts as breeding and resting sites are an impor-tant resource to badgers (e.g. Kaneko *et al.*, 2010) and suitable sett sites can be a limiting resource (Macdonald *et al.*, 2004). Male badgers are thought to have a greater investment in defending breeding setts than do females (Roper, 1992) and this suggestion is supported by Stewart *et al.*'s (1999) finding that males of higher sta-tus (i.e. large, mature, frequently copulating individu-als) put more effort into sett maintenance, such as digging and enlargement, than do females or males of lower status. These observations have led to the sug-gestion that investment in setts might encourage breeding females to forgo dispersal and/or benefit the survivorship of sired litters (Stewart *et al.*, 1999; Kaneko *et al.*, 2010). The observation that latrines are situated closer together in the vicinity of the sett, pos-sibly increasing the chances for intruders to happen across them (Roper *et al.*, 1986; Buesching *et al.*, 2016), supports this hypothesis. Frequent object marking, as well as the use of sett latrines by pregnant females dur-ing the breeding season, further indicates a potential role of latrines in advertising a commitment to defend sleeping or breeding sites (see Buesching & Macdonald, 2004). Nevertheless, trapping records indicate that, although long-term dispersal is rare in high-density populations, overnight visits and short-term excursions to other setts are frequent (Macdonald *et al.*, 2008), contradicting the hypothesis that setts are a latrine-defended resource.

Mates

Sex- and seasonal-biased differences in the use of boundary latrines by European badgers are interpreted as partially demonstrating that they function in mate defence. This is done by deterring males from entering occupied territories for mating purposes (Roper *et al.*, 1993; Stewart *et al.*, 2002). Territorial defence in badgers, in the form of overt aggression and latrine use, shows a seasonal peak in early spring, which coincides with peak mating activity (Neal, 1977; Kruuk, 1978; Roper *et al.*, 1986; Buesching & Macdonald, 2004) and there is evidence for olfactory mate-guarding (Buesching *et al.*, 2003).

In addition, badgers have a promiscuous mating system (Dugdale *et al.*, 2011), and extra-group paternity is common (Carpenter *et al.*, 2005; Dugdale *et al.*, 2007; Annavi *et al.*, 2014), while intra- (as well as inter-) sexual aggression during the mating season is low, which does not correlate with peaks in bite-wounding (Delahay *et al.*, 2006). The strength of any correlation between latrine use and mate defence is not known either (Roper *et al.*, 1986).

Information Centre

While olfactory communication can be unrelated to territoriality, the territorial affiliation of the depositor may nonetheless be inferred from the location of the mark. Anal- and sub-caudal gland secretions deposited at latrines have been shown to contain individual- and group-specific information related to the fitness of the marking individual (Gorman *et al.*, 1984; Davies *et al.*, 1988; Buesching *et al.*, 2002a,b,c; Tinnesand *et al.*, 2015). As latrines are sited preferentially in areas of high badger activity (i.e. in the vicinity of the sett), rather than evenly around the territory circumference, and they can encompass several hundred faeces (see above) and are obvious and easily detectable, the chances that conspecifics happen across them are maximized. Thus, they are ideally suited not only as centres for information exchange between members of the same, and of neighbouring social groups (Buesching & Macdonald, 2004), but also as 'notes to self'. Badgers are known to spend a lot of time in the vicinity of latrines (Tinnesand *et al.*, 2015), sniffing faecal deposits from conspecifics (Stewart *et al.*, 2002; Palphramand & White, 2007) as well as glandular scent-marks (Buesching & Macdonald, 2004; Tinnesand *et al.*, 2015). Overmarking of scent-marks, particularly from unfamiliar or reproductively active individuals (Kruuk *et al.*, 1984; Buesching & Macdonald, 2004), but also of faeces (Delahay *et al.*, 2000) is frequent. Behavioural experiments (Bodin *et al.*, 2006; Tinnesand *et al.*, 2015) showed that badgers have relatively high levels of socio-spatial awareness and use olfactory cues to aid in navigation.

Parasites

Badgers are host to a variety of gut parasites, which can be transmitted through faecal sniffing or ingestion. In Oxfordshire, coccidiosis infection, leading to impaired growth and increased mortality, is prevalent in 100% of cubs under the age of about seven months, while the greater-than-expected prevalence of co-infection with *Eimeria melis* and *Isospora melis* is consistent with a common source of infection, such as communal latrines (Anwar *et al.*, 2000; Newman *et al.*, 2001). In Portugal, over 62% of faecal samples are infected with one or several parasitic helminth species, which are most likely transmitted at latrines (Rosalino *et al.*, 2006). In the UK, badgers are implicated in the spread of bovine tuberculosis (see review in Carter *et al.*, 2007). Their socio-spatial organization and the distribution of their setts – and especially latrines – are likely to increase transmission rates and thus facilitate the intra-specific spread of this disease (Böhm *et al.*, 2008). As badgers sniff latrines (including faeces) intensively, often using their vomeronasal organ and/or licking faecal matter (Stewart *et al.*, 2002; Tinnesand *et al.*, 2015), the hypothesis that latrine use in badgers could reduce the spread of parasites and/or diseases is unlikely.

Conclusion

In conclusion, as more data from fine-scale movement analyses, often achieved with modern technology (e.g. from proximity loggers and/or GPS collars; Drewe *et al.*, 2012; Noonan *et al.*, 2014, 2015), become available, evidence is mounting that latrine lines do not represent strict 'keep out' signs akin to scent fences, but that badgers can cross these lines freely, thus trespassing into each other's territories as well as setts (Macdonald *et al.*, 2008). While a singular, independent function of latrines appears impossible to determine, and influential components of the complete function appear to vary in their relative significance in context, a purely defensive function of latrines in badgers, as traditionally assumed, seems increasingly unlikely as it fails to fully take into account the olfactory information content conveyed by latrines. Because badger latrines are situated along well-travelled paths from the sett to the boundary and along the boundary to major feeding areas (e.g. pastures with an abundance of earthworms), the accumulation of faeces in these latrines could result from badgers spending more time in these areas, and would also increase their likelihood of detection. Particularly, as latrines are often placed at conspicuous landmarks, a role in orientation (Benhamou, 1989), either as a note to self or to others, appears likely (Buesching & Macdonald, 2001).

The Meerkat

In meerkats, latrines are defined as containing at least two faeces within 1 m of each other, although typically 100 or more faeces occur in an area of 0.5–6 m². Faeces are usually deposited in specially dug pits which are ~3 cm in diameter and ~1–4 cm in depth (Jordan *et al.*, 2007) and latrines are most active during the peak breeding season (Jordan *et al.*, 2007). Each pit can contain one or several faeces, as well as urine. The surrounding bushes and short vegetation can be marked with anal gland secretions (Jordan *et al.*, 2007), especially by the dominant male of each group (Jordan *et al.*, 2007). In addition to latrine-marking, meerkats use their anal glands to mark conspecifics within their own group, particularly during inter-group encounters (i.e. allo-marking: Buesching *et al.*, 2003; Jordan, 2005).

Although latrine use by wild meerkat groups has been studied year-round, our knowledge of their spatial context is limited (e.g. Jordan *et al.*, 2007). While meerkats are distributed over much of the arid part of southern Africa, their scent-marking behaviour has only been studied in detail in one area of recovering ranchland in the Northern Cape of South Africa's southern Kalahari Desert (Jordan, 2007; Jordan *et al.*, 2007).

In this environment, meerkats deposit faeces on the surface and in shallow pits which they excavate themselves, and of which many tens can be aggregated in the same latrine. Latrine sites are located significantly closer to refuges (bolt-holes) than to random points within their ranges, and are associated with vegetation that may provide a protective function, and which likely increases the longevity of any signals contained within (Jordan *et al.*, 2007). Each group of meerkats usually shares one latrine with each of their known neighbouring groups, which probably allows efficient inter-group monitoring of surrounding land tenure, probably via faeces-matching as described in badgers (Stewart *et al.*, 2001). The remaining latrines, however, are concentrated primarily in territorial core regions comparable to hinterland latrines in badgers (Jordan *et al.*, 2007).

While bait-marking has been employed to study latrine use in badgers, meerkat scent research has relied on direct observations to determine broad-scale spatial distribution, temporal visit patterns and to investigate inter-individual variation in latrine use (Figure 7.5). This approach has shown that meerkats investigate latrines frequently and that they are more likely to overmark scent-marks deposited by opposite-sex than same-sex individuals (Jordan, 2007).

Chemical analyses of the main constituent of meerkat latrines – faeces – have yet to be undertaken, but

Figure 7.5 An individually marked and recognizable meerkat, *Suricata suricatta*, defecating in a pit at a latrine site. Wooden skewers were added by researchers to demarcate the locations of known faeces. *Source:* Photos © Neil R. Jordan.

GCMS analyses on meerkat anal gland secretions have shown that the chemical composition varies by social status (Fenkes, 2011). Since only dominant males routinely contribute anal gland secretions during latrine visits, the chemical composition of these scents is of most direct importance in this context. As in badgers, analyses suggest that anal gland secretions contain information encoding group membership, sex, reproductive status, age, and perhaps individuality (Fenkes, 2011). No research has yet been conducted on meerkat scent decay/longevity. We will now review the possible function of latrines in regard to the following six resources/hypotheses.

Territoriality

Meerkat groups with known overlapping home ranges share at least one latrine with their neighbours (Jordan *et al.*, 2007) and thus the spatial and temporal distribution of their latrines are highly likely to play a role in inter-group communication. Despite the fact that latrines are no more likely to be placed in border regions vs. core areas of the home ranges, Jordan *et al.* (2007) concluded that latrines may play a role in territory defence. As described above within the Hyaenidae (Gorman & Mills, 1984), this is because in species where home ranges have long borders and animals travel as a group, it is not economical to effectively delineate the entire border. Intruders, therefore, are likely to slip through the olfactory net. By placing scents in the vicinity of particular landscape features, like refuge holes and vegetation cover, between which intruding meerkats often commute (Jordan *et al.*, 2007) and probably know the location of (Manser & Bell, 2004), resident groups likely maximize the likelihood of intruders encountering their scent. Through the mechanism of scent-matching (Gosling & McKay, 1990), intruders may be able to match the scent of residents with any meerkats they encounter in the area, which establishes a mismatch in the subsequent potential benefits of conflicts, since residents have invested more in defending the area and, as a result, have more to lose than intruders have to gain (*sensu* scent-matching: Gosling, 1982). Thus, both the broad and the localized distribution patterns of meerkat latrines might have evolved to increase the chances that intruders encounter the scents of territory owner(s), and a non-boundary-biased distribution may not necessarily be indicative of the target receiver(s) being from within the group or resident within the territory.

Food

Unfortunately, latrine distribution has not yet been investigated in relation to food abundance and utilization in meerkats. However, the seasonal patterns of latrine use may be correlated positively with the abundance of food. Latrine visit is highest in the breeding season (October–April; Jordan *et al.*, 2007) which coincides with increased food abundance, suggesting that latrine use is not related to the defence of food. The fact that latrine visit rates correlate with encounters with intruding males (see below) further suggests that this seasonal effect is more likely to be attributable to mate defence rather than food defence. Although the possibility exists for latrine use to play a role in communicating the depletion of food sources, this hypothesis has neither been tested explicitly for meerkats nor is it suggested or suspected here.

Dens

It has been suggested, though not empirically shown, that dominant meerkats scent-mark mostly around the communal dens and bolt-holes in a group's territory (Fenkes, 2011). However, it does not seem that these scent-marks, or latrines, are very effective in den site defence. Indeed, during the breeding season, prospecting male meerkats sometimes remain at the den of the group they have been following until late in the evening, and creep down a nearby bolt-hole to begin their foray again the next morning (N.R. Jordan, personal observation).

In addition to burrows used as breeding den sites, meerkats overnight in underground burrows, and forage in the vicinity of bolt-holes, and residents apparently know the location of the closest one throughout their range (Manser & Bell, 2004). Clearly, bolt-holes are an important resource throughout meerkat territories, and so the selective position of latrines close to these features (Jordan *et al.*, 2007) may play a role in their defence or advertising their presence. However, Manser & Bell (2004) covered bolt-holes with rubber car floormats and placed olfactory cues at sites where no bolt-hole existed before. Upon playback of meerkat alarm calls through a loudspeaker, the meerkats ran to the covered sites and ignored the scents, suggesting that scents were not used to signal the location of the bolt-holes (Manser & Bell, 2004). Selective positioning of latrines close to bolt-holes may, therefore, be explained through maximizing the chance of intercepting intruders, since other meerkats will also use bolt-holes as they move through the environment. As the use of bolt-holes is ephemeral, they are unlikely to warrant defence. An intruding meerkat that places a higher priority on land tenure than escape at the approach of a predator is unlikely to pass these priorities on to future generations. Again, it must be borne in mind that any selective positioning of latrines in regard to these features may also be the result of the meerkats' own selective positioning in the vicinity of these features. As in any species, if meerkats scent-mark at similar rates regardless of their location, scent-marks will accumulate in the vicinity of features where they spend most of their time. This may seem an obvious point, but it is surprising how infrequently it is considered.

Mates

Seasonal differences in meerkat latrine use, and the location of latrines in the environment, is suggestive of a role in mate defence. Although latrine use by meerkats did not increase when resident females were sexually receptive, visits were significantly more likely during the peak breeding period, and occurred at significantly greater rates during observation periods when intruding/prospecting males were encountered (Jordan *et al.*, 2007). Interestingly, however, in translocation experiments of faeces from other groups (Figure 7.6), dominant males spent less time inspecting the samples when the resident dominant female was in oestrus than when she was pregnant, and only countermarked when the dominant female was pregnant (Mares *et al.*, 2011). Additionally, subordinate females have been described as increasing their anal-marking frequencies during inter-group interactions, when foreign-group individuals are in close vicinity (Fenkes, 2011). The spatial distribution patterns of latrines may also be argued to play a role in this regard, as threats to breeding occur from outside the range in the form of prospecting males (Young *et al.*, 2007; Spong *et al.*, 2008). As noted by Young *et al.* (2007), male meerkats from neighbouring groups frequently approach resident groups, but, due to high levels of reproductive skew, the potential fitness costs and benefits of deterring these males vary according to the sex and breeding status of residents. Although all individual meerkats in a pack visited latrines for similar durations, latrine scent-mark composition at the end of each visit was highly male-biased (Jordan, 2007).

Figure 7.6 Controlled scent presentation to wild meerkats. Direct presentations of this nature are rarely possible with wild animals, but ingenuity and technology (e.g. camera-traps) could be utilized to a greater extent in less amenable species. *Source:* Photos © Krystyna A. Golabek.

Male meerkat behaviour supported a mate-defence function, as males scent-marked at significantly higher rates than females and preferentially overmarked female scent-marks (Jordan, 2007). Intruding prospecting males represent a real threat to a resident male's reproductive success (Spong *et al.*, 2008). As such, a mate-defence strategy in meerkats may look very similar to a territory-defence strategy. Insofar as representing an economical approach to communicating with intruders, the spatial distribution of meerkat latrines may also support a mate-defence function. On a more local scale, the selective positioning of latrines close to bolt-holes is also likely to be an effective strategy for intercepting intruding reproductive rivals (prospecting males), which move through the range from bolt-hole to bolt-hole. Meerkats are also able to discriminate between resident and intruding male scent-marks, and dominant males have the strongest overall response to intruder scent-marks (Mares *et al.*, 2011). That this response does not increase with female receptiveness might be explained by the importance of maintaining control of a territory (and the breeding opportunities within it) year round. Thus, Mares *et al.* (2011) expand to suggest that although all group members may be affected by the presence of intruders, reproductive conflict may be the main reason for the stronger response of dominant males to extra-group male scent-marks.

In contrast to males, competition between female meerkats is most intense within the group and females invested heavily in scent-mark investigation, but did not selectively overmark existing scent-marks of either sex (Jordan, 2007). Monitoring of other females, particularly their reproductive status, may therefore be an important function of latrine visits for females. Importantly, rather than cooperatively contributing to territorial defence, individuals appear to participate selfishly at latrine sites, with ultimate explanations for scent-marking potentially being related to both the sex and breeding status of group members. This is likely to be true in most, if not all, species.

Information Centre

While Jordan (2005) found no evidence of kin discrimination of translocated faeces from equally familiar individuals, Le Claire *et al.* (2013) translocated anal gland secretions instead, and found that females spent more time investigating scents from unfamiliar related individuals than unfamiliar unrelated individuals. This suggests that females may use a phenotype-matching mechanism (or recognition alleles) to discriminate kinship on the basis of odour. Repeated investigation of scents at latrine sites might thus aid in reinforcement of such olfactory templates. Fenkes (2011) compared the chemical qualities of anal gland secretions from four mongoose species, including meerkats and banded mongooses, and found species-specific patterns related to social complexity. The highly gregarious meerkats and banded mongooses produced scents with significantly higher chemical diversity than those of the solitary slender mongooses, *Galerella sanguinea*, and socially flexible yellow mongooses, and thus appear to be adapted for encoding more complex chemical messages (Fenkes, 2011). As all of the above mongoose species are territorial, the additional olfactory complexity of anal gland secretions in group-living species suggests multiple signalling functions unrelated to territory defence.

Parasites

Like European badgers, meerkats spend much of their time at latrine sites sniffing faeces and other deposits (Jordan, 2007). While the concentration of faecal matter away from feeding sites may reduce the likelihood of parasite intake in those areas, the fact that groups repeatedly return to latrines results in their exposure to parasites and other sources of infection in these locations. Additionally, since multiple packs share latrines, transmission of parasites and disease (e.g. tuberculosis; Drewe, 2010) is likely increased by latrine use, although this has not yet been investigated empirically.

Conclusion

Overall, meerkat latrines appear to play a primary role in both territoriality *per se*, and in intrasexual competition, predominantly as communicative signals to reproductive rivals from outside of the resident pack.

The Banded Mongoose

Latrines of banded mongooses are scent-marking sites that contain clusters of faeces, generally >4 faeces within ~2 m^2, and frequently include many tens of

Figure 7.7 Defecation by a subadult male banded mongoose, *Mungos mungo*. *Source:* Photos © Emmanuel Do Linh San.

faeces at any one time. Such latrines almost always lack pits, and faeces are instead spread around on the surface (Figure 7.7).

Scent-marking behaviour has been investigated in wild banded mongooses using a combination of behavioural observations, experimental presentations, and chemical analyses. Scent (as opposed to latrines specifically) appears to be involved in intrasexual competition in this species, both within and between packs (e.g. Müller & Manser, 2008; Jordan *et al.*, 2011a). Packs encounter latrines at higher densities in overlapping versus exclusive areas of their home ranges, although deposition of four types of scent (anal gland secretion, urine, faeces, and cheek gland secretion) and investigation of scents in latrines is similar in overlapping versus exclusive areas (Jordan *et al.*, 2010). However, despite sharing latrine sites with neighbouring packs, banded mongooses do not discriminate between individuals from groups of equal familiarity on the basis of anal gland secretions, urine, or faeces (Jordan *et al.*, 2010). In fact, Jordan *et al.* (2010) suggest that banded mongoose scent-marking is primarily involved in communicating within social groups, a pattern which, somewhat counter-intuitively, still reflects the importance of scent in the acquisition and defence of mates. This is because, in contrast to most other territorial systems, including that of meerkats, reproductive competition is most intense within and not between banded mongoose packs (Müller & Manser, 2008).

Here we will review the possible function of banded mongoose latrines in regard to the five following resources/hypotheses.

Territoriality

Banded mongooses are territorial with interactions between packs described as occurring around territory boundaries (Cant *et al.*, 2002; Jordan *et al.*, 2010). These encounters are typically extremely hostile, and frequently lead to severe injury, with between 8% (Jordan *et al.*, 2010) and 24% (Müller & Manser, 2007) of known adult mortality reported from inter-pack fights. Despite the importance of territoriality to banded mongooses, scent-marks are deposited non-selectively throughout the home range, but are encountered more frequently in regions that overlap with the home ranges of other groups. As groups generally scent-mark at similar rates in overlapping and exclusive regions, the increased density of scent-marks in overlapping regions may simply be explained by multiple groups using these regions, and is thus not likely to be a result of selective positioning in these areas (Jordan *et al.*, 2010).

Despite the inter-pack overlap in latrine use, chemical analyses of several hundred anal gland secretions failed to elucidate a group-specific 'signature' (Jordan *et al.*, 2010); a result which is backed up by the experimental presentation of single anal gland samples from different groups of equal familiarity which recipients failed to discriminate between (Jordan *et al.*, 2010). In simulated latrine sites, however, where six or seven scats or urine were translocated from one pack and presented to another, the intensity of the response of the recipient pack depended on whether the donor pack was a neighbour or non-neighbour (Müller & Manser, 2008). In contrast to work on badgers supporting the Dear Enemy Phenomenon as described above, banded mongoose packs respond more intensively (i.e. through more inspections and vocalizations) to the scents of their neighbours than to non-neighbours, and, therefore, support the Threat-Level Hypothesis (Müller & Manser, 2007). While these experimental results suggest a role for latrines in inter-pack communication, the mechanism of discrimination or recognition in the absence of a pack-specific signature remains unknown. However, it must be borne in mind that chemical analyses of urine and faeces were not conducted, and one or both of these scent types may contain information sufficient to distinguish packs by scent.

Food

Neighbouring banded mongoose packs compete over access to food, with larger groups sometimes extending their territories at the expense of their neighbours (Müller & Manser, 2007). Such competition is most intense in the Mweya population (Queen Elizabeth National Park, Uganda) where access to human refuse is common and can even affect fecundity if not ultimate reproductive success (Otali & Gilchrist, 2004). Thus, it would be possible that latrines play a role in alleviating competition, especially as latrine sites could signal resource depletion effectively and maximize foraging efficiency for all group members including the marker (*sensu* 'note to self': Buesching & Macdonald, 2001; and the Foraging Book-Keeping Hypothesis: Henry, 1977). Faeces volume and consistency are likely to be honest signals of the type and richness of resources exploited (*sensu* Buesching & Macdonald, 2001). However, as mongooses forage throughout their ranges, hitherto it has been impossible to assess the location of latrine sites in relation to food patches. Nevertheless, counter to this suggestion, banded mongoose females' intrasexual overmarking is not related to the acquisition of resources necessary to breed and rear offspring, as the frequency with which females are approached in competition for food is unrelated to their overmarking score (Jordan *et al.*, 2011c).

Dens

Although den sites are a valuable resource in which banded mongooses raise their young (Rood, 1975), they are unlikely to constitute a restricting resource in this species. Whereas some dens are revisited frequently (Cant, 1998), many are frequently changed, sometimes on a nightly basis, hinting at the abundance of dens as a resource. No association between den sites and latrine sites has been described in banded mongooses.

Mates

Sexual selection has resulted in the elaboration of secondary sexual characteristics in many animals.

Although mammalian scent glands, secretions, and marking behaviour are commonly sexually dimorphic, these traits have received little attention in this regard, especially in comparison to avian plumage and mammalian weaponry. In this context (or rather *contest*), overmarking is of particular interest because, due to the costs of repeatedly monitoring and covering the scent-marks of rivals, it may provide an honest indication of a male's resource holding potential. Jordan *et al.* (2011a) investigated the relationship between overmarking investment and mating frequency in wild banded mongooses. They not only found a relationship between these factors, but their results also suggested that overmarking may primarily affect mating through male intrasexual competition and not by female mate choice.

Anal gland secretions of banded mongooses are sexually dimorphic, and adult mongooses are more likely to overmark the scent marks of same-sex individuals, whereas juveniles overmark apparently indiscriminately with regard to sex (Jordan *et al.*, 2011a). Such same-sex-specific patterns of overmarking within groups have not yet been described in detail in any other carnivore species, and may reflect the unique social system of banded mongooses, where intrasexual competition for reproduction within packs is intense in both sexes.

In order for overmarking to affect mate choice or access to mates, it is essential that it is possible for individuals to associate the scents of conspecifics with the particular individuals that deposited them. This is commonly referred to as 'individual recognition', and requires that (i) there are chemical characteristics of scent which are individually distinctive to the individual that produced them, and (ii) individuals actually associate the scents with the particular individual that deposited them. By combining chemical (GCMS) analyses in the laboratory with a field discrimination experiment, Jordan *et al.* (2011a) demonstrated that banded mongooses exhibit these two criteria. Male banded mongooses have a degree of individual specificity in their anal gland secretions, and complementary field experiments showed that mongooses can discriminate between scents from different individuals (Jordan *et al.*, 2011b). Additionally, observations of overmarking patterns showed a relationship between overmarking score and mating success. However, instead of females actively choosing to mate with males with high over-marking scores, direct competition between males appeared to be the mechanism regulating mating success. Males that overmarked more started to mate-guard females at a significantly younger age than males with lower overmarking scores. That mate-guarding males obtain the vast majority of matings suggests that overmarking may be an important component of intrasexual competition for mating opportunities in this species, and since most over-marking occurs at latrine sites, latrines are likely to be integral to this competition.

Compared to that of males, scent-marking by females is relatively infrequent and poorly understood but appears to be common in species where females compete for mates. Jordan *et al.* (2011c) combined chemical analyses, behavioural observations and experimental presentations in attempting to distinguish three non-mutually exclusive hypotheses for 'female intrasexual overmarking'. Though less pronounced than for male mongooses, female anal gland secretions had individually specific characteristics, and observations suggested that female intrasexual overmarking was involved in competing for males. Females with higher intrasexual overmarking scores tended to receive more mating attempts and the female with the highest overmarking score in a group was mate-guarded by males in better condition than the female with the lowest overmarking score. However, the degree of direct competition between males for access to females was not related to the female's over-marking score, and models controlling for female age and weight showed that female intrasexual overmarking score did not explain the degree of harassment received from males (Jordan *et al.*, 2011c). The authors also found no evidence to suggest that intrasexual overmarking in females was involved in reproductive suppression. Females with the highest and lowest overmarking scores in each group matured at similar ages and were mate-guarded by males for similar durations. Females that were first mate-guarded at the onset of oestrus were also equally likely to have higher or lower mean overmarking scores than other females. In addition, reproductive status affected responses to translocated scent-marks, with both sexes increasing their responses when females in the pack were in oestrus (Müller & Manser, 2007). Evidence suggests that latrines have a function in intrasexual competition for mates – and mating opportunities – within banded mongoose packs.

Parasites

Like European badgers and meerkats, banded mongooses are affected by, and therefore implicated in, the spread of tuberculosis (Alexander *et al.*, 2002). In the case of banded mongooses, a new strain of the disease, *Mycobacterium mungi*, has been identified in the Chobe district of northern Botswana (Alexander *et al.*, 2010). Unlike other *Mycobacterium* strains, *M. mungi* appears to enter by a non-respiratory route through the nasal planum. This is suggestive of environmental transmission, with contact with human faeces suspected to be the most likely source (Alexander *et al.*, 2010). Therefore, while the use of latrines (or indeed flushing toilets) by the human population might reduce the spread of this disease that is fatal to mongooses, it is unlikely that mongoose latrine use is either a response to, or effective mechanism against, the transmission of disease. Like meerkats, banded mongooses repeatedly return to these communal marking sites, and investigate the scents left behind by their conspecifics; behaviours generally not expected to reduce parasitic or pathogenic infection, but instead potentially (though not yet proven) contributing to the spread of the disease.

Conclusion

Though not completely discounting the role of latrines in territoriality in banded mongooses, the evidence suggests that a major function of scent-marking in this species is in intrasexual competition, predominantly within packs.

General Discussion

In this review, we have provided a summary of the current knowledge on small carnivore latrines in space and time, concentrating on three well-studied focal species: the European badger, the meerkat and the banded mongoose. In conducting and presenting these case studies, we have developed a multi-disciplinary approach to the study of latrines and scent function more generally, and highlighted the importance of collecting data from most of the following categories to determine the function of latrine use in any species: (i) spatial distribution patterns of latrines within the home range; (ii) temporal usage patterns; (iii) individual visit and contribution patterns; and (iv) information content of the scent-signal as well as its biological/behavioural relevance. We hope that the presented framework will be useful, particularly at the research planning stage, in stimulating the collection of data belonging to a range of different categories that will allow potential hypotheses to be tested and distinguished more clearly in the future.

Nevertheless, even in the three focal species for which we and others have acquired information from most of these suggested data types, the exact function of latrine use remains unclear. In part, this is due to the fact that despite these being amongst the most well-studied carnivore species, there are still considerable gaps in our knowledge and data collection. Thus, it is important to reaffirm a few general points that we have encountered in compiling these datasets:

First – and this is a point that has been made by others (e.g. Gosling & Roberts, 2001) but not always heeded by subsequent researchers, reviewers, and editors – it is important to emphasize that spatial data alone are almost always insufficient to allow functional interpretation. Figure 7.1 clearly illustrates the fallacy of such a unilateral approach, but it is also important to realize that, with spatial data alone, it is possible to support a territorial function to latrine use on *any* spatial distribution of latrines by imploring an argument of the economics of scent-marking in this manner. In effect, this negates the value of the spatial data *alone* in determining latrine function, although in combination with other data, spatial data can indeed add to functional interpretations. These requisite additional data may include, for example, information on the movements of intruding individuals.

Second, in comparing data collected by different researchers and across different study areas as well as across species, it is of paramount importance that we consider even slight differences in definitions of the term 'latrine'. Unfortunately, and although we are confident that the term latrine has been used relatively consistently in the cited studies, definitions vary slightly even across our three focal species. Generally,

a latrine is defined as a localized cluster of faeces, indicative of, and resulting from, the repeated visit and faecal deposition by one or more individual(s) or social group(s), and is often associated with the deposition of other scent-marks, resulting in a 'composite latrine'.

Third, given the variability in carnivore sociality and associated latrine use across their geographic range (e.g. European badgers: Johnson *et al.*, 2000), it is important to emphasize that isolated studies may not be representative of the species as a whole. For example, European badgers in Doñana National Park, Spain, tend to live in small family groups and use mostly hinterland latrines around the sett (Revilla & Palomares, 2002), while the highly gregarious badgers in the Southwest of England predominantly use border latrines, at the periphery of their group's range. Moreover, while Mediterranean badgers stay mostly within their home ranges, ranging patterns of Oxfordshire badgers indicate that they traverse across these lines of border latrines frequently and freely (Macdonald *et al.*, 2015). Studying only one of these populations could therefore lead to erroneous broad generalizations for the species as a whole and the entry of such biases into our species accounts is also possible for meerkats and banded mongooses. While meerkats and banded mongooses range over much of southern Africa (as well as East Africa for banded mongooses), in each case, their scent-marking behaviour has only been studied in detail once, in one area and in one habitat type. For meerkats, the bulk of our knowledge is derived from a study population utilizing recovering ranchland in South Africa's southern Kalahari Desert, whereas research on banded mongooses is focused on an isolated peninsula in Uganda where inter- and intra-pack dynamics may be affected by reduced dispersal opportunities, as well as access to abundant food resources in the form of human refuse (Otali & Gilchrist, 2004). As in all studies, neither may be representative of the species as a whole, and it is particularly important to bear in mind that even within species, geographic differences may affect the function of latrines (e.g. within the Hyaenidae: Gorman & Mills, 1984). Thus, these differences in the social system and habitat use within the same species highlight that in addition to the four categories of data suggested above, information should ideally be collected from different populations across a range of habitats. In short, data from a range of habitats may be as informative as data from a range of species because they would provide a richer understanding of the ecological and economic constraints of latrine use and scent-marking.

Fourth, it is important to remember that survey techniques usually fail to account for the number of individuals or groups using specific areas. For example, reports that latrines are concentrated along territory borders may neglect the possibility that several individuals utilize these areas, particularly as overlapping areas are common in the periphery of home ranges. Clearly, the presence of an increased density of latrines in areas used by more individuals is not necessarily evidence of the strategic marking of those areas. Similarly, in species that den communally but forage independently, scent aggregations could form 'passively' in proximity to dens (e.g. European badger: Buesching & Macdonald, 2004), kills or shared feeding sites (e.g. spotted hyena: Bearder & Randall, 1978; African palm civet: Charles-Dominique, 1978), or features of the landscape in general (e.g. meerkat bolt-holes: Jordan *et al.*, 2007). Hence, apparent selective positioning of latrines may be due to the increased density of individuals utilizing these areas, or the increased time spent in these areas, rather than by any active selective positioning of scents.

Finally, it is worth emphasizing that although our three focal study species and study populations are particularly well suited to the study of latrines, technological and scientific advancements may be fruitfully applied to less tractable species. Technology has particular potential value in furthering our understanding of individual behaviour at latrine sites. The use of camera-traps, genetic analyses, bait-marking and high-resolution GPS and/or RFID collars could be used either alone or in combination to advance the study of latrine function beyond a basic interpretation and understanding of latrine spatial distribution alone.

Conclusion

In summary, in this chapter, we have identified a more complete research framework for the study of latrines, which relies on a multi-disciplinary methodological approach, comprising four data categories (spatial, temporal, behavioural, and semio-chemical). As we have demonstrated, the collection of any one of these data types in isolation, however, will result in the limitation of the conclusions that can be drawn on latrine function. We have described considerable variation in functional interpretations within and across species, and in doing so, we have developed a framework for the study of latrine function which highlights potential hypotheses and, where possible, any distinguishing predictions between them. We hope that this broad framework will be of use in future studies of latrine function, and in scent-marking studies more generally.

Acknowledgements

We would like to extend our thanks to Emmanuel Do Linh San for inviting us to write this chapter and expertly guiding us through its development. We are grateful to Aliza le Roux and Michael Ferkin for useful comments. N.R. Jordan's post-graduate supervisors (Mike Cherry, Mike Cant, Marta Manser, and Tim Clutton-Brock) contributed to parts of the text while in thesis-form, and supported the meerkat and mongoose work in the field.

References

Alexander, A. & Ewer, R.F. (1959) Observations on the biology and behaviour of the smaller African polecat (*Poecilogale albinucha*). *African Wildlife* 13, 313–320.

Alexander, K.A., Pleydell, E., Williams, M.C., Lane, E.P., Nyange, J.F. & Michel, A.L. (2002) *Mycobacterium tuberculosis*: an emerging disease of free-ranging wildlife. *Emerging Infectious Diseases* 8, 598–601.

Alexander, K.A., Laver, P.N., Michel, A.L., Williams, M., van Helden, P.D., Warren, R.M. & van Pittius, N.C.G. (2010) Novel *Mycobacterium tuberculosis* complex pathogen, *M. mungi. Emerging Infectious Diseases* 16, 1296–1299.

Almeida, D., Copp, G.H., Masson, L., Miranda, R., Murai, M. & Sayer, C.D. (2012) Changes in the diet of a recovering Eurasian otter population between the 1970s and 2010. *Aquatic Conservation: Marine and Freshwater Ecosystems* 22, 26–35.

Annavi, G., Newman,C., Dugdale, H.L., Buesching, C.D., Sin, Y.W., Burke, T. & Macdonald, D.W. (2014) Neighbouring-group composition and within-group relatedness drive extra-group paternity rate in the European badger (*Meles meles*). *Journal of Evolutionary Biology* 27, 2191–2203.

Anwar, M.A., Newman, C., Macdonald, D.W., Woolhouse, M.E.J. & Kelly, D.W. (2000) Coccidiosis in the European badger (*Meles meles*) from England, an epidemiological study. *Parasitology* 120, 255–260.

Apps, P., Rafiq, K. & McNutt, J.W. (2019) Do carnivores have a world wide web of interspecific scent signals? In: *Chemical Signals in Vertebrates 14* (ed. C.D. Buesching), pp. 182–202. Springer, Cham.

Bailey, T.N. (1974). Social organization in a bobcat population. *Journal of Wildlife Management* 38, 438–446.

Barja, I. & List, R. (2006) Faecal marking behaviour in ringtails (*Bassariscus astutus*) during the non-breeding period: spatial characteristics of latrines and single faeces. *Chemoecology* 16, 219–222.

Barja, I. & de Miguel, F.J. (2010) Chemical communications in large carnivores: urine-marking frequencies in captive tigers and lions. *Polish Journal of Ecology* 58, 397–400.

Barja, I., de Miguel, F.J. & Bárcena, F. (2004) The importance of crossroads in faecal marking behaviour of the wolves (*Canis lupus*). *Naturwissenschaften* 91, 489–492.

Barja, I., Silván, G., Martínez-Fernández, L. & Illera, J.C. (2011) Physiological stress responses, fecal marking behavior, and reproduction in wild European pine martens (*Martes martes*). *Journal of Chemical Ecology* 37, 253–259.

Barocas, A., Golden, H.N., Harrington, M.W., McDonald, D.B. & Ben-David, M. (2016) Coastal latrine sites as social information hubs and drivers of river otter fission–fusion dynamics. *Animal Behaviour* 120, 103–114.

Barrientos, R. (2006) Year-round defecation pattern in wild genets (*Genetta genetta* L.) in a mountain forest (Toledo, Central Spain). *Polish Journal of Ecology* 54, 325–328.

Bartels, E. (1964) On *Paradoxurus hermaphroditus* (Horsfield 1824). *Beaufortia* 19, 193–201.

Bearder, S.K. & Randall, R.M. (1978) Use of fecal marking sites by spotted hyenas and civets. *Carnivore* 1, 32–48.

Begg, C.M., Begg, K.S., Du Toit, J.T. & Mills, M.G.L. (2003) Scent-marking behaviour of the honey badger, *Mellivora capensis* (Mustelidae), in the southern Kalahari. *Animal Behaviour* 66, 917–929.

Bekele Tsegaye, Afework Bekele & Balakrishnan, M. (2008) Scent-marking by the African civet *Civettictis civetta* in the Menagesha–Suba State Forest, Ethiopia. *Small Carnivore Conservation* 38, 29–33.

Benhamou, S. (1989) An olfactory orientation model for mammals' movements in their home ranges. *Journal of Theoretical Biology* 139, 379–388.

Bizani, M. (2014) *Diet of the Yellow Mongoose* (Cynictis penicillata) *in the Albany Thicket Biome of South Africa.* MSc thesis, University of Fort Hare, Alice.

Blomsterberg, S.E. (2016) *The Temporal Use of Latrines by Rusty-Spotted Genets (Genetta maculata* Gray 1830*) in Telperion Nature Reserve.* BSc Honours thesis, University of Pretoria.

Bodin, C., Benhamou, S., & Poulle, M.-L. (2006) What do European badgers (*Meles meles*) know about the spatial organisation of neighbouring groups? *Behavioural Processes* 72, 84–90.

Böhm, M., Palphramand, K.L., Newton-Cross, G., Hutchings, M.R. & White, P.C. (2008) Dynamic interactions among badgers: implications for sociality and disease transmission. *Journal of Animal Ecology* 77, 735–745.

Borowski, Z. (1998) Influence of weasel (*Mustela nivalis* Linnaeus, 1766) odour on spatial behaviour of root voles (*Microtus oeconomus* Pallas, 1776). *Canadian Journal of Zoology* 76, 1799–1804.

Bothma, J. du P. & Nel, J.A.J. (1980) Winter food and foraging behavior of the aardwolf (*Proteles cristatus*) in the Namib–Naukluff Park. *Madoqua* 12, 141–147.

Boydston, E.E., Morelli, T.L. & Holekamp, K.E. (2001) Sex differences in territorial behavior exhibited by the spotted hyena (Hyaenidae, *Crocuta crocuta*). *Ethology* 107, 369–385.

Brown, R.E. & Macdonald, D.W. (eds) (1985) *Social Odours in Mammals*. Clarendon Press, Oxford.

Buesching, C.D. & Macdonald, D.W. (2001) Scent-marking behaviour of the European badger (*Meles meles*): resource defence or individual advertisement? In: *Chemical Signals in Vertebrates 9* (eds A. Marchlewska-Koj, J.J. Lepri & D. Müller-Schwarze), pp. 321–327. Springer, New York.

Buesching, C.D. & Macdonald, D.W. (2004) Variations in scent-marking behaviour of European badgers *Meles meles* in the vicinity of their setts. *Acta Theriologica* 49, 235–246.

Buesching, C.D., Newman, C. & Macdonald, D.W. (2002a) Variations in colour and volume of the subcaudal gland secretion of badgers (*Meles meles*) in relation to sex, season and individual-specific parameters. *Mammalian Biology* 67, 147–156.

Buesching, C.D., Waterhouse, J.S. & Macdonald, D.W. (2002b) Gas-chromatographic analyses of the subcaudal gland secretion of the European badger (*Meles meles*). Part I: chemical differences related to individual parameters. *Journal of Chemical Ecology* 28, 41–56.

Buesching, C.D., Waterhouse, J.S. & Macdonald, D.W. (2002c) Gas-chromatographic analyses of the subcaudal gland secretion of the European badger (*Meles meles*). Part II: time-related variation in the individual-specific composition. *Journal of Chemical Ecology* 28, 57–69.

Buesching, C.D., Stopka, P. & Macdonald, D.W. (2003) The social function of allo-marking in the European badger (*Meles meles*). *Behaviour* 140, 965–980.

Buesching, C.D., Newman, C. & Macdonald, D.W. (2014) How dear are deer volunteers: the efficiency of monitoring deer using teams of volunteers to conduct pellet group counts. *Oryx* 48, 593–601.

Buesching, C.D., Newman, C., Service, K., Macdonald, D.W. & Riordan, P. (2016) Latrine marking patterns of badgers (*Meles meles*) with respect to population density and range size. *Ecosphere* 7, e01328.

Burgener, N., East, M.L., Hofer, H. & Dehnhard, M. (2008) Do spotted hyena scent marks code for clan membership? In: *Chemical Signals in Vertebrates 11* (eds J. Hurst, R.J. Beynon, S.C. Roberts & T.D. Wyatt), pp. 169–177. Springer, New York.

Camenzind, F.J. (1978) Behavioral ecology of coyotes on the national elk refuge, Jackson, Wyoming.
In: *Coyotes: Biology, Behavior, and Management* (ed. M. Bekoff), pp. 267–294. Academic press, New York.

Cant, M.A. (1998) *Communal Breeding in Banded Mongooses and Theory of Reproductive Skew*. PhD thesis, University of Cambridge.

Cant, M.A., Otali, E. & Mwanguhya, F. (2002) Fighting and mating between groups in a cooperatively breeding mammal, the banded mongoose. *Ethology* 108, 541–555.

Carpenter, P.J., Pope, L.C., Greig, C., Dawson, D.A. Rogers, L.M., Erven, K., Wilson, G.J., Delahay, R.J., Cheeseman, C.L. & Burke, T. (2005) Mating system of the Eurasian badger, *Meles meles*, in a high density population. *Molecular Ecology* 14, 273–284.

Carter, S.P., Delahay, R.J., Smith, G.C., Macdonald, D.W., Riordan, P., Etherington, T.R., Pimley, E.R., Walker, N.J. & Cheeseman, C.L. (2007) Culling-induced social perturbation in Eurasian badgers *Meles meles* and the management of TB in cattle: an analysis of a critical problem in applied ecology. *Proceedings of the Royal Society B: Biological Sciences* 274, 2769–2777.

Charles-Dominique, P. (1978) Écologie et vie sociale de *Nandinia binotata* (Carnivores, Viverridés): comparaison avec les prosimiens sympatriques du Gabon. *Revue d'Écologie – La Terre et la Vie* 32, 477–528. (In French with English summary).

Clapperton, B.K. (1989) Scent-marking behaviour of the ferret, *Mustela furo* L. *Animal Behaviour* 38, 436–446.

Clutton-Brock, T.H. (ed.) (1988) *Reproductive Success: Studies of Individual Variation in Contrasting Breeding Systems*. University of Chicago Press, Chicago.

Corbett, L.C. (1979) *Feeding Ecology and Social Organisation of Wildcats (*Felis silvestris*) and House Cats (*Felis catus*) in Scotland*. PhD thesis, University of Aberdeen.

Darden, S.K., Steffensen, L.K. & Dabelsteen, T. (2008) Information transfer among widely spaced individuals: latrines as a basis for communication networks in the swift fox? *Animal Behaviour* 75, 425–432.

Davies, J.M., Lachno, D.R. & Roper, T.J. (1988) The anal gland secretion of the European badger (*Meles meles*) and its role in social communication. *Journal of Zoology* 216, 455–463.

Dawkins, R. & Krebs, J.R. (1978) Animal signals: information or manipulation? In: *Behavioural Ecology: An Evolutionary Approach* (eds J.R. Krebs & N.B. Davies), pp. 282–309. Blackwell, Oxford.

Delahay, R.J., Brown, J.A., Mallinson, P.J., Spyvee, P.D., Handoll, D., Rogers, L.M. & Cheeseman, C.L. (2000) The use of marked bait in studies of the territorial organization of the European badger (*Meles meles*). *Mammal Review* 30, 73–87.

Delahay, R.J., Walker, N.J., Forrester, G.J., Harmsen, B., Riordan, P., Macdonald, D.W., Newman, C. & Cheeseman, C.L. (2006) Demographic correlates of bite wounding in Eurasian badgers, *Meles meles* L., in stable and perturbed populations. *Animal Behaviour* 71, 1047–1055.

Do Linh San, E., Ferrari, N. & Weber, J.-M. (2007a) Spatio-temporal ecology and density of badgers *Meles meles* in the Swiss Jura Mountains. *European Journal of Wildlife Research* 53, 265–275.

Do Linh San, E., Ferrari, N. & Weber, J.-M. (2007b) Socio-spatial organization of Eurasian badgers (*Meles meles*) in a low-density population of central Europe. *Canadian Journal of Zoology* 85, 973–984.

Do Linh San, E., Nqinana, A., Madikiza, Z.J. & Somers, M.J. (2020) Diet of the marsh mongoose around a non-permanent reservoir: response of a generalist opportunist forager to the absence of crabs. *African Zoology* 55, 240–244.

Drewe, J.A. (2010) Who infects whom? Social networks and tuberculosis transmission in wild meerkats. *Proceedings of the Royal Society B: Biological Sciences* 277, 633–642.

Drewe, J.A., Weber, N., Carter, S.P., Bearhop, S., Harrison, X.A., Dall, S.R.X., McDonald, R.A. & Delahay, R.J. (2012) Performance of proximity loggers in recording intra-and inter-species interactions: a laboratory and field-based validation study. *PLoS One* 7, e39068.

Dugdale, H.L., Macdonald, D.W., Pope, L.C. & Burke, T. (2007) Polygynandry, extra-group paternity and multiple-paternity litters in European badger (*Meles meles*) social groups. *Molecular Ecology* 16, 5294–5306.

Dugdale, H.L., Griffiths, A. & Macdonald, D.W. (2011) Polygynandrous and repeated mounting behaviour in European badgers, *Meles meles*. *Animal Behaviour* 82, 1287–1297.

Dunbar, I.F. (1977) Olfactory preferences in dogs: the response of male and female beagles to conspecific odors. *Behavioral Biology* 20, 471–481.

Eppley, T.M., Ganzhorn, J.U. & Donati, G. (2016) Latrine behaviour as a multimodal communicatory signal station in wild lemurs: the case of *Hapalemur meridionalis*. *Animal Behaviour* 111, 57–67.

Eisenberg, J.F. & Kleiman, D.G. (1972) Olfactory communication in mammals. *Annual Review of Ecology and Systematics* 3, 1–32.

Engel, T. (1998) Seeds on the roundabout – tropical forest regeneration by *Genetta rubiginosa*. *Small Carnivore Conservation* 19, 13–20.

Engel, T.R. (2000) *Seed Dispersal and Forest Regeneration in a Tropical Lowland Biocoenosis (Shimba Hills, Kenya).* PhD thesis, University of Bayreuth/Logos Verlag, Berlin.

Espírito-Santo, C., Rosalino, L.M. & Santos-Reis, M. (2007) Factors affecting the placement of common genet latrine sites in a Mediterranean landscape in Portugal. *Journal of Mammalogy* 88, 201–207.

Estes, R.D. (1991) *The Behaviour Guide to African Mammals.* University of California Press, Oakland.

Ewer, R.F. (1963) The behaviour of the meerkat, *Suricata suricatta* (Schreber). *Zeitschrift für Tierpsychologie* 20, 570–607.

Ewer, R.F. (1973) *The Carnivores.* Cornell University Press, Ithaca.

Feldman, H.N. (1994) Methods of scent marking in the domestic cat. *Canadian Journal of Zoology* 72, 1093–1099.

Fenkes, M. (2011) *Comparative Analysis of Volatile Components in the Anal Gland Secretion of Three Sympatric but Socially Different Mongoose Species* (Galerella sanguinea, Cynictis penicillata, Suricata suricatta). MSc thesis, University of Potsdam.

Fisher, J.B. (1954) Evolution and bird sociality. In: *Evolution as a Process* (eds J. Huxley, A.C. Hardy & E.B. Ford), pp. 71–83. Allen & Unwin, London.

Gilbert, K.A. (1997) Red howling monkey use of specific defecation sites as a parasite avoidance strategy. *Animal Behaviour* 54, 451–455.

Gompper, M.E., Kays, R.W., Ray, J.C., Lapoint, S.D., Bogan, D.A. & Cryan, J.R. (2006) A comparison of noninvasive techniques to survey carnivore communities in northeastern North America. *Wildlife Society Bulletin* 34, 1142–1151.

Gorman, M.L. (1976) A mechanism for individual recognition by odour in *Herpestes auropunctatus* (Carnivora: Viverridae). *Animal Behaviour* 24, 141–146.

Gorman, M.L. (1984) Scent marking and territoriality. *Acta Zoologica Fennici* 171, 49–53.

Gorman, M.L. (1990) Scent marking strategies in mammals. *Revue suisse de Zoologie* 97, 3–29.

Gorman, M.L. & Mills, M.G.L. (1984) Scent marking strategies in hyaenas (*Mammalia*). *Journal of Zoology* 202, 535–547.

Gorman, M.L. & Trowbridge, B.J. (1989) The role of odor in the social lives of carnivores. In: *Carnivore Behavior, Ecology, and Evolution. Volume 1* (ed. J.L. Gittleman), pp. 57–88. Cornell University Press, Ithaca.

Gorman, M.L., Kruuk, H. & Leitch, A. (1984) Social functions of the sub-caudal scent gland secretion of the European badger *Meles meles* (Carnivora: Mustelidae). *Journal of Zoology* 203, 549–559.

Gosling, L.M. (1981) Demarkation in a gerenuk territory: an economic approach. *Zeitschrift für Tierpsychologie* 56, 305–322.

Gosling, L.M. (1982) A reassessment of the function of scent marking in territories. *Zeitschrift für Tierpsychologie* 60, 89–118.

Gosling, L.M. (1985) *Social Behaviour in Mammals*. Chapman & Hall, New York.

Gosling, L.M. (1986) The evolution of mating strategies in male antelopes. In: *Ecological Aspects of Social Evolution: Birds and Mammals* (eds D.I. Rubenstein & R.W. Wrangham), pp. 244–281. Princeton University Press, Princeton.

Gosling, L.M. & McKay, H.V. (1990) Competitor assessment by scent matching: an experimental test. *Behavioral Ecology and Sociobiology* 26, 415–420.

Gosling, L.M. & Roberts, S.C. (2001) Testing ideas about the function of scent marks in territories from spatial patterns. *Animal Behaviour* 62, F7–F10.

Green, M.L., Monick, K., Manjerovic, M.B., Novakofski, J. & Mateus-Pinilla, N. (2015) Communication stations: cameras reveal river otter (*Lontra canadensis*) behavior and activity patterns at latrines. *Journal of Ethology* 33, 225–234.

Hammerstein, P. (1981) The role of asymmetries in animal contests. *Animal Behaviour* 29, 193–205.

Harris, R.L., Sprent, J. & Nicol, S.C. (2019) Latrines as potential communication centres in short-beaked echidnas. In: *Chemical Signals in Vertebrates 14* (ed. C.D. Buesching), pp. 13–26. Springer, Cham.

Hediger, H. (1949) Säugetier-Territorien und ihre Markierung. *Bijdragen Tot de Dierkunde* 28, 172–184.

Henry, J.D. (1977) The use of urine marking in the scavenging behavior of the red fox (*Vulpes vulpes*). *Behaviour* 61, 82–106.

Hirsch, B.T., Prange, S., Hauver, S.A. & Gehrt, S.D. (2014) Patterns of latrine use by raccoons (*Procyon lotor*) and implications for *Bayliscaris procyonis* transmission. *Journal of Wildlife Diseases* 50, 243–249.

Ikeda, H. (1984) Raccoon dog scent marking by scats and its significance in social behaviour. *Journal of Ethology* 2, 77–84.

Irwin, M.T., Samonds, K.E., Raharison, J.L. & Wright, P.C. (2004) Lemur latrines: observations of latrine behavior in wild primates and possible ecological significance. *Journal of Mammalogy* 85, 420–427.

Johnsingh, A.J.T. (1982) Reproductive and social behaviour of the dhole, *Cuon alpinus* (Canidae). *Journal of Zoology* 198, 443–463.

Johnson, R.P. (1973) Scent marking in mammals. *Animal Behaviour* 21, 521–535.

Johnson, D.D.P., Macdonald, D.W. & Dickman, A.J. (2000) An analysis and review of models of the sociobiology of the Mustelidae. *Mammal Review* 30, 171–196.

Jordaan, R.K., Somers, M.J. & McIntyre, T. (2017) Dancing to the message: African clawless otter scent marking behaviour. *Hystrix, the Italian Journal of Mammalogy* 28, 277–279.

Jordan, N.R. (2005) *Meerkat Latrines: Cooperation, Competition and Discrimination*. MSc thesis, University of Stellenbosch.

Jordan, N.R. (2007) Scent-marking investment is determined by sex and breeding status in meerkats. *Animal Behaviour* 74, 531–540.

Jordan, N.R., Cherry, M.I. & Manser, M.B. (2007) Latrine distribution and patterns of use by wild meerkats: implications for territory and mate defence. *Animal Behaviour* 73, 613–622.

Jordan, N.R., Mwanguhya, F., Kyabulima, S., Rüedi, P. & Cant, M.A. (2010) Scent marking within and between groups of wild banded mongooses. *Journal of Zoology* 280, 72–83.

Jordan, N.R., Manser, M.B., Mwanguhya, F., Kyabulima, S., Rüedi, P. & Cant, M.A. (2011a) Scent marking in wild banded mongooses: 1. Sex-specific scents and overmarking. *Animal Behaviour* 81, 31–42.

Jordan, N.R., Mwanguhya, F., Furrer, R.D., Kyabulima, S., Rüedi, P. & Cant, M.A. (2011b) Scent marking in wild banded mongooses: 2. Intrasexual overmarking and competition between males. *Animal Behaviour* 81, 43–50.

Jordan, N.R., Mwanguhya, F., Kyabulima, S., Rüedi, P., Hodge, S.J. & Cant, M.A. (2011c) Scent marking in wild banded mongooses: 3. Intrasexual overmarking in females. *Animal Behaviour* 81, 51–60.

Kaneko, Y., Suzuki, T. & Atoda, O. (2009) Latrine use in a low density Japanese badger (*Meles anakuma*) population determined by a continuous tracking system. *Mammal Study* 34, 179–186.

Kaneko, Y., Newman, C., Buesching, C.D. & Macdonald, D.W. (2010) Variations in badger (*Meles meles*) sett microclimate: differential cub survival between main and subsidiary setts, with implications for artificial sett construction. *International Journal of Ecology* 2010, 1–10.

Kaufmann, J.H. (1983) On the definitions and functions of dominance and territoriality. *Biological Reviews* 58, 1–20.

Kasper, C.B., Bastazini, V.A.G., Salvi, J. & Grillo, H.C.Z. (2008) Trophic ecology and the use of shelters and latrines by the Neotropical otter (*Lontra longicaudis*) in the Taquari Valley, Southern Brazil. *Iheringia, Série Zoologia* 98, 469–474.

Kent, L. & Tang-Martinez, Z. (2014) Evidence of individual odors and individual discrimination in the raccoon, *Procyon lotor*. *Journal of Mammalogy* 95, 1254–1262.

Kilshaw, K., Newman, C., Buesching, C.D., Bunyan, J. & Macdonald, D.W. (2009) Coordinated latrine use by European badgers, *Meles meles*: potential consequences for territory defense. *Journal of Mammalogy* 90, 1188–1198.

Kimura, R. (2001) Volatile substances in feces, urine and urine-marked feces of feral horses. *Canadian Journal of Animal Science* 81, 411–420.

King, T.W., Salom-Pérez, R., Shipley, L.A., Quigley, H.B. & Thornton, D.H. (2017) Ocelot latrines: communication centers for Neotropical mammals. *Journal of Mammalogy* 98, 106–113.

Kingdon, J. (1978) *East African Mammals*. Academic Press, London.

Kinoshita, K., Inada, S., Aramaki, Y., Seki, K., Ashida, M., Hama, N. & Kusunoki, H. (2009) Relationship between sexual behaviors and fecal estrogen levels in a female snow leopard (*Uncia uncia*) and a female

cheetah (*Acinonyx jubatus*) under captivity. *Japanese Journal of Zoo and Wildlife Medicine* 14, 59–66.

Kleiman, D.G. (1966) Scent marking in the Canidae. *Symposia of the Zoological Society of London* 18, 167–177.

Kleiman, D.G. (1972) Social behavior of the maned wolf (*Chrysocyon brachyurus*) and bush dog (*Speothos venaticus*): a study in contrast. *Journal of Mammalogy* 53, 791–806.

Krebs, J.R. (1982) Territorial defence in the great tit (*Parus major*): do residents always win? *Behavioral Ecology and Sociobiology* 11, 185–194.

Krebs, J., Ashcroft, R. & Webber, M. (1978) Song repertoires and territory defence in the great tit. *Nature* 271, 539–542.

Kruuk, H. (1972) *The Spotted Hyaena*. University of Chicago Press, Chicago.

Kruuk, H. (1978) Spatial organization and territorial behaviour of the European badger *Meles meles*. *Journal of Zoology* 184, 1–19.

Kruuk, H. (1989) *The Social Badger. Ecology and Behavior of a Group-Living Carnivore (Meles meles)*. Oxford University Press, Oxford.

Kruuk, H. (1992) Scent marking by otters (*Lutra lutra*): signaling the use of resources. *Behavioral Ecology* 3, 133–140.

Kruuk, H. & Hewson, R. (1978) Spacing and foraging of otters (*Lutra lutra*) in a marine habitat. *Journal of Zoology* 185, 205–212.

Kruuk, H. & Jarman, P.J. (1995) Latrine use by the spotted-tailed quoll (*Dasyurus maculatus*: Dasyuridae, Marsupialia) in its natural habitat. *Journal of Zoology* 236, 345–348.

Kruuk, H. & Sands, W.A. (1972) The aardwolf (*Proteles cristatus* Sparrman) 1783 [sic] as predator of termites. *African Journal of Ecology* 10, 211–227.

Kruuk, H., Gorman, M. & Leitch, A. (1984) Scent-marking with the subcaudal gland by the European badger, *Meles meles* L. *Animal Behaviour* 32, 899–907.

Lack, D. (1966) *Population Studies of Birds*. Clarendon Press, Oxford.

Le Claire, S., Nielsen, J.F., Thavarajah, N.K., Manser, M. & Clutton-Brock, T.H. (2013) Odour-based kin discrimination in the cooperatively breeding meerkat. *Biology Letters* 9, 20121054.

le Roux, A., Cherry, M.I. & Manser, M.B. (2008) The effects of population density and sociality on scent marking in the yellow mongoose. *Journal of Zoology* 275, 33–40.

Leuchtenberger, C. & Mourão, G. (2009) Scent-marking of giant otter in the southern Pantanal, Brazil. *Ethology* 115, 210–216.

Leuthold, W. (1977) African ungulates. *Zoophysiology and Ecology* 8, 1–307.

Lockie, J.D. (1966) Territory in small carnivores. *Symposia of the Zoological Society of London* 18, 143–165.

Logiudice, K. (2001) Latrine foraging strategies of two small mammals: implications for the transmission of *Baylisascaris procyonis*. *The American Midland Naturalist* 146, 369–378.

Loureiro, F., Rosalino, L.M., Macdonald, D.W. & Santos-Reis, M. (2007) Path tortuosity of Eurasian badgers (*Meles meles*) in a heterogeneous Mediterranean landscape. *Ecological Research* 22, 837–844.

Macdonald, D.W. (1978) Observations on the behaviour and ecology of the striped hyena, *Hyaena hyaena*, in Israel. *Israel Journal of Zoology* 27, 189–198.

Macdonald, D.W. (1979) The flexible social system of the golden jackal, *Canis aureus*. *Behavioral Ecology and Sociobiology* 5, 17–38.

Macdonald, D.W. (1980) Patterns of scent marking with urine and faeces amongst carnivore communities. *Symposia of the Zoological Society of London* 45, 107–139.

Macdonald, D.W. (1985) The carnivores; Order Carnivora. In: *Social Odours in Mammals* (eds R.E. Brown & D.W. Macdonald), pp. 619–722. Clarendon Press, Oxford.

Macdonald, D.W. (1987) *Running with the Fox*. Facts on File Publications, New York.

Macdonald, D.W. & Johnson, D.P.P. (2015) Patchwork planet: resource dispersion and the ecology of life. *Journal of Zoology* 295, 75–107.

Macdonald, D.W., Stewart, P.D., Johnson, P.J., Porkert, J. & Buesching, C. (2002) No evidence of social hierarchy amongst feeding badgers, *Meles meles*. *Ethology* 108, 613–628.

Macdonald, D.W., Newman, C., Dean, J., Buesching, C.D. & Johnson, P.J. (2004) The distribution of Eurasian badger, *Meles meles*, setts in a high-density area: field observations contradict the sett dispersion hypothesis. *Oikos* 106, 295–307.

Macdonald, D.W., Newman, C., Buesching, C.D. & Johnson, P.J. (2008) Male-biased movement in a high-density population of the Eurasian badger (*Meles meles*). *Journal of Mammalogy* 89, 1077–1086.

Macdonald, D.W., Newman, C. & Buesching, C.D. (2015) Badgers in the rural landscape – conservation paragon or farmland pariah? Lessons from the Wytham Badger Project. In: *Wildlife Conservation on Farmland. Volume 2: Conflict in the Countryside* (eds D.W. Macdonald & R. Feber), pp. 65–95. Oxford University Press, Oxford.

Maddock, A.H. (1988) *Resource Partitioning in a Viverrid Assemblage.* PhD thesis, University of Natal.

Manser, M.B. & Bell, M.B. (2004). Spatial representation of shelter locations in meerkats, *Suricata suricatta*. *Animal Behaviour* 68, 151–157.

Mares, R., Young, A.J., Levesque, D.L., Harrison, N. & Clutton-Brock, T.H. (2011) Responses to intruder scents in the cooperatively breeding meerkat: sex and social status differences and temporal variation. *Behavioral Ecology* 22, 594–600.

Maynard-Smith, J. & Parker, G.A. (1976) The logic of asymmetric contests. *Animal Behaviour* 24, 159–175.

Mbatyoti, A. (2010) *The Diet of the Cape Grey Mongoose* Galerella pulverulenta *in the Albany Thicket Biome (South Africa).* BSc Honours thesis, University of Fort Hare, Alice.

McColl, I. (1967) The comparative anatomy and pathology of anal glands. Arris and Gale lecture delivered at the Royal College of Surgeons of England on 25th February 1965. *Annals of the Royal College of Surgeons of England* 40 36.

McComb, K., Packer, C. & Pusey, A. (1994) Roaring and numerical assessment in contests between groups of female lions, *Panthera leo*. *Animal Behaviour* 47, 379–387.

Mech, L.D. (1977) Wolf-pack buffer zones as prey reservoirs. *Science* 198, 320–321.

Medina-Barrios, O. & Morales-Betancourt, D. (2019) Notes on the behaviour of Neotropical river otter (*Lontra longicaudis*) in Palomino River (La Guajira, Colombia). *IUCN Otter Specialist Group Bulletin* 36, 34–47.

Mertl-Millhollen, A.S. (2006) Scent marking as resource defense by female *Lemur catta*. *American Journal of Primatology* 68, 605–621.

Mills, M.G.L. & Gorman, M.L. (1987) The scent-marking behaviour of the spotted hyaena *Crocuta crocuta* in the southern Kalahari. *Journal of Zoology* 212, 483–497.

Mills, M.G.L., Gorman, M.L., & Mills, M.E. (1980) The scent marking behaviour of the brown hyaena *Hyaena brunnea*. *South African Journal of Zoology* 15, 240–248.

Molsher, R.L. (1999) *The Ecology of Feral Cats,* Felis catus*, in Open Forest in New South Wales: Interactions with Food Resources and Foxes.* PhD thesis, University of Sydney.

Moreno, R. & Giacalone, J. (2006) Ecological data obtained from latrine use by ocelots (*Leopardus pardalis*) on Barro Colorado Island, Panama. *Tecnociencia* 8, 7–21.

Moreno, R. & Giacalone, J. (2014) Use of video cameratraps to study ocelot (*Leopardus pardalis*) behavior at latrines. *Mesoamericana* 18, 55–60.

Mrubata, Z. (2018) *Comparing Latrine Use in Genets (*Genetta genetta *and* G. tigrina*) with Camera-Trapping and Scat Surveys at the Great Fish River Reserve.* BSc Honours thesis, University of Fort Hare, Alice.

Müller, C.A. & Manser, M.B. (2007) 'Nasty neighbours' rather than 'dear enemies' in a social carnivore. *Proceedings of the Royal Society B: Biological Sciences* 274, 959–965.

Müller, C.A. & Manser, M.B. (2008) Scent-marking and intrasexual competition in a cooperative carnivore with low reproductive skew. *Ethology* 114, 174–185.

Müller-Schwarze, D. & Heckman, S. (1980) The social role of scent marking in beaver (*Castor canadensis*). *Journal of Chemical Ecology* 6, 81–95.

Mykytowycz, R. (1970) The role of skin glands in mammalian communication. *Advances in Chemoreception* 1, 327–360.

Mykytowycz, R., Hesterman, E.R., Gambale, S. & Dudziński, M.L. (1976) A comparison of the effectiveness of the odors of rabbits, *Oryctolagus cuniculus*, in enhancing territorial confidence. *Journal of Chemical Ecology* 2, 13–24.

Neal, E. (1977) *The Badger.* Collins, London.

Nel, J.A.J. & Bothma, J. du P. (1983) Scent marking and midden use by aardwolves (*Proteles cristatus*) in the Namib Desert. *African Journal of Ecology* 21, 25–39.

Newman, C., Macdonald, D.W. & Anwar, M.A. (2001) Coccidiosis in the European badger, *Meles meles* in Wytham Woods: infection and consequences for growth and survival. *Parasitology* 123, 133–142.

Noonan, M.J., Markham, A., Newman, C., Trigoni, N., Buesching, C.D., Ellwood, S.A. & Macdonald, D.W. (2014) Climate and the individual: inter-annual variation in the autumnal activity of the European badger (*Meles meles*). *PLoS One* 9, e83156.

Noonan, M.J., Markham, A., Newman, C., Trigoni, N., Buesching, C.D., Ellwood, S.A. & Macdonald, D.W. (2015) A new magneto-inductive tracking technique to uncover subterranean activity: what do animals do underground? *Methods in Ecology and Evolution* 6, 510–520.

Noonan, M.J., Tinnesand, H.V., Müller, C.T., Rosell, F., Macdonald, D.W. & Buesching, C.D. (2019) Knowing me, knowing you: anal gland secretion of European badgers (*Meles meles*) codes for individuality, sex and social group membership. *Journal of Chemical Ecology* 45, 823–837.

Otali, E. & Gilchrist, J.S. (2004) The effects of refuse feeding on body condition, reproduction, and survival of banded mongooses. *Journal of Mammalogy* 85, 491–497.

Owen, D. & Pemberton, D. (2005) *Tasmanian Devil: A Unique and Threatened Animal.* Allen and Unwin, Sidney.

Ozaga, J.J. & Harger, E.M. (1966) Winter activities and feeding habits of northern Michigan coyotes. *The Journal of Wildlife Management* 30, 809–818.

Page, L.K., Swihart, R.K. & Kazacos, K.R. (1998) Raccoon latrine structure and its potential role in transmission of *Baylisascaris procyonis* to vertebrates. *The American Midland Naturalist* 140, 180–185.

Palomares, F. (1993) Faecal marking behaviour by free-ranging common genets *Genetta genetta* and Egyptian mongooses *Herpestes ichneumon* in southwestern Spain. *Zeitschrift für Säugetierkunde* 58, 225–231.

Palphramand, K.L. & White, P.C. (2007) Badgers, *Meles meles*, discriminate between neighbour, alien and self scent. *Animal Behaviour* 74, 429–436.

Panaman, R. (1981) Behaviour and ecology of free-ranging farm cats (*Felis catus* L.). *Zeitschrift für Tierpsychologie* 56, 59–673.

Peters, R.P. & Mech, L.D. (1975) Scent-marking in wolves: radio-tracking of wolf packs has provided definite evidence that olfactory sign is used for territory maintenance and may serve for other forms of communication within the pack as well. *American Scientist* 63, 628–637.

Pigozzi, G. (1990) Latrine use and the function of territoriality in the European badger, *Meles meles*, in a mediterranean coastal habitat. *Animal Behaviour* 39, 1000–1002.

Pigozzi, G. (1992) Frugivory and seed dispersal by the European badger in a Mediterranean habitat. *Journal of Mammalogy* 73, 630–639.

Piñeiro, A. & Barja, I. (2015) Evaluating the function of wildcat faecal marks in relation to the defence of favourable hunting areas. *Ethology Ecology & Evolution* 27, 161–172.

Pusey, A.E. & Packer, C. (1997) The ecology of relationships. In: *Behavioural Ecology. An Evolutionary Approach*, 4th edition (eds J.R. Krebs & N.B. Davies), pp. 254–283. Blackwell, Oxford.

Ralls, K. (1971) Mammalian scent marking. *Science* 171, 443–449.

Ralls, K. & Smith, D.A. (2004) Latrine use by San Joaquin kit foxes (*Vulpes macrotis mutica*) and coyotes (*Canis latrans*). *Western North American Naturalist* 64, 544–547.

Rasa, O.A.E. (1977) The ethology and sociology of the dwarf mongoose. *Zeitschrift für Tierpsychologie* 43, 337–406.

Raymer, J., Wiesler, D., Novotny, M., Asa, C., Seal, U.S. & Mech, L.D. (1986) Chemical scent constituents in the urine of wolf (*Canis lupus*) and their dependence on reproductive hormones. *Journal of Chemical Ecology* 12, 297–313.

Remonti, L., Balestrieri, A., Smiroldo, G. & Prigioni, C. (2011) Scent marking of key food sources in the Eurasian otter. *Annales Zoologici Fennici* 48, 287–294.

Revilla, E. & Palomares, F. (2002) Spatial organization, group living and ecological correlates in low-density populations of Eurasian badgers, *Meles meles*. *Journal of Animal Ecology* 71, 497–512.

Roberts, P.D., Somers, M.J., White, R.M. & Nel, J.A.J. (2007) Diet of the South African large-spotted genet *Genetta tigrina* (Carnivora, Viverridae) in a coastal dune forest. *Acta Theriologica* 52, 45–53.

Rodgers, T.W., Giacalone, J., Heske, E.J., Pawlikowski, N.C. & Schooley, R.L. (2015) Communal latrines act as potentially important communication centers in ocelots *Leopardus pardalis*. *Mammalian Biology* 80, 380–384.

Rood, J.P. (1975) Population dynamics and food habits of the banded mongoose. *African Journal of Ecology* 13, 89–111.

Rood, J.P. (1983) The social system of the dwarf mongoose. In: *Advances in the Study of Mammalian Behavior*, Special Publication 7 (eds J.F. Einsenberg & D.G. Kleinman), pp. 454–488. American Society of Mammalogists, Pittsburgh.

Roper, T.J. (1992) The structure and function of badger setts. *Journal of Zoology* 227, 691–694.

Roper, T.J. (2010) *Badger*. The New Naturalist Library. Collins, London.

Roper, T.J., Shepherdson, D.J. & Davies, J.M. (1986) Scent marking with faeces and anal secretion in the European badger (*Meles meles*): seasonal and spatial characteristics of latrine use in relation to territoriality. *Behaviour* 97, 94–117.

Roper, T.J., Conradt, L., Butler, J., Christian, S.E., Ostler, J. & Schmid, T.K. (1993) Territorial marking with faeces in badgers (*Meles meles*): a comparison of boundary and hinterland latrine use. *Behaviour* 127, 289–307.

Rosalino, L.M., Macdonald, D.W. & Santos-Reis, M. (2004) Spatial structure and land-cover use in a low-density Mediterranean population of Eurasian badgers. *Canadian Journal of Zoology* 82, 1493–1502.

Rosalino, L.M., Torres, J. & Santos-Reis, M. (2006) A survey of helminth infection in Eurasian badgers (*Meles meles*) in relation to their foraging behaviour in a Mediterranean environment in southwest Portugal. *European Journal of Wildlife Research* 52, 202–206.

Rosell, F., Jojola, S.M., Ingdal, K., Lassen, B.A., Swenson, J. E., Arnemo, J.M. & Zedrosser, A. (2011) Brown bears possess anal sacs and secretions may code for sex. *Journal of Zoology* 283, 143–152.

Rostain, R.R., Ben-David, M., Groves, P. & Randall, J.A. (2004) Why do river otters scent-mark? An experimental test of several hypotheses. *Animal Behaviour* 68, 703–711.

Ruibal, M., Peakall, R. & Claridge, A. (2011) Socio-seasonal changes in scent-marking habits in the carnivorous marsupial *Dasyurus maculatus* at communal latrines. *Australian Journal of Zoology* 58, 317–322.

Santos, L.B. & dos Reis, N.R. (2012) Use of shelters and marking sites by *Lontra longicaudis* (Olfers, 1818) in lotic and semilotic environments. *Biota Neotropica* 12, 199–205.

Schilling, A. (1979) Olfactory communication in prosimians. In: *The Study of Prosimian Behavior* (ed. G.A. Doyle), pp. 461–542. Academic Press, London.

Schwarzenberger, F., Möstl, E., Palme, R. & Bamberg, E. (1996) Faecal steroid analysis for non-invasive monitoring of reproductive status in farm, wild and zoo animals. *Animal Reproduction Science* 42, 515–526.

Seitz, E. (1969) Die Bedeutung geruchlicher Orientierung beim Plumplori *Nycticebus coucang* Boddaert 1785 (*Prosimii, Lorisidae*). *Zeitschrift für Tierpsychologie* 26, 73–103.

Sillero-Zubiri, C. & Gottelli, D. (1995) Diet and feeding behavior of Ethiopian wolves (*Canis simensis*). *Journal of Mammalogy* 76, 531–541.

Sillero-Zubiri, C. & Macdonald, D.W. (1998) Scent-marking and territorial behaviour of Ethiopian wolves *Canis simensis*. *Journal of Zoology* 245, 351–361.

Sillero-Zubiri, C., Gottelli, D. & Macdonald, D.W. (1996) Male philopatry, extra-pack copulations and inbreeding avoidance in Ethiopian wolves (*Canis simensis*). *Behavioral Ecology and Sociobiology* 38, 331–340.

Sliwa, A. (1996) *A Functional Analysis of Scent Marking and Mating Behaviour in the Aardwolf,* Proteles cristatus *(Sparman, 1783).* PhD thesis, University of Pretoria.

Sliwa, A. & Richardson, P.R.K. (1998) Responses of aardwolves, *Proteles cristatus*, Sparrman 1783, to translocated scent marks. *Animal Behaviour* 56, 137–146.

Sneddon, I.A. (1991) Latrine use by the European rabbit (*Oryctolagus cuniculus*). *Journal of Mammalogy* 72, 769–775.

Spong, G.F., Hodge, S.J., Young, A.J. & Clutton-Brock, T.H. (2008) Factors affecting the reproductive success of dominant male meerkats. *Molecular Ecology* 17, 2287–2299.

Stewart, P.D., Anderson, C. & Macdonald, D.W. (1997) A mechanism for passive range exclusion: evidence from the European badger (*Meles meles*). *Journal of Theoretical Biology* 184, 279–289.

Stewart, P.D., Bonesi, L. & Macdonald, D.W. (1999) Individual differences in den maintenance effort in a communally dwelling mammal: the Eurasian badger. *Animal Behaviour* 57, 153–161.

Stewart, P.D., Macdonald, D.W., Newman, C. & Cheeseman, C.L. (2001) Boundary faeces and matched advertisement in the European badger (*Meles meles*): a potential role in range exclusion. *Journal of Zoology* 255, 191–198.

Stewart, P.D., Macdonald, D.W., Newman, C. & Tattersall, F.H. (2002) Behavioural mechanisms of information transmission and reception by badgers, *Meles meles*, at latrines. *Animal Behaviour* 63, 999–1007.

Stoddart, D.M. (1980) *The Ecology of Vertebrate Olfaction.* Chapman & Hall, London.

Temeles, E.J. (1994) The role of neighbours in territorial systems: when are they 'dear enemies'? *Animal Behaviour* 47, 339–350.

Theis, K.R., Venkataraman, A., Dycus, J.A., Koonter, K.D., Schmitt-Matzen, E.N., Wagner, A.P., Holecamp, K.E. & Schmidt, T.M. (2013) Symbiotic bacteria appear to mediate hyena social odors. *Proceedings of the National Academy of Sciences of the United States of America* 110, 19832–19837.

Tinnesand, H.V., Buesching, C.D., Noonan, M.J., Newman, C., Zedrosser, A., Rosell, F. & Macdonald, D.W. (2015) Will trespassers be prosecuted or assessed according to their merits? A consilient interpretation of territoriality in a group-living carnivore, the European Badger (*Meles meles*). *PLoS One* 10, e0132432.

Tobin, M.E., Engeman, R.M. & Sugihara, R.T. (1995) Effects of mongoose odors on rat capture success. *Journal of Chemical Ecology* 21, 635–639.

Torgerson, T.J. (2014) *Latrine Site Selection and Seasonal Habitat Use of a Coastal River Otter Population.* PhD thesis, Humboldt State University.

Trapp, G.R. (1978) Comparative behavioural ecology of the ringtail and the gray fox in Southwestern Utah. *Carnivore* 1, 3–31.

Trivers, R.L. (1972) Parental investment and sexual selection. In: *Sexual Selection and the Descent of Man* (ed. B.G. Campbell), pp. 139–179. Aldine, Chicago.

Tsunoda, M., Kaneko, Y., Sako, T., Koizumi, R., Iwasaki, K., Mitsuhashi, I., Saito, M.U., Hisano, M., Newman, C., Macdonald, D.W. & Buesching, C.D. (2019) Human disturbance affects latrine-use patterns of raccoon dogs. *Journal of Wildlife Management* 83, 728–736.

Tuyttens, F.A.M., Long, B., Fawcett, T., Skinner, A. Brown, J.A., Cheeseman, C.L., Roddam, A.W. & Macdonald, D.W. (2001) Estimating group size and population density of Eurasian badgers *Meles meles* by quantifying latrine use. *Journal of Applied Ecology* 38, 1114–1121.

Verberne, G. (1976) Chemocommunication among domestic cats, mediated by the olfactory and vomeronasal senses. II. The relation between the function of Jacobson's organ (vomeronasal organ) and flehmen behaviour. *Zeitschrift für Tierpsychologie* 42, 113–128.

Verberne, G. & DeBoer, J. (1976) Chemocommunication among domestic cats, mediated by the olfactory and vomeronasal senses. I. Chemocommunication. *Zeitschrift für Tierpsychologie* 42, 86–109.

Vitale, J.D., Jordan, N.R., Gilfillan, G.D., McNutt, J.W. & Reader, T. (2020) Spatial and seasonal patterns of communal latrine use by spotted hyenas (*Crocuta crocuta*) reflect a seasonal resource defense strategy. *Behavioral Ecology and Sociobiology* 74, 1–14.

Walls, S.C., Mathis, A., Jaeger, R.G. & Gergits, W.F. (1989) Male salamanders with high-quality diets have faeces attractive to females. *Animal Behaviour* 38, 546–548.

Walther, F.R. (1978) Mapping the structure and the marking system of a territory of the Thompson's gazelle. *East African Wildlife Journal* 16, 167–176.

Ward, J.F., Macdonald, D.W. & Doncaster, C.P. (1997) Responses of foraging hedgehogs to badger odour. *Animal Behaviour* 53, 709–720.

Woodroffe, G.L. & Lawton, J.H. (1990) Patterns in the production of latrines by water voles (*Arvicola terrestris*) and their use as indices of abundance in population surveys. *Journal of Zoology* 220, 439–445.

Wrangham, R.W. (1982) Mutualism, kinship and social evolution. In: *Current Problems in Sociobiology* (ed. King's College Sociobiology Group), pp. 269–289. Cambridge University Press, Cambridge.

Yamamoto, I. (1984) Latrine utilization and feces recognition in the raccoon dog, *Nyctereutes procyonoides*. *Journal of Ethology* 2, 47–54.

Young, A.J., Spong, G. & Clutton-Brock, T.H. (2007) Subordinate male meerkats prospect for extra-group paternity: alternative reproductive tactics in a cooperative mammal. *Proceedings of the Royal Society B: Biological Sciences* 274, 1603–1609.

Zemouche, J. (2018) *Trophic Ecology of Rusty-Spotted Genet* Genetta maculata *and slender mongoose* Herpestes sanguineus *in Telperion Nature Reserve, with a Focus on Dietary Segregation as a Possible Mechanism of Coexistence.* MSc thesis, University of the Witwatersrand.

Zhang, J.X., Ni, J., Ren, X.J., Sun, L., Zhang, Z.B. & Wang, Z.W. (2003) Possible coding for recognition of sexes, individuals and species in anal gland volatiles of *Mustela eversmanni* and *M. sibirica*. *Chemical Senses* 28, 381–388.

Zhang, J.X., Liu, D., Sun, L., Wei, R., Zhang, G., Wu, H., Zhang, H. & Zhao, C. (2008) Potential chemosignals in the anogenital gland secretion of giant pandas, *Ailuropoda melanoleuca*, associated with sex and individual identity. *Journal of Chemical Ecology* 34, 398–407.

Zhou, Y., Chen, W., Kaneko, Y., Newman, C., Liao, Z., Zhu, X., Buesching, C.D., Xie, Z. & Macdonald, D.W. (2015a) Seasonal dietary shifts and food resource exploitation by the hog badger (*Arctonyx collaris*) in a Chinese subtropical forest. *European Journal of Wildlife Research* 61, 125–133.

Zhou, Y., Chen, W., Buesching, C.D., Newman, C., Kaneko, Y., Xiang, M., Nie, C., Macdonald, D.W. & Xie, Z. (2015b) Hog badger (*Arctonyx collaris*) latrine use in relation to food abundance: evidence of the scarce factor paradox. *Ecosphere* 6, 19.

Ziko, B.A. (2018) *Latrine Site Characteristics and Selection by Genets (*Genetta spp.*) in the Great Fish River Reserve.* BSc Honours thesis, University of Fort Hare, Alice.

Zuri, I., Gazit, I. & Terkel, J. (1997) Effect of scent-marking in delaying territorial invasion in the blind mole-rat *Spalax ehrenbergi*. *Behaviour* 134, 867–880.

8

Factors Affecting European Badger Movement Lengths and Propensity: Evidence of Density-Dependent Effects?

Andrew W. Byrne[1,2,3,]*, James O'Keeffe[1,3], and S. Wayne Martin[4]

[1] Centre for Veterinary Epidemiology and Risk Analysis, School of Veterinary Medicine, University College Dublin, Dublin, Ireland
[2] Veterinary Sciences Division, Bacteriology Branch, Agri-Food and Biosciences Institute, Belfast, UK
[3] Department of Agriculture, Food and the Marine, Dublin, Ireland
[4] Department of Population Medicine, Ontario Veterinary College, University of Guelph, Guelph, Canada

SUMMARY

Understanding the mechanisms underpinning animal movement patterns is one of the key goals of animal ecology. The motivation to move across populations can be driven by a number of factors, including finding new mates, reducing competition or exploiting new resources. The movement ecology of wildlife hosts of zoonotic diseases – e.g. European badger, *Meles meles*, a reservoir of bovine tuberculosis – is also important when attempting to manage spill-back infection to humans or domestic animals. We studied badger movements, using mark–recapture data (2008–2012) at a large spatial scale (755 km^2) in Ireland. We investigated both intrinsic (sex, age-class, or weight at capture) and extrinsic (territory size, group size, or population density) factors that may have affected either movement length or the propensity to move across putative territorial boundaries. We constructed several models using differing metrics of territory size and density, forming a matrix of competing models, from which we assessed similarities and differences. Older badgers tended to make shorter movements relative to other age classes. Movement length increased with greater time intervals between captures. Importantly, there was negative density-dependence with movement length; shorter movements were associated with higher-density areas. The propensity to move across putative territories varied depending on the metrics of territory configuration or badger abundance. Across models, there was a general trend toward lower movement propensity for older badgers and higher densities (or group sizes) and a higher propensity with increasing time between captures. Taken together, our data suggest that there are density-dependent mechanisms affecting movement patterns in badgers within subpopulations. Badgers in higher density areas generally exhibited shorter and less frequent movements than badgers in lower-density areas. However, overall, there was no net tendency for badgers to move into higher- or lower-density areas. These findings help us understand badger movement ecology and will have implications for understanding bovine tuberculosis dynamics across badger populations.

Keywords

Bovine tuberculosis – dispersal – Ireland – mark–recapture – *Meles meles* – movement ecology – wildlife management

Introduction

Understanding underlying mechanisms that shape the patterns of wild animal movements is one of the most enduring, and challenging, research themes in ecology (Lidicker & Stenseth, 1992; Nathan *et al.*, 2008; Clobert *et al.*, 2012). While recording animal movement patterns, and the characterization of these patterns (e.g. in terms of frequency and distance), is an essential component to understanding how animals

---

* Corresponding author.

Small Carnivores: Evolution, Ecology, Behaviour, and Conservation, First Edition. Edited by Emmanuel Do Linh San, Jun J. Sato, Jerrold L. Belant, and Michael J. Somers.

utilize their environment (*how* questions), it is limited in its ability to give a mechanistic understanding of underlying processes (*why* questions). In recent years, animal movement research has started to develop theoretical frameworks, supported by empirical exploration, to help explain why animals move – what are the 'motivations' that drive movement? An emerging transdisciplinary paradigm (Nathan *et al.*, 2008) has developed out of this thinking, 'emphasizing the need to understand the movement of living organisms . . . in the context of their internal states, traits, constraints, and interactions among themselves and with the environment' (Nathan & Giuggioli, 2013). Whereas establishing what motivations are driving movement 'choices' within individual animals proves extremely challenging, across populations internal and external *information* and intrinsic and extrinsic *traits* can be estimated (Lidicker & Stenseth, 1992; Clobert *et al.*, 2012). These can be used to make inferences about potential mechanisms that might be involved. For example, the *intrinsic motivation* to move may be greater in one sex (for example, males looking for mates), but this motivation may be modulated by *external information* about the environment (e.g. proximal dangers and more distal risks, such as unfamiliarity with foraging landscape).

Density is one key parameter that has the potential to affect movement characteristics across spatially structured populations (Benton & Bowler, 2012). When animals move across density gradients, there may be positive, negative, or combination (u-shaped) density-dependent (DD) relationships with movement parameters (Matthysen, 2005; Kim *et al.*, 2009). Positive density-dependence occurs where increasing density results in increasing emigration (Matthysen, 2005). This pattern has been recorded in many animal systems and is the most common hypothesis presented in dispersal and movement studies (Matthysen, 2005; Kim *et al.*, 2009). It occurs due to the negative effects of overcrowding (general decrease of environmental conditions) or where resources are exploited to the maximum carrying capacity. Negative density-dependence relates to decreasing emigration within increasing density. This phenomenon can occur when there are benefits to aggregating (e.g. predator vigilance) or if there are strong pressures against moving away imposed by conspecifics ('social fence'; Hestbeck, 1982)

who contribute to a type of 'social viscosity' (Byrne *et al.*, 2014b). Combination patterns have been recorded, where there is increased emigration at very high densities (e.g. due to the negative effects of overcrowding) and also at very low densities (e.g. an animal cannot find a mate; therefore, it emigrates to find one; commonly referred to as an 'Allee effect' [Courchamp *et al.*, 2008]), but lower relative emigration at moderate densities (Kim *et al.*, 2009). There may be occasions where movement could be affected by density, but not by the intrinsic gradient in density across space. For example, it is known that European badger, *Meles meles*, territories are larger in poor habitats (lower carrying capacity) such as uplands, but territories are much smaller in lowland pastoral landscapes (Feore & Montgomery, 1999). Thus, we might expect to record larger movements in the larger territories than in the smaller territories, simply by virtue of their size, but not necessarily expect any net movement bias at the interface of the two habitat types. The carrying capacity of the neighbouring environment could be an important factor in these instances, and, therefore, we could even observe movements from low- to high-density environments.

In the current study, we investigated intrinsic traits and extrinsic factors that were associated with metrics of movement (both intra- and inter-territorial movements) by European badger using mark–recapture data. European badgers are a species within the family Mustelidae and are unusual within this taxon because they exhibit nascent sociality (Macdonald, 1983; Revilla & Palomares, 2002). Badgers form 'social groups', which are composed of a variable number of individuals (from 2 to > 20) that cohabit in communal burrows (setts) and utilize a group range which is also referred to as a territory (Roper, 2010; Byrne *et al.*, 2012b). These territories, in high-density populations, can be defended through aggression and demarcated by scent marking at border latrine pits (Kruuk, 1989; Kilshaw *et al.*, 2009; Roper, 2010). In lower-density populations, territoriality appears to become more flexible and territorial boundaries may be more diffuse, without elevated levels of aggression or scent marking at borders (Hutchings *et al.*, 2002; Revilla & Palomares, 2002; Do Linh San *et al.*, 2007). Genetic studies have suggested that badgers are polygynandrous, meaning that both males and females do not mate with exclusive partners, with multiple individuals within a social group mating and breeding within a given year (Annavi *et al.*, 2014). In high-density populations, badgers generally exhibit natal philopatry, with reduced permanent dispersal propensity (Macdonald *et al.*, 2008). In lower-density populations, the social structure may be more fluid (Revilla & Palomares, 2002; Byrne *et al.*, 2014b, 2019; Gaughran *et al.*, 2018, 2019). However, even at high densities, temporary inter-group movements and extra-group mating occur frequently as a mechanism to avoid inbreeding (Macdonald *et al.*, 2008; Annavi *et al.*, 2014), though group members are usually relatively closely related (within-group relatedness; Annavi *et al.*, 2014). Badgers are omnivorous, feeding on soil invertebrates (earthworms and insect larvae), but also seasonally abundant fruits and occasionally small mammals and birds (Byrne *et al.*, 2012b). Home range size and socio-spatial organization may also be affected by the composition and configuration of these resources within the landscape (Johnson *et al.*, 2001). Due to these sources of variation, it is difficult to assess where territorial boundaries are located in lower-density populations, especially if they are unstable or changing over time (Revilla & Palomares, 2002). We used simple methods to demark 'territories' within our study population (mean density within our study population: ~1 badger km^{-2}; Byrne *et al.*, 2012a), based on the location of main setts and regular grids. Using these spatial structures, we were able to ask questions about the relative movement propensity across 'groups' and the relationship between movement length and density.

An understanding of wildlife movement is not only of ecological significance, but can also be of epidemiological importance. Wild animals can act as disease reservoirs, and understanding movement behaviour is critical to developing and evaluating various management strategies aimed at containing or reducing the spread of zoonotic infections. Badgers are an important wildlife host of bovine tuberculosis (Delahay *et al.*, 2002; Gortazar *et al.*, 2012) caused by the bacterial infection of *Mycobacterium bovis* (Corner *et al.*, 2011). Indirect evidence suggests that they are implicated in the transmission of the disease to cattle in Britain and Ireland (Tolhurst *et al.*, 2009; Biek *et al.*, 2012; Allen *et al.*, 2018; Campbell *et al.*, 2019;

Milne *et al.*, 2020) with interrelated biological factors such as badger density, population structure, social organization, group size, movement and inbreeding affecting disease incidence, prevalence, and progressed infection in badgers, and, therefore, disease transmission (Vicente *et al.*, 2007; Woodroffe *et al.*, 2009; Benton *et al.*, 2016, 2018; McDonald *et al.*, 2018; Rozins *et al.*, 2018). The ensuing animal health issues and economic hardship to farmers have brought about drastic management measures, including badger removal in targeted areas (Bourne *et al.*, 2007; Brunton *et al.*, 2017; Downs *et al.*, 2019; Martin *et al.*, 2020). Culling operations have themselves led to changes in badger behaviour, social organization and movements (Tuyttens *et al.*, 2000; Carter *et al.*, 2007; Riordan *et al.*, 2011; Ham *et al.*, 2019) with contrasting effects on the prevalence of bovine tuberculosis in cattle (Donnelly *et al.*, 2006; Woodroffe *et al.*, 2006; Corner *et al.*, 2008; Byrne *et al.*, 2015a). Altogether, the control of this chronic infectious disease is challenging and impeded by both ecological and epidemiological factors (Allen *et al.*, 2018). The results of the present study not only help inform about the movement ecology of badgers but also the management and epidemiology of bovine tuberculosis in this species.

Materials and Methods

Study Area

Badgers were captured as part of a mark–recapture study arising from a badger-vaccination trial which ran from summer 2008 to summer 2012 (see Byrne *et al.*, 2012a, 2014b). The study area covered 755 km^2 of north-western Co. Kilkenny (52.6477 °N, 7.2561 °W) in the Republic of Ireland. The area was comprised of predominantly agricultural low-altitude rich pasture land divided by an extensive hedgerow network (Byrne *et al.*, 2012a).

Badger Population and Capturing Regime

In Ireland, badgers usually form social groups (typically mixed-sex groups of three to six adults, with the young of the year) that inhabit territories (see above; Byrne *et al.*, 2012a). Badger setts are typically located within hedgerow or woodlands (Byrne *et al.*, 2012b). These setts can be categorized into 'main' setts and 'non-main' setts using combinations of the number of openings (non-main setts usually had less than three openings), the activity levels and the presence of a conspicuous spoil-heap (Byrne *et al.*, 2012b). Main setts are usually in continuous use by a badger social group and are often the location where cubs are born. Typically there is only one main sett per social group. All setts were surveyed for activity twice per year. Attempts to capture badgers were made at all active setts twice per year; sett activity was assessed using cues such as the presence of fresh spoil and bedding material (Byrne *et al.*, 2013a). Within a capturing session, traps were laid for eight nights, over a two-week period at each active sett (Byrne *et al.*, 2012a). For more information on the badger population within this study site, see Byrne *et al.* (2012a, 2014b).

Badger-capturing methods conformed to national legislation for the humane trapping of wildlife (Wildlife Act, 1976, Regulations 2003, S.l. 620 of 2003). Badgers were captured under license (1876 Cruelty to Animals Act; Irish Department of Health & Children), and the work on badgers (capturing, marking, vaccinating, and releasing) was approved by the University College Dublin animal ethics committee. Badgers were captured predominantly using wire-stopped restraints. Supplementary capture effort using baited steel wire-mesh cages was also employed (Byrne *et al.*, 2012a). Both these capturing methodologies have been shown to result in very few injuries to captured animals (Woodroffe *et al.*, 2005; Murphy *et al.*, 2009; Byrne *et al.*, 2015b). Each captured badger was anaesthetized by an on-site veterinarian. At first capture, badgers were tattooed and a microchip inserted to ensure that each badger could be uniquely identified. Badger age-class (young < 18 months, adult, or old, based on an assessment of tooth wear following Murphy *et al.*, 2010), sex and weight were recorded at each capture. 'Old' animals were categorized subjectively by having heavily worn teeth, but their precise age could not be determined.

Analyses

Badger movements were inferred from the mark–recapture histories of badgers within the study area (Byrne *et al.*, 2014b). Our outcome (dependent) variables were either movement length or movement propensity (i.e. a movement score – see below). We measured movement lengths as the Euclidean (straight line) distance between capture points (usually setts) using a Geographical Information System (GIS; ArcGIS®). Capturing can disrupt the normal movement patterns of wildlife (e.g. Schütz *et al.*, 2006; Arzamendia & Vilá, 2012; Quinn *et al.*, 2012); therefore, we ensured that there were ≥ 3 days between recaptures (Schütz *et al.*, 2006) when measuring movements (7.9% of movements occurred within this period of ≤ 3 days).

All statistical modelling was undertaken in Stata 11® (Statacorp, College Station, TX, 2009). We used a Generalized Linear Mixed Model (GLMM) to model factors affecting movement length, with badger being the random variable (movements clustering within badgers; xtreg in Stata 11®) when we used a single abundance metric for the total study period (i.e. total number of individuals per territory). We also used time-varying abundance metrics (i.e. yearly number of individuals per territory) and in these analyses, we used multi-level models with two hierarchies for year and badger identification (xtmixed in Stata®). Likelihood-ratio tests were used to compare the mixed models against their nested (non-cluster adjusted) linear-regression equivalents. The movement lengths were log-transformed in order to meet the assumptions of a linear model (Dohoo *et al.*, 2009). A skewness/kurtosis test for normality suggested that the distribution did not significantly deviate from a normal distribution after transformation ($\chi^2 = 4.53$, df = 2, $p = 0.104$).

We assessed movement propensity following a similar approach to Rogers *et al.* (1998). We created a binary outcome variable that represented a movement score. Each observation was a capture, with a score of 1 if a badger was caught at a different social-group territory than the previous capture, or a score of 0 if caught in the same territory. This binary movement score was used in logistic mixed models (xtlogit for total counts; xtmelogit for yearly counts; Stata 11®) to assess which factors influenced the propensity of badgers to move across territories. We used a stepwise backward selection procedure to arrive at parsimonious final models. Competing models were compared using Akaike's information criterion (AIC), and models with the smallest values were preferred. In order to use AIC, we implemented the models with Maximum Likelihood Estimation (MLE) during model fitting. Robust standard errors were used throughout.

We estimated simple metrics of local density in a number of ways. Firstly, we enumerated the total number of unique badgers identified during the full study period (2008–2012) in each defined 'territory'. As it is likely that this method overestimated group size (positive bias), we also enumerated the total unique badger numbers per territory captured per year (with years starting on 1 July). Finally, we estimated the minimum number alive (MNA) for each territory. We followed Byrne *et al.* (2012a) when estimating MNA, using the information on badgers captured in territories prior to, and after the capture period of interest (using yearly periods), instead of capture sessions that were used in Byrne *et al.* (2012a). We used data from the first and last years of capturing to inform the estimate for the intervening years; therefore, we could not estimate MNA for those (first and last) years. As trappability was relatively low during this study, the MNA estimates are underestimates (negative bias) of the true group size (Byrne *et al.*, 2012a; Byrne & Do Linh San, 2016). When a badger was caught in two or more territories during a year, the badger was added to both group counts because there was no reliable way to assign the 'home' territory. We converted group sizes into density estimates by dividing group size by the 'territory' area – see below for details on territory delineation.

In order to estimate local metrics of density, we had to estimate the possible configuration of territories within the study area. To do this, we created tessellations (also called Voronoi diagrams and Thiessen polygons), using the location of main setts as the 'seed' for each polygon (Hammond & McGrath, 1998). Tessellations are formed by drawing perpendicular straight lines at the halfway point between two points (setts) within an array. These lines stop when they intersect with other lines, forming a two-dimensional lattice. The resulting shapes have been used to estimate the configuration and extent of badger territories (Hammond & McGrath, 1998; Woodroffe *et al.*, 2009) and they work reasonably well in comparison with empirical data on badger territories that were

delineated using bait-marking and telemetry techniques (Doncaster & Woodroffe, 1993; but see Roper, 2010). These territories can be improved (more accurately representing the size and configuration of territories) by using the location of boundary latrines (these can indicate the frontier between two adjoining territories in high-density populations; Woodroffe *et al.*, 2009). However, we did not have data on latrine locations during the present study. Furthermore, previous research suggested that badgers make less boundary markings in lower-density populations (Hutchings *et al.*, 2002; Revilla & Palomares, 2002; Do Linh San *et al.*, 2007). Therefore, the configuration of these territories remains as a heuristic indicator of possible territory shape only. We added a constraint to the growth of tessellations (as employed by Halls *et al.*, 2001), by not allowing any polygon to extend more than 2 km from the main sett (theoretical maximum territory size without neighbours: 12.57 km^2), as there is no empirical evidence to suggest that territories extend greater distances than this in Ireland (see Byrne *et al.*, 2012b). Furthermore, due to the potential for edge effects (due to our limited knowledge of the location of main setts outside our study area), we excluded all territories that overlapped with our study area boundary. This approach has been used previously for comparable large-scale projects (Woodroffe *et al.*, 2009), and offers the best metric of the true configuration of the territories when data are limited (i.e. without tracking data or bait-marking techniques). However, Roper (2010) has criticized the unguided use of Thiessen polygons for demarking badger territories (also see Blackwell & Macdonald, 2000). Therefore, to ensure that this metric did not affect our inferences significantly, we repeated our analyses using standardized grids of 2 × 2 km and 3 × 3 km, respectively (a similar approach has been employed by Woodroffe *et al.*, 2009).

In addition to metrics of badger density, we investigated whether movement length and propensity varied depending on sex, age-class, and weight at capture (see Table 8.1 for a list of independent variables). We also explored first-order interactions among pairs of these variables. We hypothesized that the time between captures may be an important predictor of movement, as greater time differences between captures would allow badgers more time to make a dispersal attempt (and/or to move further away from the source).

Table 8.1 Independent variables used during the multivariable model building of the length of the European badger, *Meles meles*, movements (km) and the propensity for badgers to move across territories in separate models.

Predictor	Description
Life stage	Young, adult, and old (categorical)
Sex	Male = 1; Female = 0 (categorical)
Weight	Weight at capture (kg; continuous)
Period	The time between captures (years; continuous)
Order	Order of movements (first, second, etc., the movement made; categorical)
No. captured ('Group' size)	Number of individual badgers associated with a territory (grid or tessellated polygon)
Territory area	Area of tessellated territory (not in grid models due to standard size) (km^2; continuous)
Density	Counts per unit area (separate models for Σcounts km^{-2} and Σyearly counts km^{-2})

We also assessed if movement order (whether the movement was the first, second, etc., of the sequence recorded) had any effect on movement length or propensity to move.

Due to the complex set of predictors – three metrics of group size and density (total enumeration, yearly counts, and MNA) and three metrics of territory configuration (tessellations, 2×2 km and 3×3 km grids) – we arrived at a matrix of final models. Our intention with these competing models was to help understand the underlying mechanisms influencing movement parameters within this population; therefore, we highlight both the similarities and differences in the outcomes and approaches.

Finally, we wanted to see whether there was evidence of a net directional bias, either going from low-to-high, high-to-low, or moving to an area of similar density. To do this, when badgers moved between putative territories, we subtracted the original density from the subsequent density (again, using core territories). Negative values indicated badgers moving from higher to lower densities. Positive values indicated badgers moving from lower to higher densities. Values close to zero indicated movements between territories with similar densities.

Results

Number of Territories, Group Size, and Population Density Estimates

Figure 8.1 depicts the configuration of territories within the study site as generated using tessellations. There were 213 core territories (non-overlapping with study boundary) within the study population, 14 of which did not yield a badger capture over the study period (6.6%). The mean ($\pm$SD) territory area within the core was 2.74 ± 1.40 km^2 (range: 0.44–8.25 km^2). Group size varied depending on metric used, with mean group-size estimates from yearly counts being 2.26 ± 1.42 and MNA of 2.34 ± 1.35. MNA was not substantially larger than yearly counts due to modest trappability (Byrne *et al.*, 2012a; and see Byrne & Do Linh San, 2016 for discussion) and was very strongly correlated with yearly counts (Pearson's correlation: $r = 0.98$, $p < 0.001$). As these two metrics essentially contribute the same information, we explored further the simpler metric (yearly counts) during our analyses. The mean total count of badgers per territory was 4.83 ± 2.67. There was a moderate correlation between yearly counts and MNA with total counts, respectively (both $r = 0.56$, $p < 0.001$). Despite this correlation, we ran separate models with total counts and with yearly counts.

In the core area, there were 746 badger setts recorded. The mean ($\pm$SD) sett density within tessellated territories was 3.50 ± 2.30 km^2. Mean badger densities in the core area using different territory-boundary delineation techniques are presented in Table 8.2. Tessellated territories generated the greatest mean density and the largest variation in density across space (largest range); whereas the 3×3 km grid generated the lowest mean density and the least variation in density across space. In total, 474 movements were recorded during this study, with a mean movement length of 1.77 ± 2.48 km (median = 1.19 km; Figure 8.2).

Factors Influencing Badger Movement Lengths

Total Counts

Irrespective of territory type (tessellation or grid), the length of badger movements was affected significantly by age class. 'Old' badgers made significantly shorter

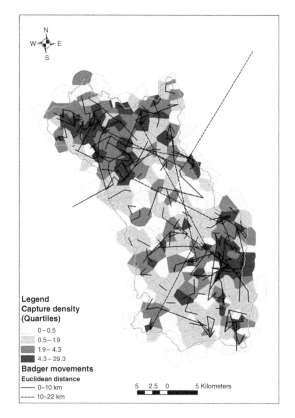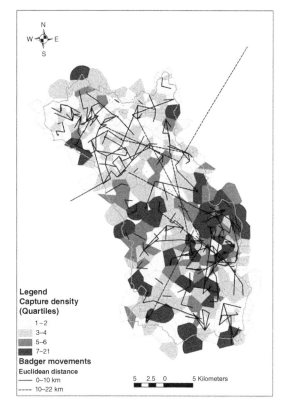

Figure 8.1   European badger, *Meles meles*, movement patterns and the configuration of tessellation-derived territories within the study area, in Kilkenny, Ireland. Metrics of badger density are presented as the total count of badger captures (left panel) and setts per unit area respectively (right panel). Territories that intersect with the study area boundary (grey line) may have edge effects and were thus discarded from the analysis.

Table 8.2   Badger densities (km^{-2}) based on yearly or total counts of badgers captured within territories of different types.

Territory type	Yearly counts		Total counts	
Core territories	Mean ± SD	Range	Mean ± SD	Range
Tessellation	1.09 ± 0.98	0.14–6.21	2.34 ± 2.01	0.14–11.61
Grid (2×2 km)	0.78 ± 0.57	0.25–2.75	2.37 ± 1.46	0.25–7.75
Grid (3×3 km)	0.58 ± 0.44	0.11–2.00	1.82 ± 1.01	0.11–4.33

movements than other badgers (mean $\beta = -0.3$, $p < 0.05$; Table 8.3), with the mean distance moved being 1.64 km (SD = 2.91 km) for old badgers versus 2.00 km (SD = 2.66 km) for other age-classes, respectively. There was no significant difference between adult and young badgers. The amount of time elapsed

between captures was associated significantly with increased movement lengths (mean per annum $\beta = 0.21$, $p < 0.01$; Figure 8.3). Movement length was also associated with movement order, but only in a model using 2×2 km grid squares ($p < 0.05$). The trend was toward longer movements during later movement

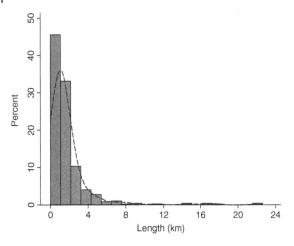

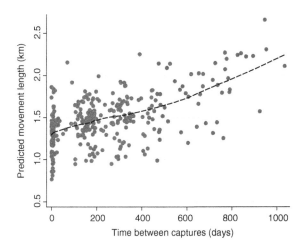

Figure 8.2 Distribution of all movement lengths recorded during a mark–recapture study of badgers from 2009 to 2012. The distribution is highly skewed, with infrequent long-distance movements, the longest of which was 22 km. The dashed line represents the kernel density of the distribution (bandwidth: 0.7 km).

Figure 8.3 Predicted relationship between badger movement length (km; y-axis) and the time period between captures (days; x-axis). The dashed line represents a locally weighted regression line (LOWESS line).

events. Finally, there was an effect of density on movement length, with shorter movements being associated with higher-density social groups (mean $\beta = -0.0863$; Figure 8.4) at the arrival point. This relationship was significant at an alpha of 0.05 for the models that used tessellations or 3×3 km^2 grid squares, whereas the relationship only approached significance using grid squares of 2×2 km^2 ($p < 0.1$; Table 8.3).

Table 8.3 Outcome from mixed models (random effects within badger) relating the length of badger movements and local metrics of badger density using the total counts of individuals captured per territory over the study period (2008–2012), in Kilkenny, Ireland. Density was calculated using tessellations and regular grids of 2×2 km and 3×3 km.

Outcome	Tessellated territories ln (length)	2 × 2 km grid ln (length)	3 × 3 km grid ln (length)
Age (old/non-old)	−0.302*	−0.387**	−0.280*
	(0.017)	(0.007)	(0.049)
Period	0.193**	0.225**	0.207**
	(0.005)	(0.005)	(0.009)
Density	−0.0881**	−0.0639*	−0.107*
	(0.001)	(0.096)	(0.041)
Order		0.136*	
		(0.049)	
Constant	0.466***	0.197	0.447***
	(<0.001)	(0.195)	(0.001)

*Significant at $p < 0.05$
**Significant at $p < 0.01$
***Significant at $p < 0.001$

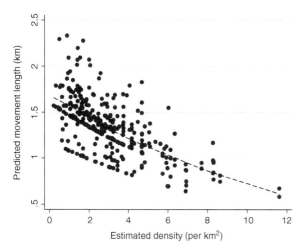

Figure 8.4 Predicted relationship between badger mean movement length (km; y-axis) and a metric of badger density (total counts per tessellated territory per km²; x-axis) at movement destination. The dashed line represents a locally weighted regression line (LOWESS line).

Yearly Counts

Very similar results were found when using yearly badger counts instead of total badger counts (see Table 8.4). Three significant factors affected movement length across territory types: age (old vs. all other groups), the period between captures, and density ($p < 0.05$). This broad agreement between modelling approaches lends strong support to the general outcomes.

Factors Influencing Badger Movement Propensity

Total Counts

There was a general difference in factors affecting movement propensity, using total counts as a metric of abundance, between tessellations and regular grid territories (Table 8.5). Using tessellations, movement propensity was affected significantly by the period of time between captures ($p < 0.001$), sex ($p = 0.001$), the size of the territory ($p = 0.042$), and 'group size' (number of unique individuals captured; $p = 0.001$). The longer the time between captures, the higher the probability of trans-territorial movements occurring ($\beta = 0.712$, OR $= 2.04$ per annum). Males had 1.95 greater odds of having a positive movement score relative to females ($\beta = 0.670$, OR $= 1.95$). The larger the group size, the lower the probability of a (immigrant) trans-territorial movement score being recorded

Table 8.4 Outcome from mixed-models (random effects at two levels: within badger and within year) relating the length of badger movements and local metrics of badger density using yearly counts of individuals captured per territory over the study period (2008–2012) in Kilkenny, Ireland. Density was calculated using tessellations and regular grids of 2 × 2 km and 3 × 3 km.

Outcome	Tessellated territories ln (length)	2 × 2 km grid ln (length)	3 × 3 km grid ln (length)
Age (old/non-old)	−0.359**	−0.380**	−0.326*
	(0.005)	(0.006)	(0.026)
Period	0.252***	0.280***	0.269***
	(0.001)	(<0.001)	(0.001)
Density	−0.154***	−0.183*	−0.310**
	(<0.001)	(0.014)	(0.005)
Constant	0.378***	0.369**	0.465***
	(<0.001)	(0.002)	(<0.001)

*Significant at $p < 0.05$
**Significant at $p < 0.01$
***Significant at $p < 0.001$

Table 8.5 Predictors of badger movement propensity with badger density based on the total count of individuals captured per territory over the study period (2008–2012) in Kilkenny, Ireland. Density was calculated using tessellations and regular grids of 2 × 2 km and 3 × 3 km.

Outcome	Tessellated territories	2 × 2 km grid	3 × 3 km grid
Age (old/non-old)		−0.713*	−0.574
		(0.038)	(0.115)
Period	0.712***	0.858***	0.813***
	(<0.001)	(<0.001)	(<0.001)
Density		−0.233*	−0.285*
		(0.012)	(0.043)
No. captured	−0.118***	[−0.058*]	[−0.032*]
	(0.001)	(0.012)	(0.043)
Territory area	−0.160*		
	(0.042)		
Sex	0.670**	0.900***	0.761**
	(0.001)	(0.001)	(0.008)
Constant	−0.863	−2.113***	−2.238***
	(0.054)	(<0.001)	(<0.001)

* Significant at $p < 0.05$
** Significant at $p < 0.01$
*** Significant at $p < 0.001$

($\beta = -0.118$, OR = 0.89 per unit increase in group size). Larger territories were associated with a significantly lower probability of a trans-territorial movement score ($\beta = -0.160$, OR = 0.85 per km²-increase in territory size).

Using regular grids to define territories resulted in age-class being a significant factor affecting movement propensity, with 'old' badgers having 0.53 times the odds of movement than other age classes (mean $\beta = -0.644$, $p = 0.077$). Similar to the tessellation models, there were greater odds of movement if a badger was male relative to females (mean $\beta = 0.831$, OR = 2.30, $p < 0.01$) and with an increasing period between captures (mean $\beta = 0.836$, OR = 2.31, $p < 0.001$). As the grid was regular in shape, there was no variation in grid size across territories. Therefore, the group size and density estimates were essentially scaled versions of each other (density was simply group size divided by four or nine, depending on the grid size – 4 or 9 km²). Both the grid models inferred an association between greater badger numbers and propensity to move ($p < 0.05$), with either

a greater social-group size or density being the explanatory mechanism. There was a general trend towards lower probability of movement into a territory with increasing density or group size ($\beta = -0.259$, OR = 0.77 for a unit increase in density; $\beta = -0.045$, OR = 0.96 per unit increase in group size).

Yearly Counts

There were different outcomes from models using yearly counts as a metric of abundance depending on how territories were defined (Table 8.6). Across territory types, there was evidence of the greater probability of movement if a badger was male (mean $\beta = 0.857$, OR = 2.35, $p < 0.01$) and with an increasing period between captures (mean $\beta = 0.825$, OR = 2.28, $p < 0.001$). In the model using tessellations as an indicator of territories, lower probabilities of trans-territorial movements were associated with larger territory sizes ($\beta = -0.216$, OR = 0.81, $p = 0.012$). Evidence in support of a relationship between age-class (old vs. all other types: $p = 0.043$) and movement

Table 8.6 Predictors of badger movement propensity with badger density based on yearly counts of individuals captured per territory over the study period (2008–2012) in Kilkenny, Ireland. Density was calculated using tessellations and regular grids of 2 × 2 km and 3 × 3 km.

Outcome	Tessellated territories	2 × 2 km grid	3 × 3 km grid
Age (old/non-old)		−0.718*	
		(0.043)	
Period	0.773***	0.872***	0.830***
	(<0.001)	(<0.001)	(<0.001)
Density		−0.347*	
		(0.040)	
Territory area	−0.216*		
	(0.012)		
Sex	0.726**	0.902**	0.942**
	(0.002)	(0.001)	(0.002)
Constant	−1.669***	−2.341***	−3.316***
	(<0.001)	(<0.001)	(<0.001)

*Significant at $p < 0.05$
**Significant at $p < 0.01$
***Significant at $p < 0.001$

propensity, and yearly count density and movement propensity ($p = 0.040$), was found only in the 2×2 km grid model. 'Old' animals exhibited lower movement propensities than other age-classes ($\beta = -0.718$, OR = 0.49, $p = 0.043$). Lower movement propensities (into territories) were recorded with higher density ($\beta = -0.347$, OR = 0.71, $p = 0.040$).

Net Movement Bias
There was no net movement bias detected during the study (Figure 8.5), with a similar number of movements from lower to higher total count density areas using tessellated territories (49.4%), as there were movements from higher to lower density areas (49.4%). The mean change in density was 0.062 badgers km^{-2} (median = 0, 95%CI: −4.467–4.440), indicating that most movements between territories did not involve a substantial change in density. Similar results were found using a 2×2 km regular grid (median = 0, mean = −0.007, 95%CI: −2.75–2.75) and a 3×3 km regular grid (median = 0.472, mean = 0.678, 95%CI: −1.556–2.722).

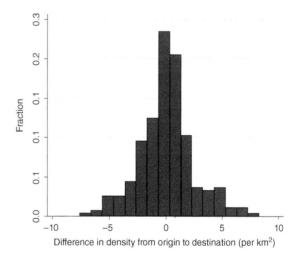

Figure 8.5 The frequency of movement types between badger density (here measured using territories defined by tessellations) at the origin and the destination of movements (mean = −0.044, 95%CI: −4.467–4.440). Negative values: badgers moving from higher to lower densities; Positive values: badgers moving from lower to higher densities; Zero values: no net difference in density between locations.

Discussion

Age-Class, Sex, and Time

We found that older badgers tend to make shorter movements and exhibit a lower propensity to cross putative territorial boundaries. This contrasts with previous studies in high-density populations where there was no significant difference amongst adult age-classes and the probability to move (e.g. in Gloucestershire or Oxford, England; Rogers *et al.*, 1998; Macdonald *et al.*, 2008). However, movements defined as 'dispersal' within an Oxford population were most likely to occur within the first year of a badger's life (Macdonald *et al.*, 2008). Age-classes were crudely estimated (young, adult, and old) during the present study due to the difficulty in aging badgers in the field (Harris *et al.*, 1992), and few young badgers were captured due to the low trappability of this age-class (Byrne *et al.*, 2012a). 'Old' in our context likely represents a smaller surviving population, which had already made attempts to disperse during earlier periods of life. Alternatively, in our study population, old badgers may represent more 'dominant' members of social groups (Revilla & Palomares, 2002), which had access to resources (mates and food) within their own territory without venturing out to neighbouring territories. Revilla & Palomares (2002) found that in a low-density population (Doñana, Spain), the oldest male badgers were 'dominant', retaining and expanding territories at the cost of subordinates. In this low-density system, 'dominant' females were reproductively active and most closely associated with the dominant males in their group, who were less likely to disperse than subordinates. It is possible that such a hierarchy exists within our study population, but further detailed research is required to test this hypothesis.

Sex was associated significantly with movement propensity, but not movement length during the present study. Males tended to make significantly more (putative) trans-territorial moves than females; this has been reported previously from this population (Byrne *et al.*, 2014b, 2019) and others (Rogers *et al.*, 1998; Revilla & Palomares, 2002; Macdonald *et al.*, 2008; but see Woodroffe *et al.*, 1993; Christian, 1994). A recent study based on our study population found that dispersal kernels (movements > 1 km) differed significantly between sexes (Byrne *et al.*, 2014b), with a higher frequency of dispersal attempts by males, but a relatively higher proportion of long-distance dispersals (LDDs) by females. We put forward a hypothesis for explaining the discordancy amongst studies with regard to sex-bias within badger populations, where movement biases varied according to the spatial scale of the study undertaken (Byrne *et al.*, 2014b). This scale-dependency may also be expanded to the temporal resolution of the movement phenomenon – as temporary movements have been shown to vary by sex, but not permanent (dispersal) movements (Macdonald *et al.*, 2008).

The period of time between captures was consistently a significant predictor of both movement length and propensity across analyses. In terms of propensity, we know that badger dispersal processes can take long periods of time and can be made up of a series of temporary exploratory movements to the recipient territory (Roper *et al.*, 2003). Therefore, increasing the time period between captures increases the prospect of a dispersal event occurring. In terms of movement length, we could interpret the relationship to indicate that badger dispersal could be thought of as a *diffusion*-type process. Badgers may make long displacements by 'hopping' across 'stepping-stone' social groups and taking up temporary residence before moving on, as opposed to a single long-distance movement. Considering the constraints for badgers to perceive their environment beyond their home range and that of their nearest neighbours (Bodin *et al.*, 2006), this 'stepping-stone mechanism' is plausible (Saura *et al.*, 2014). Saura *et al.* (2014) demonstrated that such a mechanism can be vital for the spread of species over long distances (LDDs) and, ultimately, contributes to the persistence of species across their range.

Density

We have found evidence of density affecting the propensity and lengths of badger movements. However, these patterns are not simple, with little evidence that badgers, overall, move up or down density gradients. Indeed, we found almost equal extents of up-gradient and down-gradient movements (irrespective of how we measured territories) during the study period. This might indicate that the population is currently relatively stable, without a large net flux in movement across gradients. In fact, most inter-territorial movements occurred across near-neutral densities and were mostly to neighbouring territories (mean movement length = 1.77 km). Recent analyses focusing on those movements between neighbouring territories showed that the propensity to move into a territory was negatively associated with corresponding group size; therefore, larger groups seemed to be more stable and less attractive or permeable to immigrants (Byrne *et al.*, 2019). This seems to match data from a high-density badger population (Woodchester, Gloucestershire) which demonstrated that while badgers moved from larger to smaller social groups, overall there were no social groups that were net 'exporters' (sources) or net 'importers' (sinks) (Rogers *et al.*, 1998). The probability of finding badgers that move from larger to smaller groups may have arisen simply due to larger groups having more members and, therefore, the potential for more movement (Macdonald *et al.*, 2008). However, in a different high-density population (Wytham, Oxford), there was evidence of male badgers moving from smaller to larger groups where there were more females (Macdonald *et al.*, 2008). This contrasts with data from our study population, where male movements into groups were not associated with group composition, but females exhibited a greater probability of moving into groups with a male-biased composition (Byrne *et al.*, 2019). Both studies carried out in high-density populations suggest that group composition may have been a factor in movement (Rogers *et al.*, 1998; Macdonald *et al.*, 2008), but neither explicitly assessed the effects of density as was attempted during the present study (but see Macdonald *et al.*, 2002 for other possible density-dependent effects).

If there was a strongly biased net movement toward one direction, for example, from high to low density (or from an area above its carrying capacity to an area below its carrying capacity), we expect that given enough time, the low-density area would eventually increase to its environmental carrying capacity, and, therefore, become unattractive to immigrating animals (Amarasekare, 2004). This does not mean, however, that density-dependent dispersal (as defined as moving up or down density gradients) does not occur across badger populations. Events that reduce density below carrying capacity, for example, a stochastic event like

an extreme flood or a non-stochastic event like a cull, can set up a gradient where the 'sink' population (into which net movement flows) is below the carrying capacity of the territory. This new vacant territory may constitute a valuable resource to neighbouring badgers which may cause net inward dispersal (a 'vacuum' effect; Macdonald, 1995) or the increase in area of neighbouring territories, enveloping the vacant territory and merging with any remaining group members (group 'fusion'; Roper & Lüps, 1993; Blejwas *et al.*, 2002; Revilla & Palomares, 2002; Ebensperger *et al.*, 2009; but see literature on the Resource Dispersion Hypothesis, e.g. Johnson *et al.*, 2001). Either of these mechanisms could lead to increased movement lengths and propensity from the home territory.

Our data were consistent with an overall effect of density on movement between subpopulations. Generally, in parts of our study area where density was low, we found increased trans-territorial movements (propensity) and also longer-distance movements. Longer-distance movements could arise simply as a function of badgers having larger territories in lower-density populations (Figure 8.6; Roper, 2010; Byrne *et al.*, 2012b). Indeed, there was some support for a negative relationship between movement propensity and territory size (using tessellated territories; Tables 8.5 and 8.6). This could possibly be due to badgers being recaptured away from their main sett (main

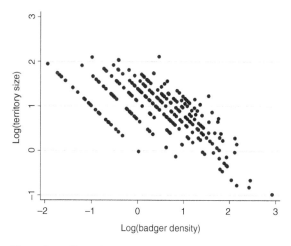

Figure 8.6 The relationship between tessellated territory size (km^2; y-axis) and estimated density (badgers km^{-2}; x-axis) based on total counts per territory.

setts exhibit the greatest probability of badger capture in this study population), but still being found within their putative home territory. Furthermore, a recent meta-analysis of data from 15 populations across Europe found a relationship between badger density and the longest-distance movements recorded (Byrne *et al.*, 2014b). Generally, longer movements are recorded where density is lower (but see Gaughran *et al.*, 2019 for very long movement distances recorded in another medium-density Irish badger population). However, some of the studies in higher-density populations may have been of too small a scale to record infrequent long-distance dispersal attempts (LDDs; Byrne *et al.*, 2014b). Higher-density populations have smaller territories packed into smaller spaces (Figure 8.6; Roper, 2010). The nearest-neighbour distances between different social-group setts are relatively close in space, indicating that the absolute distance may not be a reasonable explanation why we found less inter-territorial movements at higher densities. Instead, there may be a cost to making inter-territorial movements. If badgers exert a type of 'stronger' territoriality at higher density, there may be a greater cost to moving between territories than in lower-density populations. The Social Fence Hypothesis (Hestbeck, 1982; Matthysen, 2005) suggests, for example, that higher densities may present a greater cost of moving through populations, a type of social viscosity (Byrne *et al.*, 2014b). Moving through a neighbouring territory may increase the chances of aggression with neighbours (Kruuk, 1989; Kilshaw *et al.*, 2009; Roper, 2010); and this is more likely to happen if a badger trespasses nearer the most valuable component of a territory, the main sett. Moving across the periphery of a territory is less problematic, as there is a lower probability of meeting a neighbour the further away from the main sett (badgers can be thought of as central place foragers; Stewart *et al.*, 1997). There also may be fewer resources to defend away from the selected location of the main sett (Doncaster & Woodroffe, 1993; Roper *et al.*, 1993; but see the Passive Range Exclusion Hypothesis (Stewart *et al.*, 1997). Data from low-density populations have shown that in large territories, demarcation through group defecation is confined to close proximity of the main sett (so-called 'hinterland latrines'; Roper *et al.*, 1993; Revilla & Palomares, 2002; Do Linh San *et al.*, 2007). It may

also be difficult for badgers to adequately mark very large territories. Therefore, we could expect more fluidity in their movements across low-density populations, especially beyond the territorial core. Potentially, we could expect more aggressive interactions between badgers in low-density subpopulations, when movements cross these core areas. However, genetic evidence has shown that even in high-density populations, extra-group mating occurs regularly (Carpenter *et al.*, 2005; Dugdale *et al.*, 2007; Annavi *et al.*, 2014), and further recent research has indicated that badgers may be moving between territories at a far greater frequency than previously believed (Macdonald *et al.*, 2008; Tinnesand *et al.*, 2015; cf. Kruuk, 1989).

Our dataset is limited by temporal resolution, and, therefore, it is difficult to assess the frequency of temporary trans-territorial 'forays' (Christian, 1994; Roper *et al.*, 2003; Macdonald *et al.*, 2008). Presumably, these forays are more frequent than what was recorded during our study and could be recorded in future studies by tracking animals through radio-telemetry or Global Positioning System collars (e.g. Gaughran *et al.*, 2018, 2019). However, the relative frequency of movement distances scales with sampling effort during mark–recapture studies (see Byrne *et al.*, 2014b), allowing us to make legitimate comparisons between groups.

The variation in movements with differing densities recorded during our study may relate to a cline in socio-spatial structures found across this population. Great differences in socio-spatial structure have been observed across the badger's population range in Europe (Revilla & Palomares, 2002; Do Linh San *et al.*, 2007; Macdonald *et al.*, 2008; Roper, 2010). The variation in tessellated 'territory' size (area range: 0.44–8.25 km^2) and density (total count range: 0.14–11.61 badgers km^{-2}) within our study population makes it an important dataset to learn more about the dynamics of badger populations across a heterogeneous landscape. Future research should investigate what factors were associated with the variations in badger density within this study area, specifically investigating habitat composition and disturbance (culling) history.

A final cautionary note relates to the fact that trappability per capture session was relatively low during this study (Byrne *et al.*, 2012a), which means that the population was only sampled and not fully enumerated. Ideally, a second metric (e.g. camera-trapping) would help validate the abundance estimates (Byrne & Do Linh San, 2016). Unfortunately, a second metric was not available due to the extensive scale and enormous effort employed during the present study. Nevertheless, the same surveying effort was expended across the site and capturing effort was dictated by the presence of field signs. We know that there is a significant relationship between field signs and badger presence and capture numbers (Byrne *et al.*, 2013b; Martin *et al.*, 2017). If 'lower-density' areas were defined due to under-sampling, we would have expected that only the most frequent movements would have been recorded (i.e. within-territory movements). If 'higher-density' areas were defined due to greater sampling effort, we would expect a higher probability of sampling rarer events (e.g. the tail of the dispersal kernel; Byrne *et al.*, 2014b). As the analysis suggested the opposite outcome (more movements within lower-density areas) we are confident that our inferences relate to the biology of the system and not simply to sampling bias.

Implications for Wildlife and Disease Management

Badgers are an important wildlife reservoir of *M. bovis*, the causative agent of bovine tuberculosis (bTB) infection in cattle in Ireland and Britain (Griffin *et al.*, 2005; Byrne *et al.*, 2014c) and a spill-over host in Spain and France (Gortazar *et al.*, 2012). Badger movements have been at the centre of a controversial hypothesis (see More *et al.*, 2007 for a critique; and King *et al.*, 2007 for further discussion) that suggests that increasing territorial disruption can increase the risk to cattle herds, through a frequency-dependent mechanism based on increased contact rates (Donnelly *et al.*, 2006; Carter *et al.*, 2007; Macdonald *et al.*, 2008). Data from a large-scale study in Britain found that bTB prevalence in badgers increased over time during repeated culling activities, supporting this contention (Woodroffe *et al.*, 2006). However, data from culling activities (proactive and targeted) in Ireland has found general declines in badger bTB prevalence over time after repeated culls (Corner *et al.*, 2008; Byrne *et al.*, 2015a). Increased badger movements and the changing configuration of badger territories have been observed after disturbance in both Ireland and Britain (O'Corry-Crowe *et al.*, 1996; Tuyttens *et al.*, 2000;

Riordan *et al.*, 2011). A long-term study has found a relationship between inter-territorial movements and badger disease incidence in a high-density population (Carter *et al.*, 2007; Vicente *et al.*, 2007). Riordan *et al.* (2011) found increased roaming and bTB prevalence in culled badger groups, but postulated that the mechanism of increasing prevalence was related to stress-induced immuno-suppression, and not due to increased contacts. Irrespective of the mechanism, there was an observed transient increase in bTB risk (≤2 years) in cattle herds adjacent (≤2 km) to culling zones in Britain (Jenkins *et al.*, 2010) which was hypothesized to be caused by social perturbation of badger populations spreading infections (Donnelly *et al.*, 2006). However, such increased risk to cattle herds has not been observed in Ireland around targeted culling areas (Olea-Popelka *et al.*, 2009; White *et al.*, 2016). It should also be noted that different capturing mechanisms (wire-restraints in the Republic of Ireland and cage-traps in Britain) were employed between the two countries, which are likely to have different biases and efficiencies (Byrne *et al.*, 2012a,b). Cage-traps depend on luring badgers using baits, whereas restraints rely on going undetected. Cage-traps may suffer learned trap avoidance by badgers, or may be biased toward the capture of 'bold' individuals, whereas restraints are poor at capturing young animals (Byrne *et al.*, 2012b). These selective differences may have differing epidemiological impacts.

Given these diverging outcomes between studies, it seems that movement and density are important parameters interacting in complex ways to affect the maintenance and spread of tuberculosis in badger populations. Badger populations are generally of lower density in Ireland than in much of southern Britain (Byrne *et al.*, 2012b, 2013a, 2014a; Judge *et al.*, 2014; but note the large variation within this study population) where most culling operations took place. In low-density populations, it seems that territoriality is looser (Revilla & Palomares, 2002), with significant numbers of trans-territorial movements and longer distances traversed in the absence of culling (Byrne *et al.*, 2014b; present study). Disturbances in such low-density situations may not significantly change the structural dynamics of badger populations, at least to a great enough degree to invoke a net increase in prevalence through increasing contacts leading to transmission. Culling in low-density areas could reduce the number of infected individuals to a threshold that overcomes the net effects of frequency-dependent transmission (Potapov *et al.*, 2012). This mechanism has been described in a mathematical model by Potapov *et al.* (2012), where density-dependent 'birth or recruitment induces compensatory growth of new, healthy individuals, which has the net effect of reducing disease prevalence by dilution'. Conversely, a dramatic reduction in abundance in a high-density population may have the potential to disrupt social structure in a way that yields a net increase in bTB prevalence within culled badger populations and spread to non-culled badger populations (Donnelly *et al.*, 2006; Woodroffe *et al.*, 2006). Clearly, a broader theoretical framework is required to help explain the apparent conflicting outcomes across populations. What is important, however, is that movement is a fundamental driver of disease spread and maintenance across spatially structured populations (Cullingham *et al.*, 2008; Mundt *et al.*, 2009), and a greater understanding of movement parameters across different ecological conditions is essential to understand and design appropriate disease interventions.

Conclusion

Variation in badger movements (propensity and length) is associated with several factors, including density, sex, age-class, and the period of time between captures. There is evidence that movement patterns are different in subpopulations that are generally of 'lower' or 'higher' density, possibly representing a cline in the socio-spatial organization across our study site. This broad variation within one landscape makes our dataset important for exploring mechanisms of group dynamics. The data support the contention that a more fluid socio-spatial system is found in lower-density populations, while more structured organizations may be apparent in higher-density populations. However, there may not be a net flux of individuals moving across density gradients in undisturbed populations close to their carrying capacity.

Acknowledgements

The broader study of badgers in the study site was funded by the Department of Agriculture, Food and the Marine (http://www.agriculture.gov.ie/). Dr. Andrew Byrne was funded through a post-doctoral Research Fellowship (PDRF-L1) within the Centre for Veterinary Epidemiology and Risk Analysis during the writing of this paper (http://www.ucd.ie/cvera). We wish to acknowledge the considerable efforts of field staff, especially Richie Browne and John Cummins. The manuscript was enriched through discussions and comments from Dr. Chris Newman (University of Oxford, UK), and the comments from two reviewers.

References

Allen, A.R., Skuce, R.A. & Byrne, A.W. (2018) Bovine tuberculosis in Britain and Ireland – a perfect storm? the confluence of potential ecological and epidemiological impediments to controlling a chronic infectious disease. *Frontiers in Veterinary Science* 5, 109.

Amarasekare, P. (2004) The role of density-dependent dispersal in source–sink dynamics. *Journal of Theoretical Biology* 226, 159–168.

Annavi, G., Newman, C., Dugdale, H.L., Buesching, C.D., Sin, Y.W., Burke, T. & Macdonald, D.W. (2014) Neighbouring-group composition and within-group relatedness drive extra-group paternity rate in the European badger (*Meles meles*). *Journal of Evolutionary Biology* 27, 2191–2203.

Arzamendia, Y. & Vilá, B. (2012) Effects of capture, shearing, and release on the ecology and behavior of wild vicuñas. *The Journal of Wildlife Management* 76, 57–64.

Benton, T.G. & Bowler, D.E. (2012) Linking dispersal to spatial dynamics. In: *Dispersal Ecology and Evolution* (eds J. Clobert, M. Baguette, T.G. Benton & J.M. Bullock), pp. 251–265. Oxford University Press, Oxford.

Benton, C.H., Delahay, R.J., Robertson, A., McDonald, R.A., Wilson, A.J., Burke, T.A. & Hodgson, D. (2016)

Blood thicker than water: kinship, disease prevalence and group size drive divergent patterns of infection risk in a social mammal. *Proceedings of the Royal Society of London B* 283, 20160798.

Benton, C.H., Delahay, R.J., Smith, F.A., Robertson, A., McDonald, R.A., Young, A.J., Burke, T.A. & Hodgson, D. (2018) Inbreeding intensifies sex- and age-dependent disease in a wild mammal. *Journal of Animal Ecology* 87, 1500–1511.

Biek, R., O'Hare, A., Wright, D., Mallon, T., McCormick, C., Orton, R.J., McDowell, S.W.J., Trewby, H., Skuce, R.A. & Kao, R.R. (2012) Whole genome sequencing reveals local transmission patterns of *Mycobacterium bovis* in sympatric cattle and badger populations. *PLoS Pathogens* 8, e1003008.

Blackwell, P.G. & Macdonald, D.W. (2000) Shapes and sizes of badger territories. *Oikos* 89, 392–398.

Blejwas, K.M., Sacks, B.N., Jaeger, M.M. & McCullough, D.R. (2002) The effectiveness of selective removal of breeding coyotes in reducing sheep predation. *The Journal of Wildlife Management* 66, 451–462.

Bodin, C., Benhamou, S. & Poulle, M.-L. (2006) What do European badgers (*Meles meles*) know about the spatial organisation of neighbouring groups? *Behavioural processes* 72, 84–90.

Bourne, F.J., Donnelly, C.A., Cox, D.R., Gettinby, G., McInerney, J.P., Morrison, W.I. & Woodroffe, R. (2007) *Bovine TB: The Scientific Evidence – A Science Base for a Sustainable Policy to Control TB in Cattle. An Epidemiological Investigation into Bovine Tuberculosis.* Final report of the Independent Scientific Group on Cattle TB. Department for Environment, Food and Rural Affairs, London.

Brunton, L.A., Donnelly, C.A., O'Connor, H., Prosser, A., Ashfield, S., Ashton, A., Upton, P., Mitchell, A., Goodchild, A.V., Parry, J.E. & Downs, S.H. (2017) Assessing the effects of the first 2 years of industry-led badger culling in England on the incidence of bovine tuberculosis in cattle in 2013–2015. *Ecology and Evolution* 7, 7213–7230.

Byrne, A.W. & Do Linh San, E. (2016) A cautionary note on the use of Minimum Number Alive-derived trappability metrics in wildlife programmes, as exemplified by the case of the European Badger (*Meles meles*). *Wildlife Biology in Practice* 12, 51–57.

Byrne, A.W., O'Keeffe, J., Green, S., Sleeman, D.P., Corner, L.A.L., Gormley, E., Murphy, D., Martin, S.W.

& Davenport, J. (2012a) Population estimation and trappability of the European badger (*Meles meles*): implications for tuberculosis management. *PLoS One* 7, e50807.

Byrne, A.W., Sleeman, D.P., O'Keeffe, J. & Davenport, J. (2012b) The ecology of the European badger (*Meles meles*) in Ireland: a review. *Biology & Environment: Proceedings of the Royal Irish Academy* 112, 105–132.

Byrne, A.W., O'Keeffe, J., Sleeman, D.P., Davenport, J. & Martin, S.W. (2013a) Impact of culling on relative abundance of the European Badger (*Meles meles*) in Ireland. *European Journal of Wildlife Research* 59, 25–37.

Byrne, A.W., O'Keeffe, J., Sleeman, D., Davenport, J. & Martin, S.W. (2013b) Factors affecting European badger (*Meles meles*) capture numbers in one county in Ireland. *Preventive Veterinary Medicine* 109, 128–135.

Byrne, A.W., Acevedo, P., Green, S. & O'Keeffe, J. (2014a) Estimating badger social-group abundance in the Republic of Ireland using cross-validated species distribution modelling. *Ecological Indicators* 43, 94–102.

Byrne, A.W., Quinn, J.L., O'Keeffe, J.J., Green, S., Sleeman, D.P., Martin, S.W. & Davenport, J. (2014b) Large-scale movements in European badgers: has the tail of the movement kernel been underestimated? *Journal of Animal Ecology* 83, 991–1001.

Byrne, A.W., White, P.W., McGrath, G., O'Keeffe, J. & Martin, S.W. (2014c) Risk of tuberculosis cattle herd breakdowns in Ireland: effects of badger culling effort, density and historic large-scale interventions. *Veterinary Research* 45, 109.

Byrne, A.W., Kenny, K., Fogarty, U., O'Keeffe, J., More, S., McGrath, G., Teeling, M., Martin, S.W. & Dohoo, I. (2015a) Spatial and temporal analyses of metrics of tuberculosis infection in badgers (*Meles meles*) from the Republic of Ireland: trends in apparent prevalence. *Preventive Veterinary Medicine* 122, 345–354.

Byrne, A.W., O'Keeffe, J., Fogarty, U., Rooney, P. & Martin, S.W. (2015b) Monitoring trap-related injury status during large-scale wildlife management programs: an adaptive management approach. *European Journal of Wildlife Research* 61, 445–455.

Byrne, A.W., O'Keeffe, J., Buesching, C.D. & Newman, C. (2019) Push and pull factors driving movement in a social mammal: context dependent behavioural plasticity at the landscape scale. *Current Zoology* 65, 517–525.

Campbell, E.L., Byrne, A.W., Menzies, F.D., McBride, K.R., McCormik, C.M., Scantlebury, M. & Reid, N. (2019) Interspecific visitation of cattle and badgers to fomites: a transmission risk for bovine tuberculosis. *Ecology and Evolution* 9, 8479–8489.

Carpenter, P.J., Pope, L.C., Greig, C., Dawson, D.A., Rogers, L.M., Erven, K., Wilson, G.J., Delahay, R.J., Cheeseman, C.L. & Burke, T. (2005) Mating system of the Eurasian badger, *Meles meles*, in a high density population. *Molecular Ecology* 14, 273–284.

Carter, S.P., Delahay, R.J., Smith, G.C., Macdonald, D.W., Riordan, P., Etherington, T.R., Pimley, E.R., Walker, N.J. & Cheeseman, C.L. (2007) Culling-induced social perturbation in Eurasian badgers *Meles meles* and the management of TB in cattle: an analysis of a critical problem in applied ecology. *Proceedings of the Royal Society B: Biological Sciences* 274, 2769–2777.

Christian, S.F. (1994) Dispersal and other inter-group movements in badgers, *Meles meles. Zeitschrift für Säugetierkunde* 59, 218–223.

Clobert, J., Le Galliard, J.-F., Cote, J., Clobert, J., Baguette, M., Benton, T.G., Bullock, J.M. & Ducatez, S. (eds) (2012) *Dispersal Ecology and Evolution.* Oxford University Press, Oxford.

Courchamp, F., Berec, L. & Gascoigne, J. (2008) *Allee Effects in Ecology and Conservation.* Oxford University Press, Oxford.

Corner, L.A., Clegg, T.A., More, S.J., Williams, D.H., O'Boyle, I., Costello, E., Sleeman, D.P. & Griffin, J.M. (2008) The effect of varying levels of population control on the prevalence of tuberculosis in badgers in Ireland. *Research in Veterinary Science* 85, 238–249.

Corner, L.A., Murphy, D. & Gormley, E. (2011) *Mycobacterium bovis* infection in the Eurasian badger (*Meles meles*): the disease, pathogenesis, epidemiology and control. *Journal of Comparative Pathology* 144, 1–24.

Cullingham, C.I., Pond, B.A., Kyle, C.J., Rees, E.E., Rosatte, R.C. & White, B.N. (2008) Combining direct and indirect genetic methods to estimate dispersal for informing wildlife disease management decisions. *Molecular Ecology* 17, 4874–4886.

Delahay, R., De Leeuw, A., Barlow, A., Clifton-Hadley, R. & Cheeseman, C. (2002) The status of *Mycobacterium bovis* infection in UK wild mammals: a review. *The Veterinary Journal* 164, 90–105.

Dohoo, I.R., Martin, S.W. & Stryhn, H. (2009) *Veterinary Epidemiological Research.* 2nd edition. VER Inc., Charlottetown, Prince Edward Island.

Do Linh San, E., Ferrari, N. & Weber, J.-M. (2007) Socio-spatial organization of Eurasian badgers (*Meles meles*) in a low-density population of central Europe. *Canadian Journal of Zoology* 85, 973–984.

Doncaster, C.P. & Woodroffe, R. (1993) Den site can determine shape and size of badger territories: implications for group-living. *Oikos* 66, 88–93.

Donnelly, C.A., Woodroffe, R., Cox, D.R. Bourne, F.J., Cheeseman, C.L., Clifton-Hadley, R.S., Wei, G., Gettinby, G., Gilks, P., Jenkins, H., Johnston, W.T., Le Fevre, A.M., McInerney, J.P. & Morrison, W.I. (2006) Positive and negative effects of widespread badger culling on tuberculosis in cattle. *Nature* 439, 843–846.

Downs, S.H. Prosser, A., Ashton, A., Ashfield, S., Brunton, L.A., Brouwer, A., Upton, P., Robertson, A., Donnelly, C.A. & Parry, J.E. (2019) Assessing effects from four years of industry-led badger culling in England on the incidence of bovine tuberculosis in cattle, 2013–2017. *Scientific Reports* 9, 14666.

Dugdale, H.L., Macdonald, D.W., Pope, L.C. & Burke, T. (2007) Polygynandry, extra-group paternity and multiple-paternity litters in European badger (*Meles meles*) social groups. *Molecular Ecology* 16, 5294–5306.

Ebensperger, L.A., Chesh, A.S., Castro, R.A., Tolhuysen, L.O., Quirici, V., Burger, J.R. & Hayes, L.D. (2009) Instability rules social groups in the communal breeder rodent *Octodon degus. Ethology* 115, 540–554.

Feore, S. & Montgomery, W.I. (1999) Habitat effects on the spatial ecology of the European badger (*Meles meles*). *Journal of Zoology* 247, 537–549.

Gaughran, A., Kelly, D.J., MacWhite, T., Mullen, E., Maher, P., Good, M. & Marples, N.M. (2018). *Super-ranging.* A new ranging strategy in European badgers. *PLoS One* 13, e0191818.

Gaughran, A., MacWhite, T., Mullen, E., Maher, P., Kelly, D.J., Good, M. & Marples, N.M. (2019) Dispersal patterns in a medium-density Irish badger population: implications for understanding the dynamics of tuberculosis transmission. *Ecology and Evolution* 9, 13142–13152.

Gortazar, C., Delahay, R.J., Mcdonald, R.A., Boadella, M., Wilson, G.J., Gavier-Widen, D. & Acevedo, P.

(2012) The status of tuberculosis in European wild mammals. *Mammal Review* 42, 193–206.

Griffin, J.M., Williams, D.H., Kelly, G.E., Clegg, T.A., O'Boyle, I., Collins, J.D. & More, S.J. (2005) The impact of badger removal on the control of tuberculosis in cattle herds in Ireland. *Preventive Veterinary Medicine* 67, 237–266.

Halls, P.J., Bulling, M., White, P.C.L., Garland, L. & Harris, S. (2001) Dirichlet neighbours: revisiting Dirichlet tessellation for neighbourhood analysis. *Computers, Environment and Urban Systems* 25, 105–117.

Ham, C., Donnelly, C.A., Astley, K.L., Jackson, S.Y.B. & Woodroffe, R. (2019) Effect of culling on individual badger *Meles meles* behaviour: potential implications for bovine tuberculosis transmission. *Journal of Applied Ecology* 56, 2390–2399.

Hammond, R.F. & McGrath, G. (1998) The use of geographical information system (GIS) derived tessellations to relate badger territory to distribution patterns of soils and land use environmental habitat variables within the East Offaly badger research area. *ERAD Tuberculosis Investigation Unit, University College Dublin, Selected Papers* 1997, 14–25.

Harris, S., Cresswell, W.J. & Cheeseman, C.L. (1992) Age determination of badgers (*Meles meles*) from tooth wear: the need for a pragmatic approach. *Journal of Zoology* 228, 679–684.

Hestbeck, J.B. (1982) Population regulation of cyclic mammals: the social fence hypothesis. *Oikos* 39, 157–163.

Hutchings, M.R., Service, K.M. & Harris, S. (2002) Is population density correlated with faecal and urine scent marking in European badgers (*Meles meles*) in the UK? *Mammalian Biology* 67, 286–293.

Jenkins, H.E., Woodroffe, R. & Donnelly, C.A. (2010) The duration of the effects of repeated widespread badger culling on cattle tuberculosis following the cessation of culling. *PLoS One* 5, e9090.

Johnson, D.D.P., Macdonald, D.W., Newman, C. & Morecroft, M.D. (2001) Group size versus territory size in group-living badgers: a large-sample field test of the Resource Dispersion Hypothesis. *Oikos* 95, 265–274.

Judge, J., Wilson, G.J., Macarthur, R., Delahay, R.J. & McDonald, R.A. (2014) Density and abundance of badger social groups in England and Wales in 2011–2013. *Scientific Reports* 4, 3809.

Kilshaw, K., Newman, C., Buesching, C., Bunyan, J. & Macdonald, D. (2009) Coordinated latrine use by European badgers, *Meles meles*: potential consequences for territory defense. *Journal of Mammalogy* 90, 1188–1198.

Kim, S.Y., Torres, R. & Drummond, H. (2009) Simultaneous positive and negative density-dependent dispersal in a colonial bird species. *Ecology* 90, 230–239.

King, D., Roper, T.J., Young, D., Woolhouse, M.E.J., Collins, D.A. & Wood, P. (2007) *Bovine Tuberculosis in Cattle and Badgers: A Report by the Chief Scientific Adviser.* Department of Environment Food and Rural Affairs, London.

Kruuk, H. (1989) *The Social Badger, Ecology and Behaviour of a Group-Living Carnivore (*Meles meles*).* Oxford University Press, Oxford.

Lidicker, W.Z. Jr & Stenseth, N.C. (1992) To disperse or not to disperse: who does it and why? In: *Animal Dispersal: Small Mammals as a Model* (eds N.C. Stenseth & W.Z. Lidicker Jr), pp. 21–36. Chapman & Hall, London.

Macdonald, D.W. (1983) The ecology of carnivore social behaviour. *Nature* 301, 379–384.

Macdonald, D.W. (1995) Wildlife rabies: the implications for Britain. Unresolved questions for the control of wildlife rabies: social perturbation and interspecific interactions. In: *Rabies in a Changing World* (eds P.H. Beynon & A.T.B. Ednay), pp. 33–48. British Small Animal Veterinary Association, Cheltenham.

Macdonald, D.W., Newman, C., Stewart, P.D., Domingo-Roura, X. & Johnson, P.J. (2002) Density-dependent regulation of body mass and condition in badgers (*Meles meles*) from Wytham Woods. *Ecology* 83, 2056–2061.

Macdonald, D.W., Newman, C., Buesching, C.D. & Johnson, P.J. (2008) Male-biased movement in a high-density population of the Eurasian badger (*Meles meles*). *Journal of Mammalogy* 89, 1077–1086.

Martin, L.E.R., Byrne, A.W., O'Keeffe, J., Miller, M.A. & Olea-Popelka, F.J. (2017) Weather influences trapping success for tuberculosis management in European badgers (*Meles meles*). *European Journal of Wildlife Research* 63, 30.

Martin, S.W., O'Keeffe, J., Byrne, A.W., Rosen, L.E., White, P.W. & McGrath, G. (2020) Is moving from targeted culling to BCG-vaccination of badgers (*Meles*

meles) associated with an unacceptable increased incidence of cattle herd tuberculosis in the Republic of Ireland? A practical non-inferiority wildlife intervention study in the Republic of Ireland (2011–2017). *Preventive Veterinary Medicine* 179, 105004. doi: 10.1016/j.prevetmed.2020.105004. Epub 2020 Apr 14.

Matthysen, E. (2005) Density-dependent dispersal in birds and mammals. *Ecography* 28, 403–416.

McDonald, J.L., Robertson, A. & Silk, M.J. (2018) Wildlife disease ecology from the individual to the population: insights from a long-term study of a naturally infected European badger population. *Journal of Animal Ecology* 87, 101–112.

Milne, G., Allen, A., Graham, J., Lahuerta-Marin, A., McCormick, C., Presho, E., Reid, N., Skuce, R. & Byrne, A.W. (2020) Bovine tuberculosis breakdown duration in cattle herds: an investigation of herd, host, pathogen and wildlife risk factors. *PeerJ* 8, e8319.

More, S.J., Clegg, T.A., McGrath, G., Collins, J.D., Corner, L.A.L. & Gormley, E. (2007) Does reactive badger culling lead to an increase in tuberculosis in cattle? *Veterinary Record* 161, 208–209.

Mundt, C.C., Sackett, K.E., Wallace, L.D., Cowger, C. & Dudley, J.P. (2009) Long-distance dispersal and accelerating waves of disease: empirical relationships. *American Naturalist* 173, 456–466.

Murphy, D., O'Keeffe, J.J., Martin, S.W., Gormley, E. & Corner, L.A. (2009) An assessment of injury to European badgers (*Meles meles*) due to capture in stopped restraints. *Journal of Wildlife Diseases* 45, 481–490.

Murphy, D., Gormley, E., Costello, E., O'Meara, D. & Corner, L.A.L. (2010). The prevalence and distribution of *Mycobacterium bovis* infection in European badgers (*Meles meles*) as determined by enhanced post mortem examination and bacteriological culture. *Research in Veterinary Science* 88, 1–5.

Nathan, R. & Giuggioli, L. (2013) A milestone for movement ecology research. *Movement Ecology* 1, 1.

Nathan, R., Getz, W.M., Revilla, E., Holyoak, M., Kadmon, R., Saltz, D. & Smouse, P.E. (2008) A movement ecology paradigm for unifying organismal movement research. *Proceedings of the National Academy of Sciences* 105, 19052–19059.

O'Corry-Crowe, G., Hammond, R., Eves, J. & Hayden, T.J. (1996) The effect of reduction in badger density on the spatial organisation and activity of badgers *Meles meles* L. in relation to farms in central Ireland. *Biology and Environment: Proceedings of the Royal Irish Academy* 96b, 147–158.

Olea-Popelka, F.J., Fitzgerald, P., White, P., McGrath, G., Collins, J.D., O'Keeffe, J., Kelton, D.F., Berke, O., More, S. & Martin, S.W. (2009) Targeted badger removal and the subsequent risk of bovine tuberculosis in cattle herds in county Laois, Ireland. *Preventive Veterinary Medicine* 88, 178–184.

Potapov, A., Merrill, E. & Lewis, M.A. (2012) Wildlife disease elimination and density dependence. *Proceedings of the Royal Society B: Biological Sciences* 279, 3139–3145.

Quinn, A.C.D., Williams, D.M. & Porter, W.F. (2012) Postcapture movement rates can inform data-censoring protocols for GPS-collared animals. *Journal of Mammalogy* 93, 456–463.

Revilla, E. & Palomares, F. (2002) Spatial organization, group living and ecological correlates in low-density populations of Eurasian badgers, *Meles meles*. *Journal of Animal Ecology* 71, 497–512.

Riordan, P., Delahay, R.J., Cheeseman, C., Johnson, P.J. & Macdonald, D.W. (2011) Culling-induced changes in badger (*Meles meles*) behaviour, social organisation and the epidemiology of bovine tuberculosis. *PLoS One* 6, e28904.

Rogers, L.M., Delahay, R., Cheeseman, C.L., Langton, S., Smith, G.C. & Clifton-Hadley, R.S. (1998) Movement of badgers (*Meles meles*) in a high-density population: individual, population and disease effects. *Proceedings of the Royal Society of London B: Biological Sciences* 265, 1269–1276.

Roper, T.J. (2010) *Badger*. Collins, London.

Roper, T.J. & Lüps, P. (1993) Disruption of territorial behaviour in badgers *Meles meles*. *Zeitschrift für Säugetierkunde* 58, 252–255.

Roper, T.J., Conradt, L., Butler, J., Christian, S.E., Ostler, J. & Schmid, T.K. (1993) Territorial marking with faeces in badgers (*Meles meles*): a comparison of boundary and hinterland latrine use. *Behaviour* 127, 289–307.

Roper, T.J., Ostler, J.R. & Conradt, L. (2003) The process of dispersal in badgers *Meles meles*. *Mammal Review* 33, 314–318.

Rozins, C., Silk, M.J., Croft, D.P., Delahay, R.J., Hodgson, D.J., McDonald, R.A. & Boots, M. (2018) Social structure contains epidemics and regulates individual roles in disease transmission in a group-living mammal. *Ecology and Evolution* 8, 12044–12055.

Saura, S., Bodin, Ö. & Fortin, M.-J. (2014) Stepping stones are crucial for species' long-distance dispersal and range expansion through habitat networks. *Journal of Applied Ecology* 51, 171–182.

Schütz, K.E., Ågren, E., Amundin, M., Röken, B., Palme, R. & Mörner, T. (2006) Behavioral and physiological responses of trap-induced stress in European badgers. *The Journal of Wildlife Management* 70, 884–891.

Stewart, P.D., Anderson, C. & Macdonald, D.W. (1997) A mechanism for passive range exclusion: evidence from the European badger (*Meles meles*). *Journal of Theoretical Biology* 184, 279–289.

Tinnesand, H.V., Buesching, C.D., Noonan, M.J., Newman, C., Zedrosser, A., Rosell, F. & Macdonald, D.W. (2015) Will trespassers be prosecuted or assessed according to their merits? A consilient interpretation of territoriality in a group-living carnivore, the European badger (*Meles meles*). *PLoS One* 10, e0132432.

Tolhurst, B.A., Delahay, R.J., Walker, N.J., Ward, A.I. & Roper, T.J. (2009) Behaviour of badgers (*Meles meles*) in farm buildings: opportunities for transmission of *Mycobacterium bovis*? *Applied Animal Behaviour Science* 117, 103–113.

Tuyttens, F.A.M., Delahay, R.J., Macdonald, D.W., Cheeseman, C.L., Long, B. & Donnelly, C.A. (2000) Spatial perturbation caused by a badger (*Meles meles*) culling operation: implications for the function of territoriality and the control of bovine tuberculosis (*Mycobacterium bovis*). *Journal of Animal Ecology* 69, 815–828.

Vicente, J., Delahay, R.J., Walker, N.J. & Cheeseman, C.L. (2007) Social organization and movement influence the incidence of bovine tuberculosis in an undisturbed high-density badger *Meles meles* population. *Journal of Animal Ecology* 76, 348–360.

White, P., Martin, S.W., McGrath, G., De Jong, M., O'Keeffe, J., Olea-Popelka, F., More, S., Byrne, A. & Frankena, K. (2016) Temporal association between badger presence and herd bovine tuberculosis (bTB) risk in Co. Monaghan, Ireland. In: *The 29th World Buiatrics Congress, Dublin 2016 – Congress Proceedings* (ed. M. Doherty), p. 236. Veterinary Ireland, Dublin.

Woodroffe, R., Macdonald, D.W. & Da Silva, J. (1993) Dispersal and philopatry in the European badger, *Meles meles*. *Journal of Zoology* 237, 227–239.

Woodroffe, R., Bourne, F.J., Cox, D.R., Donnelly, C.A., Gettinby, G., McInerney, J.P. & Morrison, W.I. (2005) Welfare of badgers (*Meles meles*) subjected to culling: patterns of trap-related injury. *Animal Welfare* 14, 11–17.

Woodroffe, R., Donnelly, C.A., Jenkins, H.E., Johnston, W.T., Cox, D.R., Bourne, F.J., Cheeseman, C.L., Delahay, R.J., Clifton-Hadley, R.S., Gettinby, G., Gilks, P., Hewinson, R.G., McInerney, J.P. & Morrison, W.I. (2006) Culling and cattle controls influence tuberculosis risk for badgers. *Proceedings of the National Academy of Sciences of the United States of America* 103, 14713–14717.

Woodroffe, R., Donnelly, C.A., Wei, G., Cox, D.R., Bourne, F.J. *et al.* (2009) Social group size affects *Mycobacterium bovis* infection in European badgers (*Meles meles*). *Journal of Animal Ecology* 78, 818–827.

9

Behavioural Adaptations of Molina's Hog-Nosed Skunk to the Conversion of Natural Grasslands into Croplands in the Argentine Pampas

Diego F. Castillo* and Mauro Lucherini

Departamento de Biología, Bioquímica y Farmacia, Universidad Nacional del Sur, Bahía Blanca, Argentina
GECM, Instituto de Ciencias Biológicas y Biomédicas del Sur (INBIOSUR), Universidad Nacional del Sur (UNS) – CONICET, Bahía Blanca, Argentina

SUMMARY

Anthropogenic habitat modification is one of the most serious threats to global biodiversity, and in areas with a high urbanization level and agricultural activities, habitat loss and fragmentation are virtually inevitable. An example of this occurs in the Pampas grassland of Argentina, which is the most densely populated and most degraded region in the country. In this chapter, we explore how the behavioural ecology of the little-studied Molina's hog-nosed skunk, *Conepatus chinga*, has been affected by agricultural activities. We review the recent advances in the ecology and natural history of this mephitid and compare data collected in two grassland areas under different land uses. Information on home range characteristics, movement patterns, habitat use and selection, denning behaviour, and activity patterns was obtained by radio-tracking skunks in a protected area (7 individuals) and a cropland area (9 individuals). Feeding ecology was also studied through the analysis of faecal samples and estimation of prey abundance. Our results confirmed that *C. chinga* is mainly a solitary carnivore. The home range size is greater in males than in females, the spatial overlap is largely limited to inter-sexual dyads, and burrows are not communal. In the Pampas grassland of Argentina, this mephitid is a nocturnal, selective predator of insects, lacking clear sexual dimorphism. That *C. chinga* selected for habitat patches with natural vegetation and predictable, abundant prey, and that its activity was strictly nocturnal in unprotected croplands, suggests that prey abundance and secure denning sites are important factors affecting its behavioural ecology. We conclude that although *C. chinga* is somewhat adaptable to human-modified landscapes, the loss of native grasslands is likely affecting the abundance of its populations in the present-day Argentine Pampas.

Keywords

Activity pattern — *Conepatus chinga* — denning behaviour — Mephitidae — social behaviour — spatial ecology — telemetry

Introduction

The impact of agricultural intensification on animals has received much attention worldwide (e.g. Donald *et al.*, 2006; Jonsson *et al.*, 2012; Mastrangelo & Gavin, 2014; Habel *et al.*, 2019). However, most information about the effects of large-scale transformations on biodiversity has been from temperate grassland in North America, northern Europe, and southern Australia (Robinson & Sutherland, 2002; Brennan & Kuvlesky, 2005; Firbank *et al.*, 2008). The consequences of this phenomenon on native fauna remain poorly understood in the heavily modified South American temperate grassland areas, such as the Pampas (Medan *et al.*, 2011).

Molina's hog-nosed skunk, *Conepatus chinga*, is a small carnivore, measuring 41–71 cm in total length and weighing 1.0–2.9 kg (Redford & Eisenberg, 1992; Castillo

---

*Corresponding author.

Small Carnivores: Evolution, Ecology, Behaviour, and Conservation, First Edition. Edited by Emmanuel Do Linh San, Jun J. Sato, Jerrold L. Belant, and Michael J. Somers.

et al., 2011b). They have the black and white colouration pattern typical of many members of the Mephitidae. Its distribution range spans from southern Bolivia through Uruguay, western Paraguay to central Chile and Argentina. Previously, two species of mephitids (*C. chinga* and *Conepatus humboldtii*) had been reported in Argentina. Both were described on the basis of external characteristics (i.e. pelage colouration patterns) and size variation (Redford & Eisenberg, 1992). However, Schiaffini *et al.* (2013) in a comprehensive review, which included variations in coat colour, mitochondrial genes, and morphometrics, proposed to synonymize both currently recognized skunk species as *C. chinga*.

Despite its wide distribution, there is little information available on its ecology (Castillo *et al.*, 2012b; Kasper *et al.*, 2012). A large proportion of its geographic range is occupied by the Pampas grassland, which constitutes the most densely populated and most degraded ecological region in Argentina. In the Pampas, 90% of the original grasslands have been converted into croplands and pastures (Bertonatti & Corcuera, 2000; Medan *et al.*, 2011) and the proportion of this ecosystem under legal protection is less than 0.2%, well below the 10–12% suggested by international standards (Burkart *et al.*, 1991).

In this chapter, we provide a comprehensive synthesis of the behavioural ecology and natural history of Molina's hog-nosed skunk in the Pampas grassland of Argentina. Additionally, in order to analyze how the behavioural ecology of these skunks is affected by the agricultural activities of the Pampas, we compare data from two grassland areas with widely different levels of anthropic disturbance (Figure 9.1).

We predicted that: (i) because a more homogeneous distribution of resources should favour smaller home ranges, skunk home ranges would be larger in the landscape fragmented by agriculture; (ii) undisturbed habitats (i.e. natural and semi-natural grasslands) would be preferred by skunks and croplands would be the least used habitat; (iii) in both areas, skunk dens would be associated with habitats where their principal prey are most abundant; alternatively, if the risk of predation affects den site selection, we may expect a concentration of dens in habitats offering the greatest vegetation cover; (iv) activity patterns would differ between our two study areas, with skunks less active where and when human activities are highest.

Study Areas

The first area surveyed, Ernesto Tornquist Provincial Park (protected area: PA; 38°00′S, 62°00′W), is located in the central part of the Ventania mountain range, southern Buenos Aires province, Argentina (Figure 9.1). This park has an area of approximately 6700 ha, ranges in altitude from 450–1172 m asl and preserves one of the last fragments of the native Pampas grassland (Bertonatti & Corcuera, 2000). Due to protected area restrictions, farming activities are not allowed here. The climate is temperate with mean annual precipitation of 500–800 mm (Frangi & Bottino, 1995). The grasslands are dominated by short prairie grasses of the genus *Stipa*, *Piptochaetium*, *Briza*, and *Festuca* but patches of introduced trees and shrubs are common (Zalba & Villamil, 2002).

The second site (cropland area: CA) is located in an unprotected farming area (Estancia San Mateo; 38°37′S, 60°53′W) of the same region of Buenos Aires province (Figure 9.1). It has an area of 8100 ha and an elevation of 120–150 m asl. The climate is temperate with mean annual precipitation of 500–1000 mm (Campo de Ferreras *et al.*, 2004). In CA, 70.3% of the total area was devoted to farming activities (Castillo *et al.*, 2015). Land was mostly used for extensive livestock breeding (cattle and sheep) and intensive agriculture activities such as wheat, barley, soybean, and sunflower production. Natural grass patches were unmanaged in marginal areas such as along railroads, streams, and rocky areas.

In both areas, vehicle traffic was markedly higher during daylight than at night, with road density nearly three times higher in CA than PA (Castillo *et al.*, 2015).

Trophic Ecology

Methods

We conducted field surveys from December 1999 to July 2005 in PA, and from January 2005 to December 2008 in CA. We determined the diet of *C. chinga* in both study areas by identifying food remains in scats that were opportunistically collected. Scat analysis followed the method described by Reynolds & Aebischer (1991). Results were expressed as Frequency of

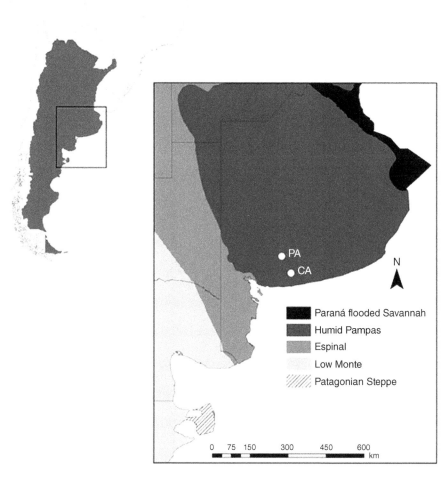

Figure 9.1 Location of the two study areas in relation to the distribution of the ecoregions of central Argentina. PA = protected area (Ernesto Tornquist Provincial Park), CA = cropland area (Estancia San Mateo).

Occurrence (FO), Numerical Frequency (NF) (Paltridge, 2002), and Percent Volume (PV) (Elmhagen *et al.*, 2000). To reduce the individual biases of these three methods, we used the Index of Relative Importance (IRI) (Home & Jhala, 2009), where IRI = (NF + PV) × FO. IRI scores for the different prey items obtained in each area were subsequently subjected to resampling with 1000 iterations using R 2.7.1 (R Development Core Team, 2008) to generate means and bias-corrected 95% confidence intervals and compare the diet between areas.

Based on the foraging habits of other skunk species and the main prey items previously found for *C. chinga* (Donadio *et al.*, 2004; Peters *et al.*, 2011), we used the abundance of invertebrates available at ground level to estimate food abundance. We placed pitfall traps in 20 × 20 m grids, each one consisting of nine plastic containers of 10 cm in diameter and 7 cm deep filled with saltwater solution and placed in pits deep enough to bury the cups up to the rim on the ground (Marrero *et al.*, 2008). In each habitat type and season, we trapped in three randomly located grids that were active for three consecutive nights. We compared the abundance of each invertebrate prey category and its use by Molina's hog-nosed skunks with Ivlev's electivity index modified by Jacobs (1974). Additionally, we calculated Bonferroni's simultaneous confidence intervals (Byers *et al.*, 1984) setting $\alpha \leq 0.05$.

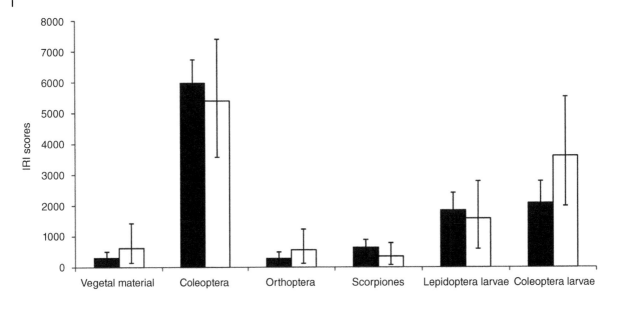

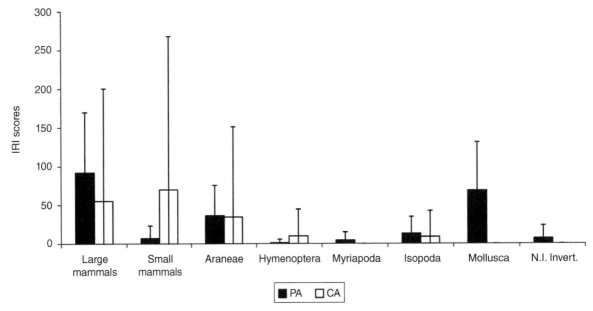

Figure 9.2 Variation in the diet of Molina's hog-nosed skunk, *Conepatus chinga*, between a protected area (PA) and a cropland area (CA) of the Argentine Pampas based on the Index of Relative Importance scores (IRI). Data are based on the analysis of faecal samples (n_{PA} = 140, n_{CA} = 27). Error bars show 95 bootstrap confidence intervals. Top chart: items with IRI > 200, bottom chart: items with IRI ≤ 200. N.I. Invert. = non-identified invertebrates.

Results

We failed to detect differences in the relative importance of most dietary items between the two study areas (Figure 9.2). However, Mollusca and Myriapoda were only found in the diet of PA skunks. In both areas, IRI scores were greatest for Coleoptera, followed by larvae of Coleoptera and Lepidoptera, then by scorpions (Figure 9.2).

In both study areas, skunks positively selected scorpions, as well as larvae of Coleoptera and Lepidoptera. Myriapoda (in PA) and Hymenoptera (CA), both poorly

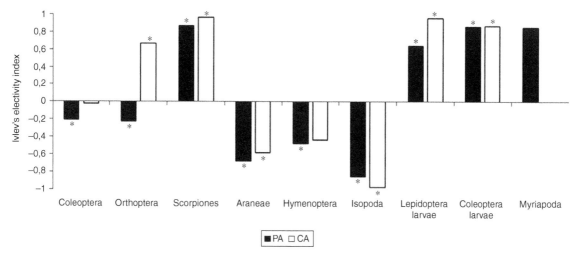

Figure 9.3 Prey selection by *Conepatus chinga* in a protected area (PA) and cropland area (CA) of the Argentine Pampas, as shown by Ivlev's electivity index. *Indicates significant positive or negative selection according to Bonferroni's simultaneous confidence intervals at $p \leqslant 0.05$.

represented in the faeces, were used proportionately to their abundance, while Araneae and Isopoda (in both areas), as well as Hymenoptera and Coleoptera (in PA) were used less than expected (Figure 9.3).

Capture and Morphometrics

Methods

During the first six years of live-trapping (from 1999 to 2005), we used mesh-wire box traps and soft-padded leg-hold traps (Victor soft catch® 1.5), but had limited success (Table 9.1). To increase capture success, we used spotlights from a vehicle to detect Molina's hog-nosed skunks and restrained them manually, using a blanket to avoid being sprayed. Restrained individuals were chemically immobilized with a combination of ketamine hydrochloride ($\bar{x} = 24.9\,\text{mg/kg}$) and xylazine ($\bar{x} = 1.9\,\text{mg/kg}$) (Castillo *et al.*, 2012a). Body weight and 12 external measurements (Table 9.3) were obtained from captured individuals.

Table 9.1 Capture effort (CEt) and capture efficiency (CEy) with box and leg-hold traps of Molina's hog-nosed skunk, *Conepatus chinga*, in a protected area (PA) and a cropland area (CA) of the Argentine Pampas.

Area	Trap type	CEt (trap nights)	No. of skunks captured	CEy (CEt/No. skunks)
PA	Box	1357	3	452.3
	Leg-hold	1278	8	159.8
	Total	**2635**	**11**	**239.5**
CA	Box	1172	—	—
	Leg-hold	1050	1	1050
	Total	**2222**	**1**	**2222**
Both areas	Box	2529	3	843
	Leg-hold	2328	9	258.7
	Total	**4857**	**12**	**404.8**

Table 9.2 Capture effort (CEt) and Capture efficiency (CEy) for manual capture of *Conepatus chinga* in a protected area (PA) and a cropland area (CA) of the Argentine Pampas.

Area	CEt (km)	No. of individuals sighted	No. of individuals captured	CEy (km/ capture)	Ratio captured/ sighted
PA	38.4	9	8	4.8	0.88
CA	431.6	13	10	43.2	0.76
Total	**470.0**	**22**	**18**	**26.1**	**0.81**

Results

We captured a total of 30 skunks (including two recaptures) during the study period in the two study areas. The overall capture effort, using both methods, was 4857 trap nights and on average, 405 trap nights were necessary to capture one skunk (Table 9.1). With similar capture effort, efficiency was almost ten times higher in PA than in CA (Table 9.1). In both study areas, leg-hold traps were thrice more efficient than box traps (Table 9.1).

Though manual capture cannot be directly compared with traps, the use of this technique increased the capture rate markedly (Table 9.2). We manually captured 18 skunks. An effort of about 5 and 43 km of searching was necessary to capture one skunk in PA and CA, respectively (Table 9.2).

Body measurements of *C. chinga* were generally similar, highlighting a lack of sexual dimorphism. Only two measurements were statistically greater in males: neck diameter and forefoot pad width (Table 9.3). Molina's hog-nosed skunks in PA were longer than in CA (Table 9.3).

Spatial Ecology

Methods

Radio-Telemetry

We fitted 16 healthy adult Molina's hog-nosed skunks with VHF radio-collars (PA: three males and four females; CA: three males and six females) equipped with activity and mortality sensors. We radio-tracked skunks on foot using portable telemetry equipment. Sampling was done using discontinuous and continuous tracking sessions (6–10 h) homogeneously throughout the day. We collected location data preferentially by the homing technique and occasionally by triangulation. We were confident that our presence did not disturb skunks because, during the day, they remained inactive in dens and in the night, they usually ignored our presence and continued with their activities.

We computed home range and core area sizes with Ranges V software (Kenward & Hodder, 1996). To estimate home range size, we calculated the 100% minimum convex polygon (100% MCP) and 95% fixed kernel (95% FK). We defined core areas as 45% FK isopleths (Reppucci *et al.*, 2009).

Spatial overlap was estimated by examining the extent of overlap among home ranges (95% FK) and core areas of neighbouring individuals. We defined two individuals as neighbours if the borders of each annual home range were at the same distance or closer than the average distance from the centre of home ranges. We calculated the percentage of the home range of animal A that was overlapped by animal B, and vice versa. Then we calculated the average overlap for selected groups of individuals.

Finally, we calculated daily movements by measuring the straight-line distance between two resting sites obtained for individual Molina's hog-nosed skunks radio-tracked across successive days.

Habitat Use and Selection

We evaluated habitat selection using compositional analysis (Aebischer *et al.*, 1993) that employs radio-collared animals as sampling units and considers all habitat types simultaneously. We used multivariate analysis of variance (MANOVA) to test the null hypothesis of no selection. Rejection of the null

Table 9.3 Morphometric measurements ($\bar{x} \pm$ SD) recorded in live-captured adults *Conepatus chinga* in a protected area (PA) and a cropland area (CA) of the Argentine Pampas. With the exception of weight (kg), measures are given in cm. Significant differences (ANOVA: $p < 0.05$) between areas or sexes are indicated in bold.

Measures	Overall	Males	Females	PA	CA	F-test
Weight	1.45±0.19	1.49±0.22	1.42±0.17	1.50±0.16	1.40±0.24	($F_{\text{sexes }1,22} = 0.28; p = 0.6$); ($F_{\text{area }1,22} = 1.1; p = 0.28$)
Total length	62.91±7.48	62.00±5.17	63.70±9.28	65.14±7.90	59.44±5.53	($F_{\text{sexes }1,20} = 1.13; p = 0.3$); **($F_{\text{area }1,20} = 4.33; p = 0.05$)**
Body length	37.38±3.13	37.10±3.95	37.60±2.33	38.75±2.86	35.56±2.60	($F_{\text{sexes }1,18} = 1,3; p = 0.26$); **($F_{\text{area }1,18} = 8.11; p = 0.01$)**
Head length	8.83±0.95	8.77±0.87	8.87±1.04	9.11±0.96	8.29±0.70	($F_{\text{sexes }1,18} = 0,43; p = 0.51$); **($F_{\text{area }1,18} = 4.24; p = 0.05$)**
Neck circumference	15.12±1.33	15.77±1.03	14.59±1.35	15.13±1.19	15.13±1.62	**($F_{\text{sexes }1,17} = 5.09; p = 0.03$)**; ($F_{\text{area }1,17} = 0.55; p = 0.46$)
Forefoot						
Total length	3.40±0.75	3.62±0.58	3.22±0.85	3.41±0.68	3.40±0.91	($F_{\text{sexes }1,13} = 1.1; p = 0.31$); ($F_{\text{area }1,13} = 0.07; p = 0.79$)
Pad length	2.28±0.78	2.40±0.58	2.19±0.94	2.08±0.64	2.56±0.92	($F_{\text{sexes }1,13} = 0.8; p = 0.38$); ($F_{\text{area }1,13} = 1.98; p = 0.18$)
Total width	2.02±0.34	2.19±0.20	1.88±0.37	2.04±0.23	2.00±0.47	($F_{\text{sexes }1,13} = 3.41; p = 0.08$); ($F_{\text{area }1,13} = 0.07; p = 0.79$)
Pad width	1.71±0.22	1.86±0.19	1.60±0.19	1.76±0.30	1.67±0.08	**($F_{\text{sexes }1,13} = 5.89; p = 0.03$)**; ($F_{\text{area }1,13} = 0.04; p = 0.89$)
Hindfoot						
Total length	3.40±0.75	4.46±0.82	4.41±1.24	4.67±0.85	4.07±1.23	($F_{\text{sexes }1, 15} = 0.00; p = 0.98$); ($F_{\text{area }1,15} = 1.41; p = 0.25$)
Pad length	3.30±1.03	3.28±0.62	3.31±1.37	3.48±0.68	3.08±1.38	($F_{\text{sexes }1, 15} = 0.07; p = 0,8$); ($F_{\text{area }1,15} = 0.67; p = 0.42$)
Total width	2.13±1.03	2.05±0.28	2.20±1.47	1.98±0.30	2.33±1.55	($F_{\text{sexes }1,15} = 0.02; p = 0.88$); ($F_{\text{area }1,15} = 0.39; p = 0.54$)
Pad width	1.81±0.52	1.78±0.20	1.84±0.69	1.76±0.17	1.89±0.79	($F_{\text{sexes }1,15} = 0.02; p = 0.88$); ($F_{\text{area }1,15} = 0.39; p = 0.54$)

hypothesis led to a series of paired Student's *t*-tests, which were used to rank habitat types from most to least preferred.

At PA habitat, analyses were conducted for seven skunks and three habitat types: rocky areas, woodland, and grassland patches (Figure 9.4). Rocky areas consisted of patches largely covered by rock outcropping with a moderate-to-high slope. The vegetation height and density in this area were strongly influenced by the presence of large numbers of feral horses (Scorolli *et al.*, 2006). Woodland was composed of introduced trees (predominantly *Pinus* sp. and *Eucalyptus* sp.). In this habitat, the density of feral horses was high too. Grassland patches were fenced areas where horses were excluded and grasses were denser and taller. At CA, analyses were conducted for nine Molina's hog-nosed skunks and we classified the following habitats: crop fields, pastures, and grassland patches (Figure 9.4). Crop fields were typically seeded, harvested, and cultivated annually with small grains (oat, wheat, and soya) or oil crops (sunflowers). Pastures consisted mostly of alfalfa and hay. Grass patches were marginal areas without management located mainly along railroads, streams, or in rocky soil.

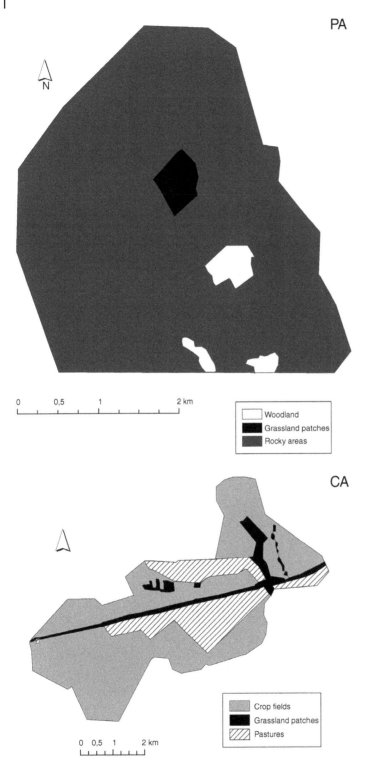

PA

Figure 9.4 Habitat types in a protected area (PA) and cropland area (CA) of the Argentine Pampas.

We examined habitat use at two of the hierarchical levels of selection described by Johnson (1980): second-order selection (the habitat composition inside 100% MCP home ranges compared to availability within the study area), and third-order selection (habitat use based on the number of locations within individual home ranges compared to habitat availability within those ranges).

Habitat Food Abundance

We performed a principal component analysis (PCA) on the invertebrate abundance data to summarize food availability for different habitat types. This analysis only included the main prey items found in the diet of *C. chinga* in the study areas: Coleoptera, Coloptera larvae, Lepidoptera larvae, and scorpions because the frequency and volume of the remaining dietary items were very small.

Results

Home Range Size

The mean ($\pm$SD) home range size (100% MCP) for 16 radio-collared Molina's hog-nosed skunks was 198 ± 109 ha and core area size (45% MCP) was 29 ± 26 ha (Table 9.4). The size of home ranges and core areas did not vary between areas (Castillo

et al., 2011b). However, in both areas, the sizes of home ranges and core areas were significantly larger for males than for females (Castillo *et al.*, 2011b).

Spatial Overlap

Mean home range overlap was greater in PA (46.9%) than in CA (21.3%) (Castillo *et al.*, 2011b). Specifically, mean overlaps between females, and between males and females, were larger in PA (Castillo *et al.*, 2011b). However, pairs of males appeared to overlap more in CA, but sample sizes (2 males in each area) were small. In both study areas, we did not detect differences in the extent of intra- and intersexual overlaps (Castillo *et al.*, 2011b).

Individual core areas also overlapped more at PA and were five times larger than at CA. At CA, core area overlap was recorded between sexes but it was absent within sexes (Castillo *et al.*, 2011b). Finally, intersexual core area overlap was significantly greater in PA than in CA (Castillo *et al.*, 2011b).

Movement Patterns

The mean distance ($\pm$SD) travelled between two consecutive resting sites was 270 ± 365 m and did not differ between areas. Movement distances were greater for males (391 ± 422 m) than females (191 ± 276 m) (Castillo *et al.*, 2011b). Finally, the mean distance travelled by

Table 9.4 Mean ($\bar{x}$) 100 and 45% minimum convex polygon (MCP) estimates and mean 95 and 45% fixed kernel (FK) estimates of home range size (ha), and corresponding standard deviations (SD), for female and male *Conepatus chinga* and all individuals combined in a protected area (PA) and a cropland area (CA) of the Argentine Pampas.

Area	Sex	100% MCP		95% FK		45% MCP		45% FK	
		$\bar{x}$	SD	$\bar{x}$	SD	$\bar{x}$	SD	$\bar{x}$	SD
PA	Females	88.3	24.9	81.2	6.9	11.6	3.3	22.2	3.3
	Males	268.1	135.6	213.1	105.3	42.1	6.1	72.8	37.7
	All individuals	165.3	125.2	137.7	93.2	24.7	16.8	43.8	34.8
CA	Females	193.2	152.7	146.5	93.6	19.5	16.9	37.5	24.1
	Males	283.6	157.6	274.4	130.7	56.3	45.3	68.6	27.4
	All individuals	223.3	151.1	189.1	117.6	31.8	32.1	47.9	28.2
Overall	Females	151.2	126.9	120.4	77.6	16.4	13.4	31.4	19.8
	Males	275.8	131.7	243.7	76.5	49.2	29.9	70.7	29.6
	All individuals	197.9	109.1	166.7	107.5	28.7	26.1	46.1	30.2

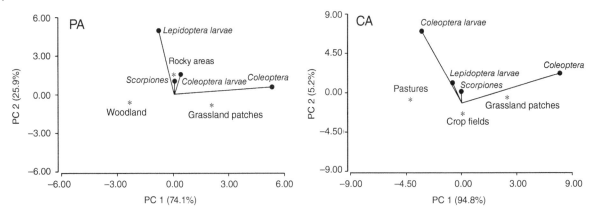

Figure 9.5 Principal component (PC) analysis performed on the abundance of the main prey (in italics) of *Conepatus chinga* in different habitat types of a protected area (PA) and a cropland area (CA) of the Argentine Pampas.

individual skunks between resting sites was correlated to their respective home range sizes (Castillo *et al.*, 2011b).

Habitat Use and Selection

In Pampas grassland, *C. chinga* used habitats non-randomly at one or both orders of resolution tested. When we studied second-order habitat selection, we found that Molina's hog-nosed skunks did not use habitats randomly. However, multiple comparison tests only detected a significant order of preference in PA, where skunks selected grassland patches over rocky areas and woodland (Castillo *et al.*, 2012b). Habitat use at the third-order level was random in PA but not in CA, where *C. chinga* showed the following order of selection: grassland patches > pastures > crop fields (Castillo *et al.*, 2012b).

Habitat Food Abundance

In both areas, the first PCA component was associated with beetle abundance (Figure 9.5). Grassland patches at PA had an abundance of beetles higher than average, while the abundance of these invertebrates in the rocky areas was close to average. Finally, woodland had the lowest abundance of Coleoptera. The second component, at PA, separated rocky areas from the remaining habitats based on greater than average abundances of Coleoptera and Lepidoptera larvae. At CA, the first component indicated that grassland patches presented a higher-than-average abundance of beetles, whereas the abundance of beetles was lower than average for pastures. The second component was associated mainly with scorpions and Coleoptera larvae. The abundance of these items was greater than average in pastures and grassland patches and lower than average in crop fields (Figure 9.5).

Population Density

Methods

Because spotlight surveys are widely accepted as a practical method of estimating the relative abundance of nocturnal animals (Sobrino *et al.*, 2009), we counted Molina's hog-nosed skunks from a vehicle (driven at a constant speed of 15 km/h) along secondary roads. Transects ranged from 8.3–16.0 km in length at PA ($n = 42$) and from 19.3–21.0 km in length at CA ($n = 40$), and were distributed homogeneously throughout the year (3–4 per month) in both areas. We used the program DISTANCE, version 6.0 (Thomas *et al.*, 2009) to estimate skunk densities in both areas based on distance sampling theory (Buckland *et al.*, 1993).

Results

In spite of the smaller sampling effort, we counted 37 skunks at PA (along 482.7 km) whereas we observed only 16 individuals at CA (along 1054 km). The estimated density of *C. chinga* by spotlighting surveys varied from 0.12–0.53 individual/km² in CA to 0.6–1.47 individual(s)/km² in PA. We obtained consistent results when skunk density was estimated on the basis of home range size (CA: 0.66 individual/km^2; PA: 1.66 individuals/km^2) (Castillo *et al.*, 2011b).

Denning Behaviour

Methods

We considered all sites where an individual skunk remained inactive during the day as occupied dens (Doty & Dowler, 2006). We located den sites via homing in on radio-collared individuals. We then categorized the dens as burrows (underground) or aboveground (rock) dens. We used the 99% fixed kernel method to build the home range contours from all locations of each skunk. Then home ranges were divided into core (delimited by the 45% kernel), middle (45–80%), and border region (the area occupied by the 19% most external locations). We calculated the densities of dens for each skunk by dividing the number of dens in any given region by its area. We calculated distances (using ArcView 9.3®) between den site locations and selected landscape features. We determined the same distances for uniformly distributed random points (generated in equal number to dens) and investigated differences between den sites and random points with Student's *t*-tests.

Results

We identified and characterized 199 ($n_{\text{males}} = 108$, $n_{\text{females}} = 91$) den sites in PA (Castillo *et al.*, 2013) and 240 ($n_{\text{males}} = 92$, $n_{\text{females}} = 148$) in CA (Castillo *et al.*, 2011a). At PA, most dens used by skunks were found in aboveground rock sites ($n = 170$, 85.4%) (Castillo *et al.*, 2013) while at CA, *C. chinga* preferred burrows ($n = 196$, 81.7%) (Castillo *et al.*, 2011a). Unlike in PA (Chi-square test of independence: $\chi^2 = 1.68$, df = 1, $p = 0.19$), we observed differences ($\chi^2 = 17.33$, df = 1, $p = 0.0001$) in the type of dens used by the two sexes at CA, with females using a higher proportion of underground shelters than males (Figure 9.6). We found no difference in the dimensions of dens entrances (height and width) between sexes or sites (Castillo, 2011). In both study areas, den density was higher in core areas than elsewhere (Table 9.5).

In both study areas, we observed relatively frequent reuse of individual resting sites by *C. chinga*, with no differences between sexes. On average, 25.1% of den sites were reused by skunks (males: 23.1%, females: 27.4%) in PA (Castillo *et al.*, 2013), and 24.1% in CA (males: 26.3%, females: 22.8%) (Castillo *et al.*, 2011a). Studied animals returned to the same den sites for a mean ($\pm$SD) of 2.6$\pm$0.4 times at PA, and 3.1$\pm$1.7 times at CA. Dens were often (PA: 42.6%; CA: 50.8%) reused on consecutive days (Castillo *et al.*, 2011a, 2013).

In both areas, den sites were located closer to grasslands and habitat edges than randomly selected points (Table 9.6). At CA, den sites were also located closer to fences than expected by chance. Contrary to other skunk species (Rosatte & Larivière, 2003), dens were not associated with streams (Table 9.6).

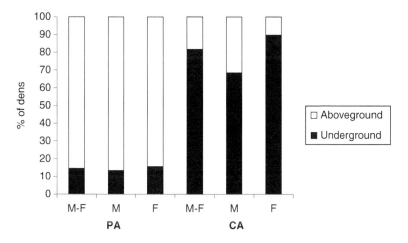

Figure 9.6 Proportion of the two types of dens used by *Conepatus chinga* in a protected area (PA) and a cropland area (CA) of the Argentine Pampas. M = males, F = females. n_{PA} = 199, n_{CA} = 240.

Table 9.5 Spatial distribution of *Conepatus chinga* dens ($n_{PA} = 199, n_{CA} = 240$) within the home ranges of radio-tagged individuals in a protected area (PA) and cropland area (CA) of the Argentine Pampas.

	Core region		Middle region		Border region	
	PA	CA	PA	CA	PA	CA
No. of dens $\pm$ SD	$14.6^a \pm 8.3$	$12.2^a \pm 5.6$	$6.6^b \pm 5.1$	$9.7^a \pm 4.1$	$7.3^b \pm 2.8$	$4.6^b \pm 3.3$
% of total dens	51.3	45.8	23.1	36.7	25.6	17.5
Density (no. dens/ha $\pm$ SD)	$0.43^a \pm 0.2$	$0.35^a \pm 0.2$	$0.15^b \pm 0.1$	$0.16^a \pm 0.1$	$0.1^b \pm 0.1$	$0.05^b \pm 0.02$

Different superscript letters indicate significant differences in home range regions (ANOVA: $p < 0.05$).
Source: Reproduced by permission of the Ecological Society of Japan and Mastozoología Neotropical.

Table 9.6 Mean distances $\pm$ SD (m) to selected landscape features associated with *Conepatus chinga* den sites (DS) and random locations (RL) in a protected area (PA) and a cropland area (CA) of the Argentine Pampas.

	PA		CA	
	DS	RL	DS	RL
Streams	178 ± 116^a	203 ± 176^a	1054 ± 555^a	1191 ± 883^a
Grasslands	305 ± 342^a	884 ± 579^b	429 ± 406^a	1097 ± 871^b
Habitat edge	132 ± 105^a	187 ± 143^b	260 ± 318^a	812 ± 773^b
Fences	—	—	87 ± 94^a	140 ± 125^b

Different superscript letters between habitat types indicate significant differences (*t*-test: $p < 0.05$).
Source: Reproduced by permission of the Ecological Society of Japan.

In both areas, *C. chinga* utilized habitats to locate den sites in a non-random manner (Castillo *et al.*, 2011a, 2013). In PA, a comparison of habitats associated with den sites and home range habitat compositions showed that rocky areas, woodland, and grassland were ranked first to third. In CA, habitats were ranked as follows: grass patches, pastures, and crop fields.

Activity Patterns

Methods

The activity was assessed as active or inactive based on two minutes' sampling of the signal from the motion sensor. This sensor was incorporated in the radio-collars (see **Spatial ecology** section) and generated different pulse rates when the skunk was inactive, and when it was moving. The percentage of time that an animal remained active during each two minutes' sample was calculated and the record was considered active if it exceeded 50%. Finally, activity was defined as the percentage of active fixes per hour.

Results

On average, Molina's hog-nosed skunks were significantly more active in PA (53.7% of records) than in CA (48.2%) (Castillo *et al.*, 2015). In both areas, females were significantly more active than males (PA: 58.3% for females, 47.9% for males; CA: 51.8% for females, 42.6% for males) (Castillo *et al.*, 2015). In both sexes and both study areas, activity was not homogeneously distributed throughout the day (Castillo *et al.*, 2015). Skunks spent most of the time

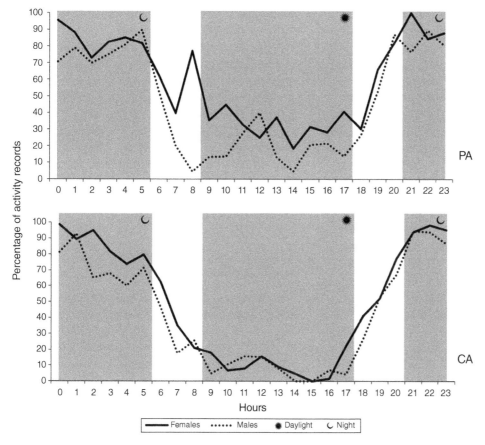

Figure 9.7 Daily activity profiles of male and female *Conepatus chinga* in a protected area (PA) and cropland area (CA) of the Argentine Pampas. The white bars correspond to sunrise and sunset periods, respectively.

active at night while the minimum proportion of active fixes was recorded during the daylight period. Intermediate values were recorded at both sunrise and sunset (Figure 9.7).

We found no differences between study areas in the proportions of activity at sunrise, sunset, and night (Castillo *et al.*, 2015). However, in PA, skunks exhibited higher activity levels during the day (Figure 9.7; Castillo *et al.*, 2015) than in CA. In the two study areas, activity peaked between 21:00/22:00 and 5:00 h when both males and females were active in more than 80% of records (Figure 9.7). In both areas, females had both higher levels and more prolonged periods of activity than males (Figure 9.7).

Discussion

Ecology of Molina's Hog-Nosed Skunk

Although the body of knowledge on *C. chinga* remains relatively small, research carried out in the last decade in two of the regions where this species is most common (central Argentina and southern Brazil) enables us to draw some substantiated inferences on a number of aspects of its ecology and natural history.

In agreement with Kasper *et al.* (2012) and Brashear *et al.* (2015) for *Conepatus leuconotus*, we conclude that intersexual differences in home range sizes are due mostly to reproductive strategies, rather than metabolic needs. In solitary carnivores, reproductive success in females is typically determined by access to trophic resources, whereas males tend to maximize their reproductive success by accessing as many females as possible (e.g. Macdonald, 1983; Sandell, 1989; Palomares *et al.*, 2017). Therefore, by increasing the size of their home ranges, males would maximize the chances of mating with a greater number of females. The high overlap in home ranges between sexes (each male overlaps with at least 2–3 females in the Argentine Pampas) supports this hypothesis. Also, the largest movements of males, compared to females, could be related to a more intense territory-patrolling behaviour and would maximize the males' probability of encountering females.

Diet composition did not differ appreciably from previous studies, supporting the data showing *C. chinga* is essentially a predator of insects, especially beetles. Similar to other skunk species, Molina's hognosed skunk has been described as an opportunistic predator (Travaini *et al.*, 1998; Zeballos *et al.*, 1998; Kasper *et al.*, 2016), capable of switching its diet depending on prey abundance. Conversely, the results inferred from our research in Pampas grassland corroborate the findings of Donadio *et al.* (2004) and Peters *et al.* (2011), who noted that *C. chinga* consumed prey in different proportions than expected based on their abundance.

We did not observe communal burrows, coparenting activities, and group foraging (Castillo, 2011). The absences of such behaviours are associated with solitary habits which are typical of many Mephitidae (Kasper *et al.*, 2012). As expected for a social system based on territorial defence, males had a higher propensity to share their home range areas with females than with other males, and home range overlap was limited to the level of core areas. This spacing pattern has also been documented for *C. chinga* in southern Brazil (Kasper *et al.*, 2012) and for *C. leuconotus* in Texas, USA (Brashear *et al.*, 2015).

Molina's hog-nosed skunks in the Pampas of central Argentina were essentially nocturnal, starting and ceasing their activities in association to sunset and sunrise, respectively. It is possible that the largely nocturnal activity exhibited by *C. chinga*, also reported by Donadio *et al.* (2001), Kasper *et al.* (2012), Tellaeche *et al.* (2014), Zúñiga *et al.* (2017), and Leuchtenberger *et al.* (2018) serves to increase the probability of encountering prey (e.g. several ground beetle species are highly vulnerable to predation during the night).

Sexual dimorphism is one of the major aspects where the data currently available on *C. chinga* are not in agreement. While an analysis with a larger sample size could detect some masked variations, it is unlikely that it would affect our conclusion that, similar to other members of the Mephitidae (Bixler & Gittleman, 2000), Molina's hog-nosed skunk in the Argentine Pampas does not show clear sexual dimorphism. This is in contrast to results obtained from studies with similar sample sizes in southern Brazil (Kasper *et al.*, 2012) and Uruguay (Redford & Eisenberg, 1992).

Effects of Land Management on Molina's Hog-Nosed Skunk

The effects of the distribution and abundance of food resources on spatial behaviour have been observed in many carnivores (Patterson & Messier, 2001; Jepsen *et al.*, 2002; Macdonald *et al.*, 2004). Although prey abundance (as well as diet composition) was similar in our two study areas, we found differences in its predictability. Whereas environments remain largely unchanged throughout the year in PA, in CA, only two habitats (grassland and pastures) were stable, while the abundance of food related to croplands varied markedly both throughout the year and between years. Contrary to our prediction, home range sizes of *C. chinga* did not vary between study sites and were almost identical to those described by Kasper *et al.* (2012) for southern Brazil, and were similar to those reported by Donadio *et al.* (2001) for the Argentine Patagonia. On the other hand, despite the limitations of the density estimators used, we estimated a lower population density of skunks at CA (supported by the low capture rate in this area). Though our conclusion could be affected by a large individual variation as well as a small sample size, we suggest that skunks in Pampas grassland are better able to tolerate changes in population density than changes in home range size. Nevertheless, female home ranges tended to be larger at CA than PA. Because females are the sex more affected by food abundance and dispersion, this result is in agreement with our first prediction that home range size would be larger in the landscape fragmented by agriculture and where resource abundance is less predictable.

As we expected (second prediction), *C. chinga* selected for grassland patches, which was evident at different spatial scales in both study areas. We speculate that this variation between areas is indicative of an adaptation by *C. chinga* to human-induced environmental changes. In general, the second-order selection is associated with key habitat patch distribution and the third-order selection is primarily related to the temporal dispersion of resources (Aebischer *et al.*, 1993; Lucherini *et al.*, 1995; Johnson *et al.*, 2000). In PA, where the habitat is relatively homogeneous and little disturbed, skunks attempted to include a large proportion of grasslands (a stable habitat with a good abundance of prey) within their home ranges while using the home range interior randomly. In the highly fragmented CA area, where the most preferred habitats are limited, it may be difficult (in terms of energetics and territory defence) for skunks to include large portions of favoured habitats within home ranges, so selection would occur primarily in the use of the different patches occurring within their home ranges.

Although food is recognized as the primary resource responsible for intraspecific variation in social organization and behaviour (Bekoff *et al.*, 1984), den sites are also an important resource for several carnivore species (Macdonald & Johnson, 2015). That dens used by *C. chinga* in our study areas were not distributed randomly, but concentrated within individual core areas and reused, indicates that dens are important. This may be more so in rural areas, where Molina's hognosed skunks are frequently injured or killed by dogs (Kasper *et al.*, 2009; Castillo & Schiaffini, 2019). In agreement with our third prediction, we found a strong selection for den sites in CA and skunks selected den sites in patches with stable and dense vegetation.

Another behavioural difference that we observed between study areas was related to the skunks' activity patterns. As predicted (fourth prediction), skunks were more active in the protected area, where the human disturbance was less. This difference was due primarily to their greater activity during daylight hours. Although we cannot rule out other factors, we speculate that *C. chinga* reduced daytime activities at CA in response to human-related disturbance, where skunks could be killed by dogs or struck by vehicles (Cunha *et al.*, 2015). This behavioural adaptation has also been reported for other carnivore species (Lucherini *et al.*, 1995; Grinder & Krausman, 2001; Tigas *et al.*, 2002; Beckmann & Berger, 2003; Wang *et al.*, 2015).

Recommendations for Conservation and Management

Studies on the effects of native grasslands' transformation on animal ecology and biodiversity (especially from regions that are generally under-represented) are of great importance for future temperate ecosystem conservation and management (Medan *et al.*, 2011). In the current Pampas landscape, native vegetation environments are restricted to roadsides, areas surrounding railways, along fences or in rocky areas where cultivation is impossible. Several results of our study indicate that these environments are also necessary for the persistence of stable *C. chinga* populations in modified Pampas ecosystems. These natural or semi-natural remnants may function as biodiversity refuges (Le Coeur *et al.*, 2002; Marshall & Moonen, 2002) and have been cited as critical for conservation in rural landscapes for birds (Renfrew & Ribic, 2001; Vickery *et al.*, 2001; Bilenca & Miñarro, 2004), rodents (Bilenca & Miñarro, 2004; Birochio, 2008), and two other carnivores occurring in the Pampas, the Pampas fox, *Lycalopex [= Pseudalopex] gymnocercus* (Luengos Vidal *et al.*, 2012), and Geoffroy's cat, *Leopardus geoffroyi* (Manfredi *et al.*, 2006; Castillo *et al.*, 2008). Therefore, the conservation of these semi-natural patches will not only benefit *C. chinga* but is likely of value to the entire carnivore community. Consistently, the association between the presence of *C. chinga* and that of native vegetation is in agreement with landscape studies in other ecoregions (Lantschner *et al.*, 2011; Caruso *et al.*, 2016).

It is well known that reduction and/or loss of connectivity of suitable habitats leads to decreased survival and reproduction rates in vertebrates (Crooks *et al.*, 2011). For appropriate management of *C. chinga* at a metapopulation scale, it is important to ensure connectivity between these patches through natural corridors, to attenuate the negative consequences of fragmentation and minimize isolation of populations of these species (Tewksbury *et al.*, 2002; Haddad *et al.*, 2015). Unfortunately, the current trend in Argentina is to convert natural habitats into croplands.

Acknowledgements

We thank E. Luengos Vidal, J. Reppucci, S. Savini, C. Manfredi, N. Caruso, M. Rodriguez, and many others who collaborated with fieldwork and in the laboratory for the success of this project. Thanks are extended to the farm owners who granted permission to work in their lands, especially the Merino family. The staff of Ernesto Tornquist Provincial Park provided logistic support. We are especially thankful to E.B. Casanave for facilitating our work and giving her full support to our research efforts. Emmanuel Do Linh San, Jerrold Belant, Michael J. Somers, Richard Dowler, and an anonymous referee greatly improved a previous version of this chapter. Our work has been financially supported by Consejo Nacional de Investigaciones Científicas y Técnicas (CONICET), Earthwatch Institute, IM40 Grant by ANPCyT, and SGCyT, UNS (PGI 24/B123, and 24B152).

References

Aebischer, N.J., Robertson, P.A. & Kenward, R.E. (1993) Compositional analysis of habitat use from animal radio-tracking data. *Ecology* 74, 1313–1325.

Beckmann, J.P. & Berger, J. (2003) Rapid ecological and behavioural changes in carnivores: the responses of black bears (*Ursus americanus*) to altered food. *Journal of Zoology* 261, 207–212.

Bekoff, M., Daniels, T.J. & Gittleman, J.L. (1984) Life history patterns and the comparative social ecology of carnivores. *Annual Review of Ecology and Systematics* 15, 191–232.

Bertonatti, C. & Corcuera, J. (2000) *Situación Ambiental Argentina*. Fundación Vida Silvestre Argentina, Buenos Aires. (In Spanish).

Bilenca, D. & Miñarro, F. (2004) *Identificacion de Áreas Valiosas del Pastizal (AVPs) en las Pampas y Campos de Argentina, Uruguay y sur de Brasil*. Fundación Vida Silvestre Argentina, Buenos Aires. (In Spanish).

Birochio, D.E. (2008) *Ecología Trófica de Lycalopex gymnocercus en la Región Pampeana: un Acercamiento Inferencial al Uso de los Recursos*. PhD Thesis, Universidad Nacional del Sur. (In spanish).

Bixler, A. & Gittleman, J.L. (2000) Variation in home range and use of habitat in the striped skunk (*Mephitis mephitis*). *Journal of Zoology* 251, 525–533.

Brashear, W.A., Ferguson, A.W., Negovetich, N.J. & Dowler, R.C. (2015) Spatial organization and home range patterns of the American hog-nosed skunk (*Conepatus leuconotus*). *The American Midland Naturalist* 174, 310–320.

Brennan, L.A. & Kuvlesky, W.P. Jr (2005) North American grassland birds: an unfolding conservation crisis? *Journal of Wildlife Management* 69, 1–13.

Buckland, S.T., Anderson, D.R., Burnham, K.P. & Laake, J.L. (1993) *Distance Sampling: Estimating Abundance of Biological Populations*. Chapman and Hall, New York.

Burkart, R., Ruiz, D., Marañal, C. & Aduea, F. (1991) *El Sistema Nacional de Áreas Naturales Protegidas de la Republica Argentina*. Asociación Parques Nacionales, Buenos Aires. (In Spanish).

Byers, C.R., Steinhorst, R.K. & Krausman, P.R. (1984) Clarification of a technique for analysis of utilization–availability data. *Journal of Wildlife Management* 48, 1050–1053.

Campo de Ferreras, A.M., Capelli de Steffens, A.M. & Diez, P.G. (2004) *El Clima del Suroeste Bonaerense*. Departamento de Geografía y Turismo, Bahía Blanca. (In Spanish).

Caruso, N.C., Lucherini, M., Fortin, D. & Casanave, E.B. (2016) Species-specific responses of carnivores to human-induced landscape changes in central Argentina. *PLoS One* 11, e0150488.

Castillo, D.F. (2011) *Ecología Espacial, Temporal y Trófica del Zorrino (Conepatus chinga) en un Área Natural y un Área de Uso Agrícola*. PhD Thesis, Universidad Nacional del Sur. (In spanish).

Castillo, D.F. & Schiaffini, M.I. (2019) *Conepatus chinga*. In: *Categorización 2019 de los Mamíferos de Argentina Según su Riesgo de Extinción. Lista Roja de los Mamíferos de Argentina*. Ministerio de Ambiente y Desarrollo Sostenible, y Sociedad Argentina para el Estudio de los Mamíferos. (In Spanish). http://cma.sarem.org.ar. Accessed on 24 May 2021.

Castillo, D.F., Luengos Vidal, E., Lucherini, M. & Casanave, E.B. (2008) First report on the Geoffroy's cat in a highly modified rural area of the Argentine Pampas. *Cat News* 49, 27–28.

Castillo, D.F., Lucherini, M. & Casanave, E.B. (2011a) Denning ecology of Molina's hog-nosed skunk in a farmland area in the Pampas grassland of Argentina. *Ecological Research* 26, 845–850.

Castillo, D.F., Lucherini, M., Luengos Vidal, E.M., Manfredi, C. & Casanave, E.B. (2011b) Spatial organization of Molina's hog-nosed skunk (*Conepatus chinga*) in two landscapes of the Pampas grassland of Argentina. *Canadian Journal of Zoology* 89, 229–238.

Castillo, D.F., Luengos Vidal, E.M., Casanave, E.B. & Lucherini, M. (2012a) Field immobilization of Molina's hog-nosed skunk (*Conepatus chinga*) using ketamine and xylazine. *Journal of Wildlife Diseases* 48, 173–175.

Castillo, D.F., Luengos Vidal, E.M., Casanave, E.B. & Lucherini, M. (2012b) Habitat selection of Molina's hog-nosed skunks in relation to prey abundance in the Pampas grassland of Argentina. *Journal of Mammalogy* 93, 716–721.

Castillo, D.F., Luengos Vidal, E.M., Caruso, N.C., Lucherini, M. & Casanave, E.B. (2013) Denning ecology of *Conepatus chinga* (Carnivora: Mephitidae) in a grassland relict of central Argentina. *Mastozoología Neotropical* 20, 373–379.

Castillo, D.F., Luengos Vidal, E.M., Caruso, N.C., Casanave, E.B. & Lucherini, M. (2015) Activity patterns of Molina's hog-nosed skunk in two areas of the Pampas grassland (Argentina) under different anthropogenic pressure. *Ethology Ecology & Evolution* 27, 379–388.

Crooks, K.R., Burdett, C.L., Theobald, D.M., Rondinini, C. & Boitani, L. (2011) Global patterns of fragmentation and connectivity of mammalian carnivore habitat. *Philosophical Transactions of the Royal Society B: Biological Sciences* 366, 2642–2651.

Cunha, G.G., Hartmann, M.T. & Hartmann, P.A. (2015) Atropelamentos de vertebrados em uma área de Pampa no sul do Brasil. *Ambiência Guarapuava* 11, 307–320. (In Portuguese).

Donadio, E., Di Martino, S., Aubone, M. & Novaro, A.J. (2001) Activity patterns, home-range, and habitat selection of the common hog-nosed skunk, *Conepatus chinga* (*Mammalia*, Mustelidae), in northwestern Patagonia. *Mammalia* 65, 49–54.

Donadio, E., Martino, S.D., Aubone, M. & Novaro, A.J. (2004) Feeding ecology of the Andean hog-nosed skunk (*Conepatus chinga*) in areas under different land use in north-western Patagonia. *Journal of Arid Environments* 56, 709–718.

Donald, P.F., Sanderson, F.J., Burfield, I.J. & van Bommel, F.P.J. (2006) Further evidence of continent-wide impacts of agricultural intensification on European farmland birds. *Agriculture Ecosystems & Environment* 116, 189–196.

Doty, J.B. & Dowler, R.C. (2006) Denning ecology in sympatric populations of skunks (*Spilogale gracilis* and *Mephitis mephitis*) in west-central Texas. *Journal of Mammalogy* 87, 131–138.

Elmhagen, B., Tannerfeldt, M., Verucci, P. & Angerbjörn, A. (2000) The arctic fox (*Alopex lagopus*): an opportunistic specialist. *Journal of Zoology* 251, 139–149.

Firbank, L.G., Petit, S., Smart, S., Blain, A. & Fuller, R.J. (2008) Assessing the impacts of agricultural intensification on biodiversity: a British perspective. *Philosophical Transactions of the Royal Society B: Biological Sciences* 363, 777–787.

Frangi, J.L. & Bottino, O.J. (1995) Comunidades vegetales de la Sierra de la Ventana, provincia de Buenos Aires, Argentina. *Revista de la Facultad de Agronomía de la Universidad de La Plata* 71, 93–133. (In Spanish).

Grinder, M.I. & Krausman, P.R. (2001) Home range, habitat use, and nocturnal activity of coyotes in an urban environment. *Journal of Wildlife Management* 65, 887–898.

Habel, J.C., Ulrich, W., Biburger, N., Seibold, S. & Schmitt, T. (2019) Agricultural intensification drives butterfly decline. *Insect Conservation and Diversity* 12, 289–295.

Haddad, N.M., Brudvig, L.A., Clobert, J., Davies, K.F., Gonzalez, A. *et al.* (2015). Habitat fragmentation and its lasting impact on Earth's ecosystems. *Science Advances* 1, e1500052.

Home, C. & Jhala, Y.V. (2009) Food habits of the Indian fox (*Vulpes bengalensis*) in Kutch, Gujarat, India. *Mammalian Biology* 74, 403–411.

Jacobs, J. (1974) Quantitative measurement of food selection. A modification of the forage ratio and Ivlev's electivity index. *Oecologia* 14, 413–417.

Jepsen, J.U., Eide, N.E., Prestrud, P. & Jacobsen, L.B. (2002) The importance of prey distribution in habitat use by arctic foxes (*Alopex lagopus*). *Canadian Journal of Zoology* 80, 418–429.

Johnson, D.H. (1980) The comparison of usage and availability measurements for evaluating resource preference. *Ecology* 61, 65–71.

Johnson, D.D.P., Macdonald, D.W. & Dickman, A.J. (2000) An analysis and review of models of the sociobiology of the Mustelidae. *Mammal Review* 30, 171–196.

Jonsson, M., Buckley, H.L., Case, B.S., Wratten, S.D., Hale, R.J. & Didham, R.K. (2012) Agricultural intensification drives landscape-context effects on host–parasitoid interactions in agroecosystems. *Journal of Applied Ecology* 49, 706–714.

Kasper, C.B., Da Fontoura-Rodriguez, M.L., Cavalcanti, G.N., De Freitas, T.R.O., Rodrigues, F.H.G., Gomez De Oliveira, T. & Ezirik, E. (2009) Recent advances in the knowledge of Molina's hog-nosed skunk *Conepatus chinga* and striped hog-nosed skunk *C. semistriatus* in South America. *Small Carnivore Conservation* 41, 25–28.

Kasper, C.B., Peters, F.B., Christoff, A.U. & De Freitas, T.R.O. (2016) Trophic relationships of sympatric small carnivores in fragmented landscapes of southern Brazil: niche overlap and potential for competition. *Mammalia* 80, 143–152.

Kasper, C.B., Soares, J.B.G. & Freitas, T.R.O. (2012) Differential patterns of home-range, net displacement and resting sites use of *Conepatus chinga* in southern Brazil. *Mammalian Biology* 77, 358–362.

Kenward, E.E. & Hodder, K.H. (1996) *RANGES V. An Analysis System for Biological Location Data*. Natural Environment Research Council, Swindon.

Lantschner, M.V., Rusch, V. & Hayes J.P. (2011) Influences of pine plantations on small mammal assemblages of the Patagonian forest–steppe ecotone. *Mammalia* 75, 249–255.

Le Coeur, D., Baudry, J., Burel, F. & Thenail, C. (2002) Why and how we should study field boundary biodiversity in an agrarian landscape context. *Agriculture, Ecosystems & Environment* 89, 23–40.

Leuchtenberger, C., De Oliveira, Ê.S., Cariolatto, L.P. & Kasper, C.B. (2018) Activity pattern of medium and large sized mammals and density estimates of *Cuniculus paca* (Rodentia: Cuniculidae) in the Brazilian Pampa. *Brazilian Journal of Biology* 78, 697–705.

Lucherini, M., Lovari, S. & Crema, G. (1995) Habitat use and ranging behaviour of the red fox (*Vulpes vulpes*) in a Mediterranean rural area: is shelter availability a key factor? *Journal of Zoology* 237, 577–591.

Luengos Vidal, E.M., Sillero-Zubiri, C., Marino, J., Casanave, E.B. & Lucherini, M. (2012) Spatial organization of the Pampas fox in a grassland relict of central Argentina: a flexible system. *Journal of Zoology* 287, 133–141.

Macdonald, D.W. (1983) The ecology of carnivore social behaviour. *Nature* 301, 379–384.

Macdonald, D.W. & Johnson, D.D.P. (2015) Patchwork planet: the resource dispersion hypothesis, society, and the ecology of life. *Journal of Zoology* 295, 75–107.

Macdonald, D.W., Creel, S. & Mills, M. (2004) Canid society. In: *The Biology and Conservation of Wild Canids* (eds D.W. Macdonald & C. Sillero-Zubiri), pp. 85–106. Oxford University Press, Oxford.

Manfredi, C., Soler, L., Lucherini, M. & Casanave, E.B. (2006) Home range and habitat use by Geoffroy's cat (*Oncifelis geoffroyi*) in a wet grassland in Argentina. *Journal of Zoology* 268, 381–387.

Marrero, H., Zalba, S. & Carpintero, D. (2008) Eficiencia relativa de distintas técnicas de captura de heterópteros terrestres en un pastizal de montaña. *BioScriba* 1, 3–9. (In Spanish).

Marshall, E. & Moonen, A. (2002) Field margins in northern Europe: their functions and interactions with agriculture. *Agriculture, Ecosystems & Environment* 89, 5–21.

Mastrangelo, M.E. & Gavin, M.C. (2014) Impacts of agricultural intensification on avian richness at multiple scales in Dry Chaco forests. *Biological Conservation* 179, 63–71.

Medan, D., Torretta, J., Hodara, K., de la Fuente, E. & Montaldo, N. (2011) Effects of agriculture expansion and intensification on the vertebrate and invertebrate diversity in the Pampas of Argentina. *Biodiversity and Conservation* 20, 3077–3100.

Palomares, F., Lucena-Pérez, M., López-Bao, J.V. & Godoy, J.A. (2017) Territoriality ensures paternity in a solitary carnivore mammal. *Scientific Reports* 7, 4494.

Paltridge, R. (2002) The diets of cats, foxes and dingoes in relation to prey availability in the Tanami Desert, Northern Territory. *Wildlife Research* 29, 389–403.

Patterson, B.R. & Messier, F. (2001) Social organization and space use of coyotes in eastern Canada relative to prey distribution and abundance. *Journal of Mammalogy* 82, 463–477.

Peters, F.B., Roth, P.R.D.O. & Christoff, A.U. (2011) Feeding habits of Molina's hog-nosed skunk, *Conepatus chinga* (Carnivora: Mephitidae) in the extreme south of Brazil. *Zoologia (Curitiba, Impresso)* 28, 193–198.

R Development Core Team (2008) *R: A Language and Environment for Statistical Computing*. Vienna, Austria. http://www.R-project.org.

Redford, K.H. & Eisenberg, J.F. (1992) *Mammals of the Neotropics, Volume 2: The Southern Cone: Chile, Argentina, Uruguay, Paraguay*. The University of Chicago Press, Chicago.

Renfrew, R. & Ribic, C. (2001) Grassland birds associated with agricultural riparian practices in southwestern Wisconsin. *Journal of Range Management* 54, 546–552.

Reppucci, J.I., Castillo, D.F., Lucherini, M., Luengos Vidal, E.M. & Casanave, E.B. (2009) Interindividual interactions of Molina's hog-nosed skunks *Conepatus chinga* in the Pampas grassland of Argentina. *Acta Theriologica* 54, 87–94.

Reynolds, J.C. & Aebischer, N.J. (1991) Comparison and quantification of carnivore diet by faecal analysis: a critique, with recommendations, based on a study of the fox *Vulpes vulpes*. *Mammal Review* 21, 97–122.

Robinson, R.A. & Sutherland, W.J. (2002) Post-war changes in arable farming and biodiversity in Great Britain. *Journal of Applied Ecology* 39, 157–176.

Rosatte, R. & Larivière, S. (2003) Skunks. In: *Wild mammals of North America: Biology, Management and Conservation* (eds G. Fledhamer, B.C. Thompson & J.A. Chapman), pp. 692–707. The Johns Hopkins University Press, Baltimore.

Sandell, M. (1989) The mating tactics and spacing patterns of solitary carnivores. In: *Carnivore Behaviour, Ecology, and Evolution* (ed. J.L. Gittleman), pp. 164–182. Cornell University Press, Ithaca.

Schiaffini, M.I., Gabrielli, M., Prevosti, F.J., Cardoso, Y.P., Castillo, D., Bo, R., Casanave, E. & Lizarralde, M. (2013) Taxonomic status of southern South American *Conepatus* (Carnivora: Mephitidae). *Zoological Journal of the Linnean Society* 167, 327–344.

Scorolli, A.L., Lopez Cazorla, A.C. & Tejera, L.A. (2006) Unusual mass mortality of feral horses during a violent rainstorn in Parque Provincial Tornquist, Argentina. *Mastozoología Neotropical* 13, 255–258.

Sobrino, R., Acevedo, P., Escudero, M., Marco, J. & Gortázar, C. (2009) Carnivore population trends in Spanish agrosystems after the reduction in food availability due to rabbit decline by rabbit haemorrhagic disease and improved waste management. *European Journal of Wildlife Research* 55, 161–165.

Tellaeche, C.G., Reppucci, J.I., Luengos Vidal, E.M. & Lucherini, M. (2014) New data on the distribution and natural history of the lesser grison (*Galictis cuja*), hog-nosed skunk (*Conepatus chinga*), and culpeo (*Lycalopex culpaeus*) in northwestern Argentina. *Mammalia* 78, 261–266.

Tewksbury, J.J., Levey, D.J., Haddad, N.M., Sargent, S., Orrock, J.L., Weldon, A., Danielson, B.J., Brinkerhoff, J., Damschen, E.I. & Townsend, P. (2002) Corridors affect plants, animals, and their interactions in fragmented landscapes. *Proceedings of the National Academy of Sciences* 99, 12923–12926.

Thomas, L., Laake, J.L., Rexstad, E., Strindberg, S., Marques, F.F.C., Buckland, S.T., Borchers, D.L., Anderson, D.R., Burnham, K.P., Burt, M.L., Hedley, S.L., Pollard, J.H., Bishop, J.R.B. & Marques, T.A. (2009) *Distance 6.0. Release 2*. Research Unit for Wildlife Population Assessment, University of St Andrews.

Tigas, L.A., Van Vuren, D.H. & Sauvajot, R.M. (2002) Behavioral responses of bobcats and coyotes to habitat fragmentation and corridors in an urban environment. *Biological Conservation* 108, 299–306.

Travaini, A., Delibes, M. & Ceballos, O. (1998) Summer foods of the Andean hog-nosed skunk (*Conepatus chinga*) in Patagonia. *Journal of Zoology* 246, 457–460.

Vickery, J.A., Tallowin, J.R., Feber, R.E., Asteraki, E.J., Atkinson, P.W., Fuller, R.J. & Brown, V.K. (2001) The management of lowland neutral grasslands in Britain: effects of agricultural practices on birds and their food resources. *Journal of Applied Ecology* 38, 647–664.

Wang, Y., Allen, M.L., & Wilmers, C.C. (2015) Mesopredator spatial and temporal responses to large predators and human development in the Santa Cruz Mountains of California. *Biological Conservation* 190, 23–33.

Zalba, S.M. & Villamil, C.B. (2002) Woody plant invasion in relictual grasslands. *Biological Invasions* 4, 55–72.

Zeballos, H., López, E. & Morales, A. (1998). Mamíferos de Chiguata, hábitat y hábitos. *Revista del Departamento Académico de Biología (BIOS)* 2, 101–114. (In Spanish).

Zúñiga, A.H., Jiménez, J.E. & de Arellano, P.R. (2017) Activity patterns in sympatric carnivores in the Nahuelbuta Mountain Range, southern-central Chile. *Mammalia* 81, 445–453.

10

Activity and Movement Patterns of Urban Stone Martens

Jan Herr[1],* and Timothy J. Roper[2]

[1] *Arrondissement Sud, Administration de la Nature et des Forêts, Leudelange, Luxembourg*
[2] *Department of Biology and Environmental Science, University of Sussex, Brighton, UK*

SUMMARY

In Europe, the stone marten, *Martes foina*, is one of the main carnivore species to inhabit urban areas. While these environments are generally resource rich, they also present a range of anthropogenic stresses, such as human persecution or road traffic, which have the potential to induce behavioural responses in urban wildlife. We radio-tracked 12 stone martens in two towns in Luxembourg in order to determine how their activity (duration of the principal activity period, nightly activity duration) and movement (nightly movement distance, movement speed, nightly range) patterns were adapted to this environment. Stone martens displayed a more strictly nocturnal lifestyle than was known from studies on this species in more rural environments. We argue this to be a behavioural adaptation to reduce the rate of potential contact with humans. In fact, during long winter nights, emergence from dens took place long after sunset and return to dens intervened mostly before traffic picked up in the mornings. Furthermore, during long nights, marten peak activity was shifted to those parts of the night when human activity was at its lowest. On the other hand, stone martens were active during the entire dark period during short summer nights. Despite presumably higher resource availability and somewhat smaller territories compared to other studies, stone marten activity duration and movement distances were similar to those recorded in forest or rural populations elsewhere. Interestingly spring mobility was more pronounced in females than would be expected. We discuss these results in the context of territorial behaviour which, in urban areas, is likely to be driven by factors such as increased perceived intruder pressure.

Keywords

Activity rhythms – Luxembourg – *Martes foina* – movement distance – telemetry – territoriality – urban wildlife

Introduction

Activity and movement patterns of animals are influenced by a variety of factors ranging from their evolutionary origin to the local abiotic and biotic environment that surrounds them. Goszczyński (1986) demonstrated the importance of phylogeny in explaining movement patterns in terrestrial mammalian predators, showing that for any given body mass mustelids generally move over greater distances than canids and felids. However, there is also a great deal of intraspecific variation in activity patterns, which may result from intersexual differences based on strong sexual dimorphism in body size (e.g. Marcelli *et al.*, 2003; Begg *et al.*, 2016) or reproductive constraints (Zalewski, 2001; Kolbe & Squires, 2007; Begg *et al.*, 2016). In addition, important biogeographical variation between populations may be explained by climatic conditions and differential availability of food resources (Kowalczyk *et al.*, 2003; Zalewski *et al.*,

---

*Corresponding author.

Small Carnivores: Evolution, Ecology, Behaviour, and Conservation, First Edition. Edited by Emmanuel Do Linh San, Jun J. Sato, Jerrold L. Belant, and Michael J. Somers.

2004), while at the individual level meteorological factors (especially temperature) and food availability may influence the time and energy animals invest in their activities (Zielinski *et al.*, 1983; Zalewski, 2000; Zalewski *et al.*, 2004; Baghli & Verhagen, 2005).

Urban habitats often provide an abundance of food, water and shelters (Adams *et al.*, 2006), and urban mammals can be expected to adapt their behaviour accordingly. Urban red foxes, *Vulpes vulpes*, and European badgers, *Meles meles*, are able to fulfil their energetic needs by foraging during shorter periods and over shorter distances than rural ones (Harris, 1982; Doncaster & Macdonald 1997; Davison *et al.*, 2009), while American black bears, *Ursus americanus*, are 36% less active in urban than in wildland areas (Beckmann & Berger, 2003). Furthermore, territory sizes in urban areas are typically smaller than in rural areas (Davison *et al.*, 2009; Herr *et al.*, 2009a) and so presumably require less effort to be patrolled. Also, it is possible that release from intraguild predation in urban areas influences the movement and activity patterns of some mesocarnivores (Crooks & Soulé, 1999; Gehrt & Prange, 2007).

Urban animals are also exposed to a variety of anthropogenic stresses, which their rural counterparts encounter to a lesser degree. Consequently, urban animals may need to adapt their behaviour in ways that allow them to mitigate these constraints (Ditchkoff *et al.*, 2006). Human activities and traffic (Mata *et al.*, 2017) are obvious factors that have to be coped with in order to successfully subsist in human-dominated environments. There are two possible ways for animals to adapt to humans: by becoming tolerant of human presence or by adjusting to human activity through temporal avoidance (i.e. urban animals become active when humans are not: Adams *et al.*, 2006). For example, bobcats, *Lynx rufus*, and coyotes, *Canis latrans*, have been shown to reduce their diurnal activity in areas with higher human activity, suggesting behavioural avoidance of humans (McClennen *et al.*, 2001; Tigas *et al.*, 2002; Riley *et al.*, 2003; Wang *et al.*, 2015; Smith *et al.*, 2018). Black bears living in the urban–wildland interface have shown comparable activity shifts (Beckmann & Berger, 2003). Similarly, urban red foxes show strong evidence of temporal adaptation of their movement patterns to avoid human disturbance, such as simple human presence (Gloor, 2002; Díaz-Ruiz *et al.*, 2016) or even very low levels of road traffic (Baker *et al.*, 2007). Higher survival has been shown for urban coyotes that

shift their peak activity towards midnight away from peak traffic levels (Murray & St Clair, 2015).

Stone martens, *Martes foina*, are commonly found in villages and urban areas (Herr *et al.*, 2009a; Dudin & Georgiev, 2016). They are generally considered as opportunistic feeders, relying mostly on fruit and on small mammals. In urbanized areas, birds seem to make up a somewhat larger proportion of the diet than elsewhere; compared to other urban carnivores, scavenging on waste and deliberate feeding by people, however, is generally considered of limited importance for stone martens (see Herr, 2008 for a review on diet).

While adaptation to urban environments might necessitate or entail changes in activity and movement patterns, this has been scarcely studied in stone martens (but see Skirnisson, 1986; Bissonette & Broekhuizen, 1995; Herrmann, 2004; Dudin & Georgiev, 2016 for limited data). By contrast, activity in stone martens inhabiting mountainous, rural and forested environments have been more thoroughly studied; initially by means of radio-telemetry (Broekhuizen, 1983; Skirnisson, 1986; Föhrenbach, 1987; Lachat Feller, 1993; Posillico *et al.*, 1995, Genovesi *et al.*, 1997; Herrmann, 2004) and more recently with camera-trapping, as part of multi-species studies (Monterroso *et al.*, 2014, 2016; Petrov *et al.*, 2016; Torretta *et al.*, 2017; Tsunoda *et al.*, 2018; Roy *et al.*, 2019).

We investigated the activity and movement patterns of stone martens in urban areas. Specifically, we determined the timing of onset and termination of their outside-the-den activities and the duration of their nightly activity period, as well as movement distances and speeds. We predicted that urban martens would shift their activities to later hours of the night by comparison with rural ones, in order to avoid human disturbance. They were also expected to be less active and to move shorter distances than rural martens, due to potentially higher resource availability in urban environments.

Materials and Methods

Study Area

The study area comprised the neighbouring towns of Bettembourg (1.6 km^2; population = 7500 inhabitants; 4700 people/km^2) and Dudelange (5.0 km^2, population = 18 300 inhabitants; 3700 people/km^2) in the south of Luxembourg (49°30′ N, 6°5′ E) (Figure 10.1).

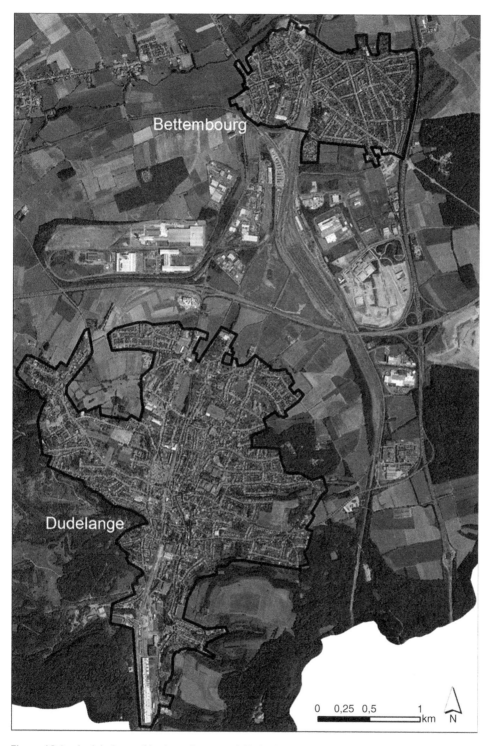

Figure 10.1 Aerial photo of both study towns delimited by the urban perimeter (black line). *Source:* Photo © Administration du Cadastre et de la Topographie, Luxembourg.

Habitat within the urban perimeter consisted mainly of residential areas characterized by private housing, gardens and a dense road network (13 km of roads/km^2). Built-up areas (all buildings and paved surfaces) covered between 42 and 45% of the total area within the urban perimeters of both towns. The surrounding habitat consisted of a patchwork of agricultural, forested, and industrial areas.

Food availability and marten diet in the study towns were not specifically determined. However, due to a high proportion of private gardens in the study towns, with the presence of numerous fruit trees, berry-producing shrubs and habitat for small mammals and passerine birds, food availability was estimated to be more favourable than in the surrounding rural landscape. Direct deliberate feeding of stone martens by people was unheard of and household waste was generally unavailable to martens due to the design of the rubbish bins.

Ambient temperature was measured with automatic Tinytalk temperature loggers (Gemini Data Loggers Ltd, Chichester, UK) that were programmed to take a temperature reading every 4 h. The loggers were placed at ground level (one within each study town) and nightly temperatures were calculated as the mean temperature from the readings taken at 20:00 h, 00:00 h, 04:00 h and 08:00 h Central European Summer Time (CEST).

Study Animals

Twelve stone martens were trapped with wire cage live-traps (81 × 23 × 23 cm; model 206, Tomahawk Live Trap Co., USA) that were covered with wooden trap covers and placed in private gardens. After a variable period of pre-baiting with chicken eggs, traps were set in the evening and checked at dawn. Trapped martens were transferred into a transparent plastic box and anaesthetized by placing a tablespoon of isoflurane into the box. As soon as the animal was motionless it was injected intramuscularly with a mix of ketamine hydrochloride (100 mg/ml) (Anesketin, Eurovet) and medetomidine hydrochloride (1 mg/ml) (Domitor, Pfizer) (ratio of 2:1 by volume) at a dose rate of 0.12 ml/kg (de Leeuw *et al.*, 2004).

Individual martens were sexed, weighed and radio-collared (Biotrack Ltd, UK, model TW-3; mass = 30 g). Age classes were defined as juvenile (<12 months),

subadult (between 12 and 24 months) and adult (> 24 months), based on tooth wear (Habermehl & Röttcher, 1967) and reproductive status (large descended testes, signs of lactation). Due to daytime road traffic, we kept trapped martens at the capture site in a holding box containing food and water and sheltered from direct sunlight, and released them the same evening after 22:00 h. Manipulation and care of animals were in accordance with the guidelines of the Animal Behaviour Society (2012).

Telemetry

Radio-tracking was carried out between June 2005 and May 2007 (Table 10.1). Tracking of focal martens was continuous and was initiated before the time of the first emergence in the evening and terminated after final retreat into a daytime den (see Herr *et al.*, 2010) the following morning, taking a locational fix every 15 min using a receiver with an attached foldable antenna (model RX-98 H, TVP Positioning AB, Sweden). We aimed to follow each marten between six and nine nights per season, defined as: summer (June–August), autumn (September–November), winter (December–February), spring (March–May) (Table 10.1). The activity was determined on the basis of fluctuations in signal strength (= active) or a steady signal (= inactive) during a 30 s period at locational fix time.

Den Emergence and Return Times

Times of emergence from and return to a den were estimated from the all-night tracking sessions but also, occasionally, on shorter observations, especially for emergence times. Emergence time was defined as the time of the first quarter-hourly fix after the marten had left the den in which it had spent the day. Return time was defined as the time of the first quarter-hourly fix after the marten had finally retreated into the den that it would subsequently spend the day in. Emergence and return times were based on direct observation of the animal exiting or entering the den or, where direct observation was impossible, on the radio-signal clearly moving away from the den or approaching it and subsequently staying there.

Table 10.1 Radio-tracking summary for 12 urban stone martens, *Martes foina*, in the towns of Bettembourg and Dudelange, Luxembourg.

Marten ID	Age[a]	Tracking period	Nights[b]	Nights/season (su,au,wi,sp)[c]		Total fixes	out/act fixes[d]
				PAP/NA/MS	NMD/NR		
Females							
F1	A	12/06/05–12/03/06	31	12,7,5,0	13,8,6,0	1033	480
F2	A	16/06/05–14/05/06	36	14,8,5,6	14,8,6,8	1203	555
F4	A	06/08/05–24/05/06	35	9,7,6,11	9,8,6,12	1275	588
F6	SA	04/05/06–06/06/06	8	0,0,0,6	0,0,0,6	259	182
F7	A	28/06/06–15/02/07	21	4,5,5,0	4,7,6,0	720	352
F8	J	01/08/06–22/08/06	5	5,0,0,0	5,0,0,0	156	97
F9	A	20/08/06–09/05/07	25	0,7,7,8	0,7,7,8	955	593
F10	A	21/08/06–07/05/07	25	0,5,7,9	0,7,7,9	962	654
F11	A	15/01/07–08/05/07	16	0,0,5,9	0,0,7,9	647	405
Males							
M1	A	09/09/05–07/03/06	19	0,11,6,0	0,12,6,0	733	411
M2	A	14/04/06–14/11/06	24	8,6,0,8	8,7,0,9	839	585
M4	A	14/06/06–01/03/07	23	8,7,6,0	8,7,7,0	879	563

[a] A = adult, SA = subadult, J = juvenile.
[b] Number of total tracking nights.
[c] Number of tracking nights considered per season for calculating various activity and movement parameters: PAP = principal activity period, NA = nightly activity, MS = movement speed, NMD = nightly movement distance, NR = nightly range.
[d] Active fixes outside the daytime den during PAP.

Principal Activity Period (PAP) and Nightly Activity (NA)

PAP (h) was only estimated for nights where both emergence and return times from and to dens were known ($n = 232$). It was defined as the time period between emergence and return. PAP also included bouts of inactivity that may have interrupted activity bouts during the night. NA (h) was defined as the actual time that a marten was active during its PAP. It was calculated as the product of the PAP duration and the proportion of active fixes during the PAP.

Activity Rhythms

Nightly activity rhythms were calculated on the basis of all fixes (Kowalczyk *et al.*, 2003), i.e. fixes from all collared individuals pooled, grouped into 1 h periods. Fixes were recorded as (i) inside the den (all fixes before the first emergence and after final retreat into a den); (ii) outside and active (active fixes during PAP); and (iii) outside and inactive (inactive fixes during PAP). We did not differentiate between locomotory and stationary activity (e.g. Lachat Feller, 1993). Activity rhythms were only established for the hours between 19:00 h and 09:00 h. When a tracking session started later than 19:00 h but before the animal had left the den, all potential fixes between 19:00 h and the actual start of the tracking were recorded as inside the den. Equally, all potential fixes between the end of the tracking period and 09:00 h were recorded as inside the den. Activity rhythms outside the PAP of martens in their daytime dens were not recorded.

Movement Patterns

Four different movement parameters were calculated following Zalewski *et al.* (2004): (i) Nightly movement distance (NMD, in km): the sum of straight-line distances between consecutive fixes taken at 15 min intervals; (ii) Movement speed (MS = NMD/NA, in km/h); (iii) Nightly range (NR, in ha): 100% minimum convex polygon (MCP100) based on all the fixes taken during a marten's PAP; (iv) Nightly range as a percentage of seasonal home range (NR/SR, in %), using the MCP95 seasonal home ranges determined by Herr *et al.* (2009a) for the calculations. Nights that were only partially completed or where too many fixes were missed were discarded from the analyses. The number of tracking nights used for calculating the different activity and movement variables is shown in Table 10.1.

Statistical Analysis

In order to avoid problems with pseudo-replication, we treated each animal as a sampling unit rather than treating each individual night as an independent sample, as has often been done for activity and movement pattern analyses (e.g. Zalewski *et al.*, 2004; Kowalczyk *et al.*, 2006). Univariate general linear models (GLM) were used to test for seasonal effects on mean seasonal values (for PAP, NA, NMD, MS, NR, NR/SR) for individual martens, which were used as blocking factors to control for inter-individual variation. *Post-hoc* pairwise comparisons were performed with a Tukey test. Due to the small number of males, these statistical analyses were only carried out for females. For individual martens, no means were calculated for seasons where they were tracked for less than four nights.

However, when analyzing emergence and return times, activity rhythms, and correlations between NA and temperature, all data were included and treated as independent samples. Chi-square tests of association were employed to investigate for potential differences in the distribution of hourly activity (proportion of outside/active fixes) between seasons. In order to test for differences in the emergence and return times, all quarter-hour intervals between 19:00h and 09:00h (the following morning) were associated with an integral number running from 1 (19:00h) to 57 (09:00h). Non-parametric tests were then run on these substitute numbers rather than on the actual times. The significance level was always set at 0.05. Statistical analyses were carried out in MINITAB®, version 14 (Minitab Inc., State College, PA, USA).

Results

Den Emergence and Return Times

The martens in both study towns were clearly nocturnal, emerging from their daytime dens after sunset and returning to the same or different dens before sunrise (Figure 10.2). On only two occasions did martens retreat into their den a few minutes after official sunrise time in summer. Both emergence and return times showed significant seasonal variation (Kruskal–Wallis test: emergence: $H = 27.24$, df = 3, $p < 0.001$; return: $H = 91.28$, df = 3, $p < 0.001$) (Table 10.2). Overall, the martens left their den sites earliest in spring, followed by summer, autumn, and winter. All seasons differed significantly from each other (Mann–Whitney tests: $p < 0.05$), apart from summer and autumn ($p = 0.144$). Return times were earliest in summer, followed by spring, autumn and winter, with all seasons differing from each other ($p < 0.001$), apart from autumn and spring ($p = 0.114$). Although there were no overall significant differences in male and female emergence times (Mann–Whitney test: $W_s = 26\,218.0$, $n_{males} = 63$, $n_{females} = 195$, $p = 0.060$), there was a trend for females to leave their dens later than males, apart from winter when males remained inside for longer (Table 10.2). Throughout the year, males returned significantly later to their den sites than females ($W_s = 22\,360.0$, $n_{males} = 65$, $n_{females} = 189$, $p < 0.001$), although spring return times were similar for both sexes (Table 10.2).

Emergence times were more closely coupled to sunset in summer and spring than in autumn and winter (Figure 10.2; Table 10.2). Although the same pattern was true for return times in relation to sunrise, it was much less pronounced. When considering only those nights where both emergence and return times were known ($n = 243$), return times were coupled significantly closer to sunrise than emergence times were to sunset in each season (Wilcoxon signed-rank tests: summer: $T = 1546.5$, $n = 67$, $p < 0.05$; autumn: $T = 1894.0$, $n = 62$, $p < 0.001$; winter: $T = 1300.0$, $n = 51$, $p < 0.001$; spring: $T = 1503.0$, $n = 61$, $p < 0.001$). This was most apparent in winter. Interestingly, there seemed to be a threshold at around 21:00h before which the martens hardly ever emerged from their den, even in late autumn and winter when sunset occurred much earlier. During the same period, there was a threshold between 08:00h and

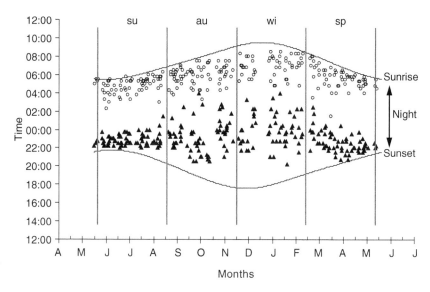

Figure 10.2 Time (Central European Summer Time = CET + 1 h) of emergence from the daytime den (▲; *n* = 258) and return to the same or another daytime den (○; *n* = 255) by 12 radio-collared urban stone martens at the start and end of their principal activity period, respectively. The lower line shows the time of sunset, the upper line the time of sunrise. The area between the two lines represents night-time hours.

Table 10.2 Median fix time (Central European Summer Time) of the first emergence from and final return to the daytime den and mean (±SD) time lag (h) between sunset to emergence and return to sunrise. $n_{(emergence)}$ = 258; $n_{(return)}$ = 255. *F* = female; *M* = male.

Season		Emergence			Return		
		F	M	F+M	F	M	F+M
Summer	Time	22:45 h	22:15 h	22:45 h	04:45 h	05:15 h	05:00 h
	Lag	1.6 ± 0.8	0.7 ± 0.3	1.4 ± 0.8	1.3 ± 0.8	0.4 ± 0.4	1.1 ± 0.8
Autumn	time	23:37 h	22:30 h	23:15 h	05:45 h	06:30 h	06:00 h
	Lag	4.6 ± 2.0	3.8 ± 1.6	4.3 ± 1.9	2.3 ± 1.1	1.5 ± 0.8	2.0 ± 1.1
Winter	Time	23:37 h	22:30 h	23:45 h	06:45 h	08:15 h	07:00 h
	Lag	5.9 ± 2.1	5.9 ± 1.5	5.9 ± 1.9	2.5 ± 1.1	1.6 ± 0.8	2.3 ± 1.1
Spring	Time	22:30 h	22:00 h	22:15 h	05:30 h	05:22 h	05:45 h
	Lag	2.4 ± 1.6	1.7 ± 1.4	2.3 ± 1.6	1.5 ± 1.1	0.8 ± 0.5	1.4 ± 1.1

08:30 h by which the martens had always returned to their dens, an hour before sunrise (Figure 10.2).

Principal Activity Period (PAP) and Nightly Activity (NA)

Even though night length varied from on average 8.4 h in summer to 15.0 h in winter, the martens did not adapt their PAP accordingly, although significant inter-individual differences in PAP were observed (univariate GLM: females – season: $F = 0.38$, df = 3.12; $p = 0.767$; individual: $F = 7.87$, df = 8.12, $p < 0.01$). During short summer nights, PAP was only about 1 h shorter than during the rest of the year where PAP remained fairly constant (Table 10.2). PAP was on average about 1 h longer for males than for females,

Table 10.3 Mean (±SD) seasonal activity and movement patterns from 12 urban stone martens.

	Season (*n* females; *n* males)			
	Summer (5;2)	Autumn (6;3)	Winter (7;2)	Spring (6;1)
Principal activity period (PAP; h)			—	—
Females	5.6±0.7	6.2±1.4	6.6±1.5	6.9±0.8
Males	6.7±0.1	7.5±0.6	7.8±1.3	7.8
Nightly activity (NA; h)		—	—	—
Females	4.8±0.6	4.9±1.4	4.7±1.4	5.8±1.2
Males	6.4±0.4	5.9±0.7	5.9±1.1	6.8
Nightly movement distance (NMD; km)			—	—
Females	2.24±0.65	2.03±0.85	2.10±0.99	4.69±2.72
Males	7.48±0.42	3.50±0.34	4.49±1.42	6.54
Movement speed (MS; km/h)			—	—
Females	0.48±0.18	0.39±0.09	0.43±0.12	0.74±0.31
Males	1.17±0.01	0.58±0.09	0.77±0.15	0.98
Nightly range (NR; ha)				
Females	11.8±9.3	13.2±8.0	13.5±8.8	19.5±10.9
Males	44.6±12.2	36.4±8.3	46.3±10.8	42.3
NR/Seasonal range (%)			—	—
Females	54.4±14.3	44.2±8.2	42.2±16.6	62.9±11.9
Males	56.1±0.4	44.6±4.7	50.9±7.2	66.2

although in autumn, winter, and spring, there was overlap between male and female PAP values.

When considering NA (i.e. the actual time that martens were active during their PAP), significant seasonal differences became apparent for females (Table 10.3; season: $F = 4.33$, df = 3.12, $p < 0.05$; individual: $F = 16.69$, df = 8.12, $p < 0.001$). Females were active for significantly longer in summer than in winter (Table 10.4), though this was not obvious from the overall seasonal mean activity values (Table 10.3). This was mostly due to F10 and F11, who exhibited generally higher activity in winter than the rest of the females, but for whom summer data were missing (Table 10.1). Male activity duration seemed to remain fairly constant across seasons and males were on average about 1 h longer active than females (Table 10.3).

There was no overall correlation between NA and nightly temperature (Pearson's product-moment correlation: $r = 0.056$, $n = 232$, $p = 0.397$). However, when

Table 10.4 Significant results for *post-hoc* multiple comparisons (Tukey test) of seasonal activity and movement variables in female urban stone martens. For meaning of variable abbreviations, see Table 10.3.

Variable	Seasons[a]	T	p
NA	su–wi	3.16	< 0.05
NMD	sp–au	3.54	< 0.05
	sp–wi	4.30	< 0.01
MS	sp–au	3.42	< 0.05
	sp–wi	3.68	< 0.05
NR	sp–wi	2.98	< 0.05
NR/SR	su–au	3.65	< 0.05
	su–wi	4.64	< 0.01
	sp–wi	3.70	< 0.05

[a] sp = spring; su = summer; au = autumn; wi = winter

seasons were considered separately, significant positive correlations were found between nightly temperature and NA in winter ($r = 0.313$, $n = 41$, $p < 0.05$) and spring ($r = 0.331$, $n = 59$, $p < 0.05$) but not in summer ($r = 0.010$, $n = 68$, $p = 0.933$) or autumn ($r = 0.072$, $n = 63$, $p = 0.574$). For each 1°C decrease in temperature, martens reduced their activity by on average 7 min in winter and 12 min in spring.

Activity Rhythms

Nocturnal activity rhythms for combined sexes showed striking differences from season to season (Figure 10.3). Summer was characterized by a rather abrupt start and end of activity outside the den with high levels above 60% throughout most of the night (23:00–04:00 h). In autumn, the onset of activity outside the den started a little earlier but increased more slowly than in summer. A peak of 80% activity was only reached relatively late, at around 03:00 h. Activity subsided gradually thereafter. In winter, the observed pattern of activity was superficially similar in shape to the one in autumn. However, there was an overall shift of activity toward the later hours of the night. A first small peak occurred around midnight at about 50%, followed by a second and higher peak 4 h later (04:00 h) at 72%. In the early morning hours, the remaining activity outside the den ceased abruptly at 07:00 h. Throughout the night, the martens spent a significant proportion of their time either inside their daytime den or outside but being inactive. The spring pattern was overall similar to the summer pattern, with a more or less bell-shaped appearance. By comparison with winter, there was a shift of peak activity back to the more central parts of the night, with high levels being maintained throughout most of the night. At the start and the end of the night, activity respectively increased and decreased less abruptly than in summer. The observed seasonal variations in nocturnal activity rhythms were statistically significant (Table 10.5).

Nightly Movement Distance (NMD)

The longest recorded NMDs were observed on two spring nights for F10 and F11 with 10.0 and 10.1 km, respectively, and on a summer night for M2 with

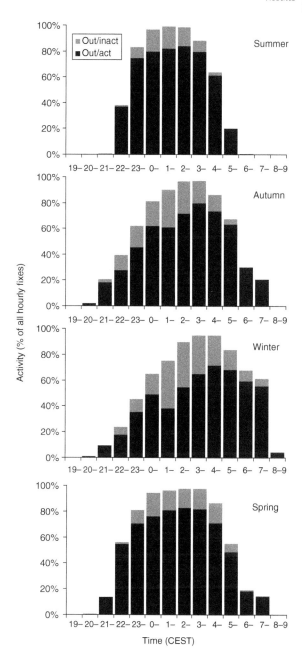

Figure 10.3 Seasonal activity rhythms of urban stone martens (both sexes combined) in Luxembourg. out/inact: inactive during the principal activity period (PAP); out/act: active during the PAP.

10.1 km. Female martens moved on average between 1.08 and 8.33 km per night depending on the season (Table 10.3). Overall, both factors, season and individual, were found to have a significant effect on female

Table 10.5 Chi-square tests of association to compare nocturnal patterns of hourly activity rates (proportion of outside/active fixes) between different seasons (sexes combined).

Seasons[a]	Time	χ^2	df	p
su–au	21:00–07:00 h	103.195	10	< 0.001
su–wi	21:00–07:00 h	200.773	10	< 0.001
su–sp	21:00–07:00 h	52.356	10	< 0.001
au–wi	21:00–07:00 h	42.082	10	< 0.001
au–sp	21:00–07:00 h	22.516	10	< 0.001
wi–sp	21:00–07:00 h	104.125	10	< 0.001

[a] sp = spring; su = summer; au = autumn; wi = winter

NMD (season: $F = 7.23$, df = 3,12, $p < 0.01$; individual: $F = 4.80$, df = 8,12, $p < 0.01$). NMD was found to be significantly higher in spring than in autumn and winter (Table 10.4), while the other seasons did not differ from each other. Males had on average higher seasonal NMDs than females, a trend that was most pronounced in summer. Males moved on average larger distances in spring and summer than in autumn and winter (Table 10.3).

Movement Speed (MS)

The speeds at which female martens moved varied significantly depending on the season (season: $F = 6.19$, df = 3,12, $p < 0.01$; individual: $F = 3.33$, df = 8,12, $p < 0.05$). They moved faster in spring than in autumn and winter (Table 10.4). This was particularly pronounced in F10 and F11 who, in spring, effectively doubled the speeds at which they travelled while they were active. Males also generally moved at greater speeds in spring and summer than in autumn and winter. In summer, the two males moved on average 2.4 times faster than the females (Table 10.3).

Nightly Range (NR)

The areas that female martens covered each night during their activities were significantly influenced by season, although a significant proportion of the observed variation was also due to inter-individual differences (season: $F = 4.91$, df = 3,12, $p < 0.05$; individual: $F = 19.19$, df = 8,12, $p < 0.001$). Spring NR was significantly larger than winter NR (Table 10.4), while other seasons did not differ significantly from each other. Across all seasons, males covered on average 2–3 times larger areas per night than females did (Table 10.3). The males M2 and M3 greatly reduced their NR from summer to autumn.

The animals covered each night on average between 14.9 and 72.3% of their MCP95 seasonal home ranges (Table 10.3). In females, NR/SR was significantly affected by season and individual (season: $F = 10.14$, df = 3,12, $p < 0.01$; individual: $F = 6.35$, df = 8,12, $p < 0.01$). Females used larger proportions of their seasonal range in summer than in autumn and winter, and in spring, they used larger proportions than in winter (Table 10.4). Males showed a similar trend and in any given season, both sexes used on average very similar proportions of their seasonal home ranges (Table 10.3).

Discussion

Activity Patterns

This study confirms the stone marten's distinctly nocturnal lifestyle. All other studies on this species based on radio-tracking have come to the same general conclusion, irrespective of the environment in which the animals lived (e.g. Broekhuizen, 1983; Skirnisson, 1986; Lachat Feller, 1993; Posillico *et al.*, 1995; Hermann, 2004; Wereszczuk & Zalewski, 2015); and similar results have been obtained through camera-trapping (Monterroso *et al.*, 2014, 2016; Dudin & Georgiev, 2016; Petrov *et al.*, 2016; Torretta *et al.*, 2017; Tsunoda *et al.*, 2018; Roy *et al.*, 2019). Nevertheless, the degree to which stone martens restrict their activities to night-time hours seems to be more pronounced in our study than has been reported elsewhere. Both Broekhuizen (1983) and Skirnisson (1986) found that emergence and return times were closely coupled to sunset and sunrise, with martens leaving the dens earlier and returning later as the nights became longer. Furthermore, during summer, they often emerged before sunset and retreated into their dens after sunrise. The fact that our urban stone martens never emerged before sunset in summer and delayed the onset of their activity in autumn and winter until after

21:00 h (20:00 h CET; winter time) suggests that they strongly avoided human activities. In fact, road and pedestrian traffic in both study towns had considerably slowed by 21:00 h (J. Herr, personal observation). Similarly, the observed return threshold at around 08:00 h (07:00 h winter time), as well as the sharp drop from very high male activity levels to very low levels between 07:00 h and 08:00 h in winter, coincided with the start of the morning rush hour (from 07:00 h to 08:00 h winter time; J. Herr, personal observation). Similar behaviour was recorded by Skirnisson (1986) and Hermann (2004) in relation to farming activities in rural villages. LaPoint (2013) showed the same responses to road traffic by urban fishers, *Pekania pennanti*. In principle, the late emergence times of our martens could result from them not needing to be active for so long in urban areas owing to possible higher food abundance, as has been suggested for other species such as the red fox and the European badger (Harris, 1982; Doncaster & Macdonald, 1997; Davison *et al.*, 2009). However, this explanation is unlikely since it is only the timing of activity onset and cessation that differs between rural and urban environments, not so much the total duration of the activity period. In other urban or semi-urban carnivores, similar shifts toward more distinctly nocturnal behaviour have also been linked to avoidance of human disturbances (Grubbs & Krausman, 2009; Gehrt & Riley, 2010; Harris *et al.*, 2010; Riley *et al.*, 2010; Díaz-Ruiz *et al.*, 2016; Smith *et al.*, 2018).

The activity rhythms of urban martens showed a general pattern of high activity throughout the night in summer and spring with generally lower levels in autumn and winter. This general pattern mirrors observations of rural and village-dwelling stone martens (Skirnisson, 1986; Lachat Feller, 1993). Temperature seems to be an important factor regulating the activity, especially during the colder seasons. Curtailing of activity with decreasing temperatures has also been observed elsewhere in stone martens (Skirnisson, 1986; Lachat Feller, 1993; Herrmann, 2004) and pine martens, *Martes martes* (Zalewski, 2000). Interestingly, urban stone martens responded very similarly to each 1°C decrease in temperature as did forest and village stone martens (Skirnisson 1986; Herrmann 2004).

The general decrease in activity in autumn and winter was associated here with a shift of peak outside-the-den activity toward later parts of the night. This activity shift has not been seen in rural stone martens (Skirnisson, 1986; Genovesi, 1993; Herrmann, 2004) but coincides with patterns observed during winter and summer weekends in stone martens inhabiting rural villages (with higher night-time human activity; Herrmann, 2004) or inhabiting predominantly developed areas in Poland (Wereszczuk & Zalewski, 2015). This suggests that during short nights, the martens used the entire dark period for their activities, but within longer nights, they preferentially used the later hours of the night with the least human disturbance. Thus, activity rhythms, as well as the emergence and return times, of urban stone martens suggest that behavioural adaptation to human activity in urbanized habitats occurs primarily by temporal avoidance rather than by tolerance.

In Switzerland, activity duration (male and female combined; 5.1–6.9 h) was similar to that of females in the present study, while in Germany, activity (6.2–8.2 h) was generally slightly higher than in urban stone martens, apart from spring (5.6 h) when urban martens were active for longer periods. However, owing to different methodology, these studies included both inside the den as well as outside activity in their estimates. One can, hence, assume that in these studies, the activity that we defined as NA (i.e. activity occurring between emergence and return) was actually similar or even lower than in the urban environment considered here. In the Netherlands, however, rural stone martens were found to be active outside the den for on average ~8 h per night irrespective of season and sex (Broekhuizen, 1983), which was similar to urban males and 1–2 h longer than urban females. However, it was not clear whether the activity levels referred to by Broekhuizen (1983) were PAP or NA.

Mobility

Few studies have collected detailed movement data on stone martens, and when these are available, they are often based on very few animals. However, Genovesi *et al.* (1997) reported NMDs for 16 forest and rural martens. Females moved on average 1.45 km and males 5.32 km in all seasons combined. They found no significant seasonal changes. Interestingly, these martens had much larger territories (M: 421 ± 231 ha, F: 217 ± 134 ha) than in the present study area (M: 112 ± 25 ha; F: 37 ± 22 ha; see Herr *et al.*, 2009a for comparison), yet urban females moved much longer distances, especially in spring. Male distances were similar in both environments. Although the limited data from other study areas and environments are somewhat conflicting, the general pattern suggests that the urban martens may have moved larger distances in relation to their home range size, especially in spring and summer, than forest and rural martens elsewhere (see Skirnisson, 1986; Lachat Feller, 1993). Movement distances may also vary depending on whether stone martens use a 'stationary' or 'roamer' home range use strategy (see Wereszczuk & Zalewski, 2019), but no relevant information is currently available in the literature.

The observed seasonal variation in nightly activity and in the various movement parameters always revealed the same general pattern insofar as spring and summer were the most active periods. In males, the evidence suggested peak mobility in summer, although spring data were only available for one male, which could have prevented us from seeing clearer trends. However, this is in line with other studies that have found male mustelids to be most active and mobile during the mating season (Jędrzejewski *et al.*, 2000; Zalewski *et al.*, 2004). The crucial period for females seemed to be spring when they moved the longest distances at the highest speeds, covering the largest areas and the largest proportions of their seasonal home ranges. There was a very clear change in behaviour from winter to spring, while other season transitions were less pronounced. Thus, the seasonal peak in female mobility did not occur during the mating season in summer as has been described for pine martens (Zalewski *et al.*, 2004), but during spring, when they usually give birth and rear their young. Incidentally, the female with the highest spring mobility values was also the only one that actually reared young during the study period. Nonetheless, these high movement values may not be solely related to the rearing and provisioning of young, since another non-rearing female showed similar movement patterns in terms of NMD, MS, and NR/SR. Both females, however, showed very strong car-patrolling behaviour (visiting, sniffing, and scent-marking cars) during this

time of the year, which is a strong indicator for territorial behaviour in this urban environment (Herr *et al.*, 2009b). The fact that females covered by far the largest proportions of their home ranges in both spring and summer also testifies to high territoriality during this period. In spring, an increasingly large proportion of marten activity took part directly on roads rather than in the gardens behind the houses (see Herr *et al.*, 2009b), which suggests that this increase in activity from winter to spring was unrelated to foraging behaviour or possible seasonal changes in food abundance.

Conclusion

The stone martens in both towns showed evidence of behavioural adaptation to human activity in urban environments. Although they lived in very close association with humans, they remained wary of them and shifted their activity so as to reduce potential contact rates with humans. This could be seen in their more pronounced nocturnal lifestyle, their late emergence from dens in winter when nights were longest, and their return to dens before traffic picked up in the mornings. They also shifted their peak activity to the later parts of the night when human activity was lowest.

Although urban habitats are generally seen as richer in resources (Adams *et al.*, 2006) and where animals exploit smaller home ranges or territories than in natural habitats, we did not observe martens to show greatly reduced activity and mobility rates as was predicted. On the contrary, there was some evidence that they were on average at least as active and mobile as their forest or rural counterparts. Spring especially seemed to be a time when females were highly mobile. This may be due to the relatively high population density of 4.7–5.8 adults/km^2 in the urban environment (Herr *et al.*, 2009a) and to the high perceived intruder pressure due to marten scent-marks being transported between territories by cars (Seiler *et al.*, 1994; Herr *et al.*, 2009b), which ultimately would lead females to be particularly territorial in urban environments. While the same can be expected for male stone martens, the small sample size did not allow us to draw clear conclusions in this respect.

Acknowledgements

We thank L. Schley, A. Krier, J. Sünnen (all Administration de la Nature et des Forêts), E. Engel (Musée National d'Histoire Naturelle), and the residents of both study towns for their cooperation. J. Herr was supported by a Bourse de formation–recherche (BFR05/042) from the Ministère de la Culture, de l'Enseignement Supérieur et de la Recherche, Luxembourg. The research complied with the Luxembourg law.

References

Adams, C.E., Lindsey, K.J. & Ash, A.J. (2006) *Urban Wildlife Management*. Taylor & Francis Group, Boca Raton.

Animal Behaviour Society (2012) Guidelines for the treatment of animals in behavioural research and teaching. *Animal Behaviour* 83, 301–309.

Baghli, A. & Verhagen, R. (2005) Activity patterns and use of resting sites by polecats in an endangered population. *Mammalia* 69, 211–222.

Baker, P.J., Dowding, C.V., Molony, S.E., White, P.C.L. & Harris, S. (2007) Activity patterns of urban red foxes (*Vulpes vulpes*) reduce the risk of traffic-induced mortality. *Behavioral Ecology* 18, 716–724.

Beckmann, J.P. & Berger, J. (2003) Rapid ecological and behavioural changes in carnivores: the responses of black bears (*Ursus americanus*) to altered food. *Journal of Zoology* 261, 207–212.

Begg, C.M., Begg, K.S., Do Linh San, E., du Toit, J.T. & Mills, M.G.L. (2016) Sexual and seasonal variation in the activity patterns and time budget of honey badgers in an arid environment. In: *Badgers: Systematics, Biology, Conservation and Research Techniques* (eds G. Proulx & E. Do Linh San), pp. 161–192. Alpha Wildlife Publications, Sherwood Park.

Bissonette, J.A. & Broekhuizen, S. (1995) *Martes* populations as indicators of habitat spatial patterns: the need for a multiscale approach. In: *Landscape Approaches in Mammalian Ecology and Conservation* (ed. W.Z. Lidicker Jr), pp. 95–121. University of Minnesota Press, Minnesota.

Broekhuizen, S. (1983) Habitat use of beech martens (*Martes foina*) in relation to landscape elements in a Dutch agricultural area. In: *Proceedings from XVI Congress of the International Union of Game Biologists* (eds M. Špenik & P. Hell), pp. 614–624. Štrebské Pleso.

Crooks, K.R. & Soulé, M.E. (1999) Mesopredator release and avifaunal extinctions in a fragmented system. *Nature* 400, 563–566.

Davison, J., Huck, M., Delahay, R.J. & Roper, J. (2009) Restricted ranging behaviour in a high-density population of urban badgers. *Journal of Zoology* 277, 45–53.

de Leeuw, A.N.S., Forrester, G.J., Spyvee, P.D., Brash, M.G.I. & Delahay, R.J. (2004) Experimental comparison of ketamine with a combination of ketamine, butorphanol and medetomidine for general anaesthesia of the Eurasian badger (*Meles meles*). *The Veterinary Journal* 167, 186–193.

Díaz-Ruiz, F., Caro, J., Delibes-Mateos, M., Arroyo, B. & Ferreras, P. (2016) Drivers of red fox (*Vulpes vulpes*) daily activity: prey availability, human disturbance or habitat structure? *Journal of Zoology* 298, 128–138.

Ditchkoff, S.S., Saalfeld, S.T. & Gibson, C.J. (2006) Animal behavior in urban ecosystems: modifications due to human-induced stress. *Urban Ecosystems* 9, 5–12.

Doncaster, C.P. & Macdonald, D.W. (1997). Activity patterns and interactions of red foxes (*Vulpes vulpes*) in Oxford city. *Journal of Zoology* 241, 73–87.

Dudin, G.S. & Georgiev, D.G. (2016) Research on the daily activity of the stone marten (*Martes foina* Erxl.) in anthropogenically influenced habitats in Bulgaria. *Journal of Bioscience and Biotechnology* 5, 259–261.

Föhrenbach, H. (1987) *Untersuchungen zur Ökologie des Steinmarders* (Martes foina, *ERXLEBEN 1777) im Alpen- und Nationalpark Berchtesgaden*. PhD thesis, Ruprecht-Karls-Universität. (In German).

Gehrt, S.D. & Prange, S. (2007) Interference competition between coyotes and raccoons: a test of the mesopredator release hypothesis. *Behavioral Ecology* 18, 204–214.

Gehrt, S.D. & Riley, S.P.D. (2010) Coyotes (*Canis latrans*). In: *Urban Carnivores: Ecology, Conflict, and Conservation* (eds S.D. Gehrt, S.P.D. Riley & B.L. Cypher), pp. 79–95. The Johns Hopkins University Press, Baltimore.

Genovesi, P. (1993) *Strategie di Sfruttamento delle Risorse e Struttura Sociale.* PhD thesis, Università di Roma "la Sapienza". (In Italian).

Genovesi, P., Sinibaldi, I. & Boitani, L. (1997) Spacing patterns and territoriality of the stone marten. *Canadian Journal of Zoology* 75, 1966–1971.

Gloor, S. (2002) *The Rise of Urban Foxes* (Vulpes vulpes) *in Switzerland and Ecological and Parasitological Aspects of a Fox Population in the Recently Colonised City of Zurich.* PhD thesis, Universität Zürich.

Goszczyński, J. (1986) Locomotor activity of terrestrial predators and its consequences. *Acta Theriologica* 31, 79–95.

Grubbs, S.E. & Krausman, P.R. (2009) Use of urban landscape by coyotes. *The Southwestern Naturalist* 54, 1–12.

Habermehl, K.H. & Röttcher, D. (1967) Die Möglichkeiten der Altersbestimmung beim Marder und Iltis. *Zeitschrift für Jagdwissenschaft* 13, 89–102. (In German).

Harris, S. (1982) Activity patterns and habitat utilization of badgers (*Meles meles*) in suburban Bristol: a radio tracking study. *Symposia of the Zoological Society of London* 49, 301–323.

Harris, S., Baker, P.J., Soulsbury, C.D. & Iossa, G. (2010) Eurasian badgers (*Meles meles*). In: *Urban Carnivores: Ecology, Conflict, and Conservation* (eds S.D. Gehrt, S.P.D. Riley & B.L. Cypher), pp. 109–119. The Johns Hopkins University Press, Baltimore.

Herr, J. (2008) *Ecology and Behaviour of Urban Stone Martens (*Martes foina*) in Luxembourg.* PhD thesis, University of Sussex.

Herr, J., Schley, L. & Roper, T.J. (2009a) Socio-spatial organization of urban stone martens. *Journal of Zoology* 277, 54–62.

Herr, J., Schley, L. & Roper, T.J. (2009b) Stone martens (*Martes foina*) and cars: investigation of a common human–wildlife conflict. *European Journal of Wildlife Research* 55, 471–477.

Herr, J., Schley, L., Engel, E. & Roper, T.J. (2010) Den preferences and denning behaviour in urban stone martens (*Martes foina*). *Mammalian Biology* 75, 138–145.

Herrmann, M. (2004) *Steinmarder in Unterschiedlichen Lebensräumen – Ressourcen, Räumliche und Soziale Organisation.* Laurenti Verlag, Bielefeld. (In German).

Jędrzejewski, W., Jędrzejewska, B., Zub, K. & Nowakowski, W.K. (2000) Activity patterns of radio-tracked weasels *Mustela nivalis* in Białowieża National Park (E Poland). *Annales Zoologici Fennici* 37, 161–168.

Kolbe, J.A. & Squires, J.R. (2007) Circadian activity patterns of Canada lynx in western Montana. *Journal of Wildlife Management* 71, 1607–1611.

Kowalczyk, R., Jędrzejewska, B. & Zalewski, A. (2003) Annual and circadian activity patterns of badgers (*Meles meles*) in Białowieża Primeval Forest (eastern Poland) compared with other Palaearctic populations. *Journal of Biogeography* 30, 463–472.

Kowalczyk, R., Zalewski, A. & Jędrzejewska, B. (2006) Daily movement and territory use by badgers *Meles meles* in Białowieża Primeval Forest, Poland. *Wildlife Biology* 12, 385–391.

Lachat Feller, N. (1993) *Eco-éthologie de la Fouine (*Martes foina *Erxleben, 1777) dans le Jura Suisse.* PhD thesis, Université de Neuchâtel. (In French).

LaPoint, S.D. (2013) *Movement Ecology of Fishers (*Pekania pennanti*) within a Semi-Urban Landscape.* PhD Thesis, Universität Konstanz.

Marcelli, M., Fusillo, R. & Boitani, L. (2003) Sexual segregation in activity patterns of European polecats (*Mustela putorius*). *Journal of Zoology* 261, 249–255.

Mata, C., Ruiz-Capillas, P. & Malo, J.E. (2017) Small-scale alterations in carnivore activity patterns close to motorways. *European Journal of Wildlife Research* 63, 64.

McClennen, N., Wigglesworth, R.R., Anderson, S.H. & Wachob, D.G. (2001) The effect of suburban and agricultural development on the activity patterns of coyotes (*Canis latrans*). *American Midland Naturalist* 146, 27–36.

Monterroso, P., Alves, P.C. & Ferreras, P. (2014) Plasticity in circadian activity patterns of mesocarnivores in Southwestern Europe: implications for species coexistence. *Behavioral Ecology and Sociobiology* 68, 1403–1417.

Monterroso, P., Rebelo, P., Alves, P.C. & Ferreras, P. (2016) Niche partitioning at the edge of the range: a multidimensional analysis with sympatric martens. *Journal of Mammalogy* 97, 928–939.

Murray, M.H. & St. Clair, C.C. (2015) Individual flexibility in nocturnal activity reduces risk of road mortality for an urban carnivore. *Behavioral Ecology* 26, 1520–1527.

Petrov, P.R., Popova, E.D. & Zlatanova, D.P. (2016) Niche partitioning among the red fox *Vulpes vulpes* (L.), stone marten *Martes foina* (Erxleben) and pine marten *Martes martes* (L.) in two mountains in Bulgaria. *Acta Zoologica Bulgarica* 68, 375–390.

Posillico, M., Serafini, P. & Lovari, S. (1995) Activity patterns of the stone marten *Martes foina* Erxleben, 1777, in relation to some environmental factors. *Hystrix* 7, 79–97.

Riley, S.P.D., Boydston, E.E., Crooks, K.R. & Lyren, L.M. (2010) Bobcats (*Lynx rufus*). In: *Urban Carnivores: Ecology, Conflict, and Conservation* (eds S.D. Gehrt, S.P.D. Riley & B.L. Cypher), pp. 121–138. The Johns Hopkins University Press, Baltimore.

Riley, S.P.D., Sauvajot, R.M., Fuller, T.K., York, E.C., Kamradt, D.A., Bromley, C. & Wayne, R.K. (2003) Effects of urbanization and habitat fragmentation on bobcats and coyotes in southern California. *Conservation Biology* 17, 566–576.

Roy, S., Ghoshal, A., Bijoor, A. & Suryawanshi, K. (2019) Distribution and activity pattern of stone marten *Martes foina* in relation to prey and predators. *Mammalian Biology* 96, 110–117.

Seiler, A., Krüger, H.-H. & Festetics, A. (1994) Reaction of a male stone marten (*Martes foina* Erxleben, 1777) to foreign feces within its territory – a field experiment. *Zeitschrift für Säugetierkunde* 59, 58–60.

Skirnisson, K. (1986) *Untersuchungen zum Raum-Zeit-System Freilebender Steinmarder (Martes foina Erxleben, 1777)*. M + K Hansa Verlag, Hamburg (In German).

Smith, J.A., Thomas, A.C., Levi, T., Wang, Y. & Wilmers, C.C. (2018) Human activity reduces niche partitioning among three widespread mesocarnivores. *Oikos* 127, 890–901.

Tigas, L.A., Van Vuren, D.H. & Sauvajot, R.M. (2002) Behavioral responses of bobcats and coyotes to habitat fragmentation and corridors in an urban environment. *Biological Conservation* 108, 299–306.

Torretta, E., Mosini, A., Piana, M., Tirozzi, P., Serafini, M., Puopolo, F. & Balestrieri, A. (2017) Time partitioning in mesocarnivore communities from different habitats of NW Italy: insights into martens' competitive abilities. *Behaviour* 154, 241–266.

Tsunoda, H., Ito, K., Peeva, S., Raichev, E. & Kaneko, Y. (2018) Spatial and temporal separation between the golden jackal and three sympatric carnivores in a human-modified landscape in central Bulgaria. *Zoology and Ecology* 28, 172–179.

Wang, Y., Allen, M.L. & Wilmers, C.C. (2015) Mesopredator spatial and temporal responses to large predators and human development in the Santa Cruz Mountains of California. *Biological Conservation* 190, 23–33.

Wereszczuk, A. & Zalewski, A. (2015) Spatial niche segregation of sympatric stone marten and pine marten – Avoidance of competition or selection of optimal habitat? *PLoS One* 10, e0139852.

Wereszczuk, A. & Zalewski, A. (2019) Does the matrix matter? Home range sizes and space use strategies in stone marten at sites with differing degrees of isolation. *Mammal Research* 64, 71–85.

Zalewski, A. (2000) Factors affecting the duration of activity by pine martens (*Martes martes*) in the Białowieża National Park, Poland. *Journal of Zoology* 251, 439–447.

Zalewski, A. (2001) Seasonal and sexual variation in diel activity rhythms of pine marten *Martes martes* in the Białowieża National Park (Poland). *Acta Theriologica* 46, 295–304.

Zalewski, A., Jędrzejewski, W. & Jędrzejewska, B. (2004) Mobility and home range use by pine martens (*Martes martes*) in a Polish primeval forest. *Ecoscience* 11, 113–122.

Zielinski, W.J., Spencer, W.D. & Barrett, R.H. (1983) Relationship between food habits and activity patterns of pine martens. *Journal of Mammalogy* 64, 387–396.

11

A 9-Year Demographic and Health Survey of a European Mink Population in Navarre (Spain): Role of the Canine Distemper Virus

Christine Fournier-Chambrillon[1],*, Juan-Carlos Ceña[2], Fermin Urra-Maya[3], Marco van de Bildt[4], M. Carmen Ferreras[5], Gloria Giralda-Carrera[6], Thijs Kuiken[4], Laëtitia Buisson[7,8], Francisco Palomares[9], and Pascal Fournier[1]

[1] *GREGE, Villandraut, France*
[2] *Ingeniero Técnico Forestal, Logroño, Spain*
[3] *Área de Biodiversidad, Gestión Ambiental de Navarra, Pamplona, Spain*
[4] *Viroscience Lab, Erasmus MC, Rotterdam, The Netherlands*
[5] *Departamento de Sanidad Animal (Anatomía Patológica), Universidad de León, León, Spain*
[6] *Servicio de Conservación de la Biodiversidad, Gobierno de Navarra, Pamplona, Spain*
[7] *CNRS, UMR 5245 EcoLab (Laboratoire Ecologie Fonctionnelle et Environnement), Toulouse, France*
[8] *INP, UPS, EcoLab, Université de Toulouse, Toulouse, France*
[9] *Departamento de Biología de la Conservación, Estación Biológica de Doñana, CSIC, Sevilla, Spain*

SUMMARY

As part of a large monitoring of the Critically Endangered European mink, *Mustela lutreola*, conducted in 2004 in Navarre, Spain, the highest densities of the western population were detected on the rivers Aragón, Arga, and Ebro. During this study, the first fatal naturally occurring canine distemper virus (CDV) infection in free-ranging European mink was detected on the 20 km downstream section of the river Arga. The European mink population of this section was then subject to a long-term demographic and CDV survey from 2004 to 2012. On average, 165 live traps were set during 10 consecutive nights generally twice per year (pre- and post-breeding periods), for a total of 13 trapping sessions. All captured mink were marked with a transponder. CDV antibodies were detected using a virus neutralization test on 126 sera collected from 79 mink from 2005 to 2012. Additionally, 81 pools of swabs collected from nose, throat, and rectum of 46 mink between 2008 and 2012 were tested for the presence of CDV-RNA using the Reverse transcriptase-polymerase chain reaction. High antibody prevalence was observed from 2005 to 2008, with in parallel a drastic decrease of the population, falling from 44 to 15 mink. Since 2009, all tested mink were seronegative and the population seemed to recover slowly but regularly, reaching 20 mink in 2012. These results strongly suggest the occurrence of a CDV epidemic as the cause of the catastrophic decline of the population studied, as years with low trapping were neither correlated to unfavourable weather conditions during sampling nor during the breeding season. This population, confronted with numerous additional factors of mortality and now immunologically naïve to CDV, remains particularly vulnerable. Strong conservation measures of this major nucleus and of all relictual nuclei of the western European mink population are, therefore, urgently needed.

Keywords

canine distemper virus − critically endangered − demography − European mink − *Mustela lutreola* − serology − Spain − survey

*Corresponding author.

Small Carnivores: Evolution, Ecology, Behaviour, and Conservation, First Edition. Edited by Emmanuel Do Linh San, Jun J. Sato, Jerrold L. Belant, and Michael J. Somers.
© 2022 John Wiley & Sons Ltd. Published 2022 by John Wiley & Sons Ltd.

Introduction

The European mink, *Mustela lutreola*, a small semiaquatic mustelid, has retracted dramatically from its former range during the past 150 years. The remaining populations are currently distributed in well-separated nuclei, in northern Spain and southwestern France, Romania, Ukraine, and Russia (Maran *et al.*, 2016). In Romania, the species still seems to be widespread (Marinov *et al.*, 2011), whereas the conservation status of the other populations is particularly alarming. In Ukraine, the European mink was recently rediscovered in deltas, but the population is considered to be highly fragmented and at the edge of extirpation. In Russia, the European mink is extinct or believed to be extinct in 40 of the 61 regions within its historical range and the current range consists of isolated distant habitat patches of different sizes. In Western Europe, the population seems to become increasingly fragmented and probably consists of small relictual nuclei (Maran *et al.*, 2016).

This decline is due to several factors whose relative roles have varied through time and acted with cumulative effect. These factors include major anthropogenic pressure (habitat loss and degradation, historical hunting, accidental trapping, vehicle collisions, and dog predation), interspecific competition with the alien invasive American mink, *Neovison vison* (also named *Mustela vison*), and infectious diseases (Fournier & Maizeret, 2003; Maran, 2007). Maran *et al.* (2016) reported that the overall number of European mink has probably suffered from at least a 90% decline since the beginning of the twentieth century. The species, listed as Endangered since 1994, was classified in 2011 as Critically Endangered (i.e. facing an extremely high risk of extinction in the wild in the near future) by *The IUCN Red List of Threatened Species* (Maran *et al.*, 2016).

In the last 20 years, numerous studies on habitat use (Zabala & Zuberogoitia, 2003; Fournier *et al.*, 2007), spatial behaviour (Zabala *et al.*, 2006; Fournier *et al.*, 2008), health and reproductive status (Fournier-Chambrillon *et al.*, 2004a,b, 2010; Philippa *et al.*, 2008; Torres *et al.*, 2008), and conservation genetics (Michaux *et al.*, 2005; Cabria *et al.*, 2011) have been conducted on the western population, including comparative studies with related European polecat, *Mustela putorius*, and feral American mink, to understand the causes of the decline and to propose conservation measures. The identification of the role of infectious diseases is of important concern for conservation strategies, especially because animals with low genetic diversity, as observed in the western population (Michaux *et al.*, 2005), could be more vulnerable to infectious diseases (O'Brien & Evermann, 1988). Recent studies in southwestern France revealed the prevalence of Aleutian disease virus (ADV), canine distemper virus (CDV), and *Leptospira*, whereas exposure to other viral pathogens appeared to be low (Fournier-Chambrillon *et al.*, 2004a; Philippa *et al.*, 2008; Moinet *et al.*, 2010).

In 2004, extensive monitoring of the European mink was conducted in Navarre, Spain (Figure 11.1) to evaluate its distribution and conservation status, using a standardized trapping method in six areas representative of the main types of streams potentially used by the species, completed with data on road collisions and indirect sampling in some areas. The results revealed the presence of the species in almost the whole fluvial network of Navarre (1070 km), except the river Esca (north-east) and the low section of the Ebro (south-east). The main populations were detected on the rivers Aragón, Arga, and on the river Ebro upstream to the confluence of the river Aragón. One adult female, captured in November 2004 on the lower Arga, showed poor body condition and nervous signs. It was transferred to a wildlife rehabilitation centre where finally it died. The necropsy revealed interstitial pneumonia, marked depletion of lymphocytes in the splenic white pulp, splenic hyalinosis, and multifocal demyelination of the white matter in the cerebellum (pons and cerebellar peduncles). Later, CDV antigen was found in different organs examined (lung, spleen, kidney, and cerebellar cortex) using immunohistochemical techniques (Sánchez-Migallón *et al.*, 2008). Unfortunately, only the prevalence of ADV antibodies was studied on the sera collected in 2004, but not the prevalence of CDV antibodies.

Distemper is an acute or subacute, highly contagious febrile disease affecting the respiratory, gastrointestinal, and central nervous systems of terrestrial carnivores. Canine distemper virus belongs to the genus *Morbillivirus* in the Paramyxoviridae family and does not survive very long outside the host in the environment. Therefore, transmission occurs mostly by close contact between infected and new hosts.

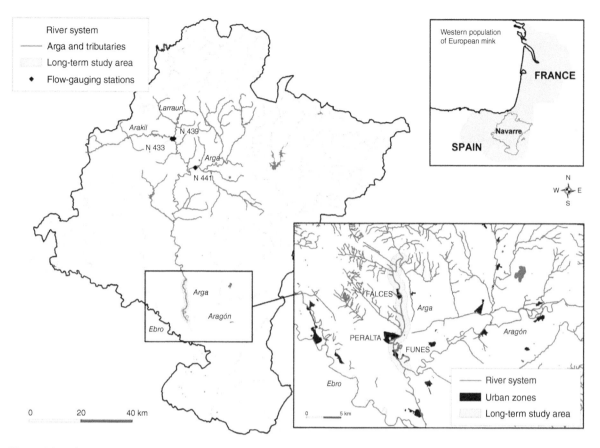

Figure 11.1 Location of the area where the long-term study on the European mink, *Mustela lutreola*, was carried out from 2004 to 2012 in Navarre, Spain, and of the flow-gauging stations upstream of the study area, from which were extracted flow data from 2003 to 2012.

The infected host excretes the virus mainly via nasal, ocular, and oral fluids and urine. The new host becomes infected through the respiratory tract, by direct (sniffing, droplet exposure) or indirect contact (breathing in virus in aerosol). Therefore, interactive behaviour with high promiscuity involves a high risk of infection transmission, and the denser the population of susceptible animals, the higher the risk of infection (Appel, 1987; Rijks, 2008). CDV has been reported worldwide in all families of terrestrial carnivores (Deem *et al.*, 2000), and mustelids are known to be among the most susceptible to CDV (Pearson & Gorham, 1987; Williams, 2001) with fatal infections reported in numerous species (Cunningham *et al.*, 2009). The effect of CDV on European mink is poorly documented, but fatal vaccine-induced distemper has been reported (Sutherland-Smith *et al.*, 1997;

Ek-Kommonen *et al.*, 2003). To our knowledge, Sánchez-Migallón *et al.* (2008) described the first fatal naturally occurring CDV infection in free-ranging European mink in our study area.

The objectives of this study were (i) to conduct a long-term demographic and CDV survey on the European mink population of the low section of the river Arga; and (ii) to assess the role of canine distemper as a cause of population decrease.

Material and Methods

Study Area

The long-term study area (Figure 11.1) is located in Navarre, Spain, on the 20 km low section of the river Arga (Municipalities of Falcès, Peralta, and Funès;

Figure 11.2 Flooded habitats of Falcès. *Source:* Photo © Juan-Carlos Ceña.

42°29′–42°44′ N, 1°78′–1°81′ W), a tributary of the river Aragón, and secondary tributary of the river Ebro. A serologic survey was also conducted during 3 years on European mink captured on the river Aragón up to ~40 km upstream to the confluence with the Arga (42°29′–42°39′ N, 1°78′–1°47′ W), and these serological results were also included in this chapter.

The long-term study area presents a total length of 25 linear km of fluvial courses and wetlands, which includes the riverbed of the river Arga and the wide network of lagoons, channels, and irrigation ditches located on its borders. The ~7 km upper section (up to the bridge of Falcès) presents mostly a natural dynamic, with wide fluvial tables interrupted by brief sections of rapids and two big active meanders showing complex spaces of river bank (Figure 11.2). This section bears a great environmental value by the presence of gallery-forest and other riparian formations that shelter a diverse fluvial fauna. In contrast, the lower section is canalized; the river bank vegetation is very scanty, being formed by a discontinuous band of new shoots of black poplar, *Populus nigra*, and some small bramble, *Rubus* spp., patches, and this sector is assumed to be used mostly as a corridor, particularly

by the semi-aquatic fauna. On the borders of the river, the complex labyrinth of former meanders connected to the old riverbed is of better quality (Figure 11.3). Indeed, since 2004, the Government of Navarre led several actions to restore and create new complexes of lagoons in this area, in order to increase the quality of habitats for the European mink, especially for breeding females. Thanks to the conservation of the former riparian woodland and the progressive increase of the hygrophilous vegetation, these extensive areas have constituted humid zones of interest (Figure 11.4), shaping an extensive network of natural protected spaces (Natural Reserves of the Government of Navarre).

Intensive Periodic Trapping

The whole study was licensed by the Navarre government. A systematic intensive trapping programme was carried out, generally twice a year, during the pre-breeding (February to April) and the post-breeding (September to November) periods, respectively. A representative sampling unit was considered to be ~20 km of the mainstream, on which live traps (60 × 15 × 15 cm)

Figure 11.3　Wet zones of 'La Muga', Peralta. *Source:* Photo © Juan-Carlos Ceña.

Figure 11.4　Open marsh of 'Soto Gil', Funès. *Source:* Photo © Juan-Carlos Ceña.

baited with fish (e.g. sardines, trouts) were set during 10 consecutive nights, along the river and around lagoons, approximately every 100 m, resulting in ~1500–2000 trap-nights depending on the local conditions to settle the traps. All individuals were marked with transponders. This methodology was considered appropriate in view of reliably comparing European mink annual densities.

In 2004, trapping sessions were conducted only on the 13 km upstream of the long-term study area, while the 7 km downstream were monitored in 2005. Data were pooled for the yearly analysis. Thereafter, from 2006 to 2012, traps were placed in the whole study area during both periods, except in spring 2006, autumn 2009, and spring 2011 (Table 11.1). On the 40 km section of the river Aragón upstream to the confluence with the Arga, a standardized intensive trapping programme was conducted only in 2004–2005. In 2006 and 2007, trapping was carried out less intensively for other purposes.

Animal Manipulation

Once captured, animals were immobilized either with an intramuscular injection of ketamine (Imalgene® 1000, Merial SAS) combined with xylazine (Rompun® 2%, Bayer) or with an intramuscular injection of medetomidine (Domitor®, Pfizer) combined with ketamine (Kétamine 500®, Virbac), and reversed with atipamezole (Antisedan®, Pfizer) (Fournier-Chambrillon *et al.*, 2003). A clinical exam was performed, including weighing and body measurements. The sex was determined and age was defined as juvenile (individuals captured in autumn, with new teeth without abrasion or tartar), subadult (individuals captured in spring, with new teeth without abrasion or tartar), and adult (animals previously marked as juveniles or subadults, otherwise individuals with teeth partly abraded and with tartar). All animals were injected with subcutaneous transponders (BackHome®, Virbac) between the shoulders. A blood sample of up to 2.5 ml was taken from the jugular vein and transferred into a plain silicone-coated glass tube (Venosafe®, Terumo).

Additionally, from 2008, swabs (Eurotubo®, Deltalab) were collected from nose, throat, and rectum, immediately placed in virus transport medium (EMEM

supplemented with glycerol, lactalbumin, penicillin, streptomycin, polymyxin B sulphate, nystatin, and gentamycin) and frozen as soon as possible in liquid nitrogen, at $< -80°C$. When the procedure was completed, each animal was placed back in the trap to recover, in a quiet and sheltered place, and was released at the capture site ~2 h after recovery.

Marked animals were also immobilized and blood sampled in case of the first capture during a new 10-night trapping session. Otherwise, they were immediately released (when recaptured during the same session).

Blood Processing, Serologic, and PCR Analyses

From spring 2005 to spring 2008, blood was centrifuged at 3000 XG for 5 min several hours after sampling in a veterinary clinic located ~70 km from the study area, and the serum was stored at $-20°C$. Because of the cytotoxicity of numerous sera due to bad conservation, from autumn 2008, the blood was centrifuged locally as soon as possible after sampling and the serum immediately stored in a cryotube (CryoTube™ Vials®, Thermo Fischer Scientific Nunc A/S) in liquid nitrogen at $< -80°C$. Transport of the sera as well as the swabs to the Vioscience Lab (Erasmus MC, Rotterdam, Netherlands) was then conducted under liquid nitrogen or dry ice via a specialized carrier.

Serum samples were tested for the presence of neutralizing antibodies against CDV using a virus neutralization (VN) test as described previously (Visser *et al.*, 1990). In brief, 50 μl of duplicate 2-log dilution series, starting at 1 : 10, of heat-inactivated serum samples were incubated with 50 μl containing 100 median tissue culture infectious dose ($TCID_{50}$) of CDV. After 1 h incubation at 37°C, 10^4 Vero cells were added to each well. After 4–6 incubation days (37°C, 5% CO_2), the plates were checked for the presence of cytopathic effect (CPE). Titers were expressed as the reciprocal of the highest serum dilution with complete inhibition of CPE. The mean from the duplicates was calculated. Samples with a mean titer of < 20 were considered negative.

Swabs were tested for the presence of *Morbillivirus* RNA. Total nucleic acids were isolated from swabs using the MagnaPure and the MagnaPure LC Total

Table 11.1 Trapping effort, detailed numbers per sex and age of the captured European mink, *Mustela lutreola*, trapping index, and rainfall and flow data, for the 13 trapping sessions conducted on the 20 km low section of the river Arga (Navarre, Spain) from 2004 to 2012. Data from 2004 and 2005 were pooled to obtain results for the whole study area, as the 13 km upstream were monitored in 2004, and the 7 km downstream in 2005.

| Year | Period | Number of traps | Number of different captured European mink | | | | | | Trapping index | Mean daily precipitation during the sampling period (L/m²) | Total precipitation during the previous breeding season (L/m²) | Mean flow during the previous breeding season (m³/s) |
			Total	Males	Females	Adults	Sub-adults	Juveniles				
2004–2005	Pre-breeding[a]	162	29	14	15	20	9	—	17.90	1.94	—	—
	Post-breeding[b]	137	26	10	16	18	—	8	18.98	1.74	81.63	15.57
	Whole year	—	44	18	26	36	—	8	—	—	—	—
2006	Pre-breeding	No trapping	—	—	—	—	—	—	—	—	—	—
	Post-breeding	174	13	7	6	5	—	8	7.47	0.95	109.55	7.23
	Whole year	—	—	—	—	—	—	—	—	—	—	—
2007	Pre-breeding	116	11	5	6	6	5	—	9.48	2.14	—	—
	Post-breeding	159	10	6	4	7	—	3	6.29	0.07	73.35	22.46
	Whole year	—	18	10	8	15	3	3	—	—	—	—
2008	Pre-breeding	202	9	4	5	6	3	—	4.46	4.47	—	—
	Post-breeding	197	7	3	4	1	—	6	3.55	0.55	187.80	40.96
	Whole year	—	15	7	8	9	—	6	—	—	—	—
2009	Pre-breeding	170	8	2	6	5	3	—	4.71	3.66	—	—
	Post-breeding	No trapping	—	—	—	—	—	—	—	—	—	—
	Whole year	—	—	—	—	—	—	—	—	—	—	—

(Continued)

Table 11.1 (Continued)

| Year | Period | Number of traps | Number of different captured European mink | | | | | | Trapping index | Mean daily precipitation during the sampling period (L/m²) | Total precipitation during the previous breeding season (L/m²) | Mean flow during the previous breeding season (m³/s) |
			Total	Males	Females	Adults	Sub-adults	Juveniles				
2010	Pre-breeding	171	8	3	5	5	3	—	4.68	0.35	—	—
	Post-breeding	171	10	5	5	5	—	5	5.85	1.06	72.05	31.69
	Whole year		14	6	8	9	—	5	—	—	—	—
2011	Pre-breeding	No trapping	—	—	—	—	—	—	—	—	—	—
	Post-breeding	170	14	5	9	6	—	8	8.24	2.60	84.85	7.07
	Whole year	—	—	—	—	—	—	—	—	—	—	—
2012	Pre-breeding	160	10	5	5	3	7	—	6.25	3.43	—	—
	Post-breeding	160	15	7	8	8	—	7	9.38	3.94	67.65	17.33
	Whole year	—	20	10	10	13	—	7	—	—	—	—

[a] February–April.
[b] October–December.

Nucleic Acid Isolation Kit (Roche Diagnostics GmbH, Mannheim, Germany) according to the manufacturer's protocols. Reverse transcriptase-polymerase chain reaction (RT-PCR) was performed to detect morbilliviral RNA using *Morbillivirus*-specific primers P1: 5'ATGTTTATGATCACAGCGGT3' and P2: 5'ATTGGGTTGCACCACTTGTC3', after first strand synthesis with specific morbilliviral primers.

Rainfall and Flow Conditions Data

As bad weather conditions as well as flooding events may influence both breeding success in mink and capture efficiency, we included weather data in the analyses. Data from 2003 to 2012 were extracted from the Navarre government website (http://www.navarra.es). The weather stations of Falcès and Funès located in the upper and lower sections of the long-term study area, respectively, were selected to obtain rainfall (Figure 11.1). The average daily precipitation (in L/m^2) of the two stations was used as mean daily precipitation for the study area. Mean daily flow data per month (in m^3/s) were extracted from the three closest gauging stations located upstream on the three tributaries converging in the study area (Figure 11.1). Given the location of these gauging stations, the sum of the flow measured in each station was considered representative of the flow in the study area.

Mean daily precipitation was calculated for each 10-night trapping session. Moreover, we assumed that bad weather and flow conditions during May and June (i.e. the period during which breeding females remain confined in their dens for parturition and lactation of newborns) may have a negative effect on breeding success. The total precipitation in May and June was thus calculated for each year. The average of the mean daily flow in May and June was also calculated for each gauging station and year, and the sum used as the flow in the study area during the breeding period. We thus assumed that a low breeding success would affect the trapping index during the subsequent post-breeding and pre-breeding periods.

Data and Statistical Analysis

Capture 'C1' was defined as the first capture of each mink during a 10-night trapping session. The trapping index during each 10-night trapping session was measured as the number of different captured mink divided by the number of traps. The total annual number of mink was calculated only when both pre-breeding and post-breeding trapping sessions occurred in the year. It was defined as the total number of different captured individuals when combining spring and autumn sessions (i.e. ≤ the sum of the data of each session because some individuals were captured twice in the same year, in both spring and autumn sessions). Data from 2004 and 2005 were pooled to obtain results for the whole study area. Age ratio was the ratio between juveniles and adults. Animals considered as subadults in spring became adults in autumn, and were, therefore, included in the age class 'adult' for yearly data.

Spearman's rank correlations were run to investigate a possible relationship between the number of different captured mink and the number of traps set during a particular session. We used chi-square tests of independence to compare the number of different captured mink, the sex-ratio and the age-ratio between years, both for annual data and specific trapping sessions (pre- or post-breeding, except for age-ratio that was only tested for post-breeding as there were no juveniles in pre-breeding session).

Cytotoxic sera in the VN test were excluded from the calculation of prevalence. For determination of annual antibody prevalence, resampled animals between spring and autumn were counted only once (i.e. the sample that tested positive in case of seroconversion). The difference of antibody prevalence between sexes was tested using a chi-square test of independence after pooling the data from Arga and Aragón from 2005 to 2008 to increase the sample size.

We ran a generalized linear mixed model (procedure GLIMMIX in SAS with a binomial distribution) on the capture data on the Arga to determine whether the presence or absence of CDV antibodies differed between sex, year, and trapping session (pre- or post-breeding). Individual identity was also included in the GLIMMIX procedure as a random factor to control for some individual being tested for CDV more than once. Captures with cytotoxic sera and without CDV VN test were excluded from the analysis.

We also performed a generalized linear model (procedure GLM in SAS) to examine whether the number of mink trapped on the Arga (i.e. the trapping index described above) was influenced by CDV prevalence of mink population, while controlling for trapping session, precipitation during the sampling period, and precipitation and flow during the previous breeding period. The trapping index was log-transformed to ensure normality.

Finally, based on the previous results (serology and multivariate analysis), two multiannual periods were identified: period 1 from 2004–2005 to 2008 during which the population was known to be exposed to CDV, and period 2 from 2009–2012, during which no CDV antibodies were detected in the population. Within each multiannual period, resampled animals were counted only once to determine the number of different mink and antibody prevalence (we used the sample that tested positive in case of seroconversion). The change in CDV antibody prevalence and possible differences in sex ratio and age ratio between both periods were tested using a chi-square test of independence. To study the temporal trend in the number of different captured mink, a Mann–Kendall trend test was performed within each period.

Spearman's rank correlations, chi-square tests, and Mann–Kendall trend tests were performed using R 2.14.2 software® (R Development Core Team, 2012). The GLIMMIX and the GLM procedures were run in SAS® 9.4 (SAS Institute Inc., 2013). For all statistical tests, $p \leq 0.05$ was considered significant (Scherrer, 1984; Sokal & Rohlf, 1995).

Results

Trapping

From 2004 to 2012, 13 intensive pre- or post-breeding 10-night trapping sessions conducted on the 20 km low section of the river Arga resulted in 170 captures 'C1' of 98 different European mink. A total of 46 males and 52 females were trapped resulting in a sex ratio not significantly different from 1:1 (Goodness-of-fit test: $\chi^2 = 0.37$, df = 1, $p = 0.544$). Thirty-five males (76%) and 25 females (48%) were recaptured at least once during the same session, and 21 males (46%) and 23 females (44%) were recaptured at least once during another session.

The mean number of traps set per 10-night session was 165 ± 23 (range: 116–202, $n = 13$) and the mean number of different captured mink per 10-night session was 13 ± 7 (range: 7–29, $n = 13$) (Table 11.1). The relationship between the number of captured mink and the number of traps was not significant (Spearman's rank correlation: $r_s = -0.53$, $p = 0.064$).

The total number of captured mink differed significantly between years, for both annually and within each trapping session (Chi-square tests of independence: annual: $\chi^2 = 27.78$, df = 4, $p < 0.001$; pre-breeding session: $\chi^2 = 26.68$, df = 5, $p < 0.001$; post-breeding session: $\chi^2 = 16.63$, df = 6, $p = 0.011$). No significant difference could be detected in sex-ratio ($p > 0.82$ in all cases) or age-ratio ($p > 0.132$ in all cases) between years, neither for the annual results, nor for pre- and post-breeding trapping sessions. On the 40 km section of the river Aragón upstream to the confluence with the Arga, 29 European mink (14 males, 15 females) were captured during the 2004–2005 standardized trapping.

CDV Survey

None of the animals sampled from 2005 to 2012 showed clinical signs of disease upon capture and sampling. However, CDV antibodies were detected every year from 2005 to 2008 (Table 11.2). Annual prevalence was high, at least 20%, with large confidence intervals because of the small number of animals sampled. Antibody titers ranged from 20 to 60. During this first period, CDV antibody prevalence was significantly higher in females than in males (9/20 vs. 3/20; Chi-square test of independence: $\chi^2 = 4.29$, df = 1, $p = 0.038$). From 2009, no antibodies to CDV were detected, and the prevalence during the second period was significantly lower than during the first period (0/32 vs. 8/24; $\chi^2 = 12.44$, df = 1, $p < 0.001$). The serologic conversion was observed in two females: they were negative in autumn 2006 but positive in autumn 2007. In addition, one of these two females was thereafter negative in spring 2009 and autumn 2010 again.

Eighty-one pools of swabs were submitted for detection of CDV-RNA. They were from 46 different European mink captured from 2008–2012, as 22 individuals were tested 2–6 times at various dates. All the 81 pools were negative, indicating that no European

Table 11.2   Annual antibody prevalence and 95% confidence interval (95% CI) to canine distemper virus (CDV) in European mink on the 20 km low section of the river Arga from 2005 to 2012, and on the 40 km section of the river Aragón upstream to the confluence with the Arga (Navarre, Spain) from 2005 to 2007, using virus neutralization (VN) test.

		Number of sera				
		Cytotoxic	Negative	Positive	Prevalence (%)	95% CI
2005	Arga	1	3	1	25.0	0.6–80.6
	Aragón	0	5	2	28.6	3.7–71.0
	Total	1	8	3	27.3	6.0–61.0
2006	Arga	3	8	1	11.1	0.3–48.2
	Aragón	1	4	2	33.3	4.3–77.2
	Total	4	12	3	20.0	4.3–48.1
2007	Arga	3	5	4	44.4	13.7–78.8
	Aragón	6	3	0	0.0	0.0–70.8
	Total	9	8	4	33.3	9.9–65.1
2008	Arga	7	6	2	25.0	3.2–65.1
2009	Arga	0	8	0	0.0	0.0–36.9
2010	Arga	0	13	0	0.0	0.0–24.7
2011	Arga	0	12	0	0.0	0.0–26.5
2012	Arga	0	18	0	0.0	0.0–18.5

mink was excreting *Morbillivirus* at the time of capture. In 15 cases, VN test was not available, in 64 cases, VN test was negative, and in two cases, VN test was positive.

Factors Affecting the Presence of CDV Antibodies and the Trapping Index

The GLIMMIX procedure was performed on 89 out of 170 captures for which a CDV serological result was available. Among the set of explanatory variables, only the year of trapping was found to explain significantly the presence of CDV antibodies ($p = 0.012$; Table 11.3).

The GLM relating the trapping index to the trapping period, the prevalence of CDV antibodies as well as rainfall and flow conditions was performed on the 12 out of 13 trapping sessions for which CDV seroprevalence had been tested. We found that none of the considered variables significantly affected the trapping index ($F = 1.25$, df = 5, $p = 0.389$).

Population Trends With and Without CDV

The number of mink was highest during the 2004–2005 pre-breeding session, with 29 different captured mink, which then decreased quickly to reach only seven mink during the 2008 post-breeding session (Table 11.1). The time series from 2004–2005 pre- to 2008 post-breeding session (period 1), during which the population was known to be exposed to CDV, showed a significant negative monotonous trend (Mann–Kendall trend test: $S = -21$, $p = 0.003$). Conversely, from 2009 pre- to 2012 post-breeding session (period 2), during which no CDV was detected in the population, a rise in the number of captured mink from 8 to 15 different individuals was observed. This positive monotonous trend was nearly significant ($S = 11$, $p = 0.051$).

The ratio of juveniles during post-breeding sessions was much higher in period 2 (20 J/10 A) than in period 1 (25 J/21 A), but this difference was not significant (Chi-square test of independence: $\chi^2 = 1.14$, df = 1, $p = 0.285$). The ratio of females during pre-breeding

Table 11.3 Results of the generalized linear mixed model (GLIMMIX) procedure conducted on the 89 captured European mink between 2004 and 2012 with available CDV serological result.

Type III tests of fixed effects

Effect	Numerator degrees of freedom	Denominator degrees of freedom	F	p
Year	1	33	7.15	0.0116
Sex	1	33	0.15	0.7054
Session	1	33	0.00	0.9854

sessions was slightly higher in period 2 (12F/8M) than in period 1 (23F/21M), but again not significant difference was found ($\chi^2 = 0.33$, df = 1, $p = 0.565$).

Discussion

In this long-term study on European mink in Navarre, Spain, we showed the presence of antibodies to CDV with high prevalence during 4 years following a fatal case of canine distemper in a free-ranging adult female captured in November 2004. We observed a simultaneous drastic decrease of population size (using our trapping index as a surrogate), from a maximum in 2004–2005 to only one-third of the original population in 2008. From 2009 to 2012, no more antibodies to CDV were detected in the captured European mink and trapping results suggested a slow but constant recovery of the population, reaching only half of the numbers recorded in 2004–2005 by 2012.

Following this fatal CDV case in November 2004, both the demographic and serologic results converge toward the hypothesis of a strong negative impact of a CDV epidemic on this probably immunologically naïve population. Indeed, all the other hypotheses which could possibly explain the decrease in trapping success were rejected. Our intensive trapping programme proved to be valuable to estimate population size or density, as no correlation was observed between the number of different captured mink and the number of traps. This finding strongly suggests that all 'trap-confident' mink were captured whatever the additional increase in trapping effort. Thus, the decline

in the number of captured mink was likely not due to a decrease in sampling effort. Second, we highlighted that years with low trapping were neither correlated to higher rainfall during sampling nor to unfavourable weather (i.e. higher precipitation and flow) conditions during the previous breeding season that would have impacted on the breeding success as hypothesized. The fact that years with low trapping were not correlated to CDV prevalence can also be easily understood. Indeed, CDV antibodies were already detected in the initially dense population, and it could not recover instantaneously from the end of the epidemic. All these findings support our hypothesis that the sharp decline in European mink population numbers probably results from the emergence of the CDV epidemic in this population from 2004.

In 2004–2005, the population density was exceptionally high in the study area, with more than 17 mink per 10 km of river, while in southwestern France, for example, mean home range lengths of this territorial species were ~13 km river in males, and ~6 km in females, i.e. ~3 mink per 10 km river (Fournier *et al.*, 2008). Given the modes of transmission of CDV (see **Introduction**), this very high density may have largely contributed to a fast spread of the virus in the population. Indeed, in 2004–2005, the number and proportion of females were particularly high (26 females vs. 18 males), and females were mainly confined to the relictual lagoons (Palomares *et al.*, 2017). Moreover, close contacts between individuals increase during the mating season, given that females can copulate with several males and conversely. In our study, we detected a significantly higher CDV antibody prevalence in females than in males, and negative to positive seroconversion was only observed in females. However, the significance of these results is unclear as they could derive either from higher infection rates in females or higher mortality in males, considering that our sample informs only about the living population.

From the detection of the CDV fatal case in November 2004 and through the period of detection of CDV antibodies in the population (i.e. period 1), the successive numbers of captured mink significantly decreased, even between pre- and post-breeding sessions conducted during the same year. These results are thus consistent with the higher transmission rate suspected during the mating season and with the known higher susceptibility of young animals to CDV (Williams, 2001; Timm *et al.*, 2009). Then, from 2009 to 2012 (period 2), no CDV antibodies were detected indicating that there was no more evidence of circulating *Morbillivirus*. Similarly, no CDV-RNA was detected in European mink captured from 2008 to 2012 showing that none was excreting *Morbillivirus* at the time of capture. During this second period, a slow but constant population recovery was observed. The last serologic results seem to support that the population has become immunologically naïve again and remains susceptible to CDV infection.

Our results, therefore, suggest that CDV cannot persist endemically in this European mink population and that an external source of virus must be at the origin of the epidemic. Unfortunately, no epidemiological investigations could be performed to determine the possible sources of the virus of the infected mink population. Relatively dense populations of susceptible hosts are usually needed to sustain CDV circulation (Williams, 2001), and domestic dogs, *Canis familiaris*, badly or not immunized, are regularly identified or suspected as the origin of CDV infections in wildlife (Ferreyra *et al.*, 2009; Gowtage-Sequeira *et al.*, 2009; Meli *et al.*, 2010; Muller *et al.*, 2011). In the rural area where this study was conducted, sheepdogs and hunting dogs are common. Many nature trails through the wet zones were also created for public awareness and are accessible to any walker. Necropsies performed on 91 mink found dead between 1999 and 2013 in the whole Navarre revealed 13 attacks by carnivores (i.e. predation without consumption) confirming regular interspecific interactions that may facilitate the

transmission of the virus (Fournier-Chambrillon *et al.*, unpublished data). Moreover, other sympatric free-ranging carnivores could be implicated as reservoir hosts in addition to domestic dogs, as suspected in other studies (Chen *et al.*, 2008; Martella *et al.*, 2010; Keller *et al.*, 2012). For instance, the presence of CDV antibodies or infection are regularly described in red fox, *Vulpes vulpes*, populations in Europe (Spain: Sobrino *et al.*, 2008; Portugal: Santos *et al.*, 2009; Italy: Nouvellet *et al.*, 2013; Switzerland: Origgi *et al.*, 2012; Luxembourg: Damien *et al.*, 2002; Germany: Frolich *et al.*, 2000; Scandinavia: Akerstedt *et al.*, 2010; Pagh *et al.*, 2018). Martella *et al.* (2010) considered that in case of a multi-host epizootic, red foxes might play a major role in CDV amplification and diffusion because of their social behaviour during the reproductive season and the sometimes long distances covered by dispersing juvenile foxes. In Central Spain, exposure to CDV was recently detected in a small sample of European wild cats, *Felis silvestris* (Candela *et al.*, 2019). Tavernier *et al.* (2012) reminded that the reservoir of CDV is a metapopulation of multiple host species, and consider that pine martens, *Martes martes*, are more likely to be a reservoir in Flanders (Belgium) than domestic dogs. In southwestern France, CDV antibody prevalence in stone martens, *Martes foina*, and European polecats were found to be high (Philippa *et al.*, 2008). Both species are sympatric with European mink in our study area, although much lower densities of polecats than mink have been observed. Nevertheless, CDV antigen was detected by immuno-histo-chemistry with intense positivity in the lung and the kidney of a European polecat victim of a road collision in August 2003, ~55 km northeast of our study area (Ferreras *et al.*, unpublished data), confirming the circulation of the virus in other free-ranging species in Navarre. Although no other data are available, the existence of a multi-host epidemic that could have spread over the country cannot be excluded, as, for instance, in Switzerland (Origgi *et al.*, 2012).

To our knowledge, the present study is the first to combine both quantitative demographic and serologic data suggesting a strong negative impact of CDV on a European mink population. It is thus of great importance for the conservation of this Critically Endangered species. Indeed, the CDV epidemic affected one of the presumably most important European mink nuclei of the western population, which remains particularly vulnerable. It appears to be immunologically naïve to CDV again, and other CDV epidemics could have dramatic consequences for the long-term survival of this population, as far as the initial high densities have not been reached again for the moment. In addition to the CDV disease, this population is confronted with many other factors of mortality limiting life expectancy, particularly road collisions and attacks by carnivores. In addition, this species gives birth only once a year to a relatively limited number of young (3.4 ± 0.9 embryos; Fournier-Chambrillon *et al.*, 2010), which probably limits the population growth rate. Therefore, it is particularly urgent to develop a panel of short- and long-term conservation measures. Currently, there is no safe and effective commercially available CDV vaccine for use in Europe in highly susceptible nondomestic species like the European mink, as the safe canarypox-based vaccine used in North America, a genetically modified organism, is not licensed for use in Europe (Philippa, 2010). The CDV-ISCOM vaccine proved to induce strong humoral immune response in the European mink (Philippa, 2007), but its large-scale production was not developed.

As far as domestic dogs could be one of the CDV reservoirs in our study area, a massive and continuous vaccination programme for the dog population should be recommended as the principle of precaution. Authorizing the use of the canarypox-based vaccine could also be of great interest as an immediate means to protect the European mink population against a new CDV epidemic. Unfortunately, the long-term demographic and CDV survey of the population has stopped, although it would have been particularly important to watch its evolution after 2012 and the possible outbreak of another CDV epidemic. Demographic and epidemiological studies on other free-ranging carnivores, at least on red foxes and martens, would also be necessary to better understand all possible routes of CDV transmission. In parallel, large-scale measures to reduce all other factors of mortality remain essential and urgent for the conservation of the European mink.

Lastly, *Morbillivirus* should be the object of continuous pronounced attention, as *Morbillivirus* diseases are re-emerging or newly emerging in recent years, with newly identified viruses or viruses crossing species barriers (Griot *et al.*, 2003), including emergence in Europe of CDV strains leading to high mortality in wild carnivores (Origgi *et al.*, 2012). Therefore, in the longer term, the development of oral vaccines for free-ranging carnivores, including mustelids, could be a relevant measure to limit these devastating epidemics. However, given the long time necessary for such developments, further research is urgently required. Long-term monitoring would be needed before applying these newly developed vaccines, particularly in highly endangered species.

Acknowledgements

This study was funded by the 'Departamento de Desarrollo Rural, Medio Ambiente y Administración Local del Gobierno de Navarra'. We would like to thank all the people involved in the field work: Alfonso Ceña Martínez, Enrique Sola Larrayoz, Gabriel Berasategui Echevarría, Uxue Itoiz Mariñelarena, Iosu Alfaro Vergarachea, and Itsaso Bidegain Garbala. We also thank Peter van Run (Erasmus MC) for his assistance in the laboratory. We are also grateful to two anonymous reviewers for their constructive comments on a previous draft of the manuscript, and to Emmanuel Do Linh San for linguistic corrections.

References

Akerstedt, J., Lillehaug, A., Larsen, I.L., Eide, N.E., Arnemo, J.M. & Handeland, K. (2010) Serosurvey for canine distemper virus, canine adenovirus, *Leptospira interrogans*, and *Toxoplasma gondii* in free-ranging canids in Scandinavia and Svalbard. *Journal of Wildlife Diseases* 46, 474–480.

Appel, M.J. (1987) Canine distemper virus. In: *Virus Infections of Carnivores* (ed. M.J. Appel), pp. 133–159. Elsevier, Amsterdam.

Cabria, M.T., Michaux, J.R., Gomez-Moliner, B.J., Skumatov, D., Maran, T., Fournier, P., Lopez De Luzuriaga, J. & Zardoya, R. (2011) Bayesian analysis of hybridization and introgression between the endangered European mink (*Mustela lutreola*) and the polecat (*Mustela putorius*). *Molecular Ecology* 20, 1176–1190.

Candela, M., Pardavilla, X., Ortega, N., Lamosa, A., Mangas, J.G. & Martínez-Carrasco, C. (2019) Canine distemper virus may affect European wild cat populations in Central Spain. *Mammalian Biology* 97, 9–12.

Chen, C.C., Pei, K.J.C., Liao, M.H. & Mortenson, J.A. (2008) Canine distemper virus in wild ferret-badgers of Taiwan. *Journal of Wildlife Diseases* 44, 440–445.

Cunningham, M.W., Shindle, D.B., Allison, A.B., Terrell, S.P., Mead, D.G. & Owen, M. (2009) Canine distemper epizootic in Everglades mink. *Journal of Wildlife Diseases* 45, 1150–1157.

Damien, B.C., Martina, B.E., Losch, S., Mossong, J., Osterhaus, A.D. & Muller, C.P. (2002) Prevalence of antibodies against canine distemper virus among red foxes in Luxembourg. *Journal of Wildlife Diseases* 38, 856–859.

Deem, S.L., Spelman, L.H., Yates, R.A. & Montali, R.J. (2000) Canine distemper in terrestrial carnivores: a review. *Journal of Zoo and Wildlife Medicine* 31, 441–451.

Ek-Kommonen, C., Rudback, E., Anttila, M., Aho, M. & Huovilainen, A. (2003) Canine distemper of vaccine origin in European mink, *Mustela lutreola* – a case report. *Veterinary Microbiology* 92, 289–293.

Ferreyra, H., Calderon, M.G., Marticorena, D., Marull, C. & Leonardo, B.C. (2009) Canine distemper infection in crab-eating fox (*Cerdocyon thous*) from Argentina. *Journal of Wildlife Diseases* 45, 1158–1162.

Fournier, P. & Maizeret, C. (2003) Status and conservation of the European Mink (*Mustela lutreola*) in France. In: *Proceedings of the International Conference on the Conservation of the European Mink,* pp. 95–102. Gobierno de la Rioja, Logroño.

Fournier, P., Maizeret, C., Jimenez, D., Chusseau, J.-P., Aulagnier, S. & Spitz, F. (2007) Habitat utilization by sympatric European mink *Mustela lutreola* and polecats *Mustela putorius* in south-western France. *Acta Theriologica* 52, 1–12.

Fournier, P., Maizeret, C., Fournier-Chambrillon, C., Ilbert, N., Aulagnier, S. & Spitz, F. (2008) Spatial behaviour of European mink *Mustela lutreola* and polecat *Mustela putorius* in south-western France. *Acta Theriologica* 53, 343–354.

Fournier-Chambrillon, C., Chusseau, J.P., Dupuch, J., Maizeret, C. & Fournier, P. (2003) Immobilization of free-ranging European mink (*Mustela lutreola*) an polecat (*Mustela putorius*) with medetomidine–ketamine and reversal by atipamezole. *Journal of Wildlife Diseases* 39, 393–399.

Fournier-Chambrillon, C., Aasted, B., Perrot, A., Pontier, D., Sauvage, F., Artois, M., Cassiède, J.-M., Chauby, X., Dal Molin, A., Simon, C. & Fournier, P. (2004a) Antibodies to Aleutian mink disease parvovirus in free-ranging European mink (*Mustela lutreola*) and other small carnivores from southwestern France. *Journal of Wildlife Diseases* 40, 394–402.

Fournier-Chambrillon, C., Berny, P.J., Coiffier, O., Barbedienne, P., Dasse, B., Delas, G., Galineau, H., Mazet, A., Pouzenc, P., Rosoux, R. & Fournier, P. (2004b) Evidence of secondary poisoning of free-ranging riparian mustelids by anticoagulant rodenticides in France: implications for conservation of European mink (*Mustela lutreola*). *Journal of Wildlife Diseases* 40, 688–695.

Fournier-Chambrillon, C., Bifolchi, A., Mazzola-Rossi, E., Sourice, S., Albaret, M., Bray, Y., Ceña, J.-C., Urra Maya, F., Agraffel, T. & Fournier, P. (2010) Reliability of stained placental scar counts in farmed American mink and application to free-ranging Mustelids. *Journal of Mammalogy* 91, 818–826.

Frolich, K., Czupalla, O., Haas, L., Hentschke, J., Dedek, J. & Fickel, J. (2000) Epizootiological investigations of canine distemper virus in free-ranging carnivores from Germany. *Veterinary Microbiology* 74, 283–292.

Gowtage-Sequeira, S., Banyard, A.C., Barrett, T., Buczkowski, H., Funk, S.M. & Cleaveland, S. (2009) Epidemiology, pathology, and genetic analysis of a canine distemper epidemic in Namibia. *Journal of Wildlife Diseases* 45, 1008–1020.

Griot, C., Vandevelde, M., Schoebesberger, M. & Zurbriggen, A. (2003) Canine distemper, a re-emerging morbillivirus with complex neuropathogenic mechanisms. *Animal Health Research Reviews* 4, 1–10.

Keller, S.M., Gabriel, M., Terio, K.A., Dubovi, E.J., VanWormer, E., Sweitzer, R., Barret, R., Thompson, C., Purcell, K. & Munson, L. (2012) Canine distemper in an isolated population of fishers (*Martes pennanti*) from California. *Journal of Wildlife Diseases* 48, 1035–1041.

Maran, T. (2007) *Conservation Biology of the European Mink,* Mustela lutreola *(Linnaeus, 1761): Decline and Causes of Extinction.* PhD thesis, Tallinn University.

Maran, T., Skumatov, D., Gomez, A., Põdra, M., Abramov, A.V. & Dinets, V. (2016) *Mustela lutreola.* The IUCN Red List of Threatened Species 2016, e.T14018A45199861. http://dx.doi.org/10.2305/IUCN. UK.2016-1.RLTS.T14018A45199861.en. Accessed on 4 January 2017.

Marinov, E.M., Kiss, J.B., Alexe, V., Doroftei, M., Doroşencu, A., Covaliov, S., Nichifor, C., Condac, M., Lupu, G., Gal, A., Iosif, N., Băcescu, G., Bucur, G., Cîrpăveche, P. & Timofei, A. (2011) *European Mink Handbook (*Mustela lutreola*) within the Danube Delta Biosphere Reserve – Romania.* Danube Delta National Institute for Research and Development, Tulcea.

Martella, V., Bianchi, A., Bertoletti, I., Pedrotti, L., Gugiatti, A., Catella, A., Cordioli, P., Lucente, M.S., Elia, G. & Buonavoglia, C. (2010) Canine distemper epizootic among red foxes, Italy, 2009. *Emerging Infectious Diseases* 16, 2007–2009.

Meli, M.L., Simmler, P., Cattori, V., Martinez, F., Vargas, A., Palomares, F., López-Bao, J.V., Simon, M.A., Lopez, G., Leon-Vizcaino, L., Hofmann-Lehmann, R. & Lutz, H. (2010) Importance of canine distemper virus (CDV) infection in free-ranging Iberian lynxes (*Lynx pardinus*). *Veterinary Microbiology* 146, 132–137.

Michaux, J.R., Hardy, O.J., Justy, F., Fournier, P., Kranz, A., Cabria, M., Davison, A., Rosoux, R. & Libois, R. (2005) Conservation genetics and population history of the threatened European mink *Mustela lutreola*, with an emphasis on the west European population. *Molecular Ecology* 14, 2373–2388.

Moinet, M., Fournier-Chambrillon, C., André-Fontaine, G., Aulagnier, S., Mesplède, A., Blanchard, B., Descarsin, V., Dumas, P., Dumas, Y., Coïc, C., Couzi, L. & Fournier, P. (2010) Leptospirosis in free-ranging endangered European mink (*Mustela lutreola*) and other small carnivores (Mustelidae, Viverridae) from southwestern France. *Journal of Wildlife Diseases* 4, 1141–1151.

Muller, A., Silva, E., Santos, N. & Thompson, G. (2011) Domestic dog origin of canine distemper virus in free-ranging wolves in Portugal as revealed by hemagglutinin gene characterization. *Journal of Wildlife Diseases* 47, 725–729.

Nouvellet, P., Donnelly, C.A., De Nardi, M., Rhodes, C.J., De Benedictis, P., Citterio, C., Obber, F., Lorenzetto, M., Dalla Pozza, M., Cauchemez, S. & Cattoli, G. (2013) Rabies and canine distemper virus epidemics in the red fox population of Northern Italy (2006–2010). *PLoS One* 8, e61588.

O'Brien, S.J. & Evermann, J.F. (1988) Interactive influence of infectious disease and genetic diversity in natural populations. *Trends in Ecology & Evolution* 3, 254–259.

Origgi, F.C., Plattet, P., Sattler, U., Robert, N., Casaubon, J., Mavrot, F., Pewsner, M., Wu, N., Giovannini, S., Oevermann, A., Stoffel, M.H., Gaschen, V., Segner, H. & Ryser-Degiorgis, M.P. (2012) Emergence of canine distemper virus strains with modified molecular signature and enhanced neuronal tropism leading to high mortality in wild carnivores. *Veterinary Pathology* 49, 913–929.

Pagh, S., Chriél, M., Madsen, A.B., Jensen, T-L.W., Elmeros, M., Asferg, T. & Hansen, M.S. (2018) Increased reproductive output of Danish red fox females following an outbreak of canine distemper. *Canid Biology & Conservation* 21(3), 12–20.

Palomares, F., López-Bao, J.V., Telletxea, G., Ceña, J.C., Fournier, P., Giralda, G. & Urra, F. (2017) Activity and home range in a recently widespread European mink population in Western Europe. *European Journal of Wildlife Research* 63, 78.

Pearson, R.C. & Gorham, J.R. (1987) Canine distemper virus. In: *Virus Infections of Carnivores* (ed. M.J. Appel), pp. 371–378. Elsevier, Amsterdam.

Philippa, J.D. (2007) *Vaccination of Non-domestic Animals Against Emerging Virus Infections.* PhD thesis, Erasmus University of Rotterdam.

Philippa, J.D. (2010) Vaccination of non-domestic carnivores: a review. In: *Transmissible Diseases Handbook.* 4th edition (eds J. Kaandorp, N. Chai & A. Bayens). European Association of Zoo and Wildlife Veterinarians. http://www.eaza.net/activities/ TDH/12%20Vaccination%20of%20Non-domestic%20 Carnivores.pdf. Accessed on 15 January 2014.

Philippa, J.D., Fournier-Chambrillon, C., Fournier, P., Schaftenaar, W., Van De Bildt, M.W., Van Herweijnen, R., Kuiken, T., Liabeuf, M., Ditcharry, S., Joubert, L., Bégnier, M. & Osterhaus, A. (2008) Serologic survey for selected viral pathogens in endangered European mink (*Mustela lutreola*) and other free-ranging mustelids in south-western France. *Journal of Wildlife Diseases* 44, 791–801.

R Development Core Team (2012) *R: A Language and Environment for Statistical Computing*. The R Foundation for Statistical Computing, Vienna.

Rijks, J.M. (2008) *Phocine Distemper Revisited. Multidisciplinary Analysis of the 2002 Phocine Distemper Virus Epidemic in the Netherlands*. PhD thesis, Erasmus University of Rotterdam.

Sánchez-Migallón, D., Carvajal, A., García-Marín, J.F., Ferreras, M.C., Pérez, V., Mitchell, M.A., Urra, F. & Ceña, J.-C. (2008) Aleutian disease serology, protein electrophoresis and pathology of the European mink (*Mustela lutreola*) from Navarra, Spain. *Journal of Zoo and Wildlife Medicine* 39, 305–313.

Santos, N., Almendra, C. & Tavares, L. (2009) Serologic survey for canine distemper virus and canine parvovirus in free-ranging wild carnivores from Portugal. *Journal of Wildlife Diseases* 45, 221–226.

Sas Institute Inc. (2013) *SAS 9.4 Guide to Software Updates*. SAS Institute Inc., Cary.

Scherrer, B. (1984) *Biostatistique*. Gaëtan Morin, Montréal. (In French).

Sobrino, R., Arnal, M.C., Luco, D.F. & Gortazar, C. (2008) Prevalence of antibodies against canine distemper virus and canine parvovirus among foxes and wolves from Spain. *Veterinary Microbiology* 126, 251–256.

Sokal, R.R. & Rohlf, F.J. (1995) *Biometry: The Principles and Practice of Statistics in Biological Research*. W.H. Freeman and Company, New York.

Sutherland-Smith, M.R., Rideout, B.A., Mikolon, A.B., Appel, M.J., Morris, P.J., Shima, A.L. & Janssen, D.J. (1997) Vaccine-induced canine distemper in European mink, *Mustela lutreola*. *Journal of Zoo and Wildlife Medicine* 28, 312–318.

Tavernier, P., Baert, K., van de Bildt, M.W.G, Kuiken, T., Cay, A.B., Maes, S., Roels, S., Gouwy, J. & van den Berge, K. (2012) A distemper outbreak in beech martens (*Martes foina*) in Flanders. *Vlaams Diergeneeskundig Tijdschrift* 81, 81–87.

Timm, S.F., Munson, L., Summers, B.A., Terio, K.A., Dubovi, E.J., Rupprecht, C.E., Kapil, S. & Garcelon, D.K. (2009) A suspected canine distemper epidemic as the cause of a catastrophic decline in Santa Catalina Island foxes (*Urocyon littoralis catalinae*). *Journal of Wildlife Diseases* 45, 333–343.

Torres, J., Miquel, J., Fournier, P., Fournier-Chambrillon, C., Liberge, M., Fons, R. & Feliu, C. (2008) Helminth communities of the autochthonous mustelids *Mustela lutreola* and *M. putorius* and the introduced *Mustela vison* in south-western France. *Journal of Helminthology* 82, 349–356.

Visser, I.K., Kumarev, V.P., Orvell, C., de Vries, P., Broeders, H.W., van de Bildt, M.W., Groen, J., Teppema, J.S., Burger, M.C., UytdeHaag, F.G. & Osterhaus, A.D.M.E. (1990) Comparison of two morbilliviruses isolated from seals during outbreaks of distemper in north west Europe and Siberia. *Archives of Virology* 111, 149–164.

Williams, E.S. (2001) Canine distemper. In: *Infectious Diseases of Wild Mammals* (eds E.S. Williams & I.K. Barker), pp. 50–59. Manson Publishing, London.

Zabala, J. & Zuberogoitia, I. (2003) Habitat use of male European mink (*Mustela lutreola*) during the activity period in south western Europe. *Zeitschrift für Jagdwissenschaft* 49, 77–81.

Zabala, J., Zuberogoitia, I. & Martinez-Climent, J.A. (2006) Factors affecting occupancy by the European mink in south-western Europe. *Mammalia* 70, 193–201.

12

Density of African Civets in a Moist Mountain Bushveld Region of South Africa

Lisa Isaacs[1], Michael J. Somers[2,3,*], and Lourens H. Swanepoel[1,4]

[1]*Centre for Wildlife Management, University of Pretoria, Pretoria, South Africa*
[2]*Eugène Marais Chair of Wildlife Management, Department of Zoology and Entomology, Mammal Research Institute, University of Pretoria, Pretoria, South Africa*
[3]*Centre for Invasion Biology, University of Pretoria, Pretoria, South Africa*
[4]*Department of Zoology, University of Venda, Thohoyandou, South Africa*

Summary

The African civet, *Civettictis civetta*, is the largest member of the Viverridae family and one of the most widely distributed mesocarnivores in Africa. Despite its wide geographic distribution, little is known about its ecology, behaviour, and conservation biology, such as abundance and density. Mesocarnivores can play important roles in ecosystem functioning and these roles may become more important, especially in areas where large carnivores are actively removed (e.g. mesocarnivore release hypothesis). In this study, we use data from a camera-trapping survey originally designed to monitor leopards, *Panthera pardus*, to report on the density of African civets across different land-use types – two conservation areas (Lapalala, Welgevonden) and one mosaic 'Farming area' consisting of hunting, ecotourism, and livestock farms – in the moist mountain bushveld region of the Waterberg Biosphere Reserve, South Africa. We fitted spatially explicit capture–recapture (secr) models, with parameter sharing, across the different sites to improve estimates. We found that the study site (and hence land use type) had a significant effect on African civet density, detection probability, and the movement parameter. Density estimates were the highest for Lapalala (8.63 ± 2.30 individuals/100 km²), followed by the Farming area (4.88 ± 1.05 individuals/100 km²) while the lowest density was detected on Welgevonden (4.43 ± 1.13 individuals/100 km²). Our results suggest that there are healthy African civet populations within the Waterberg Biosphere Reserve, but that land use might play an important role in African civet population demographics. We hypothesize that differences in African civet density might be a result of factors such as top–down regulation from large carnivores, recreational hunting, poisoning, resource provisioning, and human activity.

Keywords

African civet–camera — trapping–density–spatially explicit capture — recapture models

Introduction

The African civet, *Civettictis civetta*, is the largest member of the Viverridae family and the sole member of its genus. Females (average body mass of 11.58 kg) are generally slightly heavier than males (10.92 kg; Skinner & Chimimba, 2005). African civets are one of the most widely distributed carnivores in Africa (Ray, 2013; Do Linh San *et al.*, 2019), and in South Africa, they are found throughout the northern sections of the country (Swanepoel *et al.*, 2016). They are opportunistic omnivores with their diet including a wide variety of insects,

---

*Corresponding author.

Small Carnivores: Evolution, Ecology, Behaviour, and Conservation, First Edition. Edited by Emmanuel Do Linh San, Jun J. Sato, Jerrold L. Belant, and Michael J. Somers.

wild fruits, rodents, and carrion when available (Ray, 1995; Skinner & Chimimba, 2005; Bekele Tsegaye *et al.*, 2008; Amiard, 2014; Tadesse Habtamu *et al.*, 2017). African civets are solitary, nocturnal, and terrestrial, but are commonly found near permanent water sources (Skinner & Chimimba, 2005). They are well adapted to live among human settlements (Mateos Ersado *et al.*, 2015), occur in croplands (Williams *et al.*, 2018) and are equally found inside and outside protected areas (Ray, 1995). Despite their wide geographic distribution, little is known about their ecology, behaviour, and conservation biology. For example, home range data on African civets are limited to two studies in Ethiopia and were only based on one adult and two sub-adult individuals (Ermias Admasu *et al.*, 2004; Ayalew Berhanu *et al.*, 2013). Data on African civets, therefore, range from anecdotal reports to general studies focusing on anatomy, physiology, and general ecology (Ray, 1995; Ermias Admasu *et al.*, 2004; De Luca & Mpunga, 2005; Skinner & Chimimba, 2005; Martinoli *et al.*, 2006; Bekele Tsegaye *et al.*, 2008; Tadesse Habtamu *et al.*, 2017). Little is known about key aspects of African civet conservation biology such as abundance and density, which is needed for conservation planning.

The lack of research on African civets seems related to their solitary nocturnal nature and their low conservation status. The species is listed as Least Concern by *The IUCN Red List of Threatened Species* (Do Linh San *et al.*, 2019) and *The Red List of Mammals of South Africa, Swaziland, and Lesotho* (Swanepoel *et al.*, 2016). Therefore, it seems to have a low research priority as it actually may benefit from anthropogenic habitat modification, favouring agricultural lands and degraded forests (Ray *et al.*, 2005; Williams *et al.*, 2018). However, African civet populations may be threatened by human–wildlife conflict and commercial exploitation, since it is regularly sold as bushmeat in western and central African markets (Angelici *et al.*, 1999; Bahaa-el-din *et al.*, 2013; Ray, 2013), and kept in captivity for the extraction of African civet musk for the perfume industry (Pugh, 1998; Yilma Delelegn, 2003; Ray *et al.*, 2005). Furthermore, understanding mesocarnivore (e.g. African civet) population dynamics in mixed-use landscapes is becoming more important since large carnivores are increasingly removed from these areas which can lead to increased mesocarnivore densities (e.g. Mesopredator

Release Hypothesis; Prugh *et al.*, 2009). As such, mesocarnivores may increasingly play important roles in ecosystem function where large carnivores have been reduced in density (Miller *et al.*, 2001; Gehrt & Prange, 2006; Prugh *et al.*, 2009; Roemer *et al.*, 2009).

Since the pioneering work of Karanth (1995), camera-trapping has become an established and preferred method in monitoring elusive carnivore species (Tobler *et al.*, 2008). In comparison with large-spotted African predators (e.g. leopard, *Panthera pardus*, spotted hyena, *Crocuta crocuta*, cheetah, *Acinonyx jubatus*), mesocarnivores have only been the focus of a limited number of camera-trapping studies (Braczkowski *et al.*, 2012; Williams *et al.*, 2018). However, data from camera-trapping studies for large carnivores might offer an opportunity to study lesser-known mesocarnivores. Previous studies have shown that valuable data could be collected on one species while monitoring another simultaneously. For example, Maffei *et al.* (2005) collected data on ocelots, *Leopardus pardalis*, even though the initial focus of their study was jaguars, *Panthera onca*. Similarly, Bista *et al.* (2012) studied the large Indian civet, *Viverra zibetha*, using surveys primarily directed at tigers, *Panthera tigris*. Therefore, existing camera-trapping databases in Africa might be a rich source of data to study widely distributed mesocarnivores like the African civet.

We used data from a camera-trapping survey originally designed to monitor leopards to estimate the density of African civets across different land-use types in the Waterberg Biosphere Reserve, South Africa. We extended maximum likelihood-based spatially explicit capture–recapture models (secr) to allow for parameter sharing across the different study sites to improve parameter estimates. We used this approach to investigate the effect of the different study sites on African civet density, detection and movement parameters.

Study Site

We conducted the study on three different study sites within the Waterberg Biosphere Reserve of Limpopo province, northeastern South Africa (Figure 12.1). Camera-trapping took place between May and October 2009, which constitutes the dry season for the study

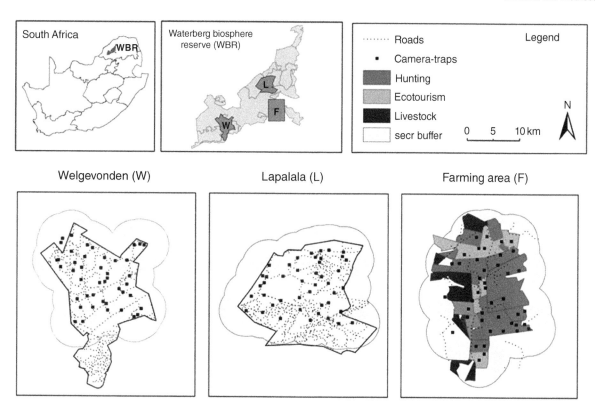

Figure 12.1 Location of Welgevonden Private Game Reserve, Lapalala Wilderness and a mosaic of hunting, ecotourism, and livestock farms ('Farming area'), northeastern South Africa and camera-trap grids used for density estimation of African civet, *Civettictis civetta*, from May to October 2009.

area. Lapalala Wilderness (hereafter Lapalala; 23°44′–23°57′ S, 28°09′–28°25′ E; surveyed from 1 June to 24 July 2009) is a privately owned reserve of 360 km², committed to conservation and environmental education. Welgevonden Private Game Reserve (hereafter 'Welgevonden'; 24°10′–24°25′ S, 27°45′–27°56′ E; surveyed from 13 May to 21 July 2009) is a 375 km² syndicated game reserve which focuses on photographic safaris, conservation, environmental education, and ecotourism. The last site of approximately 350 km² is a game and livestock farming area (hereafter 'Farming area'; 28°16′–28°30′ E, 23°58′–24°16′ S; surveyed from 4 August to 7 October 2009). The Farming area consists of seven farms dedicated to ecotourism, eight hunting game farms, and 11 livestock farms. Welgevonden had the full complement of large carnivores, while Lapalala and the Farming area had no lion, *Panthera leo*, or cheetah (Ramnanan *et al.*, 2013). Furthermore, a variety of carnivores were present at all three sites (Table 12.1). Also, all sites supported a diversity of ungulate species, including greater kudu, *Tragelaphus strepsiceros*, impala, *Aepyceros melampus*, blue wildebeest, *Connochaetes taurinus*, and Burchell's zebra, *Equus burchelli* (Isaacs *et al.*, 2013). All study sites had topographies ranging from undulating rocky hills to elevated plateaus. Rainfall is restricted to the summer season (October to March; average annual rainfall of 650–900 mm) while vegetation is classified as Waterberg moist mountain bushveld (Rutherford *et al.*, 2006).

Material and Methods

Our camera-trapping survey was initially set up to estimate leopard density (Swanepoel, 2013; Swanepoel *et al.*, 2015). We followed survey design guidelines for closed-population capture–recapture studies on large

Table 12.1 Carnivore species photographed (number of photos/100 trap days) during a camera-trapping survey in the Waterberg Biosphere Reserve, South Africa, based on an effort of 1374 trap days for Lapalala, 1606 for Welgevonden, and 840 for the Farming area. Table adapted from Ramnanan *et al.* (2013).

Species	Lapalala (2008–2010)	Welgevonden (2008–2010)	Farming area (2009)
Large carnivores	—	—	—
Wild dog, *Lycaon pictus*	0.73	0.00	0.00
Lion, *Panthera leo*	0.00	2.30	0.00
Leopard, *Panthera pardus*	3.00	3.54	2.95
Cheetah, *Acinonyx jubatus*	0.00	0.17	0.00
Spotted hyena, *Crocuta crocuta*	0.09	1.25	0.00
Brown hyena, *Hyaena brunnea*	12.63	20.83	4.26
Meso- and small carnivores	—	—	—
Blacked-back jackal, *Canis [= Lupulella] mesomelas*	2.93	7.02	4.02
Bat-eared fox, *Otocyon megalotis*	0.00	0.26	0.00
Serval, *Leptailurus serval*	0.00	1.61	0.23
Caracal, *Caracal caracal*	1.30	2.55	1.61
African wild cat, *Felis lybica cafra*	0.59	0.63	0.48
White-tailed mongoose, *Ichneumia albicauda*	0.21	0.38	0.36
Water mongoose, *Atilax paludinosus*	0.16	0.00	1.19
Banded mongoose, *Mungos mungo*	1.45	0.08	0.57
Slender mongoose, *Galerella sanguinea*	0.64	0.84	2.49
Aardwolf, *Proteles cristatus*	0.12	0.00	0.00
Honey badger, *Mellivora capensis*	0.90	2.42	4.06
Striped polecat, *Ictonyx striatus*	0.00	0.00	0.12
African civet, *Civettictis civetta*	14.95	17.83	18.81
Rusty-spotted genet, *Genetta maculata*	3.84	1.05	2.32
Small-spotted genet, *Genetta genetta*	0.36	1.08	0.00

carnivores (Karanth & Nichols, 1998). Camera-traps were set out in pairs in a grid pattern, where the size of each grid (6.25 km^2) represented approximately half the home range of the smallest home range recorded for leopards in mountainous terrain (10 km^2; Smith, 1978). This resulted in an average spacing of 1.6 km between camera-traps, which ensured that all the study sites were adequately surveyed and that all the African civets within the survey grid had a non-zero capture probability. Since the only home range data available for African civet (11.1 km^2; Ermias Admasu *et al.*, 2004) at the time our study took place was similar to the leopard home range size (10 km^2) we used to set our grid sizes, we assumed our survey design was adequate to monitor African civets. Due to a shortage of cameras, we divided each study site into three equally sized trapping blocks, with 15 camera stations in each block. Each block was surveyed for 18–22 consecutive days, before moving the camera-traps to the next block with 15 camera stations (Karanth, 1995). We considered each 24 h period as a sampling occasion (Otis *et al.*, 1978). We used infrared digital camera-traps (Moultrie I40, Moultrie Feeders Inc.) at Welgevonden, and a combination of Moultrie and flash film camera-traps (DeerCam DC100, Non Typical Inc.; Stealth Cam MC2-GV, Stealth Cam LLC; Trailmaster® TM1550, Goodson & Associates Inc.) at Lapalala and the Farming area. We programmed units to run for 24 h, with a 1 min delay between pictures for digital cameras and a 10 min delay between pictures for film cameras.

Camera-traps were positioned at either side of animal paths or vehicle roads, at a height of 45 cm, and at a 45° angle down the road. Due to a slow trigger speed for digital camera traps, we baited every camera station with rotten eggs and fermented fish. Camera stations were checked every 4–5 days to replace film, monitor battery power, and memory card space, and to replace lures.

We identified individual African civets based on distinct natural markings. The flanks of an animal are usually the most reliable body parts for successful individual identification. We pooled capture data from all three survey blocks such that day 1 of each trapping block was considered as day 1 of the survey (e.g. first sampling occasion; Nichols & Karanth, 2002). We developed capture histories for individual African civets in a standard X-matrix format.

We estimated African civet densities using full maximum likelihood spatially explicit capture–recapture models (secr; Efford, 2014) in R, version 3.0.2 (R Development Core Team, 2013). Secr combines a state and an observation model; the state model estimates the distribution of animal home ranges in the area, and the observation model relates the probability of detecting an animal at a particular camera station to the distance of that station from the centre of the animal's home range (Borchers & Efford, 2008). The detection function in secr has two parameters: (i) g0: capture probability at an animal's home range centre; and (ii) sigma (σ): function of the scale of animal movements (Gray & Prum, 2011). Since our study included three different sites, we fitted a multi-study closed spatial model which allowed sharing of the detection (g0) and scale parameter (σ) across the study sites to improve estimates (Tobler *et al.*, 2014). We were unable to identify the age or sex of African civets from photographs which prevented us from investigating the effects of these variables on estimated parameters. We, therefore, fitted a half-normal detection function by maximizing the full likelihood where we fitted models constraining density (D), σ, and g0 to vary only by study site. We did model selection based on Akaike's Information Criteria scores corrected for small sample sizes (AIC_c) and regarded the model with the lowest AIC_c score as the most parsimonious (Borchers & Efford, 2008). We further assumed that models within 2 AIC_c units of the most parsimonious model had equal support (ΔAIC_c; Burnham & Anderson, 2002). We used model-averaged estimates if there was model uncertainty (Boulanger *et al.*, 2004). We assumed a Poisson distribution of home range centres and set a 6000 m buffer width around the trapping grid.

Results

From May to October 2009, we obtained a capture rate (expressed as the number of African civet photos/100 trap nights) of 17.15 at Lapalala, 13.35 at Welgevonden, and 10.61 at the Farming area (Table 12.2), while

Table 12.2 Camera-trapping periods and effort for African civets, *Civettictis civetta*, in three study sites within a moist mountain bushveld region, South Africa, 2009.

Study area	Sampling period	Camera polygon[a]	No. of stations	Trap nights	Photos	Captures	Capture rate[b]	Individuals
Lapalala	1 June – 24 July 2009	232.87	44	723	311	124	17.15	28
Welgevonden	13 May – 21 July 2009	250.13	45	1019	303	136	13.35	18
Farming area	4 August – 7 October 2009	264.49	43	850	201	90	10.50	23
Total		747.49	132	2592	815	350	13.50	69

[a]Camera polygon was estimated as the minimum convex polygon (km^2) around camera stations
[b]Capture rate expressed as the number of African civet photos per 100 trap nights

Table 12.3 Secr parameters in the three study sites, South Africa. Models ranked based on Akaike's Information Criteria corrected for small sample size (AIC$_c$).

Model	Detectfn	K	Loglink	AIC$_c$	ΔAIC$_c$	w
D~reserve g0~reserve σ~reserve	Halfnormal	9	−1370.54	2762.13	0.00	0.81
D~1 g0~reserve σ~reserve	Halfnormal	7	−1374.71	2765.25	3.12	0.17
D~reserve g0~1 σ~reserve	Halfnormal	7	−1377.02	2768.04	7.74	0.02

D = density; g0 = detection probability; σ = sigma (movement parameter); K = number of parameters; ΔAIC$_c$ = difference between each model and model with lowest AICc; w = model weight.

Table 12.4 Model averaged parameter estimates by a spatially explicit capture–recapture model for African civets in three study sites within a moist mountain bushveld region of South Africa.

Study area	Sampling period	n	σ (m)	g0	Density	Home range (km^2)^a
Lapalala	1 June – 24 July 2009	28	1006.86 (± 56.60)	0.30 (± 0.04)	8.63 (± 2.30)	55.36
Welgevonden	13 May – 21 July 2009	18	1525.61 (± 86.90)	0.21 (± 0.03)	4.43 (± 1.13)	83.89
Farming area	4 August – 7 October 2009	23	1902.12 (± 153.45)	0.12 (± 0.03)	4.88 (± 1.05)	104.50

n = number of individual African civets; σ = sigma (movement parameter); g0 = detection probability.
^{a}Approximation of African civet home range size based on the 95% probability interval of the circular bivariate normal distribution with a radius of 2.45 × sigma (Tobler *et al.*, 2013)

rejecting unclear photos (14 encounters at Lapalala, 15 at Welgevonden, and 21 at the Farming area). Multimodel inference showed overwhelming support that sigma (movement parameter) varied between the study sites (Table 12.3). Based on model-averaged estimates, the Farming area had the highest sigma (1902 ± 153 m), followed by Welgevonden (1525 ± 87 m) and Lapalala (1007 ± 57 m; Table 12.4). These estimated sigma values suggest that African civet home range sizes were the largest in the Farming area (105 km^2), followed by Welgevonden (84 km^2), and Lapalala (55 km^2; Table 12.4). Similarly, we found support that African civet density and g0 varied between the different study sites (Table 12.3), where Lapalala had the highest African civet density (8.63 ± 2.30 individuals/100 km^2), followed by the Farming area (4.88 ± 1.05 individuals/100 km^2) and Welgevonden (4.44 ± 1.13 individuals/100 km^2; Table 12.4). African civets at Lapalala had a high detection probability (0.30 ± 0.04), while detections at Welgevonden (0.21 ± 0.03) were lower and the Farming area had the lowest detection (0.12 ± 0.03; Table 12.4).

Discussion

We demonstrated the feasibility of using an existing camera-trapping database, initially set up to monitor leopard densities, to estimate the density of the African civet. However, this success hinged on the fact that we employed relatively small camera-trapping grids and that African civet home ranges were similar to the leopard home range we used to set up the survey grids. It is unlikely that our camera-trapping data will be suitable for smaller mesocarnivores like the rusty-spotted genet, *Genetta maculata*, which has a much smaller home range size (0.06 km^2; Ikeda *et al.*, 1982 [species referred to as large-spotted genet, *G. tigrina*, in their paper]). The survey grid needs to be sufficiently large compared to the animal's home range so that all individuals have non-zero detection probabilities (Maffei & Noss, 2008). Therefore, larger grids set up to accommodate large carnivores will increase the probability of holes in the survey grid, which will violate some assumptions of closed-population capture–recapture (Foster & Harmsen, 2012).

Our study is the first to provide robust estimates of African civet densities within the moist mountain bushveld region of South Africa. While Lapalala might have a healthy population of African civets, the densities for the Farming area and Welgevonden were relatively low, given the size and generalist behaviour of African civets. These estimates are also lower than that reported for African civets in other savannah areas of South Africa (e.g. mean = 10 individuals/100 km^2 for dry savannah; Amiard, 2014), and are similar to large carnivore estimates for the same area (e.g. mean leopard density = 5.5 individuals/100 km^2; Swanepoel *et al.*, 2015).

We found that the study site had a significant effect on density, sigma, and detection probability. Since two of the study sites were private reserves, and the Farming area was dominated by game farms, we suggest that food resources and availability were similar at all three study sites, particularly because African civets are omnivorous and opportunistic scavengers (Ray, 1995; Skinner & Chimimba, 2005; Tadesse Habtamu *et al.*, 2017). As such, we hypothesize that differences in estimated parameters might result from other – not necessarily mutually exclusive – factors such as top–down regulation from large carnivores, recreational hunting, poisoning, bush encroachment, food provisioning, or human activity.

For example, large carnivores can both directly (through predation) and indirectly (by altering prey behaviour) impact the community structure of mesocarnivores (Miller *et al.*, 2001; Davis *et al.*, 2019; Prugh & Sivy, 2020). The site with the full complement of large carnivores, Welgevonden, had the lowest African civet density. We suggest that top–down regulation by large carnivores (e.g. Mesopredator Release Hypothesis; Prugh *et al.*, 2009) might play an important role at Welgevonden by keeping mesocarnivore densities lower than at the other sites. In addition, competition from other scavengers (e.g. spotted and brown hyenas, *Hyaena brunnea*, both present at Welgevonden) might contribute to the low African civet densities (e.g. interspecific competition; Yarnell *et al.*, 2013). On the other hand, large carnivore richness was similar for Lapalala and the Farming area, yet Lapalala had a higher African civet density. Because leopard densities (most abundant large carnivore in the study area; Swanepoel, 2013) were similar across all study sites, we believe that

other factors than leopard predation affect African civet density. First, bush encroachment has been found to play an important role on small carnivore diversity and abundance, where encroached areas have lower abundance and diversity (Blaum *et al.*, 2007). In the Waterberg moist bushveld, bush encroachment is among the important drivers of landscape degradation of farming areas, which might also play a role in reducing mesocarnivore abundance and diversity (Henning, 2002; Blaum *et al.*, 2007). Secondly, the low African civet density in the Farming area might be related to trophy hunting and accidental killing of mesocarnivores during problem animal control. For example, mesocarnivore abundance and diversity can be greatly affected by persecution levels (Proulx, 2021), and the Waterberg Biosphere Reserve experiences among the highest carnivore conflict rates in South Africa which can directly affect mesocarnivore densities (Swanepoel *et al.*, 2014).

The estimated home ranges based on the movement parameters are among the first estimates of African civet home ranges in South Africa. Our estimated home ranges were larger than those recorded in two previous radio-tracking studies conducted in Ethiopia (0.71–11 km^2; Ermias Admasu *et al.*, 2004; Ayalew Berhanu *et al.*, 2013). However, both studies took place in the vicinity of human settlements, which probably supplied civets with rich anthropogenic food resources and could have resulted in such small observed home ranges (Quinn & Whisson, 2005; Ayalew Berhanu *et al.*, 2013). Furthermore, Ermias Admasu *et al.*'s (2004) study was based on only one subadult African civet male for a limited time period and so the home range size could have been severely underestimated. Our estimates were also larger than that of a large Indian civet home range in Thailand (12 km^2; Rabinowitz, 1991). While we have to acknowledge that our home range estimates are only an approximation, we believe that the large values we obtained are probably a result of the same factors affecting density (e.g. top–down regulation from large carnivores, recreational hunting, poisoning, bush encroachment, food provisioning, or human activity).

Lastly, we have to consider some limitations to our modelling approaches. For example, we could have captured more males than females at our camera stations since overall male carnivores tend to have larger home ranges (Ray, 1995). The sex of an animal can affect its home range size and movement, which will affect density estimates in secr (Gray & Prum, 2011). We did not, however, include covariates such as sex in our models because it was impossible to determine the sex of animals based on photos. We also did not include age in our models for similar reasons. Because we only had access to one reference of African civet home range size, based on a single subadult male (Ermias Admasu *et al.*, 2004), at the time our project started, it is possible that our study design was inadequate and biased our results. Nonetheless, we have presented the first density estimates and other ecological parameters for the African civet in South Africa. Our results illustrated apparently viable African civet populations in mixed-use landscapes in the Waterberg Biosphere Reserve. However, land use played an important role in African civet population parameters and we highlight that densities of even an opportunistic omnivore can be quite low on non-protected farmlands. We suggest that the differences in density estimates between the three sites result from differences in large carnivore assemblages and varying land uses practised in the areas.

Acknowledgements

Funding for the project was received from the IFS (D/4984-1), Wild Foundation (2008-011), Wilson Foundation, the National Research Foundation of South Africa, and the University of Pretoria. L.H. Swanepoel was further supported by the NRF (#74819 & #88179) and M.J. Somers by the DSI-NRF Centre of Excellence for Invasion Biology and the NRF. We thank Mathias Tobler and Philippe Gaubert for their comments that improved the manuscript and Chris Sutherland for his help with secr models.

References

Amiard, P. (2014) *Ecology of the African Civet* (Civettictis civetta) *in Three Different Vegetation Types of South Africa: Study of the Population Density, the Habitat Use and the Diet.* Master thesis, University of Reims Champagne-Ardenne.

Angelici, F.M., Luiselli, L., Politano, E. & Akani, G.C. (1999) Bushmen and mammal fauna: a survey of the mammals traded in bush-meat markets of local people in the rainforests of southeastern Nigeria. *Anthropozoologica* 30, 51–58.

Ayalew Berhanu, Afework Bekele, & Balakrishnan, M. (2013) Home range and movement patterns of African civet *Civettictis civetta* in Wondo Genet, Ethiopia. *Small Carnivore Conservation* 48, 83–86.

Bahaa-el-din,L., Henschel, P., Aba'a, R., Abernethy, K., Bohm, T. *et al.* (2013) Notes on the distribution and status of small carnivores in Gabon. *Small Carnivore Conservation* 48, 19–29.

Bekele Tsegaye, Afework Bekele, & Balakrishnan, M. (2008) Feeding ecology of the African civet *Civettictis civetta* in the Menagesha–Suba State Forest, Ethiopia. *Small Carnivore Conservation* 39, 19–24.

Bista, A., Chanchani, P., Warrier, R., Mann, R., Gupta, M. & Vattakavan, J. (2012) Detection of a large Indian civet *Viverra zibetha* in camera-trap surveys in and around Dudhwa National Park in the Terai Region of North India. *Small Carnivore Conservation* 47, 54–57.

Blaum, N., Rossmanith, E., Schwager, M. & Jeltsch, F. (2007) Responses of mammalian carnivores to land use in arid savannah rangelands. *Basic and Applied Ecology* 8, 552–564.

Borchers, D.L. & Efford, M.G. (2008) Spatially explicit maximum likelihood methods for capture–recapture studies. *Biometrics* 64, 377–385.

Boulanger, J., Stenhouse, G. & Munro, R. (2004) Sources of heterogeneity bias when DNA mark–recapture methods are applied to grizzly bears *Ursus arctos* populations. *Journal of Mammalogy* 85, 618–624.

Braczkowski, A., Watson, L., Coulson, D., Lucas, J., Peiser, B. & Rossi, M. (2012) The diet of caracal *Caracal caracal*, in two areas of the southern Cape, South Africa as determined by scat analysis. *South African Journal of Wildlife Research* 42, 111–116.

Burnham, K.P. & Anderson, D.R. (2002) *Model Selection and Multimodel Inference: A Practical Information-theoretic Approach.* 2nd edition. Springer, New York.

Davis, C.L., Rich, L.N., Farris, Z.J., Kelly, M.J., Di Bitteti, M.S. *et al.* (2019) Ecological correlates of the spatial co-occurrence of sympatric mammalian carnivores worldwide. *Ecology Letters* 21, 1401–1412.

De Luca, D.W. & Mpunga, N.E. (2005) *Carnivores of the Udzungwa Mountains: Presence, Distributions and Threats*. Wildlife Conservation Society, Mbeya.

Do Linh San, E., Gaubert, P., Wondmagegne Daniel & Ray, J. (2019) *Civettictis civetta. The IUCN Red List of Threatened Species* 2019, e.T41695A147992107. https://dx.doi.org/10.2305/IUCN.UK.2019-2.RLTS.T41695A147992107.en.

Efford, M.G. (2014) *secr: Spatially Explicit Capture–Recapture Models*. R package version 2.9.0.

Ermias Admasu, Thirgood, S.J., Afework Bekele & Laurenson, M.K. (2004) A note on the spatial ecology of African civet *Civettictis civetta* and common genet *Genetta genetta* in farmland in the Ethiopian Highlands. *African Journal of Ecology* 42, 160–162.

Foster, R.J. & Harmsen, B.J. (2012) A critique of density estimation from camera-trap data. *The Journal of Wildlife Management* 76, 224–236.

Gehrt, S.D. & Prange, S. (2006) Interference competition between coyotes and raccoons: a test of the mesopredator release hypothesis. *Behavioral Ecology* 18, 204–214.

Gray, T.N.E. & Prum, S. (2011) Leopard density in post-conflict landscape, Cambodia: evidence from spatially explicit capture–recapture. *The Journal of Wildlife Management* 76, 163–169.

Henning B.J. (2002) *The Relevance of Ecosystems to Ecotourism in the Waterberg Biosphere Reserve.* PhD thesis, University of Pretoria.

Ikeda, H., Ono, Y., Baba, M., Doi, T. & Iwamoto, T. (1982) Ranging and activity patterns of three nocturnal viverrids in Omo National Park, Ethiopia. *African Journal of Ecology* 20, 179–186.

Isaacs, L., Somers, M.J. & Dalerum, F. (2013) Effects of prescribed burning and mechanical bush clearing on ungulate space use in an African savannah. *Restoration Ecology* 21, 260–266.

Karanth, K.U. (1995) Estimating tiger *Panthera tigris* populations from camera-trap data using capture–recapture models. *Biological Conservation* 71, 333–338.

Karanth, K.U. & Nichols, J.D. (1998) Estimation of tiger densities in India using photographic captures and recaptures. *Ecology* 79, 2852–2862.

Maffei, L. & Noss, A.J. (2008) How small is too small? Camera trap survey areas and density estimates for ocelots in the Bolivian Chaco. *Biotropica* 40, 71–75.

Maffei, L., Noss, A.J., Cuellar, E. & Rumiz, D.I. (2005) Ocelot *Felis pardalis* population densities, activity, and ranging behaviour in the dry forests of eastern Bolivia: data from camera trapping. *Journal of Tropical Ecology* 21, 1–6.

Martinoli, A., Preatoni, D., Galanti, V., Codipietro, P., Kilewo, M., Fernandes, C.A.R., Wauters, L.A. & Tosi, G. (2006) Species richness and habitat use of small carnivores in the Arusha National Park (Tanzania). *Biodiversity and Conservation* 15, 1729–1744.

Mateos Ersado, Zerihun Girma, Mamo Yosef & Megersa Debele (2015) Community attitude towards African Civet *Civettictis civetta* conservation in eastern sub-catchment of Lake Hawassa basin, Southern Ethiopia. *Discovery* 27, 2–7.

Miller, B., Dugelby, B., Foreman, D., Martinez del Rio, C., Noss, R., Phillips, M., Reading, R., Soulé, M.E., Terborgh, J. & Willcox, L. (2001) The importance of large carnivores to healthy ecosystems. *Endangered Species UPDATE* 18, 202–210.

Nichols, J.D. & Karanth, K.U. (2002) Statistical concepts: estimating absolute densities of tigers using capture–recapture sampling. In: *A Manual for Researchers, Managers and Conservationists in Tropical Asia* (eds K.U. Karanth & J.D. Nichols), pp. 124–137. Centre for Wildlife Studies, Bangalore.

Otis, D.L., Burnham, K.P., White, G.C. & Anderson, D.R. (1978) Statistical inference from capture data on closed animal populations. *Wildlife Monographs* 62, 1–135.

Proulx, G. (2021) How human–small carnivore conflicts can repeat themselves: examples from Western Canada. *Canadian Wildlife Biology and Management* 10, 1–24.

Prugh, L.R. & Sivy, K.J. (2020) Enemies with benefits: integrating positive and negative interactions among terrestrial carnivores. *Ecology Letters* 23, 902–918.

Prugh, L.R., Stoner, C.J., Epps, C.W., Bean, W.T., Ripple, W.J., Laliberte, A.S. & Brashares, J.S. (2009) The rise of the mesopredator. *Bioscience* 59, 779–791.

Pugh, M. (1998) *Civet Farming. An Ethiopian Investigation*. World Society for the Protection of Animals, London.

Quinn, J.H. & Whisson, D.A. (2005) The effect of anthropogenic food on the spatial behaviour of small Indian mongooses (*Herpestes javanicus*) in a subtropical rainforest. *Journal of Zoology* 267, 339–350.

R Development Core Team (2013) *R: A Language and Environment for Statistical Computing. R Foundation for Statistical Computing, Edition 3.0.2.* R Foundation for Statistical Computing, Vienna. http://www.R-project.org. Accessed on 14 July 2013.

Rabinowitz, A.R. (1991) Behaviour and movements of sympatric civet species in Huai Kha Khaeng Wildlife Sanctuary, Thailand. *Journal of Zoology* 223, 281–289.

Ramnanan, R., Swanepoel, L.H. & Somers, M.J. (2013) The diet and presence of African wild dogs (*Lycaon pictus*) on private land in the Waterberg region, South Africa. *South African Journal of Wildlife Research* 43, 68–73.

Ray, J.C. (1995) *Civettictis civetta. Mammalian Species* 488, 1–7.

Ray, J.C. (2013) *Civettictis civetta* African civet. In: *The Mammals of Africa. V. Carnivores, Pangolins, Equids and Rhinoceroses* (eds J. Kingdon & M. Hoffmann), pp. 255–259. Bloomsbury, London.

Ray, J.C., Hunter, L. & Zigouris, J. (2005) *Setting Conservation and Research Priorities for Larger African Carnivores.* WCS Working Paper No. 24. Wildlife Conservation Society, New York.

Roemer, G.W., Gompper, M.E. & Van Valkenburgh, B. (2009) The ecological role of the mammalian mesocarnivore. *Bioscience* 59, 165–173.

Rutherford, M.C., Mucina, L., Lötter, M., Bredenkamp, G.J., Smit, J.H.L. *et al.* (2006) Savannah biome. In: *The vegetation of South Africa, Lesotho and Swaziland. Strelitzia 19* (eds L. Mucina & M.C. Rutherford), pp. 438–539. South African National Biodiversity Institute (SANBI), Pretoria.

Skinner, J.D. & Chimimba, C.T. (eds) (2005) *The Mammals of the Southern African Subregion.* 3rd edition. Cambridge University Press, Cambridge.

Smith, R.M. (1978) Movement patterns and feeding behavior of the leopard in the Rhodes Matopos National Park, Rhodesia. *Arnoldia* 8, 1–16.

Swanepoel, L.H. (2013) *Viability of Leopards* Panthera pardus *(Linnaeus, 1758) in South Africa.* PhD thesis, University of Pretoria.

Swanepoel, L.H., Camacho, G., Power, R.J., Amiard, P. & Do Linh San, E. (2016) A conservation assessment of *Civettictis civetta.* In: *The Red List of Mammals of South Africa, Swaziland and Lesotho* (eds M.F. Child, L. Roxburgh, E. Do Linh San, D. Raimondo & H.T. Davies-Mostert). South African National Biodiversity Institute and Endangered Wildlife Trust, Johannesburg.

Swanepoel, L.H., Lindsey, P., Somers, M.J., van Hoven, W. & Dalerum, F. (2014) The relative importance of trophy harvest and retaliatory killing for large carnivores: a case study on South African leopards. *South African Journal of Wildlife Research* 44, 115–134.

Swanepoel, L.H., Somers, M.J. & Dalerum, F. (2015) Density of leopards (*Panthera pardus*) on protected and non-protected land in the Waterberg Biosphere, South Africa. *Wildlife Biology* 21, 263–268.

Tadesse Habtamu, Afework Bekele, Raya Ahmed, Tsegaye Gadisa, Belay Birlie, Taye Tolemariam & Berhanu Belay (2017) Diets of the African civet *Civettictis civetta* (Schreber, 1778) in selected coffee forest habitat, south-western Ethiopia. *African Journal of Ecology* 55, 573–579.

Tobler, M.W., Carrillo-Percastegui, S.E., Pitman, R.L., Mares, R. & Powell, G. (2008) An evaluation of camera traps for inventorying large- and medium-sized terrestrial mammals. *Animal Conservation* 11, 169–178.

Tobler, M.W., Carrillo-Percastegui, S.E., Zúñiga Hartley, A. & Powell, G.V.N. (2013) High jaguar densities and large population sizes in the core habitat of the southwestern Amazon. *Biological Conservation* 159, 375–381.

Tobler, M.W., Hibert, F., Debeir, L. & Richard-Hansen, C. (2014) Estimates of density and sustainable harvest of the lowland tapir *Tapirus terrestris* in the Amazon of French Guiana using a Bayesian spatially explicit capture–recapture model. *Oryx* 48, 410–419.

Williams, S.T., Maree, N., Taylor, P., Belmain, S.R., Keith, M. & Swanepoel, L.H. (2018) Predation by small mammalian carnivores in rural agro-ecosystems: an undervalued ecosystem service? *Ecosystem Services* 30, 362–371.

Yarnell, R.W., Phipps, W.L., Burgess, L.P., Ellis, J.A., Harrison, S.W.R., Dell, S., MacTavish, D., MacTavish, L.M. & Scott, D.M. (2013) The influence of large predators on the feeding ecology of two African mesocarnivores: the black-backed jackal and the brown hyaena. *South African Journal of Wildlife Research* 43, 155–166.

Yilma Delelegn (2003) Sustainable utilization of the African civet (*Civettictis civetta*) in Ethiopia. In: *2nd Pan-African Symposium on the Sustainable Use of Natural Resources in Africa* (ed. Bihini Won wa Musiti), pp. 197–207. IUCN, Gland and Cambridge.

Part IV

Interspecific Interactions and Community Ecology

13

Spatio-Temporal Overlap Between a Native and an Exotic Carnivore in Madagascar: Evidence of Spatial Exclusion

Zach J. Farris[1,], Brian D. Gerber[2], Sarah Karpanty[3], Felix Ratelolahy[4],*
Vonjy Andrianjakarivelo[4], and Marcella J. Kelly[3]

[1] *Department of Health & Exercise Science, Appalachian State University, Boone, NC, USA*
[2] *Department of Natural Resources Science, University of Rhode Island, Kingston, RI, USA*
[3] *Department of Fish & Wildlife Conservation, Virginia Tech, Blacksburg, VA, USA*
[4] *Wildlife Conservation Society Madagascar Program, WCS Makira Antseranamborondolo, Maroantsetra, Madagascar*

SUMMARY

The exclusion or local extirpation of native species by exotic or introduced carnivores is a burgeoning issue for conservation. Exotic carnivores may indeed present a serious threat as they have the potential to negatively influence and/or interact with native wildlife via exploitative or interference competition, intraguild predation and/or transmission of pathogens. So far, studies investigating co-occurrence have failed to include both a spatial and temporal component which is likely to lead to improper inference. Here, we used a novel approach to investigate the relationship between native and exotic carnivores across both space and time and provide insight on the spatial exclusion of the native spotted fanaloka, *Fossa fossana* (listed as Vulnerable by the IUCN), by the exotic small Indian civet, *Viverricula indica*, across Madagascar's eastern rainforest ecosystem. We combined both spatial (single-species and two-species occupancy analyses) and temporal (kernel density estimation) analyses to investigate potential spatio-temporal interactions across the landscape, comparing degraded and non-degraded forests. We found that the exotic Indian civet negatively influenced spotted fanaloka occupancy, which resulted in a strong decrease in occupancy across degraded forests. Further, spotted fanaloka occupancy decreased by 40% at sites where Indian civet were present, resulting in a strong lack of co-occurrence between these two species. Finally, we recorded strong spatio-temporal overlap during the nocturnal time period within degraded, patchy forests. As a result, we suggest that this reveals evidence of spatial exclusion of the spotted fanaloka. This novel approach provides a unique investigation across both space and time – allowing us to identify more accurately the precise locations where co-occurring carnivores are potentially interacting – and has wide-ranging implications for conservation managers working to address the negative impacts of exotic species on native wildlife.

Keywords

Camera-trapping — conservation — co-occurrence — degradation — invasive species — occupancy — rainforest

Introduction

Understanding how exotic species affect the range, behaviour and/or demography of threatened species is important for conservation and management as these alterations may lead to the exclusion and/or extirpation of native threatened species (Mack *et al.*, 2000; Tompkins *et al.*, 2003; Salo *et al.*, 2007; Santulli *et al.*, 2014). The presence of exotic carnivores may seriously endanger threatened, native carnivores as exotics

---

*Corresponding author.

Small Carnivores: Evolution, Ecology, Behaviour, and Conservation, First Edition. Edited by Emmanuel Do Linh San, Jun J. Sato, Jerrold L. Belant, and Michael J. Somers.

have been shown to negatively affect or interact with native wildlife via exploitative or interference competition, intraguild predation and/or transmission of pathogens (Tompkins *et al.*, 2003; Salo *et al.*, 2007; Clout & Russell, 2008; Vanak & Gompper, 2010; Santulli *et al.*, 2014). In addition, exotic carnivores have been shown to influence both the spatial and temporal activities of native carnivore species (Vanak & Gompper, 2010; Silva-Rodríguez & Sieving, 2012; Gerber *et al.*, 2012a; Santulli *et al.*, 2014; Farris *et al.*, 2015a,c, 2016). These spatio-temporal, inter-species interactions among native and exotic carnivores may alter community structure and may result in declines of native species. Consequently, the influx of exotic carnivores worldwide, and their interactions with native wildlife, is a growing management and conservation concern. Additional research on the interactions between exotic and native carnivores is needed, particularly in little-studied, biologically diverse ecosystems, such as tropical and sub-tropical regions where the ecological roles of carnivores (especially small and medium-sized ones) are only poorly understood.

Madagascar represents a top global conservation priority due to its high levels of biodiversity and endemism, as well as the numerous pressures threatening it (Myers *et al.*, 2000). In particular, Madagascar's carnivores are perhaps the most threatened yet least-studied carnivores in the world (Brooke *et al.*, 2014; Wampole *et al.*, 2021). The small body size and limited distribution of Madagascar's 10 native, extant carnivore species may explain the limited amount of research conducted on these endemic species (Brooke *et al.*, 2014). While studies on Madagascar's carnivores are limited, recent research has highlighted a number of threats resulting from anthropogenic pressures, namely habitat degradation and fragmentation, human encroachment, exotic carnivores and unsustainable hunting (Gerber *et al.*, 2012a,b; Farris *et al.*, 2014, 2015b,c, 2017a, 2020; Wampole *et al.*, 2021). In particular, Farris (2014) provided insight on how temporal overlap of Madagascar's exotic (domestic/feral dogs, *Canis familiaris*, domestic/feral/wild cats, *Felis* sp., and small Indian civet, *Viverricula indica*) and native carnivores may be related to lower occupancy of some native carnivores and lemurs across a diverse landscape ranging from intact, contiguous rainforest to degraded, fragmented forests to highly cultivated areas. However, efforts to

investigate both the spatial and temporal interactions of native and exotic carnivores, and the effects of any spatio-temporal overlap on native carnivore population parameters, are still lacking not only for Madagascar but also for co-occurring carnivores worldwide.

The native Malagasy spotted fanaloka, *Fossa fossana*, currently listed as Vulnerable by *The IUCN Red List of Threatened Species* (Hawkins, 2015), occupies lowland to mid-altitude rainforest across eastern and northern Madagascar (Goodman, 2012). Little is currently known about this small-bodied (1.3–2.1 kg) carnivore (Figure 13.1); however, recent research has highlighted a strong decrease in population density and occupancy in response to habitat degradation and fragmentation (Gerber *et al.*, 2012b; Farris *et al.*, 2015b). Spotted fanalokas have a generalist diet and utilize a wide spectrum of prey ranging from insects to small mammals; however, anecdotal accounts and opportunistic observations suggest that this carnivore may prefer wetland environments where it feeds on an assortment of aquatic species, such as fish, amphibians, crabs and others (Kerridge *et al.*, 2003; Goodman, 2012). Recent population density estimates (expressed as individuals/km^2) of spotted fanalokas from non-degraded forest (3.19 in South East Madagascar: Gerber *et al.*, 2012b; and 2.71 in North East Madagascar: Farris & Kelly, 2011) were higher compared to degraded forest (1.38 in South East Madagascar: Gerber *et al.*, 2012b). Moreover, these recent studies also highlighted the

Figure 13.1 The spotted fanaloka, *Fossa fossana*, is one of 7 to 9 Eupleridae species endemic to Madagascar. This 'Malagasy spotted civet' was once classified within the family Viverridae. *Source:* Photo © Chien C. Lee (chienclee.com).

threat of forest loss, fragmentation, human presence, exotic carnivores and hunting on spotted fanalokas across both the south-east (Gerber *et al.*, 2012a,b) and north-east (Farris & Kelly, 2011; Farris *et al.*, 2015b) regions of Madagascar. For example, Farris *et al.* (2015c) showed how spotted fanalokas had a lower probability of occupancy at sites that had high exotic carnivore and human activity, while Gerber *et al.* (2012b) found that spotted fanalokas were absent from fragmented forest where exotic carnivore activity was high.

The exotic small Indian civet (Figure 13.2) is one of Madagascar's three confirmed introduced carnivore species. While native to Bangladesh, Bhutan, Cambodia, China, India, Indonesia, Lao People's Democratic Republic, Malaysia, Myanmar, Nepal, Pakistan, Sri Lanka, Taiwan, Thailand, and Vietnam (Choudhury *et al.*, 2015), the introduction process of the small Indian civet to Madagascar remains unknown, though it may have occurred as a result of traders using civet glands for the production of perfume during the Islamic Golden Age (Goodman, 2012; Gaubert *et al.*, 2017). Presently, it is widely distributed across Madagascar, occupying degraded forests, marshland, and anthropogenic areas in every region

Figure 13.2 The small Indian civet, *Viverricula indica*, may have been brought to Madagascar by Islamic traders for the production of perfume. It is now widely distributed on the island and even occupies anthropogenic areas. The individual on the picture was found resting in an outbuilding in a village in eastern Madagascar. *Source:* Photo © Nick Garbutt (www.nickgarbutt.com).

of Madagascar excluding the dry, spiny forest of the south (Goodman, 2012). Similar to the spotted fanaloka, the small Indian civet is a small-bodied (2.0–4.0 kg) carnivore with a generalist diet ranging from insects to small mammals; however, the small Indian civet is also a confirmed lemur predator (Goodman, 2003). Recent research in eastern Madagascar has shown that small Indian civet occupancy and activity increase in degraded, fragmented forests, particularly in non-forested areas near villages (Gerber *et al.*, 2012b; Kotschwar *et al.*, 2015). More specifically, Gerber *et al.* (2012b) found Indian civet occupancy to be 0.94 (± 0.04 Standard Error [SE]) in fragmented forest, while the species was absent in contiguous forest. Similarly, Farris *et al.* (2015b) found small Indian civet occupancy to be considerably higher in degraded forest, while the species was absent in intact, contiguous forest. This prior research further indicates that small Indian civets may negatively influence native carnivore populations, including the spotted fanaloka (Gerber *et al.*, 2012b; Farris *et al.*, 2015b,c). The overlap in ecological niche (Farris *et al.*, 2015a) – particularly in diet – between the spotted fanaloka and the small Indian civet suggests that there is potential for direct competitive interactions for food or space, and if spatio-temporal overlap occurs, there is potential for intraguild predation and disease transmission as well. The above findings provide evidence of these potential interactions; however, we need more in-depth investigation of how these two carnivores interact spatially and temporally across the landscape to better understand this relationship.

In this study we investigate the spatio-temporal overlap between the native spotted fanaloka and the exotic small Indian civet, including how these interactions may influence spotted fanaloka population parameters across the eastern rainforest landscape in Madagascar. To achieve this goal, we combined occupancy estimation (both single-species and two-species analyses) and temporal activity overlap (kernel density estimation) to identify potential spatio-temporal interactions across the landscape, comparing degraded and non-degraded forests. Using these spatial and temporal analyses, we make inference on the evidence for spatial exclusion between these two carnivores.

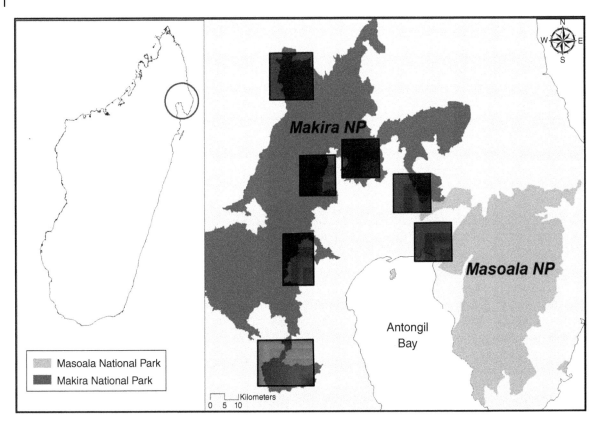

Figure 13.3 Map of the Masoala–Makira landscape including the outline of the regions in which the surveys were conducted at seven study sites. Photographic surveys were conducted from 2008 to 2012. *Source:* Map modified from Farris (2014).

Methods

Study Area

From 2008 to 2013, we conducted photographic sampling (camera-trapping) of Madagascar's carnivore community across the Masoala–Makira landscape (Figure 13.3). It consists of the Masoala National Park (240000 ha) and Makira Natural Park (372470 ha of protected area and 351037 ha of community management zone); together this landscape makes up Madagascar's largest protected area complex. Across this landscape we selected seven study sites with varying degrees of degradation and fragmentation. Human use of these forests varied from no use (no human presence) to heavy resource extraction and intense hunting pressure (Farris, 2014). These seven sites were selected as part of an ongoing, expansive carnivore and lemur survey to investigate the influence of anthropogenic pressures on these populations (Farris, 2014). Two of the sites were sampled multiple times over a six-year period providing a total of 13 surveys.

Photographic Sampling

At each of the seven study sites, we established a camera grid consisting of 18–25 camera stations spaced approximately 500 m apart, where each station had two passive remote cameras positioned on opposing sides of human (> 0.5 m wide) or game (< 0.5 m wide) trails. All cameras, either digital (Reconyx PC85 & HC500, Wisconsin, USA; Moultrie D50 & D55, Alabama, USA; Cuddeback IR, Wisconsin, USA) or film-loaded (DeerCam DC300), were placed approximately 20–30 cm off the ground and were offset to prevent mutual flash interference. We did not use bait or lure at stations. We checked cameras every 5–10 days to change memory cards and ensure proper functioning. Each camera grid was active for an average (± Standard Deviation) of 67 ± 8 days.

Habitat and Landscape Sampling

To quantify habitat features (Table 13.1), we sampled vegetation at each camera station by walking a 50 m transect in three directions (0, 120, and 240°) starting at each individual camera station. We estimated canopy height and percent cover at 10 m intervals along each transect. At 25 and 50 m on each transect, we used the point-quarter method (Pollard, 1971) to estimate tree density and basal area, recording diameter at breast height (DBH) for any stem/tree with ≥5 cm diameter. At 20 and 40 m, we established a 20 m transect running perpendicular to the established 50 m micro-habitat transect and we measured understory cover at three levels (0–0.5 m, 0.5–1 m, and 1–2 m) by holding a 2 m pole perpendicular to the ground at 1 m intervals and recording presence (1 = vegetation touching pole) or absence (0 = no vegetation touching pole) of understory cover (Davis *et al.*, 2011).

We also characterized landscape features (Table 13.1) in our study area by classifying habitat-cover types from Landsat satellite imagery (2004, 2006, and 2009) using Erdas Imagine (Intergraph Corporation, Madison, AL, USA) rainforest, degraded forest and matrix (non-forest, cultivation area). We placed a 500 m (landscape-level) buffer around individual camera stations and clipped the classified imagery for each of the resulting seven camera grid buffers (each providing an approximately 10–15 km^2 area) for analysis in the program FragStats, version 4.0 (McGarigal *et al.*, 2012). Using FragStats we created the following landscape level covariates and clipped imagery from each camera grid buffer for use in our occupancy models: (i) number of habitat patches: total number of rainforest, degraded forest, and matrix patches (based on habitat classifications from satellite imagery) within the buffer, where a patch is an area of habitat type separated from similar habitat by ≥50 m; (ii) largest patch index: the percentage of total buffered area comprised by the largest rainforest patch; (iii) landscape shape index (LSI) or the standardized measure of total edge adjusted for the size of the buffered area (McGarigal *et al.*, 2012); (iv)

percent rainforest within the buffered area; (v) percent matrix or non-forest, cultivated area within the buffered area; (vi) total rainforest core area: the sum of the core areas (accounting for edge of depth of 500 m) of each rainforest patch within the buffer; and (vii) total edge (in m/ha) (McGarigal *et al.*, 2012). Further, we calculated the average distance from each camera station to the nearest village and to the nearest forest edge using satellite imagery.

Occupancy Analysis (Single-species and Two-species)

Occupancy modelling provides estimates of the proportion of the landscape used by a target species while accounting for our inability to perfectly detect individuals (MacKenzie *et al.*, 2006). This estimation approach is useful as it is able to provide reliable estimates of occupied area for a species of interest, which may or may not be individually identifiable, and it is able to identify important relationships, via covariates, that explain patterns of habitat use across a diverse landscape. We estimated single-season, single-species occupancy (ψ) for the spotted fanaloka and the small Indian civet in the program PRESENCE, version 7.0 (Hines, 2006) using habitat and landscape variables and photographic capture rates for co-occurring exotic carnivores (domestic/feral dog and domestic/feral/wild cat) as covariates. Capture rates were calculated by dividing the total number of captures, defined as the number of photos of the target species within a 30 minute period (Di Bitetti *et al.*, 2006), by the total number of trap nights, defined as the number of nights at least one camera at a camera station was properly functioning, multiplied by 100. To evaluate the spatial co-occurrence between the spotted fanaloka and the small Indian civet, we used the two-species, single-season occupancy modelling, with the psi Ba parameterization (Richmond *et al.*, 2010), in the program PRESENCE, while also incorporating the landscape variables to determine the relationship between these two carnivores. This particular parameterization was chosen as it allows for an investigation of change in occupancy as a result of the presence of a co-occurring species. Using this modelling approach, we estimated the following parameters: occupancy of the small Indian civet (ψC), occupancy of the spotted fanaloka

Table 13.1 Station-level habitat (camera station) and landscape (500 m grid buffer) features for the seven study sites, ranked from least degraded (S01) to most degraded (S07), across the Masoala–Makira landscape.

Level	Study site	Least → Site S01	Site S02	Site S03	Level of degradation Site S04	Site S05	Most Site S06	Site S07
Habitat	TreeDen (stems ≥5 cm/ha)[a]	1200 (300)	3500 (900)	4100 (1600)	4600 (1700)	4400 (1100)	—	3000 (700)
	BA (stems ≥5 cm, m²/ha)[b]	82.00 (10.22)	57.4 (6.11)	22.85 (4.59)	73.54 (13.03)	76.54 (8.48)	—	49.85 (6.35)
	Can Ht (m)[c]	16.97 (1.95)	12.50 (0.96)	7.48 (0.67)	10.55 (1.23)	12.89 (1.08)	—	9.75 (1.27)
	% Can cover[d]	64.15 (5.58)	57.05 (4.89)	62.75 (3.17)	43.52 (6.82)	60.84 (4.09)	—	42.45 (5.14)
	% Understory cover (0–2 m)	0.50 (0.05)	0.44 (0.04)	0.53 (0.03)	0.46 (0.04)	0.44 (0.05)	—	0.52 (0.04)
Landscape	# Patches[e]	3	10	22	21	31	116	190
	Largest patch index[f]	60.38	52.33	44.88	51.30	39.90	43.72	50.36
	LSI[g]	1.04	1.34	2.12	1.95	2.02	3.11	6.76
	% Rainforest	99.94	98.89	94.48	95.19	96.87	96.06	81.07
	% Matrix[h]	0.05	0.66	4.38	0.59	0.76	0.19	4.07
	Tot core rainforest (ha)[i]	0.88	0.99	0.85	0.87	1.14	0.72	0.59
	Tot edge (m/ha)[j]	0.03	0.59	1.85	1.53	2.13	3.51	7.89
	Avg. dist. to village (km)[k]	10.96	2.80	3.33	2.08	4.82	2.71	1.45
	Avg. dist. to edge (km)[l]	1.14	0.68	0.29	0.36	0.34	0.60	0.18

[a] TreeDen = tree density averaged (Standard Error) across all camera stations ($n = 18$–25) for each study site.
[b] BA = average (Standard Error) basal area.
[c] Can Ht = average (Standard Error) canopy height.
[d] % Can cover = average (Standard Error) percentage canopy cover.
[e] # Patches = total number of rainforest, degraded forest and matrix patches within the camera grid buffer.
[f] Largest patch index = the percentage of total landscape area comprised by the largest rainforest patch.
[g] LSI = landscape shape index or the standardized measure of total edge adjusted for the size of the landscape.
[h] % Matrix = percentage matrix defined as non-forest land cover consisting of cultivation, open field or early succession.
[i] Tot core rainforest (ha) = total core area defined as the sum of the core areas within the camera grid buffer (accounting for 500 m edge depth) of each rainforest patch.
[j] Tot edge = total edge area within the camera grid buffer which lies 50 m or less from the edge of the forest.
[k] Avg. dist. to village = average distance from each camera station to the nearest village based on satellite imagery.
[l] Avg. dist. to edge = average distance from each camera station to the nearest forest edge based on satellite imagery.

when the small Indian civet is present (ψFC) and when the small Indian civet is absent (ψFc), the probability of detection for the small Indian civet (pC) and spotted fanaloka (pF) and the probability of detecting the spotted fanaloka when the small Indian civet is present (pFC) and absent (pFc). Additionally, using the equation from Richmond *et al.* (2010), we derived the species interaction factor (SIF), which is a measure of co-occurrence where individuals are said to occur independently (SIF = 1), occur together more often than if they were independent (hereafter 'co-occur'; SIF > 1) or occur together less often than if they were independent (hereafter 'lack of co-occurrence'; SIF < 1). For both single-species and two-species occupancy modelling, we conducted model selection using Akaike's Information Criterion (AIC), where the minimum AIC of a model set is the most parsimonious (i.e. represents the trade-off of model fit and complexity) model for the data (Akaike, 1973). We only report top-ranking models having a ΔAIC value < 2, and all top-ranking model output, including associated AIC values and beta (β) values can be found in Farris (2014).

Temporal Analysis

To determine the level of temporal overlap between the spotted fanaloka and the small Indian civet, we modelled captures (capture events/available hours) for each time category (day, dawn, dusk, night; Farris, 2014). Further, we estimated the probability density of temporal activity distribution for each species using a non-parametric kernel density analysis (Ridout & Linkie, 2009). This analysis allowed us to make inference from the most parsimonious model and to determine the dominant activity pattern for both carnivores. Finally, we estimated the coefficient of overlap of the probability densities using an estimator supported for small sample size [denoted Δ_1] (Ridout & Linkie, 2009).

Relative Species Interaction

To investigate the level of spatio-temporal overlap or interaction potential between spotted fanalokas and small Indian civets, we developed a novel approach to calculate the spatio-temporal value (STV), which is

the joint overlap density resulting from the kernel density analysis (temporal) and two-species occupancy SIF (spatial). More specifically, we calculated the STV by multiplying the coefficient of overlap across the diel cycle from our kernel density analysis by the SIF value across the landscape. This STV value provides a measure of overlap between the two carnivores such that a value of 0 indicates no overlap or potential interaction, and as the STV value increases, this indicates increasing overlap or increasing potential for interaction (see Farris *et al.*, 2020 for more details).

Results

Across our 13 surveys at the seven study sites, we recorded 820 photographic captures of spotted fanalokas (*n* = 5 sites) and 43 photographic captures of small Indian civets (*n* = 6 sites). The average (± SE) capture rate was 4.89 ± 1.32 for the spotted fanaloka and 0.29 ± 0.15 for the small Indian civet across the landscape. Spotted fanalokas were more likely to be captured in non-degraded forest (96% of captures), while 86% of small Indian civet captures took place in degraded forest. Spotted fanalokas had a probability of occupancy (± SE) of 0.70 ± 0.07 across the landscape, as compared to only 0.11 ± 0.04 for small Indian civets (Farris *et al.*, 2015c). Spotted fanaloka occupancy was negatively influenced by the presence of small Indian civet (β = −1.20 ± 0.52 SE; Figure 13.4) and feral cats (β = −2.65 ± 1.00). Small Indian civet probability of occupancy was negatively associated with distance to village, and thus higher small Indian civet occupancy was recorded closer to villages (β = −1.59 ± 0.87; Farris *et al.*, 2015c). Spotted fanalokas decreased in occupancy from 0.73 ± 0.08 to 0.50 ± 0.09 going from non-degraded to degraded forest, respectively, while small Indian civets increased from a naïve occupancy estimate of 0.02 in non-degraded forest (not accounting for imperfect detection due to low capture rate) to 0.47 ± 0.15 in degraded forest.

Our two-species occupancy modelling revealed that spotted fanalokas had a higher probability of occupancy when small Indian civets were not present (0.72 ± 0.31; ψFc) compared to when small Indian civets were

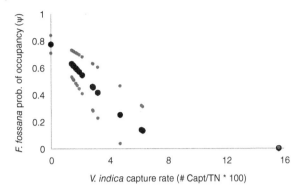

Figure 13.4 Probability of occupancy ψ (large black dots) with 95% confidence interval (CI) (small grey dots) for the spotted fanaloka in relation to small Indian civet capture rate, defined as the number of captures divided by the total number of trap nights (TN) multiplied by 100. Photographic sampling of both carnivores occurred from 2008 to 2013 in North East Madagascar.

in non-degraded, contiguous forest, which had a lower number of habitat patches. Both carnivores exhibited nocturnal activity patterns with the spotted fanaloka displaying a higher level of activity during the early morning hours (01:00–05:00h) and almost no diurnal activity compared to the small Indian civet (Farris *et al.*, 2015a). Overall, the two carnivores demonstrated a high probability of temporal overlap ($\Delta_1 = 0.80$; Figure 13.6).

We found that spotted fanalokas and small Indian civets varied in their spatio-temporal overlap with the highest level of overlap (STV > 1.50) occurring from 18:00 to 20:00h within highly patchy (> 800 habitat patches across the surveyed area) forest habitat (Figure 13.6). While temporal overlap was high from 18:00 to 24:00h across the landscape, the STV overlap value dropped considerably during this time frame in contiguous, non-patchy forest as a result of the limited presence of small Indian civets within this forest habitat (Figure 13.6).

Discussion

Our research shows that the native spotted fanaloka and the exotic small Indian civet have a high level of spatio-temporal overlap in Madagascar's eastern

present (0.33 ± 0.11; ψFC; Farris *et al.*, 2015b). This resulted in a SIF of 0.50 ± 0.26 revealing a lack of co-occurrence between these two carnivores (Figure 13.5; Farris *et al.*, 2015b). This interaction between the carnivores was best explained by the number of habitat patches within the surveyed area (Farris *et al.*, 2015b,c), such that the two carnivores were less likely to co-occur

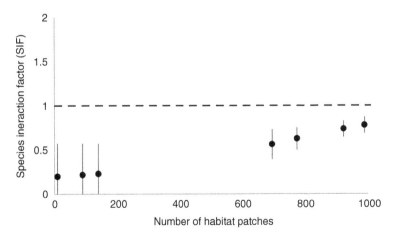

Figure 13.5 The species interaction factor (SIF) between the spotted fanaloka and the small Indian civet revealing a lack of co-occurrence between the two carnivores in relation to the number of habitat patches. SIF of 1 (dashed line) denotes independent occurrence, while SIF < 1 indicates lack of co-occurrence. Black bars indicate standard error. Photographic sampling of carnivores occurred across the Masoala–Makira landscape from 2008 to 2013.

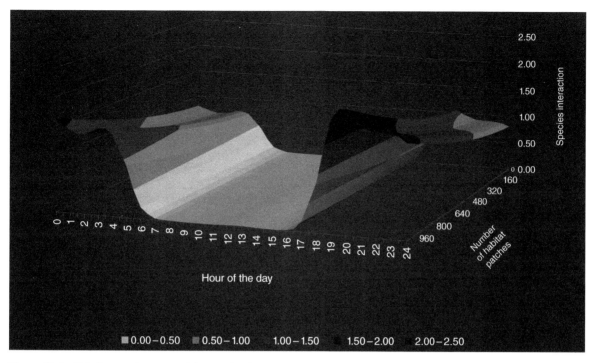

Figure 13.6 Relative spatio-temporal interaction between the spotted fanaloka and the small Indian civet revealing the level of spatial and temporal interaction (y-axis) in relation to the hour of day (x-axis) and number of habitat patches (z-axis). The level of interaction ranges from no interaction (indicated by interaction value of 0 and light colour) to strong potential interaction (indicated by interaction value of > 1.50 and dark colour). Photographic sampling of both carnivores occurred from 2008 to 2013 in North East Madagascar.

rainforest habitat during the nocturnal time period within degraded, patchy forests. We suggest that this strong spatio-temporal overlap may be leading to negative impacts of the exotic small Indian civet on the native spotted fanaloka as evidenced by a strong decrease in occupancy of the spotted fanaloka within degraded forest across the landscape when in the presence of the small Indian civet. We cannot deduce the mechanism for the decline at this time, but rather underline that the overlap exists, thus increasing the potential for competitive interactions (including predation) and for disease transmission. The strong overlap in body size, diet, and apparent preference for wetland habitat further highlights the danger in the strong spatio-temporal overlap between this native and exotic carnivore pairing. Parts of these findings are based on a novel analysis that combines spatial and temporal techniques to investigate the cumulative potential for co-occurrence, and hence potential competition, between carnivore species. Through use of

this novel modelling approach, we highlight the need for incorporating both time and space components in studies investigating co-occurring species, as failure to account for both niche dimensions may lead to improper inference and underestimates of potential competition.

The spotted fanaloka appears to be the most common carnivore (highest capture rate, highest probability of occupancy) across eastern rainforest habitat, including a high to moderate probability of occupancy across both non-degraded and degraded forests (Farris & Kelly, 2011; Gerber *et al.*, 2012b; Farris *et al.*, 2015b). The generalist diet of this small, native carnivore likely drives its ability to persist in a number of habitat types (Goodman, 2012). However, despite its generalist diet and high probability of occupancy across eastern Madagascar, this native carnivore is heavily constrained by the presence of the exotic small Indian civet in degraded forest habitat (Gerber *et al.*, 2012b; Farris *et al.*, 2015b). In particular, spotted fanalokas have a considerably higher occupancy, or wider distribution, across the landscape in the absence of small Indian civets (as was also observed by Gerber *et al.*, 2012b), which is likely to result in a higher probability of local extirpation in degraded forests where small Indian civets are present. The loss of the spotted fanaloka from rainforest habitat in Madagascar may have widespread negative impacts on this threatened ecosystem given the generalist diet and the associated ecosystem services provided by this wide-ranging, common native carnivore species. However, if non-degraded forests are effectively conserved and managed, this native carnivore, as well as additional co-occurring native wildlife, are not likely to be negatively impacted by competition with the small Indian civet, as this exotic species does not appear to utilize non-degraded forest. The negative association with small Indian civet presence and the high spatio-temporal overlap in degraded, patchy forest habitat are particularly alarming given the current forest degradation rates and increasing exotic carnivore trends in Madagascar's eastern rainforests (Harper *et al.*, 2007; Allnutt *et al.*, 2008, 2013; Gerber *et al.*, 2012a,b; Farris *et al.*, 2015b,c). As more of Madagascar's rainforest habitat is cleared and/or degraded and fragmented for cultivation and resource extraction, interactions between these two carnivores are likely to increase, resulting in further decreases in spotted fanaloka populations.

Our ongoing research on Madagascar's carnivore community in North East Madagascar highlights negative interactions between native and exotic carnivores, both spatially (i.e. dogs and Eastern falanouc, *Eupleres goudotii*, and small Indian civet and broad-striped vontsira, *Galidictis fasciata*; Farris *et al.*, 2015c) and temporally (i.e. small Indian civet and fosa, *Cryptoprocta ferox*, and dog and brown-tailed vontsira, *Salanoia concolor*; Farris *et al.*, 2015a), as well as spatio-temporally (Farris *et al.*, 2020). However, despite numerous cases of spatial or temporal overlap between native and exotic carnivores, the only exotic vs. native carnivore pairings having both spatial and temporal overlap are small Indian civet and spotted fanaloka, and small Indian civet and broad-striped vontsira (Farris *et al.*, 2020).

The small Indian civet not only poses a threat to the spotted fanaloka but has also been shown to negatively influence the spatial patterns of the Eastern falanouc, the ring-tailed vontsira, *Galidia elegans* and the broad-striped vontsira across eastern rainforest habitat in North East Madagascar (Farris *et al.*, 2015c). This introduced small carnivore also demonstrates a strong temporal overlap with the broad-striped vontsira and the fosa in North East Madagascar (Farris *et al.*, 2015a). Gerber *et al.* (2012a,b) found that the small Indian civet negatively influenced the temporal activity and occupancy of multiple native carnivores in South East Madagascar as well. Although no demographic data are available for the small Indian civet, it is likely that populations of this exotic carnivore are increasing throughout Madagascar, particularly across eastern rainforest habitat, given the widespread intensifications in forest loss and human encroachment (Ganzhorn *et al.*, 2001; Harper *et al.*, 2007). The apparent preference of the small Indian civet for habitat located near anthropogenic areas (Goodman, 2012; Gerber *et al.*, 2012b; Farris *et al.*, 2015b) greatly diminishes the probability of interactions with native carnivores in contiguous rainforest, thus placing greater emphasis on the need for intact, contiguous rainforest for effective, long-term conservation of wildlife in Madagascar. We suspect that the reason for the strong impact of the exotic small Indian civet on the native spotted fanaloka and other co-occurring native carnivores is due to the high overlap in body size, diet and/or temporal activity with these native carnivore species. The small Indian civet appears to have a similar preference for wetland habitat exhibited by both the spotted fanaloka and the Eastern falanouc, and its documented diet in South East Asia is similar to that of the spotted fanaloka (Nowak, 2005; Su & Sale, 2007). The generalist diet of the small Indian civet; its capacity to persist in heavily degraded, fragmented forests and anthropogenic areas and its overall widespread distribution throughout its native range are likely contributors to the ability of this small-bodied carnivore to invade the Madagascar landscape, exploit numerous resources, compete with native carnivores, transmit pathogens to native wildlife and spatially exclude the spotted fanaloka and other native carnivores from degraded forest sites. This hypothesis on the ability of the small Indian civet to successfully invade a wide range of Madagascar's habitats is further supported by Gantchoff *et al.* (Chapter 20, this volume) who found that preadaptation to local climate and inherent superiority hypothesis were most important for explaining successful (re-)introductions of small carnivores to new areas.

Based on our findings, we suggest that the small Indian civet poses a great threat to the persistence of the spotted fanaloka across degraded eastern rainforest habitat, emphasizing the need to protect remaining intact forest habitat, which is preferred by this species and seemingly unsuitable for competing small Indian civets. Farris *et al.* (2015b) found that across the Masoala–Makira landscape, the small Indian civet was the most consumed carnivore out of both native and exotic species. However, the current hunting pressure on small Indian civets across this region does not appear to be effectively controlling or diminishing their populations (Farris *et al.*, 2015b). As a result, conservation and management strategies should consider intensified efforts to decrease populations of small Indian civets, including trap-kill programmes surrounding anthropogenic areas, particularly within or near high-priority conservation areas. It is also believed that the widespread hunting of small Indian civets near settlements will alleviate hunting pressures on native carnivore populations throughout intact forests. These trap-kill programmes should involve live-trapping and be monitored by experienced researchers to ensure that native carnivores are not killed due to bycatch. Indeed, local hunters opportunistically capture carnivores and often confuse small Indian civets and spotted fanalokas owing to the similar coat markings of both species.

To better understand the influence of the small Indian civet on the spotted fanaloka and co-occurring native carnivores, and to aid in the development of targeted management strategies to combat the influx of exotic carnivores in Madagascar's forests, we need long-term, repeated surveys; baseline demographic information; diet analyses and data on disease and pathogen loads for both native and exotic carnivores. To date, we still lack population density estimates for small Indian civets from any location in Madagascar, despite our ability to identify individuals based on their unique coat markings. To obtain density estimates and additional demographic data, such as survival and recruitment, we require surveys across highly degraded forests and anthropogenic areas,

which we currently lack throughout Madagascar. Baseline demographic data on small Indian civets are needed to prevent the type of exotic carnivore conservation issues plaguing other areas. For example, Santulli *et al.* (2014) found that over a 12-year period the exotic American mink, *Neovison vison*, was causing the local extirpation of the native European mink, *Mustela lutreola*. Additionally, Salo *et al.* (2007), via a meta-analysis, established that exotic predators impose more intense suppression on native prey populations compared to native predators in habitats worldwide, an alarming trend that may be occurring in Madagascar's diverse ecosystems.

In this chapter, we also presented a novel approach for quantifying spatio-temporal overlap between co-occurring carnivores, which combines both spatial (two-species occupancy modelling) and temporal (kernel density estimation) analyses. This technique provides a unique investigation across both time and space, allowing researchers to identify more accurately the precise locations where co-occurring carnivores are potentially interacting. Most ecological studies on co-occurring species often address space or time individually but ideally should incorporate both niche dimensions to more thoroughly investigate their cumulative effects. Moreover, we demonstrated how studies investigating co-occurrence across only space may lead to an improper inference and underestimate potential competition, by not incorporating a temporal component.

Conclusion

We used a novel modelling method to investigate the spatio-temporal overlap between the native spotted fanaloka and the exotic small Indian civet. This approach provided strong evidence for the spatial exclusion of this native species within degraded habitat across the eastern rainforest landscape of Madagascar. We calculated the probability of occupancy for these two carnivores and demonstrated how the presence of the exotic small Indian civet negatively influences the occupancy of the native spotted fanaloka, which results in a lack of co-occurrence for these two carnivores across the landscape. We highlighted the strong spatial, temporal and diet overlap between these two carnivores and discussed the potential implications of losing this native carnivore from Madagascar's eastern rainforest habitat. We also emphasized the need for targeted management plans to address the influx and resulting negative impacts of the small Indian civet and other exotic species across Madagascar (to learn about their effects on other native carnivores, see Farris *et al.*, 2017a,b). This work expands on our ongoing research investigating the relationships among native and exotic carnivores and presents a unique framework for investigating the adverse effects of exotic species on native wildlife in general (see e.g. Rasambainarivo *et al.*, 2017; Murphy *et al.*, 2019), a burgeoning conservation issue worldwide.

Acknowledgements

This research was funded by the following organizations: Cleveland Metroparks Zoo, European Association for Zoos and Aquariums, Idea Wild, National Geographic Society-Waitts grant (#W96-10), People's Trust for Endangered Species, Virginia Tech Chapter of Sigma Xi, Virginia Tech Department of Fish & Wildlife and logistical and financial support from the Wildlife Conservation Society (WCS) Madagascar Program. We thank our Malagasy field assistants (B.L. Donah, M. Helin, V. Andrianjakarivelo and R. Wilson) and Malagasy collaborators (C.B. Beandraina, B.A. Salofo, R.C. Christian, Didice, B. Papin, Rabeson, Tobey, Cressent, J. Fernando and Sassid), our field volunteers (A. Evans, T. Nowlan, K. Miles, H. Doughty, K. Galbreath, J. Larson, C. Miller and H. Davis), our Virginia Tech data-entry volunteers and A. Murphy for her collaborative efforts. We thank the Antananarivo and Maroantsetra staff of the WCS for their logistical support, particularly C. Holmes, the Antongil Conservation staff and the Madagascar Ministry of Environment, Water, Forest, and Tourism for permitting this project (permit No 128/11 and 128/12). Lastly, we are grateful to Chien C. Lee (chienclee.com) and Nick Garbutt (www.nickgarbutt.com) for allowing us to use a photograph of spotted fanaloka and small Indian civet, respectively, to illustrate this chapter.

References

Akaike, H. (1973) Information theory and an extension of the maximum likelihood principle. In: *Second International Symposium on Information Theory*, pp. 267–281. Akademinai Kiado, Budapest, Hungary.

Allnutt, T.F., Ferrier, S., Manion, G., Powell, G.V., Ricketts, T.H., Fisher, B.L., Harper, G.J., Irwin, M.E., Kremen, C. & Labat, J.N. (2008) A method for quantifying biodiversity loss and its application to a 50-year record of deforestation across Madagascar. *Conservation Letters* 1, 173–181.

Allnutt, T.F., Asner, G.P., Golden, C.D. & Powell, G.V. (2013) Mapping recent deforestation and forest disturbance in northeastern Madagascar. *Tropical Conservation Science* 6, 1–15.

Brooke, Z.M., Bielby, J., Nambiar, K. & Carbone, C. (2014) Correlates of research effort in carnivores: body size, range size and diet matter. *PLoS One* 9, e93195.

Choudhury, A., Duckworth, J.W., Timmins, R., Chutipong, W., Willcox, D.H.A., Rahman, H., Ghimirey, Y. & Mudappa, D. (2015) *Viverricula indica*. *The IUCN Red List of Threatened Species* 2015, e. T41710A45220632. http://dx.doi.org/10.2305/IUCN.UK.2015-4.RLTS.T41710A45220632.en. Accessed on 16 December 2016.

Clout, M.N. & Russell, J.C. (2008) The invasion ecology of mammals: a global perspective. *Wildlife Research* 35, 180–184.

Davis, M., Kelly, M. & Stauffer, D. (2011) Carnivore co-existence and habitat use in the Mountain Pine Ridge Forest Reserve, Belize. *Animal Conservation* 14, 56–65.

Di Bitetti, M., Paviolo, A. & De Angelo, C. (2006) Density, habitat use and activity patterns of ocelots (*Leopardus pardalis*) in the Atlantic Forest of Misiones, Argentina. *Journal of Zoology* 270, 153–163.

Farris, Z.J. (2014) *Response of Madagascar's Endemic Carnivores to Fragmentation, Hunting, and Exotic Carnivores Across Masoala–Makira Landscape.* PhD thesis, Virginia Polytechnic Institute and State University.

Farris, Z.J. & Kelly, M.J. (2011) *Assessing Carnivore Populations Across the Makira Protected Area, Madagascar: WCS Pilot Camera Trapping Study.* Wildlife Conservation Society Madagascar Program Official Report, September 2011.

Farris, Z.J., Karpanty, S.M., Ratelolahy, F. & Kelly, M.J. (2014) Predator–primate distribution, activity, and co-occurrence in relation to habitat and human activity across fragmented and contiguous forests in Northeastern Madagascar. *International Journal of Primatology* 35, 859–880.

Farris, Z.J., Gerber, B.D., Karpanty, S.M., Murphy, A., Ratelolahy, F. & Kelly, M.J. (2015a) When carnivores roam: temporal patterns and partitioning among Madagascar's native and exotic carnivores. *Journal of Zoology* 296, 45–57.

Farris, Z.J., Golden, C., Karpanty, S., Murphy, A., Stauffer, D., Andrianjakarivelo, V., Ratelolahy, F., Holmes, C. & Kelly, M.J. (2015b) Hunting, exotic carnivores, and habitat loss: anthropogenic effects on a native carnivore community, Madagascar. *PLoS One* 10, e0136456.

Farris, Z.J., Kelly, M.J., Karpanty, S.M. & Ratelolahy, F. (2015c) Patterns of spatial co-occurrence among native and exotic carnivores in NE Madagascar. *Animal Conservation* 19, 189–198.

Farris, Z.J., Kelly, M.J., Karpanty, S. & Ratelolahy, F. (2016) Patterns of spatial co-occurrence among native and exotic carnivores in north-eastern Madagascar. *Animal Conservation* 19, 189–198.

Farris, Z.J., Kelly, M.J., Karpanty, S., Murphy, A., Ratelolahy, F., Andrianjakarivelo, V. & Holmes, C. (2017a) The times they are a changin': multi-year surveys reveal exotics replace native carnivores at a Madagascar rainforest site. *Biological Conservation* 206, 320–328.

Farris, Z.J., Gerber, B.D., Valenta, K., Rafaliarison, R., Razafimahaimodison, J.C., Larney, E., Rajaonarivelo, T., Randriana, Z., Wright, P.C. & Chapman, C.A. (2017b) Threats to a rainforest carnivore community: a multi-year assessment of occupancy and co-occurrence in Madagascar. *Biological Conservation* 210, 116–124.

Farris, Z.J., Gerber, B.D., Karpanty, S., Murphy, A., Wampole, E., Ratelolahy, F. & Kelly, M.J. (2020) Exploring and interpreting spatiotemporal interactions between native and invasive carnivores across a gradient of rainforest degradation. *Biological Invasions* 22, 2033–2047.

Gantchoff, M., Libal, N.S. & Belant, J.L. (Chapter 20, this volume) Small carnivore introductions: ecological and biological correlates of success. In: *Small Carnivores: Evolution, Ecology, Behaviour, and Conservation* (eds E. Do Linh San, J.J. Sato, J.L. Belant & M.J. Somers). Wiley–Blackwell, Oxford.

Ganzhorn, J.U., Lowry, P.P., Schatz, G.E. & Sommer, S. (2001) The biodiversity of Madagascar: one of the world's hottest hotspots on its way out. *Oryx* 35, 346–348.

Gaubert, P., Patel, R.P., Veron, G., Goodman, S.M., Willsch, M., Vasconcelos, R., Lourenço, A., Sigaud, M., Justy, F., Joshi, B.D., Fickel, J. & Wilting, A. (2017) Phylogeography of the small Indian civet and origin of introductions to western Indian Ocean islands. *Journal of Heredity* 108, 270–279.

Gerber, B.D., Karpanty, S.M. & Randrianantenaina, J. (2012a) Activity patterns of carnivores in the rain forests of Madagascar: implications for species coexistence. *Journal of Mammalogy* 93, 667–676.

Gerber, B.D., Karpanty, S.M. & Randrianantenaina, J. (2012b) The impact of forest logging and fragmentation on carnivore species composition, density and occupancy in Madagascar's rainforests. *Oryx* 46, 414–422.

Goodman, S. (2003) Predation on lemurs. In: *The Natural History of Madagascar* (eds S. Goodman & J. Benstead), pp. 1159–1186. The University of Chicago Press, Chicago.

Goodman, S.M. (2012) *Les Carnivora de Madagascar.* Association Vahatra, Antananarivo. (In French).

Harper, G.J., Steininger, M.K., Tucker, C.J., Juhn, D. & Hawkins, F. (2007) Fifty years of deforestation and forest fragmentation in Madagascar. *Environmental Conservation* 34, 325–333.

Hawkins, F. (2015) *Fossa fossana. The IUCN Red List of Threatened Species* 2015, e.T8668A45197868. http://dx.doi.org/10.2305/IUCN.UK.2015-4.RLTS.T8668A45197868.en. Accessed on 20 December 2016.

Hines, J.E. (2006) *PRESENCE 6.2 – Software to Estimate Patch Occupancy and Related Parameters.* USGS-PWRC. http://www.mbr-pwrc.usgs.gov/software/presence.html.

Kerridge, F., Ralisoamalala, R., Goodman, S. & Pasnick, S. (2003) *Fossa fossana*, Malagasy striped civet, Fanaloka. In: *The Natural History of Madagascar* (eds S. Goodman & J. Benstead), pp. 1363–1365. The University of Chicago Press, Chicago.

Kotschwar, M., Gerber, B., Karpanty, S.M., Justin, S. & Rabenahy, F. (2015) Assessing carnivore distribution from local knowledge across a human-dominated landscape in central-southeastern Madagascar. *Animal Conservation* 18, 82–91.

Mack, R.N., Simberloff, D., Mark Lonsdale, W., Evans, H., Clout, M. & Bazzaz, F.A. (2000) Biotic invasions: causes, epidemiology, global consequences, and control. *Ecological Applications* 10, 689–710.

MacKenzie, D.I., Nichols, J.D., Royle, J.A., Pollock, K.H., Bailey, L.L. & Hines, J.E. (2006) *Occupancy Estimation and Modeling: Inferring Patterns and Dynamics of Species Occurrence*. Academic Press, Burlington.

McGarigal, K., Cushman, S. & Ene, E. (2012) *FRAGSTATS v4: Spatial Pattern Analysis Program for Categorical and Continuous Maps*. University of Massachusettes, Amherst. http://www.umass.edu/landeco/research/fragstats/fragstats.html.

Murphy, A., Kelly, M.J., Karpanty, S.M., Andrianjakarivelo, V. & Farris, Z.J., (2019) Using camera traps to investigate spatial co-occurrence between exotic predators and native prey species: a case study from northeastern Madagascar. *Journal of Zoology* 307, 264–273.

Myers, N., Mittermeier, R.A., Mittermeier, C.G., Da Fonseca, G.A. & Kent, J. (2000) Biodiversity hotspots for conservation priorities. *Nature* 403, 853–858.

Nowak, R.M. (2005) *Walker's Carnivores of the World*. The Johns Hopkins University Press, Baltimore.

Pollard, J. (1971) On distance estimators of density in randomly distributed forests. *Biometrics* 27, 991–1002.

Rasambainarivo, F., Farris, Z.J., Andrianalizah, H. & Parker, P.G. (2017) Interactions between carnivores in Madagascar and the risk of disease transmission. *EcoHealth* 14, 691–703.

Richmond, O.M., Hines, J.E. & Beissinger, S.R. (2010) Two-species occupancy models: a new parameterization applied to co-occurrence of secretive rails. *Ecological Applications* 20, 2036–2046.

Ridout, M.S. & Linkie, M. (2009) Estimating overlap of daily activity patterns from camera trap data. *Journal of Agricultural, Biological, and Environmental Statistics* 14, 322–337.

Salo, P., Korpimäki, E., Banks, P.B., Nordström, M. & Dickman, C.R. (2007) Alien predators are more dangerous than native predators to prey populations. *Proceedings of the Royal Society B: Biological Sciences* 274, 1237–1243.

Santulli, G., Palazón, S., Melero, Y., Gosálbez, J. & Lambin, X. (2014) Multi-season occupancy analysis reveals large scale competitive exclusion of the critically endangered European mink by the invasive non-native American mink in Spain. *Biological Conservation* 176, 21–29.

Silva-Rodríguez, E.A. & Sieving, K.E. (2012) Domestic dogs shape the landscape-scale distribution of a threatened forest ungulate. *Biological Conservation* 150, 103–110.

Su, S. & Sale, J. (2007) Niche differentiation between common palm civet *Paradoxurus hermaphroditus* and small Indian civet *Viverricula indica* in regeneration degraded forest, Myanmar. *Small Carnivore Conservation* 36, 30–34.

Tompkins, D., White, A. & Boots, M. (2003) Ecological replacement of native red squirrels by invasive greys driven by disease. *Ecology Letters* 6, 189–196.

Vanak, A.T. & Gompper, M.E. (2010) Interference competition at the landscape level: the effect of free-ranging dogs on a native mesocarnivore. *Journal of Applied Ecology* 47, 1225–1232.

Wampole, E.M., Farris, Z.J. & Gerber, B.D. (2021) A synthesis of life-history traits, functional traits, and consequences of anthropogenic pressures on Madagascar's threatened carnivorans, Eupleridae. *Mammal Review* 51, 402–419.

14

Colonization of Agricultural Landscapes by the Pine Marten: Influence of Habitat Constraints and Interspecific Competition

Luigi Remonti[1,2], Aritz Ruiz-González[3,4,5], and Alessandro Balestrieri[1,6],*

[1] Department of Earth and Environmental Sciences, University of Pavia, Pavia, Italy

[2] Institut Agricole Régional, Aosta, Italy

[3] Department of Zoology and Animal Cell Biology, University of the Basque Country (UPV-EHU), Vitoria-Gasteiz, Spain

[4] Systematics, Biogeography and Population Dynamics Research Group, University of the Basque Country (UPV/EHU), Vitoria-Gasteiz, Spain

[5] Conservation Genetics Laboratory, National Institute for Environmental Protection and Research (ISPRA), Ozzano dell'Emilia, Italy

[6] Department of Environmental Sciences and Policy, University of Milan, Milan, Italy

SUMMARY

Ranges of species can expand or contract over time due to anthropogenic and/or environmental factors. The pine marten, *Martes martes*, occurs throughout much of Europe and is generally associated with mature coniferous and mixed forests. Nonetheless, in the past 15 years, the southward range expansion of alpine pine marten populations has reached the intensively cultivated Po-Venetian plain (Northern Italy), where residual woods primarily consist of small forest fragments within an agricultural matrix. Seasonal trapping confirmed the low richness of the small mammal community and the poor prey-base of this heavily altered habitat. Road-kills and non-invasive genetic analyses revealed that the pine marten is using the best-conserved riparian forests as colonization corridors. The species is now widely distributed throughout the northwestern portion of the River Po plain, with the river, which crosses northern Italy from the west to the east, acting as a natural barrier. In contrast, genetic sampling showed the strong range contraction of the similar-sized stone marten, *Martes foina*, which was recently considered to be widespread in the plain. Analyses of habitat selection by the pine marten in this area suggest that its distribution and abundance are influenced by both the shape and the degree of fragmentation of residual forest patches and availability of riparian corridors. The importance of river valleys as expansion routes is highlighted by the high pine marten density in riparian woods. The effects of range expansions on interspecific relationships may vary from coexistence to the displacement of indigenous species by invaders. We suggest that expansion of the pine marten into habitats where the availability of animal prey is lowered by human activities may have heightened interspecific competition for food in the guild of mammalian carnivores. Because competition among species with similar morphology and food habits can be severe, range expansion by the pine marten may have led to the decrease in stone marten numbers. As suggested by the high dietary overlap, the red fox, *Vulpes vulpes*, may also have influenced interspecific interactions among the carnivore community in the Po-Venetian plain.

Keywords

activity patterns – diet – habitat selection – niche overlap – non-invasive genetic sampling – *Martes martes* – range expansion – red fox – stone marten

* Corresponding author.

Small Carnivores: Evolution, Ecology, Behaviour, and Conservation, First Edition. Edited by Emmanuel Do Linh San, Jun J. Sato, Jerrold L. Belant, and Michael J. Somers.
© 2022 John Wiley & Sons Ltd. Published 2022 by John Wiley & Sons Ltd.

Introduction

Ranges of species can change over time, expanding or contracting according to various factors, such as landscape modification, climate change, introduction of exotics, persecution by humans or hunting restrictions and protection measures (Lubina & Levin, 1988; Lensink, 1997; Parmesan & Yohe, 2003; Morrison *et al.*, 2005). Demographic trends for several mammalian carnivores have reversed in the past 20 years, resulting in either the recolonization of areas where they had disappeared during the twentieth century (e.g. the Eurasian otter, *Lutra lutra* [Prigioni *et al.*, 2007], and the grey wolf, *Canis lupus* [Boitani, 2003], in Italy) or the occupancy of novel environments (e.g. urban red foxes, *Vulpes vulpes* [Gloor *et al.*, 2001; DeCandia *et al.*, 2019], and European badgers, *Meles meles* [Geiger *et al.*, 2018], in Switzerland).

Early attempts to model the rate of such range expansions were based on population parameters, i.e. population growth and dispersal of individuals beyond the 'wave front' of a species' current range (van den Bosch *et al.*, 1992). Recent studies have recognized that both the direction and rate of expansions may be influenced by external factors, such as the availability of suitable habitats or landscape heterogeneity (Darimont *et al.*, 2005; Veech *et al.*, 2011). In fact, when expanding in heterogeneous landscapes, individuals tend to colonize preferred habitats and avoid less-suitable ones, depending on habitat availability, landscape heterogeneity and the connectivity of suitable habitat corridors (Wilson *et al.*, 2009).

The pine marten, *Martes martes*, is distributed throughout much of Europe and northern and central Asia and has been subject to long-term decline in most regions (Proulx *et al.*, 2004). The species is generally associated with forest habitats, mainly mature coniferous and mixed forests (Proulx *et al.*, 2004; Zalewski & Jędrzejewski, 2006). Deforestation and forest fragmentation affect the distribution and density of pine martens (Brainerd *et al.*, 1995; Kurki *et al.*, 1998), which generally avoid treeless areas (Storch *et al.*, 1990; Brainerd & Rolstad, 2002; Pereboom *et al.*, 2008) and are believed to need a minimum of 2 km^2 of forested habitats to survive (Zalewski & Jędrzejewski, 2006). As a consequence, the pine marten is reported to be particularly sensitive to the effects of human activities, including habitat loss and landscape-scale habitat fragmentation (Bright, 2000; Pereboom *et al.*, 2008; Mergey *et al.*, 2011).

In contrast, recent studies have suggested that the species is more generalist in terms of habitat preferences than previously reported (Virgós *et al.*, 2012). As already observed in Mediterranean Italy (De Marinis & Masseti, 1993), these studies showed that the pine marten can colonize agricultural landscapes that contain remnant forest patches and highlighted the importance of the surrounding matrix for providing food resources, den sites and complementary habitats enhancing connectivity among residual forest patches (Pereboom *et al.*, 2008; Balestrieri *et al.*, 2010; Mergey *et al.*, 2011; Caryl *et al.*, 2012).

Throughout much of Europe, the pine marten occurs sympatrically with the closely related stone marten, *Martes foina* (Proulx *et al.*, 2004). The stone marten is synanthropic in most of its geographic range (Herr *et al.*, 2009), but also inhabits more natural landscapes, particularly in the southern part of its range (Sacchi & Meriggi, 1995; Virgós *et al.*, 2000). Recent studies have shown contrasting results about the relationship between these two species at finer spatial scales. The pine marten was the only *Martes* species present in a mountainous area of northwestern Spain (Rosellini *et al.*, 2008); however, these two martens can coexist in forest–field mosaics (Posluszny *et al.*, 2007; Ruiz-González *et al.*, 2008). The coexistence of these species probably depends on a combination of several factors, including the relative abundance of each species within the local carnivore guild and food availability. However, current knowledge of their ecological relationships in areas of sympatry is extremely limited; further studies are needed to fully understand the factors influencing this complex relationship (Virgós *et al.*, 2012).

The pine marten is currently colonizing the western portion of the intensively cultivated Po-Venetian plain in northern Italy, where the stone marten also occurs. In this chapter, we review both published and unpublished information on this expansion process, highlighting the progressive colonization of intensive agricultural areas by pine martens using river valleys as ecological corridors, the role of residual forest patches for providing resources and the role of interspecific competition in the guild of mammalian predators in shaping the current distribution of these two marten species. We used non-invasive genetic techniques to monitor pine and stone marten populations in the field (Waits & Paetkau, 2005; Schwartz & Monfort, 2008). We used the standard method of transect surveys for collecting faeces (Birks *et al.*, 2004), combined with genetic analyses of deoxyribonucleic acid (DNA) obtained from faecal samples (Ruiz-González *et al.*, 2008; Ruiz-González *et al.*, 2013a). These reliable and cost-effective techniques are particularly useful for investigating the distribution, abundance and food habits of sympatric martens, resolving information gaps, and designing effective management programmes, all of which are lacking in many European countries (Proulx *et al.*, 2004).

Habitat Conditions Within the Study Area

Since the 1950s, the intensification and modernization of agricultural techniques in Europe have caused a widespread decline in landscape diversity, due to loss of natural vegetation, fragmentation of uncultivated features, increase of field size and monocultures and widespread use of herbicides and pesticides (Matson *et al.*, 1997; Robinson & Sutherland, 2002).

During the past few decades, forest cover has increased in the Alps and Apennines in Italy, while in lowlands intensive agriculture has spread even further. The Po-Venetian plain is the largest in Italy (~46 000 km^2), and is one of the most densely populated areas in the country. The pedogenetic and micromorphological characteristics of the soils of the lower plain, crossed by the River Po (652 km in length), support high levels of agricultural productivity and are intensively managed for cattle husbandry and the production of rice, maize, wheat, sugar beets, fruit and horticultural products.

Residual forests cover < 5% of the Po-Venetian plain (Falcucci *et al.*, 2007). About 70% of forests are in the western and central plain (Camerano *et al.*, 2010) and either consist of small fragments (mean patch size = 4.5 ha; Lassini *et al.*, 2007) scattered within the

agricultural matrix or, as in most European lowlands (Coles *et al.*, 1989), occur along major rivers (Figure 14.1). The largest and best-conserved riparian forests in northern Italy are located in the valley of the River Ticino, a left-bank tributary of the River Po (Figure 14.1). The river crosses an area that is intensively cultivated for cereal crops: rice and maize fields cover about 70% of the valley, followed by poplar, *Populus* spp., plantations (~15%) and towns (~5%). The resulting ecosystem mosaic is considered a corridor that non-native grey squirrels, *Sciurus carolinensis*, may use to move between the plain and the Alps (Bertolino *et al.*, 2006).

Range Expansion by the Pine Marten in the Po-Venetian Plain

Road-Kills

Available information on the status of Italian mustelids is incomplete; however, the pine marten has traditionally been associated with deciduous and coniferous forests ranging in elevation from 1000 to 2000 m a.s.l. (e.g. Spagnesi & De Marinis, 2002). Accordingly, in the past three decades of the 1900s, the stone marten was the only marten species reported for the Po-Venetian plain (Bon *et al.*, 1995; Martinoli, 2001a,b; Mantovani, 2010). At the end of the twentieth century, road-killed pine martens began to be reported in intensively cultivated areas of the western Po plain (Sindaco, 2006; Savoldelli & Sindaco, 2008). Twenty-four records of road-killed pine martens below the 300 m a.s.l. contour, which broadly marks the upper limit of the plain (Balestrieri *et al.*, 2010), revealed an exponential growth of pine marten records in the first decade of the twenty-first century and suggested a progressive range expansion by the pine marten throughout the interior of the plain (Balestrieri *et al.*, 2010). Records occurred only in the western part of the plain and most of them were associated with a river valley, suggesting that riparian zones may serve as natural expansion corridors for pine martens. All road-kill records were evaluated by expert zoologists and, in three cases, confirmed by the genetic analysis of tissue samples, following the methods described in Ruiz-González *et al.* (2008).

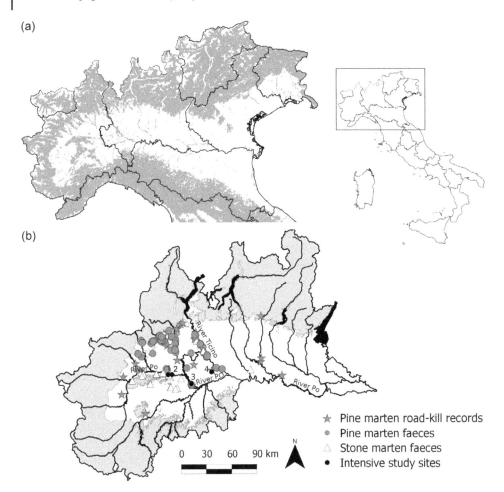

Figure 14.1 Study area: (a) distribution of forests in northern Italy (shaded area), including broad-leaved forests, coniferous forests, mixed forests and transitional woodland-shrub of the European landscape database CORINE Land Cover. *Source:* Modified from http://www.eea.europa.eu/publications/CORO-landcover; (b) distribution of major rivers, pine marten, Martes martes, and stone marten, Martes foina, records (1988–2015) and intensive study sites (1: Camino, 2: Coniolo, 3: Valenza, 4: San Massimo) in the western and central River Po plain; the shaded area is above the 300 m a.s.l. contour line, which broadly marks the upper limit of the plain. *Source:* From Balestrieri *et al.*, 2010, 2016a, Mantovani, 2010, and Remonti *et al.*, 2012.

The colonization of such apparently unsuitable habitats by the pine marten suggested a certain degree of plasticity in both its territorial (Balharry, 1993) and feeding behaviours (Jędrzejewski *et al.*, 1993; Zhou *et al.*, 2011). Nonetheless, agricultural areas could represent suboptimal habitats acting as dispersal sinks with high turnover of extinction – (re)colonization events (Kawecki, 1995; Baguette, 2004). If so, then the occurrence of pine martens in the plain would depend on the immigration of non-resident martens from surrounding areas. We used genetic analyses of faecal samples to test this hypothesis.

Non-Invasive Genetic Surveys: Species and Individual Identification

From 2007 to 2009, we surveyed linear transects for marten faeces in four study sites (Remonti *et al.*, 2012; Figure 14.1). Because the stone marten had been reported previously in all of these sites (Balestrieri *et al.*, 2010), we also searched the hay-lofts of abandoned farm buildings for its latrines (Michelat *et al.*, 2001). We used genetic techniques to identify faecal samples to species because pine marten faeces cannot be distinguished morphometrically from those of the stone marten, and can also be confused with those of other carnivores, such as the red fox and the European polecat, *Mustela putorius* (Davison *et al.*, 2002; Birks *et al.*, 2004).

Because the faeces of adult foxes are expected to be larger than those of martens (Bang & Dahlström, 1974), and a preliminary genetic survey did not identify any marten faeces larger than 10–12 mm in diameter, faecal samples larger than 15 mm were assigned to the red fox, whereas faeces between 15 and 10 mm (~10%) were discarded to optimize both the rate of success and cost-effectiveness of faeces identification. A small portion (~1 cm) of marten-like faeces <10 mm in diameter was preserved for genetic identification in 96% ethanol or by freezing; the remainder was retained for dietary analysis. DNA was isolated using the QIAamp DNA Stool Mini Kit (Qiagen).

Species Identification Using Mitochondrial DNA

The identification of faecal samples to species was accomplished using the polymerase chain reaction–restriction fragment length polymorphism (PCR–RFLP) method (Ruiz-González *et al.*, 2008), using two specifically designed primers that amplify 276 bp fragments of mitochondrial DNA (mtDNA) from both *Martes* species and from four *Mustela* species (among which only the European polecat was reported to occur in our study area); red fox mtDNA will not amplify. The simultaneous use of the restriction enzymes *Rsa*I and *Hae*III differentiates pine and stone martens from each other, and from other mustelids whose mtDNA amplifies with these primers.

Two sites ('Camino' and 'Coniolo'; Figure 14.1) contained mosaics of field and forest, extending from the bank of the River Po up to about 200 m a.s.l. Oak, *Quercus* sp., and woods covered about 25% of each area, providing patches of suitable habitat for the pine marten. Nonetheless, all 91 faecal samples genotyped from those sites were from the stone marten. In contrast, the pine marten was the dominant species in the other study sites ('Valenza' and 'San Massimo'), two flat areas covered extensively by rice and maize fields containing small, isolated forest patches. In these areas, all 109 genotyped faeces collected along linear transects were from the pine marten; however, in San Massimo, we also collected 15 stone marten faeces at a

barn (Remonti *et al.*, 2012). As reported for recovering Eurasian otters in southern Italy (Remonti *et al.*, 2008), the distribution of viable source metapopulations and the degree of resistance (permeability) within the landscape, rather than habitat suitability *per se*, may have played a major role in shaping the distribution of expanding pine marten populations.

From 2011 to 2012, the search for marten faeces was extended to the areas surrounding San Massimo, and then further upstream along a 30 km portion of the River Ticino valley, which may represent a suitable dispersal route for the pine marten from the Alps to the plain. These surveys provided 284 'marten-like' faeces, of which 177 were genotyped successfully: 165 were assigned to the pine marten (92.2%), 10 to the stone marten (5.6%) and 2 to the European polecat (1.1%; Remonti *et al.*, 2012). Stone marten samples were found inside two barns and on one transect that went through a wooded area in the northernmost part of the study area. One of these barns was the same location in which we found stone marten faeces at the periphery of the San Massimo study site in 2008 (Remonti *et al.*, 2012) and had been regularly checked for marten faeces since that time. Nonetheless, all the genotyped faeces found in the surrounding paths were from the pine marten. This pattern of stone marten distribution (i.e. extremely localized with pronounced synanthropic behaviour) suggested that the density of stone martens was much lower than that of the pine marten, and that residual populations mainly occur within human settlements (Delibes, 1983; Goszczyński *et al.*, 2007; Balestrieri *et al.*, 2019a).

Successively, to ascertain the actual level of pine marten penetration in lowland areas of north-western Italy and assess the current distribution of the stone marten, we expanded the faecal mtDNA-based survey so as to cover the whole western Po plain (Balestrieri *et al.*, 2016a).

The results showed that the pine marten was widespread throughout the northern section of the study area, with the River Po, which crosses northern Italy from the west to the east, acting as a natural barrier. In contrast, north of the River Po stone marten occurrence was negligible, suggesting that pine marten expansion coincided with a sharp reduction in stone marten range (Figure 14.1; Balestrieri *et al.*, 2016a).

Fifteen out of the 21 samples (71.5%) assigned to the pine marten through DNA sequencing belonged to the

Mediterranean phylogroup (MED), while 6 samples (28.5%) were from the Central–Northern European phylogroup (CNE). This relatively high percentage of CNE martens, which currently represents a *unicum* in southern Europe (Ruiz-González *et al.*, 2013b), is consistent with the hypothesis of an ongoing expansion of Alpine and trans-Alpine pine marten populations.

Individual Identification Using Microsatellite DNA

A sub-sample of 27 pine marten faecal samples collected from September 2008 to March 2009 in the San Massimo study site were identified to individual by analyses of microsatellite DNA.

DNA samples were genotyped at 15 variable microsatellite loci using a multiplex protocol specifically designed for the analysis of degraded faecal DNA (Ruiz-González *et al.*, 2013a). DNA quality was initially screened by amplifying each DNA sample four times at four loci using PCR (MP0188; MP0059; Gg-7; Ma-1). Only samples with > 50% positive PCRs in our initial screening were amplified four times at the remaining 11 loci. Samples with ambiguous results after four amplifications per locus or with < 50% successful amplifications across loci were removed from further analyses. Multiplex PCR products were run on an ABI 3130XL automated sequencer (Applied Biosystems, Foster City, CA) using the internal lane size standard GS500 LIZ™. Fragment analyses were conducted using the ABI software Genemapper 4.0.

Consensus genotypes from four replicates of faecal DNA samples were reconstructed using the program GIMLET, version 1.3.4 (Valière, 2002), accepting heterozygotes if both alleles occurred in at least two replicates, and accepting homozygotes if a single allele occurred in at least three replicates. To identify the number of different individuals in our data set, we grouped identical consensus genotypes using the same software. GIMLET was also used to estimate allelic dropouts (ADOs) and false alleles (FAs) (Pompanon *et al.*, 2005). ADO and FA rates were similar to those reported in similar genetic studies of these marten species (Ruiz-González *et al.*, 2013a).

To test the discriminating power of the set of 15 microsatellites, we calculated the probability of pairs of individuals having an identical multilocus genotype (P_{ID} and P_{ID-sib}) using GIMLET, version 1.3.4 (Valière, 2002). The $P_{ID-sibs}$ obtained was lower than the 0.01 threshold necessary to prevent the shadow effect (i.e. the presence of two or more individuals with the same multilocus genotype; Mills *et al.*, 2000). Thus, we are confident that this microsatellite panel can distinguish between closely related pine martens (Ruiz-González *et al.*, 2013a) and that matching genotypes represent recaptures of the same individual.

The first quality-screening test was not satisfied for six samples (22.2%), which were discarded; additional three samples were discarded after subsequent replicated genotyping of the remaining 11 loci. Full multilocus microsatellite genotypes were obtained for the remaining 18 (66.7%) samples. After a regrouping procedure, we identified six individual pine martens. The number of times each individual was detected varied from one to six ($\bar{x} = 2.83$). Three of these six individuals were identified in both September 2008 and February–March 2009, suggesting stable occupancy of the area.

To estimate pine marten abundance, we used a subsample of 14 genotyped faeces collected in an interior forest patch within the study site during three weeks in September 2008. Because the number of genetic profiles was small, population size was estimated using a rarefaction curve method, based on the function $y = ax/(b+x)$, where y is the cumulative number of individual genotypes, x is the number of samples, a is the asymptote that estimates population size and b is the rate of decline of the curve. The regression was repeated after randomization of the sample, because the order by which faeces are analyzed affects the shape of the curve and, hence, the magnitude of the asymptote (Kohn *et al.*, 1999; Wilson *et al.*, 2003; Prigioni *et al.*, 2006). The mean value of the estimated parameter a and the corresponding 95% confidence interval (CI) were calculated.

One-hundred and sixty iterations of the rarefaction analysis (Figure 14.2) resulted in an estimate of 9–11 (95% CI) pine martens in an area about 1 km² in size. The relatively high number of pine martens found at this site was not consistent with the sole occurrence of roaming, non-resident animals. Anyway, the small sizes of both the faecal samples and the area where faeces were collected did not

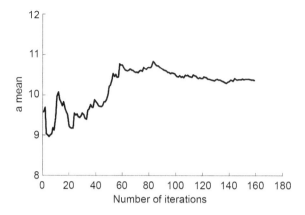

Figure 14.2 Mean values for the asymptote of the rarefaction curve (a mean) relating the number of genotyped pine marten faecal samples to the cumulative number of genotypes identified, plotted against the number of iterations.

allow for reliable comparisons with density data (0.03–1.75 individual(s)/km^2; Zalewski & Jędrzejewski, 2006) from other European areas.

To get over this limitation, we then assessed pine marten population density in a 12 km^2 large area of the River Ticino valley by applying a faecal DNA-based genetic census (Balestrieri *et al.*, 2016b). All the faecal samples identified by the PCR-RFLP method as belonging to pine marten were genotyped at 15 microsatellite loci using a multiplex protocol. We identified 15 different individual genotypes corresponding to a density of 1.83 (1.25–3.16; 95% CI) individuals/km^2, as assessed using capwire estimators. Density was among the highest ever recorded in continental Europe and similar to that reported for fragmented landscapes of Ireland (0.5–3.0 individual(s)/km^2; Mullins *et al.*, 2010; Sheehy *et al.*, 2014). The record of three family groups demonstrated that the pine marten can successfully reproduce in the study area, confirming that its population is not strictly dependent on the persistence of immigration (Balestrieri *et al.*, 2016b).

Camera-Trapping and Temporal Niche Partitioning

Over the past two decades, camera-trapping has been an important source of behavioural and ecological data for secretive species such as most mustelids (e.g. Macdonald *et al.*, 2004; Mori *et al.*, 2014; Monterroso

et al., 2014, 2016). Whilst their utility for assessing mustelid density requires further trials to be demonstrated unequivocally (Balestrieri *et al.*, 2016b), camera-traps represent a powerful method for the analysis of activity patterns (Ridout & Linkie, 2009).

We used unbaited, passive infrared camera-traps (MultiPIR SG550) for a total of 62 trap-sites and 400 trap-days, setting cameras as to record 15-s long video clips. Diel activity patterns were estimated non-parametrically through the probability density function, using Kernel density estimate (Ridout & Linkie, 2009). The pine marten showed a peak of activity between 09:00 and 11:00 h, with 51.4% of records having been collected during daylight. Nevertheless, according to Ivlev's index, it did not select any period of the diel cycle. This cathemeral pattern has been successively confirmed for other study areas of northern Italy, suggesting that pine marten's higher competitive ability with respect to the stone marten may depend on its higher ability to avoid interference competition within the local community of mainly nocturnal mammalian predators (Fonda *et al.*, 2017; Torretta *et al.*, 2017).

Food Habits and Interspecific Competition

Using standard procedures (Balestrieri *et al.*, 2011; Remonti *et al.*, 2012), we analyzed the contents of 195 genotyped pine marten faeces collected from 2008 to 2012 in the valley of the River Ticino. We investigated seasonal variation in the percent mean volume (%mV = total estimated percent volume of each food item ingested/total number of examined faeces) of primary pine marten food items using one-way analysis of similarities (ANOSIM), based on a Euclidean similarity matrix. The one-tailed significance level was computed by 10 000 permutations of group membership. Differences in the consumption of individual food categories were investigated using the Kruskal–Wallis test. All comparisons were performed using the software PAST (Hammer *et al.*, 2001).

Mammals (especially rodents) and fruit formed the bulk of the pine marten diet; birds and lagomorphs were important alternative sources of animal food (Table 14.1). The consumption of forest rodents, such as the bank vole, *Myodes glareolus*, by the pine marten at levels greater than expected based on relative

Table 14.1 Percent mean volume (%mV) of the main foods of the pine marten, *Martes martes*, in the western River Po plain, based on the analysis of 195 faeces.

	Winter (January–March)	Spring (April–June)	Summer (July–September)	Autumn (October–December)
Fruits	1.61	22.78	67.10	14.62
Insects	1.61	1.98	4.84	1.28
Birds	16.61	18.49	7.42	9.62
Mammals	71.61	52.46	20.00	72.69
Lagomorphs	20.65	19.13	14.19	2.56
Rodents	47.10	33.33	5.81	70.13
Myodes glareolus	16.13	4.29	1.61	38.33
Apodemus sp.	4.84	13.81	0	11.54

availability (Balestrieri *et al.*, 2011), confirmed the mustelid's preference for forested habitats. One-way ANOSIM revealed significant seasonal variation in the pine marten diet ($R = 0.12$, $p = 0.0001$), with either fruit (adjusted $H = 66.73$, df $= 3$, $p < 0.00001$) or rodents (adjusted $H = 37.27$, df $= 3$, $p < 0.00001$) being the most important food category during different seasons of the year. Contrary to findings from higher latitudes (Jędrzejewski *et al.*, 1993; Posluszny *et al.*, 2007), fruit was clearly the most important food during the summer months.

To evaluate the importance of exploitative competition within the guild of mammalian carnivores after colonization by the pine marten, we assessed trophic overlap between both the stone marten (Camino and Coniolo study sites, $n = 91$) and pine marten (Valenza and San Massimo study sites, $n = 109$) and the red fox ($n = 78$ and 114, respectively), which was widespread in all sites both before and after pine marten colonization (Prigioni, 2001). Because none of our study sites contained sympatric populations of pine and stone martens, we could not compare trophic niches of the two marten species directly.

Red fox and stone marten diets overlapped extensively, in terms of both food items eaten and the prevalence of major food items, whereas the red fox and the pine marten had a broad, but less pronounced, dietary overlap (Remonti *et al.*, 2012). Extensive overlaps in the diet of sympatric carnivores indicate the potential for exploitative competition, which is predicted to be intense if resources are limited (Creel, 2001). In intensively cultivated areas, reductions in habitat quality and habitat fragmentation are reportedly the primary causes of declines in biological diversity (Stoate *et al.*, 2001; Benton *et al.*, 2003; Sánchez-Zapata *et al.*, 2003). In agricultural landscapes dominated by arable lands, small mammals (the primary prey for a wide variety of predators) are mainly confined to residual forest patches and other non-cropped areas scattered in cultivated fields (Fitzgibbon, 1997); in such areas, rodent communities have lower species diversity (Millán de la Peña *et al.*, 2003) and biomass (Michel *et al.*, 2006) than those in forested habitats. The small mammal community at the Valenza study site is dominated by only two species, the bank vole and the wood mouse, *Apodemus sylvaticus* (Balestrieri *et al.*, 2017). In 2011 and 2012, the two main habitats in the study area – forest patches and cereal fields – were sampled each season using trap-strips consisting of 14–15 trapping sites spaced 20 m apart containing three different traps (Sherman, snap, and pitfall traps). Each trapping session lasted four consecutive nights. Observed and expected frequencies for both the bank vole and the wood mouse in each habitat type were compared using the χ^2 test with Yates' correction for small sample sizes. Expected frequencies were calculated based on the number of trap-nights in each habitat type.

A total of 2012 trap-nights resulted in the capture of 133 individuals of three species: the wood mouse ($n = 104$), bank vole ($n = 27$), and common shrew, *Sorex araneus* ($n = 2$). Forests were clearly selected by

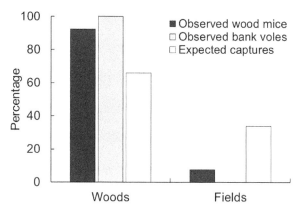

Figure 14.3 Comparison between expected and observed percentages of wood mice, *Apodemus sylvaticus*, and bank voles, *Myodes glareolus*, trapped in woods and cereal fields. Expected percentages were based on the number of trap-nights in each habitat type.

both the wood mouse ($\chi^2 = 30.8$, df = 1, $p < 0.0001$) and the bank vole ($\chi^2 = 8.2$, df = 1, $p = 0.008$). Only 6% of captured small mammals came from cultivated fields next to the wood border (Figure 14.3). We hypothesized that range expansion by the pine marten in such poor habitat conditions may have heightened interspecific competition for food among the guild of mammalian carnivores, altering the previous equilibrium and leading to the decline of the similar-sized stone marten (Remonti *et al.*, 2012). As suggested by high dietary overlap among red foxes, pine martens and stone martens in our study sites, red foxes may have influenced interspecific interactions among these carnivores.

Recently, a wide literature has demonstrated that in many species foraging is mainly driven by the need for regulating the amounts and balance of macronutrients – protein, lipids and carbohydrates – to an intake target (Kohl *et al.*, 2015). In agricultural landscapes dominated by crop cultivations, the decline of animal prey may lead to an unbalanced diet for predators, with implications for either physiology or ecology (Remonti *et al.*, 2011; Balestrieri *et al.*, 2019b). To test this hypothesis, we reviewed available data on the foods eaten by pine marten across Europe, estimated the percentage of macronutrients in each diet and then analyzed both seasonal and geographic variation in the percentage of macronutrients associated with the recorded diets. In the western Po plain, dietary

macronutrient ratios of pine marten differed from the target to an almost negligible extent. The poor prey base of the study area in terms of small rodents was compensated by the pine marten by relying on introduced Eastern cottontail, *Sylvilagus floridanus*, confirming that plasticity in feeding behaviour plays a major role in the ability of pine marten to colonize agricultural areas (Remonti *et al.*, 2016). In addition, a recent comparative analysis showed that in Europe pine and stone martens have very similar nutrient profiles when living in allopatry (Gazzola & Balestrieri, 2020). However, in sympatric areas the stone marten eats more carbohydrates (fruits) and less protein than the pine marten. This may partially result from interspecific competition, suggesting once again that the pine marten may be the superior competitor.

Influence of Habitat Characteristics on Pine Marten Abundance

We used faecal samples collected from October 2011 to June 2012 in the River Ticino valley to investigate habitat selection by pine martens (Balestrieri *et al.*, 2015). The analysis was performed at two spatial scales. At the transect scale, we assigned each sample to the habitat where it was found and compared habitat use to available habitat types using the χ^2 test and Bonferroni's CI of the proportion of use (White & Garrott, 1990). At a broader scale, we sampled 21 grid squares of 2×2 km and related pine marten marking intensity to seven habitat variables of potential importance to pine martens by a linear multiple regression (Figure 14.4).

At the transect scale, pine marten faeces were not distributed in accordance with habitat availability; rather, wooded areas were preferred and fields avoided. At the grid scale, the best model included the mean area of wooded patches, which was positively correlated with pine marten marking intensity, whereas the second model included the mean perimeter–area ratio of forest patches, which was negatively correlated with marking intensity (Balestrieri *et al.*, 2015). Thus, although the pine marten is capable of colonizing highly fragmented agricultural habitats (Pereboom *et al.*, 2008; Caryl *et al.*, 2012), both the size and shape of residual forest patches, that provide both food and cover (Zalewski & Jędrzejewski, 2006; Goszczyński

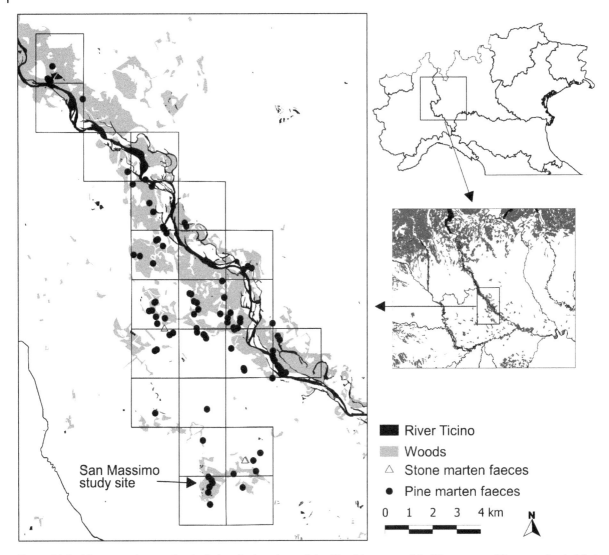

Figure 14.4 Maps at various scales depicting the locations of the 21 grid squares of 2 × 2 km surveyed for assessing habitat selection by the pine marten. *Source:* Modified from Balestrieri *et al.* (2015). Reproduced by permission of the Ecological Society of Japan.

et al., 2007; Balestrieri *et al.*, 2019a), influenced its distribution and abundance in riparian corridors.

When expanding in heterogeneous landscapes, individuals should tend to colonize preferred habitats and avoid less-suitable ones, depending on habitat availability, landscape heterogeneity and connectivity of suitable habitat corridors (Wilson *et al.*, 2009). To identify the main factors driving the colonization of lowlands by the pine marten and predict its potential south- and eastwards expansion, we collected available occurrence data of the pine marten in the western

River Po plain and related them to a set of environmental variables by developing nine different Species Distribution Models and using average ensemble predictions (Balestrieri *et al.*, 2016c).

Distance from water and distance from woods played a main role in shaping pine marten distribution, confirming the importance of riparian corridors and woods as key habitat features for this species (see also Balestrieri *et al.*, 2019a). Based on the relatively large availability of suitable areas, the pine marten may further expand in the western lowland, whilst the negligible residual wood cover of large areas in the central and eastern plain makes the habitat unsuitable for the pine marten (Balestrieri *et al.*, 2016c).

Conclusions

The pine marten has long been considered a strictly forest-dwelling species. It has been reported to be arboreal (Goszczyński *et al.*, 2007), select mature forests and avoid open areas (Brainerd & Rolstad, 2002; Zalewski & Jędrzejewski, 2006; Goszczyński *et al.*, 2007) and feed primarily on small mammals, such as the bank vole (Jędrzejewski *et al.*, 1993; Pulliainen & Ollinmäki, 1996; Sidorovich *et al.*, 2005; Balestrieri *et al.*, 2011). Moreover, historical variation in the range of the pine marten in Europe has been related to changes in forest distribution. During the maximum ice expansion of the Weichselian glaciation, disjunct pine marten populations persisted within forested glacial refugia on both the Iberian and Italian Peninsulas and in the Carpathian Mountains (Sommer & Benecke, 2004; see also Sommer & Crees, Chapter 4, this volume), which led to the emergence of a Mediterranean mtDNA phylogroup (Ruiz-González *et al.*, 2013b). Both genetic and fossil data suggest that the postglacial recolonization of central and northern Europe by pine martens was from source populations that persisted in Mediterranean peninsulas and in a cryptic central-European glacial refugium (Ruiz-González *et al.*, 2013b), as forests advanced during glacial retreat (Sommer & Benecke, 2004).

During the last few centuries, agriculture has substantially modified the environment. In montane areas of Europe, the decrease in forest cover from the expansion of low-intensity farming and livestock rearing has reversed during recent decades, following widespread agricultural abandonment (MacDonald *et al.*, 2000). In the Alps, forest cover increased by about 50% from the 1960s to 2000, replacing open pastures (Falcucci *et al.*, 2007), which would have had a positive effect on forest-dwelling species (MacDonald *et al.*, 2000). Alpine pine marten populations, which are known to have high densities and small home ranges in mature and productive forests (indicative of high carrying capacities), may have taken advantage of this increase in forest cover (Zalewski & Jędrzejewski, 2006). A

similar scenario was reported for the roe deer, *Capreolus capreolus*, a forest ungulate that has increased rapidly in density and progressively expanded into open, subalpine habitats during the twentieth century (Kaluzinski, 1974; Tellería & Virgós, 1997; Jepsen & Topping, 2004). Moreover, the ongoing recovery of the pine marten in both Ireland (O'Mahony *et al.*, 2006, 2012) and Scotland (Croose *et al.*, 2013) has been related to the expansion of commercial forestry. A southward range expansion by alpine pine martens that probably used riparian forests as dispersal corridors is supported by the presence of a high percentage of Central–Northern European pine marten haplotypes among samples from the Po plain.

The habitat characteristics of the lower River Ticino valley have not changed substantially during the last six decades, because the pressure of intensive agriculture does not allow the recovery of forested habitats. Although riparian woods are currently more widespread north of our study area, the overall percentage of woods changed from 6.1% in 1954, to 6.2% in 1980, and 5.5% in 2013 (Figure 14.5). Consequently, we hypothesize that the expansion of pine marten populations at lower elevations is driven by the saturation of alpine populations, rather than by increases in habitat suitability in recently colonized areas. Because of its behavioural flexibility, the pine marten is able to persist in heavily fragmented landscapes by exploiting isolated forest patches and wooded slopes surrounded by agricultural land (Virgós *et al.*, 2012), as was reported in rural portions of France (Pereboom *et al.*, 2008; Mergey *et al.*, 2011).

When a predator joins a stable community, either exploitative or interference competition can occur. In the Po plain, the pine marten has joined a relatively species-poor assemblage of up to five mammalian predators (Remonti *et al.*, 2012). Species that either belong to the same functional or taxonomic groups (Rosenzweig, 1966) or have similar body mass (Kelt & Brown, 1999; Campbell, 2004) are less likely to coexist. For example, expansion of the coyote, *Canis latrans*, in the midwestern United States, paralleled decreases in bobcat, *Lynx rufus*, and red fox numbers, or their shift from valley bottoms to less-productive areas (DeBow *et al.*, 1998). Within the carnivore community that occupies the Po plain, the pine and stone martens are the most similar in morphology and feeding habits

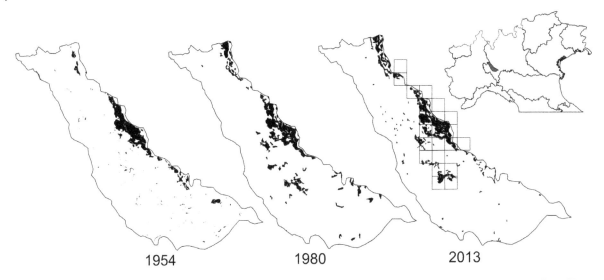

1954   1980   2013

Figure 14.5 Variation in wood cover from 1954 to 2013 in the western valley of the downstream stretch of the River Ticino (about 400 km²), based on land-cover shape files available at the institutional website of the Lombardy Region (http://www. cartografia.regione.lombardia.it/geoportale/ptk). The 2 × 2 km grid that was surveyed to assess habitat selection by the pine marten is shown in the map for 2013 (see Figure 14.4).

(Larroque *et al.*, 2015) and, thus, the most likely to compete (Powell & Zielinski, 1983). Although they have been reported to coexist in the same forest patches in field–forest mosaics in Poland and Spain (Posluszny *et al.*, 2007; Ruiz-González *et al.*, 2008), in heavily human-altered habitats the reduced availability of food resources may increase the severity of competition, leading to the decline of the species with the greatest competitive disadvantage (Powell & Zielinski, 1983).

Results from the studies reported here have helped elucidating the ecological implications of range expansions by pine martens into apparently unsuitable habitats. Our ongoing research programme involves the use of species distribution models to compare pine and stone marten habitat selection in both rural and montane sympatric areas, to assess how environmental characteristics shape the relative distribution of the two species. As knowledge on the distribution and abundance of *Martes* spp. in Alpine and sub-Alpine areas is still relatively poor, several opportunities are still available for field research.

Acknowledgements

A. Ruiz-González held a postdoctoral fellowship awarded by the Department of Education, Universities and Research of the Basque Government (Ref. DKR-2012-64). This study was partially funded by the Basque Government through the research group on 'Systematics, Biogeography and Population Dynamics' (Ref. IT317-10, GIC10/76).

References

Baguette, M. (2004) The classical metapopulation theory and the real, natural world: a critical appraisal. *Basic and Applied Ecology* 5, 213–224.

Balestrieri, A., Remonti, L., Ruiz-González, A., Gómez-Moliner, B.J., Vergara, M. & Prigioni, C. (2010) Range expansion of the pine marten (*Martes martes*) in an agricultural landscape matrix (NW Italy). *Mammalian Biology* 75, 412–419.

Balestrieri, A., Remonti, L., Ruiz-González, A., Capelli, E., Gómez-Moliner, B.J. & Prigioni, C. (2011) Food habits of genetically identified pine martens (*Martes martes*) expanding in agricultural lowlands (NW Italy). *Acta Theriologica* 56, 199–207.

Balestrieri, A., Remonti, L., Ruiz-González, A., Zenato, M., Gazzola, A., Dettori, E.E., Saino, N., Capelli, E., Gómez-Moliner, B.J., Guidali, F. & Prigioni, C. (2015)

Distribution and habitat use by pine marten *Martes martes* in a riparian corridor crossing intensively cultivated lowlands. *Ecological Research* 30, 153–162.

Balestrieri, A., Ruiz-González, A., Capelli, E., Vergara, M., Prigioni, C. & Saino, N. (2016a) Pine marten vs. stone marten in agricultural lowlands: a landscape-scale, genetic survey. *Mammal Research* 61, 327–335.

Balestrieri, A., Ruiz-González, A., Vergara, M., Capelli, E., Tirozzi, P., Alfino, S., Minuti, G., Prigioni, C. & Saino, N. (2016b) Pine marten density in lowland riparian woods: a test for the Random Encounter Model. *Mammalian Biology* 81, 439–446.

Balestrieri, A., Bogliani, G., Boano, G., Ruiz-González, A., Saino, N. & Milanesi, P. (2016c) Modelling the distribution of forest-dependent species in human-dominated landscapes: patterns for the pine marten in intensively cultivated lowlands. *PLoS One* 11, e0158203.

Balestrieri, A., Remonti, L., Morotti, L., Saino, N., Prigioni, C. & Guidali, F. (2017) Multilevel habitat preferences of *Apodemus sylvaticus* and *Clethrionomys glareolus* in an intensively cultivated agricultural landscape. *Ethology Ecology & Evolution* 29, 38–53.

Balestrieri, A., Mori, E., Menchetti, M., Ruiz-González, A. & Milanesi, P. (2019a) Far from the madding crowd: tolerance towards human disturbance shapes distribution and connectivity patterns of closely related *Martes* spp. *Population Ecology* 61, 289–299.

Balestrieri, A., Remonti, L., Saino, N. & Raubenheimer, D. (2019b) The 'omnivorous badger dilemma': towards an integration of nutrition with the dietary niche in wild mammals. *Mammal Review* 49, 324–339.

Balharry, D. (1993) Social organization in martens: an inflexible system? *Symposia of the Zoological Society of London* 65, 321–345.

Bang, P. & Dahlström, P. (1974) *Animal Tracks*. Collins, London.

Benton, T.G., Vickery, J.A. & Wilson, J.D. (2003) Farmland biodiversity: is habitat heterogeneity the key? *Trends in Ecology & Evolution* 18, 182–188.

Bertolino, S., Lurz, P.W.W. & Rushton, S.P. (2006) *Prediction of the Grey Squirrel Spread from Italy into Adjacent Countries.* Report prepared for the European Squirrel Initiative (ESI). http://www.europeansquirrelinitiative.org/Italy1.pdf. Accessed on 17 January 2013.

Birks, J.D.S., Messenger, J.E., Braithwaite, T.C., Davison, A., Brookes, R.C. & Strachan, C. (2004) Are scat surveys a reliable method for assessing distribution and population status of pine martens? In: *Martens and Fishers (*Martes*) in Human-Altered Environments: An International Perspective* (eds D.J. Harrison, A.K. Fuller & G. Proulx), pp. 235–252. Springer-Verlag, New York.

Boitani, L. (2003) Wolf conservation and recovery. In: *Wolves. Behaviour, Ecology, and Conservation* (eds L.D. Mech & L. Boitani), pp. 317–340. The University of Chicago Press, Chicago.

Bon, M., Paolucci, P., Mezzavilla, F., Battisti, R. & Vernier, E. (1995) *Atlante dei Mammiferi del Veneto*. Lavori della Società Veneziana di Scienze Naturali 21 (Suppl.). Grafic House editrice, Mestre. (In Italian).

Brainerd, S.M. & Rolstad, J. (2002) Habitat selection by Eurasian pine martens *Martes martes* in managed forests of southern boreal Scandinavia. *Wildlife Biology* 8, 289–297.

Brainerd, S.M., Helldin, J.O., Lindstrom, E.R., Rolstad, E., Rolstad, J. & Storch, I. (1995) Pine marten (*Martes martes*) selection of resting and denning sites in scandinavian managed forests. *Annales Zoologici Fennici* 32, 151–157.

Bright, P.W. (2000) Lessons from lean beasts: conservation biology of the mustelids. *Mammal Review* 30, 217–226.

Camerano, P., Grieco, C. & Terzuolo, P.G. (2010) *I Boschi Planiziali: Conoscenza, Conservazione e Valorizzazione.* Regione Piemonte e Blu edizioni, Torino. (In Italian).

Campbell, L.A. (2004) *Distribution and Habitat Associations of Mammalian Carnivores in the Central and Southern Sierra Nevada.* PhD thesis, University of California, Davis.

Caryl, F.M., Quine, C.P. & Park, K.J. (2012) Martens in the matrix: the importance of nonforested habitats for forest carnivores in fragmented landscapes. *Journal of Mammalogy* 93, 464–474.

Coles, T.F., Southey, J.M., Forbes, I. & Clough, T. (1989) River wildlife data bases and their value for sensitive environmental management. *Regulated Rivers* 4, 179–189.

Creel, S. (2001) Four factors modifying the effect of competition on carnivore population dynamics as illustrated by African wild dogs. *Conservation Biology* 15, 271–274.

Croose, E., Birks, J.D.S. & Schofield, H.W. (2013) *Expansion Zone Survey of Pine Marten (*Martes martes*) Distribution in Scotland.* Scottish Natural Heritage Commissioned Report No. 520.

Darimont, C.T., Paquet, P.C., Reimchen, T.E. & Crichton, V. (2005) Range expansion by moose into coastal temperate rainforests of British Columbia, Canada. *Diversity and Distributions* 11, 235–239.

Davison, A., Birks, J.D.S., Brookes, R.C., Braithwaite, T.C. & Messenger, J.E. (2002) On the origin of faeces: morphological versus molecular methods for surveying rare carnivores from their scats. *Journal of Zoology* 257, 141–143.

DeBow, T.M., Webster, W.M.D. & Sumner, P.W. (1998) Range expansion of the coyote, *Canis latrans* (Carnivora: Canidae), into North Carolina; with comments on some management implications. *The Journal of the Elisha Mitchell Scientific Society* 114, 113–118.

DeCandia, A.L., Brzeski, K.E., Heppenheimer, E., Caro, C.V., Camenisch, G., Wandeler, P., Driscoll, C. & vonHoldt, B.M. (2019) Urban colonization through multiple genetic lenses: the city-fox phenomenon revisited. *Ecology and Evolution* 9, 2046–2060.

Delibes, M. (1983) Interspecific competition and the habitat of the stone marten *Martes foina* (Erxleben, 1777) in Europe. *Acta Zoologica Fennica* 174, 229–231.

De Marinis, A.M. & Massetti, M. (1993) Distribution of the pine marten *Martes martes* L., 1758 (Mammalia, Carnivora) on the island of Elba, northern Tyrrhenian Sea. *Supplementi di Ricerca di Biologia della Selvaggina* 21, 263–267.

Falcucci, A., Maiorano, L. & Boitani, L. (2007) Changes in land-use/land-cover patterns in Italy and their implications for biodiversity conservation. *Landscape Ecology* 22, 617–631.

Fitzgibbon, C.D. (1997) Small mammals in farm woodlands: the effects of habitat, isolation and surrounding land-use patterns. *Journal of Applied Ecology* 34, 530–539.

Fonda, F., Torretta, E., Balestrieri, A. & Pavanello, M. (2017) Time partitioning in pine- and stone marten from the Carnic Pre-Alps (NE Italy). Poster presented at the 32nd European Mustelid Colloquium, Lyon, 15–17 November 2017.

Gazzola, A. & Balestrieri, A. (2020) Nutritional ecology provides insights into competitive interactions between closely related *Martes* species. *Mammal Review* 50, 82–90.

Geiger, M., Taucher, A.L., Gloor, S., Hegglin, D. & Bontadina, F. (2018) In the footsteps of city foxes: evidence for a rise of urban badger populations in Switzerland. *Hystrix, the Italian Journal of Zoology* 29, 236–238.

Gloor, S., Bontadina, F., Hegglin, D., Deplazes, P. & Breitenmoser, U. (2001) The rise of urban fox populations in Switzerland. *Mammalian Biology* 66, 155–164.

Goszczyński, J., Posłuszny, M., Pilot, M. & Gralak, B. (2007) Patterns of winter locomotion and foraging in two sympatric marten species: *Martes martes* and *Martes foina*. *Canadian Journal of Zoology* 85, 239–249.

Hammer, Ø., Harper, D.A.T. & Ryan, P.D. (2001) PAST: paleontological statistics software package for education and data analysis. *Palaeontologia Electronica* 4, 1–9.

Herr, J., Schley, L. & Roper, T.J. (2009) Socio-spatial organization of urban stone martens. *Journal of Zoology* 277, 54–62.

Jędrzejewski, W., Zalewski, A. & Jędrzejewska, B. (1993) Foraging by pine marten *Martes martes* in relation to food resources in Białowieża National Park, Poland. *Acta Theriologica* 38, 405–426.

Jepsen, J.U. & Topping, C.J. (2004) Modelling roe deer (*Capreolus capreolus*) in a gradient of forest fragmentation: behavioural plasticity and choice of cover. *Canadian Journal of Zoology* 82, 1528–1541.

Kaluzinski, J. (1974) The occurrence and distribution of field ecotype of roe deer in Poland. *Acta Theriologica* 19, 291–300.

Kawecki, T.J. (1995) Demography of source-sink populations and the evolution of ecological niches. *Evolutionary Ecology* 9, 38–44.

Kelt, D.A. & Brown, J.H. (1999) Community structure and assembly rules: confronting conceptual and statistical issues with data on desert rodents. In: *Ecological Assembly Rules: Perspectives, Advances and Retreats* (eds E. Weiher & P. Keddy), pp. 75–107. Cambridge University Press, Cambridge.

Kohl, K.D., Coogan, S.C.P. & Raubenheimer, D. (2015) Do wild carnivores forage for prey or for nutrients? *Bioessays* 37, 701–709.

Kohn, M., York, E.C., Kamradt, D.A., Haught, G., Sauvajot, R.M. & Wayne, R.K. (1999) Estimating population size by genotyping feces. *Proceedings of the Royal Society of London, B. Biological Sciences* 266, 657–663.

Kurki, S., Nikula, A., Helle, P. & Linden, H. (1998) Abundances of red fox and pine marten in relation to the composition of boreal forest landscapes. *Journal of Animal Ecology* 67, 874–886.

Larroque, J., Ruette, S., Vandel, J.-M. & Devillard, S. (2015) Where to sleep in a rural landscape? A comparative study of resting sites pattern in two syntopic *Martes* species. *Ecography* 38, 1129–1140.

Lassini, P., Monzani, F. & Pileri, P. (2007) A green vision for the renewal of the Lombardy landscape. In: *Europe's Living Landscapes. Essays on Exploring our Identity in the Countryside* (eds B. Pedroli, A. Van Doorn, G. De Blust, M.L. Paracchini, D. Wascher & F. Bunce), pp. 83–100. Landscape Europe, Wageningen and KNNV Publishing, Zeist.

Lensink, R. (1997) Range expansion of raptors in Britain and the Netherlands since the 1960s: testing an individual-based diffusion model. *Journal of Animal Ecology* 66, 811–826.

Lubina, J.A. & Levin, S.A. (1988) The spread of a reinvading species: range expansion in the California sea otter. *The American Naturalist* 131, 526–543.

Macdonald, D., Crabtree, J.R., Weisinger, G., Dax, T., Stamou, N., Fleury, P., Gutierrez Lazpita, J. & Gibon, A. (2000) Agricultural abandonment in mountain areas of Europe: environmental consequences and policy response. *Journal of Environmental Management* 59, 47–69.

Macdonald, D.W., Buesching, C.D., Stopka, P., Henderson, J., Ellwood, S.A. & Baker, S.E. (2004) Encounters between two sympatric carnivores: red foxes (*Vulpes vulpes*) and European badgers (*Meles meles*). *Journal of Zoology* 263, 385–392.

Mantovani, S. (2010) Recenti segnalazioni della martora, *Martes martes*, in provincia di Cremona. *Pianura* 25, 95–107.

Martinoli, A. (2001a) *Martes martes*. In: *Atlante dei Mammiferi della Lombardia* (eds C. Prigioni, M. Cantini & A. Zilio), pp. 236–238. Regione Lombardia e Università degli Studi di Pavia. (In Italian).

Martinoli, A. (2001b) *Martes foina*. In: *Atlante dei Mammiferi della Lombardia* (eds C. Prigioni, M. Cantini & A. Zilio), pp. 240–242. Regione Lombardia e Università degli Studi di Pavia. (In Italian).

Matson, P.A., Parton, W.J., Power, A.G. & Swift, M.J. (1997) Agricultural intensification and ecosystem properties. *Science* 277, 504–509.

Mergey, M., Helder, R. & Roeder, J.-J. (2011) Effect of forest fragmentation on space-use patterns in the European pine marten (*Martes martes*). *Journal of Mammalogy* 92, 328–335.

Michel, N., Burel, F. & Butet, A. (2006) How does landscape use influence small mammal diversity, abundance and biomass in hedgerow networks of farming landscapes? *Acta Oecologica* 30, 11–20.

Michelat, D., Quéré, J.P. & Giraudoux, P. (2001) Caractéristiques des gîtes utilisés par la fouine (*Martes foina*, Erxleben, 1777) dans le Haut-Doubs. *Revue suisse de Zoologie* 108, 263–274. (In French with English abstract).

Millán De La Peña, N., Butet, A., Delettre, Y., Paillat, G., Morant, P. & Burel, F. (2003) Response of the small mammal community to changes in western French agricultural landscapes. *Landscape Ecology* 18, 265–278.

Mills, L.S., Citta, J.J., Lair, K.P., Schwartz, M.K. & Tallmon, D.A. (2000) Estimating animal abundance using noninvasive DNA sampling: promise and pitfalls. *Ecological Applications* 10, 283–294.

Monterroso, P., Alves, P.C. & Ferreras, P. (2014) Plasticity in circadian activity patterns of mesocarnivores in Southwestern Europe: implications for species coexistence. *Behavioral Ecology and Sociobiology* 68, 1403–1417.

Monterroso, P., Rebelo, P., Alves, P.C. & Ferreras, P. (2016) Niche partitioning at the edge of the range: a multidimensional analysis with sympatric martens. *Journal of Mammalogy* 97, 928–939.

Mori, E., Menchetti, M. & Balestrieri, A. (2014) Interspecific den sharing: a study on European badger setts using camera traps. *Acta Ethologica* 18, 121–126.

Morrison, L.W., Korzukhin, M.D. & Porter, S.D. (2005) Predicted range expansion of the invasive fire ant, *Solenopsis invicta*, in the eastern United States based on the VEMAP global warming scenario. *Diversity and Distributions* 11, 199–204.

Mullins, J., Statham, M., Roche, T., Turner, P. & O'Reilly, C. (2010) Remotely plucked hair genotyping: a reliable and non-invasive method for censusing pine marten (*Martes martes*, L. 1758) populations. *European Journal of Wildlife Research* 56, 443–453.

O'Mahony, D., O'Reilly, C. & Turner, P. (2006) National pine marten survey of Ireland 2005. *Coford Connects, Environment* 7, 1–8.

O'Mahony, D., O'Reilly, C. & Turner, P. (2012) Pine marten (*Martes martes*) distribution and abundance in Ireland: a cross-jurisdictional analysis using non-invasive genetic survey techniques. *Mammalian Biology* 77, 351–357.

Parmesan, C. & Yohe, G. (2003) A globally coherent fingerprint of climate change impacts across natural systems. *Nature* 421, 37–42.

Pereboom, V., Mergey, M., Villerette, N., Helder, R., Gerard, J.F. & Lodé, T. (2008) Movement patterns, habitat selection, and corridor use of a typical woodland-dweller species, the European pine marten (*Martes martes*), in fragmented landscape. *Canadian Journal of Zoology* 86, 983–991.

Pompanon, F., Bonin, A., Bellemain, E. & Taberlet, P. (2005) Genotyping errors: causes, consequences and solutions. *Nature Reviews Genetics* 6, 847–859.

Posluszny, M., Pilot, M., Goszczynski, J. & Gralak, B. (2007) Diet of sympatric pine marten (*Martes martes*) and stone marten (*Martes foina*) identified by genotyping of DNA from faeces. *Annales Zoologici Fennici* 44, 269–284.

Powell, R.A. & Zielinski, W.J. (1983) Competition and coexistence in mustelid communities. *Acta Zoologica Fennica* 174, 223–227.

Prigioni, C. (2001) *Vulpes vulpes*. In: *Atlante dei Mammiferi della Lombardia* (eds C. Prigioni, M. Cantini & A. Zilio), pp. 219–222. Regione Lombardia e Università degli Studi di Pavia. (In Italian).

Prigioni, C., Remonti, L., Balestrieri, A., Sgrosso, S., Priore, G., Mucci, N. & Randi, E. (2006) Estimation of European otter (*Lutra lutra*) population size by fecal DNA typing in southern Italy. *Journal of Mammalogy* 87, 855–858.

Prigioni, C., Balestrieri, A. & Remonti, L. (2007) Decline and recovery in otter *Lutra lutra* populations in Italy. *Mammal Review* 37, 71–79.

Proulx, G., Aubry, K.B., Birks, J., Buskirk, S.W., Fortin, C., Frost, H.C., Krohn, W.B., Mayo, L., Monakhov, V., Payer, D., Saeki, M., Santos-Reis, M., Weir, R. & Zielinski, W.J. (2004) World distribution and status of the genus *Martes* in 2000. In: *Martens and Fishers (*Martes*) in Human-Altered Environments: An International Perspective* (eds D.J. Harrison, A.K. Fuller & G. Proulx), pp. 21–76. Springer-Verlag, New York.

Pulliainen, E. & Ollinmäki, P. (1996) A long-term study of the winter food niche of the pine marten *Martes martes* in northern boreal Finland. *Acta Theriologica* 41, 337–352.

Remonti, L., Prigioni, C., Balestrieri, A., Sgrosso, S. & Priore, G. (2008) Distribution of a recolonising species may not reflect habitat suitability alone: the case of the Eurasian otter (*Lutra lutra*) in southern Italy. *Wildlife Research* 35, 798–805.

Remonti, L., Balestrieri, A. & Prigioni, C. (2011) Percentage of protein, lipids, and carbohydrates in the diet of badger (*Meles meles*) populations across Europe. *Ecological Research* 26, 487–495.

Remonti, L., Balestrieri, A., Ruiz-González, A., Gómez-Moliner, B.J., Capelli, E. & Prigioni, C. (2012) Intraguild dietary overlap and its possible relationship to the coexistence of mesocarnivores in intensive agricultural habitats. *Population Ecology* 54, 521–532.

Remonti, L., Balestrieri, A., Raubenheimer, D. & Saino, N. (2016) Functional implications of omnivory for dietary nutrient balance. *Oikos* 125, 1233–1240.

Ridout, M.S. & Linkie, M. (2009) Estimating overlap of daily activity patterns from camera trap data. *Journal of Agricultural Biology and Environmental Statistics* 14, 322–337.

Robinson, R.A. & Sutherland, W.J. (2002) Post-war changes in arable farming and biodiversity in Great Britain. *Journal of Applied Ecology* 39, 157–176.

Rosellini, S., Osorio, E., Ruiz-González, A., Piñeiro, A. & Barja, I. (2008) Monitoring the small-scale distribution of sympatric European pine martens (*Martes martes*) and stone martens (*Martes foina*): a multi-evidence approach using faecal DNA analysis and camera-traps. *Wildlife Research* 35, 434–440.

Rosenzweig, M.L. (1966) Community structure in sympatric Carnivora. *Journal of Mammalogy* 47, 602–612.

Ruiz-González, A., Rubines, J., Berdión, O. & Gomez-Moliner, B.J. (2008) A non-invasive genetic method to identify the sympatric mustelids pine marten (*Martes martes*) and stone marten (*Martes foina*): preliminary distribution survey on the northern Iberian peninsula. *European Journal of Wildlife Research* 54, 253–261.

Ruiz-González, A., Madeira, M.J., Randi, E., Urra, F. & Gómez-Moliner, B.J. (2013a) Non invasive genetic sampling of sympatric marten species (*Martes martes* and *Martes foina*): assessing species and individual identification success rates on faecal DNA genotyping. *European Journal of Wildlife Research* 59, 371–386.

Ruiz-González, A., Madeira, M.J., Randi, E., Abramov, A.V., Davoli, F. & Gómez-Moliner, B.J. (2013b) Phylogeography of the forest-dwelling European pine marten (*Martes martes*): new insights into cryptic northern glacial refugia. *Biological Journal of the Linnean Society* 109, 1–18.

Sacchi, O. & Meriggi, A. (1995) Habitat requirements of the stone marten (*Martes foina*) on the Tyrrhenian slopes of the northern Apennines. *Hystrix* 7, 99–104.

Sánchez-Zapata, J.A., Carrete, M., Gravilov, A., Sklyarenko, S., Ceballos, O., Donázar, J.A. & Hiraldo, F. (2003) Land use changes and raptor conservation in steppe habitats of Eastern Kazakhstan. *Biological Conservation* 111, 71–77.

Savoldelli, P. & Sindaco, R. (2008) *Grandi e Piccoli Predatori*. Collana 'La nostra fauna'. Osservatorio Faunistico per la Fauna Selvatica, Regione Piemonte. (In Italian).

Schwartz, M.K. & Monfort, S.L. (2008) Genetic and endocrine tools for carnivore surveys. In: *Noninvasive Survey Methods for Carnivores* (eds R.A. Long, P. Mackay, W.J. Zielinski & J.C. Ray), pp. 238–262. Island Press, Washington, DC.

Sheehy, E., O'Meara, D.B., O'Reilly, C., Smart, A. & Lawton, C. (2014) A non-invasive approach to determining pine marten abundance and predation. *European Journal of Wildlife Research* 60, 223–236.

Sidorovich, V.E., Krasko, D.A. & Dyman, A.A. (2005) Landscape-related differences in diet, food supply and distribution pattern of the pine marten, *Martes martes* in the transitional mixed forest of northern Belarus. *Folia Zoologica* 54, 39–52.

Sindaco, R. (2006) Segnalazioni faunistiche piemontesi e valdostane. *Rivista Piemontese di Storia Naturale* 27, 443–459. (In Italian).

Sommer, R. & Benecke, N. (2004) Late- and Post-Glacial history of the Mustelidae in Europe. *Mammal Review* 34, 249–284.

Sommer, R.S. & Crees, J.J. (Chapter 4, this volume) Late Quaternary biogeography of small carnivores in Europe. In: *Small Carnivores: Evolution, Ecology, Behaviour, and Conservation* (eds E. Do Linh San, J.J. Sato, J.L. Belant & M.J. Somers). Wiley–Blackwell, Oxford.

Spagnesi, M. & De Marinis, A.M. (eds) (2002) *Mammiferi d'Italia*, Quaderni di Conservazione della Natura, 14. Ministero dell'Ambiente – Istituto Nazionale per la Fauna Selvatica. (In Italian).

Stoate, C., Boatman, N.D., Borralho, R.J., Carvalho, C.R., de Snoo, G.R. & Eden, P. (2001) Ecological impacts of arable intensification in Europe. *Journal of Environmental Management* 63, 337–365.

Storch, I., Lindstrom, E. & Dejounge, J. (1990) Diet and habitat selection of the pine marten in relation to competition with the red fox. *Acta Theriologica* 35, 311–320.

Tellería, J.L. & Virgós, E. (1997) Distribution of an increasing roe deer population in a fragmented Mediterranean landscape. *Ecography* 20, 247–252.

Torretta, E., Mosini, A., Piana, M., Tirozzi, P., Serafini, M., Puopolo, F., Saino, N. & Balestrieri, A. (2017) Time partitioning in mesocarnivore communities from different habitats of NW Italy: insights into martens' competitive abilities. *Behaviour* 154, 241–266.

Valière, N. (2002) GIMLET: a computer program for analysing genetic individual identification data. *Molecular Ecology Notes* 2, 377–379.

van den Bosch, F., Hengeveld, R. & Metz, J.A.J. (1992) Analysing the velocity of animal range expansion. *Journal of Biogeography* 19, 135–150.

Veech, J.A., Small, M.F. & Baccus, J.T. (2011) The effect of habitat on the range expansion of a native and an introduced bird species. *Journal of Biogeography* 38, 69–77.

Virgós, E., Recio, M.R. & Cortés, Y. (2000) Stone marten (*Martes foina*) use of different landscape types in the mountains of central Spain. *Zeitschrift für Säugetierkunde* 65, 375–379.

Virgós, E., Zalewski, A., Rosalino, L.M. & Mergey, M. (2012) Habitat ecology of genus *Martes* in Europe: a review of the evidences. In: *Biology and Conservation of Martens, Sables, and Fisher: A New Synthesis* (eds K.B. Aubry, W.J. Zielinski, M.G. Raphael, G. Proulx & S.W. Buskirk), pp. 255–266. Cornell University Press, Ithaca.

Waits, L.P. & Paetkau, D. (2005) Noninvasive genetic sampling tools for wildlife biologists: a review of applications and recommendations for accurate data collection. *Journal of Wildlife Management* 69, 1419–1433.

White, G.C. & Garrott, R.A. (1990) *Analysis of Wildlife Radio-Tracking Data*. Academic Press, San Diego.

Wilson, G.J., Frantz, A.C., Pope, L.C., Roper, T.J., Burke, T.A., Cheesan, C.L. & Delahay, R.J. (2003) Estimation of badger abundance using faecal DNA typing. *Journal of Applied Ecology* 40, 658–666.

Wilson, J.R.U., Dormontt, E.E., Prentis, P.J., Lowe, A.J. & Richardson, D.M. (2009) Something in the way you move: dispersal pathways affect invasion success. *Trends in Ecology & Evolution* 24, 136–144.

Zalewski, A. & Jędrzejewski, W. (2006) Spatial organisation and dynamics of the pine marten *Martes martes* population in Bialowieza Forest (E Poland) compared with other European woodlands. *Ecography* 29, 31–43.

Zhou, Y.-B., Newman, C., Xu, W.-T., Buesching, C.D., Zalewski, A., Kaneko, Y., Macdonald, D.W. & Xie, Z.-Q. (2011) Biogeographical variation in the diet of Holarctic martens (genus *Martes*, Mammalia: Carnivora: Mustelidae): adaptive foraging in generalists. *Journal of Biogeography* 38, 137–147.

15

Spatial and Temporal Resource Partitioning of Small Carnivores in the African Rainforest: Implications for Conservation and Management

Yoshihiro Nakashima[1], *Yuji Iwata[2], Chieko Ando[2], Chimene Nze-Nkogue[3], Eiji Inoue[2], Etienne François Akomo-Okoue[3], Pierre Philippe Mbehang Nguema[3], Thierry Diop Bineni[3], Ludovic Ngok Banak[3], Yuji Takenoshita[4], Alfred Ngomanda[3], and Juichi Yamagiwa[2]*

[1] *College of Bioresource Science, Nihon University, Fujisawa City, Japan*
[2] *Graduate School of Science, Kyoto University, Kyoto, Japan*
[3] *Institut de Recherche en Ecologie Tropicale, CENAREST, Libreville, Gabon*
[4] *Department of Children, Faculty of Child Studies, Chubu-Gakuin University, Sika, Japan*

SUMMARY

Although African rainforests harbour a high diversity of small carnivores, few studies have been conducted on these species' ecology and interspecific relations. We carried out a camera-trapping survey to examine habitat use and activity patterns of small carnivores in the Moukalaba–Doudou National Park, Gabon. The study area (~500 km^2) consists of various types of vegetation, including forest on dry soils, swamp forest, montane forest and savannah. We detected nine species of small carnivores in the study area. The seven most common carnivores were broadly classified into forest-interior species ($n = 3$), savannah/forest-edge species ($n = 3$) and aquatic-habitat species ($n = 3$), in agreement with observations by other researchers. Occupancy analysis suggested further habitat separation within the small carnivore assemblage: among the savannah/forest-edge species, African civets, *Civettictis civetta*, more often used the forest edge and less frequently entered the savannah interior compared with Egyptian mongooses, *Herpestes ichneumon*, and rusty-spotted genets, *Genetta maculata*. Among the forest-interior species, black-legged mongooses, *Bdeogale nigripes*, were more closely associated with mature secondary dry forest than were long-nosed mongooses, *Xenogale naso*, and servaline genets, *Genetta servalina*. These two forest mongoose species, with similar body size and diet, exhibited different activity patterns. However, their habitat use and activity patterns were not affected by one another's presence, indicating that they had different preferences. Our results show that most pairs of small carnivores in the Moukalaba differ in either habitat use or time of activity, which may promote their coexistence across this region. This suggests that maintenance of habitat heterogeneity may be important for the conservation of these species. The relative proportion of small carnivores over space and time may reflect the degree of degradation of the forest; therefore, long-term monitoring by using camera-traps is highly recommended.

Keywords

Activity time — camera-trapping — coexistence — Gabon — habitat use — occupancy model — savannah

Introduction

African rainforests harbour a high diversity of small carnivores with more than 10 species occurring sympatrically (Ray, 2001; Bahaa-el-din *et al.*, 2013). Despite their high diversity and potential ecological importance, few comprehensive studies have been conducted on the ecology of these small carnivores (Ray, 2001; but see: Ray 1997; Angelici *et al.*, 1999; Angelici, 2000; Ray & Sunquist, 2001; Angelici &

*Corresponding author.

Small Carnivores: Evolution, Ecology, Behaviour, and Conservation, First Edition. Edited by Emmanuel Do Linh San, Jun J. Sato, Jerrold L. Belant, and Michael J. Somers.

Luiselli, 2005; Mills *et al.*, 2019). Although most species that occur in the Congo Basin are currently listed as Least Concern on *The IUCN Red List of Threatened Species* (Do Linh San *et al.*, 2013; IUCN, 2020), they are gradually declining under increasing pressure from human activities such as clearance for agriculture, bushmeat hunting and logging (Happold, 1996; Ray, 2001; Bahaa-el-din *et al.*, 2013). In order to effectively conserve the carnivore community and properly manage its habitat, urgent research is required on these species' ecology and interspecific relations.

In accordance with community ecology principles, ecologically similar species must exhibit niche differentiation and resource partitioning to coexist (Hardin, 1960; Schoener, 1974). Three possible dimensions are generally considered for resource partitioning in animals: food, habitat and activity time (Schoener, 1974). Food partitioning may be the most important dimension in many carnivore communities, and has been associated with morphological differences among species, including body size and dental morphology (Rosenzweig, 1966; Azevedo *et al.*, 2006; Davies *et al.*, 2007). Spatial and temporal resource partitioning may also be important between similar-sized species with similar feeding habits (Palomares *et al.*, 1996; Durant, 1998; Di Bitetti *et al.*, 2009, 2010; Davis *et al.*, 2019).

Differential use of resources can occur either when focal pairs of species have different resource preferences (i.e. different species traits) or when they have similar preferences but different competitive ability (Chase *et al.*, 2002). Although such mechanisms of resource partitioning are concealed in natural conditions, and therefore difficult to evaluate, the differences may become clear if either species is experimentally or naturally eliminated from the community (Dickman, 1988; Kasparian *et al.*, 2002; Trewby *et al.*, 2008). A better understanding of the mechanisms that promote such differential resource use becomes important in the prediction of the ecological consequences of species extinction (Moreno *et al.*, 2006). Furthermore, understanding animal spatial patterns and their determinants is profoundly relevant to species conservation and management in the African rainforest, where habitat is characterized by high spatial heterogeneity, often maintained by human activities (e.g. burning, livestock grazing; Primack & Corlett, 2005).

It has been suggested that carnivore species in the African rainforest clearly exhibit resource partitioning in the dimensions of diet, habitat and activity time (Ray, 1997, 1998, 2001; Ray & Sunquist, 2001). Based on these data, Ray (2001) categorized African rainforest carnivores into forest-interior, forest-edge and aquatic-habitat species. Furthermore, Ray (1997) suggested that two forest mongooses – the marsh mongoose, *Atilax paludinosus*, and the long-nosed mongoose, *Xenogale naso* – with similar dietary and habitat preferences may exhibit temporal variation/separation (i.e. different activity patterns) (Ray, 1997). However, these results are based on preliminary data, with the exception of radio-tracked studies of the two above-mentioned forest mongooses. While resource-partitioning studies have relied heavily on radio-tracking (e.g. Haines *et al.*, Chapter 16, this volume), recent advances in camera-trapping enable further systematic evaluation of habitat use and activity patterns of African small carnivores, as exemplified by a recent research carried out in Kibale National Park, Uganda (Mills *et al.*, 2019).

In this study, we examined the habitat use and activity patterns of small carnivores using camera-trapping and investigated the possibility of spatial and temporal resource partitioning among these carnivores in the Moukalaba–Doudou National Park, Gabon. Although broad patterns of resource separation are indeed most likely (Ray, 2001), we hypothesized that finer-scale spatial and temporal separations should exist, and that such resource partitioning may facilitate coexistence of diverse carnivore species in the Moukalaba. The aim of the present study was to collect basic information on habitat use and activity patterns of small carnivores to test this hypothesis.

Materials and Methods

Study Area

Our study site was located in the north-eastern part of Moukalaba–Doudou National Park, Gabon (Figure 15.1). The Moukalaba is located on the south-western boundary of the humid tropical rainforest of the Congo Basin and the northern boundary of the 'southern Congolian forest–savannah mosaic'

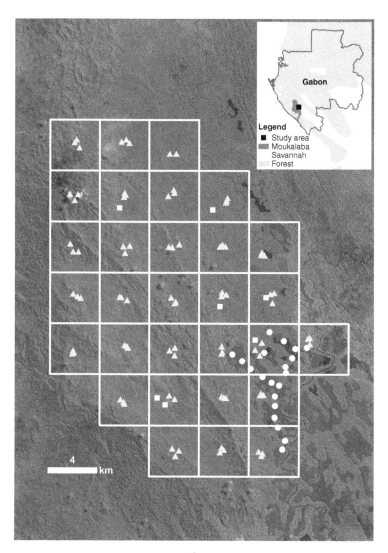

Figure 15.1 The study area (~500 km²) in Moukalaba–Doudou National Park, Gabon was divided into 30 grid cells of 4 × 4 km each. Green areas correspond to rainforest, and brown areas are savannah. Yellow triangles indicate the location of camera-traps in the forest, while white circles and squares represent the location of camera-traps in savannah and forested stream, respectively.

stretching from the southern part of the Republic of Congo. The savannah in the Moukalaba is currently maintained by annual artificial burning. We selected an area of approximately 500 km² on the western side of the Moukalaba River as the focal study area; this area contained various types of vegetation and topography. Elevation in the study area ranged from 68 to 723 m a.s.l. The south-eastern part of the study area contained savannah (Figure 15.1). Forest vegetation was categorized into low-altitude (< 200 m),

medium-altitude (200–450 m), and high-altitude (> 450 m) forest by Sosef *et al.* (2004). Low-altitude forest can further be subdivided into forest on sandy soil, swamp forest and gallery forest. The forests in this area were selectively logged from the 1960s until the 1980s. The logging intensity was moderate (logging in Gabon is selective for a few timber species), and forest has recovered well from the past damage. Annual rainfall in the study area ranged from 1582 to 1886 mm between 2004 and 2006, and mean monthly minimum and maximum temperatures ranged from 21.3 to 24.1°C and from 29.3 to 33.7°C, respectively, during this period (Takenoshita *et al.*, 2008). There are two distinct seasons: a rainy season from October to April and a dry season from May to September, with little rains during the three months in the middle of the dry season (Thibault *et al.*, 2004).

Camera-Traps

Our study consisted of two surveys. The first survey was conducted in the dry season in 2010. In the first survey, we targeted forested habitat and stratified the area into 30 grid cells of $16 \, km^2$ ($4 \times 4 \, km$; Figure 15.1). We generated three random points within 500 m circles centred on each grid cell and placed a camera (Bushnell Trophy Cam, 2010) by an animal path within 20 m of each random point; cameras were located > 100 m apart. Ninety cameras were used, mounted on trees at approximately 30 cm above the ground. The cameras were operated 24 h/day with a three-minute delay between sequential photographs and were left in the forest for three months. Three photographs were taken within 0.1 s each time a sensor was triggered, which helped us to identify species with similar appearances (e.g. the marsh mongoose and the long-nosed mongoose), as different postures of the animals were captured.

The second survey was conducted in the dry season in 2011. We targeted savannah and aquatic habitats because different species may occur in these habitats (Ray, 2001). We placed 20 cameras in savannah areas. To avoid direct sunlight, cameras were placed in trees at forest-edge areas (boundary between continuous forest and savannah; 10 cameras) and on *Nauclea latifolia* trees within savannah areas (> 100 m from forest edge; 10 cameras). These cameras were left for two months. In addition, seven cameras were positioned at a small forested stream. Cameras in this second survey were located approximately 1 km apart of each other (Figure 15.1). Camera positioning in the second survey was less systematic than during the first survey, because the exact position of each camera was determined by the presence of trees suitable for mounting cameras. In addition, the capture rates (the number of photos captured by camera-traps/the number of trap-days) in the second survey

were not comparable to those of the forest survey, because of different camera settings.

Habitat Analysis

We analyzed the first and second surveys separately because they differed in the length of the research period and in camera placement. For the first survey, we compared mean capture rate of camera-traps among vegetation types using Kruskal–Wallis tests. When we obtained significant results ($p < 0.05$), we tested the difference between each pair of vegetation types with a Scheffé's *post hoc* procedure. The data were filtered to exclude photos of the same species at the same camera within 30 minutes to make sure that the events were independent. We classified vegetation from the 30 grid cells into lowland old secondary forest (forest on dry soils in Sosef *et al.*, 2004; 14 cells), lowland swamp forest (8 cells) and montane forest (8 cells). Although Sosef *et al.* (2004) recognized gallery forest within lowland forest areas, the forests located between savannahs in our study area were swamp forest. Gallery forest areas were associated with rivers within savannah, and annual flooding of the river during the rainy season created swamp forest; therefore, we did not distinguish between swamp forest and gallery forest in this study. We used vegetation information derived from a classification based on Landsat Thematic Mapper (TM), radar and aerial imagery (Ministère des Eaux et Forêts et du Reboisement, Tecsult International, Québec Canada; provided by World Wildlife Fund Gamba Complex Programme). Details of vegetation classifications are provided in Nakashima *et al.* (2013). We compared mean capture rate between forest-edge areas (boundary between large continuous forest and savannah) and savannah-interior areas (> 100 m from forest edge) with Mann–Whitney tests.

We also applied single-season occupancy models (MacKenzie *et al.*, 2006) to evaluate the effect of habitat on occupancy rate (ψ) and detection probability (p). Since occupancy relies on independence between grids, grid-cell size should be at least similar or larger than the home range of target species. Although we did not estimate the home range size of small carnivores in the Moukalaba, our grid size ($4 \times 4 \, km$) is much larger than that predicted from body sizes

(Lindstedt *et al.*, 1986), suggesting that the assumption of occupancy modelling is well held in our study design. For data from the first survey, we used vegetation type as a covariate of occupancy rate and detection probability. To increase the detection probability for each sampling period, we combined data from the three camera-traps within each grid cell and combined sections of five-day camera-trapping periods into individual sampling occasions resulting in 18 repeated observations. Because cameras often malfunctioned due to elephant damage (20%) and moisture (11%), the total number of sampling days from the three camera-traps in each grid cell varied and was incorporated as a factor affecting detection probability. For the second survey, we used camera position (forest edge or savannah interior) as a single categorical covariate and collapsed 5 survey days into a single occasion resulting in 12 repeated observations. We used these pooled data (five days) to construct an X matrix needed for occupancy models. We did not apply occupancy analysis to the data from the stream due to the limited sample size ($n = 8$). We fitted the occupancy models with the user-contributed R (R Language and Environment for Statistical Computing) package 'unmarked' (Fiske & Chandler, 2011). We selected the most parsimonious models based on Akaike's Information Criterion corrected for small sample sizes (AIC_c; Akaike, 1987). We compared the AIC_c value of all combinations of main effects without interactions using the 'dredge' function in the R package 'MuMIn' (Bartoń, 2009). Models with $\Delta AIC_c < 2$ were considered equally parsimonious. Occupancy analysis was restricted to the three most common species in forest (long-nosed mongoose, black-legged mongoose, *Bdeogale nigripes*, and servaline genet, *Genetta servalina*) and to the three most common species in savannah/forest edge (African civet, *Civettictis civetta*, Egyptian mongoose, *Herpestes ichneumon*, and rusty-spotted genet, *Genetta maculata*).

Activity Analysis

For species captured > 10 times, we categorized into either diurnal or nocturnal species. We defined diurnal species (D) as those for which $> 90\%$ of captures occurred during daytime (06:01 to 18:00 h) and nocturnal species (N) as those for which $> 90\%$ of photos were taken during night-time (18:01 to 06:00 h), following the definition by Grassman *et al.* (2006). We also assessed animals' daily activity patterns by applying statistical methods developed by Ridout & Linkie (2009). We used Kernel density estimation or fitted trigonometric-sum distributions to the data (Fernández-Durán, 2004). This analysis was performed using the R code made available by Linkie & Ridout (2011).

Interspecific Interactions

We also examined whether the occurrence of one mongoose species affected the temporal and spatial use of other mongoose species. This analysis was restricted to two forest mongooses – the black-legged mongoose and the long-nosed mongoose – due to insufficient independent photographs for the other species and because these forest mongooses are most likely to interact due to their similar body size and feeding habits (Ray, 1997; Baker & Ray, 2013; Van Rompaey & Colyn, 2013). For spatial partitioning, we used two-species occupancy modelling to test whether their occurrences were independent or were affected by one another (MacKenzie *et al.*, 2006). However, the number of sampled grid cells ($n = 30$) was inadequate to apply a two-species occupancy model (MacKenzie *et al.*, 2006). We therefore assumed that the detection records captured by each of the three camera-traps in each grid cell were independent of one another ($n = 90$). To test the validity of this assumption, we tested for spatial auto-correlation of the capture rates of each camera using Geary's C-values for capture rate (Geary, 1954) based on 1000 permutations (we shuffled the capture rate within the vegetation types). Geary's C-values range from 0 to 2, where 1 means no spatial auto-correlation. Values < 1 indicate increasing positive spatial auto-correlation (maximum $= 0$), while values > 1 indicate increasing negative spatial auto-correlation (maximum $= 2$). This analysis showed that Geary's C-values for the black-legged and long-nosed mongooses were 0.81 ($p = 0.33$) and 0.92 ($p = 0.76$), respectively, suggesting that the detection records of the three cameras can be regarded as independent.

We fitted the two-species occupancy models using the same covariates as for the best-fit single-species

Table 15.1 Mean capture rate (number of photos/100 trap-days) of small carnivores in different habitats in Moukalaba–Doudou National Park, Gabon. *n* = number of camera-trap locations.

Common name	Species	Forest (*n* = 90)	Forest edge (*n* = 10)	Savannah (*n* = 10)	Stream (*n* = 7)
African palm civet	*Nandinia binotata*	0.02 ± 0.00			
African civet	*Civettictis civetta*	0.12 ± 0.01	3.74 ± 4.58	0.40 ± 1.26	
Central African oyan	*Poiana richardsonii*	0.02 ± 0.00			
Rusty-spotted genet	*Genetta maculata*		2.46 ± 3.61	0.80 ± 1.69	
Servaline genet	*Genetta servalina*	0.47 ± 0.01	0.54 ± 1.13		
Black-legged mongoose	*Bdeogale nigripes*	1.92 ± 0.04			
Egyptian mongoose	*Herpestes ichneumon*		1.79 ± 1.70	0.93 ± 1.41	
Long-nosed mongoose	*Xenogale naso*	0.83 ± 0.02	0.87 ± 1.41		0.21 ± 0.01
Marsh mongoose	*Atilax paludinosus*	0.05 ± 0.01			1.66 ± 0.12

model to account for habitat preference and differences in occupancy rate (ψ) between sites. Following Tobler *et al.* (2009), we further assumed that the detection probability for each species was independent. We estimated the parameter η ($\eta = \psi^{AB}/\psi^A \times \psi^B$), which defines the relationship between the occurrence of each species (MacKenzie *et al.*, 2006). Values of η that are <1 indicate species avoidance, while $\eta > 1$ indicates species attraction, and $\eta = 1$ suggests that species occur independently (MacKenzie *et al.*, 2006). To test for interaction, we compared models with species interactions using the parameter η to a model where η was set equal to 1 (MacKenzie *et al.*, 2006). We fitted the two-species occupancy models using the software PRESENCE 4.4 (Hines, 2006).

For temporal interactions, we focused on the activity patterns of the black-legged mongoose and long-nosed mongoose, because temporal partitioning appears to be particularly important for these two insectivorous species (see Results; Ray, 2001). Our results showed that the black-legged mongoose was strictly nocturnal, while the long-nosed mongoose was active during both day and night time (see Figure 15.3). Therefore, we tested whether the long-nosed mongoose increased its diurnal activity where the black-legged mongoose was more abundant. Because occupancy rate of the black-legged mongoose was higher in old secondary forest and lower in swamp forest, we compared the activity time of the long-nosed mongoose between these forests using Fisher's exact test.

Results

Habitat Analysis

Our survey effort resulted in 6379 camera-trapping days for the first survey and 1685 for the second survey (1275 in forest/savannah and 410 at small streams in forests). We detected nine small carnivore species during the surveys, and there was a distinct difference in species composition among vegetation types (Table 15.1). Only African civets were detected across forest, forest/savannah edge and savannah, while three species (black-legged mongoose, long-nosed mongoose and servaline genet) were exclusively detected within the forest. Two species (Egyptian mongoose and rusty-spotted genet) were detected only near the savannah during the second survey, while the marsh mongoose was detected primarily at the streams.

The mean capture rate of most species differed among vegetation types within forest and savannah (Figure 15.2). In forest areas, the capture rate among the vegetation types was significantly different for the black-legged mongoose (Kruskal–Wallis test: $\chi^2 = 7.00$, df = 2, $p = 0.03$) and the long-nosed mongoose ($\chi^2 = 16.91$, df = 2, $p = 0.0002$) but not for the servaline genet ($\chi^2 = 0.719$, df = 2, $p = 0.70$). The capture rate of the black-legged mongoose was significantly higher in old secondary forest than in swamp forest (Scheffé's *post hoc* test: $\chi^2 = 6.92$, $p = 0.03$), while that of the long-nosed mongoose was higher in swamp forest

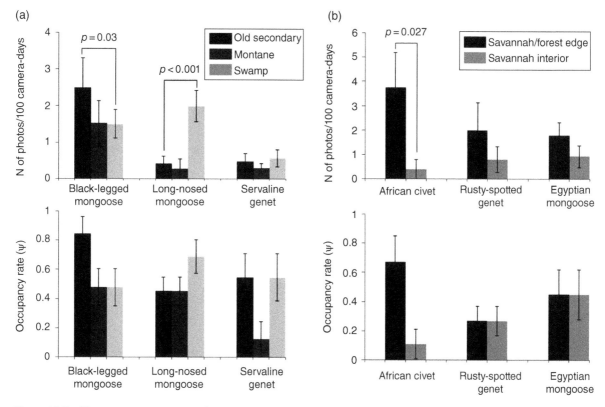

Figure 15.2 The mean capture rate and occupancy rate of (a) forest-interior species and (b) savannah/forest-edge species in Moukalaba–Doudou National Park, Gabon. Significant *p*-values from Kruskall–Wallis and Mann–Whitney tests, respectively, are indicated.

than in old secondary forest ($\chi^2 = 14.50$, $p = 0.0007$). Differences between other habitat pairs were not statistically significant.

Occupancy analysis followed a similar trend (Figure 15.2a; Table 15.2), where occupancy rate of the black-legged mongoose was higher in old secondary forest ($\psi = 0.84 \pm 0.12$ SD) than in montane or swamp forest (0.48 ± 0.13). In contrast, occupancy rate of the long-nosed mongoose was higher in swamp forest (0.69 ± 0.11) than in old secondary or montane forest (0.45 ± 0.09). Occupancy rate of the servaline genet was lower in montane forest (0.55 ± 0.16) than in old secondary or swamp forest (0.13 ± 0.12).

For the second survey, mean capture rate at the savannah edge was significantly higher than in the savannah interior for the African civet (Mann–Whitney test: $U = 25$, $p = 0.03$). There were no significant differences for the rusty-spotted genet ($U = 38$, $p = 0.28$) and the Egyptian mongoose ($U = 37$,

$p = 0.31$) (Figure 15.2b; Table 15.3). Occupancy rate of the African civet was 0.67 ± 0.18 in savannah/forest edge and 0.11 ± 0.10 in small forest patches within savannah; that of the rusty-spotted genet and the Egyptian mongoose were 0.27 ± 0.10 and 0.45 ± 0.17 across savannah/forest edge and small forest patches, respectively (Figure 15.2b).

Activity Analysis

Most small carnivores (seven out of nine species) showed nocturnal activity patterns, the exceptions being Egyptian mongooses and long-nosed mongooses (Figure 15.3; Table 15.4). The four mongoose species showed variable activity times: long-nosed mongooses were mostly diurnal but were occasionally (9.8%) detected at night, black-legged mongooses were strictly nocturnal (100%), marsh mongooses were nocturnal and Egyptian mongooses diurnal, although both

Table 15.2 Summary of occupancy model selection for small carnivores in forest interior in Moukalaba–Doudou National Park, Gabon. Only models with model weight (*w*) > 0.1 are shown. AIC$_c$ = corrected Akaike's Information Criterion.

Species	Model[a]	AIC$_c$	ΔAIC$_c$	*w*	df
Black-legged mongoose	ψ(Secondary) p(Effort)	407	0	0.206	4
	ψ(Secondary) p(.)	407.2	0.24	0.184	3
	ψ(.) p(Effort)	408.4	1.42	0.101	3
Long-nosed mongoose	ψ(Swamp) p(Effort)	212.5	0	0.31	4
	ψ(Swamp + Montane) p(Effort)	213.1	0.58	0.232	5
	ψ(Montane) p(Effort + Montane)	214.1	1.54	0.143	5
	ψ(Montane) p(Effort)	214.4	1.82	0.125	4
Servaline genet	ψ(Montane) p(Effort + Montane)	147.4	0	0.226	5
	ψ(.) p(Effort)	147.5	0.07	0.218	3
	ψ(.) p(Effort + Montane)	148.4	0.87	0.146	4
	ψ(Montane) p(Effort + Montane)	148.4	0.95	0.12	5

[a] Covariates affecting occupancy rate (ψ) and detection probability (p) are shown within parentheses. ψ(.) and p(.) mean constant occupancy and detection probabilities, respectively.

Table 15.3 Summary of occupancy model selection for small carnivores in forest edge and savannah in Moukalaba–Doudou National Park, Gabon. Only models with a model weight (*w*) > 0.1 are shown. AIC$_c$ = corrected Akaike's information criterion. See Table 15.2 for additional explanations.

Species	Model	AIC$_c$	ΔAIC$_c$	*w*	df
Egyptian mongoose	ψ(.) p(.)	139.8	0	0.471	2
	ψ(Position) p(.)	141	1.21	0.257	3
	ψ(.) p(Position)	141.4	1.59	0.213	3
African civet	ψ(Position) p(.)	109.6	0	0.688	3
	ψ(Position) p(Position)	112.7	3.08	0.148	4
	ψ(.) p(.)	113.1	3.44	0.123	2
Rusty-spotted genet	ψ(.) p(.)	82.6	0	0.362	2
	ψ(Position) p(.)	82.8	0.18	0.331	3
	ψ(.) p(Position)	83.6	1.01	0.218	3

species also exhibited some crepuscular activity (diurnal activity 18 and 78%, respectively); however, the sample sizes (*n* = 11 and *n* = 18, respectively) were too small to definitively determine the activity patterns.

Interspecific Interactions of Forest Mongooses

The two-species occupancy model for black-legged and long-nosed mongooses where η = 1 had lower support (AIC$_c$ = 646.81) than the model that considered species interactions (AIC$_c$ = 648.87), suggesting that species were distributed independently and that there was no spatial separation of species within habitat types.

Furthermore, the diurnal activity pattern of long-nosed mongooses was not affected by the occurrence of black-legged mongooses. Of 38 observations of long-nosed mongooses in old secondary forest, 30 observations (78.9%) were diurnal while 8 (21.1%)

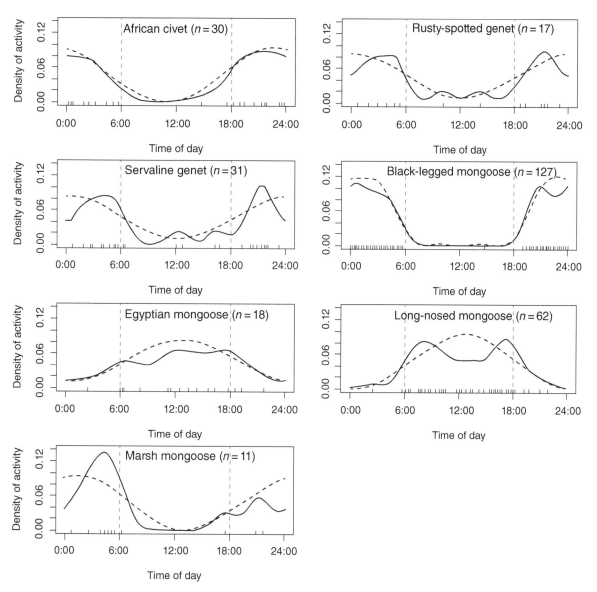

Figure 15.3 Activity patterns of seven small carnivore species in the Moukalaba–Doudou National Park, Gabon. The solid lines are Kernel density estimates, whereas the dashed lines are trigonometric-sum distributions. The short vertical lines above the x-axis indicate the times of individual photographs, and the grey-dashed vertical lines indicate the approximate time of sunrise and sunset. *n* = number of camera-trapping detection records.

were nocturnal. Of 17 observations of long-nosed mongooses in swamp forest and forest/savanna edge, 14 observations (82.4%) were diurnal and 3 (17.6%) were nocturnal (Fisher's exact test: $p > 0.05$). Therefore, long-nosed mongooses were active during the daytime regardless of the presence or absence of black-legged mongooses.

Discussion

Species Composition and Richness in the Moukalaba

We detected 9 small carnivore species in our study area, which is lower than the 12 Bahaa-el-din *et al.* (2013) predicted to exist in the area based on theoretical

Table 15.4 Different use of spatial and temporal resources among small carnivores in the Moukalaba–Doudou National Park, Gabon. N = nocturnal; D = diurnal.

Species	Forest			Forest edge		
	Terrestrial	Semi-arboreal	Arboreal	Terrestrial	Semi-arboreal	Aquatic
African palm civet			N			
African civet				N		
Central African oyan			N			
Rusty-spotted genet					N	
Servaline genet		N				
Black-legged mongoose	N					
Egyptian mongoose				D		
Long-nosed mongoose	D					
Marsh mongoose						N?[a]

[a] Activity time of marsh mongooses could not be determined due to low sample size.

geographic distribution. Indeed, we failed to detect the flat-headed cusimanse, *Crossarchus platycephalus*, the slender mongoose, *Galerella sanguinea*, and the honey badger, *Mellivora capensis*. This was expected since some of these species (e.g. *C. platycephalus*) are not widely geographically distributed, while others (e.g. *M. capensis*) have few valid records in Gabon (Bahaa-el-din *et al.*, 2013). Nonetheless, our study still confirmed the occurrence of all of Gabon's common small carnivore species and, in addition, we visually detected the presence of Congo clawless otter, *Aonyx congicus*, at a large river. Importantly, we recorded the occurrence of two additional savannah/forest-edge species (the rusty-spotted genet and the Egyptian mongoose) compared to other studies in the same area (e.g. Dzanga–Sanga National Park, Democratic Republic of the Congo, characterized by dense rainforest; Ray, 1996). In a recent study, Mills *et al.* (2019) recorded 11 species of small carnivores (excluding the larger canid and felid species) in Kibale National Park, with 7 of them (including Congo clawless otter) in common with our study. Black-legged and long-nosed mongooses do not occur in Uganda (Angelici & Do Linh San, 2015; Ray *et al.*, 2015). Contrary to us, the authors detected slender mongoose and honey badger, as well as two species that are absent from Gabon (Alexander's cusimanse,

Crossarchus alexandri) or only marginally present in the south-east (banded mongoose, *Mungos mungo*), respectively.

Resource Use and Interspecific Competition

All pairs of small carnivores – with exception of African palm civet, *Nandinia binotata*, and Central African oyan, *Poiana richardsonii* – exhibited spatial and/or temporal resource partitioning (Table 15.4). The patterns of resource use that we detected basically agreed with those reviewed by Ray (2001): in accordance with her description, the small carnivores were classified into forest-interior species, forest-edge species and aquatic-habitat species. In addition, the most ecologically similar mongooses (black-legged and long-nosed mongooses) were separated by the temporal dimension. Therefore, our results provide empirical evidence for conclusions drawn by Ray (2001) and further extend the scope for the coexistence of small carnivores in this region. Studies have been conducted outside of the Central African rainforest on habitat use and activity patterns of species investigated here, including African civet (Ikeda *et al.*, 1982; Ermias Admasu *et al.*, 2004), rusty-spotted genet (Fuller *et al.*, 1990; Roux, 2018; Ndzinisa, 2018), Egyptian mongoose (e.g. Delibes & Beltrán, 1985; Maddock & Perrin, 1993; Palomares & Delibes, 1990, 1991, 1992, 1993; Streicher *et al.*, 2020) and marsh mongoose (Maddock & Perrin, 1993; Streicher *et al.*, 2020, 2021). However, ours is one of the few systematic surveys – following those by J. C. Ray in Dzanga–Sanga National Park (Ray, 1996), F. M. Angelici in south-eastern Nigeria (Angelici *et al.*, 1999; Angelici, 2000; Angelici & Luiselli, 2005) and D. Mills in Kibale National Park (Mills *et al.*, 2019) – on temporal and spatial resource use of the entire small carnivore community in Central Africa.

Our study had two important findings. First, it seems that habitat was partitioned within the forest-interior and forest/savannah-edge species. For example, for forest-interior species, occupancy rates of black-legged mongooses were higher in mature secondary forest than in swamp or montane forest. On the other hand, servaline genets had lower occupancy rates in montane forest than swamp and mature secondary forest, while long-nosed mongooses had higher occupancy rates in swamp forest than mature secondary forest and montane forest. For the three savannah/forest-edge species, we found that African civets often used forest edge and rarely entered the savannah interior in comparison to Egyptian mongooses and rusty-spotted genets. Overall, black-legged mongoose had the strongest affinity for mature forest. Collectively these results suggest that the small carnivore species composition gradually transitioned from mature forest to savannah.

Ray & Sunquist (2001) found high food-niche overlap between black-legged and long-nosed mongooses in Dzanga–Sangha National Park. Ray (2001) suggested that these two forest mongoose species may be separated by the temporal dimension (i.e. different activity patterns) or that the superabundant or highly renewable prey may promote their sympatric occurrences. In our study, black-legged mongoose was nocturnal and long-nosed mongoose diurnal. However, resource partitioning in the temporal dimension alone is generally rare and unstable (Schoener, 1974; Kronfeld-Schor & Dayan, 2003). We therefore suggest that the slight difference in habitat use observed between both mongoose species may be important for their coexistence. For example, swamp forest has

softer soils covered by a dense mat of understory vegetation compared to dry, terrestrial forest in the Moukalaba. We observed that the long-nosed mongoose searched for its prey by digging in the soft soils of swamp forests with its snout, while the black-legged mongoose searched for prey on the soil surface (Y. Nakashima, unpublished data). These observations suggest that the two species have adapted to different feeding methods, and therefore they may be inherently separated in terms of fine-scale habitat use.

The second important finding from our study supports the theory that the detected spatial differences may be associated with the traits of these two mongoose species. If interspecific competition were prominent, resource use by the inferior competitor should be affected by that of the superior competitor. However, we did not obtain such evidence for either their spatial or their temporal resource partitioning. For example, the two-species occupancy analysis suggested that the occurrence of the species was determined independently. Furthermore, the activity patterns of the species were also not affected by one another. However, our results may still be inconclusive. This is because we compared only the activity patterns of the long-nosed mongoose in areas where the black-legged mongoose was more and less abundant. We could for example expect different patterns where the black-legged mongoose is entirely absent (i.e. competition may have been strong enough to increase the diurnal activity of long-nosed mongooses even where black-legged mongooses were less abundant). In addition, our results did not include members of the other small carnivore species, which may also affect the results. For example, marsh mongooses have been found in upland and both closed and open habitats away from water in other study sites (Rosevear *et al.*, 1974; Stuart, 1981; Baker, 1992; Skinner & Chimimba, 2005; Mills *et al.*, 2019) and may be active in morning and late afternoon in KwaZulu-Natal, South Africa (Rowe-Rowe, 1977; but see Maddock & Perrin, 1993). Therefore, the habitat use and activity patterns of marsh mongooses in the Moukalaba may be affected by other species. However, the strong consistency of our results with those of Ray (2001) suggests that spatial and temporal resource use by small carnivores was relatively stable across this region. Given that most pairs of mongooses have differing morphology or are separated broadly by habitats, it is very likely that they inherently occupy different ecological niches, which would promote their stable coexistence.

Implications for Conservation

These findings have two important implications for the conservation of small carnivores and the management of their habitats. First, as also highlighted by Mills (2019) for Uganda's rainforest, the different habitat preferences among species suggest that maintenance of habitat heterogeneity is necessary for the conservation of diverse carnivore assemblages. Second, from their different affinities for mature forest, the relative proportions of small carnivores over space and time may indicate the degree of degradation of the forest habitat and can serve as indicators of forest integrity.

Our results suggest that conservation priorities in the Moukalaba should focus on savannah/forest-edge species (Egyptian mongooses and rusty-spotted genets). In the Lopé National Park, Gabon, savannah areas have been maintained by annual, human-induced fires for approximately the past 9000 years (White, 2001). Currently, forest is advancing into savannah areas that are no longer burned (White, 2001), and a burning regime has been designed to maintain savannahs and the species that depend on them (Jeffery *et al.*, 2011). The situation is the same in the Moukalaba. Although soils in the Moukalaba savannah area consist of karst, which is much less fertile than soils of forested areas (Thibault *et al.*, 2004), forest has been invading into savannah. Comparing the current vegetation to that of Landsat images taken in 1992 confirms that the area of savannah has been diminishing in the past 20 years. This suggests that regular burning is necessary to maintain this vegetation type in the Moukalaba and also to conserve the small carnivores and other species (e.g. waterbuck, *Kobus ellipsiprymnus*, and bushbuck, *Tragelaphus scriptus*) that inhabit savannah/forest-edge

areas. Although burning is now conducted under the control of the government's National Park Agency (ANPN), savannah in remote areas has not been burnt regularly, suggesting the need to consolidate management systems for regular burning.

Our study also suggests that carnivore species should be monitored over the long term because they may be good indicators of forest condition. There was a gradient in affinity for mature forest or savannah among the small carnivore assemblages in the Moukalaba. Such habitat preferences appeared to be associated with species traits and did not vary largely across habitats. Given its high trophic status and important role in maintaining forest ecosystems (Nakashima & Sukor, 2010; Nakashima *et al.*, 2010a, 2010b), the carnivore community well reflects the conditions of the forest ecosystem as a whole. These characteristics make the carnivore community desirable as a biological indicator (Lindenmayer *et al.*, 2000). Although fewer camera-trapping studies have been conducted in lowland rainforest in Africa than in other regions (McCallum, 2012), monitoring of these small carnivores using camera-traps provides an efficient and a cost-effective method of assessing the transition of forest conditions over space and time.

Acknowledgements

We thank the Agence Nationale des Parcs Nationaux (ANPN) and the Centre National de la Recherche Scientifique et Technique (CENAREST) of Gabon for permission to conduct our research in Moukalaba–Doudou National Park. This study was financially supported by JST/JICA, SATREPS (Science and Technology Research Partnership for Sustainable Development). This research complied with PACUC regulations (Protocol 98-047 and 08-021) and the laws of the Gabonese Republic.

References

Akaike, H. (1987) Factor analysis and AIC. *Psychometrika* 52, 317–332.

Angelici, F.M. (2000) Food habits and resource partitioning of carnivores (Herpestidae, Viverridae) in the rainforests of southeastern Nigeria: preliminary results. *Revue d'Écologie – La Terre et la Vie* 55, 67–76.

Angelici, F.M. & Do Linh San, E. (2015) *Bdeogale nigripes. The IUCN Red List of Threatened Species*

2015, e.T41592A45205243. http://dx.doi.org/10.2305/ IUCN.UK.2015-4.RLTS.T41592A45205243.en. Accessed on 8 December 2020.

Angelici, F.M. & Luiselli, L. (2005) Habitat associations and dietary relationships between two genets, *Genetta maculata* and *Genetta cristata*. *Revue d'Écologie – La Terre et la Vie* 60, 341–354.

Angelici, F.M., Luiselli, L. & Politano, E. (1999) Distribution and habitat of selected carnivores (Herpestidae, Mustelidae, Viverridae) in the rainforests of southeastern Nigeria. *Zeitschrift für Säugetierkunde* 64, 116–120.

Azevedo, F.C.C., Lester, V., Gorsuch, W., Larivière, S., Wirsing, A.J. & Murray, D.L. (2006) Dietary breadth and overlap among five sympatric prairie carnivores. *Journal of Zoology* 269, 127–135.

Bahaa-el-din, L., Henschel, P., Aba'a, R., Abernethy, K., Bohm, T. *et al.* (2013) Notes on the distribution and status of small carnivores in Gabon. *Small Carnivore Conservation* 48, 19–29.

Baker, C.M. (1992) *Atilax paludinosus. Mammalian Species* 408, 1–6.

Baker, C.M. & Ray, J.C. (2013) *Atilax paludinosus* Marsh mongoose. In: *The Mammals of Africa. Volume V: Carnivores, Pangolins, Equids and Rhinoceroses* (eds J. Kingdon & M. Hoffmann), pp. 298–302. Bloomsbury, London.

Bartoń, K. (2009) *MuMIn: Multi-Model Inference*. R package version 1.7.11.

Chase, J.M., Abrams, P.A., Grover, J.P., Diehl, S., Chesson, P., Holt, R.D., Richards, S.A., Nisbet, R.M. & Case, T.J. (2002) The interaction between predation and competition: a review and synthesis. *Ecology Letters* 5, 302–315.

Davies, J.T., Meiri, S., Barraclough, T.G. & Gittleman, J.L. (2007) Species co-existence and character divergence across carnivores. *Ecology Letters* 10, 146–152.

Davis, C.L., Rich, L.N., Farris, Z.J., Kelly, M.J., Di Bitteti, M.S. *et al.* (2019) Ecological correlates of the spatial co-occurrence of sympatric mammalian carnivores worldwide. *Ecology Letters* 21, 1401–1412.

Delibes, M. & Beltrán, J.F. (1985) Activity, daily movements and home range of an Ichneumon or Egyptian mongoose (*Herpestes ichneumon*) in southern Spain. *Journal of Zoology* 207, 610–613.

Di Bitetti, M.S., Balanco, Y.E.D., Pereira, J.A., Paviolo, A. & Pérez, I.J. (2009) Time partitioning favors the coexistence of sympatric crab-eating foxes (*Cerdocyon thous*) and Pampas foxes (*Lycalopex gymnocercus*). *Journal of Mammalogy* 90, 479–490.

Di Bitetti, M.S., De Angelo, C.D., Di Blanco, Y.E. & Paviolo, A. (2010) Niche partitioning and species coexistence in a Neotropical felid assemblage. *Acta Oecologica* 36, 403–412.

Dickman, C.R. (1988) Body size, prey size, and community structure in insectivorous mammals. *Ecology* 69, 569–580.

Do Linh San, E., Ferguson, A.W., Belant, J.L., Schipper, J., Hoffmann, M., Gaubert, P., Angelici, F.M. & Somers, M.J. (2013) Conservation status, distribution and species richness of small carnivores in Africa. *Small Carnivore Conservation* 48, 4–18.

Durant, S.M. (1998) Competition refuges and coexistence: an example from Serengeti carnivores. *Journal of Animal Ecology* 67, 370–386.

Ermias Admasu, Thirgood, S.J., Afework Bekele & Laurenson, M.K. (2004) A note on the spatial ecology of African civet *Civettictis civetta* and common genet *Genetta genetta* in farmland in the Ethiopian Highlands. *African Journal of Ecology* 42, 160–162.

Fernández-Durán, J.J. (2004) Circular distributions based on nonnegative trigonometric sums. *Biometrics* 60, 499–503.

Fiske, I. & Chandler, R. (2011) Unmarked: an R package for fitting hierarchical models of wildlife occurrence and abundance. *Journal of Statistical Software* 43, 1–23.

Fuller, T., Biknevicius, A. & Kat, P. (1990) Movements and behaviour of large spotted genets (*Genetta maculata* Gray 1830) near Elmenteita Kenya (Mammalia Viverridae). *Tropical Zoology* 3, 13–19.

Grassman, L.I., Haines, A.M., Janečka, J.E. & Tewes, M.E. (2006) Activity periods of photo-captured mammals in north central Thailand. *Mammalia* 70, 306–309.

Geary, R.C. (1954) The contiguity ratio and statistical mapping. *The Incorporated Statistician* 5, 115–146.

Haines, A., Grassman, L.I. Jr & Tewes, M.E. (Chapter 16, this volume) Ecological separation and co-existence in a carnivore community in north-central Thailand. In: *Small Carnivores: Evolution, Ecology, Behaviour, and Conservation* (eds E. Do Linh San, J.J. Sato, J.L. Belant & M.J. Somers). Wiley–Blackwell, Oxford.

Happold, D.C.D. (1996) Mammals of the Guinea–Congo rain forest. *Proceedings of the Royal Society of Edinburgh B* 104, 243–284.

Hardin, G. (1960) The competitive exclusion principle. *Science* 131, 1292–1297.

Hines, J.E. (2006) *PRESENCE2 – Software to Estimate Patch Occupancy and Related Parameters*. USGS-PWRC. http://www.mbr-pwrc.usgs.gov/software/presence.html.

Ikeda, H., Ono, Y., Baba, M. & Iwamoto, T. (1982) Ranging and activity patterns of three nocturnal viverrids in Orno National Park, Ethiopia. *African Journal of Ecology* 20, 179–186.

IUCN [International Union for the Conservation of Nature] (2020) *The IUCN Red List of Threatened Species*. Version 2020-3. http://www.iucnredlist.org. Accessed on 8 December 2020.

Jeffery, K., Palla, F., Walters, G., Mendoza, I., Abernethy, K. & White, L.J. (2011) Fire management in a changing landscape: a case study from Lopé National Park, Gabon. *Parks* 20, 39–52.

Kasparian, M.A., Hellgren, E.C. & Ginger, S.M. (2002) Food habits of the Virginia opossum during raccoon removal in the Cross Timbers ecoregion, Oklahoma. *Proceedings of the Oklahoma Academy of Sciences* 82, 73–78.

Kronfeld-Schor, N. & Dayan, T. (2003) Partitioning of time as an ecological resource. *Annual Review of Ecology, Evolution, and Systematics* 34, 153–181.

Lindenmayer, D.B., Margules, C.R. & Botkin, D.B. (2000) Indicators of biodiversity for ecologically sustainable forest management. *Conservation Biology* 14, 941–950.

Lindstedt, S.L., Miller, B.J. & Buskirk, S.W. (1986) Home range, time, and body size in mammals. *Ecology* 67, 413–418.

Linkie, M. & Ridout, M.S. (2011) Assessing tiger–prey interactions in Sumatran rainforests. *Journal of Zoology* 284, 224–229.

MacKenzie, D.I., Nichols, J.D., Royle, J.A., Pollock, K.H., Bailey, L.L. & Hines, J.E. (2006) *Occupancy Estimation and Modelling: Inferring Patterns and Dynamics of Species Occurrence*. Academic Press, Burlington.

Maddock, A.H. & Perrin, M.R. (1993) Spatial and temporal ecology of an assemblage of viverrids in Natal, South Africa. *Journal of Zoology* 229, 277–287.

McCallum, J. (2012) Changing use of camera traps in mammalian field research: habitats, taxa and study types. *Mammal Review* 43: 196–206.

Mills, D., Do Linh San, E., Robinson, H., Isoke, S., Slotow, R. & Hunter, L. (2019) Competition and specialization in an African forest carnivore community. *Ecology and Evolution* 9, 10092–10108.

Moreno, R.S., Kays, R.W. & Samudio, J.R. (2006) Competitive release in diets of ocelot (*Leopardus pardalis*) and puma (*Puma concolor*) after jaguar (*Panthera onca*) decline. *Journal of Mammalogy* 87, 808–816.

Nakashima, Y. & Sukor, J.A. (2010) Importance of common palm civets (*Paradoxurus hermaphroditus*) as a long-distance disperser for large-seeded plants in degraded forests. *Tropics* 18, 221–229.

Nakashima, Y., Inoue, E., Inoue-Murayama, M. & Sukor, J.A. (2010a) Functional uniqueness of a small carnivore as seed dispersal agents: a case study of the common palm civets in the Tabin Wildlife Reserve, Sabah, Malaysia. *Oecologia* 164, 721–730.

Nakashima, Y., Inoue, E., Inoue-Murayama, M. & Sukor, J.A. (2010b) High potential of a disturbance-tolerant frugivore, the common palm civet *Paradoxurus hermaphroditus* (Viverridae), as a seed disperser for large-seeded plants. *Mammal Study* 35, 209–215.

Nakashima, Y., Iwata, Y., Ando, C., Nze-Nkogue, C.N., Inoue E., Akomo-Okoue, E.F., Nguema, P.M., Bineni, T.D., Banak, L.N., Takenoshita, Y., Ngomanda, A. & Yamagiwa, J. (2013) Assessment of landscape-scale distribution of sympatric great apes in African rainforests: concurrent use of nest and camera-trap surveys. *American Journal of Primatology* 75, 1220–1230.

Ndzinisa, G. (2018) *Activity Patterns and Home Range of Rusty-Spotted Genet (Genetta maculata) in a Low-Lying Savannah, Mbuluzi Game Reserve, Eswatini*. MSc thesis, University of Eswatini.

Palomares, F. & Delibes, M. (1990) Habitat preference of large grey mongooses *Herpestes ichneumon* in Spain. *Acta Theriologica* 35, 1–6.

Palomares, F. & Delibes, M. (1991) Assessing three methods to estimate daily activity patterns in radio-tracked mongooses. *Journal of Wildlife Management* 55, 698–700.

Palomares, F. & Delibes, M. (1992) Circadian activity patterns of free-ranging large gray mongooses, *Herpestes ichneumon*, in southwestern Spain. *Journal of Mammalogy* 73, 173–177.

Palomares, F. & Delibes, M. (1993) Key habitats for Egyptian mongooses in Doñana National Park, south-western Spain. *Journal of Applied Ecology* 30, 752–758.

Palomares, F., Ferreras, P., Fedriani, J.M. & Delibes, M. (1996) Spatial relationships between Iberian lynx and other carnivores in an area of south-western Spain. *Journal of Applied Ecology* 33, 5–13.

Primack, R.B. & Corlett, R. (2005) *Tropical Rain Forests: An Ecological and Biogeographical Comparison.* Blackwell Science, Malden.

Ray, J.C. (1996) *Resource Use Patterns Among Mongooses and Other Carnivores in a Central African Rainforest.* PhD thesis, University of Florida.

Ray, J.C. (1997) Comparative ecology of two African forest mongooses, *Herpestes naso* and *Atilax paludinosus. African Journal of Ecology* 35, 237–253.

Ray, J.C. (1998) Temporal variation of predation on rodents and shrews by small African forest carnivores. *Journal of Zoology* 244, 363–370.

Ray, J.C. (2001) Carnivore biogeography and conservation in the African forest: a community perspective. In: *African Rainforest Ecology and Conservation – An Interdisciplinary Perspective* (eds W. Weber, L.J.T. White, A. Vedder & L. Naughton-Treves), pp. 214–232. Yale University Press, New Haven.

Ray, J.C. & Sunquist, M. (2001) Trophic relations in a community of African rainforest carnivores. *Oecologia* 127, 395–408.

Ray, J., Bahaa-el-din, L., Angelici, F.M. & Do Linh San, E. (2015) *Herpestes naso. The IUCN Red List of Threatened Species* 2015, e.T41615A45207915. http://dx.doi.org/10.2305/IUCN.UK.2015-4.RLTS.T41615A45207915.en. Accessed on 8 December 2020.

Ridout, M.S. & Linkie, M. (2009) Estimating overlap of daily activity patterns from camera trap data. *Journal of Agricultural, Biological, and Environmental Statistics* 14, 322–337.

Rosenzweig, M.L. (1966) Community structure in sympatric Carnivora. *Journal of Mammalogy* 47, 602–612.

Rosevear, D.R., Parsons, R., Wolseley, P. & Shaffer, M. (1974) *The Carnivores of West Africa.* British Museum (Natural History) London, London.

Roux, R. (2018) *Spatio-Temporal Ecology of the Rusty-Spotted Genet,* Genetta maculata, *in Telperion Nature Reserve (Mpumalanga, South Africa).* MSc thesis, University of South Africa.

Rowe-Rowe, D.T. (1977) Food ecology of otters in Natal, South Africa. *Oikos* 28, 210–219.

Schoener, T.W. (1974) Resource partitioning in ecological communities. *Science* 185, 27–39.

Skinner, J.D. & Chimimba, C.T. (eds) (2005) *The Mammals of the Southern African Subregion.* Cambridge University Press, Cape Town.

Sosef, M.S.M., Issembe, Y., Bourobou, H.P.B. & Koopman, W.J.M. (2004) Botanical diversity of the Pleistocene forest refugia Mounts Doudou. In: *Monts Doudou, Gabon: Floral and Faunal Inventory with Reference to Elevational Distribution* (ed. B.L. Fisher), pp. 17–92. California Academy of Science, San Francisco.

Streicher, J.P., Ramesh, T. & Downs, C.T. (2020) Home range and core area utilisation of three co-existing mongoose species: large grey, water and white-tailed in the fragmented landscape of the KwaZulu-Natal Midlands, South Africa. *Mammalian Biology* 100, 273–283.

Streicher, J.P., Ramesh, T. & Downs, C.T. (2021) An African urban mesocarnivore: navigating the urban matrix of Durban, South Africa. *Global Ecology and Geography* 26, e01482.

Stuart, C.T. (1981) Notes on the mammalian carnivores of the Cape Province, South Africa. *Bontebok* 1, 1–58.

Takenoshita, Y., Ando, C. & Yamagiwa, J. (2008) Fruit phenology of the great ape habitat in the Moukalaba–Doudou National Park, Gabon. *African Study Monographs. Supplementary Issue* 39, 23–39.

Thibault, M., Fisher, B.L. & Goodman, S.M. (2004) Description of Mounts Doudou, Gabon, and the 2000 biological inventory of the reserve. In: *Monts Doudou, Gabon: A Floral and Faunal Inventory with Reference to Elevational Distribution* (ed. B.L. Fisher), pp. 3–16. California Academy of Science, San Francisco.

Tobler, M.W., Carrillo-Percastegui, S.E. & Powell, G. (2009) Habitat use, activity patterns and use of mineral licks by five species of ungulate in south-eastern Peru. *Journal of Tropical Ecology* 25, 261–270.

Trewby, L.D., Wilson, G.J., Delahay, R.J., Walker, N., Young, R., Davison, J., Cheeseman, C., Robertson, P.A., Gorman, M.L. & McDonald, R.A. (2008) Experimental evidence of competitive release in sympatric carnivores. *Biology Letters* 4, 170–172.

Van Rompaey, H. & Colyn, M. (2013) *Herpestes naso* Long-nosed mongoose (long-snouted mongoose). In: *The Mammals of Africa. Volume V: Carnivores, Pangolins, Equids and Rhinoceroses* (eds J. Kingdon & M. Hoffmann), pp. 295–297. Bloomsbury, London.

White, L.J.T. (2001) Forest–savanna dynamics and the origins of Marantaceae forest in central Gabon. In: *African Rainforest Ecology and Conservation – An Interdisciplinary Perspective* (eds W. Weber, L.J.T. White, A. Vedder & L. Naughton-Treves), pp. 165–182. Yale University Press, New Haven.

16

Ecological Separation and Coexistence in a Carnivore Community in North-Central Thailand

Aaron M. Haines[1], Lon I. Grassman Jr[2],*, and Michael E. Tewes[2]

[1] *Department of Biology, Applied Conservation Lab, Millersville University, Millersville, PA, USA*
[2] *Feline Research Center, Caesar Kleberg Wildlife Research Institute, Texas A&M University–Kingsville, Kingsville, TX, USA*

SUMMARY

Carnivore species are believed to exert strong competitive pressure on each other, resulting in adaptations to allow for niche separation through resource partitioning. However, factors that promote ecological separation among species in tropical forests are difficult to explain and are poorly understood because robust field studies are lacking. We examined spatial, temporal and morphological segregation between tropical carnivores in a protected forest in north-central Thailand. Sympatric spatial overlap was calculated from radio-telemetry data of 38 individuals from six species (5 yellow-throated martens, *Martes flavigula*, 20 leopard cats, *Prionailurus bengalensis*, 2 Asiatic golden cats, *Catopuma temminckii*, 4 clouded leopards, *Neofelis nebulosa*, 5 binturongs, *Arctictis binturong*, and 2 dholes, *Cuon alpinus*) in the same study area. Spatial overlap was then correlated with 14 independent variables (i.e. skull and dental morphology, body mass, habitat use and activity patterns) compared among the six species. We predicted that carnivores with differing morphology and activity patterns would exhibit more spatial overlap because these species would compete less for prey resources. Our statistical analyses indicated that lower mean carnassial length and activity patterns in closed habitat cover were significantly correlated ($p < 0.05$) with species spatial overlap. Binturongs appeared to have the greatest amount of spatial overlap with other species of carnivores, whereas dholes had the least spatial overlap; also, dholes and yellow-throated martens tended to be more active in open habitats and during diurnal time periods, whereas clouded leopards and Asiatic golden cats were more active in closed cover and were more arrhythmic in activity. Although these results provide useful information on carnivore coexistence, we recommend that future studies monitor larger sample sizes of carnivore species over the same time period to provide more robust statistical analyses. In addition, we suggest that future research on carnivore coexistence evaluates the impacts of anthropogenic activity on study results.

Keywords

competitive exclusion – morphological differences – resource partitioning – spatial overlap – spatial segregation – temporal segregation

Introduction

Gause (1934) developed the principle of competitive exclusion, which states that 'complete competitors cannot coexist'. If there are no ecological differences between species in nature, they will not be able to coexist (Hutchinson, 1978). Thus, differences between species are essential for their coexistence (Chase & Leibold, 2003), and natural selection should favour adaptations that would reduce competition among similar species through resource partitioning (Agostinho *et al.*, 2003). Carnivore species are believed to exert

---

* Corresponding author.

Small Carnivores: Evolution, Ecology, Behaviour, and Conservation, First Edition. Edited by Emmanuel Do Linh San, Jun J. Sato, Jerrold L. Belant, and Michael J. Somers.

strong competitive pressure on each other, resulting in adaptations to allow for niche separation through resource partitioning (Schoener, 1974; Di Bitetti *et al.*, 2010). Pianka (1969) highlighted that niche separation can occur with differences in morphology, spatial activity or temporal activity, which has been found among wild carnivores (Hardin, 1960; Creel *et al.*, 2001; Pfennig & Pfennig, 2005; Schuette *et al.*, 2013; de Cassia Bianchi *et al.*, 2016).

Co-occurring species might be morphologically similar because they are adapted to the same environment or morphologically dissimilar to minimize competition (Davies *et al.*, 2007). In the latter case, morphological differences may lead to trophic segregation and, therefore, it is possible that sympatric species exhibit slightly different but distinct morphology to reduce competition for food resources (Brown & Wilson, 1956). This pattern has been documented for wild felids and mustelids (Dayan & Simberloff, 1994; Davies *et al.*, 2007). In addition, spatial and temporal segregation, as well as behavioural differentiation, might lessen competition by reducing the frequency of encounters among competitor species through habitat or diel partitioning (Durant, 1998; Fedriani *et al.*, 2000; Linnell & Strand, 2000; Creel, 2001; Farlow & Pianka, 2002; Grassman *et al.*, 2006; Gómez-Ortiz *et al.*, 2015; Ramesh *et al.*, 2017). This ecological strategy has been found to be especially important for predators in tropical forests in Central and South America (Davis *et al.*, 2011; Santos *et al.*, 2019), Africa (Mills *et al.*, 2019; Nakashima *et al.*, Chapter 15, this volume) and Southeast Asia (Ngoprasert *et al.*, 2012; Lynam *et al.*, 2013; Singh & Macdonald, 2017; Haidir *et al.*, 2018; Petersen *et al.*, 2019; Pudyatmoko, 2019).

Grassman *et al.* (2006) underlined that in a diverse ecosystem such as tropical forests, the interaction among species and their ecological separation may be complex because of the sympatry of a large number of species. In tropical regions, distributions of species are predominantly limited by biotic factors such as competitive interactions (Lomolino *et al.*, 2005). However, factors that promote ecological separation among species in tropical forests are difficult to explain and are poorly understood because robust field studies are lacking (Ray & Sunquist, 2001; Sanchez-Cordero *et al.*, 2008).

Grassman *et al.* (2006) examined factors that permitted the yellow-throated marten, *Martes flavigula*, to coexist with five other carnivore species as part of a long-term radio-telemetry study on carnivore ecology in Phu Khieo Wildlife Sanctuary (PKWS), Thailand. Their objectives were to compare interspecific differences in body mass, activity pattern, habitat preference, home range size, home range overlap and spatial distribution among 5 yellow-throated martens, 20 leopard cats, *Prionailurus bengalensis*, 2 Asiatic golden cats, *Catopuma temminckii*, 2 clouded leopards, *Neofelis nebulosa*, 2 dholes, *Cuon alpinus*, and 5 binturongs, *Arctictis binturong*. Grassman *et al.* (2006) focused specifically on the yellow-throated marten and did not compare ecological separation within the larger carnivore community.

For comparison to Grassman *et al.* (2006), we reviewed the same spatial data set to correlate morphological differences, as well as spatial and temporal differences among all species studied in this carnivore community, with the goal to identify which ecological factors were most correlated with species coexistence or spatial overlap. We expanded this data set to include skull, jaw and dental morphology because Davies *et al.* (2007) found that morphological disparity, particularly carnassial tooth morphology, was important in explaining carnivore co-occurrence at regional scales. In addition, Grassman *et al.* (2006) considered ecological separation to be multi-causal among this carnivore community, and thus we combined spatial and temporal attributes to assess if combinations of these factors were correlated to species co-occurrence.

In this study, we attempted to recognize the ecological processes that allowed for carnivore coexistence in PKWS by correlating carnivore spatial overlap (i.e. degree of sympatry) with contrasts in morphological and behavioural characteristics. Our null hypothesis was that spatial overlap would be unrelated to morphological, habitat and temporal differences among these six carnivore species. Based on the observations of Davies *et al.* (2007), we predicted that carnivores with differing dental and skull morphology, as well as differing mass, would exhibit spatial overlap in range because they would most likely feed on different prey species. We also predicted that carnivore species that exhibit spatial overlap would do so via activity avoidance to reduce competition over prey resources.

Materials and Methods

Study Area

The study site at PKWS was located in north-central Thailand (16°05′–16°35′ N, 101°20′–101°55′ E; Figure 16.1) and encompassed 1560 km² of forests within the larger 4550 km² Western Issan Forest Complex (Kumsuk *et al.*, 1999). This site represented the largest protected area within the Issan region. Phu Khieo is one of the three protected areas in Thailand that does not contain a permanent human settlement (Kekule, 1999). The habitat consisted of forested hills transitioning into mountains with evergreen forest (75%), mixed deciduous forest (13%), dry dipterocarp forest (4%), bamboo (4%), grassland (3%) and a forest plantation (1%; Anonymous, 2000). Grassland communities were patchy and dominated by the Thung Kha Mang grassland (3 km²) near the sanctuary headquarters (Figure 16.2). The study area (~200 km²) was located in the north-central portion of the sanctuary. It included the Thung Kha Mang headquarters, minor walking trails, the Phrom River and several perennial streams (Figure 16.2). Site selection was based on its central location within the sanctuary, abundant carnivore sign and low tourism.

Collection of Field Data

Field data consisted of trapping, radio-collaring and radio-tracking carnivores intermittently from September 1998 through February 2003. Carnivores were trapped in box traps baited with live chickens and were sedated either with ketamine hydrochloride and xylazine hydrochloride or with tiletamine hydrochloride and zolazepam hydrochloride. Study carnivores were fitted with very high frequency (VHF) radio-collars with activity sensors. For a full description of field methods and analyses of spatial and temporal data, see Grassman *et al.* (2006).

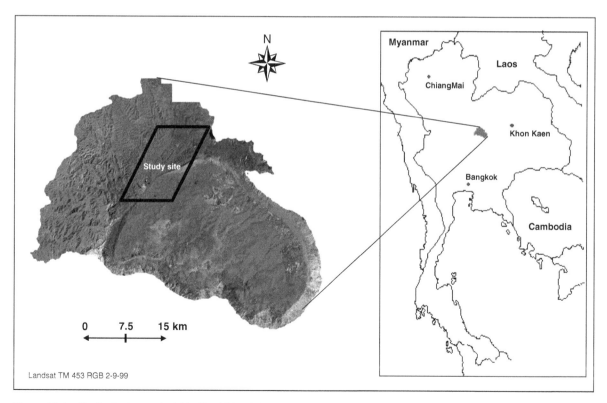

Figure 16.1 Study site located within Phu Khieo Wildlife Sanctuary (PKWS) in north-central Thailand.

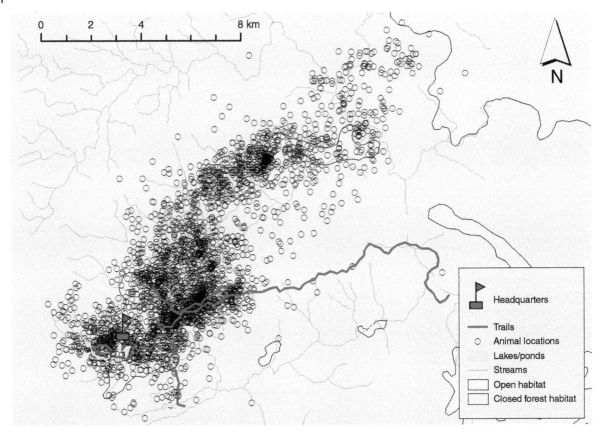

Figure 16.2   Locations of 38 individuals of 6 carnivore species radio-collared and tracked at PKWS in north-central Thailand, from 1998 to 2003, as related to habitat, water sources and trails.

We recorded activity, time of day and the type of habitat for each telemetry location. Animal locations were derived from terrestrial radio-telemetry with ≥3 bearings used for each location. Activity (e.g. locomotion) was determined by a variable radio-pulse combined with fluctuations in pulse volume. Activity levels for each radio-collared animal were recorded for each bearing during location telemetry and also intermittently during diels. Diels consisted of uninterrupted monitoring during a 24 hour period, and we assumed that 15 minutes were sufficient for independence of observations between each activity reading during diels (Grassman, 2004). Activity recorded during radio-tracking for spatial locations was separated by ≤5 minute/reading.

Because there was little variation in sunrise and sunset times in our study site throughout the year, we considered locations taken between 06:00 and 18:59 h to be diurnal and locations taken between 19:00 and 05:59 h to be nocturnal. Carnivore species with < 10% of radio-telemetry readings occurring during the nocturnal period were characterized as 'diurnal', whereas species with > 90% of active readings during the night were considered 'nocturnal'. Species with activity levels between 10 and 90% during both the day and the night were classified as 'arrhythmic' (i.e. no clear activity pattern).

Vegetation types were delineated on a 1:50 000 scale topographic map developed from SPOT satellite imagery obtained in March 1992. Habitat cover was virtually unchanged between the date of this imagery and the dates of field data collection. Two major cover types were delineated within the study site: closed forest and open forest/grassland (including abandoned orchard; Figure 16.2).

Data Analysis

Degree of Sympatry (Response Variable)

The response variable for our model was degree of sympatry (spatial overlap at the local scale). We calculated an area of use for each species to determine the degree of sympatry between coexisting carnivore species. An area of use was calculated using a 95% fixed-kernel range-size estimator with least-squared cross-validation using all locations for all individuals. This step was performed using the spatial software BIOTAS® (Ecological Software Solutions, Inc., Sacramento, California, USA). We then used BIOTAS to calculate area of overlap between each species. The degree of sympatry (DS) between two species A and B was defined as the proportion of the more restricted species' total range overlapped by the more widespread species (Chesser & Zink, 1994):

$$DS_{AB} = \frac{\text{area of overlap between species A and B}}{\text{total range size of species with the smaller range}}$$

A spatial overlap of 0 signified no range overlap, whereas a value of 1 indicated that the total range of one species entirely overlapped the other. This is similar to the relative area of common occurrence calculated by Barraclough *et al.* (1998) and Barraclough & Vogler (2000). As with most long-term radio-telemetry studies involving multiple species over a large geographic area, there were likely uncollared, sympatric cohorts whose influence could not be measured against our study animals. However, we believe that this influence was minimal given the intensive live-trapping effort (i.e. 27 928 trap-nights over four years), which likely resulted in capturing the majority of individuals in the study area.

Independent Predictor Variables of Species Coexistence

In an attempt to identify the main factors allowing for, or explaining, species spatial coexistence among this tropical forest carnivore community, we selected 11 potential predictor variables where divergence between species could be correlated with species degree of sympatry. Predictor variables were based on previous studies that identified body size, dental morphology, skull morphology and species behaviour

(e.g. habitat partitioning or temporal segregation) as good predictors of carnivore character displacement or coexistence (Dayan & Simberloff, 1996; Farlow & Pianka, 2002; Grassman *et al.*, 2006; Davies *et al.*, 2007; Lynam *et al.*, 2013; de Cassia Bianchi *et al.*, 2016). These characteristics included body mass, canine length, canine diameter, lower carnassial length and upper carnassial length, skull length, jaw length, activity patterns in closed and open habitat and activity patterns during nocturnal and diurnal time periods (Table 16.1).

The first predictor variable, body mass (kg), was recorded for each individual in this study by Grassman *et al.* (2006) (Table 16.1). Morphology on skull length and jaw length for each species was determined by taking the average of skull-length measurements reported by Lekagul & McNeeley (1977), Cohen (1978), van Valkenburgh (1985), van Valkenburgh & Ruff (1987), Wayne *et al.* (1989) and Farlow & Pianka (2002) for specimens found in Thailand (Table 16.1). Median measurements of canine length, canine diameter, lower carnassial length and upper carnassial length were recorded by Dayan & Simberloff (1996), Meiri *et al.* (2005), Davies *et al.* (2007) and provided by Shai Meiri (Tel-Aviv University, Department of Zoology, Tel Aviv, Israel) from specimens in Southeast Asia (Table 16.1). Field-based predictor variables included percentage of bearings used to triangulate locations in which the animal was active for both nocturnal and diurnal periods and for which the animal was active while in closed and open habitat cover (Table 16.1). Data for predictor variables were pooled for each species.

Variable divergence was quantified by performing contrasts in the logarithm of the variable values between species pairs (Davies *et al.*, 2007). Hence, for each species pair, the divergence in variable X between species A and B was calculated as follows:

$$\log\left(X_A\right) - \log\left(X_B\right),$$

where X_A and X_B are the variable values and $X_A > X_B$ (Davies *et al.*, 2007). These calculations of variable divergence were the predictor variables values that were correlated to species degree of sympatry.

Due to the limited number of species comparisons ($n = 15$), we converted our variable proportions to

Table 16.1 Measurements of predictor variables used to model degree of sympatry (spatial overlap) for six carnivore species radio-collared and tracked at PKWS, Thailand, from 1998 to 2003. Species are ranked based on decreasing mean body mass.

Species	Mean body mass (kg)[a]	Median upper right canine length (mm)[b]	Median canine diameter (mm)[b]	Median lower carnassial (mm)[b]	Median upper carnassial (mm)[b]	Mean jaw length (mm)[c]	Mean skull length (mm)[c]	Percentage of active bearings for locations in closed cover	Percentage of active bearings for locations in open cover	Percentage of active bearings for nocturnal locations	Percentage of active bearings for diurnal locations
Dhole, *Cuon alpinus*	16.50	20.00	9.52	20.60	19.22	130.00	165.00	50	47	24	68
Clouded leopard, *Neofelis nebulosa*	13.00	27.50	11.31	14.55	19.86	95.00	155.00	57	66	56	60
Binturong, *Arctictis binturong*	12.30	18.40	7.90	8.69	6.93	112.50	140.00	49	41	63	42
Asiatic golden cat, *Pardofelis temminckii*	10.70	22.50	7.39	12.02	15.72	73.00	135.00	61	41	48	64
Yellow-throated marten, *Martes flavigula*	2.80	10.60	4.99	9.56	8.95	67.00	100.00	57	61	28	63
Leopard cat, *Prionailurus bengalensis*	2.70	9.92	4.44	7.70	9.87	48.00	92.00	53	56	52	54

[a] Grassman et al. (2006).

[b] Dayan & Simberloff (1996), Meiri et al. (2005), Davies et al. (2007) and Shai Meiri (Tel-Aviv University, Department of Zoology, Tel Aviv, Israel) from specimens in Southeast Asia.

[c] Lekagul & McNeeley (1977), Cohen (1978), van Valkenburgh (1985), van Valkenburgh & Ruff (1987), Wayne et al. (1989) and Farlow & Pianka (2002).

percentages and tested each variable for normality using the Kolmogorov–Smirnov statistic. All variables with a p-value ≥ 0.10 were considered to have a normal distribution. All predictor variables were found to have a normal distribution, with exception of body mass. In response, we conducted non-parametric and parametric analyses for our data. We ran univariate non-parametric Spearman's rank correlation tests to identify possible significant relationships between divergence in predictor variables and species degree of sympatry. Statistical significance was based on a p-value <0.05. We also used parametric multiple-regression analysis of all predictor variables modelled against the response variable (degree of sympatry) using a best subsets regression analysis (Hocking, 1976) to identify which models produced the highest r^2 values. We then weighted these models based on Akaike's Information Criterion corrected for small sample sizes (AIC_c) and calculated ΔAIC_c to identify plausible models that explained which variables were most associated with spatial overlap between carnivores (Anderson *et al.*, 2000). We only considered models as plausible when within approximately 2 ΔAIC_c units (Burnham & Anderson, 2002). All statistical analyses were conducted using Minitab (Minitab® 16.1.1 2010, State College, PA, USA).

Results

We found some general trends in the degree of sympatry within the carnivore community. Binturongs had the greatest amount of spatial overlap with other species of carnivores, whereas dholes had the least amount of spatial overlap (Table 16.2). We also found that these species exhibited general differences in their habitat and temporal preferences when active. For example, dholes and yellow-throated martens were more active in open habitats and during diurnal time periods, whereas clouded leopards and Asiatic golden cats were more active in closed cover and were more arrhythmic in activity.

Based on Spearman's rank correlation analysis, we found that species range overlap (degree of sympatry) was significantly negatively correlated with divergence in median lower carnassial length ($r_s = -0.62$, $p = 0.01$) and positively correlated with percentage of

active bearings in closed cover habitat ($r_s = 0.56$, $p = 0.03$; Table 16.2). We did not find any other divergence contrast in predictor variables to be closely correlated to the species degree of sympatry according to Spearman's rank correlation tests ($r_s < 0.31$, $p > 0.27$) (Table 16.2). Based on parametric best subsets multiple regression analysis, we found only one plausible model with a $\Delta AIC_c < 2$. This model had an r^2-value of 0.77 and contained median lower carnassial length ($F = 28.87$, df $= 1$, $p = <0.01$), median upper carnassial length ($F = 7.36$, df $= 1$, $p = 0.02$) and percentage of active bearings for locations in closed cover ($F = 4.81$, df $= 1$, $p = 0.05$) as predictor variables that best determine species degree of sympatry. All other models had a $\Delta AIC_c > 2$, and thus were not considered plausible models in explaining spatial overlap between carnivores monitored in our study.

Both statistical tests identified carnassial length and activity in closed habitat cover as important variables predicting degree of spatial overlap of carnivores. Greater divergence in median carnassial length was correlated with low species spatial overlap, while greater divergence in activity periods in closed habitat cover was correlated with greater species spatial overlap. Thus, species with greater differences in carnassial length tended to avoid each other spatially, whereas species that exhibited different activity patterns in closed habitat cover had more spatial overlap.

Discussion

We predicted that carnivores with differing dental and skull morphology, as well as differing body mass, would exhibit spatial overlap because they would most likely feed on different prey species. Davies *et al.* (2007) found that 63% of variation in distribution range overlap among carnivore sister species (defined as pairs of species that are each other's closest extant relatives) was explained by morphological divergence of dentition, with species that differed more in carnassial tooth length having a greater geographic overlap. These authors suggested that their results are consistent with the idea that competition among sympatric and ecologically similar carnivore species drives ecological character displacement or competitive exclusion. We also found that carnassial tooth length was a

Table 16.2 Degree of sympatry (response variable) and divergence of predictor variables for each pair of carnivores radio-collared and tracked at PKWS, Thailand, 1998–2003. Spearman's rank correlation coefficients (r_s) are used to identify significant correlations ($p < 0.05$) and these are highlighted in bold.

Species pair	Degree of sympatry	Mean body mass (kg)[a]	Median upper right canine length (mm)[b]	Median canine diameter (mm)[b]	Median lower carnassial (mm)[b]	Median upper carnassial (mm)[b]	Mean jaw length (mm)[c]	Mean skull length (mm)[c]	Percentage of active bearings for locations in closed cover	Percentage of active bearings for locations in open cover	Percentage of active bearings for nocturnal locations	Percentage of active bearings for diurnal locations
Dhole/leopard cat	0.398	0.785	0.305	0.331	0.427	0.289	0.433	0.254	0.025	0.076	0.336	0.100
Dhole/yellow-throated marten	0.585	0.801	0.276	0.281	0.333	0.332	0.288	0.217	0.057	0.113	0.067	0.033
Dhole/clouded leopard	0.621	0.007	0.138	0.075	0.151	0.014	0.136	0.027	0.057	0.147	0.368	0.054
Clouded leopard/yellow-throated marten	0.649	0.794	0.414	0.355	0.182	0.346	0.152	0.190	0.000	0.034	0.301	0.021
Dhole/binturong	0.662	0.147	0.036	0.081	0.375	0.443	0.063	0.071	0.009	0.059	0.419	0.209
Yellow-throated marten/leopard cat	0.666	0.015	0.029	0.051	0.094	0.042	0.145	0.036	0.032	0.037	0.269	0.067
Clouded leopard/leopard cat	0.686	0.779	0.443	0.406	0.276	0.304	0.296	0.227	0.032	0.071	0.032	0.046
Clouded leopard/Asiatic golden cat	0.695	0.169	0.087	0.185	0.083	0.101	0.114	0.060	0.029	0.207	0.067	0.028
Clouded leopard/binturong	0.706	0.141	0.175	0.156	0.224	0.457	0.073	0.044	0.066	0.207	0.051	0.155
Asiatic golden cat/leopard cat	0.733	0.610	0.356	0.221	0.193	0.202	0.182	0.167	0.061	0.135	0.035	0.074
Dhole/Asiatic golden cat	0.736	0.176	0.051	0.110	0.234	0.087	0.251	0.087	0.086	0.059	0.301	0.026
Binturong/leopard cat	0.794	0.638	0.268	0.250	0.053	0.154	0.370	0.182	0.034	0.135	0.083	0.109
Asiatic golden cat/yellow-throated marten	0.841	0.625	0.327	0.171	0.099	0.245	0.037	0.130	0.029	0.173	0.234	0.007
Binturong/yellow-throated marten	0.869	0.654	0.240	0.200	0.041	0.111	0.225	0.146	0.066	0.173	0.352	0.176
Binturong/Asiatic golden cat	0.869	0.028	0.087	0.029	0.141	0.356	0.188	0.016	0.095	0.000	0.118	0.183
Spearman's rank correlation coefficient (r_s)		−0.200	−0.084	−0.286	**−0.617**	−0.068	−0.088	−0.304	**0.555**	0.178	−0.195	0.191

[a]Grassman et al. (2006).
[b]Dayan & Simberloff (1996), Meiri et al. (2005), Davies et al. (2007) and Shai Meiri (Tel-Aviv University, Department of Zoology, Tel Aviv, Israel) from specimens in Southeast Asia.
[c]Lekagul & McNeeley (1977), Cohen (1978), van Valkenburgh (1985), van Valkenburgh & Ruff (1987), Wayne et al. (1989) and Farlow & Pianka (2002).

good predictor of the degree of spatial overlap among carnivores in our study area, but this was a significantly negative correlation. In other words, contrary to what was predicted, carnivore species with different carnassial sizes had less spatial overlap.

This discrepancy could result from at least two different reasons. First, our study did not focus on sister species *per se*, but on several members of the local carnivore community, with many of the species pairs included in the analyses belonging to different families. While it may be expected that closely related (and therefore morphologically similar) species will exhibit spatial segregation to avoid competition at the local scale, the same may not necessarily apply to species belonging to different families (see e.g. Pudyatmoko, 2019), as partially incorporated and tested in the present study.

Second, Davies *et al.*'s (2007) statistical analyses clearly suggested that a large proportion (37%) of the variation in geographic range overlap among sister species is affected by other variables, and it is likely that variables not included in their model also play a role. At the local scale, two factors need to be taken into consideration. On the one hand, species with similar morphology may avoid competition through temporal rather than spatial segregation (see below). On the other hand, the risk of intraguild predation and interspecific killing (Palomares & Caro, 1999; Donadio & Buskirk, 2006) must be taken into account when interpreting our analyses. Indeed, Grassman *et al.* (2006) found that in our study site the yellow-throated marten avoided core ranges of the larger-toothed Asiatic golden cat and clouded leopard possibly to avoid antagonistic confrontations but showed more overlap with carnivore species of similar tooth size (e.g. leopard cat). The authors concluded that smaller-toothed carnivore species avoided spatial overlap with larger-toothed carnivore species to reduce the chance of interspecific confrontations.

The above reasoning is compatible with the results of a recent study that investigated the ecological correlates of the spatial co-occurrence of sympatric carnivores worldwide based on camera-trapping data. Davis *et al.* (2019) found that co-occurrence probabilities were greatest for pairs of carnivores that shared ecological traits such as similar body size, diet or temporal activity pattern. However, co-occurrence decreased as compared to other species pairs when the pair included a large-bodied carnivore, confirming that large carnivores play an important top-down role (Roemer *et al.*, 2009; Wallach *et al.*, 2015; Ramesh *et al.*, 2017; Zhao *et al.*, 2020). Donadio & Buskirk (2006) found that interspecific killing among carnivores occurred with greater frequency and intensity when a larger species was 2–5.4 times the mass of the smaller victim. As body sizes become more similar, killing interactions decrease due to high risk of injury among carnivores of the same size (Donadio & Buskirk, 2006). We found no correlation between interspecific spatial overlap and divergence in body mass. Interestingly, several species had high spatial overlap with the binturong (Table 16.2), which was the third largest species by body mass but has relatively small carnassials (Table 16.1). Binturongs feed primarily on figs (Grassman *et al.*, 2005c), thus many species of smaller body mass, but relatively similar carnassial tooth size (Table 16.1), did not avoid binturongs probably because they did not compete for food resources and/ or did not incur any risks of being killed.

Pessino *et al.* (2001) found that during periods of low prey abundance pumas, *Puma concolor*, consumed more small wild felids. Therefore, at small spatial scales smaller carnivore species may reduce spatial overlap with much larger carnivore species to avoid interspecific mortality. If smaller carnivore species avoid overlap with much larger carnivore species, then how do carnivore species of similar size coexist? Davies *et al.* (2007) suggested that direct interspecific interactions may be reduced with different preferences in microhabitats, activity patterns and food. Thus, we predicted that species with spatial overlap would exhibit avoidance via diel activity patterns in order to avoid competition over prey resources. Our results indeed suggested that species that were active in closed cover overlapped more with species that were inactive in closed cover. Within the felid community in Sumatra, species of similar size and with similar-sized prey were also shown to exhibit temporal avoidance (Sunarto *et al.*, 2015), as notably observed for the diurnal Asiatic golden cat and nocturnal Sunda clouded leopard, *Neofelis diardi* (Haidir *et al.*, 2018). At three study sites in South America (Argentina, Ecuador and Suriname), two morphologically similar small cats also had contrasting activity patterns: the jaguarundi, *Herpailurus [= Puma] yagouaroundi*, was diurnal, while the margay, *Leopardus wiedii*, was exclusively nocturnal (Di Bitetti *et al.*, 2010; Santos *et al.*, 2019). More generally, several authors suggested that temporal habitat segregation may help maintain diverse predator communities in tropical forest habitats (Davis *et al.*, 2011; Lynam *et al.*, 2013; Singh & Macdonald, 2017; Mills *et al.*, 2019; Marinho *et al.*, 2020; Nakashima *et al.*, Chapter 15, this volume).

In comparison, other researchers have found that predators maintain coexistence via prey and fine-scale habitat partitioning. For example, several studies showed that leopards, *Panthera pardus*, may coexist with tigers, *Panthera tigris*, by avoiding places where tigers prefer to hunt and rest, while also partitioning prey based on their size (Karanth & Sunquist, 1995; Sunquist & Sunquist, 2002; Ngoprasert *et al.*, 2012; Pokheral & Wegge, 2019; cf. Li *et al.*, 2019). Similarly, in South America, Schaller & Crawshaw (1980), Scognamillo *et al.* (2003) and Haines (2006) concluded that the jaguar, *Panthera onca*, and the puma exhibit mutual avoidance through fine-scale habitat separation and food habits. In addition, Horne *et al.* (2009) found that ocelots, *Leopardus pardalis*, and bobcats, *Lynx rufus*, exhibited fine-scale habitat separation within their sympatric ranges. Hence, habitat partitioning seems to be another consistent form of resource partitioning among carnivores of similar size and morphology (Farlow & Pianka, 2002).

In our study site the majority of the habitat consisted of closed forest cover (Figure 16.2), and we did not define fine-scale habitat differences. However, we did find that species that overlapped spatially exhibited different activity patterns in closed cover habitat. This suggests an interaction between resource partitioning strategies, rather than one dominant strategy. Thus, other combinations of resource partitioning may be involved among carnivores to reduce interspecific competition. For example, only the dhole appeared to somewhat avoid closed forest cover and hunt as cooperative packs to kill larger prey in open habitat during diurnal periods (Grassman *et al.*, 2005b); however, several dhole locations occurred in closed forest, mainly during inactive periods at night. In contrast, the clouded leopard was mainly found in closed forest habitat, but at night it travelled into more open habitat areas (Grassman *et al.*, 2005c). Some other carnivore species increased their activity at night in closed forest

habitat, such as the binturong (Grassman *et al.*, 2005a), whereas other species such as the Asiatic golden cat became inactive during nocturnal periods but were more active during crepuscular and diurnal periods in closed cover (Grassman *et al.*, 2005c).

In Southwest China, Bu *et al.* (2016) evaluated spatial co-occurrence and activity patterns of five meso-carnivore species, masked palm civet, *Paguma larvata*, leopard cat, greater hog badger, *Arctonyx collaris*, yellow-throated marten and Siberian weasel, *Mustela sibirica*. Only the masked palm civet and greater hog badger avoided each other, while other species pairs occurred independently of each other. With regard to diel activity, masked palm civet, greater hog badger and leopard cat were primarily nocturnal and crepuscular; yellow-throated marten was diurnal; and the Siberian weasel was arrhythmic for most of the year. Overall, the diel activity patterns recorded for the small carnivores we studied in PKWS broadly matched those obtained for many of these species elsewhere in Southeast Asia (Lynam *et al.*, 2013; Sunarto *et al.*, 2015; Bu *et al.*, 2016; Chutipong *et al.*, 2017; Singh & Macdonald, 2017; Mukherjee *et al.*, 2019; Petersen *et al.*, 2019). This suggests that the temporal niches of these carnivores are the results of pre-existing adaptations, with local differences due to slightly dissimilar community compositions, prey base and contemporary competitive effects.

Conclusion and Future Research

To summarize, we found that carnivore species with smaller carnassials avoided spatial overlap with larger carnassial carnivore species, and carnivore species that did overlap spatially partitioned their activity periods within closed cover habitat. These findings are similar to those noted in some other studies; however, our analyses are preliminary and we recommend that future research efforts expand upon this study. For example, the sample size of animals monitored was small, with less than six individuals being tracked for each species, with the exception of the leopard cat ($n = 20$). In addition, there were several other carnivore species in this community that were not analyzed including tiger, leopard, Asiatic black bear, *Ursus thibetanus*, marbled cat, *Pardofelis marmorata*, golden jackal, *Canis aureus*, back-striped weasel, *Mustela strigidorsa*, large-toothed ferret-badger, *Melogale personata*, Javan mongoose, *Urva javanica*, and five species of civets (Grassman, 2004). To validate the results of this study, we recommend that a larger sample size of individuals representing a greater diversity of carnivore species be monitored during the same time period to provide a more robust sample size for statistical analyses. The number of individuals and species trapped, radio-collared and monitored at the same time would require a large cost in equipment and man hours by researchers. However, PKWS provides an ideal study area for such a large-scale study to take place. Until a telemetry study is conducted, an alternative could be a large-scale camera-trap project to assess spatial and temporal overlap of carnivore species, as recently done elsewhere in Thailand (Chutipong *et al.*, 2017; Petersen *et al.*, 2019).

Phu Khieo is a remote sanctuary with no permanent human settlement and minimal anthropogenic activity, although poaching does occur sporadically (Grassman *et al.*, 2006). Large remote sanctuaries with minimal human activities may provide carnivores more opportunities for resource partitioning to avoid direct competition, and thus support a diverse carnivore community while experiencing lower anthropogenic-caused mortality compared to other sanctuaries. Competition and impacts from humans can cause significant declines in carnivore diversity and abundance because of direct persecution, human and vehicular traffic, poaching of prey species, spread of disease, habitat alteration and fragmentation with growing agriculture, livestock and development (Haines, 2006; Rogala *et al.*, 2011; Bevins *et al.*, 2012; Gubbi *et al.*, 2012; Bu *et al.*, 2016). Because expanding anthropogenic activity would certainly exacerbate such negative effects, we recommend that future analyses of carnivore community ecology that report the level of carnivore coexistence also incorporate the level of human activity in the study areas. This would help in determining how human activity impacts carnivore diversity and the ability of various species therein to coexist.

Acknowledgements

Royal Forest Department staff is thanked for logistical support during the field study. Shai Meiri is thanked for providing specimen data, and Kitti Kreetiyutanont and the late Surapon Poomkhonsan are gratefully acknowledged for their assistance with field work. This study was supported by the Bosack and Kruger Foundation through the Cat Action Treasury and the Caesar Kleberg Wildlife Research Institute at Texas A&M University–Kingsville. Additional support was provided by Sierra Endangered Cat Haven, Hexagon Farm, Parco Faunistica La Torbiera, Columbus Zoo, Point Defiance Zoo and Mountain View Farms Conservation Breeding Centre. Research permission was granted by the National Research Council of Thailand (#0004.3/0301) and the Royal Forest Department of Thailand. Research methodology was approved by the TAMUK Institutional Animal Care and Use Committee (#2003-8-12). This is publication #15-105 of the Caesar Kleberg Wildlife Research Institute.

References

Agostinho, C.S., Hahn, N.S. & Marques, E.E. (2003) Patterns of food resource use by two congeneric species of piranhas (*Serrasalmus*) on the Upper Paraná River floodplain. *Brazilian Journal of Biology* 63, 177–182.

Anderson, D.R., Burnham, K.P. & Thompson, W.L. (2000) Null hypothesis testing: problems, prevalence, and an alternative. *Journal of Wildlife Management* 64, 912–923.

Anonymous (2000) Basic physical and biological information of wildlife sanctuaries of Thailand. Unpublished report. GIS Sub-division, Wildlife Conservation Division, Natural Resources Conservation Office, Royal Forest Department, Bangkok, Thailand.

Barraclough, T.G. & Vogler, A.P. (2000) Detecting the geographical pattern of speciation from species-level phylogenies. *American Naturalist* 155, 419–434.

Barraclough, T.G., Vogler, A.P. & Harvey, P.H. (1998) Revealing the factors that promote speciation. *Philosophical Transactions of the Royal Society London B, Biological Sciences* 353, 241–249.

Bevins, S.N., Carver, S., Boydston, E.E., Lyren, L.M., Alldredge, M., Logan, K.A., Riley, S.P.D., Fisher, R.N.,

Vickers, T.W., Boyce, W., Salman, M., Lappin, M.R., Crooks, K.R. & VandeWoude, S. (2012) Three pathogens in sympatric populations of pumas, bobcats, and domestic cats: implications for infectious disease transmission. *PLoS One* 7, e31403.

Brown, W. & Wilson, E. (1956) Character displacement. *Systematic Biology* 5, 49–64.

Bu, H., Wang, F., McShea, W.J., Lu, Z., Wang, D. & Li, S. (2016) Spatial co-occurrence and activity patterns of mesocarnivores in the temperate forests of southwest China. *PLoS One* 11, e0164271.

Burnham, K.P. & Anderson, D.R. (2002) *Model Selection and Multimodel Inference: A Practical Information-Theoretic Approach*. 2nd edition. Springer, New York.

Chase, J.M. & Leibold, M.A. (2003) *Ecological Niches: Linking Classical and Contemporary Approaches*. The University of Chicago Press, Chicago.

Chesser, R.T. & Zink, R.M. (1994) Modes of speciation in birds: a test of Lynch's method. *Evolution* 48, 490–497.

Chutipong, W., Steinmetz, R., Savini, T. & Gale, G.A. (2017) Assessing resource and predator effects on habitat use of tropical small carnivores. *Mammal Research* 62, 21–36.

Cohen, J.A. (1978) *Cuon alpinus. Mammalian Species* 100, 1–3.

Creel, S. (2001) Four factors modifying the effect of competition on carnivore population dynamics as illustrated by African wild dogs. *Conservation Biology* 15, 271–274.

Creel, S., Spong, G. & Creel, N. (2001) Interspecific competition and the population biology of extinction-prone carnivores. In: *Carnivore Conservation* (eds J.L. Gittleman, S.M. Funk, D. Macdonald & R.K. Wayne), pp. 35–60. Cambridge University Press, Cambridge.

Davis, M.L., Kelly, M.J. & Stauffer, D.F. (2011) Carnivore co-existence and habitat use in the Mountain Pine Ridge Forest Reserve, Belize. *Animal Conservation* 14, 1–10.

Davis, C.L., Rich, L.N., Farris, Z.J., Kelly, M.J., Di Bitteti, M.S. *et al.* (2019) Ecological correlates of the spatial co-occurrence of sympatric mammalian carnivores worldwide. *Ecology Letters* 21, 1401–1412.

Davies, J.T., Meiri, S., Barraclough, T.G. & Gittleman, J.L. (2007) Species co-existence and character divergence across carnivores. *Ecology Letters* 10, 146–152.

Dayan, T. & Simberloff, D. (1994) Character displacement, sexual dimorphism, and morphological variation among British and Irish mustelids. *Ecology* 75, 1063–1073.

Dayan, T. & Simberloff, D. (1996) Patterns of size separation in carnivore communities. In: *Carnivore Behavior, Ecology, and Evolution. Volume 1* (ed. J.L. Gittleman), pp. 243–266. Cornell University Press, Ithaca.

de Cassia Bianchi, R., Olifiers, N., Gompper, M.E. & Mourão, G. (2016) Niche partitioning among mesocarnivores in a Brazilian wetland. *PLoS One* 11, e0162893.

Di Bitetti, M.S., De Angelo, C.D., Di Blanco, Y.E. & Paviolo, A. (2010) Niche partitioning and species coexistence in a neotropical felid assemblage. *Acta Oecologia* 36, 403–412.

Donadio, E. & Buskirk, S.W. (2006) Diet, morphology, and interspecific killing in Carnivora. *American Naturalist* 167, 524–536.

Durant, S. (1998) Competition refuges and coexistence: an example from Serengeti carnivores. *Journal of Animal Ecology* 67, 370–386.

Farlow, J.A. & Pianka, E.R. (2002) Body size overlap, habitat partitioning and living space requirements of terrestrial vertebrate predators: implications for the paleoecology of large theropod dinosaurs. *Historical Biology* 16, 21–40.

Fedriani, J., Fuller, T. & Sauvajot, R. (2000) Competition and intraguild predation among three sympatric carnivores. *Oecologia* 125, 258–270.

Gause, G.F. (1934) *The Struggle for Existence*. Hafner Publishing Company, London.

Gómez-Ortiz, Y., Monroy-Vilchis, O. & Mendoza-Martínez, G.D. (2015) Feeding interactions in an assemblage of terrestrial carnivores in central Mexico. *Zoological Studies* 54, 16.

Grassman, L.I. Jr (2004) *Comparative Ecology of Sympatric Felids in Phu Khieo Wildlife Sanctuary, Thailand*. PhD dissertation, Texas A&M University–Kingsville.

Grassman, L.I. Jr, Tewes, M.E. & Silvy, N.J. (2005a) Ranging, habitat use and activity patterns of binturong *Arctictis binturong* and yellow-throated marten *Martes flavigula* in north-central Thailand. *Wildlife Biology* 11, 49–57.

Grassman, L.I. Jr, Tewes, M.E., Silvy, N.J. & Kreetiyutanont, K. (2005b) Spatial ecology and diet of the dhole *Cuon alpinus* (Canidae, Carnivora) in north central Thailand. *Mammalia* 69, 11–20.

Grassman, L.I. Jr, Tewes, M.E., Silvy, N.J. & Kreetiyutanont, K. (2005c) Ecology of three sympatric felids in a mixed evergreen forest in north-central Thailand. *Journal of Mammalogy* 86, 29–38.

Grassman, L.I. Jr, Janečka, J.E. & Tewes, M.E. (2006) Ecological separation of *Martes flavigula* with five sympatric mesocarnivores in north-central Thailand. In: *Martes in Carnivore Communities. Proceedings of the 4th International Martes Symposium* (eds M. Santos-Reis, J.D.S. Birks, E.C. O'Doherty & G. Proulx), pp. 63–76. Alpha Wildlife Publications, Sherwood Park.

Gubbi, S., Poornesha, H.C. & Madhusudan, M.D. (2012) Impact of vehicular traffic on the use of highway edges by large mammals in a South Indian wildlife reserve. *Current Science* 102, 1047–1051.

Haidir, I.A., Macdonald, D.W. & Linkie, M. (2018) Assessing the spatiotemporal interactions of mesopredators in Sumatra's tropical rainforest. *PLoS One* 13, e0202876.

Haines, A.M. (2006) Is there competition between jaguars (*Panthera onca*) and pumas (*Puma concolor*)? *Acta Zoologica Sinica* 52, 1142–1147.

Hardin, G. (1960) The competitive exclusion principle. *Science* 131, 1292–1297.

Hocking, R.R. (1976) A biometrics invited paper: the analysis and selection of variables in linear regression. *Biometrics* 32, 1–49.

Horne, J., Haines, A.M., Tewes, M.E. & Laack, L.L. (2009) Habitat partitioning of sympatric ocelots and bobcats: implications for recovery of ocelots in southern Texas. *Southwestern Naturalist* 54, 119–126.

Hutchinson, G. (1978) *An Introduction to Population Ecology*. Yale University Press, New Haven.

Karanth, K.U. & Sunquist, M.E. (1995) Prey selection by tiger, leopard, and dhole in tropical forests. *Journal of Animal Ecology* 64, 439–450.

Kekule, B. (1999) *Wildlife in the Kingdom of Thailand*. WKT Publishers, Bangkok.

Kumsuk, M., Kreetiyutanont, K., Suyannakorn, V. & Sanguanyat, N. (1999) *Diversity of Wildlife Vertebrates in Phu Khieo Wildlife Sanctuary, Chaiyaphum Province*. Wildlife Conservation Division, Royal Forest Department, Bangkok, Thailand.

Lekagul, B. & McNeeley, J. (1977) *Mammals of Thailand*. Association for the Conservation of Wildlife, Bangkok.

Li, Z., Wang, T., Smith, J.L.D., Feng, R., Feng, L., Mou, P. & Ge, J. (2019) Coexistence of two sympatric flagship carnivores in the human-dominated forest landscapes of Northeast Asia. *Landscape Ecology* 34, 291–305.

Linnell, J.D.C. & Strand, O. (2000) Interference interactions, co-existence and conservation of mammalian carnivores. *Diversity and Distributions* 6, 169–176.

Lomolino, M., Riddle, B.R., Whittaker, R.J. & Brown, J.H. (2005) *Biogeography*. Sinauer, Sunderland.

Lynam, A.J., Jenks, K.E., Tantipisanuh, N., Chutipong, W., Ngoprasert, D. *et al.* (2013) Terrestrial activity patterns of wild cats from camera-trapping. *The Raffles Bulletin of Zoology* 61, 407–415.

Marinho, P.H., Fonseca, C.R., Sarmento, P., Fonseca, C. & Venticinque, E.M. (2020) Temporal niche overlap among mesocarnivores in a Caatinga dry forest. *European Journal of Wildlife Research* 66, 34.

Meiri, S., Dayan, T. & Simberloff, D. (2005) Variability and correlations in carnivore crania and dentition. *Functional Ecology* 19, 337–343.

Mills, D., Do Linh San, E., Robinson, H., Isoke, S., Slotow, R. & Hunter, L. (2019) Competition and specialization in an African forest carnivore community. *Ecology and Evolution* 9, 10092–10108.

Mukherjee, S., Singh, P., Silva, A., Ri, C., Kakati, K., Borah, B., Tapi, T., Kadur, S., Choudhary, P., Srikant, S., Nadig, S., Navya, R., Björklund, M. & Ramakrishnan, U. (2019) Activity patterns of the small and medium felid (Mammalia: Carnivora: Felidae) guild in northeastern India. *Journal of Threatened Taxa* 11, 13432–13447.

Nakashima, Y., Iwata, Y., Ando, C., Nze-Nkogue, C., Inoue, E., Akomo-Okoue, E.F., Mbehang Nguema, P., Diop Bineni, T., Ngok Banak, L., Takenoshita, Y., Ngomanda, A. & Yamagiwa, J. (Chapter 15, this volume) Spatial and temporal resource partitioning of small carnivores in the African rainforest: implications for conservation and management. In: *Small Carnivores: Evolution, Ecology, Behaviour, and Conservation* (eds E. Do Linh San, J.J. Sato, J.L. Belant & M.J. Somers) Wiley–Blackwell, Oxford.

Ngoprasert, D., Lynam, A.J., Sukmasuang, R., Tantipisanuh, N., Chutipong, W. *et al.* (2012) Occurrence of three felids across a network of protected areas in Thailand: prey, intraguild, and habitat associations. *Biotropica* 44, 810–817.

Palomares, F. & Caro, T.M. (1999) Interspecific killing among mammalian carnivores. *American Naturalist* 153, 492–508.

Pessino, M.E.M., Sarasola, J.H., Wander, C. & Besoky, N. (2001) Respuesta a largo plazo del puma (*Puma concolor*) a una declinación poblacional de la vizcacha (*Lagostomus maximus*) en el desierto del Monte, Argentina. *Ecologia Austral* 11, 61–67. (In Spanish).

Petersen, W., Savini, T., Steinmetz, R. & Ngoprasert, D. (2019) Periodic resource scarcity and potential for interspecific competition influences distribution of small carnivores in a seasonally dry tropical forest fragment. *Mammalian Biology* 95, 112–122.

Pfennig, K. & Pfennig, D. (2005) Character displacement as the "best of a bad situation": fitness trade-offs resulting from selection to minimize resource and mate competition. *Evolution* 59, 2200–2208.

Pianka, E.R. (1969) Sympatry of desert lizards (*Ctenotus*) in western Australia. *Ecology* 50, 1012–1030.

Pokheral, C.P. & Wegge, P. (2019) Coexisting large carnivores: spatial relationships of tigers and leopards and their prey in a prey-rich area in lowland Nepal. *Ecoscience* 26, 1–9.

Pudyatmoko, S. (2019) Spatiotemporal inter-predator and predator–prey interactions of mammalian species in a tropical savannah and deciduous forest in Indonesia. *Mammal Research* 64, 191–202.

Ramesh, T., Kalle, R., Downs, C.T. (2017) Staying safe from top predators: patterns of co-occurrence and inter-predator interactions. *Behavioral Ecology and Sociobiology* 71, 41.

Ray, J.C. & Sunquist, M.E. (2001) Trophic relations in a community of African rainforest carnivores. *Oecologia* 127, 395–408.

Roemer, G., Gompper, M.E. & Van Valkenburgh, B. (2009) The ecological role of the mammalian mesocarnivore. *BioScience* 59, 165–173.

Rogala, J.K., Hebblewhite, M., Whittington, J., White, C.A., Coleshill, J. & Musiani, M. (2011) Human activity differentially redistributes large mammals in the Canadian Rockies national parks. *Ecology and Society* 16, 16.

Sanchez-Cordero, V., Stockwell, D., Sarkar, S., Liu, H., Stephens, C.R. & Gimenez, J. (2008) Competitive interactions between felid species may limit the southern distribution of bobcats *Lynx rufus*. *Ecography* 31, 757–764.

Santos, F., Carbone, C., Wearn, O.R., Rowcliffe, J.M., Espinosa, S. *et al.* (2019) Prey availability and temporal partitioning modulate felid coexistence in Neotropical forests. *PLoS One* 14, e0213671.

Schaller, G.B. & Crawshaw, P.G. (1980) Movement patterns of jaguar. *Biotropica* 12, 161–168.

Schoener, T.W. (1974) Resource partitioning in ecological communities. *Science* 185, 27–39.

Schuette, P., Wagner, A.P., Wagner, M.E. & Creel, S. (2013) Occupancy patterns and niche partitioning within a diverse carnivore community exposed to anthropogenic pressures. *Biological Conservation* 158, 301–312.

Scognamillo, D.M., Maxit, I.E., Sunquist, M.E. & Polisar, J. (2003) Coexistence of jaguar (*Panther onca*) and puma (*Puma concolor*) in a mosaic landscape in the Venezuelan llanos. *Journal of Zoology* 259, 269–279.

Singh, P. & Macdonald, D.W. (2017) Populations and activity patterns of clouded leopards and marbled cats in Dampa Tiger Reserve, India. *Journal of Mammalogy* 98, 1453–1462.

Sunarto, S., Kelly, M.J., Parakkasi, K. & Hutajulu, M.B. (2015) Cat co-existence in central Sumatra: ecological characteristics, spatial and temporal overlap, and implications for management. *Journal of Zoology* 296, 104–115.

Sunquist, F. & Sunquist, M. (2002) *Tiger Moon – Tracking the Great Cats of Nepal*. The University of Chicago Press, Chicago.

van Valkenburgh, B. (1985) Locomotor diversity within past and present guilds of large predatory mammals. *Paleobiology* 11, 406–428.

van Valkenburgh, B. & Ruff, C.B. (1987) Canine tooth strength and killing behaviour in large carnivores. *Journal of Zoology* 212, 379–397.

Wallach, A.D., Izhaki, I., Toms, J.D., Ripple, W.J. & Shanas, U. (2015) What is an apex predator? *Oikos* 124, 1453–1461.

Wayne, R.K., van Valkenburgh, B., Kat, P.W., Fuller, T.K., Johnson, W.E. & O'Brien, S.J. (1989) Genetic and morphological divergence among sympatric canids. *Journal of Heredity* 80, 447–454.

Zhao, G., Yang, H., Xie, B., Gong, Y., Ge, J. & Feng, L. (2020) Spatio-temporal coexistence of sympatric mesocarnivores with a single apex carnivore in a fine-scale landscape. *Global Ecology and Conservation* 21, e00897.

17

Interactions Between Honey Badgers and Other Predators in the Southern Kalahari: Intraguild Predation and Facilitation

Colleen M. Begg[1], Keith S. Begg[1], Emmanuel Do Linh San[2], Johan T. du Toit[3,4] and Michael G.L. Mills[5]

[1]*Niassa Carnivore Project, The Ratel Trust, Rondebosch, South Africa*

[2]*Department of Zoology and Entomology, University of Fort Hare, Alice, South Africa*

[3]*Department of Wildland Resources, Utah State University, Logan, UT, USA*

[4]*Department of Zoology and Entomology, Mammal Research Institute, University of Pretoria, Pretoria, South Africa*

[5]*School of Biology and Environmental Sciences, Faculty of Agriculture and Natural Sciences, University of Mpumalanga, Nelspruit, South Africa*

SUMMARY

Relationships and interactions among predators are multifaceted and intricate and they affect the fitness and survival of individuals. We followed and watched nine habituated honey badgers, *Mellivora capensis*, during > 5800 h over a 42-month period to investigate their direct interactions with sympatric carnivorous mammals and birds in the southern Kalahari, South Africa. We recorded foraging associations between honey badgers and seven other species (two mammals, five birds), most commonly facultative commensalistic or 'producer–scrounger' interactions between honey badgers and pale chanting goshawks, *Melierax canorus*, and black-backed jackals, *Canis [= Lupulella] mesomelas*. The goshawks and jackals benefited from increased hunting opportunities and intake rate. In addition, goshawks showed increased strike success and an expanded prey base when hunting with honey badgers compared to hunting alone in similar habitat in the Little Karoo. Overall honey badgers did not show any significant differences in digging success, intake rate, or predator vigilance when foraging in association compared to foraging alone. The only exception relates to the jackal–badger association, which resulted in a significant decrease (5% of their prey overall) in the amount of prey caught above ground by honey badgers. This form of kleptoparasitism by jackals may have costs for honey badgers in the cold-dry season, when prey availability is low and the foraging association is most common. Based on our field observations and previously published dietary analyses, we recorded or inferred antagonistic interactions between honey badgers and 12 other carnivore species. The outcomes of interspecific aggression (i.e. interference competition) could be predicted from relative body size and were largely asymmetrical. Intraguild predation was common and honey badgers preyed or attempted to prey on all mammalian carnivores smaller than themselves, as well as the young of medium-sized carnivores. Lions, *Panthera leo*, leopards, *Panthera pardus*, and probably spotted hyenas, *Crocuta crocuta*, preyed on honey badger adults and cubs, and cubs were killed by black-backed jackals. The web of interactions observed to date in the taxocenosis of Kalahari carnivores is complex and we encourage further investigations to help better understand how interactions between carnivores shape the whole community structure, both in pristine and altered ecosystems.

Keywords

Black-backed jackal – commensalism – foraging associations – interference competition – intraguild predation – kleptoparasitism – pale chanting-goshawk – 'producer–scrounger' interactions – ratel – taxocenosis

Small Carnivores: Evolution, Ecology, Behaviour, and Conservation, First Edition. Edited by Emmanuel Do Linh San, Jun J. Sato, Jerrold L. Belant, and Michael J. Somers.

Introduction

Relationships among predators are complex and can involve exploitative and interference competition (Cooper, 1991; Mills & Biggs, 1993; Creel *et al.*, 2001; Tannerfeldt *et al.*, 2002; St-Pierre *et al.*, 2006). Although there is extensive diet overlap between some sympatric mammalian carnivores (e.g. Avenant & Nel, 1997; Azevedo *et al.*, 2006; Vogel *et al.*, 2019), evidence of exploitative competition remains elusive (Cupples *et al.*, 2011; Remonti *et al.*, 2012). In contrast, direct interspecific aggression, the most obvious form of interference competition, occurs between a wide variety of species (e.g. Mills & Biggs, 1993; Palomares & Caro, 1999; Fedriani *et al.*, 2000; Macdonald & Thom, 2001; Hunter & Caro, 2008). Aggressive interactions can influence the fitness, abundance, and distribution of the subordinate competitors (Linnell & Strand, 2000; Berger & Gese, 2007), and can even be lethal (Palomares & Caro, 1999; Donadio & Buskirk, 2006). A special case of interspecific killing is intraguild predation, that is when two species compete for prey and one species also preys on the other (Polis & Holt, 1992; Holt & Polis, 1997; Lourenço *et al.*, 2014). Interference competition also includes interspecific feeding associations where an individual of one species intentionally approaches an individual of another species to gain some foraging advantage (Dean & Macdonald, 1981; Packer & Ruttan, 1988; Ellis *et al.*, 1993).

In Africa, interspecific interactions (or avoidance) between large carnivores have been intensively studied over the past five decades (e.g. Kruuk, 1972; Schaller, 1972; Mills, 1990; Mills & Biggs, 1993; Caro, 1994; Mills & Gorman, 1997; Durant, 1998; Creel *et al.*, 2001; Hayward & Slotow, 2009; Swanson *et al.*, 2016; Dröge *et al.*, 2017; Mugerwa *et al.*, 2017; Rafiq *et al.*, 2020; amongst many others). In contrast, a deeper interest in mesocarnivore interactions has emerged only recently (Loveridge & Macdonald, 2002; Do Linh San & Somers, 2006; Kamler *et al.*, 2013; Bagniewska & Kamler, 2014; de Satgé *et al.*, 2017; Mills *et al.*, 2019; Easter *et al.*, 2020; Nakashima *et al.*, Chapter 15, this volume). So far, only two camera-trap studies have focused on evaluating the potential interactions among syntopic small, medium-sized and large carnivores (Schuette *et al.*, 2013; Ramesh *et al.*, 2017). This is surprising considering that the average African

carnivore shares food resources with 22 other carnivore species and is vulnerable to predation by 15 of them (Caro & Stoner, 2003). In addition, interspecific competition has been argued to play an important role in shaping carnivore communities (Linnell & Strand, 2000; Donadio & Buskirk, 2006). This lack of studies of whole carnivore taxocenoses constitutes a clear gap in our knowledge of predator inter-relations and biological communities as a whole.

Honey badgers, *Mellivora capensis*, are medium-sized (females: 6.2 kg; males: 9.2 kg; Begg, 2001) generalist carnivores. Their life history (Begg *et al.*, 2005a), foraging ecology (Begg *et al.*, 2003a; Gil-Sánchez *et al.*, 2020), socio-spatial organization (Begg *et al.*, 2005b), habitat use (Kheswa *et al.*, 2018; Chatterjee *et al.*, 2020; Sharifi *et al.*, 2020) and behaviour (Begg *et al.*, 2003b; Begg *et al.*, 2016a; Allen *et al.*, 2018; Chatterjee *et al.*, 2020) have recently been investigated in detail. Although they can be persecuted in some areas of their wide distribution range in Africa and Asia, their global conservation status is classified as Least Concern by the International Union for the Conservation of Nature (Do Linh San *et al.*, 2016); and this categorization also applies for the Southern African region (Begg *et al.*, 2016c).

Honey badgers are of particular interest as there are accounts of a foraging association between them and black-backed jackals, *Canis [= Lupulella] mesomelas*, and pale chanting goshawks, *Melierax canorus* (Cooper, 1976; Mills *et al.*, 1984; Borello & Borello, 1986; Nelson & Nelson, 1987; Paxton, 1988; Lombard, 1989). It is generally agreed that jackals and goshawks catch fleeing rodents that escape while a honey badger is digging (Mills *et al.*, 1984; Dean *et al.*, 1990), but the possible benefits or costs to the honey badger are unclear. It has been proposed that goshawks indicate the presence of rodent burrows to honey badgers (Cooper, 1976; Dean & Macdonald, 1981; Borello & Borello, 1986) in a possible example of facultative mutualism in which individuals of each associating species gain a foraging advantage, although each individual can forage alone (Rasa, 1983). Alternatively, goshawks – but also jackals – might steal food from the honey badger, i.e. kleptoparasitism (Cooper, 1991; Caro, 1994; Gorman *et al.*, 1998; Creel *et al.*, 2001), or the association may have no negative or positive effect on the honey badger, i.e. commensalism

(Ellis *et al.*, 1993). While commensalism and kleptoparasitism are relatively common, the relationship between common dwarf mongooses, *Helogale parvula*, and eastern yellow-billed hornbills, *Tockus flavirostris*, is the only verified example of a facultative mutualistic foraging association between two predators (Rasa, 1983). In this association, the hornbills feed on insects flushed by the mongooses, and give warnings of avian predators, including those relevant only to the mongooses (Rasa, 1983; Kemp, 1995).

Negative interactions such as predation (here particularly intrataxocenosis or intraguild predation), competition and kleptoparasitism can adversely affect the spatial use (Wilson *et al.*, 2010), activity patterns (Arjo & Pletscher, 1999), as well as foraging and vigilance behaviours (Garvey *et al.*, 2015, 2016) of the victim, but more importantly, its population growth rate (Laurenson, 1995; Carbone *et al.*, 1997; Creel *et al.*, 2001) and energetic intake (Cooper, 1991). In contrast, a mutualistic or commensalistic foraging association may increase the fitness of one or both species through increased energetic returns, increased breeding success, and/or increased vigilance (Rasa, 1983).

In this chapter, we investigate direct interactions between honey badgers and other predators in the southern Kalahari and assess the direct effects of these interactions. Besides raptors, 18 other mammalian carnivore species share this semi-arid environment with honey badgers, and so we expected that these predators interact with honey badgers in a variety of ways. We gave particular attention to foraging associations between honey badgers, pale chanting goshawks, and black-backed jackals.

Materials and Methods

Study Area

We carried out field work from July 1996 until December 1999 in the South African part (area: 9600 km^2) of the Kgalagadi Transfrontier Park (KTP). This semi-desert environment belongs to the Kalahari Duneveld Bioregion (Rutherford *et al.*, 2006) and is a very open savannah of grey camel-thorn, *Acacia haemotoxylon,* common camel-thorn, *Acacia erioloba*, and desert grasses. We conducted the study mainly in the central dune area, which is characterized by medium to high dunes on reddish sands where *A. haemotoxylon* appears in a shrub-like form with occasional *A. erioloba* and shepherd's trees, *Boscia albitrunca*. Dune areas are interspersed with slightly undulating open plains, with similar plant composition but with no *B. albitrunca* trees, and pans and yellowish sands, which support shrub veld of three-thorn, *Rhigozum trichotomum*, and *Monechma* spp. (Van Rooyen *et al.*, 2001).

The study area falls between the 200 and 250 mm rainfall isohyets and is characterized by low, irregular annual rainfall (Mills & Retief, 1984). Rainfall variability has a major effect on the vegetation of the KTP (Leistner, 1967) and large variations in floristic composition, basal cover, and density can take place over medium- or short-term periods (van Rooyen *et al.*, 1984). Three seasons are distinguished into: the hot-wet season (January to April) when the mean monthly temperature is ~20 °C or higher and when 70% of the rain falls; the cold-dry season (May to August) when the mean monthly temperature is below 20 °C and rainfall is rare; and the hot–dry season (September to December) when the mean monthly temperature is ~20 °C and usually not more than 20% of the rain falls (Mills & Retief, 1984). Temperature extremes are 30–40 °C by day during the hot seasons and −5 to 5 °C at night during the cold-dry season when ground frost is common (Mills, 1977; Mills & Retief, 1984).

Data Collection

We habituated nine radio-implanted, adult honey badgers (five females with one cub each, four males) to the research vehicle until we could follow them without any obvious influence on their foraging behaviour. Detailed capture, radio-marking and habituation techniques are presented in Begg *et al.* (2016b). We followed selected animals continuously for 91 observation periods ranging from 1–12 days ($\bar{x} = 4$ days), with an additional 57 short observation periods (<24 h) ranging from 45 min to 20 h. Overall, we spent 5811 h following and watching these honey badgers.

Honey badgers and other species were observed from the roof of a vehicle, approximately 10–30 m away, depending on visibility and grass height. We used a spotlight for night observations. During continuous observations, we timed some activities to the nearest minute with a stopwatch and recorded the success of each digging attempt, the location of prey capture (in a hole or above ground), and the prey type/species. Detailed analyses of the diet and foraging behaviour of honey badgers in the KTP were published in Begg et al. (2003a), while information on activity patterns is presented in Begg et al. (2016a).

For the purpose of our study, we divided mammalian carnivores present in the KTP into three size classes: small (<1 kg), medium (1–12 kg), and large (>12 kg). We assessed the relative abundance of medium and large mammalian carnivores through spotlight counts and spoor transects. We conducted spotlight counts along roads in the dunes (370 km; 18.4 h) and rivers (565 km; 24.5 h) in the KTP during February–March 1996 from an hour after sunset until 23:30 h. We monitored a fixed-length spoor transect of 30 km along a dusty road through the central study area in the early morning at regular intervals during the study period ($n = 20$). We identified the spoor of each carnivore that had crossed the road during the previous night with the aid of a Bushman tracker. We could only express data as the presence or absence of spoor from each species on the transect, as it was often difficult to distinguish individual tracks and, therefore, possible multiple crossings of the same individual(s) over the transect.

We obtained basic data on the activity schedules and diets of mammalian carnivores and associating birds from the literature, where possible from the KTP or similar semi-arid habitats, i.e. small- to medium-sized carnivores (except the canids): Mills et al. (1984); Skinner & Smithers (1990); bat-eared fox, *Otocyon megalotis*: Nel (1990); Cape fox, *Vulpes chama*: Nel (1984); black-backed jackal: Ferguson (1980); Nel (1984); Ferguson et al. (1988); African wild cat, *Felis lybica cafra*: Herbst (2009); Herbst & Mills (2010); large carnivores: Mills (1990). For the small-spotted genet, *Genetta genetta*, we obtained data on prey species from a stomach analysis study conducted in Botswana (Skinner & Smithers, 1990). We used visual observations of hunting behaviour in the Little Karoo, South Africa, for the pale chanting goshawk (Malan & Crowe, 1997) and pellet analysis data for the owls (Steyn, 1982).

Data Analysis

After Minta et al. (1992), we considered animals to be interacting when the attention of one was focused on the other. Aggressive interactions ended when one individual was killed, or when neither individual was focused on the other. We considered a foraging association to begin when the associating species appeared to be following a foraging honey badger's movements and it ended when this individual lost interest in the interaction and the honey badger systematically moved off out of view.

We used both qualitative and quantitative information to assess interactions. For each interaction, we recorded each species' response to the other's presence, assuming that behaviour that initiates or maintains the association is evidence that the net outcome for the behaving animal is likely to be neutral or positive, while behaviour that tends to avoid or terminate the association is evidence that the net outcome is likely to be negative (Minta et al., 1992). Although animals may sometimes be obliged to tolerate associates that inflict a cost, e.g. when the cost of getting rid of them is higher than the cost they inflict, we could not pick up and quantify such instances during our study. However, during foraging associations, in particular, we described any behaviour or vocalizations that suggested aggression or disadvantages, or conversely, coordinated hunting or non-hunting advantages (e.g. increased predator vigilance) of the association for either individual.

We expressed results as the percentage of time we observed honey badgers interacting with each of the other species, i.e. the total time we observed the two species together as a percentage of the total time we observed active honey badgers. For species that only associated with honey badgers during the day, we calculated this metric using only the number of hours honey badgers were active during the day. We divided interaction periods into spot observations (<5 min), where the associating animal was obviously disturbed by the vehicle and moved off almost immediately, and sample observations of >5 min. We calculated the frequency of occurrence of an association as the percentage of observations where an associating species was with a foraging honey badger at the start of an observation period. This is termed 'initial sightings' in the text.

For pale chanting goshawks and black-backed jackals, we recorded the prey category or species caught, the strike (goshawk) or capture (jackal) success (percentage of successful attempts), and we calculated the intake rate (g/h) when foraging with a honey badger. For goshawks, we also calculated the prey capture rate (number of successful strikes per hour). Because all pale chanting goshawks in the KTP are likely to associate with honey badgers, we compared these metrics with data from the literature for hunting alone in similar habitat in the Little Karoo (Malan, 1998). We, however, acknowledge that any differences between these datasets may also be linked to slight inter-habitat differences and inter-annual variations. No data are available on the capture success and prey capture rate of black-backed jackals when hunting alone in similar habitat and thus no direct comparisons could be made. We also compared the digging success (percentage of digging events that resulted in capture) and intake rate (g/h) of honey badgers when foraging with and without black-backed jackals and pale chanting goshawks in attendance. For the calculation of intake rate of all focal predators, we used the average prey mass data calculated and listed by Begg *et al.* (2003a). We used non-parametric chi-squared analysis to compare seasonal differences in the absolute frequency of occurrence of associations, and Fisher's exact test to analyse 2×2 contingency tables (Zar, 1999). We ran two-sample, two-sided *t*-tests to compare the intake rate and digging success of honey badgers when foraging with and without associating species; we arcsine transformed proportions before running *t*-tests (Zar, 1999).

Results

Overview and Relative Abundance of Medium to Large Carnivores

Honey badgers interacted directly with five bird species and 14 of the 18 other carnivore species that occur in the KTP (Figure 17.6). No observations of direct encounters with African wild dog, *Lycaon pictus*, Caracal, *Caracal caracal*, small-spotted genet, and banded mongoose, *Mungos mungo*, were made during the study period. We could classify the interspecific interactions we observed into three categories: foraging associations, aggressive (predator–prey), and neutral interactions (Table 17.1).

The results of the spoor and spotlight counts show that black-backed jackals were the most common species among medium to large mammalian carnivores in the KTP (Table 17.2) and they were the carnivores that interacted most commonly with honey badgers (Table 17.1). We only observed small-spotted genets during the river spotlight count and this is likely to be due to the low availability of trees in the dune areas (Table 17.2). Because we only recorded species as present or absent on the spoor transects, this method underestimated the relative abundance of common species, as well as group-living species (i.e. bat-eared fox and lion, *Panthera leo*). However, spoor transects were more successful than spotlight counts at detecting the presence of, and locating, honey badgers. The lack of success at locating honey badgers during spotlight counts may be due to their small eyes (poor eyeshine) and habit of moving away from a disturbance with their heads low and seldom looking back.

Foraging Associations

We observed seven species (two mammals and five birds) following foraging honey badgers. The most common associations were between honey badgers and pale chanting goshawks and black-backed jackals. On 41 occasions, jackals and goshawks were associated with honey badgers at the same time (Figure 17.1), and as many as three goshawks and two jackals followed a single honey badger. Other associating species included the African wild cat, three owl species (barn owl, *Tyto alba*, marsh owl, *Asio capensis*, and spotted eagle-owl, *Bubo africanus*) and one passerine (ant-eating chat, *Myrmecocichla formicivora*).

Honey badgers catch more than 80% of their prey by digging (Begg *et al.*, 2003b). Small mammals (<100 g) and small reptiles (<100 g) were the most common prey and contributed ~80% of the individual prey eaten in all seasons (females: 79.4%; males: 82.6%; Begg *et al.*, 2003a). When digging for small mammals and reptiles, honey badgers caught 55% of the prey items in a hole, but 45% of the prey items fled above ground (Table 17.3) and it is these fleeing prey items that are available for capture by associating species.

Table 17.1 Type and frequency of interspecific interactions observed between honey badgers, *Mellivora capensis*, and 14 other mammalian carnivore species in the southern Kalahari from direct observations and tracking spoor.

Species	Body mass (kg)	Interactions with honey badgers		
		Category	Type	Number
Small				
Slender mongoose, *Galerella [= Herpestes] sanguinea*	0.37–0.79	Aggressive	Predation attempts by honey badgers on adults and juveniles	1
Yellow mongoose, *Cynictis penicillata*	0.44–0.9	Aggressive	Predation attempts by honey badgers on adults and juveniles	1
Meerkat, *Suricata suricatta*	0.62–0.97	Aggressive	Predation attempts by honey badgers on adults and juveniles	3
Striped polecat, *Ictonyx striatus*	0.4–1.4 (♀) 0.7–1.5 (♂)	Aggressive	Predation attempts by honey badgers on adults and juveniles	3
Medium				
Cape fox, *Vulpes chama*	2–3.3	Aggressive	Predation attempts by honey badgers on juveniles	5
			Aggressive defence of pups by adult Cape foxes	26
Bat-eared fox, *Otocyon megalotis*	3.4–5.4	Aggressive	Predation attempts by honey badgers on juveniles	2
			Aggressive defence of pups by adult bat-eared foxes	3
African wild cat, *Felis lybica cafra*	2–5.8 (♀) 2–7.7 (♂)	Aggressive	Predation attempts by honey badgers on adults and juveniles	2
		Foraging	Foraging association	8
Black-backed jackal, *Canis [= Lupulella] mesomelas*	5.9–10 (♀) 6.4–11.1 (♂)	Aggressive	Predation attempts by honey badgers on jackal pups	19
			Predation attempts by jackals on honey badger cubs	3
		Foraging	Foraging association	137
Aardwolf, *Proteles cristatus*	7.7–14	Aggressive	Aggressive display by honey badgers	7
			Aggressive display by aardwolves	3
Large				
Brown hyena, *Parahyaena brunnea*	28–47.5 (♀) 35–49.5 (♂)	Neutral	Scavenge honey badger carcass	1
			Neutral	9
Leopard, *Panthera pardus*	17–42 (♀) 20–90[a] (♂)	Aggressive	Predator of adult and juvenile honey badgers	1
			Threat display by honey badgers	2
			Avoidance by honey badgers	1
Cheetah, *Acinonyx jubatus*	21–51 (♀) 29–64 (♂)	Neutral	Neutral	1
Spotted hyena, *Crocuta crocuta*	56–86 (♀) 49–79 (♂)	Aggressive	Threat display by honey badgers	1
			Avoidance by honey badgers	2
Lion, *Panthera leo*	110–168 (♀) 150–272 (♂)	Aggressive	Predator of honey badger (adults and juveniles)	3
			Threat display by honey badgers	1
			Avoidance by honey badgers	4

Interactions are ranked in ascending order of the body mass of associating species based on data from Hunter & Barrett (2018). Comparatively, honey badger body mass varies from 6.2–13.6 kg (♀) and 7.7–14.5 kg (♂).

[a] Such extreme body masses for male leopards have probably not been recorded in the Kalahari.

Table 17.2 Relative abundance of medium and large carnivores in the central dune area of the South African part of the Kgalagadi Transfrontier Park as determined from spotlight counts in dune and river habitat and a repeated spoor transect (*n* = 20) of 34 km through the study area.

| Carnivore species | Spotlight transects (936 km) | | | | Spoor transects | |
| | Dunes (380 km) | | Dry river (556 km) | | | |
	#/100 km	PO	#/100 km	PO	Present[a]	PO
Lion	2.4	14.3	2	3.7	3	15
Spotted hyena	0	0	0.9	1.5	1	5
Cheetah	0.5	3.2	0.9	1.5	2	10
Leopard	0	0	0.2	0.3	2	10
Brown hyena	0	0	0.9	1.5	15	75
Caracal	0.8	4.8	0.5	0.9	17	85
Aardwolf	0.3	1.5	0	0	3	15
Honey badger	0	0	0	0	7	35
Black-backed jackal	5.3	31.7	21.4	42.8	20	100
African wild cat	1.6	9.5	1.8	3	16	80
Bat-eared fox	3.7	22.2	16.7	28.6	1	5
Cape fox	2.1	12.7	7.7	13.2	20	100
Small-spotted genet	0	0	1.6	2.8	0	0

Species are ranked in descending order of body mass. PO = percentage occurrence.
[a] Refers to the number of transects for which a species was recorded as present.

Males and females differed significantly in the proportion of successful attempts when digging prey items in holes (Fisher's exact test: *p* = 0.0023; Table 17.3) and as a result, more prey escaped when females were digging. Therefore, female honey badgers are likely to be the most productive sex for associating species. Honey badgers captured 16.5% of the prey that escaped from the digging hole by chasing them above ground, but this represented only 11.8% of the total small mammal and small reptile prey caught overall.

Honey Badgers and Pale Chanting Goshawks

Goshawks were in attendance at 36% of the initial daylight sightings of honey badgers (*n* = 319). Both adult and subadult goshawks followed honey badgers, with up to six individuals in attendance at one time (Figure 17.2). Goshawks associated with honey badgers for 111 h, that is 15.8% of the time honey badgers foraged during the day (Table 17.4). Interaction periods varied from 2 to 366 min (*n* = 194) with 66 spot observations (< 5 min) and 128 sample observations with a mean duration of 52 min. The foraging association was more common in the cold-dry season than the hot-dry or hot-wet seasons (Chi-square test of independence: $\chi^2 = 26.3$, df = 2, *p* < 0.01) and more common with female honey badgers (46% of 163 initial sightings) than males (25% of 156 sightings; Fisher's exact test: *p* = 0.0001).

Of the 71 prey items caught by goshawks when foraging with honey badgers, 39% were small reptiles (barking gecko, *Ptenopus garrulus*: *n* = 26; western three-striped skink, *Trachylepis [= Mabuya] occidentalis*: *n* = 2) and 61% were small mammals (hairy-footed gerbil, *Gerbillurus paeba*: *n* = 20; Brants's gerbil, *Gerbilliscus [= Tatera] brantsii*: *n* = 4; arid four-striped grass mouse, *Rhabdomys bechuanae* [previously *Rhabdomys pumilio*]: *n* = 19).

Figure 17.1 In the KTP, black-backed jackals, *Canis [= Lupullela] mesomelas*, and pale chanting-goshawks, *Melierax canorus*, often use a 'sit-and-wait' strategy and closely monitor the foraging activities of honey badgers, *Mellivora capensis*. Here, a honey badger just dragged a Cape cobra, *Naja nivea*, from a hole and is now giving the snake a fatal head bite; a jackal and a goshawk are alert and ready to seize the opportunity to grab any leftovers, or even steal the dead prey. *Source:* Photo © Peet van Schalkwyk.

Figure 17.2 Three pale chanting goshawks perched on low bushes in the central dune area of the KTP, with the 'purpose' of capturing small rodents and reptiles flushed from their underground refuges through the digging activity of a honey badger. *Source:* Photo © Mario Fazekas.

Table 17.3 Position of prey capture by female (*n* = 236) and male (*n* = 400) honey badgers when digging for small mammals (< 100 g) and small reptiles (< 100 g), showing honey badger digging success when foraging alone, as well as the percentage of prey that escaped the digging hole and were therefore potentially available for capture above ground by associating predators.

	Females		Males		Overall	
Outcome of digging event	**AF**	**%**	**AF**	**%**	**AF**	**%**
Prey caught in digging hole by a honey badger (A)	112	47.5	240	60.0	352	55.3
Prey escaped the hole but was then caught above ground by a honey badger (B)	18	7.6	29	7.3	47	7.4
Prey that escaped capture (C)	106	44.9	131	32.7	237	37.3
Prey available for associating species (B + C)	124	52.5	160	40.0	284	44.7

AF = absolute frequency; % = percentage frequency.

The strike success of goshawks when hunting in association with honey badgers was 58.4% (*n* = 95 strikes), compared to 10–14% when hunting alone or 21–25% when hunting with conspecifics in a similar environment (Malan, 1998). The prey capture rate when hunting with honey badgers in the KTP (0.89 ± 1.05 strike/h, *n* = 34 observation periods) was also higher than the prey capture rate of adult goshawks hunting alone (0.15 ± 0.24 strike/h, *n* = 84; Malan, 1998), but this difference was not significant (two-sample, two-sided *t*-test: *p* > 0.05). Overall, goshawks caught 61% of the prey that fled a honey badger's digging (Table 17.5) and the mean intake rate of goshawks foraging with honey badgers was 22 g/h. The percentage of prey that fled from holes and was subsequently caught above ground by honey badgers decreased from about 17% (7.4/44.7 in Table 17.3) to 13% (Table 17.5) when pale chanting goshawks were present, but this decrease was not significant (Fisher's exact test: *p* > 0.05) and represented a loss of

Table 17.4 Seasonal differences in the number of hours and relative percentage of time black-backed jackals, *Canis [= Lupullela] mesomelas*, and pale chanting goshawks, *Melierax canorus*, were observed with honey badgers.

Category	Seasons			
	Hot-wet	Cold-dry	Hot-dry	Overall
A. Pale chanting goshawk				
Number of initial daylight sightings of honey badgers[a]	79	110	130	319
Number of hours honey badgers were observed active (day)	184	220	314	718
Percentage of observation time goshawks were with honey badgers	11.3%	23.0%	12.3%	15.8%
B. Black-backed jackal				
Number of initial sightings of honey badgers (day + night)[a]	121	136	169	426
Number of hours honey badgers were observed active (day + night)	276	596	796	1668
Percentage of observation time jackals were with honey badgers	5.5%	8.3%	4.1%	5.9%

The data for each season are pooled over the 42-month study period (1996–1999).
[a] Initial sightings refer to the start of a honey badger observation period.

Table 17.5 Percentage of available prey[a] items caught by honey badgers, pale chanting goshawks, and black-backed jackals when foraging in association.

Outcome of digging event	Honey badger + goshawk (%)	Honey badger + jackal (%)
Escaped prey caught by associating species	60.6	68.9
Escaped prey caught by honey badger	12.8	4.9
Escaped prey not caught	26.6	26.2

[a] Available prey refers to prey items that fled above ground while honey badgers were digging.

less than 2% of prey items. There was also no difference between the digging success or intake rate of a honey badger hunting alone or in association with a goshawk (two-sided, two-sample *t*-test, $p > 0.05$; Table 17.6).

A goshawk generally initiated an interaction by flying in and perching near a foraging honey badger. However, on three occasions, a goshawk flew in and perched within 20 m of a resting burrow and waited at least 1 h before a honey badger emerged. Goshawks live in family groups and are strictly territorial (Malan & Crowe, 1996). Although we did not know where the territorial boundaries of the goshawks were, our observations suggested that individual goshawks stopped following a honey badger when a territory boundary was reached.

Goshawks frequently followed foraging honey badgers from high perches (commonly *B. albitrunca* trees) at least 100 m away, while also hunting independently. Once a honey badger began to dig, the goshawk would fly in and either stand on the ground within 1–2 m of the honey badger and follow on foot, or would perch alongside on low shrubs (Figure 17.2).

When foraging with a honey badger, goshawks made a variety of calls ranging from a shrill shriek when striking at a prey item, to a loud, high-pitched cheeping when they saw a prey item, and a quieter, continuous cheeping or 'murmuring' when perched. A similar quieter, continuous cheeping was used between group members when no honey badger was present, particularly in response to a prey item, and its function may be to inform group members of intent (G. Malan, personal communication). The loud, high-pitched cheeping heard in this study may simply be a louder form of the low continuous cheeping.

On seven occasions, the focal honey badger looked up on hearing vocalizations from goshawks, particularly the high-pitched cheeping and on four occasions, approached the goshawk to investigate. A honey badger caught a prey item as a result of moving toward a goshawk on only one occasion. On this occasion, the

Table 17.6 Average digging success and intake rate of female and male honey badgers foraging alone for small mammals (<100 g) and small reptiles (<100 g), compared with foraging in association with pale chanting goshawks and black-backed jackals.

	Females $\bar{x}$ (SE; n)			Males $\bar{x}$ (SE; n)		
	Alone	With goshawk(s)	With jackal(s)	Alone	With goshawk(s)	With jackal(s)
Digging success[a] (%)	43 (1.4; 156)	44 (3.4; 20)	41 (2.4; 14)	46 (1.4; 128)	49 (3.4; 17)	52 (8.5; 18)
Intake rate (g/h)	48 (6; 84)	30 (6; 10)	66 (12; 13)	60 (6; 109)	60 (6; 24)	90 (18; 11)

The differences in digging success and intake rate in the three scenarios were not significant. Note that digging success is neither related to the time period nor to the type of prey. Data with jackals are partly related to night-time foraging when the digging rate was higher and heavier prey (rodents) were dug out, therefore leading to a higher – albeit non-significant – intake rate than with the exclusively diurnal goshawks. SE = standard error; n = sample size.

[a] Digging success was calculated as the percentage of digging events that had a successful outcome (prey was caught) per observation period. The term 'capture success' was avoided here, as honey badgers would sometimes manage to capture fleeing prey while they were above ground (see Table 17.3).

honey badger had entered a resting burrow but on hearing the nearby excited cheeping of the goshawk perched on the ground 2–3 m from the hole, it re-emerged and approached the goshawk, which was standing at a rodent burrow. The honey badger began to dig and caught a hairy-footed gerbil. Given how rarely this was seen, it is unlikely that goshawks consistently aid the foraging efforts of honey badgers by showing them where to dig. On only one occasion was a goshawk heard to give an alarm call in the presence of a honey badger, and the honey badger did not respond. For radio-tagging, we ambushed honey badgers on foot and caught them in hand nets (n = 66; Begg *et al.*, 2016b). Goshawks were in attendance at 21 of these capture events. On all occasions, the goshawks flew off before the honey badger displayed any behaviour suggesting awareness of danger.

Honey Badgers and Black-Backed Jackals

Black-backed jackals were in attendance at 16% of the initial sightings of honey badgers (n = 426; night and day). From one to four jackals were observed with a honey badger at one time. Jackals were observed with honey badgers for 236 h during 156 observation periods ranging in duration from 2 to 847 min ($\bar{x}$ = 110 min). This included 27 spot observations (<5 min) and 129 sample observations (5–847 min). Unlike associations with goshawks, the relationship between honey badgers and jackals was not simply a foraging one, and we could divide the behaviour of the jackals

during sample observations into foraging (n = 94 events, 98 h of observation), resting (n = 24 events, 135 h) and aggressive interactions (n = 12 events, 3 h). We report foraging and resting behaviour in this section, while aggressive interactions – including those that took place during foraging bouts – are presented separately below.

In all instances, the jackal initiated the interaction. Jackals travelled close (2–10 m) behind, ahead, or alongside a foraging honey badger and stopped when the honey badger stopped to dig. On a few occasions, the honey badger appeared to be following the jackal for a short distance. For 6% of the time that honey badgers were foraging, they had a jackal in attendance (Table 17.4). Jackals were more frequently seen with honey badgers in the cold-dry season than the hot-dry or hot-wet seasons (Chi-square test of independence: χ^2 = 8.62, df = 2, p < 0.05) and more frequently observed with female honey badgers (22% of 223 initial sightings) than males (11% of 203 sightings; Fisher's exact test: p = 0.0026). The percentage of time jackals were observed with foraging honey badgers during each hour of the diel period ranged from 2.5–11.5%, with no visually discernible difference between day and night (Figure 17.3). On 24 occasions, jackals rested and even slept close (2–15 m) to a honey badger's resting burrow for periods ranging from 48 to 847 min ($\bar{x}$ = 385 min), waiting until the honey badger emerged and then associated with it during its next foraging bout.

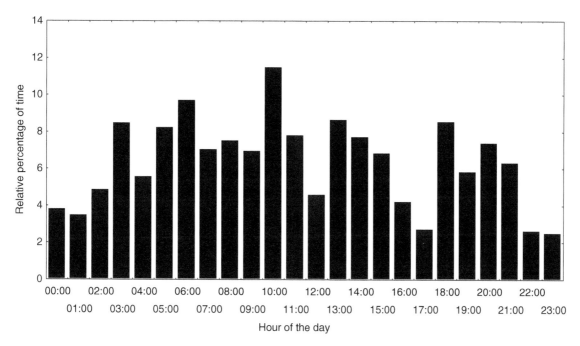

Figure 17.3 Percentage of time that foraging honey badgers had black-backed jackals associating with them during each hour of the day, averaged over the study period (1996–1999).

Jackals were observed to catch 69% of the prey that fled from digging honey badgers (Table 17.5). All of the 111 prey caught by jackals foraging with honey badgers were murids: *G. paeba* (41.5%), *G. brantsii* (28.8%), and *R. bechuanae* (29.7%). On average, jackals caught 296 g/h when foraging with a honey badger. In an extreme case, during a period of 212 min, a jackal caught 17 murids (715 g) that fled a digging honey badger. The proportion of prey items caught above ground by honey badgers decreased significantly (Fisher's exact test: $p < 0.01$) from about 17% without jackals to 5% (Table 17.5) when jackals were present. However, this corresponded to only a 5% decrease in the number of prey caught. There was no difference in the digging success or intake rate of honey badgers foraging with and without a jackal (two-sided, two-sample t-test, $p > 0.05$; Table 17.6) and there was little evidence to suggest that honey badgers and jackals actively cooperated when foraging in association, although, on several occasions, the presence of another predator waiting at possible rodent escape holes may have influenced the outcome.

A jackal gave a predator alarm call on two occasions when with a honey badger and on both occasions, the honey badger stopped its activities to listen, but then continued foraging. During 52 honey badger captures with hand nets (Begg *et al.*, 2016b), at least one jackal was with the honey badger on nine occasions. However, the jackal(s) were never heard to warn the honey badger of danger and the honey badger did not appear to take any notice when the jackal ran off at the approach of the capture team.

Other Associating Species

Groups of up to eight ant-eating chats followed foraging honey badgers on 65 occasions for periods ranging from 2 to 130 min ($\bar{x} = 20$ min), corresponding to 3% of the time that honey badgers were foraging during the day. They flew in from as far as 200 m and followed the honey badger for distances up to 500 m from their nesting and roosting sites in the roofs of aardvark, *Orycteropus afer*, burrows. The chats landed and stood on the ground within 1–2 m of a digging honey badger, gleaning insects that were disturbed by the honey badger's digging. On one occasion, a chat caught and ate a barking gecko that escaped while the honey badger was digging. The honey badger ignored the chats during these associations, although both adults and chicks are prey items (Begg *et al.*, 2003a), and there were no obvious benefits or costs to the honey badger from this association.

Individuals of three species of owl (spotted eagle-owl: *n* = 7; barn owl: *n* = 5; marsh owl: *n* = 3) followed foraging honey badgers at night on 15 occasions. While the owls were not observed to catch prey, on five occasions, they circled and swooped over a digging honey badger before landing on the ground within 5 m of it and following its foraging path. Similarly, African wild cats followed honey badgers (two females, two males) at night on eight occasions but all interactions were of short duration (3–22 min). In each instance, the cats intently watched the digging of the honey badger from less than 5 m away, but were nervous of the vehicle. During these specific interactions, honey badgers never displayed any aggression or interest toward the African wild cats, although both kittens and adults have been recorded as prey items (Begg *et al.*, 2003a). It is possible that these species associate when undisturbed, as their diets overlap considerably (honey badger: Begg *et al.*, 2003a; African wild cat: Herbst & Mills, 2010).

Predation and Aggressive Interactions

During the course of the Honey Badger Project in general (Begg, 2001), we observed predation attempts by honey badgers on the adults and juveniles of the smaller carnivores, namely yellow mongoose, *Cynictis penicillata*, slender mongoose, *Galerella [= Herpestes] sanguinea*, meerkat, *Suricata suricatta*, striped polecat, *Ictonyx striatus*, and African wild cat; scat analyses showed that all are the prey of honey badgers, as are young Cape foxes and bat-eared foxes (Table 17.1; Begg *et al.*, 2003a). Adult Cape foxes, bat-eared foxes, and black-backed jackals were observed to bite and chase honey badgers that were close to, or raiding, dens (Table 17.1). On three occasions, an aardwolf, *Proteles cristatus*, gave a threat display (raising its mane and vocalizing with a hoarse bark) toward a honey badger that entered a den, and it is likely that young aardwolves are the prey of honey badgers.

On 14 occasions, a jackal bit a honey badger adult and/or cub during a foraging association and all these brief attacks appeared unprovoked. Adults were bitten on the rump whilst they were digging. Cubs were only

bitten when their mothers were occupied. In all instances, the cubs vocalized, at which the adult females rushed at the jackal and the jackal backed off a few meters. On 42 occasions, a honey badger rushed aggressively at the jackal that was following it. These chases were less than 5 m and were frequently accompanied by a short rattle/growl by the honey badger; they occurred in response to a jackal being in the way, too close, or while both species were trying to catch a prey item that escaped above ground. On no occasion did a honey badger catch a jackal and the only effect of these chases was that the space between the jackal and the honey badger increased temporarily.

Nineteen longer (5–48 min) aggressive interactions that were not associated with foraging were initiated by jackals. These interactions consisted of repeated chasing, biting, and harassing of a honey badger until it moved off. These interactions occurred predominantly (68%) in the hot-dry season, with six (32%) observations in the cold-dry season and none in the hot-wet season. The hot-dry season (Sep–Dec) coincides with the breeding season of jackals in the KTP (Ferguson, 1980) and on two occasions, jackal dens with cubs were known to be within 100 m of the interaction. Honey badgers were not observed to eat jackal pups in the KTP, but this has been observed elsewhere (see **Discussion**).

In the KTP, honey badgers are killed by lions (Figure 17.4) and leopards, *Panthera pardus* (Figure 17.5), and possibly spotted hyenas, *Crocuta crocuta* (Table 17.1). Of seven adult male honey badgers killed, three were thought to have been killed by lions due to the presence of lion spoor around the remains and skull puncture wounds. The thick, loose skin of the honey badger is thought to provide some protection against predator bites and enables a honey badger to twist around and bite the attacker. This was supported by observations of a female leopard, which fought with a 6 kg honey badger for 52 min before the leopard was able to deliver a killing bite to the throat (Figure 17.5). In addition, signs (spoor, teeth marks on radio-implants) suggested that two other badger females were also killed by a large predator. Three honey badger cubs were thought (from spoor and teeth marks) to have been killed by large predators and on two of these occasions, both the mother and the cub died. On four occasions, a female honey badger

Figure 17.4 (a) A honey badger female trying to fend off a young male lion, *Panthera leo*, with a characteristic threatening rattle-roar and rushing movement toward the predator.
(b) Despite the honey badger's bravery, the young male lion will subsequently kill both the female and, as illustrated here, her young. *Source:* Photos © Shane Saunders.

(a)

(b)

Figure 17.5 This female leopard, *Panthera pardus*, fought with a 6 kg honey badger for 52 min before it was able to deliver a killing bite to the throat. *Source:* Photo © Colleen Begg.

with a <1-month-old cub in a den was observed to aggressively chase a jackal over 80–100 m away from the den. On two of these occasions, the jackal put its head and shoulders into the den burrow while the female was foraging within 10 m of the den. This behaviour suggests that black-backed jackals might prey on honey badger cubs.

Honey badgers avoided interactions with large predators (Table 17.1). On seven occasions, honey badgers were observed to intensively smell the ground and grass stalks around the fresh tracks of a lion ($n = 4$), leopard ($n = 1$), and spotted hyena ($n = 2$). In all cases, after smelling the spoor, the honey badger changed direction and in one case, on smelling fresh lion spoor, a male honey badger bolted into a nearby burrow and did not come out until the following day.

When avoidance was not possible, a honey badger's response to a large predator was one of 'fight' rather than 'flight'. Its formidable close-quarters defence consists of a threatening rattle-roar, pilo-erection, the release of scent from anal glands, and a rushing movement toward the predator (Figure 17.4a). This defence was observed to be successful at warding off leopard ($n = 2$; one observation was made by D. and C. Hughes, National Geographic filmmakers, South Africa, personal communication), lion ($n = 1$), and spotted hyena ($n = 1$). During the spotted hyena interaction, a male honey badger initially laid flat and motionless in the grass on smelling three hyenas. When the hyenas were within 2–3 m of the honey badger, he stood up, rushed at the hyenas, released the scent, vocalized, and then ran off. The hyenas made no further attempt to pursue the honey badger.

Visual observations suggested that neither brown hyenas, *Parahyaena brunnea*, nor cheetahs, *Acinonyx jubatus*, are predators of honey badgers, although, on two occasions, honey badger carcasses were found in brown hyena dens (Table 17.1). On the eight occasions that honey badgers and brown hyenas were observed to interact, the honey badger was not seen to use the threat display. On all occasions, the brown hyena approached the honey badger and then changed direction and moved off.

Discussion

Foraging Associations

Like most mammalian carnivores, honey badgers are solitary and have not been recorded to hunt cooperatively with conspecifics, except occasionally as mother–cub pairs (Begg *et al.*, 2003a). Yet honey badgers were observed to forage in association with individuals of seven other species (two mammals, five birds) in the KTP. Two of the associating species (pale chanting-goshawk and ant-eating chat) are exclusively diurnal (Maclean, 1985), while the owls (Steyn, 1982) and African wild cat (Herbst, 2009) are primarily nocturnal. Black-backed jackals and honey badgers are active during day and night (Ferguson, 1980; Begg *et al.*, 2016b).

Ant-eating chats predominantly catch insects disturbed while the honey badger is digging for vertebrate prey, and so this relationship appears to be one of commensalism, which has been reported in a wide variety of birds and mammals (Dean & Macdonald, 1981). Relationships between honey badgers and pale chanting goshawks and black-backed jackals (and possibly the three owl species and African wild cat) are more complicated as associating individuals appear to catch the prey items that the honey badger is digging for. As these are generally small mammals and small reptiles, only the successful hunter can feed and there is no potential for food sharing (Packer & Ruttan, 1988).

There is little doubt that the associating individuals benefit directly from the digging efforts of honey badgers. In goshawks, in particular, the strike success and prey capture rate appear to be substantially higher when they are hunting for similar prey with a honey badger than when hunting alone or in conspecific groups in similar habitat (Malan & Crowe, 1996). The association also extends the prey base of the exclusively diurnal goshawk to include nocturnal prey species that are normally unavailable to it when it hunts alone (i.e. rodents such as *P. garrulus, G. brantsii*, and *G. paeba*; Malan & Crowe, 1996). Foraging with a honey badger may increase the reproductive and survival fitness of goshawks, particularly since they have been observed to take food back to the nest after foraging with a honey badger (C.M. Begg & K.S. Begg, personal observation).

Whereas many species show rigidity in their foraging strategies (Bouskila, 1998), jackals switched from their typical active hunting strategy (Ferguson, 1980) to a 'sit-and-wait' strategy (Figure 17.1) when foraging with a honey badger and this is likely to decrease the search and handling costs for the jackals. In addition, the association is likely to provide increased opportunities for the jackals to catch prey above ground. Ferguson (1980) observed that when jackals hunt alone, mice and small reptiles frequently escape into holes in the ground and are thereafter unavailable to a jackal, as in only one case was a jackal successful at digging them out. An investigation into a similar foraging association between coyotes, *Canis latrans*, and American badgers, *Taxidea taxus*, when hunting Uinta ground squirrels, *Urocitellus armatus*, showed that the association benefited the coyote with an increased consumption rate (Minta *et al.*, 1992), but possible foraging benefits to the American badger were unclear. Other species like swift foxes, *Vulpes velox*, and hawks have also been reported to hunt near excavating badgers (Wauer & Egbert, 1977; Devers *et al.*, 2004; Ausband & Ausband, 2006), but these associations are likely rare.

Current theory suggests that mutualism is best viewed as reciprocal exploitations that nonetheless provide net benefits to each partner rather than as reciprocal beneficial relationships (Herre *et al.*, 1999). It has been shown that the associating individuals benefit from the association by exploiting the prey that escapes while a honey badger is digging, but there is no evidence to suggest that a honey badger benefits from increased overall capture success (i.e. through digging and capturing prey that fled from the hole) and intake rate or decreased search time through cooperative hunting. While honey badgers are aware of associating individuals and react to their behaviour, this seldom results in their own capture of a prey item. Unlike the foraging association between eastern yellow-billed hornbills and common dwarf mongooses

(Rasa, 1983), there is also no evidence that either pale chanting goshawks or black-backed jackals warn honey badgers of potential predators.

Many interactions between animals can usefully be regarded as 'producer–scrounger' relationships where individuals of one species (scrounger) use the behavioural investment of another (producer) to obtain a limited resource (Barnard & Sibley, 1981; King *et al.*, 2009). Scroungers reduce their costs of exploiting the resource (in this case, food) by letting the producers invest the necessary time and energy in foraging and then usurping the results of their efforts. Producers can maximize their food intake by staying far away from potential scroungers, who, in turn, can maximize their intake by staying near potential producers (Flynn & Giraldeau, 1998; Giraldeau & Mottley, 1998). In large carnivores, kleptoparasitism or food stealing is fairly common and usually involves one carnivore scavenging prey from the other (Creel *et al.*, 2001). In this study, the associating species ('scroungers') always initiated the interaction and both jackals and goshawks were observed to wait at a resting burrow for a honey badger to emerge. However, honey badgers did not appear to avoid the associating species (i.e. by going into a hole, or shifting their activity schedule) and were rarely aggressive toward the associating individuals. This might partly be because, contrarily to what is observed in large carnivores, the honey badgers' associates only deprive them of potential prey rather than actual resources in the form of killed prey. As a result of the low success of honey badgers chasing and capturing prey above ground, the majority of prey items that escaped during digging are unavailable to honey badgers even when associating individuals are not present. In addition, compared to the kleptoparasitism in large carnivores where a group of females or subadult lions might lose almost 20% (Cooper, 1991) and cheetahs 9.2% (Caro, 1994) of the edible portion of their kills to spotted hyena, honey badgers lose less than 5% of their overall potential prey to jackals and only 2% to goshawks. In Etosha National Park (Namibia), Gorta (2020) recently observed a honey badger kleptoparasitizing a yellow mongoose killed by a black-backed jackal. Based on the context of his observations, the author assumed that a pair of honey badgers, tightly followed by a jackal, had discovered a burrow and sought to prey on the two yellow mongooses that lived in there. Contrary to the jackal, the honey badgers failed to kill one of the two fleeing mongooses. Gorta (2020) suggested that in general, defence, reclamation, or even theft of prey by honey badgers may represent an opportunistic trade-off of reward for effort. Hence, when preys are small, such as rodents and geckos, the reward for trying to steal a prey that was killed by a competitor is likely not worth the energetic investment and possible injury risk.

In the KTP, our data show that both black-backed jackals and pale chanting goshawks were more commonly seen with female honey badgers than with males. This may be due to the lower digging success of females and the higher number of prey that escape them, and as a result, females may be more productive for associating species to follow. In addition, black-backed jackals and pale-chanting goshawks may be more likely to randomly encounter foraging females than males, as females spend 97% of their active time above ground foraging, while males spend 19% of their active time engaged in non-foraging activities, that is scent marking, long-distance trotting and male–male interactions (Begg *et al.*, 2003a, 2016a).

It is unlikely that the small decrease in the number of prey caught above ground (and related intake) by a honey badger during an association is detrimental to the honey badger's fitness, and these associations are most likely to be an example of facultative commensalism. However, both jackals and goshawks most frequently follow honey badgers in the cold-dry season, possibly because honey badgers are more diurnal during this season (Begg *et al.*, 2016a). This is the 'lean' season for honey badgers in the KTP (Begg *et al.*, 2003b) and may also be the 'lean' season for jackals and goshawks since they take similar prey. It is during this season that the associations, particularly with jackals, are likely to have the greatest negative effect on honey badgers.

Both goshawks and jackals frequently hunt with conspecifics and are likely to benefit from social learning, and this may facilitate the spread of the association within a population. Sasvari & Hegyi (1998) have shown that tits, *Parus* spp., are able to learn from conspecifics that they can find food near other animals. These associations would be reinforced by regular interactions but are likely to be sensitive to anthropogenic disturbance. For instance, persecution may result in honey badgers becoming exclusively nocturnal (C.M. Begg & K.S. Begg, personal observation) and this will cause the loss of their association with goshawks. Berger (1999) suggests that desensitization in interspecific responsiveness can occur in less than ten generations.

Dark chanting goshawks, *Melierax metabates*, and black-backed jackals following honey badgers in wooded, mesic habitats in the Lowveld of South Africa (I. Thomas, personal communication; P. Chadwick, personal communication) and Ethiopian wolves, *Canis simensis*, following honey badgers in the Bale Mountains of Ethiopia (Sillero-Zubiri, 1996), show that these interspecific foraging associations are not limited to arid open habitats where they are easily seen. Similar hunting associations have also been observed between pale chanting goshawks and slender mongooses (Dean & Macdonald, 1981), black-footed cats, *Felis nigripes*, and marsh owls (Sliwa, 1994), pied kingfishers, *Ceryle rudis*, and Cape clawless otters, *Aonyx capensis* (Boshoff, 1978), and African marsh harriers, *Circus ranivorus*, and Cape grey mongooses, *Galerella [= Herpestes] pulverulenta* (Lombard, 1989). It is likely that these also represent facultative commensalism and the result of opportunistic predators keying in to the opportunities provided by the hunting efforts of other species.

Predation and Aggressive Defence

In the KTP, the majority of predation events or attempts and defensive threat displays observed between the honey badger and other species were asymmetrical (one species being the aggressor; Palomares & Caro, 1999) and could largely be predicted by relative body size (Donadio & Buskirk, 2006). All small carnivores weighing <1 kg are the prey of the honey badger, as are the young of medium-sized carnivores (bat-eared fox, Cape fox, African wild cat and, probably, aardwolf and black-backed jackal) that are in the same size class. Interactions between honey badgers and black-backed jackals appear to be symmetrical – each species attempts (and likely sometimes succeeds) to kill the other's young – with aggression initiated by either species. A honey badger has been recorded killing jackal pups in a den in Etosha National Park (Hancock, 1999) and Dragesco-Joffé (1993)

reports two golden jackals, *Canis aureus* (but now regarded as African golden wolves, *Canis lupaster*), killing a honey badger in northern Niger. Although no direct encounters with caracal and small-spotted genet were observed during the study, based on body size/mass differences and interactions recorded with similar-sized carnivores, honey badgers may potentially prey on genets and their cubs, as well as on young caracals. Similarly, adult caracals could occasionally prey on honey badger cubs. African wild dogs and banded mongooses are rare – and, probably, only transients – in the KTP, but here too, body size/mass asymmetry and, probably, group size effect, should determine the outcome of potential direct interactions between honey badgers and each of these two species where sympatric.

There appears to be a large dietary overlap between medium-sized mammalian carnivores, medium-sized raptors, and large snakes in the Kalahari, with murids the predominant prey items (honey badger: Begg *et al.*, 2003b; African wild cat: Skinner & Smithers, 1990; Herbst & Mills, 2010; black-backed jackal: Ferguson, 1980; Nel, 1984; Cape fox: Nel, 1984; medium-sized raptors: Steyn, 1982; large snakes: Sprawl & Branch, 1995). Differences in hunting behaviour and habitat and prey selection probably reduce the ecological overlap between these species (Simberloff & Dayan, 1991), but indirect exploitative competition for food could be an important regulatory factor for these predators (including honey badgers). The dietary overlap between mesocarnivores and the observed or inferred predation on these species by honey badgers in the KTP highlights the presence of intraguild predation, which is defined as 'the killing and eating of species that use similar, potentially limited resources and are therefore potential competitors' (Polis & Holt, 1992; Holt & Polis, 1997).

Interactions between honey badgers and other medium-sized carnivores commonly involve biting and this may have important consequences for the spread of diseases. Honey badgers are vectors of rabies (Bingham *et al.*, 1997) and have been associated with outbreaks of rabies amongst black-backed jackals (J. Bingham, personal communication). A faecal sample from the Kalahari also indicated infection with feline panleukopenia virus (Steinel *et al.*, 2000) and canine distemper has also been implicated in honey badger deaths (McKenzie, 1993; Kingdon, 1989; L. Hunter, Panthera, USA, personal communication).

Due to their formidable threat display, strength, and aggressiveness, and despite their small size, honey badgers are commonly reported to have no enemies except humans and lions (Estes, 1992; Skinner & Chimimba, 2005). A recent camera-trapping study carried out in the grass plains and open woodlands of the Serengeti, Tanzania, showed that honey badgers were not avoiding larger carnivores such as lions and spotted hyenas, neither temporally nor spatially, but were instead potentially seeking out similar habitats and niches (Allen *et al.*, 2018). This partially contrasts with other camera-trapping studies conducted in KwaZula-Natal, South Africa, in a mosaic of open habitats and woodlands. There the detection probability of honey badgers at camera sites where leopards and spotted hyenas had been detected was higher in closed than open habitat (Ramesh *et al.*, 2017). This suggests that although exploiting the same sites as larger carnivores, honey badgers feel safer in closed habitats and/or avoid direct interactions in the more open habitats. Indeed, our results and other studies showed that honey badgers, besides by lions, are also sometimes preyed on by leopards, African wild dogs and, possibly, spotted hyenas (Turnbull-Kemp, 1967; Pienaar, 1969; Pienaar *et al.*, 1987; Bailey, 1993). Honey badgers appear to be particularly vulnerable to predation because they are easily surprised, especially while digging, and are relatively slow runners (Begg *et al.*, 2016b). To offset this, they generally avoid the direct presence of large carnivores and, when surprised at close quarters, perform the above-mentioned and frequently successful threat display. Brown hyenas and cheetahs appeared to be cautious of confrontation with honey badgers and Owens & Owens (1978) and Mills (1990) reported brown hyenas chasing honey badgers, but aborting the attack in the face of the honey badger's threat display. A record of a honey badger dragging a small cheetah cub in its mouth in the Kruger National Park (M. Allsopp, personal communication) and circumstantial evidence from the KTP (Mills & Mills, 2017) suggest that honey badgers may sometimes kill the young of the larger predators.

Eaton (1976) suggested that the honey badger's defensive attributes and striking colouration have resulted in Batesian mimicry by cheetah cubs of adult honey badgers. The long, white dorsal hair of infant cheetahs might indeed mimic the appearance from above of honey badger adults and thus protect the cheetah cubs from predation. The predation of honey badgers by large predators lessens the likelihood that the long hair on cheetah cubs serves to protect them from large mammalian predators, as lions and spotted hyenas are also major predators of cheetah cubs (Laurenson, 1995). However, the mimicry may be intended for mesocarnivores and/or aerial predators; and, in fact, no records of aerial predators killing a honey badger could be found. As emphasized by Hunter & Caro (2008), over evolutionary time, species that co-occur with a large number of competitors would likely evolve morphological and/or behavioural adaptations to reduce intraguild competition and decrease predation risk. In American carnivores, conspicuous species with both contrasting facial and body colouration co-occur with more potential predators than less striking species, suggesting that intrataxocenosis predation is an evolutionary driver of contrasting coat colouration in carnivores. Mephitidae (skunks) are potentially under greater predation pressure than less conspicuous carnivores, but they reduce predation risk by producing and spraying noxious anal-gland secretions to potential predators. Similarly, honey badgers, who share their environment with a large number of carnivore species, have evolved a threatening rattle-roar, pilo-erection, release of anal-gland secretions and rushing movements to deter predators. In addition, their characteristic black and white colouration might be a warning to larger would-be predators (e.g. Newman *et al.*, 2005; Stankowich *et al.*, 2011, 2014).

Conclusion

Honey badgers were observed to interact with at least 14 of the 18 other species of terrestrial carnivores that occur in the southern Kalahari, and with 5 predatory birds (Figure 17.6). As could be expected, interspecific aggressive interactions (predation vs. defensive threat display) were determined by relative body size and were largely asymmetrical. Attempts of intraguild or intrataxocenosis predation varied both in frequency and success depending on the pair of species involved.

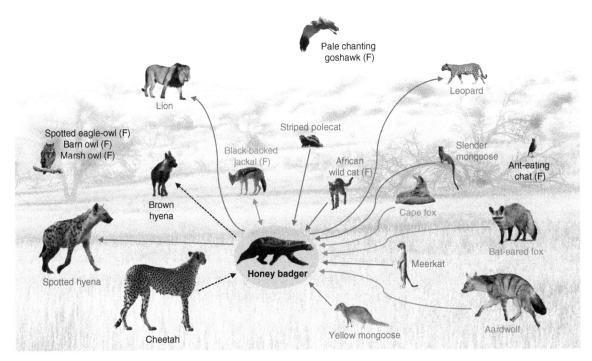

Figure 17.6 Schematic summary of the interactions – observed directly, deduced from dietary analyses or inferred from observations in other areas – of the honey badger with 19 other predator species (14 mammals, 5 birds) in the Kalahari Transfrontier Park, South Africa. Species involved in foraging associations with honey badgers are denoted with '(F)'. Aggressive interactions are represented with different colour codes: red = predators of adult and juvenile honey badgers; blue = prey (adults and juveniles) of honey badgers; green = prey (juveniles only) of honey badgers. Among the latter category, black-backed jackals similarly prey on juvenile honey badgers. It is likely that honey badgers may occasionally kill the young of larger predators such as cheetahs, while brown hyenas may perhaps kill juvenile honey badgers (dashed black arrows). No direct encounters between honey badgers and caracals or small-spotted genets were observed during the study period. Scientific names of predators and detailed information on interspecific interactions are provided in Table 17.1 and/or in the text. *Source:* Photos © Emmanuel Do Linh San, with exception of striped polecat (Johan and Estelle van Rooyen).

All species of carnivores smaller than honey badgers were prey items, as were the young of medium-sized carnivores. Large carnivores, with the notable exception of cheetahs and brown hyenas, preyed on honey badger adults and cubs, and cubs were also killed by black-backed jackals. Foraging associations between honey badgers and seven other species (two mammals, five birds) were recorded, most commonly commensalistic or 'producer–scrounger' interactions between honey badgers and black-backed jackals and pale chanting goshawks.

The web of interactions observed to date in the community of Kalahari predators is complex, and considering the likely high predation pressure put by large carnivores on smaller species, it is not surprising that the honey badger evolved a contrasting facial and body colouration and the associated defensive display and weaponry. The latter seems to be so effective and deterrent that, depending on the large carnivore species involved, either successful predation attempts or even direct attacks on honey badgers have rarely been witnessed. Interspecific interactions between the smaller carnivore species appear to be more diverse, frequent, and intricate, but these have largely been neglected due to the difficulties in obtaining visual observation of these interactions compared to those between the larger carnivores. Further research in both pristine and altered ecosystems is strongly encouraged, as comparative studies could help better understand how human-induced modifications of carnivore taxocenoses (especially the removal of large predators) may not only affect interspecific interactions between carnivores, but also impact on community structure as a whole.

Acknowledgements

This research was generously funded by the Carnivore Conservation Group of the Endangered Wildlife Trust (Johannesburg), the National Research Foundation of South Africa, and the University of Pretoria. M.G.L. Mills was supported by the Tony and Lisette Lewis Foundation. We thank South African National Parks for permission to carry out the study and aerial tracking support in the Kgalagadi Transfrontier Park, as well as Douw Grobler, Pete Morkel, Mike Kock, and Erick Verreynne for veterinary assistance. We remain indebted to the late Klaas Kruiper for finding and interpreting the tracks of honey badgers. Mario Fezekas, Shane Saunders, Johan and Estelle van Rooyen, and Peet van Schalkwyk graciously supplied photographs to illustrate this chapter. Lastly, we are obliged to Peter Apps and David Eads whose critical and pertinent comments helped us significantly improve this chapter.

References

Allen, M.L., Peterson, B. & Krofel, M. (2018) No respect for apex carnivores: distribution and activity patterns of honey badgers in the Serengeti. *Mammalian Biology* 89, 90–94.

Arjo, W.M. & Pletscher, D.H. (1999) Behavioral responses of coyotes to wolf recolonization in northwestern Montana. *Canadian Journal of Zoology* 77, 1919–1927.

Ausband, D.E. & Ausband, E.A. (2006) Observations of interactions between swift fox and American badger. *Prairie Naturalist* 38, 63–64.

Avenant, N.L. & Nel, J.A.J. (1997) Prey use by four syntopic carnivores in a strandveld ecosystem. *South African Journal of Wildlife Research* 27, 86–93.

Azevedo, F.C.C., Lester, V., Gorsuch, W., Larivière, S., Wirsing, A.J. & Murray, D.L. (2006) Dietary breadth and overlap among five sympatric prairie carnivores. *Journal of Zoology* 269, 127–135.

Bagniewska, J.M. & Kamler, J.F. (2014) Do black-backed jackals affect numbers of smaller carnivores and prey? *African Journal of Ecology* 52, 564–567.

Bailey, T.N. (1993) *The African Leopard: Ecology and Behaviour of a Solitary Felid*. Columbia University press, New York.

Barnard, C.J. & Sibley, R.M. (1981) Producers and scroungers: a general model and its application to captive flocks of house sparrows. *Animal Behaviour* 29, 543–550.

Begg, C.M. (2001) *Feeding Ecology and Social Organisation of Honey Badgers (*Mellivora capensis*) in the Southern Kalahari.* PhD thesis, University of Pretoria.

Begg, C.M., Begg, K.S., du Toit, J.T. & Mills, M.G.L. (2003a) Sexual and seasonal variation in the diet and foraging behaviour of a sexually dimorphic carnivore, the honey badger (*Mellivora capensis*). *Journal of Zoology* 260, 301–316.

Begg, C.M., Begg, K.S., du Toit, J.T. & Mills, M.G.L. (2003b) Scent-marking behaviour of the honey badger, *Mellivora capensis* (Mustelidae) in the southern Kalahari. *Animal Behaviour* 66, 917–929.

Begg, C.M., Begg, K.S., du Toit, J.T. & Mills, M.G.M. (2005a) Life-history variables of an atypical mustelid, the honey badger *Mellivora capensis*. *Journal of Zoology* 265, 17–22.

Begg, C.M., Begg, K.S., du Toit, J.T. & Mills, M.G.L. (2005b) Spatial organization of the honey badger *Mellivora capensis* in the southern Kalahari: home-range size and movement patterns. *Journal of Zoology* 265, 23–35.

Begg, C.M., Begg, K.S., Do Linh San, E., du Toit, J.T. & Mills, M.G.L. (2016a) Sexual and seasonal variation in the activity patterns and time budget of honey badgers in an arid environment. In: *Badgers: Systematics, Biology, Conservation and Research Techniques* (eds G. Proulx & E. Do Linh San), pp. 161–192. Alpha Wildlife Publications, Sherwood Park.

Begg, C.M., Begg, K.S., Do Linh San, E., du Toit, J.T. & Mills, M.G.L. (2016b) An evaluation of techniques used for the capture, immobilization, marking, and habituation of honey badgers (*Mellivora capensis*). In: *Badgers: Systematics, Biology, Conservation and Research Techniques* (eds G. Proulx & E. Do Linh San), pp. 339–362. Alpha Wildlife Publications, Sherwood Park.

Begg, C.M., Begg, K.S., Power, R.J., van der Merwe, D., Camacho, G., Cowell, C., Isham, J. & Do Linh San, E.

(2016c) A conservation assessment of *Mellivora capensis*. In: *The Red List of Mammals of South Africa, Swaziland and Lesotho* (eds M.F. Child, L. Roxburgh, E. Do Linh San, D. Raimondo & H.T. Davies-Mostert). South African National Biodiversity Institute and Endangered Wildlife Trust, Johannesburg.

Berger, J. (1999) Anthropogenic extinction of top carnivores and interspecific animal behaviour: implications of the rapid decoupling of a web involving wolves, bears, moose and ravens. *Proceedings of the Royal Society of London* 266, 2261–2267.

Berger, K. & Gese, E.M. (2007) Does interference competition with wolves limit the distribution and abundance of coyotes? *Journal of Animal Ecology* 76, 1075–1085.

Bingham, J., Schumacher, C.L., Aubert, M.F.A., Hill, F.W.G. & Aubert, A. (1997) Innocuity studies of SAG-2 oral rabies vaccine in various Zimbabwean wild non-target species. *Vaccine* 15, 937–943.

Borello, W. & Borello, R. (1986) Chanting goshawks foraging with honey badger. *Babbler* 12, 25.

Boshoff, A.F. (1978) Possible pied kingfisher-clawless otter commensalism. *Ostrich* 49, 89.

Bouskila, A. (1998) Should a predator forage actively or wait in ambush? Predator–prey encounter probabilities for animals which do not move randomly. In: *International Conference on Foraging Behaviour, 21–24 July*. University of California, Santa Cruz.

Carbone, C., du Toit, J.T. & Gordon, I.J. (1997) Feeding success in African wild dogs: does kleptoparasitism by spotted hyaenas influence hunting group size. *Journal of Animal Ecology* 66, 318–326.

Caro, T. (1994) *Cheetah of the Serengeti Plains: Group Living in an Asocial Species*. University of Chicago Press, Chicago.

Caro, T. & Stoner, C.J. (2003) The potential for interspecific competition among African carnivores. *Biological Conservation* 110, 67–75.

Chatterjee, N., Nigam, P. & Habib, B. (2020) Population estimate, habitat-use and activity patterns of the Honey badger in a dry deciduous forest of central India. *Frontiers in Ecology and Evolution* 8, 585256.

Cooper, R.L. (1976) Strange hunting companions. *Custos* June/July, 42–44.

Cooper, S.M. (1991) Optimal hunting group size: the need for lions to defend their kills against loss to spotted hyaenas. *African Journal of Ecology* 29, 130–136.

Creel, S., Spong, G. & Creel, N. (2001) Interspecific competition and the population biology of extinction-prone carnivores. In: *Carnivore Conservation* (eds J.L. Gittleman, S.M. Funk, D. Macdonald & R.K. Wayne), pp. 35–60. Cambridge University Press, Cambridge.

Cupples, J.B., Crowther, M.S., Story, G. & Letnic, M. (2011) Dietary overlap and prey selectivity among sympatric carnivores: could dingoes suppress foxes through competition for prey? *Journal of Mammalogy* 92, 590–600.

Dean, W.R.J. & Macdonald, I.A.W. (1981) A review of African birds feeding in association with mammals. *Ostrich* 52, 135–155.

Dean, W.R.J., Siegfried, W.R. & Macdonald, I.A.W. (1990) The fallacy, fact and fate of guiding behaviour in the greater honeyguide. *Conservation Biology* 4, 99–101.

de Satgé, J., Teichman, K. & Cristescu, B. (2017) Competition and coexistence in a small carnivore guild. *Oecologia* 184, 873–884.

Devers, P.K., Koenen, K. & Krausman, P.R. (2004) Interspecific interactions between badgers and red-tailed hawks in the Sonoran Desert, southwestern Arizona. *Southwestern Naturalist* 49, 109–111.

Do Linh San, E. & Somers, M.J. (2006) Mongooses on the move: an apparent case of interspecific cooperative vigilance between carnivores. *South African Journal of Wildlife Research* 36, 201–203.

Do Linh San, E., Begg, C., Begg, K. & Abramov, A. (2016) *Mellivora capensis. The IUCN Red List of Threatened Species* 2016, e.T41629A45210107. http://dx.doi.org/10.2305/IUCN.UK.2016-1.RLTS.T41629A45210107.en. Accessed on 27 November 2017.

Donadio, E. & Buskirk, S.W. (2006) Diet, morphology, and interspecific killing in carnivora. *American Naturalist* 167, 524–536.

Dragesco-Joffé, A. (1993) *La Vie Sauvage au Sahara*. Delachaux & Niestlé, Neuchâtel – Paris.

Dröge, E., Creel, S., Becker, M.S. & M'Soka, J. (2017) Spatial and temporal avoidance of risk within a large carnivore guild. *Ecology and Evolution* 7, 189–199.

Durant, S. (1998) Competition refuges and coexistence: an example from Serengeti carnivores. *Journal of Animal Ecology* 67, 370–386.

Easter, T., Bouley, P. & Carter, N. (2020) Intraguild dynamics of understudied carnivores in a human-altered landscape. *Ecology and Evolution* 10, 5476–5488.

Eaton, R.L (1976) A possible case of mimicry in larger mammals. *Evolution* 30, 853–856.

Ellis, D.H., Bednarz, J.C., Smith, D.G. & Flemming, S.P. (1993) Social foraging classes in raptorial birds. *Bioscience* 43, 14–20.

Estes, R.D. (1992) *The Behaviour Guide to African Mammals*. University of California Press, Los Angeles.

Fedriani, J., Fuller, T. & Sauvajot, R. (2000) Competition and intraguild predation among three sympatric carnivores. *Oecologia* 125, 258–270.

Ferguson, J.W.H. (1980) *Die Ekologie van die Rooijackals* Canis mesomelas *Schreber 1778 met Speciale Verwysing na Bewegings en Sosiale Organisasie.* MSc thesis, University of Pretoria. [In Afrikaans].

Ferguson, J.W.H., Galpin, J.S. & De Wet, M.J. (1988) Factors affecting the activity patterns of black-backed jackals *Canis mesomelas. Journal of Zoology* 214, 55–69.

Flynn, R. & Giraldeau, L.-A. (1998) Producer-scrounger tactics affect the geometry of foraging. In: *International Conference on Foraging Behaviour, 21–24 July*. University of California, Santa Cruz.

Garvey, P.M., Glenn, S.S. & Pech, P.R. (2015) Foraging ermine avoid risk: behavioural responses of a mesopredator to its interspecific competitors in a mammalian guild. *Biological Invasions* 17, 1771–1783.

Garvey, P.M., Glenn, S.S. & Pech, P.R. (2016) Dominant predator odour triggers caution and eavesdropping behaviour in a mammalian mesopredator. *Behavioral Ecology and Sociobiology* 70, 481–492.

Gil-Sánchez, J.M., Herrera-Sánchez, F.J., Rodríguez-Siles, J., Sáez, J.M. & Díaz-Portero, M.A. (2020) Feeding specialization of honey badgers in the Sahara Desert: a trial of life in a hard environment. *Diversity* 12, 59.

Giraldeau, L.-A. & Mottley, K. (1998) Experimental evidence of equilibrium producer scrounger solutions to producer-scrounger games. In: *International Conference on Foraging Behaviour, 21–24 July*. University of California, Santa Cruz.

Gorman, M.L., Mills, M.G.L., Raath, J.P. & Speakman, J.R. (1998) High hunting costs make African wild dogs vulnerable to kleptoparasitism by hyaenas. *Nature* 391, 479–481.

Gorta, S.B.Z. (2020) What goes around comes around: complex competitive interactions between two widespread southern African mesopredators. *Canid Biology & Conservation* 22(2), 8–10.

Hancock, D.A. (1999) Blackjack, high stakes. *Timbila* 1, 60–68.

Hayward, M.W. & Slotow, R. (2009) Temporal partitioning of activity in large African carnivores: tests of multiple hypotheses. *South African Journal of Wildlife Research* 39, 109–125.

Herbst, M. (2009) *Behavioural Ecology and Population Genetics of the African Wild Cat,* Felis silvestris *Forster 1980, in the Southern Kalahari.* PhD thesis, University of Pretoria.

Herbst, M. & Mills, M.G.L. (2010) The feeding habits of the Southern African wildcat, a facultative trophic specialist, in the southern Kalahari (Kgalagadi Transfrontier Park, South Africa/Botswana). *Journal of Zoology* 280, 403–413.

Herre, E.A., Knowlton, N., Mueller, U.G. & Rehner, S.A. (1999) The evolution of mutualisms: exploring the paths between conflict and cooperation. *Trends in Ecology and Evolution* 14, 49–53.

Hunter, L. & Barrett, P. (2018) *A Field Guide to the Carnivores of the World*. 2nd edition. Bloomsbury Publishing, London.

Hunter, J. & Caro, T. (2008) Intraspecific competition and predation in American carnivore families. *Ethology Ecology & Evolution* 20, 295–324.

Holt, R.D. & Polis, G.A. (1997) A theoretical framework for intraguild predation. *American Naturalist* 149, 745–764.

Kamler, J.F., Stenkewitz, U. & Macdonald, D.W. (2013) Lethal and sublethal effects of black-backed jackals on Cape foxes and bat-eared foxes. *Journal of Mammalogy* 94, 295–306.

Kemp, A. (1995) *The Hornbills*. Oxford University Press, Oxford.

Kheswa, E.Z.Y., Tharmalingam, R., Kalle, R. & Downs, C.T. (2018) Habitat use by honey badgers and the influence of predators in iSimangaliso Wetland Park, South Africa. *Mammalian Biology* 90, 22–29.

King, A.J., Isaac, N.J. & Cowlishaw, G. (2009) Ecological, social, and reproductive factors shape producer–scrounger dynamics in baboons. *Behavioral Ecology* 20, 1039–1049.

Kingdon, J. (1989) *East African Mammals: Carnivores. 3A*. University of Chicago Press, Chicago, pp. 87–103.

Kruuk, H. (1972) *The Spotted Hyaena*. University of Chicago Press, Chicago.

Laurenson, M.K. (1995) Implications of high offspring mortality for cheetah population dynamics. In: *Serengeti II: Dynamics, Management and Conservation of an Ecosystem* (eds A.R.E. Sinclair & P. Arcese), pp. 385–399. University of Chicago Press, Chicago.

Leistner, O.A. (1967) The plant ecology of the southern Kalahari. *Memoirs of the Botanical Survey of South Africa* 38, 1–172.

Linnell, J.D.C. & Strand, O. (2000) Interference interactions, co-existence and conservation of mammalian carnivores. *Diversity and Distributions* 6, 169–176.

Lombard, A.P.F. (1989) Associations between raptors and small carnivores. *Gabar* 4, 1.

Lourenço, R., Penteriani, V., Rabaca, J.E. & Korpimaki, E. (2014) Lethal interactions among vertebrate top predators: a review of concepts, assumptions and terminology. *Biological Reviews* 89, 270–283.

Loveridge, A.J., & MacDonald, D.W. (2002) Habitat ecology of two sympatric species of jackals in Zimbabwe. *Journal of Mammalogy* 83, 599–607.

Macdonald, D.W. & Thom, M.D. (2001) Alien carnivores: unwelcome experiments in ecological theory. In: *Carnivore Conservation Carnivore Conservation* (eds J.L. Gittleman, S.M. Funk, D. Macdonald & R.K. Wayne), pp. 93–122. Cambridge University Press, Cambridge.

Maclean, G.L. (1985) *Roberts' Birds of Southern Africa*. John Voelcker Bird Book Fund, Cape Town.

Malan, G. & Crowe, T.M. (1996) The diet and conservation of monogamous and polyandrous pale chanting goshawks in the Little Karoo, South Africa. *South African Journal of Wildlife Research* 26, 1–10.

Malan, G. & Crowe, T.M. (1997) Perch availability and ground cover: factors that may constitute suitable hunting conditions for pale chanting goshawk families. *South African Journal of Zoology* 32, 14–20.

Malan, G. (1998) Solitary and social hunting in pale chanting-goshawk (*Melierax canorus*) families: why follow both strategies. *Journal of Raptor Research* 32, 195–201.

McKenzie, A.A. (1993) *The Capture and Care Manual*. Wildlife Decision Support Services and South African Veterinary Foundation, Pretoria.

Mills, M.G.L. (1977) *Diet and Foraging Behaviour of the Brown Hyaena,* Hyaena brunnea *Thunberg 1820 in the Southern Kalahari*. MSc thesis, University of Pretoria.

Mills, M.G.L. (1990) *Kalahari Hyaenas: The Comparative Behavioural Ecology of Two Species*. Unwin Hyman Ltd, London.

Mills, M.G.L. & Biggs, H.C. (1993) Prey apportionment and related ecological relationships between large carnivores in Kruger National Park. *Symposia of the Zoological Society of London* 65, 253–268.

Mills, M.G.L. & Gorman, M.L. (1997) Factors affecting the density and distribution of wild dogs in the Kruger National Park. *Conservation Biology* 11, 1397–1406.

Mills, M.G.L. & Mills, M.E.J. (2017) *Kalahari Cheetahs: Adaptations to an Arid Region*. Oxford University Press, Oxford.

Mills, M.G.L. & Retief, P.F. (1984) The response of ungulates to rainfall along the riverbeds of the southern Kalahari, 1972–1982. *Koedoe* Supplement (1984), 129–142.

Mills, M.G.L., Nel, J.A.J. & Bothma, J. du P. (1984) Notes on some smaller carnivores from the Kalahari Gemsbok National Park. *Koedoe* Supplement (1984), 221–229.

Mills, D., Do Linh San, E., Robinson, H., Isoke, S., Slotow, R. & Hunter, L. (2019) Competition and specialization in an African forest carnivore community. *Ecology and Evolution* 9, 10092–10108.

Minta, S.C., Minta, K. & Lott, D.F. (1992) Hunting associations between badgers (*Taxidea taxus*) and coyotes (*Canis latrans*). *Journal of Mammalogy* 73, 814–820.

Mugerwa, B., du Preez, B., Tallents, L.A., Loveridge, A.J. & Macdonald, D.W. (2017) Increased foraging success or competitor avoidance? Diel activity of sympatric large carnivores. *Journal of Mammalogy* 98, 1443–1452.

Nakashima, Y., Iwata, Y., Ando, C., Nze-Nkogue, C., Inoue, E., Akomo-Okoue, E.F., Mbehang Nguema, P., Diop Bineni, T., Ngok Banak, L., Takenoshita, Y.,

Ngomanda, A. & Yamagiwa, J. (Chapter 15, this volume) Spatial and temporal resource partitioning of small carnivores in the African rainforest: implications for conservation and management. In: *Small Carnivores: Evolution, Ecology, Behaviour, and Conservation* (eds E. Do Linh San, J.J. Sato, J.L. Belant & M.J. Somers). Wiley-Blackwell, Oxford.

Nel, J.A.J. (1984) Behavioural ecology of canids in the south-western Kalahari. *Koedoe* Supplement (1984), 195–221.

Nel, J.A.J. (1990) Foraging and feeding of bat-eared foxes *Otocyon megalotis* in the southwestern Kalahari. *Koedoe* 33, 9–15.

Nelson, R. & Nelson, J. (1987) Observations of a honey badger and chanting goshawks at Nxai pan. *Babbler* 14, 18–20.

Newman, C., Buesching, C.D. & Wolff, J.O. (2005) The function of facial masks in "midguild" carnivores. *Oikos* 108, 623–633.

Owens, M.J. & Owens, D.D. (1978) Feeding ecology and its influence on social organization in brown hyaenas (*Hyaena brunnea*, Thunberg) of the central Kalahari Desert. *East African Wildlife Journal* 16, 113–135.

Packer, C. & Ruttan, L. (1988) The evolution of cooperative hunting. *American Naturalist* 132, 159–198.

Palomares, F. & Caro, T.M. (1999) Interspecific killing among mammalian carnivores. *American Naturalist* 153, 492–508.

Paxton, M. (1988) Foraging associations between pale chanting goshawks, honey badgers and slender mongoose. *Gabar* 3, 82–84.

Pienaar, U. de V. (1969) Predator–prey relationships amongst the larger mammals of the Kruger National Park. *Koedoe* 12, 108–176.

Pienaar, U. de V., Joubert, S.C.J., Hall Martin, A., de Graaf, G. & Rautenbach, I.L. (1987) *The Mammals of the Kruger National Park*. Struik Publishers, Cape Town, pp. 73–74.

Polis, G.A. & Holt, R.D. (1992) Intraguild predation: the dynamics of complex trophic interactions. *Trends in Ecology & Evolution* 7, 151–154.

Rafiq, K., Hayward, M.W., Wilson, A.M., Meloro, C., Jordan, N.R., Wich, S.A., McNutt, J.W. & Golabek, K.A. (2020) Spatial and temporal overlaps between leopards (*Panthera pardus*) and their competitors in the African large predator guild. *Journal of Zoology* 311, 246–259.

Ramesh, T., Kalle, R. & Downs, C.T. (2017) Staying safe from top predators: patterns of co-occurrence and inter-predator interactions. *Behavioral Ecology and Sociobiology* 71, 41.

Rasa, O.A.E. (1983) Dwarf mongoose and hornbill mutualism in the Taru Desert, Kenya. *Behavioral Ecology and Sociobiology* 12, 181–190.

Remonti, L., Balestrieri, A., Ruiz-González, A., Gomez-Moliner, B.J., Capelli, E. & Prigioni, C. (2012) Intraguild dietary overlap and its possible relationship to the coexistence of mesocarnivores in intensive agricultural habitats. *Population Ecology* 54, 521–532.

Rutherford, M.C., Mucina, L., Lötter, M.C., Bredenkamp, G.J., Smith, J.H.L. *et al.* (2006) Savannah Biome. In: *The Vegetation of South Africa, Lesotho and Swaziland. Strelitzia 19* (eds L. Mucina & M.C. Rutherford), pp. 438–538. South African National Biodiversity Institute, Pretoria.

Sasvari, L. & Hegyi, Z. (1998) How mixed-species foraging flocks develop in response to benefits from observational learning. *Animal Behaviour* 55, 1461–1469.

Schaller, G.B. (1972) *The Serengeti Lion: A Study of Predator Prey Relations*. University of Chicago Press, Chicago.

Schuette, P., Wagner, A.P., Wagner, M.E. & Creel, S. (2013) Occupancy patterns and niche partitioning within a diverse carnivore community exposed to anthropogenic pressures. *Biological Conservation* 158, 301–312.

Sharifi, H., Malekian, M. & Shahnaseri, G. (2020) Habitat selection of honey badgers: are they at the risk of an ecological trap? *Hystrix* 31, 131–136.

Sillero-Zubiri, C. (1996) Records of honey badger, *Mellivora capensis* (Carnivora, Mustelidae), in afroalpine habitat. *Mammalia* 60, 323–325.

Simberloff, D. & Dayan, T. (1991) Guilds and the structure of ecological communities. *Annual Review of Ecology and Systematics* 22, 115–143.

Skinner, J.D. & Smithers, R.H.N. (1990) *The Mammals of the Southern African Subregion*. 2nd edition. University of Pretoria, Pretoria.

Skinner, J.D. & Chimimba, C.T. (eds) (2005) *The Mammals of the Southern African Subregion*. 3rd edition. Cambridge University Press, Cambridge.

Sliwa, A. (1994) Marsh owl associating with black-footed cat. *Gabar* 9, 23.

Sprawl, S. & Branch, B. (1995) *The Dangerous Snakes of Africa*. Southern Book Publishers, Halfway House.

Stankowich, T., Caro, T. & Cox, M. (2011) Bold coloration and the evolution of aposematism in terrestrial carnivores. *Evolution* 65, 3090–3099.

Stankowich, T., Haverkamp, P.J. & Caro, T. (2014) Ecological drivers of antipredator defenses in carnivores. *Evolution* 68, 1415–1425.

Steinel, A., Munson, L., Van Vuuren, M. & Truyen, U. (2000) Genetic characteristics of feline parvovirus sequences from various carnivores. *Journal of General Virology* 81, 345–350.

Steyn, P. (1982) *Birds of Prey of Southern Africa*. David Philip, Cape Town.

St-Pierre, C., Ouellet, J.-P. & Crête, M. (2006) Do competitive intraguild interactions affect space and habitat use by small carnivores in a forested landscape? *Ecography* 29, 487–496.

Swanson, A., Arnold, T., Kosmala, M., Forester, J. & Packer, C. (2016) In the absence of a "landscape of fear": how lions, hyenas, and cheetahs coexist. *Ecology and Evolution* 6, 8534–8545.

Tannerfeldt, M., Elmhagen, B. & Angerbjörn, A. (2002) Exclusion by interference competition? The relationship between red and arctic foxes. *Oecologia* 132, 213–220.

Turnbull-Kemp, P. (1967) *The Leopard*. Howard Timmins, Cape Town.

van Rooyen, N., van Rensburg, D.J., Theron, G. K. & Bothma, J. du P. (1984) A preliminary report on the dynamics of the vegetation of the Kalahari Gemsbok National Park. *Koedoe* Supplement (1984), 83–102.

van Rooyen, N., Bezuidenhout, H. & de Kock E. (2001) *Flowering Plants of the Kalahari Dunes*. Ekotrust, Pretoria.

Vogel, J.T., Somers, M.J. & Venter, J.A. (2019) Niche overlap and dietary resource partitioning in an African large carnivore guild. *Journal of Zoology* 309, 212–223.

Wauer, R.H. & Egbert, J. (1977) Interactions between a Harris' hawk and a badger. *Western Birds* 8, 155.

Wilson, B.R., Blankenship, T.L., Hooten, M.B. & Shivik, J.A. (2010) Prey-mediated avoidance of an intraguild predator by its intraguild prey. *Oecologia* 164, 921–929.

Zar, J.H. (1999) *Biostatistical Analysis*. Prentice-Hall, New Jersey.

18

Seed Dispersal by Mesocarnivores: Importance and Functional Uniqueness in a Changing World

Yoshihiro Nakashima[1],* and Emmanuel Do Linh San[2]

[1] College of Bioresource Science, Nihon University, Fujisawa City, Japan
[2] Department of Zoology and Entomology, University of Fort Hare, Alice, South Africa

SUMMARY

Seed dispersal may be a major ecological role of mesocarnivores. However, the general features of seed dispersal by mesocarnivores are not well elucidated. Here, we review the published literature regarding frugivory and seed dispersal by mesocarnivores and briefly summarize the features of seed dispersal. In particular, we focus on the characteristics of seed dispersal by mesocarnivores during a series of events after fruit have been picked up for consumption. We notably identified a minimum of 73 'basic' seed dispersal studies carried out in 26 countries across the globe and focusing on 42 mesocarnivore species from 7 families which ingested the seeds of >700 plant species. The review suggests that seed dispersal by mesocarnivores has important and unique impacts on seed fates. Mesocarnivores regularly swallow seeds (including large seeds), transport these seeds beyond the crown of the mother plant, and subsequently defecate viable seeds; this implies that mesocarnivores have a role as *legitimate* seed dispersers. A subset of studies on 24 mesocarnivore species from five families indicated that gut passage predominantly has a positive or neutral effect on seed germination rate and time. On the other hand, mesocarnivores regularly deposit seeds at 'open sites' (where the vegetation cover is sparse or completely absent) often far from parent trees, which may be hostile to most plant species; this implies that mesocarnivores may be *inefficient* seed dispersers. However, seed dispersal by mesocarnivores may still be indispensable, particularly in fragmented landscapes, as the tendency of mesocarnivores to deposit seeds in specific microhabitats may promote vegetation recovery at artificially created small or narrow open sites. Moreover, long-distance seed dispersal by mesocarnivores is essential to the persistence and recovery of plant populations, and genetic diversity in fragmented landscapes. These unique features of seed dispersal by mesocarnivores are strongly associated with their morphological and behavioural characteristics (e.g. dental morphology, short gut length, and faeces deposition at specific sites). Therefore, a fundamental understanding of mesocarnivore biology is essential to elucidate the importance and functional uniqueness of seed dispersal by mesocarnivores in a changing world.

Keywords

Carnivora — forest fragmentation — frugivory — seed fate — seed handling — vegetation recovery

Introduction

The majority of members of the order Carnivora can be categorized into small- or medium-sized carnivores (< 15 kg), often collectively termed 'mesocarnivores' (Roemer *et al.*, 2009). Some typical representatives are foxes (Canidae), martens and badgers (Mustelidae), genets and civets (Viverridae) or raccoons (Procyonidae). Depending on various types of food resources, mesocarnivores inhabit a wide range of

---

* Corresponding author.

Small Carnivores: Evolution, Ecology, Behaviour, and Conservation, First Edition. Edited by Emmanuel Do Linh San, Jun J. Sato, Jerrold L. Belant, and Michael J. Somers.
© 2022 John Wiley & Sons Ltd. Published 2022 by John Wiley & Sons Ltd.

habitats at relatively high density (Wilson & Mittermeier, 2009). Moreover, in communities where larger carnivores are naturally absent or artificially eliminated, mesocarnivores may constitute the highest trophic levels (Crooks & Soulé, 1999; Ritchie & Johnson, 2009). Therefore, mesocarnivores have a potentially important impact on the community structure of lower trophic levels, and may ultimately influence ecosystem structure and function (Roemer *et al.*, 2009). Nonetheless, relatively little attention has been paid to the ecological roles of mesocarnivores compared with those of larger carnivores – e.g. large cats (Felidae), wolves (Canidae) – which are known to have strong predation-driven direct effects or fear-driven indirect effects (Ripple & Beschta, 2004). Given that mesocarnivores have the ability to thrive in diverse habitats, often including human-modified landscapes, their ecological roles are likely to become increasingly important (Roemer *et al.*, 2009).

Seed dispersal may be a major ecological role of mesocarnivores. With the exception of secondary/ indirect seed dispersal or diplochory (Vander Wall & Longland, 2004) through predation on frugivores (Nogales *et al.*, 1996; Sarasola *et al.*, 2016), larger apex carnivores generally do not perform this function (Carbone *et al.*, 1999). On the other hand, many mesocarnivores have long been known to routinely ingest large quantities of fleshy fruits (Ridley, 1930; McIntosh, 1963; Buskirk & MacDonald, 1984). Indeed, with the exception of the Felidae, the majority of species belonging to the order Carnivora often consume fruits (Wilson & Mittermeier, 2009). However, systematic surveys of the roles of carnivores in seed dispersal only started a few decades ago – from the late 1980s to the early 1990s (Debussche & Isenmann, 1989; Herrera, 1989; Bustamante *et al.*, 1992; Pigozzi, 1992; Chávez-Ramírez & Slack, 1993; Willson, 1993; Schupp *et al.*, 1997). Since then, a number of studies have shown that mesocarnivores regularly transport viable seeds of many plant species, in a wide variety of habitats. Nonetheless, most of these studies have been based on records of plant species and/or on tests of germinability of seeds contained in faeces; other contributory processes have rarely been considered. Researchers may have assumed that the importance of mesocarnivores in seed dispersal was inferior to that of other more specialized frugivores, such as birds

(Clergeau, 1992; Jordano & Schupp, 2000; Pérez-Méndez & Rodríguez, 2018; Godínez-Alvarez *et al.*, 2020), bats (Fleming & Heithaus, 1981; Lopez & Vaughan, 2004; Carvalho-Ricardo *et al.*, 2014; Saldaña-Vázquez *et al.*, 2019), and primates (Chapman, 1989; Russo & Chapman, 2011; Fuzessy *et al.*, 2016; Sengupta *et al.*, 2020).

Nowadays, the situation is changing considerably. Recently, an increasing number of studies have indicated that seed dispersal by mesocarnivores may be much more important for plant population and community dynamics than was previously thought. Mesocarnivores may disperse seeds in a different manner from other sympatric frugivores, and may therefore contribute differently to plant population and community dynamics (Martínez *et al.*, 2008; Fedriani & Delibes, 2009a; Nakashima *et al.*, 2010a), and gene flow (Jordano *et al.*, 2007; Tsunamoto *et al.*, 2020). Interestingly, such a unique method of seed dispersal seems to be well associated with the morphological and behavioural characteristics of carnivores and may be a consequence of their non-specialized features for frugivory. Seed dispersal by mesocarnivores may become increasingly important in human-modified landscapes (Matías *et al.*, 2010; López-Bao & González-Varo, 2011; Escribano-Ávila *et al.*, 2012; Suárez-Esteban *et al.*, 2013a,b; Escribano-Ávila, 2019; Salgueiro *et al.*, 2020) and natural habitats lacking other mammalian frugivores (e.g. alpine or subalpine zone; Otani, 2005). Therefore, elucidation of the general features of seed dispersal by mesocarnivores will not only provide a valuable insight into plant population and community dynamics, but may also be crucial for the effective management of mesocarnivore habitats.

Here, we review the published literature regarding frugivory and seed dispersal by mesocarnivores, and briefly summarize the unique features of seed dispersal. In particular, we focus on the characteristics of seed dispersal by mesocarnivores during a series of events after fruit have been picked up for consumption. Throughout this review, we use the terms 'fruit(s)' and 'seeds' in their ecological, and not their anatomical, sense. We do not discuss the characteristics of seed dispersal by mesocarnivores before fruit processing, e.g. the numbers or types of plant species selected for consumption (and potential dispersal). Regarding this aspect, we refer readers to other relevant published literature (Debussche & Isenmann, 1989; Herrera, 1989; Willson, 1993; Mudappa *et al.*, 2010; David *et al.*, 2015; Kurek, 2015; Hisano & Deguchi, 2018; Koike & Masaki, 2019). The plants consumed by the different mesocarnivores reviewed here are however listed in Appendix 18.1. Diets and behaviours vary considerably among mesocarnivore species, and therefore different species may play distinct and even diverse roles in seed dispersal. However, in this review, we aim to highlight the characteristics that are widely shared among mesocarnivores and to reveal how these characteristics differ from those of other frugivorous taxa. In addition, we discuss how these characteristics are associated with the morphological and behavioural traits of mesocarnivores. Finally, based on our existing knowledge, we consider the importance of the unique seed dispersal by mesocarnivores for plant population and community dynamics. In this chapter, we focus on endozoochorous seed dispersal by mesocarnivores, although they may also be potential vectors for epizoochorous seed dispersal (Hovstad *et al.*, 2009). Lastly, we do not deal with secondary seed dispersal, and, more specifically, diploendozoochory, a mechanism which has recently been reviewed for carnivorous predators (including birds of prey), and its potential ecological significance highlighted (Hämäläinen *et al.*, 2017).

Mesocarnivore Species Studied and Geographic Distribution of Seed Dispersal Studies

We performed a search in Web of Science, Google Scholar, and ResearchGate for papers published until December 2020, using 'seed dispersal' AND ('carnivores' OR 'mesocarnivores') as keywords. We then used cross-referencing to identify other studies that could not be detected through our keyword search. This process allowed us to identify a minimum of 73 studies (see list and study attributes in Appendix 18.1) that dealt with at least some basic aspects of seed dispersal by mesocarnivores, such as the plant species dispersed, the viability of ingested seeds, effects on germination, dispersal distances, and/or characteristics of deposition sites. These studies almost equally took place in Europe, Asia, and in the Americas (Figure 18.1), in a total of 26 countries (Figure 18.2). Very few studies were carried out in Africa and Oceania. Although seven families from the order Carnivora are represented, a majority of studies focused on Canidae and Mustelidae, and to a lesser extent, Viverridae (Figure 18.3). They involved 42 mesocarnivore species and > 700 plant species (Appendix 18.1). The red fox, *Vulpes vulpes*, is by far the most studied species, followed by the European

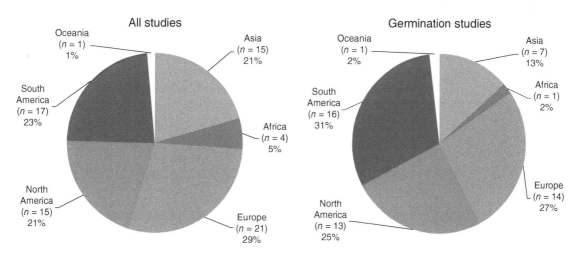

Figure 18.1 Proportional geographic distribution of the 73 'basic' seed dispersal studies identified in the present review (left pie chart), as well as of the subset of 52 studies that compared germination of non-ingested seeds vs. seeds ingested by mesocarnivores (right pie chart). Studies focusing purely on dietary ecology or advanced spatial patterns of seed dispersal were not included in this analysis.

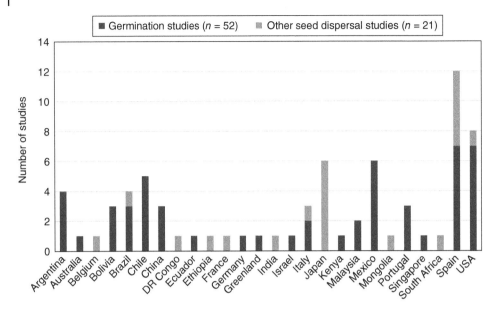

Figure 18.2 Distribution of seed dispersal studies on mesocarnivores based on the country (*n* = 26) where the research was carried out.

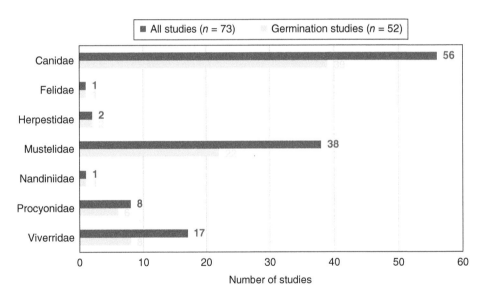

Figure 18.3 Number of seed dispersal studies focusing on at least one species of mesocarnivores (*n* = 42) identified in the present review and grouped based on family.

badger, *Meles meles*, several marten species, *Martes* spp., the culpeo fox, *Lycalopex culpaeus*, and the coyote, *Canis latrans* (Figure 18.4). The above analysis does not include the numerous studies that purely focused on dietary ecology (even if they revealed the frugivorous habits of the studied species), nor a limited number of landscape-scale studies that dealt predominantly with advanced spatial aspects of seed dispersal (seed shadows, dispersal kernels, and effects of landscape structure on seed dispersal and vice versa). The latter studies are discussed in the text where relevant.

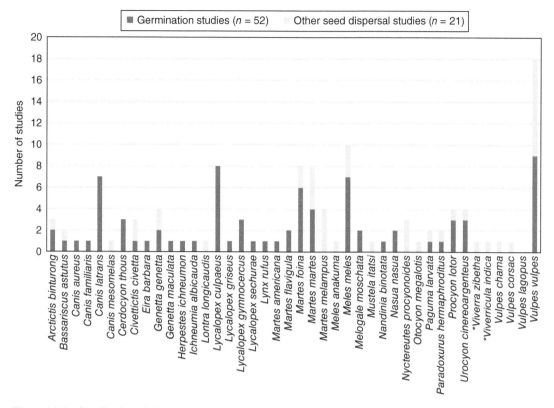

Figure 18.4 Distribution of the number of seed dispersal studies on mesocarnivores based on the focal species (*n* = 42). Note that some studies included several mesocarnivore species. The two species indicated with an asterisk (*) were present in one of the study areas, but it is unclear whether their faeces were effectively part of the samples investigated by the authors.

Seed Handling by Mesocarnivores – the Initial Cause of Unique Seed Dispersal

To evaluate the characteristics of a particular animal species as a seed dispersal agent, it is critically important to observe how the animal handles seeds. Seed handling by frugivores not only determines the immediate mortality of seeds at the fruiting tree, but also influences seed fates during many subsequent processes. For example, when animals discard seeds beneath parent trees, seeds may suffer from high density-dependent mortality or from the higher conspecific competition (Janzen, 1970; Connell, 1978). On the other hand, when animals swallow seeds and transport them in their gut, seeds may colonize new areas located at a distance from the parent trees,

thereby establishing a new population (Howe & Smallwood, 1982).

The way in which frugivores handle seeds is also of considerable importance to the frugivores themselves (Corlett, 2011) because the seeds represent a non-valuable nuisance or may even be harmful. Seeds often contain large amounts of toxic secondary compounds and are dangerous for frugivores to eat (Janzen, 1971). Additionally, frugivores may not be willing to swallow seeds because the seeds represent unwanted ballast, occupying precious gut space and increasing the body weight (Lambert, 1999). Removing seeds from fruit is time-consuming and may decrease feeding efficiency (Corlett, 2011). Therefore, the way in which frugivores handle seeds is closely associated with their feeding strategies. In general, frugivores tend to avoid swallowing larger seeds because larger seeds are typically

(a)   (b)

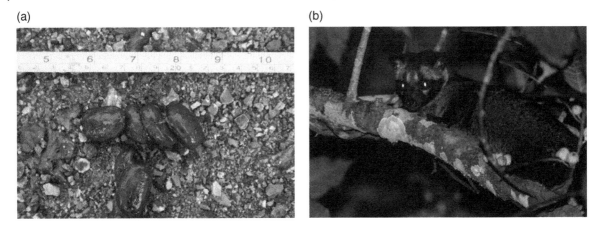

Figure 18.5 The faeces of mesocarnivores often contain disproportionally large seeds with respect to their body size. (a) This photo shows viable seeds (28.6 × 20.3 mm) of *Aglaia grandis* defecated by (b) the common palm civet, *Paradoxurus hermaphroditus* (body mass < 3 kg) in north Borneo. *Source:* Photos © Yoshihiro Nakashima (a), Chien C. Lee @ chienclee.com (b).

easier and less costly to remove from fleshy parts; and frugivores tend to regurgitate these seeds in case they swallowed them (Corlett, 1998, 2017). Many frugivores exhibit morphological and behavioural adaptations to effectively discriminate seeds from fleshy parts (Corlett, 1998, 2017). For example, cercopithecine monkeys store fruits in their cheek pouches, where they separate the seeds from the fleshy parts and spit out seeds with a width greater than ~2 mm (Corlett & Lucas, 1990; Lucas & Corlett, 1998).

By contrast, carnivores do not (or cannot) handle or manipulate seeds with their paws. Additionally, the mouths of carnivores are unsuitable for delicate seed handling. Few studies have included direct observations of seed handlings by mesocarnivores; however, the faeces of mesocarnivores contain many large intact seeds (Zhou *et al.*, 2008b; Nakashima *et al.*, 2010b). For example, Nakashima *et al.* (2010b) showed that the faeces of the common palm civet, *Paradoxurus hermaphroditus*, a relatively small carnivore (< 3 kg) that inhabits rainforests in north Borneo, frequently contained large intact seeds (Figure 18.5). The maximum width (i.e. the second longest axis) of seeds swallowed was 20.3 mm (*Aglaia grandis* [Meliaceae]); this width equalled or exceeded the maximum seed size swallowed by larger-bodied frugivorous mammals in Borneo (Nakashima *et al.*, 2010b). Escribano-Ávila *et al.* (2013) showed that seeds of Spanish juniper, *Juniperus thurifera* (Cupressaceae), dispersed by two

mesocarnivores – the red fox and the stone marten, *Martes foina* – in temperate forests were larger than the seeds dispersed by thrushes, *Turdus* spp., in Mediterranean woodlands of central Spain, possibly because these birds actively selected smaller seeds in the available pool size. Interestingly, seeds dispersed by carnivores had higher survival rates at the seedling stage than did seeds dispersed by birds (60% vs. 30%). In this system, the effects of seed size on seed germination and early survival may be greater than the density-dependent effects on seed mortality. Therefore, swallowing large seeds may have an important influence on subsequent seed fates.

Seed dispersal by mesocarnivores may not be restricted by the morphological characteristics of certain fruits as is the case for birds. In Hube Houhe National Nature Reserve in central China, Zhou *et al.* (2013) surveyed seed dispersal of oriental raisin tree, *Hovenia dulcis* (Rhamnaceae), which produces drupe fruit consisting anatomically of an outer fibroid exocarp and a mesocarp surrounding a shell of hardened endocarp containing the seed; these fruit develop at the end of edible fleshy pedunculate fruit stalks, which swell to form a type of accessory fruit (infructescence). The authors reported that two species of herbivores (Chinese serow, *Capricornis milneedwardsii*, and Chinese goral, *Naemorhedus griseus*) and four species of carnivores (masked palm civet, *Paguma larvata*, yellow-throated marten, *Martes flavigula*, Chinese ferret badger, *Melogale moschata*, and Asiatic black bear, *Ursus thibetanus*) swallowed the edible peduncles and drupe fruit whole, and dispersed the seeds. On the other hand, seven bird species frequently pecked only the peduncles and did not remove the drupe fruit containing the seeds. Lima *et al.* (2015) similarly report on the contribution of mesocarnivores to seed dispersal of the introduced *H. dulcis* in southern Brazil.

Some species of mesocarnivores do exhibit adaptive behaviours to offset seed costs. For example, in Khao Yai National Park, Thailand, the small-toothed palm civet, *Arctogalidia trivirgata*, squeezes fluid from the juicy fruit of *Ficus hispida* (Moraceae) and then discards the residue – including most of the seeds – from its mouth (Duckworth & Nettelbeck, 2008). Low & Vogrinc (2017) also observed masked palm civets crush palm fruits between their molars and swallow only the juice and pulp and drop the residue. In addition, seeds contained in the faeces of the Chinese ferret badger (Zhou *et al.*, 2008b) and the red fox (Traba *et al.*, 2006) were significantly smaller than seeds removed from fruits by researchers, suggesting that these mesocarnivores remove larger seeds before swallowing. Detailed observations using video traps at fruiting trees, or by observing captive individuals, are required to reveal the precise seed-handling behaviour by these and other mesocarnivores. However, the range of plants dispersed by mesocarnivores is clearly less strictly determined by seed size or morphological characteristics of fruits than is that of most smaller or more specialized frugivores.

Due to morphological constraints, other mesocarnivores generally swallow relatively large seeds proximately but this may be part of a feeding strategy for fruits within those constraints. Frugivorous primates are reported to use two trade-off strategies for fruits (Lambert, 1999). The first strategy is to maximize digestibility *per capita* by having a long period of digestion in the gut. This strategy, which is adopted by cercopithecine monkeys, may effectively result in the acquisition of energy and nutrition from limited amounts of food resources; however, it places constraints on the amount of fruit that an animal can consume (based on limited gut space), and also requires investment to dissolve toxic secondary compounds of fruits. The second strategy is to consume larger quantities of food that pass through the gut rapidly so that only readily digestible nutrients (e.g. carboxylates) are absorbed and the rest is defecated in an undigested form. This strategy may result in the acquisition of a limited amount of energy and nutrition *per capita*. However, it enables an increased food intake relative to the availability and does not require an investment to dissolve toxic secondary compounds. This second strategy is adopted by hominoid primates in which the larger body size may compensate for the apparently less-efficient feeding strategy. Importantly, the costs of swallowing seeds may be extremely large in the first strategy, but much smaller in the second one. Indeed, as mentioned earlier, cercopithecine monkeys strictly reject seeds with a width greater than ~2 mm, whereas hominoid primates swallow much larger seeds (width of > 20 mm). Mesocarnivores may adopt a similar strategy to that of hominoid primates. Despite having a much smaller body size, mesocarnivores can supplement their frugivorous diets with other rapidly digestible high-quality food, such as small animals.

Other studies have shown that mesocarnivores rarely destroy seeds, either intentionally or accidentally, during mastication. For example, Herrera (1989) examined more than 1500 scat samples deposited by carnivores in Mediterranean habitats in southeastern Spain and found that only 0.89% of the seeds contained in the scats were visibly damaged (broken or cracked). Similarly, Perea *et al.* (2013) collected 1596 faecal samples deposited by six mammals – European rabbit, *Oryctolagus cuniculus*, European badger, red fox, red deer, *Cervus elaphus*, fallow deer, *Dama dama*, and wild boar, *Sus scrofa* – inhabiting Mediterranean woodlands of southwestern Spain, and showed that high percentages of seeds contained in the scats of European badgers (86.5%) and red foxes (96.0%) were apparently viable, whereas seeds contained in the pellets of rabbits, deer, and boars were more frequently damaged. The structure of the sectorial or bunodont cheek teeth may minimize damage to seeds during mastication in mesocarnivores, whilst some frugivorous and/or herbivorous species of other taxa exhibit adaptations in dental morphology, to more effectively chew their foods (Van Valkenburgh, 1989; Evans *et al.*, 2007), and these adaptations may damage large seeds (Nakashima *et al.*, 2008; Colon & Campos-Arceiz, 2013). In conclusion, the seeds of fruits consumed by mesocarnivores are generally swallowed without being damaged (see also main findings of studies summarized in Appendix 18.1), and therefore remain viable at the earliest stage of seed dispersal.

Effect of Seed Swallowing on Subsequent Seed Fates

In comparison with seed rejection or regurgitation, seed dispersal via swallowing has very different impacts on seed fates; moreover, the effects on the plant may be positive or negative. Importantly, when seeds are swallowed by animals, the direction and magnitude of these effects are completely dependent on animal physiological (e.g. digestion intensity) and behavioural characteristics. Seed-swallowing frugivores influence seed fates in at least four different ways. Firstly, seed germinability may be increased or decreased by passage through the animal gut. Secondly, dispersal distances may be increased because of longer seed retention time in the animal body. Thirdly, animals defecate at specific sites and, therefore, seed survival and growth may be positively or negatively affected by the microhabitat characteristics of these deposition sites. Fourthly, seeds swallowed by animals are deposited as relatively large faecal clumps, and the density of these clumps may affect plant survival and growth. In the following subsections, we briefly summarize our existing knowledge on these four aspects of seed dispersal by mesocarnivores and examine whether each aspect has positive or negative impacts on seed fates.

Effects of Seed Swallowing on Seed Germinability

It is well recognized that passage through the animal gut has an important impact on seed germinability (reviewed in Traveset, 1998; Traveset & Verdú, 2002; Traveset *et al.*, 2007, 2008). The fleshy parts of fruits contain inhibitors (e.g. organics acids, ethylene) of seed germination, and, therefore, seed germinability can be enhanced by complete removal of the fleshy parts from seeds during gut processing (i.e. deinhibition effect: Samuels & Levey, 2005; Robertson *et al.*, 2006). On the other hand, the removal of fruit pulp may make seeds more vulnerable to seed predators (Fedriani & Delibes, 2011). Further, passage through

the animal gut may alter the chemical and physical structure (e.g. thickness) of the seed coat or endocarp, and this scarification may positively or negatively affect seed viability (Traveset, 1998; Traveset *et al.*, 2007). Lastly, faecal material surrounding the seeds may either have no effect on germination, act as a fertilizer and enhance germination rate and seedling growth, or, in contrast, have toxic effects that decrease seedling survival (Malo & Suárez, 1995; Traveset *et al.*, 2001a, 2007).

Regarding mammals, a global study on the effects of gut passage on seed germination indicated that elephants, primates, and new world marsupials act as important enhancers of seed germination (Torres *et al.*, 2020). In the case of the 23 species of carnivores (among them 18 mesocarnivores) included in size effects analyses, no overall significant effect was found. At the family level, no effects were found for Viverridae, Mustelidae, and Procyonidae, but the sample size used in the analyses was small. The more numerous studies on Canidae and Ursidae revealed overall negative and positive effects on germination, respectively.

Among the 'basic' seed dispersal studies identified in the present review, 52 (71%) compared the germinability of seeds contained in mesocarnivore faeces with that of seeds removed from fruiting trees and/or collected on the ground. These germination experiments focused on a subset of 31 mesocarnivore species (from 7 families) and >400 dispersed plant species in 18 countries, with here very few studies taking place not only in Africa and Oceania, but also in Asia (Figures 18.1–18.4; Appendix 18.1). Of the 48 studies that statistically tested the effects of mesocarnivores on germination rate (percentage) and/or germination time, the overall results were as follows: positive (17%), negative (6%), neutral (19%), and mixed effects (58%). However, these results, especially the mixed effects category, are blurred by the inclusion of several mesocarnivore and plants species in the majority of studies. A detailed analysis based on summed test effects for each mesocarnivore species indicates that the direction and magnitude of the effects of gut passage differed markedly among mesocarnivore (Table 18.1) and plant species (Appendix 18.1). Overall, however, 58% of the mesocarnivore species ($n = 24$) had a positive effect on germination, 17% had no effect (neutral), and 25% had a negative effect. At the family level, Canidae,

Table 18.1 Effects of passage through the gut of various mesocarnivore species on the germination rate/percentage (GR) and germination time (GT) of the seeds of a wide range of plant species (see Appendix 18.1 for the names of the plant species involved). Test effects are indicated as follows: + = positive; 0 = neutral; − = negative; DNG = did not germinate (neutral[a]). For germination time, positive and negative effects mean shorter and longer germination times, respectively. For English names and/or selected attributes of the focal mesocarnivores, including geographic distribution, see Table 18.2 and Appendix A.

Family and species	No. plant species[b]	No. effects tested	+ GR	0 GR	− GR	DNG	+ GT	0 GT	− GT	Weighted overall numerical effect[c]	Overall effect
Canidae											
Canis aureus	1	3	1		1		1			0.33	Positive
Canis familiaris	1	1			1					−1.00	Negative
Canis latrans	12 (14)	17	2	6	4	3	1	1		−0.06	Negative
Cerdocyon thous	3	5	1	2	1		1			0.20	Positive
Lycalopex culpaeus	7 (9)	12	3	3	2			2	2	−0.08	Negative
Lycalopex griseus	1	1		1						0.00	Neutral
Lycalopex gymnocercus	4	6	3	1			1			0.50	Positive
Lycalopex sechurae	3	3	3							1.00	Positive
Urocyon cinereoargenteus	9	15	2	5	1		6	1		0.47	Positive
Vulpes lagopus	14	14	5		9					−0.29	Negative
Vulpes vulpes	9	12	4	4	1		3			0.50	Positive
Undetermined	1	1	1							1.00	Positive
Family total	**60 (69)**	**90**	**25**	**22**	**20**	**3**	**13**	**4**	**3**	**0.17**	**Positive**
Family percentage/ average			35.7%	31.4%	28.6%	4.3%	65.0%	20.0%	15.0%	0.21	Positive
Felidae											
Lynx rufus	1	1	1							1.00	Positive
Mustelidae											
Eira barbara	1	2	1		1					0.00	Neutral
Martes americana	3	3		1	1	1				−0.33	Negative
Martes flavigula	5 (6)	6	4		2					0.33	Positive
Martes foina	4	8	1	3			1	3		0.25	Positive
Martes foina/martes	4	5	1	3				1		0.20	Positive
Martes martes	1	1		1						0.00	Neutral

(Continued)

Table 18.1 (Continued)

Family and species	No. plant species[b]	No. effects tested	+ GR	0 GR	– GR	DNG	+ GT	0 GT	– GT	Weighted overall numerical effect[c]	Overall effect
Meles meles	4 (5)	8	4		1		2	1		0.63	Positive
Melogale moschata	3 (4)	3	1	1	1	1[d]				0.00	Negative
Family total	**21 (28)**	**36**	**12**	**9**	**6**	**1**	**3**	**5**	**0**	**0.25**	**Positive**
Family percentage/average			42.9%	32.1%	21.4%	3.6%	37.5%	62.5%	0.0%	0.13	Positive
Procyonidae											
Nasua nasua	7	15	3	5			1	6		0.27	Positive
Procyon lotor	2 (3)	4	1	1	1		1			0.25	Positive
Family total	**9 (10)**	**19**	**4**	**6**	**1**	**0**	**2**	**6**	**0**	**0.26**	**Positive**
Family percentage/average			36.4%	54.5%	9.1%	0.0%	25.0%	75.0%	0.0%	0.26	Positive
Viverridae											
Arctictis binturong	6	12	2	3	1		6			0.58	Positive
Genetta genetta	1	1	1							1.00	Positive
Paguma larvata	1	1		1						0.00	Neutral
Family total	**8**	**14**	**3**	**4**	**1**	**0**	**6**	**0**	**0**	**0.57**	**Positive**
Family percentage/average			37.5%	50.0%	12.5%	0.0%	100.0%	0.0%	0.0%	0.53	Positive
Undetermined/mixed species	14	14	2	7	4			1		–0.14	Negative
GRAND TOTAL	**100 (130)**	**174**	**47**	**48**	**32**	**4**	**24**	**16**	**3**	**0.21**	**Positive**
OVERALL PERCENTAGE/ AVERAGE			35.9%	36.6%	24.4%	3.1%	55.8%	37.2%	7.0%	0.21	Positive
			Positive	Neutral	Negative	Neutral	Positive	Neutral	Negative		

[a]These were regarded as neutral effects as both seeds ingested by mesocarnivores and seeds directly collected from the fruit did not germinate.

[b]This corresponds to the number of unique plant species tested. When different, the total number of plant species used in all tests across the different studies is given in parentheses; this includes some plant species tested in two or more studies on the same or different mesocarnivore species.

[c]This index varies between –1 (fully negative effect) and +1 (fully positive effect). It corresponds to the total germination effect for a species/family divided by the total number of effects tested for the taxon. To calculate the total germination effect, all test effects were summed; positive test effects were allocated a score of +1, negative effects a score of –1, and neutral effects a score of 0.

[d]This value was excluded from the family- and order-level analyses as the authors did not indicate whether the seeds directly collected from the fruit germinated or not.

Mustelidae, Procyonidae, and Viverridae all had an overall positive effect on germination (Table 18.1). The single felid species studied (bobcat, *Lynx rufus*) also had a positive effect. In total, 36% and 40% of the tests ($n = 131$) highlighted a positive or neutral effect on germination rate, respectively. Similarly, tests ($n = 43$) mostly revealed a positive (59%) or neutral (37%) effect on germination time. The same trends were observed at the family level (Table 18.1).

As an example of differing effects of gut passage on different plant species, Schaumann & Heinken (2002) showed that ingestion by martens, *Martes* spp., significantly improved the germination rate of European blueberry, *Vaccinium myrtillus* (Ericaceae), seeds compared with non-ingested seeds (80.7% vs. 36.2%), but had no significant effect on the seed germination rate of the bramble, *Rubus* spp. (Rosaceae). The authors examined the structures of the seed coat by using scanning electron microscopy, and observed that ingested *V. myrtillus* seeds showed a clearly visible abrasion on the seed testa; on the other hand, there was no visible abrasion on the drupes of *Rubus* spp. Numerous other examples are presented in Appendix 18.1.

Besides seed characteristics (e.g. size, structure, and thickness of seed coat), the effects of gut passage may also vary according to the type and quality of food consumed by mesocarnivores because these factors have important impacts on gut passage time and degrees of acidity within the gut and in scats (Traveset, 1998). Given that most mesocarnivores are omnivorous, and consume a wide variety of food resources, the effects of food consumed on seed germinability may be relatively large; however, empirical support for this theory is lacking.

Nonetheless, in comparison with other frugivores, the effects of mesocarnivores on seed germinability may be locally moderate (Traveset *et al.*, 2001b; Zhou *et al.*, 2013; see also Traveset & Willson, 1997). In some cases, the effects of gut passage may be too weak to promote germination (Rosalino *et al.*, 2010). Traveset *et al.* (2001b) showed that germination of the myrtle, *Myrtus communis* (Myrtaceae), was improved by passage through the gut of birds under outdoor conditions in Mediterranean woodlands of northwestern Mallorca. On the other hand, passage through the digestive tract of pine marten, *Martes martes*, had no effect on seed germinability. In central China,

Zhou *et al.* (2013) planted mammal-dispersed *Hovenia dulcis* seeds that had previously been ingested by two species of herbivores and four species of carnivores (see the previous section), and compared their germination rates with those of control seeds obtained directly from wild plants. The authors found that seeds ingested by the two herbivore species and by the carnivore *Martes flavigula* had significantly higher germination rates than did control seeds; on the other hand, the germination rates of seeds consumed by the remaining three carnivore species (*Paguma larvata*, *Melogale moschata*, and *Ursus thibetanus*) were similar to those of control seeds.

Relatively weak effects on seed germinability may be attributable to weak gut processing. Mesocarnivores have short and simple gastrointestinal tracts (Stevens & Hume, 1998; McGrosky *et al.*, 2016), and, therefore, relatively short gut passage times (Table 18.2). However, gut passage time varies considerably within a single individual and may be affected more markedly by dietary contents (Silva *et al.*, 2005) than by seed characteristics (Graae *et al.*, 2004; Varela & Bucher, 2006; Zhou *et al.*, 2008b; Tsuji *et al.*, 2011a). Graae *et al.* (2004) clearly detected negative impacts of a longer passage time on seed germinability in the Arctic fox, *Vulpes lagopus*, in western Greenland. The authors showed that this species had a relatively long gut passage time (mean passage time varied from 16.2 to 25.5 h depending on the plant species), possibly because they were fed animal meat (dead chicken and raw fish), which typically increases the gut passage time (Silva *et al.*, 2005). Graae *et al.* (2004) further investigated the germinability of ingested vs. non-ingested (control) seeds for 14 plant species and showed that the germinability was inhibited in nine species, not affected in three species, and favoured by gut passage in two species. In order to remain viable, seeds of most species must pass through the gut within 12–24 h. Interestingly, the germinability of seeds of three plant species (*Oxyria digyna*, *Sibbaldia procumbens*, and *Silene acaulis*) was favoured when the gut passage time was < 10 h; further, for most species, the germinability decreased as the gut passage time increased. To sum, the gut passage times of mesocarnivores vary according to species and dietary content; nonetheless, mesocarnivores rarely prevent seed germination of most plant species.

Table 18.2   Gut passage times of mesocarnivores consuming fruits. See Appendix A for selected attributes (including geographic distribution) of the mesocarnivores listed below.

Scientific name	English name	Range (h)	References
Canidae			
Cerdocyon thous	Crab-eating fox	4.5–13	Varela & Bucher (2006)
Lycalopex culpaeus	Culpeo fox	12–18	Silva *et al.* (2005)
Lycalopex gymnocercus	Pampas fox	5.5–8.1	Varela & Bucher (2006)
Vulpes[a] lagopus	Arctic fox	4–48	Graae *et al.* (2004)
Vulpes vulpes	Red fox	5–10	Szuman & Skrzydlewski (1962), Artois *et al.* (1987)
Vulpes vulpes	Red fox	17–48	Grünewald *et al.* (2010)
Herpestidae			
Ichneumia albicauda	White-tailed mongoose	< 24–84	Engel (2000)
Mustelidae			
Martes americana	American marten	3.8–10.3	Hickey *et al.* (1999)
Martes melampus	Japanese marten	0.6–1.8	Tsuji *et al.* (2011a)
Martes melampus	Japanese marten	3.82–4.85[b] 3.02–9.18[c]	Tsuji *et al.* (2015)
Martes melampus	Japanese marten	0.6–51.7	Tsuji *et al.* (2016)
Meles meles	European badger	8–15	Grünewald *et al.* (2010)
Melogale moschata	Chinese ferret badger	0.8–5.9	Zhou *et al.* (2008b)
Nandiniidae			
Nandinia binotata	African palm civet	2–3	Charles-Dominique (1978)
Nandinia binotata	African palm civet	2–19	Engel (2000)
Procyonidae			
Nasua nasua	South American coati	2–3	Alves-Costa & Eterovick (2007)
Potos flavus	Kinkajou	0.7–5.6	Lambert *et al.* (2014)
Viverridae			
Arctictis binturong	Binturong	3.3–9.3	Lambert *et al.* (2014)
Arctictis binturong	Binturong	9–57	Colon & Campos-Arceiz (2013)
Civettictis civetta	African civet	< 24–48	Engel (2000)
Genetta maculata[d]	Rusty-spotted genet	12–24	Engel (1998)
Paradoxurus hermaphroditus	Common palm civet	0.7–7.2	Nakashima & Sukor (2010); Y. Nakashima (unpublished data)

[a] Referred to as *Alopex lagopus* in the original publication.
[b] Range of mean retention time for large seeds (spinach).
[c] Range of mean retention time for small seeds (kiwi fruit).
[d] Referred to as *Genetta rubiginosa* in the original publication.

Recently, Soltani *et al.* (2018) performed a meta-analysis of 76 publications focusing mostly on vertebrate frugivores to determine the effects of seed size, the class of dormancy (*sensu* Baskin & Baskin, 2004, 2014), as well as their interaction on the germination of defecated seeds. The authors found that overall germination of both medium and large seeds increased by 18% after gut passage, whereas germination of small seeds decreased by 8%. Germination of physically dormant, water-impermeable seeds increased by 69% compared with control seeds, but the magnitude of increase was higher for large than for medium and small seeds. Germination of non-dormant, physiologically dormant, and morphologically/morphophysiologically dormant seeds (all water-permeable) significantly decreased by 40%, 18%, and 14%, respectively, after gut passage. However, analyses performed with individual seed size categories and dormancy classes indicated that gut passage decreased or did not change the germination rate of non-dormant seeds, and increased germination of medium-sized physiologically dormant and morphologically/morphophysiologically dormant seeds more than it did for large and small seeds. Although this meta-analysis did not discriminate between frugivore taxa, it suggests that seed dormancy type plays an important role. Hence, it is recommended that both seed size and dormancy class be included in future studies investigating the effect of mesocarnivore gut passage on germination.

Effects of Seed Swallowing on Dispersal Distance

For seeds swallowed by animals, dispersal distance is a function of gut passage time, and animal speed and movement patterns. Many carnivores typically have omnivorous diets and search widely for food resources that are often scattered over a wide distance and in small amounts. Thus, carnivores constantly move around, and can potentially transport swallowed seeds over long distances. In previous studies, the dispersal kernel (the probability density function of dispersed seeds relative to distance from maternal plants; Levin *et al.*, 2003) was estimated by using three main methods: (i) combining the information on animal movement speed (estimated through radio-telemetry) with information on gut passage time (seed retention in the gut based on observations of captive individuals) (Hickey *et al.*, 1999; Nakashima & Sukor, 2010; Tsuji *et al.*, 2016); (ii) applying DNA-based genotyping techniques to seeds retrieved from animal faeces and also to maternal candidate plants (Jordano *et al.*, 2007; Gelmi-Candusso *et al.*, 2019); and (iii) searching colour-coded seed mimics embedded in experimental fruits offered to carnivores at feeding stations (González-Varo *et al.*, 2013; Sakamoto & Takatsuki, 2015; Herrera *et al.*, 2016; Mise *et al.*, 2016; González-Varo *et al.*, 2017). In addition, some studies have determined the distance from the nearest seed source (Pendje, 1994; Fedriani & Delibes, 2009b). More recently, an innovative approach using the oxygen isotope ratio of seeds has been developed to estimate vertical seed dispersal distance, which plays a key role in plant escape and/or expansion under climate change (Naoe *et al.*, 2016a,b, 2019).

These studies revealed that some mesocarnivore species routinely disperse seeds over distances of more than several hundred metres. In temperate coniferous forests of southeast Alaska, Hickey *et al.* (1999) used radio-tracking data and gut passage times to estimate that the American marten, *Martes americana*, transported seeds of the Alaskan blueberry, *Vaccinium alaskaense*, and the salmonberry, *Rubus spectabilis*, over a modal (22%) distance of 501–1000 m. The median distances for seeds of these two plant species were 492 and 519 m, respectively. The maximum distance was > 3500 m. In the forests of the northern base of Mt Fuji, central Japan, Tsuji *et al.* (2016) used the same approach and estimated that Japanese martens, *M. melampus*, frequently dispersed seeds over distances of 501–1000 m, with a maximum of 4001–5001 m. In rainforests of north Borneo, Nakashima & Sukor (2010) showed that common palm civets carried rambutan, *Nephelium lappaceum* (Sapindaceae), seeds over an average distance of 216 m. The maximum dispersal distance was > 800 m.

The shape of dispersal kernels may vary considerably within sympatric carnivores. In woodlands of northwestern Spain, González-Varo *et al.* (2013) detected a distinct dispersal kernel between two carnivore species by using colour-coded seed mimics: multimodality and long-distance dispersal for the red fox (median = 1101 m, maximum = 2846 m); and unimodality and short-distance dispersal for the pine marten (median = 260 m, maximum = 1233 m). This finding likely reflects differences in body mass and home range size of these two species. However, differences in landscape structure may also affect seed dispersal distances and kernels. In Central Portugal, using the same methodology, Herrera *et al.* (2016) found much shorter dispersal distances for the red fox, but these differed between a site with spatially continuous and abundant forest cover (median = 614 m, maximum = 965 m) and a site with spatially aggregated and scarcer forest cover (median/maximum = 816 m). Dispersal distances for stone marten were similar to those found for pine marten in Spain, but here again differences were found between habitats with homogeneous (median = 477 m, maximum = 1117 m) and sparse forest cover (median = 322 m, maximum = 944 m).

Differences in seed dispersal distances between mesocarnivores and types of landscape structure, together with species-specific microhabitat characteristics of deposition sites (see below) may result in different two-dimensional spatial distribution patterns of deposited seeds. Fedriani & Delibes (2009b) showed that the spatial patterns of deposited seeds of the Iberian pear, *Pyrus bourgaeana* (Rosaceae), differed among mammalian dispersers: the patterns for European badger (Figure 18.6) faeces and dispersed seeds were clearly clustered at small spatial scales, whereas those for wild boar and red fox faeces were relatively scattered across a 72 ha plot. These three mammals may differentially contribute to the aggregated spatial patterns of adult *P. bourgaeana* trees. Boars tended to release seeds in the vicinity of adult trees, thereby contributing to the maintenance and enlargement of existing tree clusters. Badgers deposited faeces and seeds in a highly clumped pattern; however, unlike boars, they delivered these clumps at a long distance from *P. bourgaeana* neighbourhoods, and were, therefore, more likely to create new tree clusters. The strong tree aggregation of *P. bourgaeana* was likely the result of a single or several non-exclusive processes, such as spatial patterning of seed delivery by dispersers and/or seedling establishment beneath mother trees. These distinct seed 'rains', contributing differently to adult plant distribution patterns, may be a common phenomenon; nonetheless, few studies have quantified the effects of seed deposition patterns on the spatial patterns of adult trees.

Long-distance seed dispersal by carnivores may also play important roles in plant escape and/or expansion under climate change (González-Varo *et al.*, 2017). In central Japan, Naoe *et al.* (2016b) showed using a stable oxygen isotope that the Asiatic black bear disperse seeds of the Korean hill cherry, *Prunus verecunda*, over several hundred metres vertically and that the dispersal direction is heavily biased toward the mountain tops. Asiatic black bears climbed the mountain following spring-to-summer plant-fruiting phenology, resulting in biased seed dispersal. Similar trends were found

(a)

(b)

Figure 18.6 (a) The European badger, *Meles meles*, defecates in small pits that may locally accumulate and form conspicuous latrines. (b) Detail of a dung pit containing several faeces revealing notably the consumption of plum, *Prunus domestica*, and maize, *Zea mays*, kernels. *Source:* Photos © Denis-Richard Blackbourn (a), Emmanuel Do Linh San (b).

in the smaller Japanese marten, but expectedly with shorter dispersal distances (mean ± SD: 461 ± 415 m), probably reflecting their smaller home range due to smaller body size and ranging behaviour. The dispersal distances achieved by mesocarnivores may be much longer than those achieved by other sympatric frugivores; however, few direct comparisons with sympatric frugivores have been conducted. In Mediterranean scrubland in Spain, Jordano *et al.* (2007) estimated the dispersal kernel of the St Lucie cherry, *Prunus mahaleb* (Rosaceae), generated by birds (small- and medium-sized passerines) and mesocarnivores (red fox, stone marten, and European badger), by using genetic markers and maternal tree assignments. The authors observed that small passerines mainly dispersed seeds over short distances (50% of passerines dispersed seeds at a distance of < 51 m from the source trees), whereas mesocarnivores and medium-sized birds mainly dispersed seeds over long distances (50% of mammals and medium-sized birds dispersed seeds at distances of > 495 and > 110 m, respectively, from the source trees). Corlett (2009) estimated frugivore assemblages in Southeast Asia based on home range size and determined five categories (0–10 m, 10–100 m, 100–1000 m, 1–10 km, and > 10 km) with regard to the maximum seed dispersal distance routinely achieved. Carnivores, including civets, martens, and bears, were categorized into the 1–10 km group. The routine seed dispersal distances for these species were shorter than those for fruit pigeons (*Ducula* spp., *Ptilinopus* spp.), large fruit bats (*Acerodon* spp., *Pteropus* spp., *Rousettus* spp.) (tiny seeds), the Asian elephant, *Elephas maximus*, and rhinoceroses, *Rhinoceros* spp. (categorized into the > 10 km group), but longer than those for most frugivores.

Effects of Seed Swallowing on Seed Deposition Sites

The microhabitats of seed deposition sites may represent another distinct characteristic of seed dispersal by carnivores. Mesocarnivores use faeces for intraspecific and interspecific communications (i.e. scent-marking: Macdonald, 1980; Gorman & Trowbridge, 1989; Hutchings & White, 2000), and therefore often deposit faeces and seeds as small clumps at sites having specific characteristics. Some species repeatedly deposit faeces at precisely the same locations, and also construct 'latrines' at specific microhabitats (Page *et al.*, 1998; Espírito-Santo *et al.*, 2007; Jordan *et al.*, 2007; for a review, see Buesching & Jordan, Chapter 7, this volume). Thus, by delivering seeds to very different microhabitats from those of deposition sites of other frugivores, mesocarnivores may have different effects on post-dispersal seed fates.

An increasing number of studies have evaluated the microhabitat and habitat characteristics of seed deposition sites (e.g. Jordan *et al.*, 2007; Martínez *et al.*, 2008; Fedriani & Delibes, 2009b; Escribano-Ávila *et al.*, 2012, 2013; Celedón-Neghme *et al.*, 2013; Peredo *et al.*, 2013; Suárez-Esteban *et al.*, 2013a,b; García-Cervigón *et al.*, 2018; and see numerous other references in Appendix 18.1). The detailed characteristics of faecal deposition sites may vary among species, and even within the same species (Engel, 2000; Espírito-Santo *et al.*, 2007), depending on habitat availability and intraspecific variation in animal habits; nonetheless, mesocarnivores habitually deposit faeces at 'open sites' where the vegetation cover is sparse or completely absent. For example, Peredo *et al.* (2013) systematically examined the faeces deposited by four carnivore species/genera (*V. vulpes*, *M. meles*, *Martes* spp., and *Mustela* spp.) in woodland pastures of northern Spain. The authors reported that most faeces containing seeds occurred in open microhabitats – such as pasture, heathland, or rocky ground – rather than beneath the forest canopy. As a consequence of the unusual habits of mesocarnivores, they may deliver seeds to microhabitats that differ from those 'selected' by other sympatric frugivores. Nearby, in northwest Spain, Martínez *et al.* (2008) examined 158 samples of faeces deposited by several species of carnivores, and, in particular, *V. vulpes* and *M. meles*. The faeces were located in specific microhabitats, such as rocks (25.8%), paths (23.7%), pastures (17.5%), beneath yew, *Taxus baccata*, trees (14.4%), latrines (7.2%), and beneath other trees and shrubs. On the other hand, sympatric frugivorous birds (e.g. *Turdus* spp.) disperse seeds of individual plants in smaller clumps, particularly in covered microhabitats. Similarly, Jordano *et al.* (2007) reported large interspecific variation among dispersers, in the seed dispersal distance, as seen above, but also in the microhabitats of deposition sites. Small-sized birds dispersed *P. mahaleb* seeds mainly beneath the shrub canopies of *P. mahaleb* and other fleshy-fruited trees or shrubs, whereas mesocarnivores (*V. vulpes*, *M. meles*, and *M. foina*) deposited seeds preferentially at open sites, namely rocky soils and open ground with sparse woody vegetation or grass cover. Conversely, medium-sized birds such as the carrion crow, *Corvus corone*, and the mistle thrush, *Turdus viscivorus*, dispersed seeds mainly in open areas and beneath pine trees.

The microhabitats of deposition sites directly influence post-dispersal seed fates. Few studies have demonstrated positive effects of non-random dispersal by mesocarnivores on seed fates. In rainforests of north Borneo, Nakashima *et al.* (2010a) showed that common palm civets non-randomly dispersed seeds to sites characterized by low stem density and canopy cover, such as the banks of small rivers, rain-flow paths, abandoned trails, and treefall gaps. Seeds of the mali-mali, *Leea aculeata* (Leeaceae), an early-successional shrub, that were dispersed by civets to the banks of rivers and gaps showed significantly higher survival and growth rates than did those dispersed to rain-flow paths or abandoned trails. Seeds that were dispersed by pig-tailed macaques, *Macaca nemestrina*, or to random locations showed low survival rates. The effects of civets on seed fates were not straightforward; however, in comparison with macaque or random dispersal, civets significantly enhanced the survival and growth of *L. aculeata* seeds after one year.

The direction and magnitude of the effects on seed fates may ultimately depend on the specific plant's requirements (Silva *et al.*, 2005) and also on habitat characteristics (e.g. soil moisture, temperature); however, seed deposition sites of mesocarnivores typically seem to be hostile to most plant species (Bustamante *et al.*, 1992; Silva *et al.*, 2005; Zhou *et al.*, 2008b; Rost *et al.*, 2012). In northeast India, Chakravarthy & Ratnam (2015) showed that civets (possibly common palm civets) deposited most seeds of *Vitex glabrata* (> 90%) onto canopy branches and fallen logs, while the majority of *Prunus ceylanica* seeds (> 70%) were deposited on the forest floor. For both species, seeds deposited on logs experienced higher seed predation than seeds defecated on the forest floor, especially when local seed densities were high. Further, seed viability of *P. ceylanica* was

significantly lower on logs than on the forest floor. For both tree species, unlike the case of *L. aculeata* in north Borneo (see above), civets did not disperse seeds to sites where seeds experienced either low predation or high survival. In central Chile shrub lands, Bustamante *et al.* (1992) observed that 71% of the Chilean acorn, *Cryptocarya alba* (Lauraceae), seeds contained in 75% of faecal samples from the culpeo fox, *Lycalopex culpaeus* (previously *Dusicyon/ Pseudalopex culpaeus*), were deposited in open habitats, rather than under shrubs. Further, the germination rate was lower in gravity-fallen seeds (22.7%) than in fox-dispersed seeds (77.3%), possibly as a result of higher mortality caused by desiccation and predation. Fedriani & Delibes (2009b) studied *Pyrus bourgaeana*, whose seeds are mainly dispersed by three species of mammals (*M. meles*, *V. vulpes*, and *S. scrofa*), and suggested that microhabitats may not have strong impacts on seed germination and early survival: red foxes and wild boars more frequently dispersed seeds to open microsites (81.8% and 44.6%, respectively) than beneath *Pistacia* (Anacardiaceae) shrubs (4.5% and 21.4%, respectively); on the other hand, European badgers dispersed seeds beneath *Pistacia* shrubs (42.9%), as well to open microsites (26.4%). Seed-predating rodents are mainly active beneath *Pistacia* shrubs; however, seed survival rates did not differ between open microsites and *Pistacia* shrubs when seeds were transported far from the parent tree. Nonetheless, they also suggested that conditions beneath *Pistacia* shrubs (e.g. soil temperature, moisture) are more favourable for subsequent seedling survival than are those occurring in open microsites. Therefore, non-random seeds dispersed by red foxes may have negative impacts on long-term survival, while seed dispersal by badgers may be favourable to the Iberian pear (see also Fedriani & Delibes, 2009a; Fedriani *et al.*, 2010; Perea *et al.*, 2013; Fedriani & Wiegand, 2014). The suitability of a microhabitat for germination and early survival may also be dependent on seed selection, handling, and gut passage effects (Escribano-Ávila *et al.*, 2013). Nonetheless, 'open sites' are typically created at locations where vegetation recovery is difficult or frequently disturbed, and therefore non-random seed dispersal by mesocarnivores should tentatively be regarded as low-quality seed dispersal.

Effects of Seed Swallowing on Seed Density at Deposition Sites

Seed dispersal via swallowing affects the number of seeds at deposition sites, and this may decrease the quality of seed dispersal by mesocarnivores. Animals that spit out or regurgitate seeds typically deposit them individually or as small clumps; on the other hand, animals that defecate seeds deposit them as large clumps (scatter dispersal vs. clump dispersal; Howe, 1989). Several hundred seeds may be contained in faeces, even those deposited by smaller carnivores (e.g. Santos *et al.*, 1999; Jordano *et al.*, 2007; Martínez *et al.*, 2008; Rosalino & Santos-Reis, 2009; López-Bao & González-Varo, 2011; Escribano-Ávila *et al.*, 2012; Perea *et al.*, 2013; and numerous other references listed in Appendix 18.1). High numbers of seeds in faeces may cause high conspecific competition and density-dependent mortality (e.g. infection by pathogens; Howe, 1989); further, the scent of faecal material may attract seed predators and secondary dispersers, such as rodents (e.g. rats) and dung beetles, thereby affecting seed survival rates (Janzen, 1982; Chapman, 1989; Vander Wall & Longland, 2004). Seed density and secondary dispersal by dung beetles have been reported to influence the fates of seeds contained in the faeces of large carnivores, such as bears (Koike *et al.*, 2012a,b).

Clump dispersal may have minor effects on the overall population dynamics of plants; however, regarding seed dispersal by mesocarnivores forming latrines, the effects cannot be negligible. Density-dependent mortality and seed predation by rodents are likely to be severe at latrines, although faecal materials may cover the seeds, thereby protecting them from insect predators (Fragoso *et al.*, 2003). In Shimba Hills, Kenya, Engel (2000) reported that seeds deposited at an African civet, *Civettictis civetta*, latrine were regularly buried by dung beetles (up to 5 cm in depth) and/or ant activities. Although effects of seed burial may or may not be advantageous for the plants, depending on growth strategy, seed size, and energy reserves, seeds may have lower risk of seed predation. LoGiudice & Ostfeld (2002) showed that seed removal from latrines of the northern raccoon, *Procyon lotor*, was significantly decreased by the addition of alternative food. This finding implies that during periods of high food availability, raccoon latrines represent safe sites for tree recruitment. The potential for seed germination and seedling establishment at latrines may ultimately be context dependent. Nonetheless, with the exception of abandoned latrines, repeated seed dispersal to precisely the same site is unlikely to favour plants' dispersal (Engel, 1998, 2000).

Potential Role of Carnivores as Important Seed Dispersers

Importance for Colonization

The present review suggests that seed dispersal by mesocarnivores has important and contrasting impacts on seed fates. Mesocarnivores regularly swallow seeds (including large seeds), transport these seeds beyond the crown of the mother plant, and, subsequently, defecate viable seeds; this implies that mesocarnivores have a role as *legitimate* seed dispersers, following the terminology of Herrera (1989). However, mesocarnivores regularly deposit seeds at 'open sites', which may be hostile to most plant species. This implies that mesocarnivores may be *inefficient* seed dispersers, following the terminology of Reid (1989), who defined *disperser efficiency* as 'the probability that a seed dispersed by the vector will lodge in a safe site and germinate'. Thus, mesocarnivores can be regarded as legitimate but inefficient seed dispersers. However, in the long term, seed dispersal by mesocarnivores may make an important contribution to plant populations. In general, long-distance seed dispersal is essential to maintaining genetic diversity, and also increasing the opportunity for plants to colonize suitable new habitats (Cain *et al.*, 2000; Nathan *et al.*, 2008; Schurr *et al.*, 2009; Jordano, 2017). In particular, long-distance seed dispersal may be crucial in fragmented landscapes because populations inhabiting such landscapes are more vulnerable to habitat deterioration and/or stochastic extinction, and therefore must frequently escape to suitable new fragments (Bacles *et al.*, 2006). Seed dispersal by mesocarnivores may be highly beneficial in this respect. Mesocarnivores can persist in fragmented landscapes (Corlett, 1998; Crooks, 2002; Grünewald *et al.*, 2010) and flexibly use different habitats, by moving across several fragments

(Cavallini & Lovari, 1994; Herrera *et al.*, 2016). A number of studies have empirically demonstrated that mesocarnivores regularly transport seeds from one fragment to another (Cavallini & Lovari, 1994) or across different vegetation types (Rabinowitz, 1991; Nakashima & Sukor, 2010).

In addition, mesocarnivores may contribute to plant colonization of open habitats or to vegetation recovery between fragments (e.g. abandoned land, burnt areas, and quarries) (Rost *et al.*, 2012; Escribano-Ávila *et al.*, 2013, 2014; Peredo *et al.*, 2013; García-Cervigón *et al.*, 2018; Fedriani *et al.*, 2018; Pereira *et al.*, 2019; Salgueiro *et al.*, 2020), although their dispersal service may also be jeopardized by strong landscape degradation (Cancio *et al.*, 2017). In comparison with other frugivores, mesocarnivores may be more effective in promoting vegetation recovery. For example, in a Mediterranean environment of central Spain, Escribano-Ávila *et al.* (2012) reported that red fox and stone marten played a greater contribution to the recolonization of *Juniperus thurifera*, on abandoned agricultural land than did sympatric frugivorous birds, such as *Turdus* spp. Thrushes dispersed large quantities of seeds in new areas, but seed deposition seemed to be limited to the patch where other fleshy-fruited trees were abundant. In contrast, carnivores dispersed seeds to all of the investigated habitats, even patches lacking juniper trees, suggesting that the carnivores may be responsible for the arrival and establishment of the first trees. Newly recruited isolated trees attract other species of frugivores, such as birds, and can function as dispersal foci to promote seed accumulation, thereby leading to rapid vegetation recovery. Escribano-Ávila *et al.* (2014) found that thrushes were the main contributors to *J. thurifera* recruitment in woodlands (73%), leading to population growth, but with a reduced impact on the colonization of old fields where carnivores contributed to 80% of recruitment (42% red fox, 38% stone marten). Therefore, there may be functional disparities between seed dispersal by carnivores and by thrushes, and their services may be complementary. Other studies have shown that red fox and European badger may defecate a large proportion of viable seeds in vacant habitats such as quarries (Salgueiro *et al.*, 2020), while both red fox and stone marten are able to disperse seeds to burnt areas located far (> 1 km) from forested habitats (Rost *et al.*, 2012). Therefore, long-distance seed dispersal by mesocarnivores may contribute to vegetation recovery at open sites located far from forested areas.

The tendency of mesocarnivores to deposit seeds at specific microhabitats may also promote vegetation recovery at artificially created small or narrow open sites. In Mediterranean scrublands of southwestern Spain, Suárez-Esteban *et al.* (2013a) systematically examined faeces deposited by carnivores (*V. vulpes* and *M. meles*), ungulates, and wild rabbits along a gravel road and in adjacent scrubland, and compared the abundance and diversity of seeds from 13 plant species, contained in faeces collected from the two sites. The authors observed that carnivores deposited 5.5 times more faeces and 3.5 times more seeds along the gravel road than in the scrubland. Carnivores dispersed a higher number of seeds than did sympatric rabbits and ungulates, but deposited the smallest amount of faeces. Suárez-Esteban *et al.* (2013b) further showed that the densities of seedlings and adult fleshy-fruited shrubs dispersed by foxes and rabbits along the road verge were 2.7 times higher than in the adjacent scrubland. The density of bird-dispersed shrubs was twice higher along the road verge than in the scrubland; on the other hand, shrubs dispersed by ungulates and European badgers showed similar densities in both habitats, as did rockroses (Cistaceae). In this ecosystem, red foxes (Figure 18.7) have a particularly important role as seed dispersers to promote vegetation recovery; however, seedling mortality may be higher, possibly because of herbivory, competition, and water stress during the Mediterranean summer. Therefore, the ability of mesocarnivores to deposit faeces in open habitats may effectively lead to the restoration of degraded habitats.

Dispersal of Non-Native Plants

On the other hand, seed dispersal by mesocarnivores may have negative effects on the native structure and diversity of plant communities. Mesocarnivores often eat fruits of exotic plants (Clevenger, 1996; Bermejo & Guitián, 2000; Borchert *et al.*, 2008; Zhou *et al.*, 2008c; Padrón *et al.*, 2011; Nakashima & Sukor, 2013;

Figure 18.7 The red fox, *Vulpes vulpes*, plays a legitimate role in the endozoochorous seed dispersal of numerous fleshy-fruited plant species. Although it may not necessarily be an efficient disperser because faeces are often deposited at open sites, this behaviour may locally contribute to the restoration of degraded habitats. *Source:* Photo © Denis-Richard Blackbourn.

López-Bao *et al.*, 2015; Amodeo *et al.*, 2017; Bravo *et al.*, 2019; Pereira *et al.*, 2019; and see other references in Appendix 18.1). In particular, they show a strong preference for fruits cultivated for human consumption, and disperse a high number of the seeds (Silverstein, 2005; Borchert *et al.*, 2008; Twigg *et al.*, 2009; Zhou *et al.*, 2008c; López-Bao *et al.*, 2015; Acosta-Rojas *et al.*, 2019; Bravo *et al.*, 2019; Spennemann, 2020); this may be because the fruit characteristics of the cultivated plants (e.g. crop size, nutritional content, pulp ratio to seeds) are in accord with their fruit preferences. Seed-dispersing mesocarnivores are highly favourable for some of these exotic plants because open sites with high light intensity may be suitable for survival and growth; moreover, long-distance dispersal can enable rapid expansion of

populations. Silva *et al.* (2005) observed that the seeds of an alien shrub, the Peruvian pepper, *Schinus molle* (Anacardiaceae), represented 98% of fruits consumed by the native *Lycalopex culpaeus*. Culpeo foxes were legitimate and efficient dispersers of this alien shrub, and > 41% of seeds were deposited at safe microsites, such as on the bottoms of ravines and at wet microsites. Most cultivated plants for human consumption are not invasive; however, they may reduce wild fruit consumption by mesocarnivores, thereby disrupting seed dispersal (López-Bao & González-Varo, 2011; Rost *et al.*, 2012). To effectively control the expansion of introduced plant populations, further detailed studies of seed dispersal by mesocarnivores are required, as highlighted by López-Bao & González-Varo (2011).

Seed Dispersal by Introduced Carnivores

If artificially introduced to new habitats, particularly in isolated islands, mesocarnivores may disrupt native interactions between fruits and frugivores. They may eradicate native seed dispersers such as lizards, and reduce dispersal opportunities for native plants; however, they may partly replace the role played by native frugivores (Traveset, 1995; Riera *et al.*, 2002; Celedón-Neghme *et al.*, 2013). Celedón-Neghme *et al.* (2013) surveyed seed dispersal of a native shrub, the spurge olive, *Cneorum tricoccon* (Rutaceae; previously Cneoraceae), in Dragonera and Mallorca islands. In Dragonera, the native frugivorous Lilford's wall lizard, *Podarcis lilfordi* (Lacertidae), persists at high density, whereas in Mallorca, where the pine marten was introduced, *P. lilfordi* became extinct. The authors observed that the pine marten partially substituted for the role of *P. lilfordi* in seed dispersal, but differed from the native species in many other aspects. This mustelid was absent from the coast and was mainly restricted to pine forests in mountainous areas, which comprise unfavourable habitats for *C. tricoccon*. Consequently, seed dispersal in coastal shrublands was lacking, leading to seriously limited recruitments (Traveset *et al.*, 2012). The alien marten may have different evolutionary consequences on seed traits than the native lizard. The probability of being dispersed by the marten increased with seed size whilst the opposite occurred for seeds dispersed by the lizard. This among-disperser variation in the selection regimes helped create a geographical variation for seed size of the Mediterranean relict *Cneorum tricoccon* (Traveset *et al.*, 2019). The effects of introduced mesocarnivores may be more drastic and observable on islands, but their impacts on fruit–frugivore interactions may not be restricted to islands.

Conclusion and Future Research

In conclusion, mesocarnivores may play unique and important roles in seed dispersal, based on their ability to deposit seeds at open sites and promote long-distance dispersal, particularly in degraded landscapes. These unique characteristics of seed dispersal by mesocarnivores are strongly associated with their morphological and behavioural characteristics (e.g. dental morphology, short gut length, faeces deposition at specific sites). Moreover, their unique contribution may be the consequence of their non-specialized features for frugivory. In general, fruit–frugivore relationships are evolutionarily diffused and the seeds of a given plant species are typically dispersed by a diverse range of species assemblages (Jordano, 1987, 2014). Therefore, the importance of seed dispersal by small carnivores highlights the evolutionary complexity and non-deterministic nature of fruit–frugivore interactions.

Our understanding of the functions of seed dispersal by mesocarnivores is limited. Many mechanisms of seed dispersal by carnivores remain unexplored, and therefore our review has several important limitations. Firstly, we have focused on the common characteristics widely shared among mesocarnivores; however, given that mesocarnivores differ in terms of diet, body size, behaviours, and activity patterns, it is expected that the dispersal quality will vary considerably among species, as suggested by the overview provided in Appendix 18.1. For example, González-Varo *et al.* (2013) clearly detected a large variation in seed dispersal kernel between the red fox and the pine marten in Mediterranean forests. Further, a series of studies by Fedriani and coworkers suggested that the European badger may be a more efficient disperser than other mammals, such as the red fox and the pine marten. However, with the exceptions of the degree of frugivory and effects of gut passage on seed germinability, few studies have evaluated functional disparity among mesocarnivores. Secondly, although published systematic surveys on the dispersal quality of mesocarnivores were conducted in several countries worldwide (see Appendix 18.1), very few studies have been carried out in Africa, and to a lesser extent, in Asia. The quality of seed dispersal by mesocarnivores is highly dependent on habitat characteristics (e.g. soil moisture and temperature), and therefore the results of investigations in specific habitats (e.g. semi-arid to arid habitats) may not accurately represent the roles of mesocarnivores in other environments (e.g. rainforests and wooded savannahs).

Further systematic surveys are required to evaluate the importance of seed dispersal by mesocarnivores. The characteristics of seed dispersal by animals are ultimately a consequence of their feeding strategies and associated behaviours. Therefore, a fundamental understanding of mesocarnivore biology is essential to elucidate the importance and functional uniqueness of seed dispersal by mesocarnivores in a changing world.

Acknowledgements

We are grateful to Chien C. Lee (chienclee.com) and Denis-Richard Blackbourn for graciously supplying animal photographs. Thomas Engel and Elias Soltani kindly provided some complementary information on their seed dispersal studies. Two anonymous referees are thanked for their comments on an earlier draft of this chapter.

References

Abiyu Abrham, Demel Teketay, Glatzel, G. & Gratzer, G. (2015) Tree seed dispersal by African civets in the Afromontane Highlands: too long a latrine to be effective for tree population dynamics. *African Journal of Ecology* 53, 588–591.

Acosta-Rojas, D.C., Jiménez-Franco, M.V., Zapata-Pérez, V.M., De la Rúa, P. & Martínez-López, V. (2019) An integrative approach to discern the seed dispersal role of frugivorous guilds in a Mediterranean semiarid priority habitat. *PeerJ* 7, e7609.

Alves-Costa, C.P. & Eterovick, P.C. (2007) Seed dispersal services by coatis (*Nasua nasua*, Procyonidae) and their redundancy with other frugivores in southeastern Brazil. *Acta Oecologica* 32, 77–92.

Amodeo, M.R., Vázquez, M.B. & Zalba, S.M. (2017) Generalist dispersers promote germination of an alien fleshy-fruited tree invading natural grasslands. *PLoS One* 12, e0172423.

Armenta-Méndez, L., Gallo-Reynoso, J.P., Wilder, B.T., Ortega-Nieblas, M.M. & Barba-Acuña, I. (2020) The role of wild canids in the seed dispersal of *Washingtonia robusta* (Arecaceae) in Sonoran Desert oases. *Revista Mexicana de Biodiversidad* 91, e913129.

Aronne, G. & Russo, D. (1997) Carnivorous mammals as seed dispersers of *Myrtus communis* (Myrtaceae) in the Mediterranean shrublands. *Plant Biosystems – An International Journal Dealing with All Aspects of Plant Biology* 131, 189–195.

Artois, M., Lemaire, M., George, J., Demerson, J. & Jacquemet, J. (1987) [Fox faeces as an index of feeding ecology and behaviour]. *Cahiers d'Ethologie appliquée* 7, 275–286. (In French with English summary).

Bacles, C.F., Lowe, A.J. & Ennos, R.A. (2006) Effective seed dispersal across a fragmented landscape. *Science* 311, 628–628.

Baskin, J.M. & Baskin, C.C. (2004) A classification system for seed dormancy. *Seed Science Research* 14, 1–16.

Baskin, C.C. & Baskin, J.M. (2014) *Seeds: Ecology, Biogeography, and Evolution of Dormancy and Germination.* 2nd edition. Elsevier/Academic Press, San Diego.

Bermejo, T. & Guitián, J. (2000) Fruit consumption by foxes and martens in NW Spain in autumn: a comparison of natural and agricultural areas. *Folia Zoologica* 49, 89–92.

Borchert, M., Davis, F.W. & Kreitler, J. (2008) Carnivore use of an avocado orchard in southern California. *California Fish and Game* 94, 61–74.

Bravo, S.P., Berrondo M.O. & Cueto, V.R. (2019) Are small abandoned plantations a threat for protected areas in Andean forests? The potential invasion of non-native cultivated species. *Acta Oecologica* 95, 128–134.

Brunner, H., Harris, R. & Amor, R. (1976) A note on the dispersal of seeds of blackberry (*Rubus procerus* PJ Muell.) by foxes and emus. *Weed Research* 16, 171–173.

Buesching, C.D. & Jordan, N.R. (Chapter 7, this volume) The function of carnivore latrines: review, case studies and a research framework for hypothesis-testing. In: *Small Carnivores: Evolution, Ecology, Behaviour, and Conservation* (eds E. Do Linh San, J.J. Sato, J.L. Belant & M.J. Somers), Wiley–Blackwell, Oxford.

Buskirk, S.W. & MacDonald, S.O. (1984) Seasonal food habits of marten in south-central Alaska. *Canadian Journal of Zoology* 62, 944–950.

Bustamante, R.O., Simonetti, J.A. & Mella, J.E. (1992) Are foxes legitimate and efficient seed dispersers? A field test. *Acta Oecologica* 13, 203–208.

Cain, M.L., Milligan, B.G. & Strand, A.E. (2000) Long-distance seed dispersal in plant populations. *American Journal of Botany* 87, 1217–1227.

Campos, C. & Ojeda, R. (1997) Dispersal and germination of *Prosopis flexuosa* (Fabaceae) seeds by desert mammals in Argentina. *Journal of Arid Environments* 35, 707–714.

Cancio, I., González-Robles, A., Bastida, J.M., Manzaneda, A.J., Salido, T. & Rey, P.J. (2016) Habitat loss exacerbates regional extinction risk of the keystone semiarid shrub *Ziziphus lotus* through collapsing the seed dispersal service by foxes (*Vulpes vulpes*). *Biodiversity and Conservation* 25, 693–709.

Cancio, I., González-Robles, A., Bastida, J.M., Isla, J., Manzaneda, A.J., Salido, T. & Rey, P.J. (2017) Landscape degradation affects red fox (*Vulpes vulpes*) diet and its ecosystem services in the threatened *Ziziphus lotus* scrubland habitats of semiarid Spain. *Journal of Arid Environments* 145, 24–34.

Carbone, C., Mace, G.M., Roberts, S.C. & Macdonald, D.W. (1999) Energetic constraints on the diet of terrestrial carnivores. *Nature* 402, 286–288.

Carvalho-Ricardo, M.C., Uieda, W., Fonseca, R.C.B. & Rossi, M.N. (2014) Frugivory and the effects of ingestion by bats on the seed germination of three pioneering plants. *Acta Oecologica* 55, 51–57.

Castro, S.A., Silva, S.I., Meserve, P.L., Gutiérrez, J.R., Contreras, L.C. & Jaksic, F.M. (1994) [Frugivory and seed dispersal by culpeo fox (*Pseudalopex culpaeus*) in Fray Jorge National Park (IV Region, Chile)]. *Revista Chilena de Historia Natural* 67, 169–176. (In Spanish with English abstract).

Cavallini, P. & Lovari, S. (1994) Home range, habitat selection and activity of the red fox in a Mediterranean coastal ecotone. *Acta Theriologica* 39, 279–279.

Celedón-Neghme, C., Traveset, A. & Calviño-Cancela, M. (2013) Contrasting patterns of seed dispersal between alien mammals and native lizards in a declining plant species. *Plant Ecology* 214, 657–667.

Chakravarthy, D. & Ratnam, J. (2015) Seed dispersal of *Vitex glabrata* and *Prunus ceylanica* by civets (Viverridae) in Pakke Tiger Reserve, north-east India: spatial patterns and post-dispersal seed fates. *Tropical Conservation Science* 8, 491–504.

Chapman, C.A. (1989) Primate seed dispersal: the fate of dispersed seeds. *Biotropica* 21, 148–154.

Charles-Dominique, P. (1978) Écologie et vie sociale de *Nandinia binotata* (Carnivores, Viverridés): comparaison avec les prosimiens sympatriques du Gabon. *Revue d'Écologie – La Terre et la Vie* 32, 477–528. (In French with English summary).

Chávez-Ramírez, F. & Slack, R.D. (1993) Carnivore fruit-use and seed dispersal of two selected plant species of the Edwards Plateau, Texas. *The Southwestern Naturalist* 38, 141–145.

Clergeau, P. (1992) The effect of birds on seed germination of fleshy-fruited plants in temperate farmland. *Acta Oecologica* 13, 679–686.

Clevenger, A.P. (1996) Frugivory of *Martes martes* and *Genetta genetta* in an insular Mediterranean habitat. *Revue d'Écologie* 51, 19–28.

Colon, C.P. & Campos-Arceiz, A. (2013) The impact of gut passage by binturongs (*Arctictis Binturong*) on seed germination. *Raffles Bulletin of Zoology* 61, 657–667.

Connell, J.H. (1978) Diversity in tropical rain forests and coral reefs. *Science* 199, 1302–1310.

Corlett, R.T. (1998) Frugivory and seed dispersal by vertebrates in the Oriental (Indomalayan) Region. *Biological Reviews* 73, 413–448.

Corlett, R.T. (2009) Seed dispersal distances and plant migration potential in tropical East Asia. *Biotropica* 41, 592–598.

Corlett, R.T. (2011) How to be a frugivore (in a changing world). *Acta Oecologica* 37, 674–681.

Corlett, R.T. (2017) Frugivory and seed dispersal by vertebrates in tropical and subtropical Asia: an update. *Global Ecology and Conservation* 11, 1–22.

Corlett, R. & Lucas, P. (1990) Alternative seed-handling strategies in primates: seed-spitting by long-tailed macaques (*Macaca fascicularis*). *Oecologia* 82, 166–171.

Crooks, K.R. (2002) Relative sensitivities of mammalian carnivores to habitat fragmentation. *Conservation Biology* 16, 488–502.

Crooks, K.R. & Soulé, M.E. (1999) Mesopredator release and avifaunal extinctions in a fragmented system. *Nature* 400, 563–566.

Cypher, B.L. & Cypher, E.A. (1999) Germination rates of tree seeds ingested by coyotes and raccoons. *The American Midland Naturalist* 142, 71–76.

David, J.P., Manakadan, R. & Ganesh, T. (2015) Frugivory and seed dispersal by birds and mammals in the coastal tropical dry evergreen forests of southern India: a review. *Tropical Ecology* 56, 41–55.

Debussche, M. & Isenmann, P. (1989) Fleshy fruit characters and the choices of bird and mammal seed dispersers in a Mediterranean region. *Oikos* 56, 327–338.

D'hondt, B., Vansteenbrugge, L., Van Den Berge, K., Bastiaens, J. & Hoffmann, M. (2011) Scat analysis reveals a wide set of plant species to be potentially dispersed by foxes. *Plant Ecology and Evolution* 144, 106–110.

Duckworth, J.W. & Nettelbeck, A.R. (2008) Observations of small toothed palm civets *Arctogalidia trivirgata* in Khao Yai National Park, Thailand, with notes on feeding technique. *Natural History Bulletin of the Siam Society* 55, 187–192.

Engel, T. (1998) Seeds on the roundabout – tropical forest regeneration by *Genetta rubiginosa*. *Small Carnivore Conservation* 19, 13–20.

Engel, T.R. (2000) *Seed Dispersal and Forest Regeneration in a Tropical Lowland Biocoenosis (Shimba Hills, Kenya).* PhD thesis, University of Bayreuth/Logos Verlag, Berlin.

Escribano-Ávila, G., Sanz-Pérez, V., Pías, B., Virgós, E., Escudero, A. & Valladares, F. (2012) Colonization of abandoned land by *Juniperus thurifera* is mediated by the interaction of a diverse dispersal assemblage and environmental heterogeneity. *PLoS One* 7, e46993.

Escribano-Ávila, G., Pías, B., Sanz-Pérez, V., Virgós, E., Escudero, A. & Valladares, F. (2013) Spanish juniper gain expansion opportunities by counting on a functionally diverse dispersal assemblage community. *Ecology and Evolution* 3, 3751–3763.

Escribano-Ávila, G., Calviño-Cancela, M., Pías, B., Virgós, E., Valladares, F. & Escudero, A. (2014) Diverse guilds provide complementary dispersal services in a woodland expansion process after land abandonment. *Journal of Applied Ecology* 51, 1701–1711.

Escribano-Ávila, G. (2019) Non-specialized frugivores as key seed dispersers in dry disturbed environments: an example with a generalist neotropical mesocarnivore. *Journal of Arid Environments* 167, 18–25.

Espírito-Santo, C., Rosalino, L.M. & Santos-Reis, M. (2007) Factors affecting the placement of common genet latrine sites in a Mediterranean landscape in Portugal. *Journal of Mammalogy* 88, 201–207.

Evans, A.R., Wilson, G.P., Fortelius, M. & Jernvall, J. (2007) High-level similarity of dentitions in carnivorans and rodents. *Nature* 445, 78–81.

Farris, E., Canopoli, L., Cucca, E., Landi, S., Maccioni, A. & Filigheddu, R. (2017) Foxes provide a direct dispersal service to Phoenician junipers in Mediterranean coastal environments: ecological and evolutionary implications. *Plant Ecology and Evolution* 150, 117–128.

Fedriani, J.M. & Delibes, M. (2009a) Functional diversity in fruit-frugivore interactions: a field experiment with Mediterranean mammals. *Ecography* 32, 983–992.

Fedriani, J.M. & Delibes, M. (2009b) Seed dispersal in the Iberian pear, *Pyrus bourgaeana*: a role for infrequent mutualists. *Ecoscience* 16, 311–321.

Fedriani, J.M. & Delibes, M. (2011) Dangerous liaisons disperse the Mediterranean dwarf palm: fleshy-pulp defensive role against seed predators. *Ecology* 92, 304–315.

Fedriani, J.M. & Wiegand, T. (2014) Hierarchical mechanisms of spatially contagious seed dispersal in complex seed–disperser networks. *Ecology* 95, 514–526.

Fedriani, J.M., Wiegand, T. & Delibes, M. (2010) Spatial pattern of adult trees and the mammal-generated seed rain in the Iberian pear. *Ecography* 33, 545–555.

Fedriani, J.M., Wiegand, T., Ayllón, D., Palomares, F., Suárez-Esteban, A. & Grimm, V. (2018) Assisting seed dispersers to restore oldfields: an individual-based model of the interactions among badgers, foxes and Iberian pear trees. *Journal of Applied Ecology* 55, 600–611.

Fleming, T.H. & Heithaus, E.R. (1981) Frugivorous bats, seed shadows, and the structure of a tropical forest. *Biotropica* 18, 307–318.

Fragoso, J.M., Silvius, K.M. & Correa, J.A. (2003) Long-distance seed dispersal by tapirs increases seed survival and aggregates tropical trees. *Ecology* 84, 1998–2006.

Fuzessy, L.F., Cornelissen, T.G., Janson, C. & Silveira, F.A.O. (2016) How do primates affect seed germination? A meta-analysis of gut passage effects on neotropical plants. *Oikos* 125, 1069–1080.

García-Cervigón, A.I., Żywiec, M., Delibes, M., Suárez-Esteban, A., Perea, R. & Fedriani, J.M. (2018) Microsites of seed arrival: spatio–temporal variations in complex seed-disperser networks. *Oikos* 127, 1001–1013.

Gelmi-Candusso, T.A., Bialozyt, R., Slana, D., Zárate Gómez, R., Heymann, E.W. & Heer, K. (2019) Estimating seed dispersal distance: a comparison of methods using animal movement and plant genetic data on two primate-dispersed Neotropical plant species. *Ecology and Evolution* 9, 8965–8977.

Godínez-Alvarez, H., Ríos-Casanovas, L. & Peco, B. (2020) Are large frugivorous birds better seed dispersers than medium- and small-sized ones? Effect of body mass on seed dispersal effectiveness. *Ecology and Evolution* 10, 6136–6143.

González-Varo, J.P., López-Bao, J.V. & Guitián, J. (2013) Functional diversity among seed dispersal kernels generated by carnivorous mammals. *Journal of Animal Ecology* 82, 562–571.

González-Varo, J.P., López-Bao, J.V. & Guitián, J. (2017) Seed dispersers help plants to escape global warming. *Oikos* 126, 1600–1606.

Gorman, M.L. & Trowbridge, B.J. (1989) The role of odor in the social lives of carnivores. In: *Carnivore Behavior, Ecology, and Evolution* (ed. J.L. Gittleman), pp. 57–88. Cornell University Press, Ithaca.

Graae, B.J., Pagh, S. & Bruun, H.H. (2004) An experimental evaluation of the Arctic fox (*Alopex Iagopus*) as a seed disperser. *Arctic, Antarctic, and Alpine Research* 36, 468–473.

Grünewald, C., Breitbach, N. & Böhning-Gaese, K. (2010) Tree visitation and seed dispersal of wild cherries by terrestrial mammals along a human land-use gradient. *Basic and Applied Ecology* 11, 532–541.

Guitián, J. & Munilla, I. (2010) Responses of mammal dispersers to fruit availability: rowan (*Sorbus aucuparia*) and carnivores in mountain habitats of northern Spain. *Acta Oecologica* 36, 242–247.

Hämäläinen, A., Broadley, K., Droghini, A., Haines, J.A., Lamb, C.T., Boutin, S. & Gilbert, S. (2017) The ecological significance of secondary seed dispersal by carnivores. *Ecosphere* 8, e01685.

Herrera, C.M. (1989) Frugivory and seed dispersal by carnivorous mammals, and associated fruit characteristics, in undisturbed Mediterranean habitats. *Oikos*, 55, 250–262.

Herrera, J.M., Sá Teixeira, I., Rodríguez-Pérez, J. & Mira, A. (2016) Landscape structure shapes carnivore-mediated seed dispersal kernels. *Landscape Ecology* 31, 731–743.

Hickey, J.R., Flynn, R.W., Buskirk, S.W., Gerow, K.G. & Willson, M.F. (1999) An evaluation of a mammalian predator, *Martes americana*, as a disperser of seeds. *Oikos*, 87, 499–508.

Hisano, M. & Deguchi, S. (2018) Reviewing frugivory characteristics of the Japanese marten (*Martes melampus*). *Zoology and Ecology* 28, 10–20.

Hovstad, K., Borvik, S. & Ohlson, M. (2009) Epizoochorous seed dispersal in relation to seed availability – an experiment with a red fox dummy. *Journal of Vegetation Science* 20, 455–464.

Howe, H.F. (1989) Scatter- and clump-dispersal and seedling demography: hypothesis and implications. *Oecologia* 79, 417–426.

Howe, H.F. & Smallwood, J. (1982) Ecology of seed dispersal. *Annual Review of Ecology and Systematics*, 13, 201–228.

Hutchings, M.R. & White, P.C. (2000) Mustelid scent-marking in managed ecosystems: implications for population management. *Mammal Review* 30, 157–169.

Janzen, D.H. (1970) Herbivores and the number of tree species in tropical forests. *American Naturalist* 104, 501–528.

Janzen, D.H. (1971) Seed predation by animals. *Annual Review of Ecology and Systematics* 2, 465–492.

Janzen, D.H. (1982) Removal of seeds from horse dung by tropical rodents: influence of habitat and amount of dung. *Ecology*, 63, 1887–1900.

Jordan, N.R., Cherry, M.I. & Manser, M.B. (2007) Latrine distribution and patterns of use by wild meerkats: implications for territory and mate defence. *Animal Behaviour* 73, 613–622.

Jordano, P. (1987) Patterns of mutualistic interactions in pollination and seed dispersal: connectance, dependence asymmetries, and coevolution. *American Naturalist* 129, 657–677.

Jordano, P. (2014) Fruits and frugivory. In: *Seeds: The Ecology of Regeneration in Plant Communities*. 3rd edition (ed. R.S. Gallagher), pp. 105–156. Commonwealth Agricultural Bureaux International, Wallingford.

Jordano, P. (2017) What is long-distance dispersal? And a taxonomy of dispersal events. *Journal of Ecology* 105, 75–84.

Jordano, P. & Schupp, E.W. (2000) Seed disperser effectiveness: the quantity component and patterns of seed rain for *Prunus mahaleb*. *Ecological Monographs* 70, 591–615.

Jordano, P., Garcia, C., Godoy, J.A. & García-Castaño, J.L. (2007) Differential contribution of frugivores to complex seed dispersal patterns. *Proceedings of the National Academy of Sciences USA* 104, 3278–3282.

Kamler, J.F., Klare, U. & Macdonald, D.W. (2020) Seed dispersal potential of jackals and foxes in semi-arid habitats of South Africa. *Journal of Arid Environments* 183, 104284.

Kato, S., Nasu, Y. & Hayashida, M. (2000) Seed dispersal and fruit morphology of endozoochorous plants dispersed by raccoon dogs, *Nyctereutes procyonoides*. *Tohoku Journal of Forest Science* 5, 9–15. (In Japanese with English summary).

Koike, S. & Masaki, T. (2019) Characteristics of fruits consumed by mammalian frugivores in Japanese temperate forest. *Ecological Research* 34, 246–254.

Koike, S., Morimoto, H., Goto, Y., Kozakai, C. & Yamazaki, K. (2008) Frugivory of carnivores and seed dispersal of fleshy fruits in cool-temperate deciduous forests. *Journal of Forest Research* 13, 215–222.

Koike, S., Morimoto, H., Kozakai, C., Arimoto, I., Soga, M., Yamazaki, K. & Koganezawa, M. (2012a) The role of dung beetles as a secondary seed disperser after dispersal by frugivore mammals in a temperate deciduous forest. *Acta Oecologica* 41, 74–81.

Koike, S., Morimoto, H., Kozakai, C., Arimoto, I., Yamazaki, K., Iwaoka, M., Soga, M. & Koganezawa, M. (2012b) Seed removal and survival in Asiatic black bear *Ursus thibetanus* faeces: effect of rodents as secondary seed dispersers. *Wildlife Biology* 18, 24–34.

Kurek, P. (2015) Consumption of fleshy fruit: are central European carnivores really less frugivorous than southern European carnivores? *Mammalian Biology* 80, 527–534.

Lambert, J.E. (1999) Seed handling in chimpanzees (*Pan troglodytes*) and redtail monkeys (*Cercopithecus ascanius*): implications for understanding hominoid and cercopithecine fruit-processing strategies and seed dispersal. *American Journal of Physical Anthropology* 109, 365–386.

Lambert, J.E., Fellner, V., McKenney, E. & Hartstone-Rose, A. (2014) Binturong (*Arctictis binturong*) and Kinkajou (*Potos flavus*) digestive strategy: implications for interpreting frugivory in Carnivora and Primates. *PLoS One* 9, e105415.

León-Lobos, P.M. & Kalin-Arroyo, M.T. (1994) [Germination of *Lithrea caustica* (Mol.) H. and *Aristotelia chilensis* (Anacardiaceae) seeds dispersed by *Pseudalopex* spp. (Canidae) in the Chilean Matorral]. *Revista Chilena de Historia Natural* 67, 59–64. (In Spanish with English abstract).

Levin, S.A., Muller-Landau, H.C., Nathan, R. & Chave, J. (2003) The ecology and evolution of seed dispersal: a theoretical perspective. *Annual Review of Ecology, Evolution, and Systematics* 34, 575–604.

Lima, R.E.M., Dechoum, S. & Castellani, T.T. (2015) Native seed dispersers may promote the spread of the invasive Japanese raisin tree (*Hovenia dulcis* Thunb.) in seasonal deciduous forest in southern Brazil. *Tropical Conservation Science* 8, 846–862.

LoGiudice, K. & Ostfeld, R. (2002) Interactions between mammals and trees: predation on mammal-dispersed seeds and the effect of ambient food. *Oecologia* 130, 420–425.

Lopez, J.E. & Vaughan, C. (2004) Observations on the role of frugivorous bats as seed dispersers in Costa Rican secondary humid forests. *Acta Chiroptera* 6, 111–119.

López-Bao, J.V. & González-Varo, J.P. (2011) Frugivory and spatial patterns of seed deposition by carnivorous mammals in anthropogenic landscapes: a multi-scale approach. *PLoS One* 6(1), e14569.

López-Bao, J.V., González-Varo, J.P. & Guitián J. (2015) Mutualistic relationships under landscape change: carnivorous mammals and plants after 30 years of land abandonment. *Basic and Applied Ecology* 16, 152–161.

Low, M.R. & Vogrinc, P.N. (2017) An alternate form of seed spitting for Masked Palm Civet *Paguma larvata*. *Southeast Asia Vertebrate Records* 2017, 7–8.

Lucas, P.W. & Corlett, R.T. (1998) Seed dispersal by long-tailed macaques. *American Journal of Primatology* 45, 29–44.

Macdonald, D.W. (1980) Patterns of scent marking with urine and faeces amongst carnivore communities. *Symposia of the Zoological Society of London* 45, 107–139.

Maldonado, D.E., Pacheco, L.F. & Saavedra, L.V. (2014) [Legitimacy of algorrobo (*Prosopis flexuosa*, Fabaceae) seed dispersal by Andean fox (*Lycalopex culpaeus*,

Canidae) in the valley of La Paz (Bolivia)]. *Ecologia en Bolivia* 49, 93–97. (In Spanish).

Maldonado, D.E., Loayza, A.P., Garcia, E. & Pacheco, L.F. (2018) Qualitative aspects of the effectiveness of Culpeo foxes (*Lycalopex culpaeus*) as dispersers of *Prosopis alba* (Fabaceae) in a Bolivian dry valley. *Acta Oecologica* 87, 29–33.

Malo, J.E. & Suárez, F. (1995) Establishment of pasture species on cattle dung: the role of endozoochorous seeds. *Journal of Vegetation Science* 6, 169–174.

Mandujano, M.C., Golubov, J. & Montana, C. (1997) Dormancy and endozoochorous dispersal of *Opuntia rastrera* seeds in the southern Chihuahuan Desert. *Journal of Arid Environments* 36, 259–266.

Martínez, I., García, D. & Obeso, J.R. (2008) Differential seed dispersal patterns generated by a common assemblage of vertebrate frugivores in 3 fleshy-fruited trees. *Ecoscience* 15, 189–199.

Matías, L., Zamora, R., Mendoza, I. & Hódar, J.A. (2010) Seed dispersal patterns by large frugivorous mammals in a degraded mosaic landscape. *Restoration Ecology* 18, 619–627.

McGrosky, A., Navarrete, A., Isler, K., Langer, P. & Clauss, M. (2016) Gross intestinal morphometry and allometry in Carnivora. *European Journal of Wildlife Research* 62, 395–405.

McIntosh, D. (1963) Food of the fox in the Canberra district. *Wildlife Research* 8, 1–20.

Mise, Y., Yamazaki, K., Soga, M. & Koike, S. (2016) Comparing methods of acquiring mammalian endozoochorous seed dispersal distance distributions. *Ecological Research* 31, 881–889.

Morales-Paredes, C., Valdivia, C.E. & Sade, S. (2015) [Frugivory by native (*Lycalopex* spp.) and allochthonous (*Canis lupus familiaris*) canids reduces the seed germination of litre (*Lithrea caustica*) in central Chile]. *Bosque* 36, 481–486. (In Spanish with English abstract).

Mudappa, D., Kumar, A. & Chellam, R. (2010) Diet and fruit choice of the brown palm civet *Paradoxurus jerdoni*, a viverrid endemic to the Western Ghats rainforest, India. *Tropical Conservation Science* 3, 282–300.

Murdoch, J.D., Buyandelger, S. & Cypher, B.L. (2009) Patterns of seed occurrence in corsac and red fox diets in Mongolia. *Journal of Arid Environments* 73, 381–384.

Nakabayashi, M., Inoue, Y., Ahmad, A.H. & Izawa, M. (2019) Limited directed seed dispersal in the canopy as one of the determinants of the low hemi-epiphytic figs' recruitments in Bornean rainforests. *PLoS One* 14, e0217590.

Nakashima, Y. & Sukor, J.A. (2010) Importance of common palm civets (*Paradoxurus hermaphroditus*) as a long-distance disperser for large-seeded plants in degraded forests. *Tropics* 18, 221–229.

Nakashima, Y. & Sukor, J.A. (2013) Space use, habitat selection, and day-beds of the common palm civet (*Paradoxurus hermaphroditus*) in human-modified habitats in Sabah, Borneo. *Jounal of Mammalogy* 94, 1169–1178.

Nakashima, Y., Lagan, P. & Kitayama, K. (2008) A study of fruit–frugivore interactions in two species of durian (*Durio*, Bombacaceae) in Sabah, Malaysia. *Biotropica* 40, 255–258.

Nakashima, Y., Inoue, E., Inoue-Murayama, M. & Sukor, J.A. (2010a) Functional uniqueness of a small carnivore as seed dispersal agents: a case study of the common palm civets in the Tabin Wildlife Reserve, Sabah, Malaysia. *Oecologia* 164, 721–730.

Nakashima, Y., Inoue, E., Inoue-Murayama, M. & Sukor, J.A. (2010b) High potential of a disturbance-tolerant frugivore, the common palm civet *Paradoxurus hermaphroditus* (Viverridae), as a seed disperser for large-seeded plants. *Mammal Study* 35, 209–215.

Naoe, S., Tayasu I., Masaki, T. & Koike, S. (2016a) Negative correlation between altitudes and oxygen isotope ratios of seeds: exploring its applicability to assess vertical seed dispersal. *Ecology and Evolution* 6, 6817–6823.

Naoe, S., Tayasu, I., Sakai, Y., Masaki, T., Kobayashi, K., Nakajima, A., Sato, Y., Yamazaki, K., Kiyokawa, K. & Koike, S. (2016b) Mountain-climbing bears protect cherry species from global warming through vertical seed dispersal. *Current Biology* 26, R315–R316.

Naoe, S., Tayasu, I., Sakai, Y., Masaki, T., Kobayashi, K., Nakajima, A., Sato, Y., Yamazaki, K., Kiyokawa, H. & Koike, S. (2019) Downhill seed dispersal by temperate mammals: a potential threat to plant escape from global warming. *Scientific Reports* 9, 1–11.

Nathan, R., Schurr, F.M., Spiegel, O., Steinitz, O., Trakhtenbrot, A. & Tsoar, A. (2008) Mechanisms of long-distance seed dispersal. *Trends in Ecology & Evolution* 23, 638–647.

Nogales, M., Medina, F.M. & Valido, A. (1996) Indirect seed dispersal by the feral cats *Felis catus* in island ecosystems (Canary Islands). *Ecography* 19, 3–6.

Ortiz, P.L., Arista, M. & Talavera, S. (1995) Germination ecology of *Ceratonia silique* L. (Caesalpiniaceae), a Mediterranean tree. *Flora* 190, 89–95.

Otani, T. (2002) Seed dispersal by Japanese marten *Martes melampus* in the subalpine shrubland of northern Japan. *Ecological Research* 17, 29–38.

Otani, T. (2005) [Characteristics of middle-size mammals as a seed disperser of fleshy-plants]. *Nagoya University Forest Science Research* 24, 7–43. (In Japanese with English summary).

Padrón, B., Nogales, M., Traveset, A., Vila, M., Martínez-Abraín, A., Padilla, D.P. & Marrero, P. (2011) Integration of invasive *Opuntia* spp. by native and alien seed dispersers in the Mediterranean area and the Canary Islands. *Biological Invasions* 13, 831–844.

Page, L.K., Swihart, R.K. & Kazacos, K.R. (1998) Raccoon latrine structure and its potential role in transmission of *Baylisascaris procyonis* to vertebrates. *The American Midland Naturalist* 140, 180–185.

Paulino-Neto, H., Nakano-Oliveira, E., De Assis Jardim, M.M. & Vasconcellos-Neto, J. (2016) Frugivory in *Bromelia balansae* (Bromeliaceae): the effect of seed passage through the digestive system of potential seed dispersers on germination in an Atlantic Rainforest, Brazil. *Journal of Ecosystem and Ecography* 6(4), 224.

Pendje, G. (1994) La frugivorie de *Civettictis civetta* (Schreiber) et son rôle dans la dispersion des graines au Mayombe. *Revue d'Écologie – La Terre et la Vie* 49, 107–116. (In French with English abstract).

Perea, R., Delibes, M., Polko, M., Suárez-Esteban, A. & Fedriani, J.M. (2013) Context-dependent fruit–frugivore interactions: partner identities and spatio-temporal variations. *Oikos* 122, 943–951.

Peredo, A., Martínez, D., Rodríguez-Pérez, J. & García, D. (2013) Mammalian seed dispersal in Cantabrian woodland pastures: network structure and response to forest loss. *Basic and Applied Ecology* 14, 378–386.

Pereira, L., Vasques, A., Maia, P., Ramos Pereira, M.J., Fonseca, C. & Matos, M. (2019) Native and exotic seed dispersal by the stone marten (*Martes foina*): implications for the regeneration of a relict climactic forest in central Portugal. *Integrative Zoology* 14, 280–292.

Pérez-Méndez, N. & Rodríguez, A. (2018) Raptors as seed dispersers. In: *Birds of Prey* (eds J.H. Sarasola, J. Grande & J. Negro), pp. 139–158. Springer, Cham.

Petersen, J.L., Ueckert, D.N., Taylor, C.A. & Shaffer, K.R. (2005) Germination of redberry juniper (*Juniperus pinchotii*) seed in western Texas. *Texas Journal of Agriculture and Natural Research* 18, 28–30.

Pigozzi, G. (1992) Frugivory and seed dispersal by the European badger in a Mediterranean habitat. *Journal of Mammalogy* 73, 630–639.

Quadros, J. & Monteiro-Filho, E.L. (2000) Fruit occurrence in the diet of the Neotropical otter, *Lontra longicaudis* in southern Brazilian Atlantic forest and its implication for seed dispersion. *Mastozoología Neotropical* 7, 33–36.

Rabinowitz, A. (1991) Behaviour and movements of sympatric civet species in Huai Kha Khaeng Wildlife Sanctuary, Thailand. *Journal of Zoology* 223, 281–298.

Rebein, M., Davis, C.N., Abad, H., Stone, T., Sol, J.D., Skinner, N. & Moran, M.D. (2017) Seed dispersal of *Diospyros virginiana* in the past and the present: evidence for a generalist evolutionary strategy. *Ecology and Evolution* 7, 4035–4043.

Reid, N. (1989) Dispersal of mistletoes by honeyeaters and flowerpeckers: components of seed dispersal quality. *Ecology* 70, 137–145.

Rico-Guzmán, E., Terrones Contreras, B., Cantó Corchado, J.L. & Bonet Jornet, A. (2012) Frugivore carnivores: preferences and contribution to seed dispersal of red fox *Vulpes vulpes* (Linnaeus, 1758) and stone marten *Martes foina* (Erxleben, 1777) in Carrascal de la Font Roja Natural Park (Alicante, Spain). *Galemys* 24, 25–33.

Ridley, H.N. (1930) *The Dispersal of Seeds Throughout the World*. Reeve, Ashford.

Riera, N., Traveset, A. & García, O. (2002) Breakage of mutualisms by exotic species: the case of *Cneorum tricoccon* L. in the Balearic Islands (Western Mediterranean Sea). *Journal of Biogeography* 29, 713–719.

Ripple, W.J. & Beschta, R.L. (2004) Wolves and the ecology of fear: can predation risk structure ecosystems? *BioScience* 54, 755–766.

Ritchie, E.G. & Johnson, C.N. (2009) Predator interactions, mesopredator release and biodiversity conservation. *Ecology Letters* 12, 982–998.

Rivadeneira-Canedo, C. (2008) Study of the Andean bear (*Tremarctos ornatus*) as a legitimate seed disperser and elements of its diet in the region of Apolobamba-Bolivia. *Ecología en Bolivia* 43, 29–40. (In Spanish with English abstract).

Robertson, A.W., Trass, A., Ladley, J.J. and Kelly, D. (2006) Assessing the benefits of frugivory for seed germination: the importance of the deinhibition effect. *Functional Ecology* 20, 58–66.

Roehm, K. & Moran, M.D. (2013) Is the coyote (*Canis latrans*) a potential seed disperser for the American persimmon (*Diospyros virginiana*)? *The American Midland Naturalist* 169, 416–421.

Roemer, G.W., Gompper, M.E. & Van Valkenburgh, B. (2009) The ecological role of the mammalian mesocarnivore. *BioScience* 59, 165–173.

Rosalino, L.M. & Santos-Reis, M. (2009) Fruit consumption by carnivores in Mediterranean Europe. *Mammal Review* 39, 67–78.

Rosalino, L.M., Rosa, S. & Santos-Reis, M. (2010) The role of carnivores as Mediterranean seed dispersers. *Annales Zoologici Fennici*, 47:195–205.

Rost, J., Pons, P. & Bas, J.M. (2012) Seed dispersal by carnivorous mammals into burnt forests: an opportunity for non-indigenous and cultivated plant species. *Basic and Applied Ecology* 13, 623–630.

Rubalcava-Castillo, F.A., Sosa-Ramírez, J., Luna-Ruíz, J.J., Valdivia-Flores, A.G., Díaz-Núñez, V., Íñiguez-Dávalos, L.I. (2020) Endozoochorous dispersal of forest seeds by carnivorous mammals in Sierra Fría, Aguascalientes, Mexico. *Ecology and Evolution* 10, 2991–3003.

Russo, S.E. & Chapman, C.A. (2011) Primate seed dispersal: linking behavioral ecology with forest community structure. *Primates in Perspective* 2, 523–534.

Sakamoto, Y. & Takatsuki, S. (2015) Seeds recovered from the droppings at latrines of the raccoon dog (*Nyctereutes procyonoides viverrinus*): the possibility of seed dispersal. *Zoological Science* 32, 157–162.

Saldaña-Vázquez, R.A., Castaño, J.H., Baldwin, J. & Pérez-Torres, J. (2019) Does seed ingestion by bats enhance germination? A new meta-analysis 15 years later. *Mammal Review* 49, 201–209.

Salgueiro, V., Silva, C., Eufrázio, S., Salgueiro, P.A. & Vaz, P.G. (2020) Endozoochory of a dry-fruited tree aids quarry passive restoration and seed soaking further increases seedling emergence. *Restoration Ecology* 28, 668–678.

Samuels, I.A. & Levey, D.J. (2005) Effects of gut passage on seed germination: do experiments answer the questions they ask? *Functional Ecology* 19, 365–368.

Santos, T., Telleria, J.L. & Virgós, E. (1999) Dispersal of Spanish juniper *Juniperus thurifera* by birds and mammals in a fragmented landscape. *Ecography* 22, 193–204.

Sarasola, J.H., Zanón-Martínez, J.I., Costán, A.S. & Ripple, W.J. (2016) Hypercarnivorous apex predator could provide ecosystem services by dispersing seeds. *Scientific Reports* 6, 19647.

Schaumann, F. & Heinken, T. (2002) Endozoochorous seed dispersal by martens (*Martes foina, M. martes*) in two woodland habitats. *Flora* 197, 370–378.

Schupp, E.W., Gómez, J.M., Jiménez, J.E. & Fuentes, M. (1997) Dispersal of *Juniperus occidentalis* (western juniper) seeds by frugivorous mammals on Juniper Mountain, southeastern Oregon. *The Great Basin Naturalist* 57, 74–78.

Schurr, F.M., Spiegel, O., Steinitz, O., Trakhtenbrot, A., Tsoar, A. & Nathan, R. (2009) Long-distance seed dispersal. *Annual Plant Reviews* 38, 204–237.

Sengupta, A., Gazagne, E., Albert-Daviaud, A., Tsuji, Y. & Radhakrishna, S. (2020) Reliability of macaques as seed dispersers. *American Journal of Primatology* 82(5), e23115.

Silva, S.I., Bozinovic, F. & Jaksic, F.M. (2005) Frugivory and seed dispersal by foxes in relation to mammalian prey abundance in a semiarid thornscrub. *Austral Ecology* 30, 739–746.

Silverstein, R.P. (2005) Germination of native and exotic plant seeds dispersed by coyotes (*Canis latrans*) in southern California. *The Southwestern Naturalist* 50, 472–478.

Soltani, E., Baskin, C.C., Baskin, J.M., Heshmati, S. & Mirfazeli, M.S. (2018) A meta-analysis of the effects of frugivory (endozoochory) on seed germination: role of seed size and kind of dormancy. *Plant Ecology* 219, 1283–1294.

Spennemann, D.H.R. (2020) The role of canids in the dispersal of commercial and ornamental palm species. *Mammal Research* 66, 57–74.

Stevens, C.E. & Hume, I.D. (1998) Contributions of microbes in vertebrate gastrointestinal tract to production and conservation of nutrients. *Physiological Reviews* 78, 393–427.

Suárez-Esteban, A., Delibes, M. & Fedriani, J.M. (2013a) Barriers or corridors? The overlooked role of unpaved roads in endozoochorous seed dispersal. *Journal of Applied Ecology* 50, 767–774.

Suárez-Esteban, A., Delibes, M. & Fedriani, J.M. (2013b) Unpaved road verges as hotspots of fleshy-fruited shrub recruitment and establishment. *Biological Conservation* 167, 50–56.

Szuman, J. & Skrzydlewski, A. (1962) Über die Durchgangszeit des Futters durch den Magen-Darm-Kanal beim Blaufuchs. *Archives of Animal Nutrition* 12, 1–4. [In German].

Torres, D.A., Castaño, J.H. & Carranza-Quiceno, J.A. (2020) Global patterns of seed germination after ingestion by mammals. *Mammal Review* 28, 668–678.

Traba, J., Sagrario, A., Herranz, J. & Clamagirand, M.C. (2006) Red fox (*Vulpes vulpes* L.) favour seed dispersal, germination and seedling survival of Mediterranean hackberry (*Celtis australis* L.). *Acta Oecologica* 30, 39–45.

Traveset, A. (1995) Seed dispersal of *Cneorum tricoccon* L. (Cneoraceae) by lizards and mammals in the Balearic Islands. *Acta Oecologica*, 16, 171–178.

Traveset, A. (1998) Effect of seed passage through vertebrate frugivores' guts on germination: a review. *Perspectives in Plant Ecology, Evolution and Systematics* 1, 151–190.

Traveset, A. & Verdú, M. (2002). A meta-analysis of the effect of gut treatment on seed germination. In: *Seed Dispersal and Frugivory: Ecology, Evolution and Conservation* (eds D.J. Levey, W.R. Silva & M. Galetti), pp. 339–350. Commonwealth Agricultural Bureaux International, Wallingford.

Traveset, A. & Willson, M.F. (1997) Effect of birds and bears on seed germination of fleshy-fruited plants in temperate rainforests of southeast Alaska. *Oikos*, 80, 89–95.

Traveset, A., Bermejo, T. & Willson, M. (2001a) Effect of manure composition on seedling emergence and growth of two common shrub species of Southeast Alaska. *Plant Ecology* 155, 29–34.

Traveset, A., Riera, N. & Mas, R.E. (2001b) Ecology of fruit-colour polymorphism in *Myrtus communis* and differential effects of birds and mammals on seed germination and seedling growth. *Journal of Ecology* 89, 749–760.

Traveset, A., Robertson, A.W. & Rodríguez-Pérez, J. (2007) A review on the role of endozoochory in seed germination. In: *Seed Dispersal: Theory and its Application in a Changing World* (eds A.J. Dennis, E.W. Schupp, R.J. Green & D.A. Westcott), pp. 78–103. Commonwealth Agricultural Bureaux International, Wallingford.

Traveset, A., Rodríguez-Pérez, J. & Pías, B. (2008) Seed trait changes in dispersers' guts and consequences for germination and seedling growth. *Ecology* 89, 95–106.

Traveset, A., González-Varo, J.P. & Valido, A. (2012) Long-term demographic consequences of a seed dispersal disruption. *Proceedings of the Royal Society of London B: Biological Sciences* 279, 3298–3303.

Traveset, A., Escribano-Avila, G., Gómez, J.M. & Valido, A. (2019) Conflicting selection on *Cneorum tricoccon* (Rutaceae) seed size caused by native and alien seed dispersers. *Evolution* 73, 2204–2215.

Tsuji, Y., Shiraishi, T. & Miura, S. (2011a) Gastrointestinal passage time of seeds ingested by captive Japanese martens *Martes melampus*. *Acta Theriologica* 56, 353–357.

Tsuji, Y., Tatewaki, T. & Kanda, E. (2011b) Endozoochorous seed dispersal by sympatric mustelids, *Martes melampus* and *Mustela itatsi*, in western Tokyo, central Japan. *Mammalian Biology* 76, 628–633.

Tsuji, Y., Miura, S., Kotoge, T., Shiraishi, T. & Murai, H. (2015) Effects of food intake on digesta passage time in captive Japanese martens (*Martes melampus*) and implications for endozoochorous seed dispersal. *Mammal Study* 40, 13–18.

Tsuji, Y., Okumura, T., Kitahara, M. & Jiang, Z. (2016) Estimated seed shadow generated by Japanese martens (*Martes melampus*): comparison with forest-dwelling animals in Japan. *Zoological Science* 33, 352–357.

Tsunamoto, Y., Naoe, S., Masaki, T. & Isagi, Y. (2020) Different contributions of birds and mammals to seed dispersal of a fleshy-fruited tree. *Basic and Applied Ecology* 43, 66–75.

Twigg, L.E., Lowe, T.J. & Martin, G.R. (2009) The presence and implications of viable seed in the faeces of invasive free-ranging European rabbits and red foxes. *Pacific Conservation Biology* 15, 158–170.

Vander Wall, S.B. & Longland, W.S. (2004) Diplochory: are two seed dispersers better than one? *Trends in Ecology & Evolution* 19, 155–161.

Van Valkenburgh, B. (1989) Carnivore dental adaptations and diet: a study of trophic diversity within guilds. In: *Carnivore Behavior, Ecology, and Evolution. Volume 1* (ed. J.L. Gittleman), pp. 410–436. Cornell University Press, Ithaca.

Varela, O. & Bucher, E. (2006) Passage time, viability, and germination of seeds ingested by foxes. *Journal of Arid Environments* 67, 566–578.

Vasconcellos-Neto, J., Albuquerque, L.B. & Silva, W. R. (2009) Seed dispersal of *Solanum thomasiifolium* Sendtner (Solanaceae) in the Linhares Forest, Espírito Santo state, Brazil. *Acta Botanica Brasilica* 23, 1171–1179.

Vázquez-Yanes, C. & Orozco-Segovia, A. (1986) Dispersal of seeds by animals: effect on light controlled dormancy in *Cecropia obtusifolia*. In: *Frugivores and Seed Dispersal* (eds A. Estrada & T.H. Fleming), pp. 71–77. Dr W. Junk Publishers, Dordrecht.

Vergara-Tabares, D.L., Whitworth-Hulse, J.I. & Funes, G. (2018) Germination response of *Lithraea molleoides* seeds is similar after passage through the guts of several avian and a single mammalian disperser. *Botany* 96, 485–490.

Villalobos-Escalante, A., Buenrostro-Silva, A. & Sánchez-de la Vega, G. (2014) [Gray fox (*Urocyon cinereoargenteus*) diet and their contribution to seed dispersal on the coast of Oaxaca, Mexico]. *Therya* 5, 355–363. (In Spanish with English summary).

Willson, M.F. (1993) Mammals as seed-dispersal mutualists in North America. *Oikos* 65, 159–176.

Wilson, D.E. & Mittermeier, R.A. (eds) (2009) *Handbook of the Mammals of the World. Volume 1. Carnivores.* Lynx Edicions, Barcelona.

Wilson, J.A. & Thomas, B. (1999) Diet and seed dispersal efficiency of the gray fox (*Urocyon cinereoragenteus*) in Chaparral. *Bulletin of the Southern California Academy of Sciences* 98, 119–126.

Yasumoto, Y. & Takatsuki, S. (2015) The Japanese marten favors *Actinidia arguta*, a forest edge liane as a directed seed disperser. *Zoological Science* 32, 255–259.

Zarco-Mendoza, P., Ríos-Casanova, L. & Godínez-Álvarez, H. (2018) Dispersal and germination of seeds ingested by carnivores in the Zapotitlan de las Salinas Valley, Mexico. *Polibotánica* 46, 139–147.

Zhou, Y.B., Slade, E., Newman, C., Wang, X.M. & Zhang, S.Y. (2008a) Frugivory and seed dispersal by the yellow-throated marten, *Martes flavigula*, in a subtropical forest of China. *Journal of Tropical Ecology* 24, 219–223.

Zhou, Y.B., Zhang, L., Kaneko, Y., Newman, C. & Wang, X.M. (2008b) Frugivory and seed dispersal by a small carnivore, the Chinese ferret-badger, *Melogale moschata*, in a fragmented subtropical forest of central China. *Forest Ecology and Management* 255, 1595–1603.

Zhou, Y.B., Zhang, J., Slade, E., Zhang, L., Palomares, F., Chen, J., Wang, X. & Zhang, S. (2008c) Dietary shifts in relation to fruit availability among masked palm civets (*Paguma larvata*) in central China. *Journal of Mammalogy* 89, 435–447.

Zhou, Y.B., Newman, C., Xie, Z. & Macdonald, D.W. (2013) Peduncles elicit large-mammal endozoochory in a dry-fruited plant. *Annals of Botany* 112, 85–93.

Zidon, R., Leschner, H., Motro, U. & Saltz, D. (2017) Endozoochory by the Persian fallow deer (*Dama mesopotamica*) reintroduced in Israel: species richness and germination success. *Israel Journal of Ecology and Evolution* 63, 28–34.

Appendix 18.1 Overview[a] of native and exotic plant seed dispersal by mesocarnivores and resulting seed germination (when investigated). Detailed explanations are provided at the bottom of the table. For English names and/or selected attributes of focal mesocarnivores, including geographic distribution, see Table 18.2 and Appendix A.

Continent and country (region)	Mesocarnivore species studied	Plant species studied and/or dispersed[b]	Diet	Dep. sites	Exotic seeds	Germ. tests	Germ. effects	Main findings (Reference[s])
AFRICA								
Democratic Republic of the Congo (Mayombe district)	*Civettictis civetta*	*Antiaris welwitschii* *Dacryodes edulis* *Elaeis guineensis* *Ganophyllum giganteum* *Pycnanthus angolensis* *Staudtia stipitata* *Trilepisium madagascariense* Two undetermined species	Y_f	Y	N	N	NT	African civets dispersed the seeds of nine plant species. Small seeds (diameter <1 cm) were dispersed through endozoochory, whereas larger ones were dispersed through synzoochory. The average minimum dispersal distance was ~40 m from the parent tree. Seed germination rates and seedling mortality rates varied widely among species (Pendje, 1994).
Ethiopia (Tara Gedam)	*Civettictis civetta*	*Cordia africana* *Mimusops kummel* *Olea europaea* subsp. *cuspidata*	N	Y	N	N	NT	African civets were shown to be specialized frugivores, as out of 143 tree species present in Tara Gedam forest, the seeds of only three species were found in civet latrines (civetries). *Mimusops kummel* had particularly high germination rates after endozoochorous seed dispersal (Abrham Abiyu *et al.*, 2015).
Kenya (Shimba Hills National Reserve)	*Civettictis civetta* *Genetta maculata*[c] *Ichneumia albicauda* *Nandinia binotata*	At least 118 plant species	Y	Y	Y	Y	NT	African civets dispersed diaspores of at least 108 plant species (including the exotic *Thevetia peruviana*) to numerous latrine sites. Most seeds were viable and germinated well. Rusty-spotted genet and white-tailed mongoose scats contained seeds from 41 and 5 different plant species, respectively. When trapped, palm civet defecated *Vismia orientalis* seeds and ate various other wild fruits in feeding experiments (Engel, 1998, 2000).
South Africa (Benfontein and Rooipoort nature reserves)	*Canis mesomelas* *Otocyon megalotis* *Vulpes chama*	*Diospyros lycioides* *Prosopis* spp. *Ziziphus mucronota* *Grewia flava*	Y_f	Y	Y	N	NT	Black-backed jackals had the highest seed dispersal potential: they consumed all the listed fruit species, exhibited medium densities, had a relatively large seed shadow, and mostly selected good germination sites. Bat-eared foxes had high consumption of fruit per area, but their seed dispersal potential was low due to their small seed shadow and poor germination sites. Cape foxes had the largest seed shadow, but their seed dispersal potential was low because of low fruit consumption, low density, and poor germination sites (Kamler *et al.*, 2020).

(Continued)

Appendix 18.1 (Continued)

Continent and country (region)	Mesocarnivore species studied	Plant species studied and/or dispersed[b]	Diet	Dep. sites	Exotic seeds	Germ. tests	Germ. effects	Main findings (Reference[s])
ASIA								
China (Hubei Houhe)	*Martes flavigula*	*Celtis biondii* [+$_{gr}$] *Diospyros lotus* [+$_{gr}$] *Hovenia dulcis* [+$_{gr}$] *Kadsura longipedunculata* [−$_{gr}$] *Sorbus hemsleyi* [−$_{gr}$]	Y$_f$	N	Y	Y	Mix	Seeds from a total of 13 plant species were found in 43% of scats investigated. Yellow-throated martens selected multi-seeded fruit species with > 50% fruit pulp, and preferred black, khaki and yellow fruits. Seed-bearing faeces peaked during the fruiting season of preferred plant species (Zhou et al., 2008a).
China (Hubei Houhe)	*Melogale moschata*	*Actinidia chinensis* *Clematoclethra scandens* [dng] *Dendrobenthamia capitata* *Dendrobenthamia japonica* *Diospyros lotus* [−$_{gr}$] *Hovenia dulcis* [0$_{gr}$] *Prunus salicina* One unidentified species	N	Y	Y	Y	Mix	Chinese ferret badgers selected and dispersed the seeds of at least eight fleshy-fruited and seed-pulp rich plant species. Defecated seeds were viable, but faeces were mostly deposited at open sites that were unfavourable for seed germination and seedling establishment (Zhou et al., 2008b).
China (Hubei Houhe)	*Martes flavigula* (*Mf*) *Melogale moschata* (*Mm*) *Paguma larvata* (*Pl*) Three other mammal species	*Hovenia dulcis* [*Mf*: +$_{gr}$; *Mm*: +$_{gr}^{d}$; *Pl*: 0$_{gr}$]	N	Y	N	Y	Mix	Raisin tree seeds occurred in 15–28% of faecal samples from the three small carnivore species studied. The percentage of damaged seeds varied from 0.2 to 2.5% only (Zhou et al., 2013).
India (Pakke Tiger Reserve)	Undetermined civet species among *Arctictis binturong*, *Paguma larvata*, *Paradoxurus hermaphroditus*, *Viverra zibetha*, *Viverricula indica*	*Prunus ceylanica* *Vitex glabrata*	N	Y	N	N	NT	For both tree species, civets neither dispersed seeds far from fruiting trees (< 50 m) nor to sites (tree branches, forest floor, fallen logs) where seeds experienced either low predation or high survival. This suggests that while civets were legitimate dispersers, they were not especially efficient (Chakravarthy & Ratnam, 2015).
Israel (Soreq Valley Nature Reserve)	*Canis aureus palaestina* One other mammal species	*Ceratonia siliqua* [+/−$_{gr}$ +$_{gt}$]	N	N	N	Y	Mix	Carob seeds ingested by golden jackals had a higher germination rate than seeds within intact pods or seeds manually exposed by the researchers, but a lower germination rate than naturally exposed seeds, collected under trees. Digested seeds, however, germinated faster than seeds from all control treatments (Zidon et al., 2017).

Location	Carnivore species	Plant species					Notes
Japan (Yamagata Prefecture)	*Nyctereutes procyonoides*	*Actidinia arguta* *Hovenia dulcis* *Malus tchonoskii* *Trillium smallii* 31 other plant species	Y$_f$	N	N	NT	Seeds from a total of 35 plant species (19 with fleshy fruits) were found in ~91% of scats investigated. Some seedlings of the main plant species were found on some faecal pile sites. Fruits eaten by raccoon dogs were large and dull in colour, and were sweet or had a strong scent (Kato *et al.*, 2000).
Japan (Ohu mountain range)	*Martes melampus*	11 plant species	Y$_f$	N	N	NT	Japanese martens dispersed seeds of three herbaceous and eight woody plant species. The latter included a vine species which was dispersed over a relatively long distance. Species with larger and heavier fruit were selected (Otani, 2002).
Japan (Okutama)	*Martes melampus* *Meles anakuma*[e] *Nyctereutes procyonoides* *Vulpes vulpes* One large carnivore species	*Actinidia* spp. *Akebia* spp. *Prunus grayana* *Prunus jamasakura* *Prunus verecunda* *Rubus* spp. And 13 other fleshy-fruited plant taxa	Y$_f$	N	N	NT	Japanese martens consumed 12 fleshy-fruited plant species, badgers 9, raccoon dogs 10, and red foxes 8. Six of these plant species were dispersed by the five carnivores studied. The seeds of fleshy-fruited species found in all carnivore scats were largely intact, whereas no intact acorns or nuts were recovered (Koike *et al.*, 2008).
Japan (Bonbori Forest Path, western Tokyo)	*Martes melampus* *Mustela itatsi*	*Cinnamonum camphora* *Citrus junos* *Ilex macropoda* *Morus bombycis* *Physalis alkelengi* *Rubus* sp. *Stachyurus praecox* 21 other plant species	Y$_f$	N	N	NT	Seeds were found in 81% and 56% of *M. melampus* and *M. itatsi* faeces, respectively. Japanese martens dispersed seeds of five herbaceous and 23 woody plant species. Among those, Japanese weasels dispersed one herbaceous species and 16 woody plant species. Almost all seeds found within faecal samples were intact (Tsuji *et al.*, 2011b).
Japan (Hinode Town)	*Nyctereutes procyonoides*	*Eurya japonica* (52.6%) *Rubus* spp. (17.4%) *Solanum nigrum* (16.0%) 47 other plant taxa	Y$_f$	Y	N	NT	Seeds of 50 plant taxa (96% with fleshy fruits) were recovered from raccoon dog droppings collected over a year (see proportions next to plant species names). Most seeds (43.5%) were deposited within only 50 m of the foraging area. However, the seeds of forest plants can be dispersed to open areas (Sakamoto & Takatsuki, 2015).
Japan (Bonbori Valley, Akiruno City)	*Martes melampus*	*Actinidia arguta*	N	Y	N	NT	Japanese martens preferentially ate the fruits of forest edge plants (~96%), and more frequently that of *A. arguta*. The density of marten faeces was also higher at forest edges than inside the forest, suggesting that Japanese martens function as directed seed dispersers of *A. arguta* (Yasumoto & Takatsuki, 2015).

(Continued)

Appendix 18.1 (Continued)

Continent and country (region)	Mesocarnivore species studied	Plant species studied and/or dispersed[b]	Diet	Dep. sites	Exotic seeds	Germ. tests	Germ. effects	Main findings (Reference[s])
Malaysia (Tabin Wildlife Reserve, Sabah)	*Paradoxurus hermaphroditus*	*Endospermum diadenum*, *Leea aculeata*, 28 other plant species	Y	N	N	Y	NT	Seeds from a total of 30 plant species (mostly trees and vines) were found in 92% of scats investigated. Seeds of all the planted species were viable and successfully germinated (Nakashima *et al.*, 2010a).
Malaysia (Danum Valley and Maliau Basin, Sabah)	*Arctictis binturong*, One bird and one primate species	*Ficus forstenii* $[-_{gr}, +_{gr}]$, *Ficus punctata* $[+_{gr}, +_{gr}]$, *Ficus stupenda* $[0_{gr}, +_{gr}]$, Other *Ficus* species	N	Y	N	Y	Mix	Binturongs deposited faeces at specific microsites in the canopy. This directed dispersal of hemi-epiphytic figs was shown to be more reliable than scattering faeces from the air (hornbills) or upper canopy (gibbons) (Nakabayashi *et al.*, 2019).
Mongolia (Ikh Nart Nature Reserve)	*Vulpes corsac*, *Vulpes vulpes*	*Allium polyrhizum*, *Amygdalus peduncalata*, *Asparagus gobicus*, *Corispermum mongolicum*, *Tribulus terrestris*, Eight other plant species	Y	N	N	N	NT	Fruits represented an important component of the diet, especially during winter, and both corsac and red foxes may facilitate seed dispersal of some plant species (Murdoch *et al.*, 2009).
Singapore (Zoo Night Safari)	*Arctictis binturong*	*Carica papaya* $[0_{gr}, +_{gr}]$, *Dimocarpus longan* $[-_{gr}, +_{gr}]$, *Manilkara zapota* $[0_{gr}, +_{gr}]$	N	N	Y	Y	Mix	Germination rate for digested seeds of the three plant species varied from 19–36%. Germination took place ~3 days faster after gut passage, which may increase seed survival. Binturongs may be efficient seed dispersal agents for some fruit species (Colon & Campos-Arceiz, 2013).
EUROPE								
Belgium (Flanders)	*Vulpes vulpes*	77 plant taxa, including woody fleshy-fruited and dry-fruited species, as well as herbaceous and graminoid taxa	N	N	N	N	NT	Seeds were found in 57% of red fox scats, although seed number per scat was generally low (<10); 82% of the seeds belonged to woody species with drupes or berries. *Rubus* was the most abundant taxon (64%). Many species consumed were cultivated plants. Inadvertent intake was suspected for most of the dry-fruited species (e.g. through the manipulation of prey) (D'hondt *et al.*, 2011).
France (Montpellier)	*Genetta genetta*, *Martes foina*, *Meles meles*, *Vulpes vulpes*	65 fleshy-fruited native plant species from 25 families	N	N	N	N	NT	Overall 32% of all studied plant species were dispersed by mammals, with the red fox and the stone marten dispersing no less than 91% cf those (Debussche & Isenmann, 1989).
Germany (Brandenburg)	*Martes foina*, *Martes martes*	*Vaccinium myrtillus* $[+_{gr}]$, *Rubus caesius/fruticosus* $[0_{gr}]$, *Rubus idaeus* $[0_{gr}]$, Nine other plant species	N	N	N	Y	Mix	Endozochorous seeds originated mostly from wild fleshy-fruited species (i.e. few domestic fruits). Ten of the 12 dispersed plant species germinated from scat samples. No endozoochorous transport of abundant Liliaceae species with toxic berries was recorded (Schaumann & Heinken, 2002).

Country (region)	Animal species	Plant species						Description
Italy (Tuscany)	Meles meles	Juniperus oxycedrus (41.6%), Arbutus unedo (12.8%), Rubus ulmifolius (10.3%), Crataegus monogyna (7.4%), Vitis vinifera (3.9%), Olea europaea (3.5%)	Y_f	Y	N	N	NT	Seeds from 10 fleshy-fruited plant species were recovered in European badger faeces (see proportions next to species names). On average <1% of the seeds ingested and examined were damaged. The seeds of the six species most frequently dispersed by badgers were not generally deposited under their respective fruiting plants. Faeces were majoritarily (> 96%) deposited in small holes <20 cm deep (dung pits) that may favour seed survival and germination (Pigozzi, 1992).
Italy (Tyrrhenian coast)	Martes spp. (M), Vulpes vulpes (Vv)	Myrtus communis [M: 0_{gt}, 0_{gr}; Vv: $+_{gr}$, $+_{gt}$]	Y	N	N	Y	Mix	Seeds from scats of the two mesocarnivores studied did not show any evident signs of damage due to gut passage (Aronne & Russo, 1997).
Italy (Sardinia)	Vulpes vulpes ichnusae	Juniperus phoenica subsp. turbinata [0_{gr}], Juniperus oxycedrus subsp. macrocarpa	N	Y	N	Y	0	Sardinian foxes dispersed the seeds from I. phoenica subsp. turbinata but not that of J. oxycedrus subsp. macrocarpa. From 30–100 seeds per hectare were dispersed per day, with 80–90% of dung released on dwarf plants, mainly Helichrysum italicum subsp. microphyllum, which positively affected the survival of emerged seedlings (Farris et al., 2017).
Portugal (Serra de Grândola)	Genetta genetta (Gg), Herpestes ichneumon, Martes martes, Meles meles (Mm), Vulpes vulpes (Vv)	Arbutus unedo, Ficus carica, Olea europaea [Gg & Mm: $+_{gr}$; Vv: $-_{gr}$], Pyrus bourgaeana [Mm & Vv: $+_{gr}$], Rubus sp., Vitis vinifera	N	Y	Y	Y	Mix	European badger, red fox and common genet had a significant positive effect on the germination rate of seeds from at least one fleshy-fruited plant species. Stone marten and Egyptian mongoose, however, had deleterious effects on most seeds. Seed size was correlated with seed survival, germination rate and germination time (Rosalino et al., 2010).
Portugal (Bussaco National Forest)	Martes foina	Arbutus unedo [0_{gr}, 0_{gt}], Celtis australis [0_{gr}, 0_{gt}], Prunus laurocerasus [0_{gr}, 0_{gt}], Rubus ulmifolius [$+_{gr}$, $+_{gt}$]	N	Y	Y	Y	Mix	Stone martens contributed to forest generation and gene flow by dispersing native plant seeds. Although the germination of P. laurocerasus was not enhanced, the preference of stone marten for its fruit may potentially contribute to the proliferation of this invasive species (Pereira et al., 2019).
Portugal (Arrábida Natural Park)	Meles meles, Vulpes vulpes	Ceratonia siliqua [0_{gr}], Arbutus unedo, Myrtus communis	N	N	N	Y	0	Endozoochorous carob, C. siliqua, seedlings experienced higher mortality rates than control seedlings. However, the net result for the plant can still be the colonization of vacant habitats by a large proportion of viable seeds. The later carob seeds were sown over the fruit-ripening season, the faster the seedlings emerged. Water soaking increased the germination rate by 6.5 times (Salgueiro et al., 2020).

(Continued)

Appendix 18.1 (Continued)

Continent and country (region)	Mesocarnivore species studied	Plant species studied and/or dispersed[b]	Diet	Dep. sites	Exotic seeds	Germ. tests	Germ. effects	Main findings (Reference[s])
Spain (Sierra de Cazorla)	*Martes martes* *Meles meles* *Vulpes vulpes*	27 plant species	N	Y	N	N	NT	Seeds from 40% of the fleshy-fruited plants occurring in the study region were recovered from carnivore faeces. On average <1% of the seeds defecated were damaged (Herrera, 1989).
Spain (Sierra de Grazalema)	*Vulpes vulpes*	*Ceratonia siliqua* [0_{gr}]	N	N	N	Y	0	> 30% of red fox scats contained carob seeds. Red foxes carried out long-distance dispersal of these seeds without damaging the embryo, but germination rate did not increase through gut passage. Carob seeds have physical dormancy afforded by their hard seed coat (Ortiz et al., 1995).
Spain (Balearic Islands)	*Genetta genetta* *Martes martes*	*Arbutus unedo* *Ceratonia siliqua* *Chamaerops humilis* *Citrus* sp. *Cneorum tricoccon* *Ficus carica* *Juniperus phoenicea* Three other plant species	Y	N	Y	N	NT	Nine species of fleshy fruits were eaten by pine marten, while four species were by common genet. Seeds of cultivated fruits were more prevalent than wild-native fruits for both species on two of three islands investigated (Clevenger, 1996).
Spain (Mallorca)	*Martes martes* Eight bird species	*Myrtus communis* [0_{gr}]	N	N	N	Y	0	Passage of myrtle seeds through the digestive tract of pine martens did not increase germination rate in outdoor conditions, contrarily to what was observed with different bird species. These differences may be explained by the different seed retention times in the gut or the chemical composition of the food ingested along with the seeds. Interestingly, no differences between treatments were observed in a germination chamber or in a greenhouse (Traveset et al., 2001a).
Spain (Doñana)	*Meles meles* (Mm) *Vulpes vulpes* (Vv) Two other mammal species	*Corema album* [Vv: $+_{gr}$, $+_{gt}$] *Pyrus bourgaeana* [Mm: $+_{gr}$, 0_{gt}.] *Rubus ulmifolius* [Mm: $+_{gr}$, $+_{gt}$; Vv: 0_{gr}, $+_{gt}$]	N	N	N	Y	Mix	Gut passage altered germination rate, but the magnitude and direction of such effects varied according to plant and disperser species. It also increased the asynchrony of germination in R. ulmifolius and P. bourgaeana. Removal from the mother plant similarly increased asynchrony in R. ulmifolius, hence likely enhancing plant fitness in unpredictable environments (Fedriani & Delibes, 2009a).
Spain (Cantabrian Range)	*Martes martes* *Vulpes vulpes*	*Sorbus aucuparia*	N	N	N	N	NT	The diet of martens and foxes closely tracked interannual variations in rowan fruit availability. Specifically, total crop size was correlated with the frequency of occurrence and the proportion of rowan by volume in faeces. Both carnivore species tended to visit the trees that exhibited a higher density of fallen fruits under the canopy (Guitián & Munilla, 2010).

Location	Animal species	Plant species						Notes
Spain (Sierra Nevada)	*Martes martes* *Vulpes vulpes* One other mammal species	16 plant species	N	Y	N	Y	N	NT	Seeds from 16 woody species (and some agrarian species) were recorded, which represent more than half of the total fleshy-fruited woody species available. Seeds showed a high viability rate for all dispersed species, irrespective of the mammal disperser. No differences in species composition of dispersed seeds were recorded between various landscape units or in the seed density between degraded habitats (Matias *et al.*, 2010).
Spain (Doñana)	*Meles meles*	*Chamaerops humilis* [−gr, +gr]	N	Y	N	Y	Y	Mix	The ripe fruit pulp of Mediterranean dwarf palm was proposed to play a defensive role due to a significantly lower seed survival and a much higher seed predation by invertebrates for badger-ingested than for control seeds. However, early-emerged seedlings came from badger-ingested seeds, suggesting an inhibitory function of fruit pulp. Seedling survival for badger-ingested seeds was higher away from than beneath conspecific plants (Fedriani & Delibes, 2011).
Spain (Guadalajara)	*Martes foina* *Meles meles* Several bird species	*Juniperus thurifera*	N	Y	N	Y	Y	NT	Germination rate of Spanish juniper for seeds dispersed by stone marten and red fox was 11.5%, and the maximum rate was reached on shrubs (16%). Mortality at the seedling stage was 40%. Overall a seed dispersed by these mesocarnivores had a probability of recruitment of 6.5%. Both species were therefore regarded as generalist high-quality but opportunistic dispersers (Escribano-Ávila *et al.*, 2013).
Spain (Catalonia)	*Martes foina* *Meles meles* *Vulpes vulpes*	*Celtis australis* [+gr] 16 other plant species consumed but not included in germination experiments	N	Y	Y	Y	Y	+	The mesocarnivores studied dispersed seeds into burnt areas. Digested seeds mostly belonged to non-indigenous species, predominantly the Mediterranean hackberry. Gut passage improved the germination of hackberry seeds, but the survival rate of the seedlings was very low (Rost *et al.*, 2012).

(Continued)

Continent and country (region)	Mesocarnivore species studied	Plant species studied and/or dispersed[b]	Diet	Dep. sites	Exotic seeds	Germ. tests	Germ. effects	Main findings (Reference[s])
Spain (Carrascal de la Font Roja Natural Park)	*Martes foina* *Vulpes vulpes*	*Amelanchier ovalis* *Celtis australis* *Ficus carica* *Juniperus phoenica* *Olea europaea* *Prunus avium* *Rhamnus alaternus* *Rubus ulmifolius*	Y	N	N	N	NT	Seeds were found from summer to winter in about a third of the scats analyzed. Red fox exploited more domesticated species than stone marten, although the latter demonstrated a strong preference for wild cherries. Both carnivores were suggested to play a relevant role in the long-distance seed dispersal of *R. alaternus* (Rico-Guzmán *et al.*, 2012).
Spain (17 sites in SE Spain)	*Vulpes vulpes* (*Vv*) *Genetta genetta* *Martes foina* *Meles meles* Two other vertebrate species	*Ziziphus lotus* [*Vv*: +$_{gr}$]	N	Y	N	Y	+	In comparison to other dispersers, red fox was responsible for the mobilization of 87% of jujube seeds to distinct points in the studied habitat remnants. Almost all seeds dispersed by red fox were viable. Overall, passage through the red fox's gut positively affected the germination rate of jujube seeds (Cancio *et al.*, 2016).
Greenland (Disko)	*Vulpes*[f] *lagopus*	*Cerastium alpinum* [+$_{gr}$] *Oxyria digyna* [+$_{gr}$][g] *Sibbaldia procumbens* [+$_{gr}$][g] *Silene acaulis* [+$_{gr}$][g] *Stellaria longipes* [+$_{gr}$] *Chamaenerion latifolium* [−$_{gr}$] *Gnaphalium norvegicum* [−$_{gr}$] *Luzula Parviflora* [−$_{gr}$] *Papaver radicatum* [−$_{gr}$] *Polygonum viviparum* (*bulbils*) [−$_{gr}$] *Ranunculus hyperboreus* [−$_{gr}$] *Salix glauca* ssp. *callicarpaea* [−$_{gr}$] *Saxifraga cernua* (*bulbils*) [−$_{gr}$] *Veronica alpina* [−$_{gr}$] Eight other plant species	N	N	N	Y	Mix	No significant differences were detected in gut passage times for seeds of 22 common plant species with different morphology. Plants with adaptations to wind dispersal appeared particularly vulnerable to gut passage. Arctic foxes provided long-distance dispersal of seeds lacking morphological adaptations to dispersal, but these generally needed to be defecated within 12 h to remain viable (Graae *et al.*, 2004).

Location	Animal species	Plant species						Notes
Mexico (Veracruz)	*Eira barbara*, Two other mammal species	*Cecropia obtusifolia* [+/−$_{gr}$]	N	N	N	Y	Mix	Under white light conditions, seeds ingested by tayras exhibited a lower germination rate (36%) than seeds from the soil seed bank (50%) or picked up from the tree (86%). However, under dark conditions (mimicking understorey conditions in the tropical forest), the germination rate of digested seeds (20%) was higher than that of both seeds from the soil seed bank (10%) or collected from the infructescence (0%) (Vázquez-Yanes & Orozco-Segovia, 1986).
Mexico (Mapimi Biosphere Reserve, Chihuahuan Desert)	*Canis latrans*, Four other vertebrate species	*Opuntia rastrera* [−$_{gr}$]	N	N	N	Y	−	Prickly-pear seeds ingested by coyotes had a lower germination rate than control seeds from ripe fruit collected on trees. Germination steadily increased with ageing of seeds, implying the presence of primary dormancy (embryo immaturity) (Mandujano et al., 1997).
Mexico (Coast of Oaxaca)	*Urocyon cinereoargenteus*	*Acacia cornigera* [−$_{gr}$, +$_{gl}$], *Byrsonima crassifolia* [+$_{gr}$, +$_{gl}$], *Comocladia engleriana*, *Ehretia tinifolia* [0$_{gr}$], *Guazuma ulmifolia* [0$_{gr}$, +$_{gl}$], *Ficus* sp., 'Mountain nanche'·^{h} (Malpighiaceae) [+$_{gr}$, +$_{gl}$]	Y	N	N	Y	Mix	Seeds belonging to a total of seven plant species were found in 75% of grey fox faeces. Gut passage generally improved germination rate, and germination occurred faster compared to control seeds (Villalobos-Escalante et al., 2014).
Mexico (Zapotitlán de la Salinas Valley)	Undetermined	*Myrtillocactus geometrizans* [+$_{gr}$], *Vallesia glabra* [+$_{gr}$], *Agonandra conzattii* [0$_{gr}$], *Castela tortuosa* [0$_{gr}$], *Chiococca* sp. [0$_{gr}$], *Lantana camara* [0$_{gr}$], *Mammillaria carnea* [0$_{gr}$], *Prosopis laevigata* [0$_{gr}$], *Neobuxbaumia tetetzo* [−$_{gr}$], *Opuntia pilifera* [−$_{gr}$], *Stenocereus pruinosus* [−$_{gr}$], *Stenocereus stellatus* [−$_{gr}$], Six other plant species	N	N	Y	Y	Mix	Seeds from 18 plant species were found in carnivore scats. On average only 3% of digested seeds were damaged. Gut passage had mixed effects on germination rate. Scats were dropped more frequently than expected in rocky areas and less frequently than expected in open and canopy areas (Zarco-Mendoza et al., 2018).

(Continued)

Appendix 18.1   (Continued)

Continent and country (region)	Mesocarnivore species studied	Plant species studied and/or dispersed[b]	Diet	Dep. sites	Exotic seeds	Germ. tests	Germ. effects	Main findings (Reference[s])
Mexico (Barjitas Canyon, Sonoran Desert)	*Canis latrans* *Urocyon cinereoargenteus*	*Washingtonia robusta* [+$_{gr}$]	N	Y	N	Y	+	Mexican fan palm seeds in scats had a higher germination rate (94%) than those dispersed directly from mother plants (55%). Seed deposition sites by the canids were likely suitable locations for colonization (Armenta-Méndez *et al.*, 2020).
Mexico (Sierra Fría)	*Bassariscus astutus* (Cl) *Canis latrans* (Cl) *Lynx rufus* (Lr) *Urocyon cinereoargenteus*	*Arctostaphylos pungens* [dng] *Juniperus deppeana* [Cl & Lr: +$_{gr}$]	N	Y	N	Y	Mix	Between 28% and 93% of the studied carnivores' scats contained seeds of the target plant species. Endozoochory and diploendozoochory enhanced the viability of the seeds, except in those of *A. pungens* dispersed by coyote. Most of the seeds of *J. deppeana* from the canopy presented perforations in their embryos caused by insects. Both the digested and canopy seeds of *A. pungens* failed to germinate, probably because they have a very hard coat that prevented the entry of water necessary to trigger germination (Rubalcava-Castillo *et al.*, 2020).
USA (Texas)	*Bassariscus astutus* *Procyon lotor* *Urocyon cinereoargenteus* *Vulpes vulpes*[f]	*Diospyros texana* *Juniperus ashei* *Juniper pinchotii* *Opuntia leptocaulis*	Y	Y	N	N	NT	Large numbers of seeds were present in faeces and few were destroyed through mastication and digestion. The four carnivores studied were considered legitimate dispersal agents of Ashe juniper and Texas persimmon. Dispersal efficiency was however regarded as low due to the clumped pattern of faeces and unsuitable locations for plant establishment (Chávez-Ramírez & Slack, 1993).
USA (Illinois)	*Canis latrans* (Cl) *Procyon lotor* (Pl)	*Asimina triloba* [Cl: 0$_{gr}$] *Celtis occidentalis* [Pl: −$_{gr}$] *Diospyros virginiana* [Cl: −$_{gr}$; Pl: +$_{gr}$] *Prunus americana* [Cl: −$_{gr}$]	N	N	N	Y	Mix	Ingestion improved germination only for persimmon, *D. virginiana*, seeds consumed by northern raccoons, but the studied tree species may benefit from decreased parental competition through seed dispersal by coyotes and raccoons. Among seeds ingested by coyotes, germination rates of persimmon were significantly higher when seeds were protected by undigested fruit pulp or intact seed sheaths, thus reducing the exposure of seeds to gastrointestinal enzymes (Cypher & Cypher, 1999).
USA (Alaska)	*Martes americana*	*Rubus spectabilis* [dng] *Vaccinium alaskaense* [−$_{gr}$] *Vaccinium ovalifolium* [0$_{gr}$]	N	N	N	Y	Mix	Median gut-passage time was 4.1 h for *V. alaskaense* and 4.6 h for *R. spectabilis*. Although digested seeds did not germinate at a higher rate than undigested seeds, gut passsage improved germination over fruit fall alone. Estimated median dispersal distances for both berry species were ~500 m (Hickey *et al.*, 1999).

Location	Animal species	Plant species						Notes
USA (California)	*Urocyon cinereoargenteus*	*Arctostaphylos glandulosa* (68%; dry fruit) *Rhamnus californica* (37%) [0_{gr}, $+_{gt}$] *Rhamnus illicifolia* (11%) [0_{gr}, $+_{gt}$] *Heteromeles abutifolia* (14%) [0_{gr}, 0_{gt}]	Y	N	N	Y	Mix	Grey foxes consumed fruits from four plant species (see proportions next to species names) and seeds occurred in 70% of scats. Days to germination for ingested seeds averaged 22–52 days depending on the fleshy-fruited plant species (Wilson & Thomas, 1999).
USA (Texas)	*Procyon lotor* Two bird species	*Juniperus pinchotii*	N	N	N	Y	NT	Germination of redberry juniper seeds ingested by northern raccoons (and birds) tended to be greater than that of untreated, hand-harvested seeds, but similar to that of hand-harvested seeds that were cool/moist stratified and mechanically scarified. This suggests that germination inhibitors were removed during gut passage and seeds provided with a scarification treatment similar to mechanical scarification by the researchers (Petersen *et al.*, 2005).
USA (California)	*Canis latrans*	*Opuntia littoralis* [$+_{gr}$] *Annona cherimola* [0_{gr}] *Citrus* sp. [0_{gr}] *Pyracantha* sp. [0_{gr}] *Heteromeles arbutifolia* [$-_{gr}$] *Arctostaphylos* sp. [dng] *Xylococcus bicolor* [dng] 31 other plant species	Y	Y	N	Y	Mix	Seeds from 38 plant species were found in coyote scats. Seeds from 7 native species and 11 exotic plant species germinated. Although coyotes dispersed viable seeds of exotic plants, none were considered invasive (Silverstein, 2005).
USA (Arkansas)	*Canis latrans*	*Diospyros virginiana* [0_{gr}, 0_{gt}]	N	N	N	Y	0	Although no difference was recorded for germination rate and time, seedlings produced by seeds artificially removed from persimmon fruit had greater survival than those resulting from seeds ingested by coyotes or contained in intact fruit (Roehm & Moran, 2013).
USA (Arkansas)	*Canis latrans (Cl) Procyon lotor (Pl)* Three other mammal species	*Diospyros virginiana* [*Cl*: 0_{gr} $+_{gt}$; *Pl*: 0_{gr}, $+_{gt}$]	N	N	N	N	Mix	Germination rate was not affected by gut passage, but time to sprout decreased and seedling quality increased. In the wild, a seed predator, *Odocoileus virginianus*, was frequently detected at fruiting trees, while potential seed dispersers were rarely detected (Rebein *et al.*, 2017).

(Continued)

Appendix 18.1 (Continued)

Continent and country (region)	Mesocarnivore species studied	Plant species studied and/or dispersed[b]	Diet	Dep. sites	Exotic seeds	Germ. tests	Germ. effects	Main findings (Reference[s])
OCEANIA								
Australia (NE Victoria)	*Vulpes vulpes* Other mammal species	*Rubus fruticosus* [0_{gr}]	N	N	N	Y	0	Blackberry seeds were found in up to 89% of monthly faecal samples. Passage of these seeds through the digestive tract of red fox maintained their viability and did not alter the germination rate. The presence of viable seeds in faeces was expected because the hard endocarp protects the embryo of blackberry seeds from damage (Brunner *et al.*, 1976).
SOUTH AMERICA								
Argentina (Mendoza)	*Lycalopex[j] griseus* Other mammal species	*Prosopis flexuosa* [0_{gr}]	N	N	N	Y	0	Passage of *P. flexuosa* seeds through the digestive tract of South American grey fox maintained seed viability and did not alter germination rate (Campos & Ojeda, 1997).
Argentina (Chaco)	*Cerdocyon thous* (*Ct*) *Lycalopex[j] gymnocercus* (*Lg*)	*Acacia aroma* [*Ct*: 0_{gr}; *Lg*: $+_{gr}$] *Celtis tala* *Syagrus romanzoffiana* *Ziziphus mistol* [*Lg*: $+_{gr}$]	N	N	N	Y	Mix	Passage through the digestive tract of both fox species did not affect the survival of seeds from the four focal plant species. Germination rate was improved in the case of ingestion by Pampas fox (Varela & Bucher, 2006).
Argentina (Sierras Grandes)	*Lycalopex gymnocercus* Three bird species	*Lithraea molleoides* [$+_{gr}$, $+_{gt}$]	N	N	N	Y	+	Passage of *L. molleoides* seeds through the digestive tract of Pampas fox maintained seed viability, increased germination rate and decreased the germination time (Vergara-Tabares *et al.*, 2018).
Argentina (Ernesto Tornquist Provincial Park)	*Lycalopex gymnocercus* Four bird and one ant species	*Prunus mahaleb* [0_{gr}, $-_{gt}$]	N	N	Y	Y	Mix	In the laboratory endocarp scarification enhanced germination, while vestiges of pulp on the stones had inhibitory effects. The proportion of surface covered with pulp vestiges was high (50–75%) for stones found in Pampas fox's faeces, and as a result, the germination rate was low and similar to that of intact fruits and hand-peeled stones (Amodeo *et al.*, 2017).
Bolivia (Madidi National Park)	*Lycalopex[k] culpaeus* One other carnivore species and 'birds'	*Gaultheria vaccinioides* [0_{gr}, 0_{gt}] Two other plant species	N	N	N	Y	0	Passage of *G. vaccinioides* seeds through the digestive tract of culpeo fox maintained seed viability but did not alter germination rate and time (Rivadeneira-Canedo, 2008).

Country (Region)	Disperser	Plant species					Notes
Bolivia (Mecapaca)	Lycalopex culpaeus	Prosopis flexuosa [0_{gt}]	N	N	Y	0	Passage of P. flexuosa seeds through the digestive tract of culpeo fox maintained seed viability but did not accelerate germination time (Maldonado et al., 2014).
Bolivia (La Paz)	Lycalopex culpaeus	Prosopis alba [0_{gr}, $-_{gt}$]	N	Y	Y	Mix	Culpeo foxes defecated viable seeds, but these were deposited in open areas rather than under woody vegetation. Probability of germination did not change after gut passage, but germination time decreased. Dispersal services of this fox were deemed inefficient (Maldonado et al., 2018).
Brazil (Santa Catarina State)	Lontra longicaudis	Manilkara subsericea Marlierea tomentosa Pouteria lasiocarpa	Y	N	N	NT	Of 202 scats, six (3%) presented fruit remains consisting of three abundant species. Seeds of M. tomentosa within the scats had germination ability (Quadros & Monteiro-Filho, 2000).
Brazil (Minas Gerais)	Nasua nasua	Casearia lasiophylla [0_{gr}, 0_{gt}] Cecropia pachystachya [0_{gr}, 0_{gt}] Ficus obtusifolia [0_{gr}, 0_{gt}] Guazuma ulmifolia [0_{gr}, 0_{gt}] Lithraea molleoides [0_{gr}, 0_{gt}] Myrcia guajavaefolia [0_{gr}, 0_{gt}] 47 other plant species	N	N	Y	Mix	Seeds were found in 55% of faecal samples and coatis consumed fruits of 53 plant species. Gut passage did not affect germination rate or time of five of the six tested species. Coatis may provide dispersal services and promote gene flow in defaunated forest fragments (Alves-Costa & Eterovick, 2007).
Brazil (Linhares Forest)	Cerdocyon thous Ten bird and one lizard species	Solanum thomasiifolium [$-_{gt}$]	N	Y	Y	–	Passage of S. thomasiifolium seeds through the digestive tract of crab-eating fox maintained seed viability but decreased germination rate compared to control seeds collected directly from fruit. Nevertheless, crab-eating fox contributed to the dispersal of 19% of S. thomasiifolium seeds over wide areas (Vasconcellos-Neto et al., 2009).
Brazil (Serra do Japi Ecological Reserve)	Cerdocyon thous Nasua nasua One other mammal species	Bromelia balansae [Ct: $+/0_{gr}$ $+_{gt}$; Nn: $+/+_{gr}$, $+_{gt}$]	N	N	Y	+	The seeds of B. balansae remained intact and viable after passing through the digestive tract of both mesocarnivore species. Germination rate was higher compared to control seeds with pulp (both species) and without pulp (coatis only), while germination time was faster for both species compared to control seeds with pulp (Paulino-Neto et al., 2016).

(Continued)

Appendix 18.1 (Continued)

Continent and country (region)	Mesocarnivore species studied	Plant species studied and/or dispersed[b]	Diet	Dep. sites	Exotic seeds	Germ. tests	Germ. effects	Main findings (Reference[s])
Chile (El Pangue)	*Lycalopex*[k] *culpaeus*	*Cryptocaria alba* [+gr]	N	Y	N	Y	+	Culpeo foxes deposited peumo seeds more often in unprotected habitats than under shrubs and were therefore regarded as legitimate but inefficient seed dispersers (Bustamante et al., 1992).
Chile (Fray Jorge National Park)	*Lycalopex*[k] *culpaeus*	*Schinus molle* (82%) [+gr], *Schinus polygamus* (10%), *Porlieria chilensis* (7%), Five other plant taxa (<1%)	Yf	Y	Y	Y	+	Culpeo foxes exhibited selective fruit consumption (see seed proportions from scats next to plant species names) and defecated seeds at microsites where the successful establishment of seedlings was possible. Fruit consumption was predominant when small mammal density was <10 individuals/ha. Gut passage did not affect the viability of pepper seeds and increased germination rate (Castro et al., 1994).
Chile (Río Clarillo National Reserve)	*Lycalopex*[k] *culpaeus*	*Lithrea caustica* (72.6%) [0gr, −gt], *Aristotelia chilensis* (24.8%), *Cryptocaria alba* (2.6%)	Yf	N	N	Y	Mix	Culpeo foxes defecated viable seeds (see seed proportions from scats next to plant species names), but ingestion delayed the germination of *L. caustica* seeds. It was hypothesized that these seeds may have a laxative effect, which could potentially reduce dispersal distances (León-Lobos & Kalin-Arroyo, 1994).
Chile (Las Chinchillas National Reserve)	*Lycalopex*[k] *culpaeus*	*Schinus molle* (98%) [+gr], *Porlieria chilensis* (<0.01%) [−gr], Other plant species (<2%)	Y	Y	Y	Y	Mix	Fruit consumption was predominant when small mammal density was <6 individuals/ha. Gut passage did not affect the viability of pepper seeds and increased the germination rate by 50%. Over 41% of seeds were deposited at safe microsites and there was germination but no establishment in the field. Overall, culpeo foxes were regarded as legitimate, efficient, and effective[m] dispersers of the alien *S. molle* shrub (Silva et al., 2005).
Chile (Río Clarillo National Reserve)	*Lycalopex*[k] *culpaeus* (Lc), *Canis familiaris* (Cf)	*Lithrea caustica* [Lc: −gr; Cf: −gr]	N	N	N	Y	−	The post-germination viability of seeds did not differ significantly between the two canids, suggesting that dogs can surrogate the seed disperser role of culpeo foxes. However, both species seem to be ineffective[m] seed dispersers of litter because gut passage decreases the germination capability of this fleshy-fruited tree (Morales-Paredes et al., 2015).

Country (site)	Mesocarnivore	Plant species	Diet	Dep. sites	Exotic seeds	Germ. tests	Germ. effects	Notes
Ecuador (Arenillas Ecological Reserve and Reserva la Ceiba)	Lycalopex sechurae	Cordia lutea [+$_{gr}$] Ficus spp. Coccoloba ruiziana [+$_{gr}$] Ziziphus thrysiflora [+$_{gr}$] Malphigia emarginata Prosopis sp. Vigna sp. Ipomoea sp. Phaseolus vulgaris	N	N	N	Y	+	The Sechuran fox dispersed the seeds of at least nine plant species, with Cordia lutea (54%) and Ficus spp. (31%) found in a large portion of faecal samples. Gut passage increased germination rate compared to control seeds and fruits for three plant species tested. Overall the deinhibition (through pulp removal) effect size was larger than that of scarification, with some interspecific variation (Escribano-Ávila, 2019).

Diet = whether (Y) or not (N) the study also provides detailed data on the proportional composition and/or seasonal variations of the diet of the focal mesocarnivore species (Y_f indicates that the study only focused on fruit components); **Dep. sites** = whether (Y) or not (N) the study provides information on the microhabitat and/or broad habitat characteristics of seed deposition sites; **Exotic seeds** = whether (Y) or not (N) exotic/alien/invasive plants were also dispersed, based partly on the authors' narratives or as far as known; **Germ. tests** = whether (Y) or not (N) germinability of ingested/defecated seeds was tested vs. control seeds or between different frugivores for at least one plant species; **Germ. effects** = overall effect of gut passage on seed germination (+: positive; 0: no effect; –: negative; Mix: mixed, i.e. any combination of two or more contrasting results; NT: not tested or not tested statistically). Specific effects (+, 0, –; dng = did not germinate) for corresponding plants are provided in square brackets and refer to germination rate/percentage (gr) or germination time (gt). For germination time, positive and negative effects mean shorter and longer germination times, respectively. Note that so-called e.g. 'positive' effects on seed germination may not be necessarily positive ecologically (see e.g. Engel, 2000).

[a] Considering the major seed dispersal studies dealing with the plants dispersed, the viability of ingested seeds, effects on germination, dispersal distances, and/or characteristics of deposition sites, but excluding studies purely focusing on dietary ecology (even if including frugivory) or advanced spatial aspects of dispersal (seed shadows, dispersal kernels, effects of landscape structure on seed dispersal and vice versa).

[b] Depending on the context, plant species are listed alphabetically, based on germination effects (+, 0, –) or ranked by decreasing contribution to the diet when proportions of seeds in scats (faeces) were evaluated.

[c] Referred to as Genetta rubiginosa in the original publication.

[d] The authors reported no effect, but the corresponding figure suggests otherwise.

[e] Referred to as Meles meles in the original publication.

[f] Referred to as Vulpes fulva in the original publication.

[g] Only for gut passages shorter than 10 h.

[h] The authors could not determine the exact species.

[i] Referred to as Pseudalopex griseus in the original publication.

[j] Referred to as Pseudalopex gymnocercus in the original publication.

[k] Referred to as Pseudalopex culpaeus in the original publication.

[l] In this study, the authors did not statistically compare germination times *per se*, but showed that similar germination rates were reached significantly later with ingested seeds.

[m] These authors regarded seed dispersal as effective/ineffective when the seeds in faeces had a higher/lower germination rate than those taken directly from parental plants. Following this definition, many of the mesocarnivores listed in Appendix 18.1 could be regarded as effective dispersers of at least some plants, though they may not necessarily be efficient (sensu Reid, 1989; i.e. seeds may not lodge in a safe site and germinate). This differs from the definition of Reid (1989), who regarded disperser effectiveness as 'the proportion of seedlings in a population that any one seed vector is responsible for disseminating'. It is therefore recommended to clearly define 'effectiveness' when using this key term in seed dispersal studies.

19

Ecology and Conservation of Southeast Asian Civets (Viverridae) and Mongooses (Herpestidae)

Andrew P. Jennings[1], and Géraldine Veron[2]*

[1] *Small Carnivores – Research and Conservation, Portland, ME, USA*
[2] *Institut de Systématique, Evolution, Biodiversité (ISYEB), Muséum National d'Histoire Naturelle, CNRS, Sorbonne Université, EPHE, Université des Antilles, Paris, France*

SUMMARY

Southeast Asia supports 13 civet and 4 mongoose species that are poorly known and some are threatened with extinction. We investigated the ecology and distribution of several species, using radio-telemetry, ecological niche modelling, and camera-trapping. On Buton Island, Sulawesi, we obtained radio-telemetry data from eight Malay civets, *Viverra tangalunga*. The mean home range size for both sexes was 70 ha, with a mean intrasexual overlap of 8% for males and 0% for females. In contrast, seven Malay civets radio-tracked on Peninsular Malaysia had a mean home range size of 143 ha, but the intrasexual overlap was similar (15% for males and 0% for females); the home range of each male overlapped that of one or two females. At both study sites, Malay civets were mainly nocturnal, and all daytime rest sites were within dense ground cover. On Peninsular Malaysia, we also obtained radio-telemetry data from five short-tailed mongooses, *Urva brachyura*. The mean home range size of males (233 ha) was significantly larger than that of females (132 ha). Females had almost exclusive home ranges but male ranges overlapped that of more than one female. Short-tailed mongooses were diurnal: mean activity during the day was 85%, compared to 6% at night. Our ecological niche modelling studies showed that the distributions of the large Indian civet, *Viverra zibetha*, and the crab-eating mongoose, *Urva urva*, were similar throughout mainland Southeast Asia; they are both found over a broad elevation range, and occur primarily in evergreen forest. The large-spotted civet, *Viverra megaspila*, occurs in lowland areas across northern Southeast Asia, and is most frequently found in deciduous forest. The Malay civet and the short-tailed mongoose are found primarily in lowland evergreen forest. They both occur south of the Thai–Malaysian border in Malaysia and Indonesia, while the Malay civet is also present in the Philippines. The small Indian civet, *Viverricula indica*, and the Javan mongoose, *Urva javanica*, are both found on mainland Southeast Asia and parts of Indonesia; they mainly occur at lower elevations, and appear to have no preference for forest type. The collared mongoose, *Urva semitorquata*, is found mainly on Borneo and may occur more frequently at higher elevations and in disturbed evergreen forests. The banded civet, *Hemigalus derbyanus*, occurs principally in lowland evergreen forest in southern Myanmar/Thailand, Peninsular Malaysia, Sumatra, Borneo, and the Mentawai Islands. Hose's civet, *Diplogale hosei*, is found in evergreen forest across the higher elevation regions of Borneo. On Sumatra, we set up camera-traps in two oil palm plantations and analyzed the data using occupancy modelling. From 3164 camera-trap days, we detected only three small carnivores: the leopard cat, *Prionailurus bengalensis*, the common palm civet, *Paradoxurus hermaphroditus*, and the Malay civet. The common palm civet had a high occupancy value and was found deep within the oil palm, whereas the Malay civet had low occupancy and detection probability values and was only detected near the edge. No covariate affected common palm civet occupancy, but the distance from the plantation edge did influence its detection probability. Malay civet occupancy was influenced by distance from the plantation edge and detection probability was affected by distance from the primary forest. Forest-dependent civet and mongoose species may be threatened by forest loss, degradation, and fragmentation. Other threats include hunting and the wildlife trade.

Keywords

Camera-trapping — ecological niche modelling — occupancy modelling — radio-telemetry

*Corresponding author.

Small Carnivores: Evolution, Ecology, Behaviour, and Conservation, First Edition. Edited by Emmanuel Do Linh San, Jun J. Sato, Jerrold L. Belant, and Michael J. Somers.

Introduction

Small carnivores (≤ 21 kg) play an important role in ecosystems as predators of small vertebrates and invertebrates, and as seed dispersers (Grassman *et al.*, 2005; Rajaratnam *et al.*, 2007; Nakashima *et al.*, 2010; Nakashima & Do Linh San, Chapter 18, this volume), but are often under-appreciated, overlooked, and under-researched. The Viverridae and Herpestidae are two poorly known families of small carnivore species that are mainly found in Africa and Asia (Gilchrist *et al.*, 2009; Jennings & Veron, 2009, 2019). The Viverridae comprises 17 species of genets and oyans (that mostly occur in Africa), and 17 species of civets (which all occur in Asia, except for the African civet, *Civettictis civetta*) (Jennings & Veron, 2009). Viverrids are solitary and generally live in forests, although some species are found in open habitats, such as savannah and grassland (Jennings & Veron, 2009). The Herpestidae consists of 25 African mongoose species and 9 Asian species (Jennings & Veron, 2019; Veron *et al.*, Chapter 3, this volume). This is an ecologically and behaviourally diverse family that occupies a broad range of habitats, from open savannah to dense rainforest, and displays a wide range of social behaviour, from being solitary to living in groups (Gilchrist *et al.*, 2009; Jennings & Veron, 2019).

Although Southeast Asia represents only 1% of the earth's land surface, it harbours 13% of the world's mammal species and has the highest number of threatened mammals (Schipper *et al.*, 2008). This region supports 13 civet species: Owston's civet, *Chrotogale owstoni*, otter civet, *Cynogale bennettii*, Hose's civet, *Diplogale hosei*, banded civet, *Hemigalus derbyanus*, binturong, *Arctictis binturong*, small-toothed palm civet, *Arctogalidia trivirgata*, Sulawesi palm civet, *Macrogalidia musschenbroekii*, masked palm civet, *Paguma larvata*, common palm civet, *Paradoxurus hermaphroditus* (for updated taxonomy, see Veron *et al.*, 2015b), large-spotted civet, *Viverra megaspila*, Malay civet, *Viverra tangalunga*, large Indian civet, *Viverra zibetha*, and small Indian civet, *Viverricula indica*; and 4 mongoose species: short-tailed mongoose, *Urva brachyura*, Javan mongoose, *Urva javanica*, collared mongoose, *Urva semitorquata*, and crab-eating mongoose, *Urva urva* (Corbet & Hill, 1992; Wozencraft, 2005; Gilchrist *et al.*, 2009;

Jennings & Veron, 2009, 2019; Patou *et al.*, 2009; Veron *et al.*, 2015a, Chapter 3, this volume). Despite this extraordinary richness of Southeast Asian viverrids and herpestids, the ecology of these two groups of small carnivores has not been thoroughly investigated (Schreiber *et al.*, 1989; Gilchrist *et al.*, 2009; Jennings & Veron, 2009, 2019), and even though several species are considered to be threatened with extinction (IUCN, 2021), the conservation status of viverrids and herpestids is currently difficult to assess due to our limited knowledge of their distributions, habitat requirements, and population trends. Without this information, it is difficult to predict extinction risks and implement effective conservation strategies to ensure their survival (Purvis *et al.*, 2000). Furthermore, good conservation planning also requires knowledge about the potential areas where a species may survive, and a greater understanding of why it only occurs in certain regions (Thorn *et al.*, 2009; Jackson & Robertson, 2011; Rondinini *et al.*, 2011). Several factors, such as interspecific competition, biogeography, and changes in climate and vegetation, can affect the distribution patterns of species (Creel, 2001; Meijaard, 2003; Patou, 2008), but the impact of these on small carnivores are poorly understood due to a paucity of studies. Many small carnivore species are endangered due to habitat disturbance (Schreiber *et al.*, 1989; Meijaard & Sheil, 2008), but the effects of habitat loss, fragmentation, and degradation on small carnivore populations are largely unknown (Creel, 2001). A greater understanding of their habitat requirements is therefore needed in order to mitigate the harmful effects of human disturbance on small carnivore diversity in Asian tropical forests. Species persistence may also be influenced by other human-induced factors, such as hunting (Corlett, 2007; Thorn *et al.*, 2009). Thus, identifying and assessing the level of anthropogenic threats is vital for developing conservation strategies.

Radio-telemetry studies can provide important information on home ranges, spatial organization, activity patterns, and habitat use (Kenward, 2000). Ecological niche modelling is a useful tool for predicting the distributions of poorly known species in remote and inaccessible regions and can aid conservation planning by highlighting potential unknown populations and key areas for fieldwork and conservation initiatives (Peterson *et al.*, 2006; Thorn *et al.*, 2009; Wilting *et al.*, 2010a; Jackson & Robertson, 2011). This presence-only modelling approach uses the environmental characteristics of known distribution points to assess the suitability of regions where currently no records of a species exist (Elith *et al.*, 2006). Modelling geographical distributions can also produce valuable ecological information, and may highlight instances where other factors, such as historical causes, biotic interactions or absence of key resources, have played a role in restricting a species' range (Phillips *et al.*, 2006; Hirzel & Le Lay, 2008). Camera-trapping studies provide important species detection records that can be used to determine the occupancy of species within different habitats and regions (Ancrenaz *et al.*, 2012).

We investigated the ecology and distributions of several civet and mongoose species within Southeast Asia, using radio-telemetry, ecological niche modelling, and camera-trapping. We radio-tracked the Malay civet on Peninsular Malaysia and Buton Island, Sulawesi, and the short-tailed mongoose on Peninsular Malaysia (Jennings *et al.*, 2006, 2010a,b). Few telemetry studies have been conducted on the Malay civet (Macdonald & Wise, 1979; Nozaki *et al.*, 1994; Colon, 2002; Evans *et al.*, 2021) and the short-tailed mongoose had never been radio-tracked before. We used ecological niche modelling and niche analyses for six civet and four mongoose species: large-spotted civet, Malay civet, large Indian civet, small Indian civet, banded civet, Hose's civet, short-tailed mongoose, Javan mongoose, collared mongoose, and crab-eating mongoose (Jennings & Veron, 2011; Jennings *et al.*, 2013). In central Sumatra, we set up camera-traps within two oil palm plantations in order to determine the impact of this non-native, human-modified habitat on small carnivore species (Naim *et al.*, 2012; Jennings *et al.*, 2015). In this chapter, we review the ecological findings of our studies and discuss the conservation implications.

Study Areas

We carried out a radio-telemetry project on the Malay civet (Figure 19.1) within the Kakenauwe Forest Reserve, central Buton Island, Sulawesi, from June to September in 2001, 2002, and 2003 (see Jennings

Figure 19.1 Malay civet, *Viverra tangalunga*, photographed in Maliau Basin, Sabah, Malaysia. *Source:* Photo © Chien C. Lee (chienclee.com).

Figure 19.2 Short-tailed mongoose, *Urva brachyura*, photographed in Danum Valley, Sabah, Malaysia. *Source:* Photo © Chien C. Lee (chienclee.com).

et al., 2006). Buton Island lies off the southeast coast of Sulawesi and is approximately 100 km long and 42 km wide. It has a tropical monsoon climate, with a dry season from June to September. The reserve comprised lowland forest on karst coral limestone (with some evidence of disturbance from selective logging and rattan collection), and was surrounded by agricultural crops. The elevation ranged from 40 to 360 m. The forested area encompassed a 1 km^2 study grid and a number of trails that ran through the reserve to the adjacent farmland areas.

Similarly, we conducted a radio-telemetry project on the Malay civet and the short-tailed mongoose (Figure 19.2) within Krau Wildlife Reserve, central Peninsular Malaysia, over five field seasons from 2004 to 2007 (see Jennings *et al.*, 2010a,b). The study area was along the southeastern edge of the reserve, near the village of Jenderak Selatan, and comprised lowland forest and adjacent plantations. Some logging had occurred along the edge of the reserve during the last 20–30 years, and the adjacent plantations consisted mainly of young oil palm, *Elaeis guineensis*, with some banana and rubber crops. The elevation ranged from 40 to 160 m. Throughout the forest were several small rivers and streams, a network of trails, and a 1 km^2 study grid. There were several dirt roads and small patches of remnant forest within the plantation areas.

Lastly, we undertook a camera-trapping study within two mature oil palm plantations in Riau Province, central Sumatra, from March 2012 to April 2013 (see Naim *et al.*, 2012; Jennings *et al.*, 2015). At both sites, the elevation ranged from 10 to 90 m. Each estate had an extensive grid network of dirt roads, flood-control ditches, and housing areas for oil palm workers. Oil palm trees were planted from 1986 to 2002. Both estates had a sparse understorey that consisted mainly of ferns, and the surrounding habitat comprised a mosaic of disturbed forest, farmland (including other oil palm plantations), and open areas (villages, bareland, and waterbodies).

Methods

The methodology is outlined below (for further details, see: Jennings *et al.*, 2006, 2010a,b, 2013, 2015; Jennings & Veron, 2011; Naim *et al.*, 2012).

Radio-Telemetry

We opportunistically set wire-cage traps of various sizes on the ground within the forest, often at the base of large trees or alongside logs, and close to human and animal trails. We placed traps at least 200 m apart and left them in place for a minimum of 10 days before moving them to new locations. We covered each trap with leaves and woody debris, and leaf litter was spread across the trap floor. Various combinations of meat and fruit were used as bait. Traps were left open continuously and checked twice a day. We restrained a captured animal within the trap and injected it with an anaesthetic drug. The anaesthetized animal was then weighed, measured, sexed and aged, and coloured plastic tags were clipped onto both ears. Adult individuals were fitted with an appropriately sized radio-collar.

We tracked radio-collared individuals on foot, using a receiver and antenna. At least three bearings from marked positions along the trail network were taken from each tagged animal (with a maximum of five minutes between successive bearings), or individuals were located using a 'box' signal (without the antenna and cable), at which point the GPS position was recorded. We determined an animal's activity based upon signal integrity: 'active' if there was signal fluctuation, and 'inactive' if the signal was steady. Continuous monitoring of an animal's activity (every 15 minutes) was undertaken over 8–12 hours' periods, and inactive animals were located to investigate rest sites.

We entered compass bearings into the program LOAS 3.0.3 (Ecological Software Solutions LLC, Hungary, www.ecostats.com) to generate location fixes, and all fixes were then imported into RANGES 6 (Version 1.04, Anatrack Ltd., UK, www.anatrack.com) for home range analysis. We calculated home range sizes using the minimum convex polygon method (MCP) using 95% of all fixes. Activity data were grouped into one-hour blocks over a 24-hour time period and expressed as the percentage of active fixes. We estimated minimum animal daily displacements by measuring the linear distance between consecutive 24-hour radio locations.

Ecological Niche Modelling

We compiled different types of occurrence records (camera-trap detections, museum specimens, etc.) from several sources (field researchers, publications, museums, and online databases). Additional records from the same location or within 1 km were not used in the analyses. All records were given an accuracy code based on the precision of the location – AC 1: location recorded using a GPS unit; AC 2: location determined using accurate maps and detailed field information; AC 3: only a description of the locality recorded; AC 4: reported record. The precision of AC 1 and AC 2 records was less than ± 250 m, and up to several kilometres for AC 3 and AC 4 records.

We modelled with two environmental layers, habitat and elevation, as these have key distribution and conservation implications for mammal species. For habitat, we used the 2000 Global Land Cover map or the 2010 Land Cover map for Southeast Asia. A Digital Elevation Model was used as an elevation layer. We only used records AC 1 and AC 2 for the modelling, as the precision of the localities were within the resolution of the environmental layers and there was a good temporal correspondence between the habitat layer and these records. For each species, the GIS layers were clipped to the extent of occurrence, resampled to the same cell size, and then entered with the occurrence data into the program Maxent 3.3.3k (www.cs.princeton.edu/~schapire/maxent). The area under the curve (AUC) of the receiver operating characteristic plot was used as a measure of model performance, and the outputs were projected in ArcView for interpretation. For species with a small sample size (less than 25), we also used a jacknife validation methodology.

To examine the habitat and elevation niche preferences of each species, all AC 1 and AC 2 records were plotted in ArcView and overlaid with the habitat and elevation layers. We then extracted the habitat type and elevation at each detection point, and, if available, we double-checked these with the information recorded in the field. Niche preferences were then defined as the frequency of occurrence within each habitat and elevation category, and we used Pianka's pairwise test to calculate the niche overlaps between species – Pianka's index varies between 0 (total separation) and 1 (total overlap).

For the banded civet and Hose's civet, we also assessed the loss of suitable habitat since historical times (early 1800s) by analyzing the 2010 land cover status at both historically and recently recorded localities (1992 onward). To accommodate for the uncertainties associated with specimen locations, a 5 km radius buffer zone was created around the position of all records. Within the buffer zones, we compared the total percentage area of each habitat type between historical and recent records. To determine the proportion of their predicted distributions that is under protected areas, we first created binary maps of presence/absence using Maxent. We then overlaid these maps with a GIS layer of protected areas and performed calculations in ArcView to determine the percentage of each distribution that lies under these protected areas.

Camera-Trapping

Within two oil palm estates on central Sumatra, we opportunistically deployed camera-traps (Reconyx and Bushnell) at 18 sites, with a minimum distance of 1 km between sites. Camera-traps were mounted on palm trees at 20–39 cm above the ground and were checked each month.

The total number of trap days for each camera-trap site was calculated based on the number of days that camera-traps operated. Each photograph was identified to species and rated as a dependent or independent event, using a time-to-independence criterion of one hour. We calculated the minimum distance of each camera-trap site to the edge of the oil palm plantation and to the edge of the nearest extensive area of lowland forest, using ArcView and the 2010 Land Cover map for Southeast Asia.

Each camera-trap was treated as an individual sampling unit. A sampling occasion of 10 days was used to construct a detection history for each sample site consisting of a row of 0 (no detection) and 1 (detection). The dataset was analyzed using PRESENCE, version 6.2 (Hines, 2006) to derive occupancy estimates ψ (the proportion of an area where the species was analyzed as present) and detectability p (the probability that a species was detected when present). We employed single-species, single-season models, with two site-level covariates: (i) minimum distance of each camera-trap site to the nearest edge of the oil palm; (ii) minimum distance of each camera-trap site from the nearest edge of extensive lowland forest. These two covariates were standardized by converting them to Z values. For each species, we compared the simplest occupancy model, where both occupancy ψ and detection probability p were constant, $\psi(.)$, $p(.)$, with the models in which occupancy and detection probability were either constant or a function of the two covariates (both individually and in combination). Akaike's Information Criterion values were corrected for small sample sizes (AIC_c) and these were used to rank each model (the lowest value representing the top-ranking model). Akaike's weights (w) were also adjusted to account for small sample sizes and these provided a strength of evidence for each model. The probability of observing a test statistic $\geq \chi^2$ was calculated based on 1000 parametric bootstraps (*p*-value) and this was used as an assessment of model fit. We used model averaging to estimate model parameters when no single model contained the majority of support ($w > 0.90$), by using the estimated values for occupancy ψ, detection probability p, and Akaike's weights *w*, from the top-ranking models with $\Delta AIC_c < 7$.

Results

The results below are taken from Jennings *et al.* (2006, 2010a,b, 2013, 2015), Jennings & Veron (2011), and Naim *et al.* (2012).

Radio-Telemetry of the Malay Civet

On Buton Island, Sulawesi, there was no significant difference between the mean home range sizes of males (86 ha; $n = 4$) and females (50 ha; $n = 3$), and the mean home range size of both sexes was 70 ha (Table 19.1; Figure 19.3). Mean intrasexual range overlap was significantly higher in males (8%) than in females (0%). Mean overall activity was the same for both sexes (78%). Malay civets were significantly more active at night (94%) than during the day (58%) (Figure 19.4). During the night, there was a drop in activity from 02:00 to 04:00 h, and in the daytime, there was a peak in activity from 12:00 to 14:00 h. The mean minimum distance covered during a 24-hour period was longer in males (415 m) than in females (286 m), but the difference was not statistically significant.

Table 19.1 Home range sizes (95% Minimum Convex Polygons, MCP) of Malay civets, *Viverra tangalunga*, and short-tailed mongooses, *Urva brachyura*, radio-tracked on Buton Island, Sulawesi, and in Krau Wildlife Reserve, Peninsular (Pen.) Malaysia.

Species	Study site	ID	Sex	Age	95% MCP (ha)	No. of locations
Viverra tangalunga	Buton Island	M01	Male	Adult	37	34
Viverra tangalunga	Buton Island	M03	Male	Adult	189	28
Viverra tangalunga	Buton Island	M04	Male	Adult	43	59
Viverra tangalunga	Buton Island	M05	Male	Adult	76	31
Viverra tangalunga	Buton Island	F01	Female	Y. Adult	13	35
Viverra tangalunga	Buton Island	F02	Female	Adult	60	30
Viverra tangalunga	Buton Island	F03	Female	Adult	24	27
Viverra tangalunga	Buton Island	F10	Female	Adult	66	34
Viverra tangalunga	Pen. Malaysia	M1	Male	Adult	160	33
Viverra tangalunga	Pen. Malaysia	M2	Male	Adult	148	31
Viverra tangalunga	Pen. Malaysia	M3[a]	Male	Adult	23	6
Viverra tangalunga	Pen. Malaysia	M4	Male	Adult	78	34
Viverra tangalunga	Pen. Malaysia	F1	Female	Adult	142	59
Viverra tangalunga	Pen. Malaysia	F2	Female	Adult	185	24
Viverra tangalunga	Pen. Malaysia	F3[a]	Female	Adult	39	10
Urva brachyura	Pen. Malaysia	M1	Male	Adult	250	31
Urva brachyura	Pen. Malaysia	M2	Male	Adult	224	33
Urva brachyura	Pen. Malaysia	M3	Male	Adult	224	31
Urva brachyura	Pen. Malaysia	F1	Female	Adult	115	33
Urva brachyura	Pen. Malaysia	F2	Female	Adult	149	32

[a] Home range did not reach an asymptote.

Y. = young.

Source: From Jennings *et al.* (2006, 2010a,b). Reproduced by permission of the Zoological Society of London, Deutsche Gesellschaft für Säugetierkunde, and Walter de Gruyter GmbH.

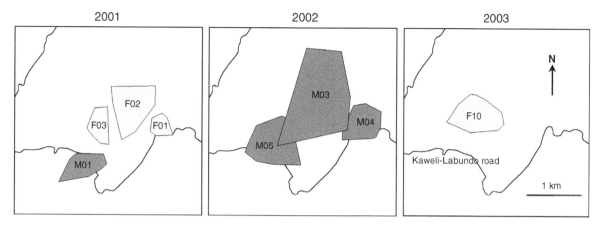

Figure 19.3 Home range polygons (95% MCP) of eight Malay civets, *Viverra tangalunga* (four males, four females), radio-tracked on Buton Island, Sulawesi, in 2001, 2002, and 2003. *Source:* From Jennings *et al.* (2006). Reproduced by permission of the Zoological Society of London.

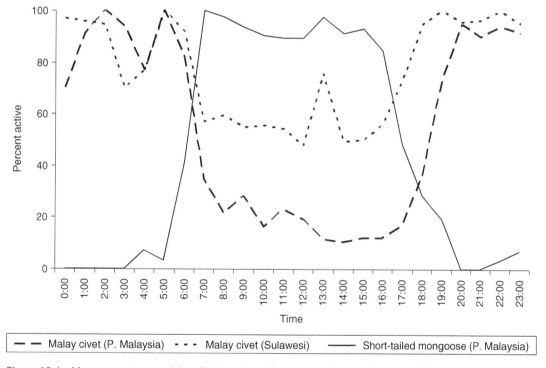

Figure 19.4 Mean percentage activity of Malay civets, *Viverra tangalunga*, and short-tailed mongooses, *Urva brachyura*, radio-tracked on Buton Island, Sulawesi, and Peninsular (P.) Malaysia, from 2001 to 2007. *Source:* Modified from Jennings *et al.* (2006, 2010a,b). Reproduced by permission of the Zoological Society of London, Deutsche Gesellschaft für Säugetierkunde, and Walter de Gruyter GmbH.

Six rest sites were found for one male and two females. All were situated at ground level and associated with some form of cover such as logs, dense brush pile, or thick herbaceous vegetation. Each rest site was accessible from all directions and none were communal.

On Peninsular Malaysia, there was no significant difference between the mean home range sizes of males (129 ha; $n = 3$) and females (164 ha; $n = 2$), and the mean home range size of both sexes was 143 ha (Table 19.1; Figure 19.5). The mean intra-gender range overlap was significantly higher in males (15%) than in females (0%).

2005

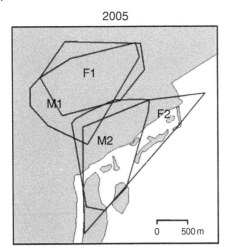

2006

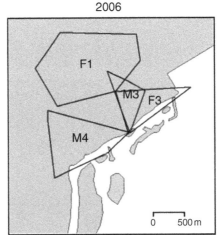

Figure 19.5 Home range polygons (95% MCP) of seven Malay civets, *Viverra tangalunga* (four males, three females), radio-tracked on Peninsular Malaysia in 2005 and 2006. Grey = lowland forest; white = plantation. *Source:* From Jennings *et al.* (2010a). Reproduced by permission of the Deutsche Gesellschaft für Säugetierkunde.

There were extensive range overlaps between M1 and F1 and between M2 and F2 (Figure 19.5). Mean overall activity was similar in males (54%) and females (51%). Malay civets were significantly more active at night (88%) than during the day (20%) (Figure 19.4). Radio-tracked individuals became very active around dusk. Activity remained high during the night, although there were some periods of rest, particularly between 23:00 and 01:00 h and 03:00 and 05:00 h. Activity decreased rapidly after dawn and reached its lowest level shortly after midday. All tagged civets were located throughout the lowland forest; however, five individuals were occasionally found in the adjacent plantation during the night (Figure 19.5). Within the oil palm plantation, the maximum distance an individual was located from the adjacent forest was 597 m. All daytime locations were within dense ground cover either in the lowland forest or within small remnant forest patches in the plantation area. Two den sites were found for one female: both sites were on the ground, within patches of large tree falls and dense foliage. These were suspected to be maternal den sites because they were used every day over a three- to four-week period.

Radio-Telemetry of the Short-Tailed Mongoose

On Peninsular Malaysia, mean home range size was significantly larger in males (233 ha; $n = 3$) than in females (132 ha; $n = 2$) (Table 19.1; Figure 19.6). The mean overlap between the two female home ranges was 4%. Mean overall activity was similar in males (44%) and females (46%). Short-tailed mongooses showed a diurnal activity pattern: mean activity during the day was 85%, compared to 6% at night (Figure 19.4). During the night, short-tailed mongooses were inactive, except for occasional and very brief stirrings at the rest site. They became fully active shortly before or during the dawn period. During the morning, there was a gradual decrease in activity from 07:00 to 12:00 h, after which it peaked again. Activity decreased rapidly as dusk approached and completely ceased soon after it became dark. The mean minimum distance covered during a 24-hour period for males (842 m) was significantly larger than for females (468 m). Trapping data revealed that 78% of capture sites were < 20 m from the edge of small streams. All collared mongooses were radio-tracked within the lowland forest, except for one male, which occasionally ventured into the adjacent plantation area; however, he was never found more than 230 m from the edge of the forest, and only moved within remnant forest patches or travelled along vegetated drainage ditches within the plantation area. Rest site locations were within the lowland forest and none were in the same place. One female rest site was inside a hollow tree log.

2005   2006   2007

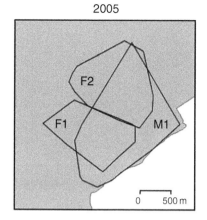
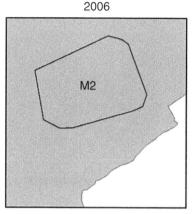
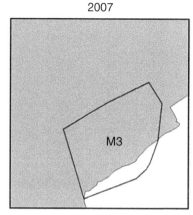

Figure 19.6   Home range polygons (95% MCP) of five short-tailed mongooses *Urva brachyura* (three males, two females), radio-tracked on Peninsular Malaysia in 2005, 2006, and 2007. Grey = lowland forest; white = plantation. *Source:* From Jennings *et al.* (2010b). Reproduced by permission of Walter de Gruyter GmbH.

Ecological Niche Modelling

Table 19.2 presents the number of records within each accuracy category, the date ranges for record groups, and the *n* values for the modelling and niche analyses. Figures 19.7–19.10 show the predicted distributions for each species, based on our Maxent modelling. The distribution models were judged to have performed well based on the high AUC values and the jacknife model testing for species with small sample sizes resulted in highly significant models.

The large Indian civet is found in mainland Southeast Asia (Cambodia, Laos, Myanmar, Peninsular Malaysia, Thailand, and Vietnam) and Singapore (Figure 19.7). Large Indian civets primarily occur in evergreen forest/scrub (84%), but are also found in degraded forest (8%), deciduous forest/scrub (7%), and

Table 19.2   Number of independent records within each accuracy category (AC 1 to AC 4), date ranges for records (AC 1 and AC 2) and (AC 3 and AC 4), and *n* values for habitat + elevation (Hab + Elv) modelling and niche analyses.

	AC 1 and AC 2			AC 3 and AC 4			Hab + Elv model/niche
	1	2	Date range	3	4	Date range	*n* = 1 + 2
V. zibetha	285	9	1993–2008	106	10	1886–2007	294
V. megaspila	61	4	1993–2008	27	3	1860–2006	65
V. tangalunga	153	145	1992–2008	199	7	1828–2007	298
Vi. indica	62	7	1993–2009	114	10	1845–2002	69
H. derbyanus	109	10	1990–2012	95	1	1838–2011	119
D. hosei	22	5	1983–2012	12	0	1891–1962	27
U. urva	83	10	1998–2008	42	23	1843–2007	93
U. brachyura	49	33	1998–2009	77	1	1884–2007	82
U. semitorquata	20	2	1998–2009	21	0	1883–2005	22
U. javanica	13	3	1997–2009	92	3	1834–2004	16

AC 1 = exact location recorded using a GPS unit; AC 2 = exact location determined using accurate maps/information; AC 3 = only a description of the locality recorded; AC 4 = reported records. *V.* = *Viverra*; *Vi.* = *Viverricula*; *H.* = *Hemigalus*; *D.* = *Diplogale*; *U.* = *Urva*. *Source:* From Jennings & Veron (2011) and Jennings *et al.* (2013). Reproduced by permission of Oxford University Press and Walter de Gruyter GmbH.

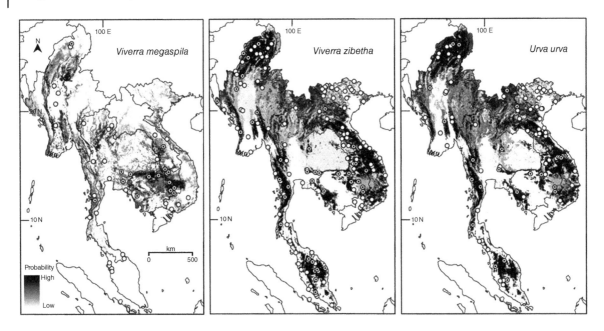

Figure 19.7 Predicted distributions for the large-spotted civet, *Viverra megaspila*, the large Indian civet, *Viverra zibetha*, and the crab-eating mongoose, *Urva urva*, on mainland Southeast Asia, based on the habitat + elevation modelling. Potential distributions are shown in grey shading, with the darker colours indicating higher probabilities of occurrence. Records AC 1 and AC 2 are shown as filled-in dots ◉, and records AC 3 and AC 4 as empty dots ○. *Source:* From Jennings & Veron (2011). Reproduced by permission of Oxford University Press.

in plantations (0.3%) (Figure 19.11). They are found over a wide elevation range, with a decreasing frequency from 0 to 2100 m (79% from 0 to 900 m, and 21% above 900 m) (Figure 19.10).

The large-spotted civet is found on mainland Southeast Asia: Cambodia, Laos, Myanmar, northwest Peninsular Malaysia, Thailand, and Vietnam (Figure 19.7). The predicted distribution is patchy and the highest probabilities of occupancy are in Cambodia and Myanmar. Large-spotted civets occur most frequently in deciduous forest/scrub (65%), followed by evergreen forest/scrub (31%), and degraded forest (5%) (Figure 19.11). Deciduous forest across northern Southeast Asia includes the mixed deciduous forests of central Myanmar and northern Thailand, and the dry dipterocarp forests in southern Laos, Cambodia, and southern Vietnam. Large-spotted civets have not been recorded at elevations higher than 600 m and most records are below 300 m (Figure 19.12).

The Malay civet is found on Borneo, Peninsular Malaysia, the Philippines (Bohol, Busuanga, Culion, Guimaras, Leyte, Lubang, Luzon, Mindanao, Mindoro,

Negro, Palawan, Panay, Samar, Sibuyan, and Siquijor Islands), Singapore, Sumatra, Sulawesi, and several other Indonesian Islands (Ambon, Bacan, Bangka, Bauwal, Billiton, Bintan, Bunguran, Buru, Halmahera, Karimata, Kundur, Laut, Lingga, Natuna, Panebangan, Rupat, Sangihe, Seram, Siao, Ternate, and Watubela) (Figure 19.8). Malay civets primarily occur in evergreen forest (80%), but are also found in degraded forest (10%), plantations (9%), and evergreen scrub (0.7%) (Figure 19.11). Although they have been recorded up to 2100 m, they are mainly found at elevations between 0 and 600 m (98%) (Figure 19.12).

The small Indian civet is found on mainland Southeast Asia (Cambodia, Laos, Myanmar, Peninsular Malaysia, Thailand, and Vietnam), northern Sumatra, Java, and several other Indonesian islands (Bali, Bawean, Bintan, Kangean, Lombok, Panaitan, and Sumbawa) (Figure 19.9). Small Indian civets occur with a similar frequency in evergreen forest/scrub (48%) and deciduous forest/scrub (43%), and are also found in degraded forest (9%) (Figure 19.11). They have been recorded at elevations up to 1500 m, but 88% of records are below 600 m (Figure 19.12).

Figure 19.8 Predicted distributions for the Malay civet, *Viverra tangalunga*, the short-tailed mongoose, *Urva brachyura*, and the collared mongoose, *Urva semitorquata*, within Southeast Asia, based on the habitat + elevation modelling. Potential distributions are shown in grey shading, with the darker colours indicating higher probabilities of occurrence. Records AC 1 and AC 2 are shown as filled-in dots ◉ , and records AC 3 and AC 4 as empty dots ◯. See **Discussion** regarding the status of the collared mongoose on Sumatra and the presence of the short-tailed mongoose and collared mongoose in the Philippines. *Source:* From Jennings & Veron (2011). Reproduced by permission of Oxford University Press.

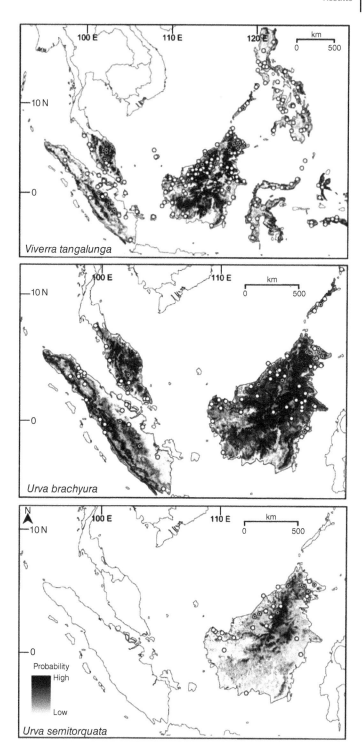

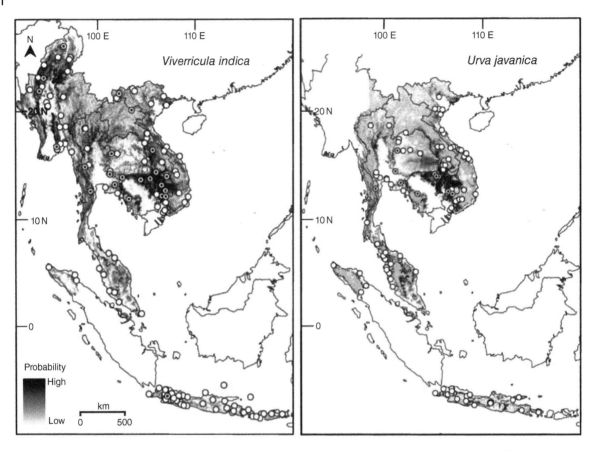

Figure 19.9 Predicted distributions for the small Indian civet, *Viverricula indica*, and the Javan mongoose, *Urva javanica*, within Southeast Asia, based on the habitat + elevation modelling. Potential distributions are shown in grey shading, with the darker colours indicating higher probabilities of occurrence. Records AC 1 and AC 2 are shown as filled-in dots ◉, and records AC 3 and AC 4 as empty dots ○. *Source:* From Jennings & Veron (2011). Reproduced by permission of Oxford University Press.

The crab-eating mongoose is found in mainland Southeast Asia: Cambodia, Laos, Myanmar, Peninsular Malaysia, Thailand, and Vietnam (Figure 19.7). Crab-eating mongooses primarily occur in evergreen forest/scrub (81%), but are also found in degraded forest (11%) and deciduous forest/scrub (9%) (Figure 19.11). They are found over a wide elevation range, with a decreasing frequency from 0 to 2100 m (74% from 0 to 900 m, and 26% above 900 m) (Figure 19.12).

The short-tailed mongoose is found on Borneo, Peninsular Malaysia, and was thought to occur in the Philippines (Palawan and Busanga Island; but see **Discussion**), and Sumatra; one specimen has been reported from southern peninsular Thailand (Figure 19.8). Short-tailed mongooses primarily occur in evergreen forest (82%), but they are also found in plantations (9%),

degraded forest (6%), and evergreen scrub (2%) (Figure 19.11). Although they have been recorded up to 1500 m, short-tailed mongooses are mainly found at elevations between 0 and 600 m (98%) (Figure 19.12).

The collared mongoose is found on Borneo (Figure 19.8); its status in Sumatra is uncertain (see **Discussion**). Collared mongooses mainly occur in evergreen/degraded forest (96%), but are also found in evergreen scrub (5%) (Figure 19.11). Most records are at elevations at 300–600 m (Figure 19.12). However, there are very limited data for this species.

The Javan mongoose is found in mainland Southeast Asia (Cambodia, Laos, Myanmar, Peninsular Malaysia, Thailand, and Vietnam), northern Sumatra, Java, and other Indonesian islands (Bali, Panaitan, and Madura) (Figure 19.9). Javan mongooses occur with a similar

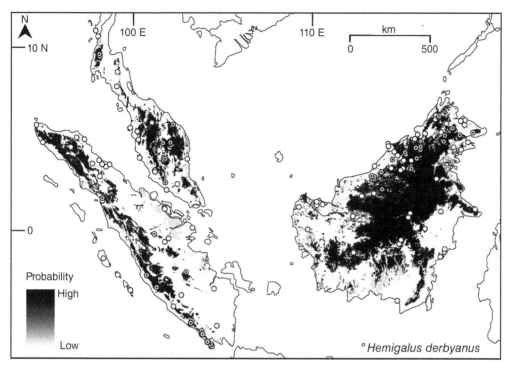

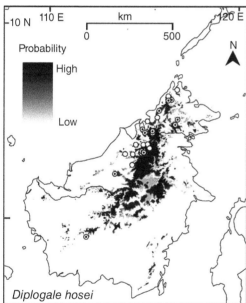

Figure 19.10 Predicted distributions for the banded civet, *Hemigalus derbyanus*, and Hose's civet, *Diplogale hosei*, within Southeast Asia, based on the habitat + elevation modelling. The predicted distribution is shown in grey shading, with darker colours indicating more suitable areas. Records AC 1 and AC 2 are shown as filled-in dots ◉, and records AC 3 and AC 4 as empty dots ○. *Source:* From Jennings *et al.* (2013). Reproduced by permission of Walter de Gruyter GmbH.

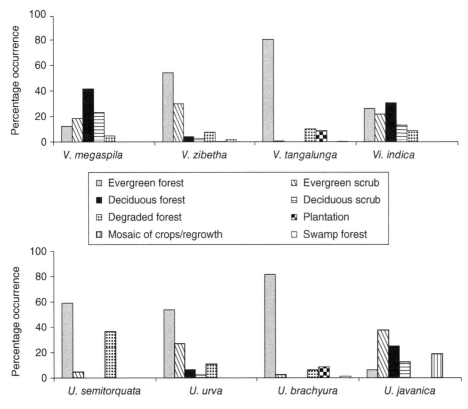

Figure 19.11 Percentage occurrence of eight civet and mongoose species in different habitat categories. *V.* = *Viverra*; *Vi.* = *Viverricula*; *U.* = *Urva. Source:* From Jennings & Veron (2011). Reproduced by permission of Oxford University Press.

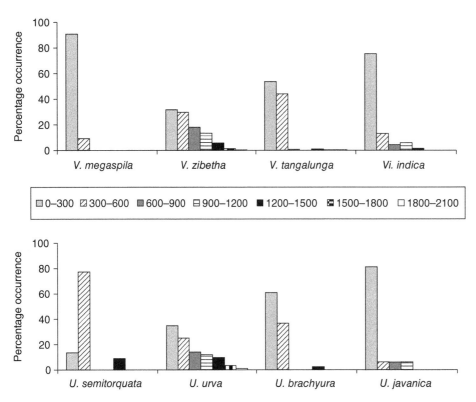

Figure 19.12 Percentage occurrence of eight civet and mongoose species at different elevations, within 300 m bandwidths. *V.* = *Viverra*; *Vi.* = *Viverricula*; *U.* = *Urva. Source:* From Jennings & Veron (2011). Reproduced by permission of Oxford University Press.

frequency in evergreen forest/scrub (44%) and deciduous forest/scrub (38%), and are also found in mosaic habitat (19%) (Figure 19.11). They have been recorded at elevations up to 1200 m, but most records are below 300 m (Figure 19.12).

The banded civet is found in southern Myanmar and Thailand, Peninsular Malaysia, and in Sumatra, Borneo, and the Mentawai Islands (Siberut, Sipora, and South Pagai; Figure 19.10), and principally occurs in evergreen forest (99%). This species has been recorded up to 1575 m, but was mainly found at elevations below 900 m (83%; Figure 19.13).

Hose's civets have only been found in evergreen forest on Borneo (Figure 19.10), and 67% of records were above 900 m (Figure 19.13).

The niche overlaps for habitat and elevation for eight species of civet and mongoose species are listed in Table 19.3. The large Indian civet and the large-spotted civet are sympatric on mainland Southeast Asia. Their elevation niches extensively overlap, particularly at low altitudes, but their habitat niche overlap is much lower: the large-spotted civet is more frequently found in deciduous forest than evergreen forest, in contrast to the large Indian civet, which primarily occurs in evergreen forest. The large Indian civet and the Malay civet have allopatric distributions, except on Peninsular Malaysia. Their niches

extensively overlap in both habitat and elevation, particularly in evergreen forest at low altitudes. The large-spotted civet and Malay civet have allopatric distributions. Their elevation niches extensively overlap as both species primarily occur in lowland areas, but their habitat niche overlap is low: the Malay civet does not occur in deciduous forest and is found more frequently in evergreen forest than the large-spotted civet. The small Indian civet is sympatric with the large Indian civet and the large-spotted civet on mainland Southeast Asia, and its distribution overlaps the Malay civet on Peninsular Malaysia and northern Sumatra. Its habitat and elevation niches extensively overlap among the three *Viverra* species, particularly with the large-spotted civet.

The crab-eating mongoose and the short-tailed mongoose have allopatric distributions, except on Peninsular Malaysia. Their niches extensively overlap in both habitat and elevation, particularly in evergreen forests at low altitudes. The short-tailed mongoose and the collared mongoose are sympatric on Borneo. Their niches extensively overlap for habitat, but their elevation niche overlap is lower: the collared mongoose occurs more frequently at higher elevations and in disturbed forests. The Javan mongoose is sympatric with the crab-eating mongoose and the short-tailed mongoose on mainland Southeast Asia. Between all three

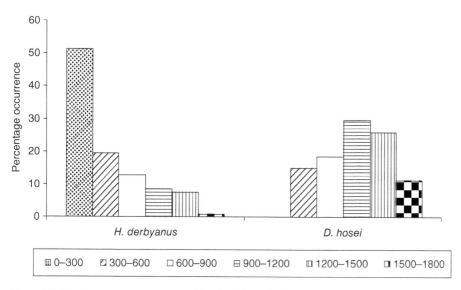

Figure 19.13 Percentage occurrence of banded civet, *Hemigalus derbyanus*, and Hose's civet, *Diplogale hosei*, at different elevations, within 300 m bandwidths. *Source:* From Jennings *et al.* (2013). Reproduced by permission of Walter de Gruyter GmbH.

Table 19.3 Niche overlap matrix for eight civet and mongoose species within Southeast Asia, calculated using Pianka's pairwise test: Pianka's index varies between 0 (total separation) and 1 (total overlap).

	V. megaspila	*V. tangalunga*	*Vi. indica*	*U. urva*	*U. brachyura*	*U. semitorquata*	*U. javanica*
V. zibetha	0.45	0.87	0.75	0.99	0.88	0.82	0.51
	0.70	0.89	0.78	0.99	0.87	0.71	0.73
V. megaspila		0.24	0.91	0.47	0.24	0.26	0.78
		0.83	0.99	0.78	0.90	0.27	0.99
V. tangalunga			0.56	0.88	0.99	0.90	0.12
			0.87	0.90	0.99	0.75	0.82
Vi. indica				0.77	0.56	0.58	0.77
				0.84	0.93	0.34	0.99
U. urva					0.89	0.86	0.49
					0.90	0.66	0.80
U. brachyura						0.88	0.14
						0.65	0.89
U. semitorquata							0.15
							0.25

For each species pair, the upper figure is the niche overlap for habitat and the lower figure the niche overlap for elevation. Niches were defined as the frequency of occurrence within each habitat and elevation category, using records AC 1 and AC 2. *V.* = *Viverra*; *Vi.* = *Viverricula*; *U.* = *Urva*. *Source:* From Jennings & Veron (2011). Reproduced by permission of Oxford University Press.

species, the habitat niche overlaps are fairly low, but their elevation niche overlaps are high.

The habitat niches of the banded civet and Hose's civet overlap extensively (Pianka's index = 0.9), as both species primarily occur in evergreen forests. The elevation niche overlap is lower (Pianka's index = 0.4): the banded civet is principally a lowland species, whereas Hose's civet is mainly found in montane regions. In 2010, the percentage of evergreen forests around the locations of historical records was 55% lower around recent records for the banded civet and was 21% lower for Hose's civet (Figure 19.14). The proportion of predicted presence within protected areas is 24% for the banded civet and 39% for Hose's civet.

Camera-Trapping in Oil Palm Plantations

A total of 3164 camera-trap days was completed and 318 independent photographs of three small carnivore species were recorded: common palm civet, Malay civet, and leopard cat, *Prionailurus bengalensis* (Jennings *et al.,* 2015). The latter is now regarded as a separate species on the Sunda islands, namely Sunda leopard cat, *Prionailurus javanensis* (Luo *et al.,* 2014; Patel *et al.,* 2017) (Figure 19.15). All three small carnivore species were detected in Estate A, but only two in Estate B, in which the Malay civet was not detected (Table 19.4). All independent camera-trap detections of the common palm civet and Malay civet were during the night/crepuscular period.

The common palm civet was found up to 3.6 km from the edge of the plantation, whereas the Malay civet was only detected in the oil palm < 1 km from the surrounding habitat (Table 19.4). The maximum distance that the Malay civet was detected from an extensive area of lowland forest was 32 km, while it was 46 km for the common palm civet (Table 19.4).

For both civet species, no single model contained the majority of support, but the combined Akaike weights for the two highest-ranked models was > 0.73 for the common palm civet, and > 0.61 for the Malay civet (Table 19.5). For the common palm civet, neither covariate had a strong influence on occupancy; however, the probability of detection was primarily a

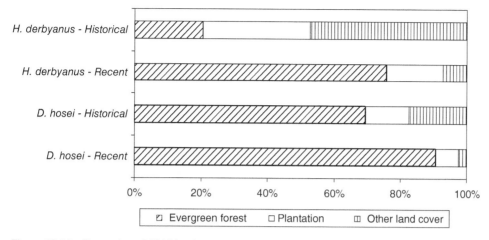

Figure 19.14 Proportion of 2010 land cover classes within 5 km radius buffers around the localities of historical and recent records for banded civet, *Hemigalus derbyanus*, and Hose's civet, *Diplogale hosei*. *Source:* From Jennings *et al.* (2013). Reproduced by permission of Walter de Gruyter GmbH.

Figure 19.15 Populations of the leopard cat, *Prionailurus bengalensis*, living on the Sunda islands (Sumatra, Java, Bali, and Borneo) and in some Philippine islands are now regarded as belonging to a separate species, namely Sunda leopard cat, *P. javanensis*. The individual on the picture was photographed in Deramakot, Sabah, Malaysia. *Source:* Photo © Chien C. Lee (chienclee.com).

function of distance from lowland forest, and to a lesser extent, distance from the plantation edge (Table 19.5). For the Malay civet, occupancy was influenced by distance from the plantation edge, and distance from primary forest had a strong effect on the detection probability (Table 19.5).

Model fit was reasonably good for the Malay civet, but poor for the common palm civet (Table 19.5). Based on model-averaged parameters (from the top-ranking models with $\Delta AIC_c < 7$), the common palm civet has higher occupancy and detection probability values than the Malay civet (Table 19.6).

Discussion

Spatio-Temporal Ecology of Malay Civet and Short-Tailed Mongoose

A wide range of home range sizes has now been documented for the Malay civet (Buton Island: 24–189 ha; Borneo: 27–283 ha; and Peninsular Malaysia: 23–185 ha; Macdonald & Wise, 1979; Nozaki *et al.*, 1994; Colon, 2002; Jennings *et al.*, 2006, 2010a; Evans *et al.*, 2021), with mean range sizes (minimum convex polygons) largest on Peninsular Malaysia, intermediate on Borneo, and smallest on Sulawesi (Jennings *et al.*, 2010a). McLoughlin & Ferguson (2000) suggested that at the population level, habitat productivity or food availability may be the primary determinant of home range sizes, and interspecific competition, habitat utilization, and diet may be contributory factors (St-Pierre *et al.*, 2006). We have insufficient data on the diet of the Malay civet and food availability in the different study areas to determine how these factors affect home range sizes across its range, but available data suggest that Borneo and Sulawesi have lower primary productivity than Peninsular Malaysia (O'Donovan, 2001; Meiri *et al.*, 2008). Colon (2002) found that the population

Table 19.4 Details of camera-trap sites and summary of photographic data obtained for small carnivores during the surveying period (1 March 2012–2 April 2013), within two mature oil palm estates, central Sumatra: camera-trap site name; camera-trap site coordinates (decimal degrees); minimum distance of each camera-trap to the nearest edge of the oil palm habitat (kilometres); minimum distance of each camera-trap from the nearest edge of primary forest (kilometres); number of camera-trap days at each site; number of independent photographs of each small carnivore species.

Estate	Site	Site coordinates (dd)		Dist. to edge (km)	Dist. from forest (km)	Trap days	No. of independent photos		
		Long.	Lat.				LC	CPC	MC
A	LI001	101.2124	0.9128	0.1	45.6	235	20	19	0
A	LI002	101.1873	0.9173	2.8	43.7	243	10	3	0
A	LI003	101.2173	0.9200	0.2	45.1	199	31	4	0
A	LI004	101.2162	0.9648	0.9	40.8	122	36	4	0
A	LI005	101.2231	0.9152	0.1	46.0	33	2	0	0
A	LI006	101.1885	0.9236	2.6	43.3	120	0	0	0
A	LI007	101.2113	0.9452	1.6	42.4	132	11	0	0
A	LI008	101.1876	0.9482	3.4	40.8	71	1	1	0
A	LI009	101.1794	0.9605	3.6	39.2	145	12	4	0
A	LI010	101.1647	0.9546	1.8	39.0	61	6	0	0
A	LI011	101.2014	0.9540	2.5	41.0	113	2	1	0
					Total	**1474**	**131**	**36**	**0**
B	RA001	101.1075	0.5632	1.2	36.0	335	7	2	0
B	RA002	101.0564	0.5520	0.1	31.9	391	45	2	1
B	RA003	101.0361	0.5228	0.9	28.0	280	19	2	4
B	RA004	101.0499	0.5377	0.1	30.2	246	27	1	0
B	RA005	101.1063	0.5314	3.2	33.0	146	1	0	0
B	RA006	101.0389	0.5383	1.1	29.6	146	29	1	0
B	RA007	101.1209	0.5337	3.6	34.3	146	5	0	0
					Total	**1690**	**133**	**8**	**5**

LC = leopard cat, *Prionailurus bengalensis* (now regarded as Sunda leopard cat, *P. javanensis*); CPC = common palm civet, *Paradoxurus hermaphroditus*; MC = Malay civet, *Viverra tangalunga*. *Source:* From Jennings *et al.* (2015). Reproduced by permission of Springer Science+Business Media.

density of Malay civets in a logged forest was lower than that in an unlogged forest and suggested that this may be due to lower fruit availability in the logged forest. Sulawesi has an impoverished mammalian carnivore guild comprising just three civet species (Sulawesi palm civet, Malay civet, and common palm civet; Veron, 2001; Veron *et al.*, 2014; Hunowu *et al.*, 2020), and the lack of interspecific interference or competition for food from other mammalian carnivores on Buton (compared to Borneo, where 22 carnivore species are present; Corbet & Hill, 1992) may result in an

increased prey base available for Malay civets on this island, allowing both sexes to meet their daily energy requirements within smaller home ranges. However, other animal species present on Buton may be potential competitors for food. On both Peninsular Malaysia and Sulawesi, we found a low intrasexual overlap of home ranges, and each male's range overlapped one or two females (Jennings *et al.*, 2006, 2010a). Colon (2002), however, found considerable home range overlap within both sexes on Borneo and concluded that the Malay civet was not territorial. Among solitary

Table 19.5 Summary of the top-ranking occupancy models (ΔAIC$_c$ < 7) for the common palm civet and Malay civet detected within two oil palm plantations in central Sumatra.

Species	Model	K	ΔAIC$_c$	w	p
Common palm civet	ψ(.), p(forest)	3	0.00	0.382	0.010
	ψ(.), p(edge + forest)	4	0.15	0.354	0.068
	ψ(forest), p(forest)	4	2.05	0.137	0.011
	ψ(edge), p(forest)	4	2.41	0.114	0.019
	ψ(edge + forest), p(edge + forest)	6	6.84	0.012	0.051
Malay civet	ψ(.), p(forest)	3	0.00	0.440	0.314
	ψ(edge), p(forest)	4	1.85	0.174	0.344
	ψ(.), p(edge + forest)	4	3.17	0.090	0.288
	ψ(forest), p(forest)	4	3.28	0.085	0.275
	ψ(forest), p(.)	3	4.14	0.055	0.194
	ψ(edge), p(edge)	4	4.31	0.051	0.220
	ψ(edge + forest), p(.)	4	5.18	0.033	0.128
	ψ(forest), p(edge)	4	5.66	0.026	0.306
	ψ(.), p(.)	2	6.35	0.018	0.025

K is the number of parameters in the model; ΔAIC$_c$ is the difference in AIC$_c$ values between each model and the lowest AIC$_c$ model; w is Akaike's weight; the p-value is an assessment of model fit and is the probability of observing a test statistic $\geq \chi^2$ based on 1000 parametric bootstraps. The covariates were: minimum distance of each camera-trap site to the nearest edge of the oil palm habitat (edge), and minimum distance of each camera-trap site from the nearest edge of primary forest (forest). *Source:* From Jennings *et al.* (2015). Reproduced by permission of Springer Science+Business Media.

Table 19.6 Naïve occupancy, model-averaged occupancy, and detection probability estimations for the two civet species found within two oil palm plantations in central Sumatra: naïve occupancy is the percentage of sampling sites at which a species was recorded.

Species	Naïve occupancy	Occupancy (SE)	Detection probability (SE)
Common palm civet	0.667	0.824 (0.141)	0.160 (0.023)
Malay civet	0.111	0.286 (0.152)	0.049 (0.028)

SE = standard error.
Source: From Jennings *et al.* (2015). Reproduced by permission of Springer Science+Business Media.

carnivores, exclusive home ranges are expected only when food resources are stable and evenly distributed (Sandell, 1989). Joshi *et al.* (1995) indeed found that home range overlap in common palm civets was minimal when food was abundant and uniform, and increased when food was most clumped. Colon (2002) expected food abundance and distribution in rainforests to vary in space and time and suggested that this accounted for the overlapping home ranges of Malay civets in Borneo. Perhaps differences in both diet and food availability accounted for the different spatial organization that we observed in Peninsular Malaysia and Sulawesi. Sandell (1989) also argued that spatial organization in female solitary carnivores is primarily determined by food resources, whereas males are influenced by both food and females. We would therefore expect female home ranges to be more closely correlated with food availability and habitat productivity than male ranges. This may explain why the mean female home range size of Malay civets on Borneo and Sulawesi were significantly smaller than on Peninsular Malaysia, whereas mean male ranges, although also smaller on these two islands, were not significantly different (Jennings *et al.*, 2010a). Our telemetry studies showed that the Malay civet is mainly nocturnal on Peninsular Malaysia and Sulawesi, although civets on Sulawesi had significantly higher diurnal activity, possibly due to differences in competitors or food resources between the two areas.

For the short-tailed mongoose, it is difficult to make comparisons between our home range size estimations and those of similar species, as there have been few studies of rainforest mongooses. The home range of a female crab-eating mongoose in China was at least 100 ha (Wang & Fuller, 2001). In the Central African Republic, the mean home range size of male long-nosed mongooses, *Xenogale naso* (42 ha; Ray, 1997), is much smaller than that of male short-tailed mongooses (233 ha). The home ranges of marsh mongooses, *Atilax paludinosus*, were generally linear in shape due to their fidelity to watercourses: one radio-collared male travelled up and down a stream ~5 km in length; its adjusted home range size was estimated at 54 ha (Ray, 1997). The home ranges of short-tailed mongooses are thus larger than those of other similar-sized rainforest mongooses. No dietary studies have been undertaken on the short-tailed mongoose to be able to determine the influence of diet and food availability on home range sizes, but we found that they were often located close to small rivers and streams, where presumably they were foraging.

Socio-spatial organization in the Herpestidae is variable, with several African mongoose species forming social groups (Gilchrist *et al.*, 2009). Our telemetry study on the short-tailed mongoose showed that it is a solitary, territorial species: females had exclusive home ranges, and males had larger home ranges than females and overlapped the range of more than one female (Jennings *et al.*, 2010b). In Africa, home ranges of male white-tailed mongooses, *Ichneumia albicauda*, were also found to be generally larger than those of females; males had exclusive home ranges, and male ranges overlapped a single female's home range (Ermias Admasu *et al.*, 2004). However, Ray (1997) found a high degree of range overlap between long-nosed mongoose individuals in streamside habitat, and Cavallini & Nel (1990) found that the home ranges of the Cape grey mongoose, *Galerella pulverulenta*, overlapped widely, both within and between sexes. Field studies on the Egyptian mongoose, *Herpestes ichneumon* (Ben Yaacov & Yom-Tov, 1983; Palomares & Delibes, 1993), and small Indian mongoose, *Urva auropunctata* (Gorman, 1979; Nellis & Everard, 1983; Roy *et al.*, 2002; Hays & Conant, 2003), have shown that there is flexibility in their social system, depending on the local circumstances. Abundant food resources have been suggested as an important factor that may facilitate group formation in mongooses (see Rood, 1986, 1989).

Our telemetry study showed that the short-tailed mongoose is diurnal (Jennings *et al.*, 2010b). Although little is known about the other Asian mongoose species, they are also believed to be mainly diurnal (Gilchrist *et al.*, 2009; Ross *et al.*, 2017), except for the Indian brown mongoose *Urva fusca* (Mudappa, 2001). The activity patterns of small carnivores are influenced by a number of factors including daily temperatures, limitations of the visual system, social behaviour, interference from competitors, and predation pressure (Palomares & Caro, 1999; Zielinski, 2000). Foraging activity is often related to food availability, and most species time their hunting to coincide with the movements of their prey (Zielinski, 2000). Unfortunately, we have insufficient information on Southeast Asian civets and mongooses to be able to relate these various factors to their activity patterns.

Distribution and Conservation of Southeast Asian Civets and Mongooses

Distribution Patterns

Our modelling and niche studies have produced predicted distribution maps for six civet and four mongoose species, revealed their potential habitat/elevation preferences and indicated areas within Southeast Asia where the presence of some viverrids and herpestids is uncertain (Jennings & Veron, 2011; Jennings *et al.*, 2013). For instance, the status of the small Indian civet and the Javan mongoose on Sumatra is unclear, with only a small number of museum records reported from the northern part of the island. Our analyses for these two species might have underestimated their predicted presence in open areas since most of their occurrence data used in the modelling were from forested habitat, whereas both species are reported to occur in more open habitats, such as scrubland and grasslands (Wells, 1989; Austin & Tewes, 1999). Thus, further field studies are needed to determine their exact habitat preferences.

The status of the collared mongoose on Sumatra is uncertain, with few records reported for this species on this island: two specimens from the Ophir District and one from Soekadana, and three recent camera-trap photographs (Patou *et al.*, 2009; Ross *et al.*, 2012; Holden & Meijaard, 2013). It is unclear whether this is an overlooked and under-recorded rare native species on Sumatra, or one that was possibly introduced (Veron *et al.*, 2015a). Little is known about the collared mongoose, and it is classified as Near Threatened by *The IUCN Red List of Threatened Species* (Mathai *et al.*, 2015a). A recent phylogenetic study (Veron *et al.*, 2015a, Chapter 3, this volume) has found that mongooses from Palawan Island (in the Philippines), which had been thought to be short-tailed mongooses, do not cluster with the other populations of this species, but are closer to the collared mongoose.

On Java, only four Malay civet and four short-tailed mongoose specimens have been reported; none of these had any precise locality information, and some were zoo animals. As no other records exist, it is unlikely that Java is a part of the native range of these two species.

On the Mentawai Islands, specimens of the banded civet have been collected from Sipora and South Pagai islands, but its presence on Siberut Island needs to be confirmed (Schreiber *et al.*, 1989; C. Abegg, D. Boussarie & M. Quinten, personal communications). The banded civet is not known to occur on any other small islands, which suggests that this species may have not been transported around Southeast Asia by humans, as have some other viverrid species (Jennings & Veron, 2011; Veron *et al.* 2014), or that ecological or biogeographical factors restricted its presence on other Southeast Asian islands.

As for the large-spotted civet, several authors had previously included all of Peninsular Malaysia in its distribution (e.g. Flower, 1900; Corbet & Hill, 1992; Francis, 2008). On closer inspection, it appears that Flower (1900) considered *V. tangalunga* (Malay civet) and *V. megaspila* (large-spotted civet) as synonyms, and thus he was actually referring to the Malay civet for Peninsular Malaysia. Although there is a large-spotted civet specimen in the London Natural History Museum labelled from Malacca, on the southwest coast of Peninsular Malaysia, this place was a centre of trade at the time (Wells, 1989). The only confirmed records of the large-spotted civet in this region are in northwest Peninsular Malaysia (Kedah, Penang, and Perak States); no recent camera-trapping surveys across central or southern Peninsular Malaysia have detected this species. Thus, it appears that the southern range limit of this species is close to the Thai–Malaysian border (Jennings & Veron, 2011), which has very important conservation implications for this threatened civet, as its range is less extensive than previously considered (Duckworth *et al.*, 2008), and which has been updated for the IUCN Red List assessment (Timmins *et al.*, 2016). Furthermore, our niche analyses suggest that the large-spotted civet is more restricted to tropical lowland deciduous forest than was previously thought (Duckworth *et al.*, 2008). In eastern Cambodia, Gray *et al.* (2010) found that the large-spotted civet was camera-trapped more frequently in deciduous dipterocarp forest than in mixed deciduous and semi-evergreen forests.

Hose's civet is a threatened species that occurs only on Borneo (Mathai *et al.*, 2015b), for which there are only a few museum specimens and a small number of sightings and camera-trap records (Figure 19.16); hence, its distribution across this island was poorly known. Our distribution modelling predicted that it mainly occurs across the higher elevation regions of Borneo (Jennings *et al.*, 2013), which is also supported by the habitat suitability modelling of Mathai *et al.* (2016, 2019). The lowest recorded elevation for Hose's civet is at 325 m (Samejima & Semiadi, 2012). There are two recent, unconfirmed reports from the lowland forests of the Lower Kinabatangan Wildlife Sanctuary (elevations below 300 m) in northeast Sabah: one possible sighting around 2005 (Mohd Azlan, personal communication), and another in April 2009 (Anonymous, 2009; R. Munds, personal communication). However, our investigations revealed that these are doubtful records, and we surmise that Hose's civet is unlikely to occur in this lowland area of northeast Sabah. As for other outlier

Figure 19.16 Hose's civet, *Diplogale hosei*, photographed on 8 February 2020 at an elevation of ~1500 m in Gunung Mulu, Sarawak, Malaysia. Although Mulu National Park is located within the predicted range of this species, this is the first record of Hose's civet in the park. *Source:* Photo © Chien C. Lee (chienclee.com).

records, a published photograph of a possible Hose's civet camera-trapped in Lanjak Entimau Wildlife Sanctuary, central Sarawak (Van Rompaey & Azlan, 2004), could instead be a masked palm civet or a yellow-throated marten, *Martes flavigula* (Jennings *et al.*, 2013).

Our predicted distribution maps can help guide future surveying and conservation priorities. For example, we predicted that Hose's civet has a limited distribution across the higher elevations of Borneo (Jennings *et al.*, 2013). As most of the records were from Brunei, Sarawak, and Sabah, it is imperative that field surveys are undertaken in the montane regions of Kalimantan to determine the true extent of this species' distribution. If the range of Hose's civet is even more restricted than the predictions suggest, then this would have major implications on its conservation status; carnivore species with small geographic ranges

have a high risk of extinction (Purvis *et al.*, 2000; Cardillo *et al.*, 2004).

Interspecific Competition and Niche Differentiation

Interspecific competition is one of several factors that could account for our predicted distribution patterns (Case & Gilpin, 1974). Similar species may avoid competition by having different geographical ranges (Caro & Stoner, 2003) or by using a variety of mechanisms that allow them to co-exist (Ross *et al.*, 2017). For instance, several studies have shown that sympatric carnivore species tend to have different morphologies, diets, spatial preferences, or activity patterns (Chuang & Lee, 1997; Davies *et al.*, 2007; Chen *et al.*, 2009; Di Bitetti *et al.*, 2009). Closely related, morphologically similar species commonly share ecological requirements, so there is often some degree of niche differentiation to allow them to co-exist or a species' distribution may be restricted by related species (Anderson *et al.*, 2002; Lucherini *et al.*, 2009; Di Bitetti *et al.*, 2010; Jennings & Veron, 2011). The three *Viverra* civet species are closely related and morphologically similar to one another (Jennings & Veron, 2009). Although large-spotted and large Indian civets are sympatric on mainland Southeast Asia, our analyses suggest that habitat partitioning occurs between these two species, as the large-spotted civet is found more frequently in deciduous forest, whereas the large Indian civet mainly inhabits the evergreen forest. Furthermore, the large-spotted civet is restricted to lower altitudes, whereas the large Indian civet can occur over a wider elevation range. So, perhaps interspecific competition between these two species is also minimized through differences in elevation preferences. It appears that the Malay civet avoids interspecific competition with the large-spotted civet through geographical separation. Banded and Hose's civets are closely related and both primarily occur in evergreen forest, yet they may separate spatially along an elevation gradient: our predicted distribution modelling found that the banded civet mainly occurs in lowland areas and Hose's civet is mostly found at higher elevations. Ross *et al.* (2016b) modelled the suitability of habitats on Borneo for the banded civet and found that it had some preference for lowland forest. Mathai *et al.* (2016, 2019) conducted similar modelling for Hose's civet and predicted that it was restricted to the mountainous interior of Borneo. However, as there is a broad overlap of elevations at which each species can occur, and both species have been recorded at the same locations on Borneo, this elevation niche pattern does not conclusively demonstrate competitive exclusion between these two civets. In addition, the banded civet is found in Sundaic regions where Hose's civet is absent, which suggests that competitive interactions with other species might also explain its niche characteristics.

Asian mongoose species are closely related and morphologically similar to one another, although the Javan mongoose is more distantly related to the other Southeast Asian mongooses (Patou *et al.*, 2009). Short-tailed and collared mongooses are sympatric on Borneo. Our ecological niche modelling results showed that the collared mongoose had a higher predicted occurrence at elevations between 300 and 600 m and occurred more frequently in evergreen scrub than the short-tailed mongoose. In Gunung Gading National Park on Borneo, Mohd-Azlan *et al.* (2020) only recorded the collared mongoose in hill and submontane forest from 300 to 700 m. The habitat suitability modelling of Hon *et al.* (2016) predicted that both lowland and upland forests on Borneo were the most suitable for the collared mongoose, and Duckworth *et al.* (2016) found that 90% of the Borneo records of the short-tailed mongoose were from below 600 m. These findings indicate that although there is considerable overlap between these two species, there may be some macrohabitat and elevation partitioning that allows them to co-exist; there may also be differences in their microhabitat preferences, diet, or activity patterns in areas where they co-occur (Wilting *et al.*, 2010b; Ross *et al.*, 2017). However, there are limited data for the collared mongoose and short-tailed mongoose and further field studies are needed to investigate interspecific competition between these two mongoose species.

Short-tailed and crab-eating mongooses are both found in evergreen forest, but their distributions are allopatric (except on Peninsular Malaysia), which suggests that interspecific competition between these two species is mainly avoided through geographical separation. The Javan mongoose is sympatric with crab-eating and short-tailed mongooses on mainland Southeast Asia. However, the Javan mongoose appears to be a forest generalist and may occur more frequently in open habitat, or possibly has other niche preferences that would allow it to co-exist with the other mongoose species.

Our modelling studies also revealed that three pairs of civets and mongooses have similar distribution patterns and habitat/elevation preferences. Although there are differences in body size between these pairs of viverrids and herpestids, there is potential for interspecific competition for food resources between these two groups (Chuang & Lee, 1997; Davies *et al.*, 2007). However, each of the civet species in our study is primarily nocturnal (Jennings & Veron, 2009), whereas each of the mongoose species appears to be diurnal (Gilchrist *et al.*, 2009; Ross *et al.*, 2017), which suggests that differences in activity patterns may reduce interspecific competition between these two taxonomic groups.

Despite the clear patterns presented above, these do not conclusively demonstrate competitive exclusion; for instance, geographical separations might have resulted from historical causes. Competitive interactions with other Southeast Asian carnivore species also need to be taken into consideration. To confirm any interspecific competition, field studies in contact zones between related species are necessary in order to determine their microhabitat usage, behaviour, and food habits. For example, in Peninsular Malaysia, where the Malay civet and the short-tailed mongoose co-occur with the large Indian civet and the crab-eating mongoose, based on our analyses, we predict that the latter two species are restricted to marginal forested habitat, particularly at higher elevations. Unfortunately, we could not gather sufficient data for these species on Peninsular Malaysia to reliably test this prediction, but within Krau Wildlife Reserve, where both the crab-eating and short-tailed mongooses have been recorded, the crab-eating mongoose has only been detected at high elevations (above 640 m), whereas the short-tailed mongoose is only found in lowland areas.

The Possible Role of Biogeography

Biogeography is another factor that could account for our predicted distribution patterns. In Southeast Asia, environmental fluctuations in the last few million years have shaped today's biogeographic patterns and regional biodiversity (Woodruff, 2010). On the Thai peninsula, the Isthmus of Kra (~10°N) has traditionally been recognized as a major transition zone between the Indochinese and Sundaic zoogeographic subregions (Meijaard, 2009; Woodruff & Turner, 2009). However, the range limits of mammals cluster in northern Peninsular Malaysia (5°N) and 800 km further north (14°N), where the peninsula joins the main continent (Woodruff & Turner, 2009). There are no obvious geophysical barriers separating these two areas, but Woodruff & Turner (2009) have suggested that repeated, rapid sea-level changes in the last 5 million years resulted in compressed species populations and local extirpations. The large-spotted, Malay, and banded civets and the short-tailed mongoose have range limits in southern Thailand/northern Peninsular Malaysia; it is thus possible that the above scenario could have restricted the distributions of these species, although other factors, such as interspecific competition or climate, may have played more important roles. Why some species that occur in both northern and southern areas of Southeast Asia, such as the small Indian civet, do not appear to have been impacted by these hypothesized peninsular events is not clear. Meijaard (2009) suggested that the distinction between the Indochinese and Sundaic faunas may be maintained by ecology rather than by localized extinction patterns. The distribution of the large-spotted civet, for example, may simply be due to its habitat preference for tropical deciduous forest, which mainly occurs across northern Southeast Asia.

At the Last Glacial Maximum (around 20 000 years ago), a lowering of the sea level by ~120 m exposed a land mass about the size of Europe on the Sunda Shelf: the large islands of Sumatra, Borneo, and Java were connected to mainland Asia, along with smaller islands (Woodruff, 2010). The climate might have been drier, colder, and more seasonal, and a continuous north-south corridor of open savannah vegetation might have existed through this continent, from Peninsular Malaysia to Java (Meijaard, 2003; Bird et al., 2005; Cannon et al., 2009). Tropical forest was mainly restricted to the mountainous regions of Sumatra and Borneo, and other small areas (Meijaard, 2003), which would have acted as refugias for forest-dependent species, such as the Malay civet and the short-tailed mongoose. Some models, however, predict a more extensive coverage of evergreen forest across the Sundaland at the Last Glacial Maximum (Cannon et al., 2009), and gallery forests could have occupied valley areas throughout the region (Bird et al., 2005), which could have allowed the dispersal of forest civet and mongoose species over the Sunda Shelf (Veron et al., 2014). A savannah corridor through the Sundaland would have provided a dispersal route from Peninsular Malaysia to Java for species that favoured more open habitats, such as the small Indian civet and the Javan mongoose. This savannah corridor would have separated forested areas to the west and east and served as a barrier to the dispersal of forest-dependent species between Sumatra and Borneo (Meijaard, 2003; Bird et al., 2005); the large rivers that dissected the exposed Sunda Shelf may also have acted as dispersal barriers

(Meijaard, 2003). Phylogenetic studies have found little genetic divergence between short-tailed mongooses from Peninsular Malaysia and Sumatra, while those from Borneo are more genetically distinct (Patou *et al.*, 2009; Veron *et al.*, 2015a, Chapter 3, this volume). Other molecular studies have also shown a strong differentiation of Bornean mammal populations (e.g. Gorog *et al.*, 2004; Patou *et al.*, 2008), suggesting that dispersal movements of forest-dependent species between Borneo and the rest of the Sunda Shelf were restricted during the last glacial period. Around 10 000 years ago, higher sea levels started to sever land connections and separate the islands of the Sundaland (Corlett, 2009). This allowed for further genetic and morphological divergences to occur between these isolated populations.

Human Transportation

Humans have moved civet and mongoose species from one area to another for a variety of reasons. For instance, terrestrial civets are farmed in some parts of Asia for the production of civetone, which is used in the perfume industry and traditional medicine (Jennings & Veron, 2009). Also, several viverrid and herpestid species are traded in local and international markets for food and as pets (Corlett, 2007; Shepherd, 2008; Jennings & Veron, 2019), or may be introduced to agricultural areas in order to catch rodents (Gilchrist *et al.*, 2009; Jennings & Veron, 2009, 2019; Louppe *et al.*, 2021). These translocations might account for the rare occurrence of some species on islands on which they may not be native (e.g. the Malay civet on Java, and the small Indian civet and the Javan mongoose on Sumatra). Human introductions may also account for the presence of the small Indian civet on some Indonesian islands (such as Lombok and Sumbawa), and the Malay civet on several islands in the Philippines and Wallacea region (Musser, 1987; Reis & Garong, 2001; Veron, 2001; Veron *et al.*, 2014), as these areas were separated from the Sundaland by deep-water channels and no land bridges have ever existed between them (van den Bergh *et al.*, 2001). However, introductions of these species are difficult to verify, as there are virtually no fossil or historical records, but molecular studies may shed some light on these events (see Veron *et al.*, 2014).

The Impacts of Deforestation, Habitat Fragmentation, and Modification

Southeast Asia has the highest rate of deforestation of any tropical region (1–2% per year), and if present levels continue unabated, this region could lose up to three-quarters of its original forest cover by 2100 and 13–85% of its biodiversity (Sodhi *et al.*, 2010a). Deforestation and land conversion have caused range contractions in many mammal species in Southeast Asia (Sodhi *et al.*, 2004, 2010a; Corlett, 2007), which could account for the absence of some species in certain areas (e.g. on Java where extensive habitat changes have occurred since humans colonized this island). Our studies have suggested that several civet and mongoose species are dependent on forested habitat and could, therefore, be threatened with extinction as a direct result of deforestation. For banded and Hose's civets, the percentage of evergreen forest in 2010 was much lower around the localities of historical records than recent records (particularly for the banded civet), which indicates that a large amount of suitable forested habitat has disappeared since the 1800s (Jennings *et al.*, 2013). Lowland forest is particularly vulnerable and disappearing fast (Meijaard & Sheil, 2008), and tropical montane cloud forests are also being destroyed at alarming rates (Peh *et al.*, 2011).

It is poorly known how Southeast Asian civet and mongoose species are impacted by logging activities, which is a major cause of forest degradation throughout Southeast Asia (Meijaard & Sheil, 2008; Sodhi *et al.*, 2010a,b). Heydon & Bulloh (1996) found that the density of civet species was significantly lower in logged forest than in primary forest. Meijaard & Sheil (2008) suggested that logging likely affects civet species negatively. The analyses of Meijaard *et al.* (2008) found that a species' phylogenetic age best predicts its sensitivity to timber harvest and thus suggested that several civet species could be severely impacted by logging. Since strictly protected areas cannot conserve the full biological diversity found within tropical forests, the fate of many species depends on what happens to forests outside protected areas. Also, forest cover is declining even within national parks and forest reserves due to illegal logging (Meijaard & Sheil, 2008). Therefore, we need to gather information on the sensitivity of civet and mongoose species to forest degradation in order to better understand what measures could be implemented to mitigate any detrimental impacts from logging.

Habitat fragmentation influences the abundance, movements, and persistence of many species, and forest carnivores are particularly vulnerable to local extinction in fragmented landscapes (Crooks, 2002; Michalski & Peres, 2005; Charles & Ang, 2010). Unfortunately, we have almost no data for assessing the impacts and extinction risks of forest fragmentation on Southeast Asian civet and mongoose species. Their dispersal abilities and metapopulation dynamics are unknown. For instance, Hose's civet has a patchy distribution across montane regions, and the viability of this metapopulation may be greatly affected by this species' ability to disperse through lower altitude habitats (Mathai *et al.*, 2016). The presence of roads throughout a forested landscape fragments major forest complexes, increases the probability of direct mortality due to vehicular traffic, and allows greater human access for logging and hunting (Meijaard & Sheil, 2008; Sodhi *et al.*, 2010b), but what impacts roads might have on civet and mongoose populations is currently unknown.

For some civet species, we calculated the amount of predicted distribution that is currently under protected areas. In order to increase this level of protection further, protected forests could be expanded to incorporate a greater proportion of each species' predicted range. However, since many human-related factors (e.g. agricultural and logging activities) would limit the expansion of protected areas, field studies are urgently needed to provide information on what would be the most effective means of increasing the level of habitat protection for Southeast Asian civet and mongoose species. For instance, protecting forested corridors between isolated forests could be given a high priority; yet we do not know the dispersal capabilities of these species through such habitat features, or the optimal conditions that may facilitate the movement of each species from one forest area to another.

The impact of human-modified habitats, such as farmland and oil palm plantations, on Southeast Asian civet and mongoose species is poorly known. Several forest-dependent species may have little or no tolerance to anthropogenic changes to a forested landscape.

In Indonesia and Malaysia, oil palm cultivation is one of several threats to tropical biodiversity due to the conversion of native forests to oil palm plantations (Sodhi *et al.*, 2010b; Wilcove & Koh, 2010). The area of mature oil palm plantations in Sumatra, for example, reached a total of 4.7 million ha by 2010, representing 10% of its total land area (Gunarso *et al.*, 2013). Although the sample size for our camera-trapping study within two oil palm plantations was limited, our preliminary results did produce useful information. We only detected three small carnivore species: leopard cat, common palm civet, and Malay civet (Naim *et al.*, 2012; Jennings *et al.*, 2015), which suggests that oil palm habitat supports few small carnivores, especially compared to the 23 species that have been recorded in natural forests across central Sumatra (Holden, 2006; Veron *et al.*, 2006; Maddox *et al.*, 2007; Wilting *et al.*, 2010a; Jennings & Veron, 2011; Jennings *et al.*, 2013). Similarly, in a logged forest and oil palm landscape in Jambi province, central Sumatra, only four small carnivore species (among 14 found in the area) were detected within the oil palm plantation (Maddox *et al.*, 2007).

We found that the common palm civet had a high occupancy value within oil palm, and was recorded deep within the plantations (up to at least 3.6 km from the edge; Jennings *et al.*, 2015). This civet species has been found in an array of human-modified habitats (Jennings & Veron, 2009). It is nocturnal, mainly arboreal and frugivorous, but also eats small vertebrates and invertebrates (Jennings & Veron, 2009). Within an oil palm landscape on Borneo, oil palm fruits were a major component of the diet of common palm civets, followed by millipedes, rodents, and insects (Nakashima *et al.*, 2013). Thus, it appears that the common palm civet can tolerate human modifications to a forested landscape and can exploit abundant food resources within oil palm plantations (oil palm fruit and rodents).

In contrast, our camera-trapping study showed that the Malay civet has a low occupancy and detection probability within oil palm and that the probability of detecting the Malay civet within oil palm habitat was affected by the proximity of extensive areas of lowland forest (Jennings *et al.*, 2015). Differences in the surrounding habitat mosaic might also explain the presence of the Malay civet in one oil palm estate and its absence in the other. The habitat suitability modelling of Ross *et al.* (2016a) showed that Malay civets on Borneo were mainly associated with lowland forest, and avoided anthropogenic habitats, such as plantations and mixed crops. On Borneo, Evans *et al.* (2021) found that several radio-collared Malay civets foraged in oil palm plantations, but all tagged individuals utilized forests. They concluded that oil palm plantations did not pose an inhospitable matrix for Malay civets, but oil palm agriculture is a less suitable habitat than remnant forests, and that proximity measures and forest structure influenced their spatial behaviours (Evans *et al.*, 2021).

On Borneo, the density of Malay civets was found to be 57% higher in an unlogged forest than in logged forest (Colon, 2002), which suggests that forest disturbance has a negative impact on this species. Heydon & Bulloh (1996) also found a marked decrease in civet density in logged forests, and Meijaard *et al.* (2008) suggested that the Malay civet is a logging-intolerant species. These findings imply that undisturbed lowland forests may act as source populations of Malay civets and that other areas could be sink habitats in which lower densities exist. Colon (2002) suggested that the ability of this species to survive in disturbed habitats might be influenced by the proximity of undisturbed forests that serves as a biological reservoir of individuals. Thus, it appears that the presence and quality of forests surrounding oil palm plantations are important factors for Malay civets within an agricultural landscape.

Our occupancy analyses have also suggested that the distance from the edge of the oil palm plantation affected the occupancy of Malay civets within oil palm. In an oil palm plantation on Peninsular Malaysia, the maximum distance that radio-collared Malay civets were located from the adjacent forest was ~600 m; tagged civets were only found in the oil palm during the night, and daytime rest sites were either in the adjacent forest or within small remnant forest patches in the plantation (Jennings *et al.*, 2010a). Evans *et al.* (2021) found that Malay civets on Borneo used more intensely habitats that were close to oil palm plantation edges and taller tree canopies. Eng (2011) found that high canopy cover, undergrowth, and logs were the most important microhabitat features for the

Malay civet within a planted forest. Radio-tagged Malay civets have been found resting during the day within dense ground cover or large logs (Colon, 2002; Jennings *et al.*, 2010a). These findings suggest that due to the sparse ground cover within oil plantations, there might be a lack of suitable daytime rest sites for Malay civets, which forces individuals to leave plantations after foraging at night. This would restrict how far a Malay civet might penetrate into oil palm to forage for food.

In contrast, common palm civets occur extensively across oil palm plantations (at least during the night), and they have been found resting in oil palm during the day (Nakashima *et al.*, 2013). Common palm civets are generally smaller than Malay civets and are arboreal (Jennings & Veron, 2009), which perhaps enables them to use rest sites that are unsuitable or unavailable for Malay civets. However, our study found that the probability of detection of common palm civets was affected by distance from the edge of the oil palm habitat. Nakashima *et al.* (2013) located radio-collared common palm civets less often in the oil palm plantation during the daytime than would be expected based on habitat availability, and individuals often returned to the adjacent forest to rest during the day. Within the oil palm plantation, common palm civets were found resting in oil palm trees with specific characteristics that differed from the surrounding trees; selected trees were covered with thick leaves and dense ferns, and were often used repeatedly (Nakashima *et al.*, 2013). These findings suggest that suitable rest sites within oil palm plantations are also a limiting factor for the common palm civet, and that dense cover is an important feature for this species within oil palm.

Continuous forests, forest fragments, and forested corridors appear to be important habitat features for tropical civet and mongoose species within oil palm-dominated landscapes, and conserving these will be beneficial for them (Maddox *et al.*, 2007). Within oil palm plantations, providing suitable rest/den sites and corridors (e.g. forested river buffer zones) might encourage civet and mongoose species from surrounding forests to utilize oil palm plantations, or at least allow them to disperse through oil palm to more suitable habitats. For example, allowing patches of dense ground vegetation to grow throughout the plantation, using dead oil palm branches to create large dense piles, and providing large rotting logs, could potentially provide suitable daytime rest and den sites for terrestrial species.

Hunting, Trade, and Other Threats

Throughout Southeast Asia, small carnivores are hunted for their meat to supply local and international markets and are also targeted for the wildlife trade in traditional medicines, skins, bones, and pets (Corlett, 2007; Shepherd, 2008; Shepherd & Shepherd, 2010; Jennings & Veron, 2019). There is also a growing demand for civet coffee (also known as 'Kopi Luwak'), a luxury coffee produced from coffee cherries that have been eaten and partially digested by civets, particularly common palm civets. Traditionally made using scat collected from the wild, the trend for 'caged' civet coffee, where live civets are taken from the wild and housed in captive conditions, is increasing (Carder *et al.*, 2016). Little is known about the extent of hunting and wildlife trade of civet and mongoose species, or the impact that these might have on populations. Even if a species is not specifically targeted, many hunting methods (such as wire-snares) are unselective and non-targeted animals are often taken (Corlett, 2007; A. Jennings & G. Veron, personal observations). Within oil palm plantations, small carnivore species are killed by illegal hunting and feral dogs (Azhar *et al.*, 2013). On Peninsular Malaysia and Buton Island, we found that there was a high turnover of Malay civets between years, particularly those individuals that used plantation areas. This suggests that Malay civets have a high mortality rate in human-modified environments. Even though the Malay civet and the short-tailed mongoose are protected species on Peninsular Malaysia, snare-trapping and hunting with guns were known to occur within our study area, and we did find evidence that both species were being killed. Other threats to civet and mongoose species include pesticides that are applied in plantations, as these contaminate groundwater systems and their prey (through bioaccumulation), which can eventually lead to adverse health problems or death in carnivores that forage in these areas (Elliott *et al.*, 1999; Melgar *et al.*, 2008; Jennings *et al.*, 2010a,b). Clearly, there needs to be increased monitoring of the wildlife trade and the level of hunting pressures, and greater enforcement of wildlife trade and hunting regulations.

Finally, the possible impacts of climate changes on civet and mongoose species is largely unknown (Jennings & Veron, 2019). Seymour *et al.* (2017) found that the body mass and survival of Malay civets increased following El Niño events, which indicates that variability in climate can have a considerable effect on the dynamics of rainforest vertebrate populations.

Conclusion

Our studies have produced important natural history and ecological information for several Southeast Asian civet and mongoose species. For example, our radio-telemetry results confirmed that the Malay civet is mainly nocturnal and the short-tailed mongoose is diurnal and that both species inhabit lowland forest. Our ecological niche modelling of several viverrids and herpestids has highlighted the areas with the highest probabilities of occurrence, thereby indicating key localities for the long-term conservation of threatened species and where further research activities should be prioritized. Our camera-trapping study has shown that oil palm cultivation has a severe impact on the diversity of small carnivore species within central Sumatra and that the current compatibility of oil palm plantations with small carnivore richness is low. However, little is still known about the ecology and conservation status of Southeast Asian civet and mongoose species, and in order to confirm our results, and further explore the mechanisms that may be responsible for their distribution and niche patterns, further field studies are needed to gather more data on their distribution, abundance, and ecology. For instance, there is insufficient occurrence data to investigate spatial or temporal changes in ecological niches, and too little is currently known about the natural history and ecology of civets and mongooses to determine how other biotic factors (such as predation and disease) or the presence of key resources (e.g. den sites) have played a role in determining their distribution patterns. Finally, our findings have important implications for the conservation of these small carnivore species. Tropical forests are vulnerable and are fast disappearing, and those species that primarily occur in this habitat could be threatened with extinction.

Acknowledgements

We would like to thank our collaborators: Zubaid Akbar, John Mathai, Jedediah Brodie, Anthony Giordano, Mohd Naim, Andreas Dwi Advento, Anak Agung Ketut Aryawan Sudharto Ps, Jean-Pierre Caliman, and Aude Verwilghen. We also thank our field assistants and students: Azilah, Suriani Bt Salihan, Ainil Hawa Bt Sazali, Ummul Nazrah Abdul Rahman, Hafiza A Hamid; and we thank the Economic Planning Unit and the Department of Wildlife and National Parks of Malaysia (PERHILITAN) for permission to work in Krau Wildlife Reserve. We are indebted to the many contributors of civet and mongoose records (acknowledged in Jennings & Veron, 2011 and Jennings *et al.*, 2013). We are grateful to Chien C. Lee (chienclee.com) for providing some stunning photographs to illustrate this chapter. These projects were financially supported by the Royal Geographical Society, Muséum National d'Histoire Naturelle and French Ministry of Research (Programme Pluriformation 'Evolution et Structure des Ecosystèmes'), French National Centre for Scientific Research (CNRS), Société d'Encouragement pour la Conservation des Animaux Sauvages (SECAS), Le Parc des Félins, International Society for Endangered Cats (Canada), Société des Amis du Muséum National d'Histoire Naturelle et du Jardin des Plantes, Malaysia Airlines, SMART Research Institute, and CIRAD.

References

Ancrenaz, M., Hearn, A.J., Ross, J., Sollmann, R. & Wilting, A. (2012) *Handbook for Wildlife Monitoring Using Camera-Traps*. BBEC II Secretariat, Natural Resources Office, Chief Minister's Department, Kota Kinabalu, Sabah.

Anderson, R.P., Peterson, A.T. & Gomez-Laverde, M. (2002) Using niche-based GIS modeling to test geographic predictions of competitive exclusion and competitive release in South American pocket mice. *Oikos* 98, 3–16.

Anonymous (2009) Nocturnal works continues. . . *Jungle Times* (Independent Newspaper of the Danau Girang Field Centre) 7, 1.

Austin, S.C. & Tewes, M.E. (1999) Observations of viverrids, mustelids and herpestids in Khao Yai NP, Thailand. *Small Carnivore Conservation* 21, 13–15.

Azhar, B., Lindenmayer, D., Wood, J., Fischer, J., Manning, A., McElhinny, C. & Zakaria, M. (2013) Contribution of illegal hunting, culling of pest species, road accidents and feral dogs to biodiversity loss in established oil-palm landscapes. *Wildlife Research* 40, 1–9.

Ben Yaacov, R. & Yom-Tov, Y. (1983) On the biology of the Egyptian mongoose, *Herpestes ichneumon*, in Israel. *Zeitschrift für Säugetierkunde* 48, 34–45.

Bird, M.I., Taylor, D. & Hunt, C. (2005) Environments of insular Southeast Asia during the Last Glacial Period: a savannah corridor in Sundaland? *Quaternary Science Reviews* 24, 2228–2242.

Cannon, C.H., Morley, R.J. & Bush, A.B.G. (2009) The current refugial rainforests of Sundaland are unrepresentative of their biogeographic past and highly vulnerable to disturbance. *Proceedings of the National Academy of Sciences of the United States of America* 106, 11188–11193.

Carder, G., Proctor, H., Schmidt-Burbach, J. & D'Cruze, N. (2016) The animal welfare implications of civet coffee tourism in Bali. *Animal Welfare* 25, 199–205.

Cardillo, M., Purvis, A., Sechrest, W., Gittleman, J.L., Bielby, J. & Mace, G.M. (2004) Human population density and extinction risk in the world's carnivores. *PLoS Biology* 2, e197.

Caro, T. & Stoner, C. (2003) The potential for interspecific competition among African carnivores. *Biological Conservation* 110, 67–75.

Case, T.J. & Gilpin, M.E. (1974) Interference competition and niche theory. *Proceedings of the National Academy of Sciences of the United States of America* 71, 3073–3077.

Cavallini, P. & Nel, J.A.J. (1990) Ranging behaviour of the Cape grey mongoose *Galerella pulverulenta* in a coastal area. *Journal of Zoology* 222, 353–362.

Charles, J.K. & Ang, B.B. (2010) Non-volant small mammal community responses to fragmentation of kerangas forests in Brunei Darussalam. *Biodiversity and Conservation* 19, 543–561.

Chen, M.T., Tewes, M.E., Pei, K.J. & Grassman, L.I. (2009) Activity patterns and habitat use of sympatric small carnivores in southern Taiwan. *Mammalia* 73, 20–26.

Chuang, S.A. & Lee, L.L. (1997) Food habits of three carnivore species (*Viverricula indica, Herpestes urva* and *Melogale moschata*) in Fushan Forest, northern Taiwan. *Journal of Zoology* 243, 71–79.

Colon, C.P. (2002) Ranging behaviour and activity of the Malay civet (*Viverra tangalunga*) in a logged and an unlogged forest in Danum Valley, East Malaysia. *Journal of Zoology* 257, 473–485.

Corbet, G.B. & Hill, J.E. (1992) *The Mammals of the Indomalayan Region: A Systematic Review*. Oxford University Press, Oxford.

Corlett, R.T. (2007) The impact of hunting on the mammalian fauna of tropical Asian forests. *Biotropica* 39, 292–303.

Corlett, R.T. (2009) *The Ecology of Tropical East Asia*. Oxford University Press, New York.

Creel, S. (2001) Four factors modifying the effect of competition on carnivore population dynamics as illustrated by African wild dogs. *Conservation Biology* 15, 271–274.

Crooks, K. (2002) Relative sensitivities of mammalian carnivores to habitat fragmentation. *Conservation Biology* 16, 488–502.

Davies, T., Meiri, S., Barraclough, T. & Gittleman, J. (2007) Species co-existence and character divergence across carnivores. *Ecology Letters* 10, 146–152.

Di Bitetti, M.S., Di Blanco, Y.E., Pereira, J.A., Paviolo, A. & Perez, I.J. (2009) Time partitioning favors the coexistence of sympatric crab-eating foxes (*Cerdocyon thous*) and pampas foxes (*Lycalopex gymnocercus*). *Journal of Mammalogy* 90, 479–490.

Di Bitetti, M.S., De Angelo, C.D., Di Blanco, Y.E. & Paviolo, A. (2010) Niche partitioning and species coexistence in a Neotropical felid assemblage. *Acta Oecologica* 36, 403–412.

Duckworth, J.W., Timmins, R.J., Olsson, A., Roberton, S., Kanchanasaka, B., Zaw, T., Jennings, A.P. & Veron, G. (2008) *Viverra megaspila. The IUCN Red List of Threatened Species*. Version 2014.3. http://www.iucnredlist.org. Accessed on 23 April 2015.

Duckworth, J.W., Hearn, A.J., Ross, J., Samejima, H., Mohamed, A. *et al.* (2016) Predicted distribution of short-tailed mongoose *Herpestes brachyurus* (Mammalia: Carnivora: Herpestidae) on Borneo. *Raffles Bulletin of Zoology*, Supplement 33, 132–141.

Elith, J., Graham, C.H., Anderson, R.P., Dudik, M., Ferrier, S. *et al.* (2006) Novel methods improve prediction of species' distributions from occurrence data. *Ecography* 29, 129–151.

Elliott, J.E., Wilson, L.K., Henny, C.J., Harris, M.L. & Norstrom, R.J. (1999) Chlorinated hydrocarbons in livers of American mink (*Mustela vison*) and river otter (*Lutra canadensis*) from the Columbia and Fraser River basins, 1990–1992. *Environmental Monitoring and Assessment* 57, 229–252.

Eng, I.S. (2011) *Microhabitat Analysis and Population Parameters of Small Carnivores in Sarawak Planted Forest, with Emphasis on the Malay Civet* Viverra tangalunga. MSc dissertation, Universiti Tunku Abdul Rahman, Kuala Kumpur.

Ermias Admasu, Thirgood, S.J., Afework Bekele & Laurenson, M.K. (2004) Spatial ecology of white-tailed mongoose in farmland in the Ethiopian highlands. *African Journal of Ecology* 42, 153–159.

Evans, M.N., Müller, C.T., Kille, P., Asner, G.P., Guerrero-Sanchez, S., Abu Bakar, M.S. & Goossens, B. (2021) Space-use patterns of Malay civets (*Viverra tangalunga*) persisting within a landscape fragmented by oil palm plantations. *Landscape Ecology* 36, 915–930.

Flower, S.S. (1900) On the mammalia of Siam and the Malay peninsula. *Proceedings of the Zoological Society of London* 1900, 306–379.

Francis, C.M. (2008) *A Field Guide to the Mammals of South-East Asia*. New Holland Publishers, London.

Gilchrist, J.S., Jennings, A.P., Veron, G. & Cavallini, P. (2009) Family Herpestidae (Mongooses). In: *Handbook of Mammals of the World. Volume 1. Carnivores* (eds D.E. Wilson & R.A. Mittermeier), pp. 262–328. Lynx Edicions, Barcelona.

Gorman, M.L. (1979) Dispersion and foraging of the small Indian mongoose *Herpestes auropunctatus* (Carnivora: Viverridae) relative to the evolution of social viverrids. *Journal of Zoology* 187, 65–73.

Gorog, A.J., Sinaga, M.H. & Engstrom, M.D. (2004) Vicariance or dispersal? Historical biogeography of three Sunda shelf murine rodents (*Maxomys surifer, Leopoldamys sabanus* and *Maxomys whiteheadi*). *Biological Journal of the Linnean Society* 91, 91–109.

Grassman, L.I., Tewes, M.E., Silvy, N.J. & Kreetiyutanont, K. (2005) Spatial organization and diet of the leopard cat (*Prionailurus bengalensis*) in northcentral Thailand. *Journal of Zoology* 266, 45–54.

Gray, T.N.E., Pin, C. & Pin, C. (2010) Status and ecology of large-spotted civet *Viverra megaspila* in eastern Cambodia. *Small Carnivore Conservation* 43, 12–15.

Gunarso, P., Hartoyo, M.E., Agus, F. & Killeen, T.J. (2013) Oil palm and land use change in Indonesia, Malaysia and Papua New Guinea. Reports from the Technical Panels of the 2nd Greenhouse Gas Working Group of the Roundtable on Sustainable Palm Oil, Kuala Lumpur.

Hays, W. & Conant, S. (2003) Male social activity in the small Indian mongoose *Herpestes javanicus*. *Acta Theriologica* 48, 485–494.

Heydon, M.J. & Bulloh, P. (1996) The impact of selective logging on sympatric civet species in Borneo. *Oryx* 30, 31–36.

Hines, J.E. (2006) *PRESENCE2 – Software to Estimate Patch Occupancy and Related Parameters.* USGS-PWRC. http://www.mbr-pwrc.usgs.gov/software/presence.html.

Hirzel, A.H. & Le Lay, G. (2008) Habitat suitability modelling and niche theory. *Journal of Applied Ecology* 45, 1372–1381.

Holden, J. (2006) Small carnivores in central Sumatra. *Small Carnivore Conservation* 34 & 35, 35–38.

Holden, J. & Meijaard, E. (2013) An orange-coloured collared mongoose *Herpestes semitorquatus* from Aceh, Sumatra, Indonesia. *Small Carnivore Conservation* 47, 26–29.

Hon, J., Hearn, A. J., Ross, J., Mohamed, A., Alfred, R. *et al.* (2016) Predicted distribution of the collared mongoose *Herpestes semitorquatus* (Mammalia: Carnivora: Herpestidae) on Borneo. *Raffles Bulletin of Zoology*, Supplement 33, 142–148.

Hunowu, I., Patandung, A., Pusparini, W., Danismend, I., Cahyana, A., Abdullah, S., Johnson, C.L., Hilser, H., Rahasia, R., Gawina, J. & Linkie, M. (2020) New insights into Sulawesi's apex predator: the Sulawesi civet *Macrogalidia musschenbroekii*. *Oryx* 54, 878–881.

IUCN [International Union for Conservation of Nature] (2021) *The IUCN Red List of Threatened Species.* Version 2021-1. https://www.iucnredlist.org. Accessed on 11 April 2021.

Jackson, C.R. & Robertson, M.P. (2011) Predicting the potential distribution of an endangered cryptic subterranean mammal from few occurrence records. *Journal for Nature Conservation* 19, 87–94.

Jennings, A.P. & Veron, G. (2009) Family Viverridae (Civets, genets, and oyans). In: *Handbook of Mammals of the World. Volume 1. Carnivores* (eds D.E. Wilson & R.A. Mittermeier), pp. 174–232. Lynx Edicions, Barcelona.

Jennings, A.P. & Veron, G. (2011) Predicted distributions and ecological niches of 8 civets and mongoose species in Southeast Asia. *Journal of Mammalogy* 92, 316–327.

Jennings, A.P. & Veron, G. (2019) *Mongooses of the World.* Whittles Publishing, Dunbeath.

Jennings, A.P., Seymour, A.S. & Dunstone, N. (2006) Ranging behaviour, spatial organization and activity of the Malay civet (*Viverra tangalunga*) on Buton Island, Sulawesi. *Journal of Zoology* 268, 63–71.

Jennings, A.P., Zubaid, A. & Veron, G. (2010a) Ranging behaviour, activity, habitat use, and morphology of the Malay civet (*Viverra tangalunga*) on Peninsular Malaysia and comparison with studies on Borneo and Sulawesi. *Mammalian Biology* 75, 437–446.

Jennings, A.P., Zubaid A. & Veron, G. (2010b) Home-ranges, movements and activity of the short-tailed mongoose (*Herpestes brachyurus*) on Peninsular Malaysia. *Mammalia* 74, 43–50.

Jennings, A.P., Mathai, J., Brodie, J., Giordano, A.J. & Veron, G. (2013) Predicted distributions and conservation status of two threatened small carnivores: banded civet and Hose's civet. *Mammalia* 77, 261–271.

Jennings, A.P., Naim, M., Advento, A.D., Aryawan, A.A.K., Sudharto, Ps, Caliman, J.-P., Verwilghen, A. & Veron, G. (2015) Diversity and occupancy of small carnivores within oil palm plantations in central Sumatra, Indonesia. *Mammal Research* 60, 181–188.

Joshi, A.R., Smith, J.L.D. & Cuthbert, F.J. (1995) Influence of food distribution and predation pressure on spacing behavior in palm civets. *Journal of Mammalogy* 76, 1205–1212.

Kenward, R.E. (2000) *A Manual for Wildlife Radio Tagging.* Academic Press, London.

Louppe, V., Lalis, A., Abdelkrim, J., Baron, J., Bed'Hom, B., Becker, A.A.M.J., Catzeflis, F., Lorvelec, O., Zieger, U., Veron, G. (2021) Dispersal history of a globally introduced carnivore, the small Indian mongoose *Urva auropunctata*, with an emphasis on the Caribbean region. *Biological Invasions* 23, 2573–2590.

Lucherini, M., Reppucci, J.I., Walker, R.S., Lilian Villalba, M., Wurstten, A., Gallardo, G., Iriarte, A., Villalobos, R. & Perovic, P. (2009) Activity pattern segregation of carnivores in the high Andes. *Journal of Mammalogy* 90, 1404–1409.

Luo, S.-J., Zhang, Y., Johnson, W.E., Miao, L., Martelli, P., Antunes, A., Smith, J.L.D. & O'Brien, S.J. (2014) Sympatric Asian felid phylogeography reveals a major Indochinese–Sundaic divergence. *Molecular Ecology* 23, 2072–2092.

Macdonald, D.W. & Wise, M.J. (1979) Notes on the behavior of the Malay civet *Viverrra tangalunga* Gray. *Sarawak Museum Journal* 48, 295–299.

Maddox, T., Priatna, D., Gemita, E. & Salampessy, A. (2007) *The Conservation of Tigers and Other Wildlife in Oil Palm Plantations. Jambi Province, Sumatra, Indonesia*. ZSL Conservation Report No. 7. The Zoological Society of London, London.

Mathai, J., Hearn, A., Brodie, J., Wilting, A., Duckworth, J.W., Ross, J., Holden, J., Gemita, E. & Hon, J. (2015a) *Herpestes semitorquatus. The IUCN Red List of Threatened Species* 2015, e.T41616A45208027. http://dx.doi.org/10.2305/IUCN.UK.2015-2.RLTS. T18711A21938411.en. Accessed on 11 April 2021.

Mathai, J., Duckworth, J.W., Wilting, A., Hearn, A. & Brodie, J. (2015b) *Diplogale hosei. The IUCN Red List of Threatened Species* 2015, e.T6635A45197564. https://dx.doi.org/10.2305/IUCN.UK.2015-4.RLTS. T6635A45197564.en. Accessed on 11 April 2021.

Mathai, J., Brodie, J., Giordano, A.J., Alfred, R., Belant, J.L., Kramer-Schadt, S. & Wilting, A. (2016) Predicted distribution of Hose's civet *Diplogale hosei* (Mammalia: Carnivora: Viverridae) on Borneo. *Raffles Bulletin of Zoology*, Supplement 33, 118–125.

Mathai, J., Niedballa, J., Radchuk, V., Sollmann, R., Heckmann, I., Brodie, J., Struebig, M., Hearn, A.J., Ross, R., Macdonald, D.W., Hon, J. & Wilting, A. (2019) Identifying refuges for Borneo's elusive Hose's civet. *Global Ecology and Conservation* 17, e00531.

McLoughlin, P.D. & Ferguson, S.H. (2000) A hierarchical pattern of limiting factors helps explain variation in home-range size. *Ecoscience* 7, 123–130.

Meijaard, E. (2003) Mammals of south-east Asian islands and their Late Pleistocene environments. *Journal of Biogeography* 30, 1245–1257.

Meijaard, E. (2009) Solving mammalian riddles along the Indochinese–Sundaic zoogeographic transition: new insights from mammalian biogeography. *Journal of Biogeography* 36, 801–802.

Meijaard, E. & Sheil, D. (2008) The persistence and conservation of Borneo's mammals in lowland rain forests managed for timber: observations, overviews and opportunities. *Ecological Research* 23, 21–34.

Meijaard, E., Sheil, D., Marshall, A.J. & Nasi, R. (2008) Phylogenetic age is positively correlated with sensitivity to timber harvest in Bornean mammals. *Biotropica* 40, 76–85.

Meiri, S., Meijaard, E., Wich, S.A., Groves, C.P. & Helgen, K.M. (2008) Mammals of Borneo – small size on a large island. *Journal of Biogeography* 35, 1087–1094.

Melgar, C., Geissen, V., Cram, S., Sokolov, M., Bastidas, P., Suarez, L.E.R., Ramos, F.J.Q. & Sanchez, A.J. (2008) Pollutants in drainage channels following long-term application of Mancozeb to banana plantations in southeastern Mexico. *Journal of Plant Nutrition and Soil Science* 171, 597–604.

Michalski, F. & Peres, C.A. (2005) Anthropogenic determinants of primate and carnivore local extinctions in a fragmented forest landscape of southern Amazonia. *Biological Conservation* 124, 383–396.

Mohd-Azlan, J., Kaicheen, S.S., Lok, L. & Brodie, J.F. (2020) Diversity and size-structured persistence of tropical carnivores in a small, isolated protected area. *Mammalia* 84, 34–40.

Mudappa, D. (2001) *Ecology of the Brown Palm Civet* Paradoxurus jerdoni *in the Tropical Forests of the Western Ghats, India*. PhD thesis, Bharathiar University, Coimbator.

Musser, G.G. (1987) The mammals of Sulawesi. In: *Biogeographical Evolution of the Malay Archipelago* (ed. T.C. Whitmore), pp. 74–95. Clarendon Press, Oxford.

Naim, M., Advento, A.D., Sudharto P.S., Jennings A.P., Veron, G., Verwilghen A. & Caliman, J.P. (2012) The presence and distribution of small carnivores in oil palm plantation and their role in controlling rat damage: preliminary results from a camera trapping study. Paper presented at the 4th IOPRI–MPOB International Seminar: Existing and Emerging Pests and Diseases of Oil Palm. Advances in Research and Management, Bandung, Indonesia, 13–14 December 2012.

Nakashima, Y. & Do Linh San, E. (Chapter 18, this volume) Seed dispersal by mesocarnivores: importance and functional uniqueness in a changing world. In: *Small Carnivores: Evolution, Ecology, Behaviour, and Conservation* (eds E. Do Linh San, J.J. Sato, J.L. Belant & M.J. Somers). Wiley-Blackwell, Oxford.

Nakashima, Y., Inoue, E., Inoue-Murayama, M. & Sukor, J.R.A. (2010) Functional uniqueness of a small carnivore as seed dispersal agents: a case study of the common palm civets in the Tabin Wildlife Reserve, Sabah, Malaysia. *Oecologia* 164, 721–730.

Nakashima, Y., Nakabayashi, M. & Sukor, J.A. (2013) Space use, habitat selection, and day-beds of the common palm civet (*Paradoxurus hermaphroditus*) in human-modified habitats in Sabah, Borneo. *Journal of Mammalogy* 94, 1169–1178.

Nellis, D.W. & Everard, C.O.R. (1983) The biology of the mongoose in the Caribbean. *Studies on Fauna of Curacao and Caribbean Islands* 64, 1–162.

Nozaki, E., Azuma, S., Sasaki, H. & Torii, H. (1994) Home-range of the Malay civet (*Viverra tangalunga*) in Teluk Kaba area, Kutai National Park, East Kalimantan, Indonesia. *Kyoto University Overseas Report Special Number*, 85–94.

O'Donovan, G. (2001) The botanical and ecological status of the Kakenauwe and Lambusanga Nature Reserves on Buton Island, *Sulawesi*. Unpublished report to Operation Wallacea, Old Bolingbroke, Spilsby.

Palomares, F. & Caro, T.M. (1999) Interspecific killing among mammalian carnivores. *The American Naturalist* 153, 492–508.

Palomares, F. & Delibes, M. (1993) Social organization in the Egyptian mongoose – group size, spatial behavior and inter-individual contacts in adults. *Animal Behaviour* 45, 917–925.

Patel, R.P., Wutke, S., Lenz, D., Mukherjee, S., Ramakrishnan, U., Veron, G., Fickel, J., Wilting, A. & Förster, D.W. (2017) Genetic structure and phylogeography of the leopard cat (*Prionailurus bengalensis*) inferred from mitochondrial genomes. *Journal of Heredity* 108, 349–360.

Patou, M.-L. 2008. *Systématique et Biogéographie des Herpestidae et Viverridae (Mammalia, Carnivora) en Asie.* PhD thesis, Muséum National d'Histoire Naturelle, Paris. (In French with English summary).

Patou, M.-L., Debruyne, R., Jennings, A.P., Zubaid, A., Rovie-Ryan, J.J. & Veron, G. (2008) Phylogenetic relationships of the Asian palm civets (Hemigalinae and Paradoxurinae, Viverridae, Carnivora). *Molecular Phylogenetics and Evolution* 47, 883–892.

Patou, M.-L., Mclenachan, P.A., Morley, C.G., Couloux, A., Jennings, A.P. & Veron, G. (2009) Molecular phylogeny of the Herpestidae (Mammalia, Carnivora) with a special emphasis on the Asian *Herpestes. Molecular Phylogenetics and Evolution* 53, 69–80.

Peh, K.S.-H., Soh, M.C.K., Sodhi, N.S., Laurance, W.F., Ong, D.J. & Clements, R. (2011) Up in the clouds: is sustainable use of tropical montane cloud forests possible in Malaysia? *Bioscience* 61, 27–38.

Peterson, A.T., Sanchez-Cordero, V., Martinez-Meyer, E. & Navarro-Siguenza, A.G. (2006) Tracking population extirpations via melding ecological niche modeling with land-cover information. *Ecological Modelling* 195, 229–236.

Phillips, S.J., Anderson, R.P. & Schapire, R.E. (2006) Maximum entropy modeling of species geographic distributions. *Ecological Modelling* 190, 231–259.

Purvis, A., Gittleman, J.L., Cowlishaw, G. & Mace, G.M. (2000) Predicting extinction risk in declining species. *Proceedings of the Royal Society of London B: Biological Sciences* 267, 1947–1952.

Rajaratnam, R., Sunquist, M., Rajaratnam, L. & Ambu, L. (2007) Diet and habitat selection of the leopard cat (*Prionailurus bengalensis*) in an agricultural landscape in Sabah, Malaysian Borneo. *Journal of Tropical Ecology* 23, 209–217.

Ray, J.C. (1997) Comparative ecology of two African forest mongooses, *Herpestes naso* and *Atilax paludinosus. African Journal of Ecology* 35, 237–253.

Reis, K.R. & Garong, A.M. (2001) Late Quaternary terrestrial vertebrates from Palawan Island, Philippines. *Palaeogeography Palaeoclimatology Palaeoecology* 171, 409–421.

Rondinini, C., Di Marco, M., Chiozza, F., Santulli, G., Baisero, D., Visconti, P., Hoffmann, M., Schipper, J., Stuart, S.N., Tognelli, M.F., Amori, G., Falcucci, A., Maiorano L. & Boitani, L. (2011) Global habitat suitability models of terrestrial mammals. *Philosophical Transactions of the Royal Society Series B* 366, 2633–2641.

Rood, J.P. (1986) Ecology and social evolution in the mongooses. In: *Ecological Aspects of Social Evolution: Birds and Mammals* (eds D.I Rubenstein & R.W. Wrangham), pp. 131–152. Princeton University Press, New Jersey.

Rood, J.P. (1989) Male associations in a solitary mongoose. *Animal Behaviour* 38, 725–728.

Ross, J., Gemita, E., Hearn, A.J. & Macdonald, D.W. (2012) The occurrence of reddish-orange mongooses *Herpestes* in the Greater Sundas and the potential for their field confusion with Malay Weasel *Mustela nudipes*. *Small Carnivore Conservation* 46, 8–11.

Ross, J., Hearn, A.J., Macdonald, D.W., Alfred, R., Cheyne, S.M. *et al.* (2016a) Predicted distribution of the Malay civet *Viverra tangalunga* (Mammalia: Carnivora: Viverridae) on Borneo. *Raffles Bulletin of Zoology*, Supplement 33, 78–83.

Ross, J., Hearn, A.J., Macdonald, D.W., Semiadi, G., Alfred, R. *et al.* (2016b) Predicted distribution of the banded civet *Hemigalus derbyanus* (Mammalia: Carnivora: Viverridae) on Borneo. *Raffles Bulletin of Zoology*, Supplement 33, 111–117.

Ross, J., Hearn, A.J. & Macdonald, D.W. (2017) The Bornean carnivore community: lessons from a little-known guild. In: *Biology and Conservation of Musteloids* (eds D.W. Macdonald, C. Newman & L.A. Harrington), pp. 326–339. Oxford University Press, Oxford.

Roy, S.S., Jones C.G. & Harris, S. (2002) An ecological basis for control of the mongoose *Herpestes javanicus* in Mauritius: is eradication possible? In: *Turning the Tide: The Eradication of Invasive Species* (eds C.R. Vietch & M.N. Clout), pp. 266–273. IUCN SSC Invasive Species Specialist Group, IUCN, Gland and Cambridge.

Samejima, H. & Semiadi, G. (2012) First record of Hose's Civet *Diplogale hosei* from Indonesia and other records of other carnivores in the Schwaner Mountains, Central Kalimantan, Indonesia. *Small Carnivore Conservation* 46, 1–7.

Sandell, M. (1989) The mating tactics and spacing patterns of solitary carnivores. In: *Carnivore Behavior, Ecology, and Evolution. Volume 1* (ed. J.L. Gittleman), pp. 164–182. Cornell University Press, Ithaca.

Schipper, J., Chanson, J.S., Chiozza, F., Cox, N.A., Hoffmann, M. *et al.* (2008) The status of the world's land and marine mammals: diversity, threat, and knowledge. *Science* 322, 225–230.

Schreiber, A., Wirth, R., Riffel, M. & Van Rompaey, H. (1989) *Weasels, Civets, Mongooses, and Their Relatives: An Action Plan for the Conservation of Mustelids and Viverrids*. IUCN/SSC Mustelid and Viverrid Specialist Group, Gland.

Seymour, A.S., Tarrant, M.R., Gerber, B.D., Sharp, A., Woollam, J. & Cox, R. (2017) Effects of El Niño on the population dynamics of the Malay civet east of the Wallace line. *Journal of Zoology* 303, 120–128.

Shepherd, C.R. (2008) Civets in trade in Medan, North Sumatra, Indonesia (1997–2001) with notes on legal protection. *Small Carnivore Conservation* 38, 34–36.

Shepherd, C.R. & Shepherd, L.A. (2010) The trade in Viverridae and Prionodontidae in Peninsular Malaysia with notes on conservation and legislation. *Small Carnivore Conservation* 42, 27–29.

Sodhi, N.S., Koh, L.P., Brook, B.W. & Ng, P.K.L. (2004) Southeast Asian biodiversity: an impending disaster. *Trends in Ecology & Evolution* 19, 654–660.

Sodhi, N.S., Posa, M.R.C., Lee, T.M., Bickford, D., Koh, L.P. & Brook, B.W. (2010a) The state and conservation of Southeast Asian biodiversity. *Biodiversity and Conservation* 19, 317–328.

Sodhi, N.S., Koh, L.P., Clements, R., Wanger, T.C., Hill, J.K., Hamer, K.C., Clough, Y., Tscharntke, T., Posa, M.R.C. & Lee, T.M. (2010b) Conserving Southeast Asian forest biodiversity in human-modified landscapes. *Biological Conservation* 143, 2375–2384.

St-Pierre, C., Ouellet, J-P. & Crête, M. (2006) Do competitive intraguild interactions affect space and habitat use by small carnivores in a forested landscape? *Ecography* 29, 487–496.

Thorn, J.S., Nijman, V., Smith, D. & Nekaris, K.A.I. (2009) Ecological niche modelling as a technique for assessing threats and setting conservation priorities for Asian slow lorises (primates: *Nycticebus*). *Diversity and Distributions* 15, 289–298.

Timmins, R., Duckworth, J.W., WWF-Malaysia, Roberton, S., Gray, T.N.E., Willcox, D.H.A., Chutipong, W. & Long, B. (2016) *Viverra megaspila. The IUCN Red List of Threatened Species* 2016, e.T41707A45220097. https://dx.doi.org/10.2305/IUCN.UK.2016-1.RLTS.T41707A45220097.en. Accessed on 11 April 2021.

van Den Bergh, G.D., De Vos, J. & Sondaar, P.Y. (2001) The Late Quaternary palaeogeography of mammal evolution in the Indonesian Archipelago. *Palaeogeography, Palaeoclimatology, Palaeoecology* 171, 385–408.

Van Rompaey, H. & Azlan, M. (2004) Hose's Civet, *Diplogale hosei. Small Carnivore Conservation* 30, 18–19.

Veron, G. (2001) The palm civets of Sulawesi. *Small Carnivore Conservation* 24, 13–14.

Veron, G., Gaubert, P., Franklin, N., Jennings A.P. & Grassman, L.I. (2006) A reassessment of the distribution and taxonomy of the Endangered otter civet *Cynogale bennettii* (Carnivora: Viverridae) of South-east Asia. *Oryx* 40, 42–49.

Veron, G., Willsch, M., Dacosta, V., Patou, M.L., Seymour, A., Bonillo, C., Couloux, A., Wong, S.T., Jennings, A.P., Fickel, J. & Wilting, A. (2014) The distribution of the Malay civet *Viverra tangalunga* (Carnivora: Viverridae) across Southeast Asia: natural or human-mediated dispersal? *Zoological Journal of the Linnean Society* 170, 917–932.

Veron, G., Patou, M.L., Debruyne, R., Couloux, A., Fernandez, D.A.P., Wong, S.T., Fuchs, J. & Jennings, A.P. (2015a) Systematics of the Southeast Asian mongooses (Herpestidae, Carnivora) – solving the mystery of the elusive collared mongoose and Palawan mongoose. *Zoological Journal of the Linnean Society* 173, 236–248.

Veron G., Patou, M.-L., Tóth, M., Goonatilake, M. & Jennings, A.P. (2015b) How many species of *Paradoxurus* civets are there? New insights from India and Sri Lanka. *Journal of Zoological Systematics and Evolutionary Research* 53, 161–174.

Veron, M., Patou, M.L. & Jennings, A.P. (Chapter 3, this volume) Systematics and evolution of the mongooses (Herpestidae, Carnivora). In: *Small Carnivores: Evolution, Ecology, Behaviour, and Conservation* (eds E. Do Linh San, J.J. Sato, J.L. Belant & M.J. Somers). Wiley-Blackwell, Oxford.

Wang, H. & Fuller, T.K. (2001) Notes on the ecology of sympatric small carnivores in southeastern China. *Mammalian Biology* 66, 251–255.

Wells, D.R. (1989) Notes on the distribution and taxonomy of Peninsular Malaysian mongooses (*Herpestes*). *Natural History Bulletin of the Siam Society* 37, 87–97.

Wilcove, D.S. & Koh, L.P. (2010) Addressing the threats to biodiversity from oil-palm agriculture. *Biodiversity and Conservation* 19, 999–1007.

Wilting, A., Cord, A., Hearn, A.J., Hesse, D., Mohamed, A. *et al.* (2010a) Modelling the species distribution of flat-headed cats (*Prionailurus planiceps*), an endangered South-East Asian small felid. *PLoS One* 5, e9612.

Wilting, A., Samejima, H. & Mohamed, A. (2010b) Diversity of Bornean viverrids and other small carnivores in Deramakot Forest Reserve, Sabah, Malaysia. *Small Carnivore Conservation* 42, 10–13.

Woodruff, D.S. (2010) Biogeography and conservation in Southeast Asia: how 2.7 million years of repeated environmental fluctuations affect today's patterns and the future of the remaining refugial-phase biodiversity. *Biodiversity and Conservation* 19, 919–941.

Woodruff, D.S. & Turner, L.M. (2009) The Indochinese–Sundaic zoogeographic transition: a description and analysis of terrestrial mammal species distributions. *Journal of Biogeography* 36, 803–821.

Wozencraft, W.C. (2005) Order Carnivora. In: *Mammal Species of the World: A Taxonomic and Geographic Reference*. 3rd edition (eds D.E. Wilson & D.M. Reeder), pp 532–628. The Johns Hopkins University Press, Baltimore.

Zielinski, W.J. (2000) Weasels and martens – carnivores in northern latitudes. In: *Activity Patterns in Small Mammals* (eds S.S. Halle & N.C. Stenseth), pp. 95–118. Ecological Studies, Vol. 141. Springer-Verlag, Berlin and Heidelberg.

Part V

Interactions with People and Conservation

20

Small Carnivore Introductions: Ecological and Biological Correlates of Success

Mariela G. Gantchoff[1],*, Nathan S. Libal[2], and Jerrold L. Belant[3]

[1] *Global Wildlife Conservation Center, State University of New York College of Environmental Science and Forestry, Syracuse, NY, USA*
[2] *Carnivore Ecology Laboratory, Mississippi State University, Mississippi State, MS, USA*
[3] *Department of Fisheries and Wildlife, Michigan State University, East Lansing, MI, USA*

SUMMARY

Successful species introductions are not homogeneously distributed over the globe, which points to the need to understand why some have succeeded, yet others failed. We summarized information on small carnivore introductions worldwide and assessed whether introduction outcomes (success or failure) supported one or more of the following hypotheses: climate-matching, propagule pressure, inherent superiority, island susceptibility and Darwin's naturalization hypotheses. Using the literature, we summarized: number of individuals released, mean body size, mean litter size, consumer type, latitude difference, ecoregions difference, congener presence, and mainland or island release. We generated generalized linear models and ranked them using Akaike's Information Criterion and Akaike's weights. We identified 253 documented introduction events of 24 species from five families, with two thirds of them involving the northern raccoon, *Procyon lotor*, the American mink, *Neovison vison*, and the small Indian mongoose, *Urva [= Herpestes] auropunctata*. Overall introduction success was high, with a success rate > 70% for four of the five represented families. We found support for climate-matching, inherent superiority, and Darwin's naturalization hypotheses. Likelihood of success increased with matching climatic conditions that allow survival, a greater body size together with a smaller litter size, a carnivorous diet, and the absence of congeners in the area of introduction. Islands were not more susceptible than the mainland, and the number of individuals introduced did not influence success. As biological invasions become increasingly widespread, understanding the biological and environmental factors affecting introduction success is important for conservation and management.

Keywords

Darwin's naturalization — inherent superiority — invasive — island susceptibility — exotic — preadaptation to climate — propagule pressure

Introduction

Biological invasions are one of many drivers of ecosystem degradation, along with habitat transformation and exploitation, environmental pollution, and climate change (Pyšek & Richardson, 2010). Invasions can adversely affect populations, communities, and food webs; disturbance regimes; biogeochemical processes; physical structure of the environment; and, in some cases, have resulted in the creation of novel

*Corresponding author.

Small Carnivores: Evolution, Ecology, Behaviour, and Conservation, First Edition. Edited by Emmanuel Do Linh San, Jun J. Sato, Jerrold L. Belant, and Michael J. Somers.
© 2022 John Wiley & Sons Ltd. Published 2022 by John Wiley & Sons Ltd.

communities (Davies, 2009). Although less common, invasive species may also have positive effects in a community, acting as novel seed dispersers, prey, or pollinators (Goodenough, 2010). Few ecosystems in the world are free of introduced species, and an increasing proportion of habitats are becoming dominated by them (Pyšek & Richardson, 2010; Seebens *et al.*, 2018).

Successful species' introductions are not homogeneously distributed worldwide, and appear concentrated on islands and in temperate areas (Vitousek *et al.*, 1997), which points to the need to understand why some introductions succeed, yet others do not (Blackburn & Duncan, 2001). Forsyth *et al.* (2004) analyzed mammal introductions to mainland Australia and found that success was associated with previous establishment success elsewhere, and climate similarity between introduced and native ranges. Successfully established reptiles and amphibians around the world also showed greater climate similarity to their native range relative to failed species' introductions (Bomford *et al.*, 2008). The success of bird introductions worldwide depended on the particular combination of species and location, such as geographical range size and the similarity between latitudes of origin and introduction (Blackburn & Duncan, 2001). For introduced mammals in New Zealand, Australia and Great Britain, the number of release events and climate similarity had the strongest influence on establishment outcomes (Bomford *et al.*, 2009). In a broader-scale analysis, Hayes & Barry (2008) examined 24 studies that identified correlates of establishment success across six animal groups (birds, mammals, fish, reptiles, amphibians, and invertebrates) and found that only three relevant characteristics consistently influenced establishment success across taxa: climate or habitat match, establishment success elsewhere, and propagule pressure.

Research on biological invasions has resulted in numerous hypotheses supporting possible mechanisms of introduction success (Lowry *et al.*, 2012). In the Climate-Matching Hypothesis (Williamson *et al.*, 1986), an introduced species can successfully establish if the environmental conditions of the introduced range are within the species' environmental tolerances (i.e. preadaptation to climate). In contrast, the Propagule Pressure Hypothesis (Williamson, 1996) states that success is the result of a large number of propagules being introduced to the novel environment. The Inherent Superiority Hypothesis (invasion potential; Di Castri, 1989) states that successful introduced species possess intrinsic abilities or traits that facilitate the establishment, such as high reproductive output or broad tolerances. Shifting attention from the particular species to the environment, islands are considered more susceptible to invasions because native species have often evolved in the absence of competition, herbivory, parasitism, or predation (Elton, 1958; Courchamp *et al.*, 2003). Lastly, Darwin stated that 'As the species of the same genus usually have, though by no means invariably, much similarity in habits and constitution, and always in structure, the struggle will generally be more severe between them' which has been termed Darwin's Naturalization Hypothesis (Rejmánek, 1996), and states that introductions should be less successful if congeners are present in the new region (Darwin, 1859). These hypotheses have been tested several times and some taxa are disproportionately represented in the literature of biological invasions (e.g. terrestrial plants; Lowry *et al.*, 2012). However, to our knowledge, these hypotheses have not been tested in the context of global small carnivore mammal introductions.

Mammals were among the first organisms to be introduced by humans as livestock or companions, and recently many species have been intentionally introduced as sporting animals, for novelty reasons, or to control pests (Clout & Russel, 2008). Overall, mammals are among the best-documented introduced taxa (Lever, 1985; Myers, 1986; Thompson, 1922; Long, 2003; Jeschke & Strayer, 2005). Introduced mammals can increase rapidly in abundance, become widely dispersed, and naturalize so thoroughly into the ecosystem that eradication or control can be difficult and expensive (Simberloff & Rejmánek, 2010). While there is substantial variation in the success of introductions among mammalian families, carnivores are among the most successful (Clout & Russel, 2008). In particular, small carnivores are the most widely introduced and successful carnivores after the domestic dog, *Canis familiaris*, and domestic cat, *Felis catus* (Clout & Russel, 2008; Gaubert, 2016). Consequences of carnivore introductions include competition with native species, interbreeding, predation, and disease propagation (Gittleman *et al.*, 2001).

Our objectives were to summarize information on small carnivore introductions worldwide and to estimate the ecological and biological correlates of broad-scale small carnivore introduction success. For the first objective, we compiled available data on species, the timing of introductions, number of individuals, location of introduction, and life-history traits for all introduced species. To address the second objective, we assessed whether correlates of introduction success supported one or more of the following hypotheses: climate-matching, propagule pressure, inherent superiority, island susceptibility, and Darwin's naturalization hypotheses.

Methods

We expanded on Long's (2003) review of worldwide mammal invasions using a literature review. We performed the search with Scopus and Web of Science™ databases for articles published from 2003 to April 2019. Similar to Jeschke *et al.* (2012), our search terms included all small carnivore genus names from the families Ailuridae, Eupleridae, Herpestidae, Mephitidae, Mustelidae, Nandiniidae, Prionodontidae, Procyonidae, and Viverridae (IUCN SSC SCSG, 2013), as well as a general search term to restrict results to invasive species (i.e. 'genera' AND [invasive OR invader OR alien OR exotic OR ruderal OR non-native OR introduced OR naturaliz*]).

We refined our results to biological sciences and then screened the title and abstract of each article to evaluate whether the study was relevant. If necessary, we also reviewed the full text of the articles. We defined relevant studies as those concerned with the description of an introduction event involving one or more of our species of interest. We then categorized introduction events by taxa, release year, release location, and success. We defined the term 'introduction' as the anthropogenic movement (intentional or accidental) of individuals of a species outside their native range, and 'success' as the establishment of a self-sustaining population following introduction.

We first defined variables of interest corresponding to our five hypotheses (Table 20.1) and, using the literature, we summarized the following information for each introduction: success or failure, number of

Table 20.1 Covariates used to test five leading invasion susceptibility hypotheses on small carnivore introductions worldwide.

Hypotheses	Variables
Climate-matching hypothesis	Latitude difference
	Ecoregion difference
Propagule pressure hypothesis	Individuals released
Inherent superiority hypothesis	Mean body size
	Mean litter size
	Consumer type
Island susceptibility hypothesis	Island/Mainland
Darwin's naturalization hypothesis	Congener presence

individuals released, mean body size, mean litter size, consumer type (i.e. carnivorous or omnivorous diet), latitude difference, ecoregion (i.e. large areas with geographically distinct assemblages of species, natural communities, and environmental conditions; WWF, 2013), congener presence, and mainland or island release. For biological correlates, we defined the number released as the number of individuals released at an introduction site (i.e. propagule pressure), which has been implicated as a major factor influencing the introduction success of many taxa (Collauti *et al.*, 2006). For mean body size, we used the average size of an adult of a given species. Organism size is correlated with several important ecological characteristics, including home range size and diet (Damuth, 1981; Reiss, 1988; Fleming, 1991). We defined mean litter size as the average number (rounded up to the nearest unit) of young per litter of a given species. High reproductive output has been implicated as an important factor influencing introduction success, particularly among plants (Mason *et al.*, 2008). We categorized consumer type as whether the species has a generally carnivorous (e.g. Mustelidae) or omnivorous (e.g. Procyonidae) diet. Broader niche breadth may give a species an advantage when introduced to new environments (Vazquez, 2006). For all life history metrics, we first consulted *Mammalian Species*

(American Society of Mammalogists) accounts, and if an account did not exist for the species of interest, we conducted a literature search for field studies or review articles.

For ecological correlates, we defined latitude difference as the difference between the northern or southern latitudinal boundary of a species' native range nearest to the location of introduction. We defined ecoregion difference as a binomial response for whether the introduction location was in the same ecoregion as the native range of the species. We included these correlates as increased introduction success is expected when climate is similar between native and novel ranges (Williamson, 1996; Mack *et al.*, 2000). Several recent ecological niche modelling studies have emerged that investigate the bioclimatic conditions of introduced species (e.g. Peterson, 2003; Thuiller *et al.*, 2005), but such approach is impractical in a worldwide analysis such as this study, with numerous species and introduction events, and unknown sources within the native range for most introductions.

For each introduction event, we also determined the presence of congeners using IUCN geographic range maps (IUCN, 2019). Indeed, Darwin's Naturalization Hypothesis predicts a greater probability of establishment when a congener is absent due to reduced competition (Darwin, 1859; Chesson, 2000; Hubbell, 2001). We then determined whether an introduction occurred on a continent (mainland) or an island. Introductions on islands may be more successful due to reduced biotic resistance (Elton, 1958; Carlquist, 1965; Wilson, 1965).

We evaluated each of the five hypotheses using generalized linear models in program R, version 3.6.0 (R Development Core Team, 2019) and variables specific to each hypothesis (Table 20.1), using introduction success as the independent variable coded as 0 (failure) or 1 (success). Before inclusion in models, we tested for multi-collinearity among independent variables using the Pearson product-moment correlation coefficient (r) and retained those variables with pairwise $|r| < 0.70$ (Dormann *et al.*, 2013). For each hypothesis, we modelled all variable combinations without interaction terms and ranked candidate models using Akaike's Information Criterion adjusted for small sample sizes (AIC_c), and Akaike's weights (w). All models within 2 AIC_c units of the top-ranked

model were considered competing models. To avoid overparameterization, we chose the simplest model (lowest AIC_c value for degrees of freedom) within competing models as the 'best-fit model', a compromise between simplicity and explanatory power (Richards *et al.*, 2011). For the best-fit model, we estimated parameter coefficients, standard errors, and 95% confidence intervals. We used R, version 3.6.0, for all statistical analyses.

Results

We identified 253 documented introduction events in 66 countries that occurred since at least 206 B.C., involving 24 species from five families (Appendix 20.1). Overall success for small carnivore introductions was high (79%), and the success rate was > 70% for four of the five represented families (Figure 20.1). Of the 253 introduction events, 169 (67%) involved three species: the small Indian mongoose, *Urva auropunctata*, the northern raccoon, *Procyon lotor*, or the American mink, *Neovison vison*. These events occurred during two pulses, the first from 1870 to 1890 that consisted largely of introductions of the small Indian mongoose to islands to control rats and pit vipers (Long, 2003). The second pulse, from 1930 to 1960, was driven by introductions of the northern raccoon and the American mink to the former USSR for the fur industry (Long, 2003) (Figure 20.2). The sable, *Martes zibellina*, has undergone reintroductions and restocking (Mel'chinov, 1958; Ishida *et al.*, 2013; Monakhov, 2016), but not introductions to non-native range. The American marten, *Martes americana*, was believed to have been introduced to Prince of Wales Island, Alaska, through the release of 10 animals in the early 1930s, but recent studies have provided evidence that this species may have naturally colonized the island during the Holocene (Pauli *et al.*, 2015). Therefore, we did not include *M. zibellina* nor *M. americana* in our summaries or analyses. Glatston (1994) mentioned that some people reported the presence (and by inference the introduction) of the crab-eating raccoon, *Procyon cancrivorus*, in Guadeloupe. However, without further evidence, we concur with the author that these reports were likely due to the misidentification of the locally introduced species, *P. lotor* (see below).

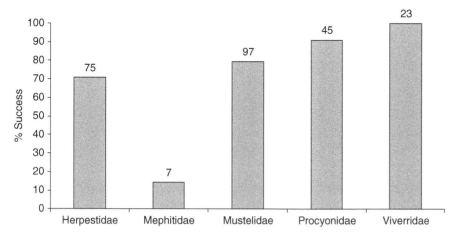

Figure 20.1 Percentage of successful introductions worldwide by small carnivore family. The number of introductions (n_{total} = 247; 6 introductions not included due to unknown outcome) is indicated above bars.

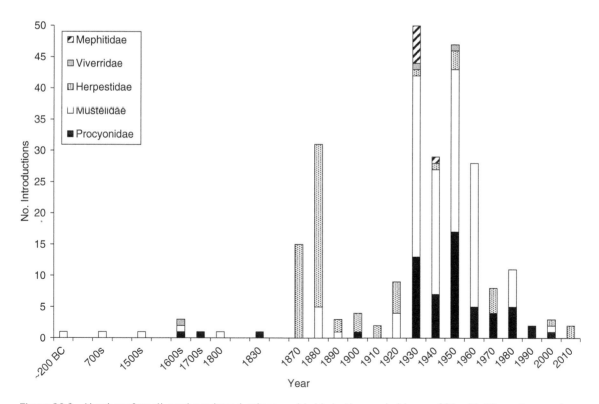

Figure 20.2 Number of small carnivore introductions worldwide by time period (n_{total} = 206, with 18 events spanning over 2 to 6 decades; 47 introductions not included due to unknown dates).

The crab-eating raccoon is however present on Trinidad (Bacon, 1970), but we could not find any evidence that it was introduced. Helgen *et al.* (2008) suggested that it may occur there naturally.

Herpestidae

Small Indian mongoose, *Urva auropunctata* (previously *Herpestes auropunctatus*) – The small Indian mongoose is an omnivorous species native to Iraq eastward through southern China (Jennings & Veron, 2009). The Javan mongoose, *Urva javanica*, and the small Indian mongoose were considered conspecific by some authors, but recent molecular studies have shown that they are separate species (Veron *et al.*, 2007, Chapter 3, this volume; Veron & Jennings, 2017). As a consequence, information on the small Indian mongoose has often been published under the name *Herpestes javanicus*. The small Indian mongoose has an average weight of 0.4 kg and a litter size of two or three young (Nellis, 1989). This is the most-introduced small carnivore (*n* = 66 events; Appendix 20.1) and is included on the IUCN list of the world's 100 worst invasive alien species (Lowe *et al.*, 2000). It has been successfully introduced to control pest populations throughout most of the Caribbean, in northern South America, Mafia Island (Tanzania), Comoros, Mauritius, Fiji, Hawaii, Japan, Croatia, and Bosnia and Herzegovina (Tinker, 1938; Hinton & Dunn, 1967; Coblentz & Coblentz, 1985; Cheke, 1987; Hoagland *et al.*, 1989; Tvrtković & Kryštufek, 1990; Simberloff *et al.*, 2000; Watari *et al.*, 2008, 2011; Barun *et al.*, 2011, 2013; Gaubert, 2016; Berentsen *et al.*, 2018; Louppe *et al.*, 2021). The species naturally expanded from Croatia along the coast to Montenegro, as well as from Eastern Herzegovina inside mainland (Ćirović *et al.*, 2011; Ćirović & Toholj, 2015). The reported introduction and/or current presence of the species in Macedonia, French Guiana, Colombia, Panama, Costa Rica, Bermuda, Bahamas, and Ambon Island (Indonesia) (Tvrtković & Kryštufek, 1990; Barun et al., 2011; CABI, 2021) require further investigation and confirmation. The introduction of the species to Dominica, Honduras and Venezuela failed (CABI, n.d.; Milne & Milne, 1962; Tvrtković & Kryštufek, 1990). In the 1950s, in Italy, a few individuals purchased from a local zoo were released and though they

successfully established for a few decades, they ultimately went extinct (Gaubert, 2016). Three small Indian mongooses were also caught or killed in 1977 on Dodge Island, Florida (Nellis et al., 1978), but since then no further sightings were reported. These individuals had probably been transported from a West Indian port in a fruit shipment. In the Pacific Ocean, similar accidental transportations of a limited number of mongooses to Tonga, Upolu (Samoa), New Caledonia and Kauai (Hawaii) were recently recorded (Barun *et al.*, 2011; Cranwell, 2016; Wostenberg *et al.*, 2019). However, it seems that the small Indian mongoose could not establish successfully on any of these islands (cf. Duffy *et al.*, 2015 for Kauai).

Indian grey mongoose, *Urva edwardsii* (previously *Herpestes edwarsii*) – The Indian grey mongoose is an omnivorous species with a native range that extends from Saudi Arabia east through India (Jennings & Veron, 2009). The average weight for this species is about 1.3 kg and the typical litter size is two to four young (Long, 2003). Repeated attempts to introduce large numbers of mongooses, mostly the Indian grey mongoose, to control rabbits in Australia and New Zealand in the late 1800s failed (Peacock & Abbott, 2010; King, 2019). A few individuals were released to control adders on private property in the Lazio region of Italy in 1952 and persisted for several decades (Gaubert & Zenatello, 2009). This species was also introduced in Peninsular Malaysia apparently without success (Tate, 1947).

Indian brown mongoose, *Urva fusca* (previously *Herpestes fuscus*) – This species is omnivorous and categorized as Least Concern although its native range is restricted to small areas in southern India and western Sri Lanka (Mudappa & Jathanna, 2015). Average weight is 2 kg and the normal litter size is two to four young (Gilchrist *et al.*, 2009; Jennings *et al.*, 2010b). It has become established in Fiji, likely from the release or escape of two private zoo specimens brought there in the late 1970s (Veron *et al.*, 2010).

Meerkat, *Suricata suricatta* – The meerkat is an omnivorous species native to Namibia, Botswana, and South Africa (Jordan & Do Linh San, 2015). Average weight is 0.7 kg and the typical litter size is three young (Van Staaden, 1994). A few individuals were apparently introduced to mainland Australia in 1876 to combat rapidly increasing rabbit populations, but the release was unsuccessful (Peacock & Abbott, 2010).

Mephitidae

Striped skunk, *Mephitis mephitis* – This omnivorous species is native to North America. Its range includes all of the United States (US) and central Canada (Helgen & Reid, 2016). Typical weight is around 2 kg with litter sizes of five to seven young (Wade-Smith & Verts, 1982). Introductions were intended for fur farming (Long, 2003). This species has been introduced at least seven times but introductions were unsuccessful, except on Prince Edward Island, Canada (around 1940) (Banfield, 1977). Failed events occurred in the 1930s and include introductions to Russia, Ukraine, and Kyrgyzstan (de Vos *et al.*, 1956; Bromlei, 1959; Yanushevich, 1963; Pavlov *et al.*, 1973).

Mustelidae

Beech marten, *Martes foina* – The beech marten is native to Eurasia. Its native range includes most of Europe (except Nordic countries and the UK), northern Middle East, China, and certain neighbouring countries, such as Afghanistan and Kyrgyzstan (Abramov *et al.*, 2016b). Average weight is 1.5 kg and the usual litter size is five young (Heptner & Naumov, 2002). It was introduced to new areas on two occasions: Wisconsin, USA (1950–1960), where it was successful, and Ibiza Island, Spain (early twentieth century), where it did not succeed (Long, 1995; Spanish Ministry of Environment, 2006).

European pine marten, *Martes martes* – The European pine marten has a native range including much of Europe, eastern Russia, and the northern Middle East (Herrero *et al.*, 2016). The average weight for his species is 1.4 kg and the typical litter size is four young (Heptner & Naumov, 2002). It has been introduced at least four times. It was successfully introduced in Mallorca and Menorca Islands (Spain), unsuccessfully (1962) in Kyrgyzstan, and successfully (1980s) to the Isle of Mull in the UK (Pavlov *et al.*, 1973; Clevenger, 1993; Solow *et al.*, 2013; Valenzuela & Alcover, 2015).

Japanese marten, *Martes melampus* – The Japanese marten is native to Japan (Honshu, Shikoku, Kyushu) (Abramov *et al.*, 2015). The average weight for this species is 1.2 kg and the typical litter size is two young (Masuda, 2009; Hunter & Barret, 2018). It has been

introduced within Japan twice: Hokkaido in the 1940s for fur farming, and Sado Island for possibly the same reason (Masuda, 2009). It has been recorded in the southern Korean Peninsula but there are no locality details to prove a wild origin and no native population has been confirmed (Won & Smith, 1999; Abe *et al.*, 2005).

Stoat, *Mustela erminea* – The stoat has a wide native range across the entire northern hemisphere. In North America, it ranges from the central US to Greenland. In Eurasia, from east to west, it ranges from Spain to Japan, and from south to north, it occupies northern Spain, Italy, Romania, Kazakhstan, and China through northern Russia (Reid *et al.*, 2016a). The average weight is 0.17 kg with an average litter size of six young (King, 1983). It has been introduced to new areas at least five times, all of them islands. Four introductions were in Northern Europe: Denmark (1980), Netherlands (Terschelling Island, 1931), and the UK (1680 and 1800s) and one in New Zealand (1885) (Fitter, 1959; Lever, 1985; King, 1990, 2017a). New Zealand individuals were imported from London, UK (King, 2017b). Introductions to Denmark, Netherlands, and New Zealand were successful, while the Whalshay, UK introduction failed.

Japanese weasel, *Mustela itatsi* – The Japanese weasel is a mustelid native to central and southern Japan (Abramov, 1999; Kaneko *et al.*, 2016). The average weight is 0.4 kg and the usual litter size is seven young (Heptner & Naumov, 2002). Reliable documentation exists for two successful introductions in Japan: Miyake-jima Island (1986) and Hokkaido Island (1880s) (Inukai, 1934; Hamao & Higuchi, 2013). The Japanese weasel was also introduced to about 50 small islands in Japan to control rats and mice (Shiraishi, 1982); however, more information is lacking. Because of conflicting evidence, Masuda and Watanabe (2009) argue that it is difficult to identify the islands where this species is native. There is evidence for an introduction to Sakhalin Island, Russia, in 1932, but there are no reliable records since 1980 (Tumanov, 2009).

European mink, *Mustela lutreola* – The European mink is unique among small carnivores in that it is the only introduced species that is Critically Endangered in its native range (Maran *et al.*, 2016). The historical range included most of Europe, but the species is now restricted to parts of Spain, France, Romania, and some areas of the European part of Russia (Maran *et al.*, 2016). The average weight is 0.45 kg with litter size typically four or five young (Youngman, 1990). It was introduced to the Île du chat in 1956, an isolated southern Atlantic island belonging to France, which involved three individuals and was unsuccessful, and Kuril Islands, Russia (1983; unknown number introduced) which appears successful (Lésel & Derenne, 1975; Schreiber *et al.*, 1989).

Least weasel, *Mustela nivalis* – The least weasel occurs across most of the northern hemisphere. In North America, it occurs from the central US to northern Canada and Alaska. It occupies almost all of Europe and northern Morocco. In Asia, it is widespread, occurring from the Arctic to Turkey, Iran, Kyrgyzstan, China, and northern Japan (McDonald *et al.*, 2016). The average weight is 0.08 kg and the average litter size is four or five young (Sheffield & King, 1994). It was brought to Mallorca Island (Spain) in ancient times, where it was successful (about 200 B.C.; see Valenzuela & Alcover, 2013). It was also successful in New Zealand, where individuals brought from London, UK (along with stoats) were introduced in 1890 (Gibb & Flux, 1973; King, 2017a,b). Introductions failed in the Netherlands (Terschelling Island, 1931) and Australia (1885) (de Vos *et al.*, 1956; Hinton & Dunn, 1967). It was introduced to São Tomé Island after European colonization, although the current status is unknown (Dutton, 1994). All introductions were on islands, except Australia.

European polecat, *Mustela putorius* – The European polecat's native range includes central Europe and southern Sweden and Finland, to eastern Russia and Ukraine (Skumatov *et al.*, 2016). The average weight is 0.73 kg and usual litter size is five young (Blandford, 1987). It has been introduced outside its native range at least nine times, though release numbers are mostly unknown. They were brought to Australia in 1885 to control rabbits, but failed to survive (de Vos *et al.*, 1956). New Zealand underwent two introductions; one consisted of thousands of individuals in the 1880s that was successful, and the second introduction of unknown numbers in 1960, which failed (Marshall, 1963; Flux & Fullagar, 1992). It was also introduced to Jamaica for rat control (date unknown), which also failed (Milne & Milne, 1962). Individuals were brought to the Canary Islands, Spain, in the 1500s, where they successfully colonized (Medina & Martín, 2010). Two

introductions occurred in Europe in the 1900s (the UK and Ireland, both failed), one in Japan (1930, failed), and one in Russia (1940, succeeded) (Kaburaki, 1940; Lavrov, 1946; Flux & Fullagar, 1992).

Siberian weasel, *Mustela sibirica* – The native range of the Siberian weasel includes central and southern Russia and central and western China (Abramov *et al.*, 2016a). The average weight is 0.4 kg and the usual litter size is seven young (Heptner & Naumov, 2002). It has been introduced at least six times outside its natural range, mostly in western Asia (Long, 2003). Numbers released are mostly unknown. Japan underwent three successful introductions in 1930, 1948, and a large release of 1600 individuals in 1966 (Kuroda, 1955; Long, 2003). One Japanese introduction failed, in the Ryukyu Islands from 1954 to 1958 that consisted of 2000 individuals (Hayashi, 1981). Siberian weasels were also introduced unsuccessfully in Kyrgyzstan (1941) and western Russia (Sakhalin Island, 1937) for fur hunting (Lavrov, 1946; Kuroda, 1955).

American mink, *Neovison vison* – The American mink is native to North America, occurring from the central US to northern Canada (Reid *et al.*, 2016b). The typical weight is 1 kg and the litter size is four on average (Larivière, 1999). This small carnivore species has been widely introduced with at least 60 introductions, 54 of them successful. It was mainly exported from North America to other parts of the world for fur farms in the early- to mid-1900s. More than half of these introductions ($n = 34$) occurred in the former USSR (Azerbaijan, Belarus, Kazakhstan, Kyrgyzstan, and Russia) (Berger, 1962; Samusenko, 1962; Beishebaev, 1963 *in* Long, 2003; Popov, 1964; Izmailov, 1968 *in* Long, 2003; Benkovsky, 1971; Pavlov *et al.*, 1973; Savenkov, 1987 *in* Long, 2003; Sinitsyn, 1987 *in* Long, 2003; Long, 2003). Release numbers typically ranged from 60 to 700 individuals, but large releases of 10 000–11 000 individuals happened twice in Russia (Long, 2003). Seven releases occurred in southern Chile and Argentina, 18 in Europe (e.g. France, Spain, Sweden, and the UK), and one in Japan (de Vos *et al.*, 1956; Thompson, 1962; Gerell, 1967a,b; Litjens, 1980; Lever, 1985; Smal, 1988; Jaksic *et al.*, 2002; Brzezinski & Marzec, 2003; Long, 2003; Bonesi & Palazon, 2007). Island introductions occurred in Britain, Isles of Lewis and Mull (Scotland), Iceland, Ireland, Japan, and Urup (part of Kuril Islands, Russia) and Sakhalin (Russia).

Procyonidae

Coati, *Nasua nasua* – The coati is a South American omnivorous species with a native range from northern Uruguay and Argentina to Colombia and Venezuela (Emmons & Helgen, 2016). Mean weight is about 3.8 kg, and litter size is from three to four young (Gompper & Decker, 1998). It has been introduced successfully outside its native range on two occasions; Juan Fernandez Islands Group (Robinson Crusoe Island), Chile in 1935, and Mallorca Island, Spain in the 1990s (Eisenberg, 1989; Mayol *et al.*, 2009). Numbers released were two to eight individuals, but it appears that undocumented releases also occurred (Long, 2003).

Northern raccoon, *Procyon lotor* – The northern raccoon is an omnivorous species with an extensive native distribution from Panama to southern Canada (Timm *et al.*, 2016). The average weight is 5.5 kg and litter size is typically three to four young (Lotze & Anderson, 1979). It has been involved in at least 43 introduction attempts, all in the 1900s for fur farms, except in the Caribbean, where it was introduced in the seventeenth and nineteenth centuries from North American populations (reasons unknown, see Pons *et al.* 1999; Helgen & Wilson, 2002, 2003). The numbers of individuals introduced ranged from 4 to 500, but usually from 20 to 90. Six introductions were in Europe, including Spain, Germany, and Belarus (Niethammer, 1963; Aliev & Sanderson, 1966; Corbet, 1966; cf. Fischer *et al.*, 2015 for Germany), where the species' range is expanding and projected to continue to do so (Salgado, 2018; Louppe *et al.*, 2019). Nineteen introductions were in Asia, including the former USSR (Azerbaijan, Kazakhstan, Kyrgyzstan, Russia, Ukraine, and Uzbekistan) (Aliev & Sanderson, 1966; Naumov, 1972; Pavlov *et al.*, 1973; Baker, 1986; Gineyev, 1987 *in* Long, 2003). Asian introductions were mostly from western European fur farms, but naturalization in Japan took place through both zoo animals that escaped and intentional release or escape of raccoons from breeding cages (Ikeda *et al.*, 2004). Six introductions took place in Alaska (USA) and Canada, and one in New Zealand (Cameron, 1950 *in* Long, 2003; Wodzicki, 1950; Burris, 1965 *in* Long, 2003; Carl & Guiguet, 1972). Nine introductions occurred in the Caribbean (three separate islands in the Bahamas as well as in Guadeloupe, Barbados, Martinique, and Saint Martin/Sint Maarten) from North American individuals (Sherman, 1954; McKinley, 1959; Husson, 1960; Lazell, 1981; Moutou, 1987; Bénito-Espinal, 1990; Bon Saint Côme & Tanasi, 1994; Lorvelec *et al.*, 2007; Helgen *et al.*, 2008; Louppe *et al.* 2020). Introductions were unsuccessful in only two known attempts: New Zealand and Petrov Island, Russia. Helgen & Wilson (2002, 2003) mentioned the possibility that the northern raccoon was introduced unsuccessfully to Jamaica, but robust evidence is lacking. It is however possible that *P. lotor* was introduced to Tres Marías Islands, Mexico, not long ago. Although Helgen & Wilson (2005) noted that insular raccoon specimens differed morphologically from the subspecies *P. lotor hernandezii* found on the Mexican mainland, they suggested that differences are not large enough to continue treating the Tres Marías raccoon as a separate species, namely *P. insularis* (cf. Dinets, 2012). It is possible that the same could apply to the case of the pygmy raccoon, *P. pygmaeus* from Cozumel Island, Mexico (Louppe *et al.*, 2020; cf. McFadden *et al.*, 2008), in which case *P. lotor* may have been introduced by the Mayas from 100 A.D. onward (Cuarón *et al.*, 2004).

Viverridae

African civet, *Civettictis civetta* – The African civet is omnivorous and its native to sub-Saharan Africa, reaching northern Namibia, Botswana, and South Africa in the south, and excluding Somalia in the east. It is the largest viverrid in Africa, weighing between 7 and 20 kg, and litter size is two to three young on average (Hunter & Barret, 2018). It was introduced to São Tomé Island by European colonizers, likely to control rodents and possibly to exploit its musk (Dutton, 1994).

Common genet, *Genetta genetta* – The common genet is omnivorous and its native range includes portions of northern, central, and southern Africa. The northern area includes Morocco and Algeria, the central area ranges from Mauritania and Senegal to Somalia and Kenya, and the southern area is from Angola and Zimbabwe to South Africa. It also occurs in south-west Yemen and Saudi Arabia (Gaubert, 2016). Its average weight is 1.9 kg and usual litter size is two to three young (Larivière & Calzada, 2001). It was successfully introduced to the Cataluña and Andalucía regions in Spain, as well in the Balearic

Islands (Mallorca, Ibiza, and Cabrera) from North African populations by early traders or invaders (Gaubert *et al.*, 2009, 2011, 2015a; Delibes *et al.*, 2019).

Masked palm civet, *Paguma larvata* – The masked palm civet is an omnivorous species with a native range from Nepal to central China in the north to Sumatra and Borneo in the south (Torri, 2009; Jennings & Veron, 2011). The average weight is 4 kg with a litter size of one to four young (Long, 2003). Unkown numbers were introduced successfully to Japan in the 1930–1940s for the fur industry (Udagawa, 1954 *in* Long, 2003; Torri, 2009), from Taiwan (Masuda *et al.*, 2010), and possibly other areas (Inoue *et al.*, 2012).

Common palm civet, *Paradoxurus hermaphroditus* – The common palm civet is an omnivorous species with a native range encompassing South and Southeast Asia; from India to the Philippines from east to west, and from southern China to Borneo from north to south (Patou *et al.*, 2010). The average weight is 3.2 kg and the usual litter size is from two to five young (Long, 2003). They were introduced successfully (unknown numbers or date) in the Aru Islands, Moluccas, Indonesia, apparently for rodent control (Lever, 1985), and in the Lesser Sunda Islands and Sulawesi (Schreiber *et al.*, 1989; Corlett, 2010; Patou *et al.*, 2010).

Malay civet, *Viverra tangalunga* – The Malay civet is omnivorous with a range including Peninsular Malaysia, Singapore, Sumatra, Borneo, the Philippines, Sulawesi, and the Moluccas (Jennings & Veron, 2011). The average weight is about 5 kg and the usual litter size is two to three young (Long, 2003; Jennings *et al.*, 2010a). The literature indicates that it has been successfully introduced at least four times on some Indonesian islands, although numbers released and dates of release appear unknown, except on Buru Island where it established in the mid-1900s (Laurie & Hill, 1954; Flannery, 1995). Based on a molecular study, Veron *et al.* (2014) suggested that humans transported the Malay civet across Wallace's Line to Sulawesi and the Moluccas, and to the Philippines and the Natuna Islands; however, natural dispersal could not be rejected.

Large Indian civet, *Viverra zibetha* – The large Indian civet is the largest invasive species within the small carnivore families. Its range extends from the Himalayan mountains in Nepal to southern China,

and south to Cambodia, Peninsular Malaysia and Singapore (Jennings & Veron, 2011). The average weight is 8.5 kg and the typical litter size is between one and four young (Long, 2003). Although the large Indian civet has been listed as Least Concern by the IUCN (Timmins *et al.*, 2016), local declines driven largely by the loss and degradation of its tropical forest habitat, and hunting, have been recorded (Jennings & Veron, 2009). The only registered introduction occurred on the Andaman Islands, India, and was successful, although the introduction date and numbers released are unknown (Lever, 1985).

Small Indian civet, *Viverricula indica* – The small Indian civet's native range extends from India to southern China, and south through Peninsular Malaysia to Java (Jennings & Veron, 2011). The average weight is 2.9 kg and the usual litter size is two to five young (Long, 2003). This species is used for both perfume production and rodent control. It was introduced successfully to at least eight islands: Sumbawa (Indonesia), Socotra (Yemen), Mafia (Tanzania), Pemba (Tanzania), Zanzibar (Tanzania), Mayotte (France), Comoros, and Madagascar (Laurie & Hill, 1954; Pakenham, 1984; Haltenorth & Diller, 1994; Louette, 1999; Kock & Stanley, 2009; Goodman, 2012; Caceres & Decalf, 2015; Gaubert *et al.*, 2017; and see Farris *et al.*, Chapter 13, this volume). Both the possible introduction and establishment of this species in the north of Sumatra remain unclear (Choudhury *et al.*, 2015).

Correlates of Success

Generalized linear mixed models suggest that small carnivore introduction success offered support for inherent superiority, climate-matching, and Darwin's naturalization hypotheses. The best-fit model for inherent superiority included the effect of body size, litter size, and consumer type (Tables 20.2 and 20.3). Percentage success decreased by an estimate of 7.6% with each additional individual in the typical litter (Figure 20.3). The best-performing model for climate-matching included latitude difference (Table 20.2). Distance between native range boundary and latitude of the introduction site had a negative effect on introduction success. For Darwin's naturalization, we found the presence of a congener in the area of introduction

Table 20.2   Competing models for each of the five invasion susceptibility hypotheses tested on small carnivore mammal introductions worldwide. *K* = number of parameters, ΔAIC_c = difference in AIC_c (Akaike's Information Criterion corrected for small sample size) score between a model and the best-supported model, and *w* = Akaike's weights. Best-fit models (lowest K within top models; top models have $\Delta AIC_c < 2$) are shown in bold.

Hypotheses	Models	*K*	ΔAIC_c	*w*
Climate-matching	**Latitude difference**	3	0	0.68
	Latitude difference + Biogeographic region	4	1.54	0.31
	Null	2	15.16	<0.01
	Biogeographic region	3	17.18	<0.01
Propagule pressure	**Null**	2	0	0.66
	Propagule pressure	3	1.36	0.34
Inherent superiority	**Body size + Litter size + Consumer type**	8	0	0.488
	Body size + Litter size	6	4.52	0.246
	Litter size	5	9.42	0.161
	Consumer type + Litter size	7	11.15	0.105
	Body size	2	29.14	<0.01
	Consumer type	3	29.60	<0.01
	Body size + Consumer type	4	30.05	<0.01
	Null	1	30.80	<0.01
Island susceptibility	**Null**	2	0	0.71
	Island/Mainland	3	1.83	0.29
Darwin's naturalization	**Congener presence**	3	0	0.92
	Null	2	5.05	0.08

Table 20.3   Parameter estimates of the best-fit models (see Table 20.2) for small carnivore invasion success worldwide. Scaled parameter estimates, standard errors (SE), and 95% confidence intervals (CI) are shown.

Hypotheses	Parameters	Estimates	SE	95% CI
Climate-matching	Latitude difference	−0.11	0.02	−0.15 to −0.06
Inherent superiority	Body size	0.10	0.03	0.05 to 0.16
	Litter size	−0.12	0.38	−0.23 to −0.09
	Consumer type	0.17	0.06	0.04 to 0.30
Darwin's naturalization	Congener	−0.06	0.02	−0.01 to −0.11

decreased the probability of introduction success (Table 20.3). The percentage of successful introductions decreased from an average of 84 to 57% when a congener was present in the area. The percentage of successful introductions for the mainland was 82% and for the islands 84%. We found no support for propagule pressure or island susceptibility hypotheses.

Discussion

Mammals and other vertebrate taxa have a low probability of arriving at a new location on their own; however, once introduced, mammals have a high likelihood (~50%) of establishment and spread (Jeschke & Strayer, 2005). Mammal introductions around the world

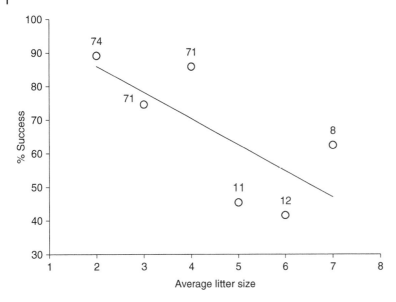

Figure 20.3 Percentage introduction success by average small carnivore litter size (rounded down to the nearest whole number). The number of introductions (n_{total} = 247; 6 introductions not included due to unknown outcome) by litter size is indicated above respective points.

peaked in the late 1800s when societies introduced valuable and charismatic species from their home areas (primarily Europe and North America) to new locations (Kegel, 2001; Long, 2003; Jeschke & Strayer, 2005). Accordingly, we found that most small carnivore introductions occurred during the late 1800s and early 1900s, primarily (67%) introductions of the small Indian mongoose for pest control, and the northern raccoon and American mink for the fur industry (Long, 2003). The success of small carnivore species introductions among families was high (> 70%), except for Mephitidae (14%).

Increasing latitude difference between native range and introduction location decreased the probability of small carnivore introduction success. Climate-matching is one of the most frequently supported hypotheses for explaining successful introductions in vertebrates (e.g. Blackburn & Duncan, 2001; Forsyth *et al.*, 2004; Bomford *et al.*, 2008). Organisms typically are adapted to a range of conditions and are not usually able to adapt rapidly to conditions beyond this range (Hayes & Barry, 2008). Changes in latitude affect the angle of the Sun's radiation, which affects temperature and climate; whereas ecoregion represents the distinct climatic, geological, and evolutionary history of species (Lomolino *et al.*, 2010). Small carnivore

species were dependent on matching climatic conditions (i.e. latitude), but not on the particular biota and environment they evolved in (i.e. ecoregion), suggesting considerable ecological flexibility. Species adapted to use a variety of resources with the ability to persist in diverse ecosystems have the potential to occupy larger areas and maintain higher population densities (Hutchinson, 1957).

The Inherent Superiority Hypothesis also received support, suggesting that successfully introduced small carnivore species possess competitively advantageous traits. Specifically, larger body size was associated with a greater probability of introduction success, which contrasts with Collauti *et al.* (2006) who found body size was similar between successful and unsuccessful introductions in several taxa. Body size can enhance the survival of terrestrial vertebrates during periods of resource shortages in more extreme seasonal environments (Boyce, 1979; Murphy, 1985; Zeveloff & Boyce, 1988). Homeostasis requires that large mammals consume more food, but at less frequent intervals, than small mammals; hence larger body sizes would be favoured when there is a need for fasting during resource shortages (Millar & Hickling, 1990). When introduced outside their native range, animals face a novel environment and may be naïve to resource distribution (Marable *et al.*, 2012). As a result, fasting is likely to happen during the first stages of colonization, which is consistent with larger carnivores having a greater probability of successful introductions. There is conflicting evidence regarding the relationship between body size and litter size in mammals. On a broad scale, larger mammalian species typically produce smaller litters at less frequent intervals (Harvey & Read, 1988). However, there is also substantial evidence of mammalian groups that do not show a significant relationship between the number of offspring per litter and adult body weight (Leitch *et al.*, 1959; Sacher & Staffeidt, 1974; Millar, 1977), a pattern we observed for the small carnivore species in this study. In contrast to our expectations, greater litter size appeared to have a negative effect on success. Nursing a larger number of individuals may require greater foraging effort and offspring defence may also be costly (Koskela *et al.*, 2000). In addition, small mammals appear to have a trade-off between the number and size of offspring (e.g. Kaufman & Kaufman, 1987). Overall, smaller litters may result in increased survival and long-term fecundity for introduced females, particularly in the context of adapting to novel environments. Finally, consumer type suggested that species with a carnivorous diet were more successful than the ones with an omnivorous diet. This failed to support the Niche-Breadth Hypothesis, which states that species with broader niches have greater invading success because they are more likely to find the necessary resources or environmental conditions (Ricciardi & Rasmussen, 1998; Lockwood *et al.*, 1999). This idea, although attractive, has produced conflicting evidence and its importance is not yet clear (Vazquez, 2006).

We found no support for islands being more susceptible to small carnivore introductions. A classic hypothesis in ecology is that following the introduction of new species, islands are easier to invade than mainland sites (Elton, 1958; Carlquist, 1965; MacDonald & Cooper, 1995). This originates from ecological theory stating that species-rich biotas better resist invasion than species-poor biotas (MacArthur, 1955; Elton, 1958; Case, 1991; Herben, 2005), and also from early observations that island communities often include a greater proportion of invasive species than mainland communities (Elton, 1958; Atkinson, 1989). Thus, colonizing predators, such as small carnivores, should seemingly benefit from this higher susceptibility of islands. Interestingly, we found that islands were not more vulnerable to small carnivore introductions, similar to more recent studies testing this prediction in other taxa (Sol, 2000; Blackburn & Duncan, 2001; Cassey, 2003). In cases where differences between islands and mainland introductions were observed, success appears linked to anthropogenic disturbances rather than natural causes (e.g. Yiming *et al.*, 2006).

Propagule pressure has been consistently associated with introduction success (e.g. Collauti *et al.*, 2006; Simberloff, 2009; Cassey *et al.*, 2018). Nonetheless, we found no support for the Propagule Pressure Hypothesis with respect to small carnivores. Releases of large numbers of individuals may allow invading populations to offset decreases in survival or reproduction caused by environmental or stochastic events (Lockwood *et al.*, 2005). In a review of 49 studies of vertebrates, insects, and plants, the number of released individuals was positively associated with introduction success (Hayes & Barry, 2008). Two meta-analyses have found that propagule pressure is positively associated with the establishment stage, suggesting that established species are introduced in greater numbers and/or more often (Collauti *et al.*, 2006, Cassey *et al.*, 2018). In an experimental setting, propagule pressure was positively associated with the introduction success of a freshwater fish invader, but only after factors including weather, resources, and competition were controlled for (Britton & Gozland, 2013). Despite this support, Moulton *et al.* (2013) were critical of the Propagule Pressure Hypothesis, arguing that the historical record is frequently inaccurate and summing propagules from multiple introduction events to represent the founding population creates a bias in favour of the Propagule Pressure Hypothesis. Moulton *et al.* (2013) further demonstrated that the number of individuals released alone did not predict introduction outcomes for passerine birds. Species-level traits (Blackburn *et al.*, 2009) and/or site level factors could play an important role in introduction success. Numbers of individuals released would seem to be of less importance if the introduced individuals cannot find sufficient food, encounter high risk, or if the new environment is climatically hostile (Moulton *et al.*, 2013).

Darwin (1859) stated that non-native species are more likely to naturalize when they belong to genera with no native species in the region. This scenario involves niche differentiation and niche gap filling by invaders to be the main drivers of introduction success (Thuiller *et al.*, 2010). Interestingly, we found evidence for small carnivore introductions to be slightly less successful when congeners were present in the introduction area. At small spatial scales, invaders taxonomically similar to native species are considered less likely to coexist because of competitive exclusion; however, at larger spatial scales, these invaders may be able to co-occur through neutral processes and dispersal limitation (Chesson, 2000; Hubbell, 2001). Thuiller *et al.* (2010) suggested that patterns observable at continental scales do not likely reflect the outcome of species interactions but instead reflect environmental filtering, regional heterogeneity and species dispersal abilities. It is increasingly acknowledged that support for Darwin's Naturalization Hypothesis will vary depending on the spatial, temporal, and phylogenetic scale considered (Diez *et al.*, 2008; Proches *et al.*, 2008; Carboni *et al.*, 2013; Pellock *et al.*, 2013). Other studies have failed to find evidence supporting taxonomic relatedness predicting introduction success (Ricciardi & Mottiar, 2006; Escobedo *et al.*, 2011) or provided only weak support (Van Wilgen & Richardson, 2011). Biotic interactions, especially competition and combined effects of enemies and mutualists are likely important determinants of establishment success (Pellock *et al.*, 2013), and once established, the continued success of invasive species may be dependent on their innate traits.

Conclusion

When considering human–carnivore conflicts, attention is typically directed toward large predators (Mladenoff *et al.*, 1997; Liu *et al.*, 2001). However, small carnivores have long been involved in competition with humans over game species, crops, apiaries, and fish stocks (Treves & Karanth, 2003), particularly after the removal of large predators (Reynolds & Tapper, 1996). Yet conflicts between small carnivores and humans are not limited to competition; for example, small carnivores are vulnerable to human-induced habitat degradation, land-use change, and exploitation (Gehrt, 2004). Also, small carnivore introductions have the potential to adversely affect native communities (Long, 2003). We found that successful introductions of small carnivore species worldwide were dependent on matching climatic conditions that allow survival, and that a greater body size, smaller litter size, and a carnivorous diet, increased chances of introduction success. In addition, the presence of a congener in the area of introduction appeared to decrease the probability of success. As previously suggested, islands were not more susceptible to successful introductions by small carnivores, and the number of individuals introduced did not have any significant impact on success. As biological invasions become increasingly frequent and widespread (Pyšek & Richardson, 2010; Seebens *et al.*, 2018), understanding the biological and environmental factors affecting their success is critically important for conservation and resource management.

Acknowledgements

We thank Luigi Boitani, Sugoto Roy, and two anonymous reviewers for their suggestions and comments, which improved this manuscript. Alexei Abramov provided additional helpful comments and kindly error-checked Russian titles and references. We also thank Emmanuel Do Linh San for his help, suggestions, and editorial assistance.

References

Abe, S. (1991) [Establishment in the wild of the mongoose (*Herpestes* sp.) on Amami-oshima Island]. *Mammalian Science (Honyurui Kagaku)* 31, 23–36. (In Japanese with English abstract).

Abe, H., Ishii, N., Ito, T., Kaneko, Y., Maeda, K., Miura, S. & Yoneda, M. (2005) *A Guide to the Mammals of Japan*. Tokai University Press, Kanagawa.

Abramov, A.V. (1999) The taxonomic status of the Japanese weasel *Mustela itatsi* (Carnivora, Mustelidae). *Zooloicheskii Zhurnal* 79, 80–88.

Abramov, A.V., Kaneko, Y. & Masuda, R. (2015) *Martes melampus. The IUCN Red List of Threatened Species* 2015, e.T41650A45213228. http://dx.doi.org/10.2305/IUCN.UK.2015-4.RLTS.T41650A45213228.en. Accessed on 2 January 2017.

Abramov, A.V., Duckworth, J.W., Choudhury, A., Chutipong, W., Timmins, R.J., Ghimirey, Y., Chan, B. & Dinets, V. (2016a) *Mustela sibirica. The IUCN Red List of Threatened Species* 2016, e.T41659A45214744. http://dx.doi.org/0.2305/IUCN.UK.2016-1.RLTS.T41659A45214744.en. Accessed on 2 January 2017.

Abramov, A.V., Kranz, A., Herrero, J., Choudhury, A. & Maran, T. (2016b) *Martes foina. The IUCN Red List of Threatened Species* 2016, e.T29672A45202514. http://dx.doi.org/10.2305/IUCN.UK.2016-1.RLTS.T29672A45202514.en. Accessed on 2 January 2017.

Agetsuma, Y.Y. (2004) [Process of establishing an introduced raccoon (*Procyon lotor*) population in Aichi and Gifu Prefectures, Japan: policy for managing threats posed by introduced raccoons]. *Mammalian Science (Honyurui Kagaku)* 44, 147–160. (In Japanese with English abstract).

Aliev, F.F. (1955) Acclimatization and the economic acclimatization of the raccoon in Azerbaijan. *Academy of Sciences of Azerbaijan SSR Report* 8, 571–578.

Aliev, F.F. & Sanderson, G.C. (1966) Distribution and status of the raccoon in the Soviet Union. *Journal of Wildlife Management* 30, 497–502.

Ando, S. & Kajiura, K. (1985) [Status of raccoons in Gifu Prefecture]. *Bulletin of the Gifu Prefectural Museum* 6, 23–30. (In Japanese).

Atkinson, I. (1989) Introduced animals and extinction. In: *Conservation for the Twenty-First Century* (eds D. Western & M. Pearl), pp. 54–69. Oxford University Press, New York.

Bacon, P.R. (1970) *The Ecology of the Caroni Swamp*, Special Publication. Central Statistical Office, Trinidad.

Baker, S.J. (1986) Irresponsible introductions and reintroductions of animals into Europe with particular reference to Britain. *International Zoo Yearbook* 24, 200–205.

Banfield, A.W.F. (1977) *The Mammals of Canada*. University of Toronto Press, Toronto and Buffalo.

Barun, A., Hanson, C.C., Campbell, K.J. & Simberloff, D. (2011) A review of small Indian mongoose management and eradications on islands. In: *Island Invasives: Eradication and Management* (eds C.R. Veitch, M.N. Clout & D.R. Towns), pp. 17–25. IUCN, Gland.

Barun, A., Niemiller, M.L., Fitzpatrick, B.M., Fordyce, J.A. & Simberloff, D. (2013) Can genetic data confirm or refute historical records? The island invasion of the small Indian mongoose (*Herpestes auropunctatus*). *Biological Invasions* 15, 2243–2251.

Beishebaev, K. (1963) [Data on numbers of the mink (*Lutreola vison* Brisson) in nut and fruit forests of southern Kirgizia]. *Izvestiya Akademii Nauk Kirgizskoi SSR, Seriya Biologicheskikh Nauk* 5, 5–10. (In Russian).

Bénito-Espinal, E. (1990) *La Grande Encyclopédie de la Caraïbe. Tome 2: La Faune*. Sanoli, Milan. (In French).

Benkovsky, L.M. (1971) [Migration of *Mustela (Mustela) vison* Schr. in the Sakhalin Region]. *Vestnik Zoologii* 31, 19–25. (In Russian with English summary).

Berentsen, A.R., Pitt, W.C. & Sugihara, R.T. (2018) Ecology of the small Indian mongoose (*Herpestes auropunctatus*) in North America. In: *Ecology and Management of Terrestrial Vertebrate Invasive Species in the United States* (eds W.C. Pitt, J.C. Beasley & G.W. Witmer), pp. 251–267. CRC Press, Boca Raton.

Berger, N.M. (1962) [American mink adaptation to new environmental conditions in Siberia and its practical use]. *Voprosky Ekologii* 6, 22–23. (In Russian).

Blackburn, T. & Duncan, R. (2001) Determinants of establishment success in introduced birds. *Nature* 414, 195–197.

Blackburn, T.M., Cassey, P. & Lockwood, J.L. (2009) The role of species traits in the establishment success of exotic birds. *Global Change Biology* 15, 2852–2860.

Blandford, P.R.S. (1987) Biology of the polecat *Mustela putorius*: a literature review. *Mammal Review* 17, 155–198.

Bomford, M., Kraus, F., Barry, S. & Lawrence, E. (2008) Predicting establishment success for alien reptiles and amphibians: a role for climate matching. *Biological Invasions* 11, 713–724.

Bomford, M., Darbyshire, R.O. & Randall, L. (2009) Determinants of establishment success for introduced exotic mammals. *Wildlife Research* 36, 192–202.

Bon Saint Côme, M. & Tanasi, M. (1994) *Le Raccoon en Martinique*. Rapport, Office National de la Chasse Fort-de-France.

Bonesi, L. & Palazon, S. (2007) The American mink in Europe: status, impacts, and control. *Biological Conservation* 134, 470–483.

Borroto-Páez, R. (2009) Invasive mammals in Cuba: an overview. *Biological Invasions* 11, 2279–2290.

Boyce, M.S. (1979) Seasonality and patterns of natural selection for life histories. *American Naturalist* 114, 569–583.

Britton, J.R. & Gozland, R.E. (2013) How many founders for a biological invasion? Predicting introduction outcomes from propagule pressure. *Ecology* 94, 2558–2566.

Bromlei, G.F. (1959) [Attempt of acclimatization of skunk (*Mephitis mephitis*) in south of Primorsky Krai]. *Soobshcheniya Dal'nevostochnogo Filiala Sibiriskogo Otdeleniya Akademii Nauk SSSR* 11, 121–123. (In Russian).

Brzezinski, M. & Marzec, M. (2003) The origin, dispersal and distribution of the American mink *Mustela vison* in Poland. *Acta Theriologica* 48, 505–514.

Buckley, D.J., Sleeman, D.P. & Murphy J. (2007). Feral ferrets *Mustela putorius furo* L. in Ireland. *The Irish Naturalists' Journal* 28, 356–360.

Burris, O.E. (1965) No title. In: *Proceedings of the 45th Annual Conference of the Western Association of State Game and Fish Commissioners*, pp. 93–104.

CABI [Center for Agriculture and Bioscience International] (n.d.). *Compendium Record*. CABI, Wallingford.

CABI [Center for Agriculture and Bioscience International] (2021) *Herpestes auropunctatus*. In: *Invasive Species Compendium*. CABI, Wallingford. https://www.cabi.org/isc. Accessed on 12 July 2021.

Caceres, S. & Decalf, G. (2015) *Stratégie de Lutte contre les Espèces Animales Invasives à Mayotte. 2015–2020*. Rapport ONCFS/DEAL, Mayotte.

Carboni, M., Münkemüller, T., Gallien, L., Lavergne, S., Acosta, A. & Thuiller, W. (2013) Darwin's naturalization hypothesis: scale matters in coastal plant communities. *Ecography* 36, 560–568.

Carl, G.C. & Guiguet, C.J. (1972) Alien animals in British Columbia. *British Columbia Provincial Museum Handbook* 14, 1–103.

Carlquist, S. (1965) *Island Life*. Natural History Press, New York.

Case, T.J. (1991) Invasion resistance, species build-up and community collapse in metapopulation models with interspecies competition. *Biological Journal of the Linnean Society* 42, 239–266.

Case, T.J. & Bolger, D.T. (1991) The role of introduced species in shaping the distribution and abundance of island reptiles. *Evolutionary Ecology* 5, 272–290.

Cassey, P. (2003) A comparative analysis of the relative success of introduced land birds on islands. *Evolutionary Ecology Research* 5, 1011–1021.

Cassey, P., Delean, S., Lockwood, J.L., Sadowski, J. & Blackburn, T.M. (2018) Dissecting the null model for biological invasions: a meta-analysis of the propagule pressure effect. *PLoS Biology* 16, e2005987.

Cheke, A. (1987) An ecological history of the Mascarene Islands, with particular reference to extinctions and introductions of land vertebrates. In: *Studies of Mascarene Island Birds* (ed. A.W. Diamond), pp. 5–89. Cambridge University Press, Cambridge.

Chesson, P. (2000) Mechanisms of maintenance of species diversity. *Annual Review of Ecology and Systematics* 31, 343–366.

Choudhury, A., Duckworth, J.W., Timmins, R., Chutipong, W., Willcox, D.H.A., Rahman, H., Ghimirey, Y. & Mudappa, D. (2015) *Viverricula indica*. *The IUCN Red List of Threatened Species* 2015: e. T41710A45220632. http://dx.doi.org/10.2305/IUCN. UK.2015-4.RLTS.T41710A45220632.en. Accessed on 2 January 2017.

Ćirović, D. & Toholj, D. (2015) Distribution of Small Indian Mongoose (*Herpestes auropunctatus*) in the Eastern Herzegovina spreading inside mainland. *Balkan Journal of Wildlife Research* 2, 33–37.

Ćirović, D., Raković, M., Milenković M. & Paunović, M. (2011) Small Indian Mongoose *Herpestes auropunctatus* (Herpestidae, Carnivora): an invasive species in Montenegro. *Biological Invasions* 13, 393–399.

Clevenger, A.P. (1993) The European pine marten *Martes martes* in the Balearic Islands, Spain. *Mammal Review* 23, 65–72.

Clout, M. & Russel, J. (2008) The invasion ecology of mammals: a global perspective. *Wildlife Research* 35, 180–184.

Coblentz, B.E. & Coblentz, B.A, (1985) Control of the Indian Mongoose *Herpestes auropunctatus* on St John, US Virgin Islands. *Biological Conservation* 33, 281–288.

Collauti, R.I., Grigorovich, I.A. & MacIsaac, H.J. (2006) Propagule pressure: a null model for biological invasions. *Biological Invasions* 8, 1023–1037.

Corbet, G.B. (1966) *The Terrestrial Mammals of Western Europe*. G.T. Foulis & Company, London.

Corbet, G.B. & Harris, S. (1991) *A Handbook of British Mammals*. 3rd edition. Blackwell, Oxford.

Corlett, R.T. (2010) Invasive aliens on tropical East Asian islands. *Biodiversity and Conservation* 19, 411–423.

Courchamp, F., Chapuis, J.-L. & Pascal, M. (2003) Mammal invaders on islands: impact, control and control impact. *Biological Reviews* 78, 347–383.

Cranwell, S, (2016) *Mongoose on the loose in Tonga*. BirdLife International, Cambridge. http://www.birdlife.org/pacific/news/mongoose-loose-tonga. Accessed on 12 July 2021.

Cuarón, A.D., Martínez-Morales, M.A., Mcfadden, K.W., Valenzuela, D. & Gompper, M.E. (2004) The status of dwarf carnivores on Cozumel Island, Mexico. *Biodiversity and Conservation* 13, 317–331.

Cuthbert, J.H. (1973) The origin and distribution of feral mink in Scotland. *Mammal Review* 3, 97–103.

Damuth, J. (1981) Home range, home range overlap, and species energy use among herbivorous mammals. *Biological Journal of the Linnean Society* 15, 185–193.

Darwin, C. (1859) *On the Origin of Species by Means of Natural Selection*. John Murray, London.

Davies, M. (2009) *Invasion Biology*. Oxford University Press, New York.

Delibes, M., Centeno-Cuadros, A., Muxart, V., Delibes, G., Ramos-Fernández, J. & Morales, A. (2019) New insights into the introduction of the common genet, *Genetta genetta* (L.) in Europe. *Archaeological and Anthropological Sciences* 11, 531–539.

de Vos, A., Manville, R.H. & VanGelder, R.G. (1956) Introduced mammals and their influence on native biota. *Zoologica* 41, 163–194.

Di Castri, E. (1989) History of biological invasions with special emphasis on the old world. In: *Biological Invasions: A Global Perspective* (eds J.A. Drake, H.A. Mooney, E. Di Castri, R.H. Groves, E.J. Kruger, M. Reimanek & M. Williamson), pp. 1–30. John Wiley, Chichester.

Diez, J.M., Sullivan, J.J., Hulme, P.E., Edwards, G. & Duncan, R.P. (2008) Darwin's naturalization conundrum: dissecting taxonomic patterns of species invasions. *Ecology Letters* 11, 674–681.

Dinets, V. (2012) Notes on Cozumel raccoon (*Procyon pygmaeus*) and Tres Marias raccoon (*P. insularis*). *Small Carnivore Conservation* 47, 14–16.

Dormann, C.F., Elith, J., Bacher, S., Buchmann, C., Carl, G., Carré, G., Marquéz, J.R.G., Gruber, B., Lafourcade, B., Leitão, P.J. & Münkemüller, T. (2013) Collinearity: a review of methods to deal with it and a simulation study evaluating their performance. *Ecography* 36, 27–46.

Duffy, D.C., Elliott, D.D., Hart, G.M., Gundersen, K., Aguon-Kona, J., Bartlett, R., Fujikawa, J., Gmelin, P., Javier, C., Kaneholani, L., Keanini, T., Kona, J., Parish, J., Penniman, J.F. & Works, A. (2015) Has the Small Indian Mongoose become established on Kaua'i Island, Hawai'i? *Pacific Science* 69, 559–565.

Dutton, J. (1994) Introduced mammals in São Tomé and Príncipe: possible threats to biodiversity. *Biodiversity and Conservation* 3, 927–938.

Eisenberg, J.F. (1989) *Mammals of the Neotropics. Volume 1. The Northern Neotropics*. University of Chicago Press, Chicago.

Elton, C.S. (1958) *The Ecology of Invasions by Animals and Plants*. Methuen and Company, London.

Emmons, L. & Helgen, K. (2016) *Nasua nasua. The IUCN Red List of Threatened Species* 2016, e. T41684A45216227. http://dx.doi.org/10.2305/IUCN.UK.2016-1.RLTS.T41684A45216227.en. Accessed on 2 January 2017

Escobedo, V.M., Aranda, J.E. & Castro, S.A. (2011) Hipótesis de Naturalización de Darwin evaluada en la flora exótica de Chile continental. *Revista Chilena de Historia Natural* 84, 543–552. (In Spanish with English abstract).

Espeut, W.B. (1882) On the aclimitization of the Indian mungoos in Jamaica. *Proceedings of the Zoological Society of London* 1882, 712–714.

Farris, Z.J., Gerber, B.D., Kelly, M.J., Ratelolahy, F. & Andrianjakarivelo, V. (Chapter 13, this volume) Spatio-temporal overlap between a native and an exotic carnivore in Madagascar: evidence of spatial exclusion. In: *Small Carnivores: Evolution, Ecology, Behaviour, and Conservation* (eds E. Do Linh San, J.J. Sato, J.L. Belant & M.J. Somers), Wiley–Blackwell, Oxford.

Fischer, M.L., Hochkirch, A., Heddergott, M., Schulze, C., Anheyer-Behmenburg, H.E., Lang, J., Michler, F.-U., Hohmann, U., Ansorge, H., Hoffmann, L.,

Klein, R. & Frantz, A.C. (2015) Historical invasion records can be misleading: genetic evidence for multiple introductions of invasive raccoons (*Procyon lotor*) in Germany. *PLoS One* 10, e0125441.

Fitter, R.S.R. (1959) *The Ark in Our Midst*. Collins, London.

Flannery, T. (1995) *Mammals of New Guinea*. Carina, Queensland.

Fleming, T.H. (1991) The relationship between body size, diet, and habitat use in frugivorous bats, genus *Carollia* (Phyllostomidae). *Journal of Mammalogy* 72, 493–501.

Flux, J.E.C. & Fullagar, P.J. (1992) World distribution of the rabbit *Oryctolagus cuniculus* on islands. *Mammal Review* 22, 151–205.

Forsyth, D.M., Duncan, R.P., Bomford, M. & Moore, G. (2004) Climatic suitability, life-history traits, introduction effort and the establishment and spread of introduced mammals in Australia. *Conservation Biology* 18, 557–569.

Garcia, J.T., Garcia, F.J., Alda, F., Gonzalez, J.L., Aramburu, M.J., Cortes, Y., Prieto, B., Pliego, B., Perez, M., Herrera, J. & Garcia-Roman, L. (2012) Recent invasion and reproduction of the raccoon (*Procyon lotor*) in Spain. *Biological Invasions* 14, 1305–1310.

Gaubert, P. (2016) Fate of the mongooses and the genet (Carnivora) in Mediterranean Europe: none native, all invasive? In: *Problematic Wildlife: A Cross-Disciplinary Approach* (ed. F.M. Angelici), pp. 109–122. Springer, New York.

Gaubert, P. & Zenatello, M. (2009) Ancient DNA perspective on the failed introduction of mongooses in Italy during the XXth century. *Journal of Zoology* 279, 262–269.

Gaubert, P., Del Cerro, I., Godoy, J.A. & Palomares, F. (2009) Early phases of a successful invasion: mitochondrial phylogeography of the common genet (*Genetta genetta*) within the Mediterranean Basin. *Biological Invasions* 11, 523–546.

Gaubert, P., Machordom, A., Morales, A., López-Bao, J.V., Veron, G., Amin, M., Barros, T., Basuony, M., Djagoun, C.A.M.S., Do Linh San, E., Fonseca, C., Geffen, E., Ozkurt, S.O., Cruaud, C., Couloux, A. & Palomares, F. (2011) Comparative phylogeography of two African carnivorans presumably introduced into Europe: disentangling natural versus human-mediated dispersal across the Strait of Gibraltar. *Journal of Biogeography* 38, 341–358.

Gaubert, P., Carvalho, F., Camps, D. & Do Linh San, E. (2015a) *Genetta genetta. The IUCN Red List of Threatened Species* 2015, e.T41698A45218636. http://dx.doi.org/10.2305/IUCN.UK.2015-4.RLTS.T41698A45218636.en. Accessed on 2 January 2017.

Gaubert, P., Del Cerro, I., Centeno-Cuadros, A., Palomares, F., Fournier, P., Fonseca, C., Paillat, J.-P. & Godoy, J.A. (2015b) Tracing historical introductions in the Mediterranean Basin: the success story of the common genet (*Genetta genetta*) in Europe. *Biological Invasions* 17, 1897–1913.

Gaubert, P., Patel, R.P., Veron, G., Goodman, S.M., Willsch, M., Vasconcelos, R., Lourenço, A., Sigaud, M., Justy, F., Joshi, B.D., Fickel, J. & Wilting, A. (2017) Phylogeography of the small Indian civet and origin of introductions to western Indian Ocean islands. *Journal of Heredity* 108: 270–279.

Gehrt, S. (2004) Ecology and management of striped skunks, raccoons, and coyotes in urban landscapes. In: *People and Predators: From Conflict to Coexistence* (eds N. Fascione, A. Delach & M.E. Smith), pp. 81–104. Island Press, Washington, DC.

Gerell, R. (1967a) Food selection in relation to habitat in mink (*Mustela vison* Schreber) in Sweden. *Oikos* 18, 233–246.

Gerell, R. (1967b) Dispersal and acclimatization of the mink (*Mustela vison* Schreber) in Sweden. *Viltrevy* 5, 1–38.

Gibb, J.A. & Flux, J.E.C. (1973) Mammals. In: *The Natural History of New Zealand* (ed. G.R. Williams), pp. 334–371. A.H. and A.W. Reed, Wellington.

Gilchrist, J.S., Jennings, A.P., Veron, G. & Cavallini, P. (2009) Family Herpestidae. In: *Handbook of the Mammals of the World. Volume 1. Carnivores* (eds D. Wilson & R.A. Mittermeier), pp. 262–328. Lynx Edicions, Barcelona.

Gittleman, J.L., Funk, S.M., Macdonald, D. & Wayne, R.K. (eds) (2001) *Carnivore Conservation*. Cambridge University Press, Cambridge.

Glatston, A.R. (ed.) (1994) *The Red Panda, Olingos, Coatis, Raccoons, and their Relatives: Status Survey and Conservation Action Plan for Procyonids and Ailurids*. IUCN/SSC Mustelid, Viverrid, and Procyonid Specialist Group, IUCN, Gland.

Gompper, M.E. & Decker, D.M. (1998) *Nasua nasua. Mammalian Species* 580, 1–9.

Goodenough, A.E. (2010) Are the ecological impacts of alien species misrepresented? A review of the "native good, alien bad" philosophy. *Community Ecology* 11, 13–21.

Goodman, S. (2012) *Les Carnivora de Madagascar.* Association Vahatra, Antananarivo. (In French).

Gorman, M.L. (1975) The diet of feral *Herpestes auropunctatus* in the Fijian Islands. *Journal of Zoology* 175, 273–278.

Hall, E.R. (1981) *The Mammals of North America. Volumes 1 & 2.* Wiley, New York.

Haltenorth, T. & Diller, H. (1994) *A Field Guide to the Mammals of Africa Including Madagascar.* Harper/Collins, London.

Hamao, S. & Higuchi, H. (2013) Effect of introduced Japanese weasels (*Mustela itatsi*) on the nest height of Japanese bush warblers (*Horornis diphone*) on Miyake-jima Island, Japan. *Wilson Journal of Ornithology* 125, 426–429.

Harvey, P.H. & Read, A.F. (1988) How and why do mammalian life histories vary? In: *Evolution of Life Histories of Mammals: Theory and Pattern* (ed. M.S. Boyce), pp. 213–232. Yale University Press, New Haven.

Hayashi, Y. (1981) The Amami rabbit. In: *Proceedings of the World Lagomorph Conference* (eds K. Myers & C.D. McInness), pp. 926–927. University of Guelph, Guelph.

Hayes, K. & Barry, S. (2008) Are there any consistent predictors of invasion success? *Biological Invasions* 10, 483–506.

Hays, W.S. & Conant, S. (2007) Biology and impacts of Pacific Island invasive species. 1. A worldwide review of effects of the small Indian mongoose, *Herpestes javanicus* (Carnivora: Herpestidae). *Pacific Science* 61, 3–16.

Helgen, K. & Reid, F. (2016) *Mephitis mephitis. The IUCN Red List of Threatened Species* 2016, e. T41635A45211301. http://dx.doi.org/10.2305/IUCN.UK.2016-1.RLTS.T41635A45211301.en. Accessed on 2 January 2017.

Helgen, K.M. & Wilson, D.E. (2002) The history of the raccoons of the West Indies. *Journal of Barbados Museum and Historical Society* 48, 1–11.

Helgen, K.M. & Wilson, D.E. (2003) Taxonomic status and conservation relevance of the raccoons (*Procyon* spp.) of the West Indies. *Journal of Zoology* 259, 69–76.

Helgen, K.M. & Wilson, D.E. (2005) A systematic and zoogeographic overview of the raccoons of Mexico and Central America. In: *Contribuciones Mastozoologicas: en Homenaje a Bernardo Villa* (eds V. Sanchez-Cordero & R.A. Medellin), pp. 221–236. Instituto de Biología e Instituto de Ecología, UNAM, Mexico.

Helgen, K.M., Maldonado, J.E., Wilson, D.E. & Buckner, S.D. (2008) Molecular confirmation of the origin and invasive status of West Indian raccoons. *Journal of Mammalogy* 89, 282–291.

Heptner, V.G. & Naumov, N.P. (eds) (2002) *Mammals of the Soviet Union. Volume II. Part 1b. Carnivora (Weasels; Additional species).* Amerind Publishing Co., New Delhi.

Herben, T. (2005) Species pool size and invasibility of island communities: a null model of sampling effects. *Ecology Letters* 8, 909–917.

Herrero, J., Kranz, A., Skumatov, D., Abramov, A.V., Maran, T. & Monakhov, V.G. (2016) *Martes martes. The IUCN Red List of Threatened Species* 2016, e. T12848A45199169. http://dx.doi.org/10.2305/IUCN.UK.2016-1.RLTS.T12848A45199169.en. Accessed on 2 January 2017.

Hinton, H.E. & Dunn A.S. (1967) *Mongooses; Their Natural History and Behaviour.* University of California Press, Berkeley.

Hoagland, D.B., Horst, G.R. & Kilpatrick, C.W. (1989) Biogeography and population biology of the mongoose in the West Indies. In: *Biogeography of the West Indies: Past, Present and Future* (ed. C.A. Woods), pp. 611–634. Sandhill Crane Press, Florida Museum of Natural History, Gainesville.

Horst, G.R., Hoagland, D.B. & Kilpatrick, C.W. (2001) The mongoose in the West Indies: the biogeography and population biology of an introduced species. In: *Biogeography of the West Indies: Patterns and Perspectives* (eds C.A. Woods & F.E. Sergile), pp. 409–424. CRC Press, Boca Raton.

Hubbell, S.P. (2001) *A Unified Neutral Theory of Biodiversity and Biogeography.* Princeton University Press, Princeton.

Hunter, L. & Barrett, P. (2018) *A Field Guide to the Carnivores of the World.* 2nd edition. Bloomsbury Publishing, London.

Husson, A.M. (1960) *De Zoogdieren van de Nederlandse Antillen. Mammals of the Netherlands Antilles.* Natuurwetenschappelijke Werkgroep Nederlandse Antillen, Curaçao, West Indies. (In Dutch with English summary).

Husson, A.M. (1978) *The Mammals of Suriname.* Zoölogische Monographieën van het Rijksmuseum van Natuurlijke Historie. Brill, Leiden.

Hutchinson, G.E. (1957) Concluding remarks. *Cold Spring Harbor Symposium on Quantitative Biology* 22, 415–457.

Ikeda, T., Asano, M., Matoba, Y. & Abe, G. (2004) Present status of invasive alien raccoon and its impact in Japan. *Global Environmental Research* 8, 125–131.

Inoue, T., Kaneko, Y., Yamazaki, K., Anezaki, T., Yachimori, S., Ochiai, K., Lin, L.K., Pei, K.J.C., Chen, Y.J., Chang, S.W. & Masuda, R. (2012) Genetic population structure of the masked palm civet *Paguma larvata*, (Carnivora: Viverridae) in Japan, revealed from analysis of newly identified compound microsatellites. *Conservation Genetics* 13, 1095–1107.

Inukai, T. (1934) Invasion of the Japanese weasel into Hokkaido and its use as predator. *Botany and Zoology* 2, 1309–1317.

Ishida, K., Sato, J.J., Kinoshita, G., Hosoda, T., Kryukov, A.P. & Suzuki, H. (2013) Evolutionary history of the sable (*Martes zibellina brachyura*) on Hokkaido inferred from mitochondrial Cytb and nuclear Mc1r and Tcf25 gene sequences. *Acta Theriologica* 58, 13–24.

IUCN [International Union for Conservation of Nature] (2019) *The IUCN Red List of Threatened Species. Version 2019-1.* https://www.iucnredlist.org. Accessed on 18 June 2019.

IUCN SSC ISSG [Invasive Species Specialist Group of the IUCN Species Survival Commission] (2018) Global Invasive Species Database (GISD). http://www.issg.org/database/welcome. Accessed on 20 February 2021.

IUCN SSC SCSG [Small Carnivore Specialist Group of the IUCN Species Survival Commission] (2013) http://www.smallcarnivoreconservation.org/home. Accessed on 10 November 2013.

Izmailov, I.V. (1968) [Some questions of transformation of Buryatia fauna]. *Uchenye Zapiski Buryatskogo Gosudarstvennogo Pedagogicheskogo Instituta* 31, 3–17. (In Russian).

Jaksic, F.M., Agustín Iriarte, J., Jiménez, J.E. & Martínez, D.R. (2002) Invaders without frontiers: cross-border invasions of exotic mammals. *Biological Invasions* 4, 157–173.

Jennings, A.P. & Veron, G. (2009) Family Viverridae (Civets, genets and oyans). In: *Handbook of the Mammals of the World. Volume 1. Carnivores* (eds D.E. Wilson & R.A. Mittermeier), pp. 174–233. Lynx, Barcelona.

Jennings, A. & Veron, G. (2011) Predicted distributions and ecological niches of 8 civet and mongoose species in Southeast Asia. *Journal of Mammalogy* 92, 316–327.

Jennings, A.P., Zubaid, A. & Veron, G. (2010a) Ranging behaviour, activity, habitat use, and morphology of the Malay civet (*Viverra tangalunga*) on Peninsular Malaysia and comparison with studies on Borneo and Sulawesi. *Mammalian Biology* 75, 437–446.

Jennings, A.P., Zubaid, A. & Veron, G. (2010b) Home ranges, movements and activity of the short-tailed mongoose (*Herpestes brachyurus*) on Peninsular Malaysia. *Mammalia* 74, 43–50.

Jeschke, J.M. & Strayer, D.L. (2005) Invasion success of vertebrates in Europe and North America. *Proceedings of the National Academy of Sciences of the United States of America* 102, 7198–7202.

Jeschke, J.M., Aparicio, L.G., Haider, S., Heger, T., Lortie, C.J., Pyšek, P. & Strayer, D.L. (2012) Support for major hypotheses in invasion biology is uneven and declining. *Neobiota* 14, 1–20.

Jordan, N.R. & Do Linh San, E. (2015) *Suricata suricatta. The IUCN Red List of Threatened Species* 2015, e. T41624A45209377. http://dx.doi.org/10.2305/IUCN. UK.2015-4.RLTS.T41624A45209377.en. Accessed on 2 January 2017.

Kaburaki, T. (1940) Further notes on the effect of some exotic animals upon the fauna of Japan. In: *Proceedings of the 5th Pacific Science Congress, Victoria and Vancouver*, pp. 229–230.

Kaneko, Y., Masuda, R. & Abramov, A.V. (2016) *Mustela itatsi. The IUCN Red List of Threatened Species* 2016, e.T41656A45214163. http://dx.doi.org/10.2305/IUCN. UK.2016-1.RLTS.T41656A45214163.en. Accessed on 2 January 2017.

Kaufman, D.W. & Kaufman, G.A. (1987) Reproduction by *Peromyscus polionotus*: number, size, and survival of offspring. *Journal of Mammalogy* 68, 275–280.

Kegel, B. (2001) *Die Ameise als Tramp: Von Biologischen Invasionen*. Heyne, Munich. (In German).

King, C.M. (1983) *Mustela erminea. Mammalian Species* 195, 1–8.

King, C.M. (1990) Stoat. In: *The Handbook of New Zealand Mammals* (ed. C.M. King), pp. 288–312. Oxford University Press, Auckland.

King, C.M. (2017a) Liberation and spread of stoats (*Mustela erminea*) and weasels (*M. nivalis*) in New Zealand, 1883–1920. *New Zealand Journal of Ecology* 41, 163–177.

King, C.M. (2017b) Pandora's box down-under: origins and numbers of mustelids transported to New Zealand for biological control of rabbits. *Biological Invasions* 19, 1811–1823.

King, C.M. (2017c) The chronology of a sad historical misjudgement: the introductions of rabbits and ferrets in nineteenth-century New Zealand. *International Review of Environmental History* 3, 139–173.

King, C.M. (2019) Failed proposals to import the mongoose, pine marten, Patagonian fox and other exotic predators into New Zealand. *Journal of the Royal Society of New Zealand* 49, 3–15.

King, C.M. & Moors, P.J. (1979) On co-existence, foraging strategy and the biogeography of weasels and stoats (*Mustela nivalis* and *M. erminea*) in Britain. *Oecologia* 39, 129–150.

Kirisa, I.D. (1972–1974) [*Acclimatization of the Hunting Industry in Animals and Birds in the USSR*]. 2 Volumes. Kirov.

Kock, D. & Stanley, W.T. (2009) Mammals of Mafia Island, Tanzania. *Mammalia* 73, 339–352.

Koskela, E., Juutistenaho, P., Mappes, T. & Oksanen, T.A. (2000) Offspring defense in relation to litter size and age: experiment in the bank vole *Clethrionomys glareolus*. *Evolutionary Ecology* 14, 99–109.

Kryštufek, B. & Tvrtković, N. (1992) New information on the introduction into Europe of the small Indian mongoose, *Herpestes auropunctatus*. *Small Carnivore Conservation* 7, 16.

Kuroda, N. (1955) The present status of the introduced mammals in Japan. *Journal of the Mammalogical Society of Japan* 1, 13–18.

Larivière, S. (1999) *Mustela vison. Mammalian Species* 608, 1–9.

Larivière, S. & Calzada, J. (2001) *Genetta genetta. Mammalian Species* 680, 1–6.

Laurie, E.M.O. & Hill, J.E. (1954) *List of Land Mammals of New Guinea, Celebes and Adjacent Islands 1758–1952*. British Museum of Natural History, London.

Lavov, M.A. (1962) [Fauna of game mammals of Kuril Islands]. *Problemy Zoologicheskikh Issledovanii v Sibiri*, pp. 149–150. Gorno-Altaiskoe knizhnoe izdatelstvo, Gorno-Altaisk. (In Russian).

Lavrov, N.P. (1941) [Methods for forecasting population changes in the ermine (*Mustela erminea* L.)]. *Trudy Tsentral'noi Laboratorii Biologii i Ochotnichego Promysla* 5, 60–77. (In Russian).

Lavrov, N.P. (1946) [*Acclimatization and Re-acclimatization of Fur-game Mammals in the USSR*]. Zagotizdat, Moscow. (In Russian).

Lazell, J.D. Jr (1981) Field and taxonomic studies of tropical American raccoons. *National Geographic Society Research Reports* 13, 381–385.

Leitch, I., Hytten, F.E. & Billewicz, W.Z. (1959) The maternal and neonatal weights of some mammalia. *Proceedings of the Zoological Society of London* 133, 11–28.

Lésel, R. & Derenne, P. (1975) Introducing animals to Iles Kerguelen. *Polar Record* 17, 485–494.

Lever, C. (1985) *Naturalized Mammals of the World*. Longman House, Essex.

Litjens, B.E.J. (1980) De beverrat, *Myocastor coypus*, in Nederland. *Lutra* 23, 43–53.

Liu, J., Linderman, M., Ouyang, Z., An, L., Yang, J. & Zhang, H. (2001) Ecological degradation in protected areas: the case of Wolong Nature Reserve for giant pandas. *Science* 292, 98–101.

Lockwood, J.L., Moulton, M.P. & Balent, K.L. (1999) Introduced avifaunas as natural experiments in community assembly. In: *Ecological Assembly Rules: Perspectives, Advances, Retreats* (eds E. Weiher & P. Keddy), pp. 108–129. Cambridge University Press, Cambridge.

Lockwood, J.L., Cassey, P. & Blackburn, T. (2005) The role of propagule pressure in explaining species invasions. *Trends in Ecology & Evolution* 20, 225–228.

Lomolino, M.V., Riddle, B.R., Whittaker, R.J. & Brown, J.H. (2010) *Biogeography*. Sinauer, Sunderland.

Long, C.A. (1995) Stone marten (*Martes foina*) in southeast Wisconsin, USA. *Small Carnivore Conservation* 13, 14.

Long, J.L. (2003) *Introduced Mammals of the World. Their History, Distribution and Influence*. CABI Publishing, Wallingford.

Lorvelec, O., Delloue, X., Pascal, M. & Mège, S. (2004) Impacts des mammifères allochtones sur quelques espèces autochtones de l'Îlet Fajou (Réserve Naturelle du Grand Cul-de-Sac Marin, Guadeloupe), établis à l'issue d'une tentative d'éradication. *Revue d'Écologie (La Terre et la Vie)* 59, 293–307. (In French with English summary).

Lorvelec, O., Pascal, M., Delloue, X. & Chapuis, J.L. (2007) Les mammifères terrestres non volants des Antilles françaises et l'introduction récente d'un écureuil. *Revue d'Écologie (La Terre et la Vie)* 62, 295–314. (In French with English summary).

Lorvelec, O., Yvon, T. & Lenoble, A. (2021) Histoire de la petite mangouste indienne *Urva auropunctata* (Hodgson, 1836) dans les Antilles: une introduction aux conséquences sociétales et écologiques majeures. *Anthropozoologica* 56, 1–22. (In French with English Abstract).

Lotze, J.H. & Anderson, S. (1979) *Procyon lotor*. *Mammalian Species* 119, 1–8.

Louette, M. (ed.) (1999) *La Faune Terrestre de Mayotte*. Annales Sciences Zoologiques Vol. 284. Musée Royal de l'Afrique Centrale, Tervuren. (In French with English Summary).

Louette, M., Meirte, D. & Jocqué, R. (eds) (2004) *La Faune Terrestre de l'Archipel des Comores*. Studies in Afrotropical Zoology Vol. 293. Musée Royal de l'Afrique Centrale, Tervuren. (In French with English Summary).

Louppe, V., Leroy, B., Herrel, A. & Veron, G. (2019) Current and future climatic regions favourable for a globally introduced wild carnivore, the raccoon *Procyon lotor*. *Scientific Reports* 9, 9174.

Louppe, V., Baron, J., Pons, J.-M. & Veron, G. (2020) New insights on the geographical origins of the Caribbean raccoons. *Journal of Zoological Systematics and Evolutionary Research* 58, 1303–1322.

Louppe, V., Lalis, A., Abdelkrim, J., Baron, J., Bed'Hom, B., Becker, A.A.M.J., Catzeflis, F., Lorvelec, O., Ziegler, U. & Veron, G. (2021) Dispersal history of a globally introduced carnivore, the small Indian mongoose *Urva auropunctata*, with an emphasis on the Caribbean region. *Biological Invasions* 23, 2573–2590.

Lowe, S., Browne, M., Boudjelas, S. & De Poorter, M. (2000) *100 of the World's Worst Invasive Alien Species a Selection from the Global Invasive Species Database*. IUCN, Gland.

Lowry, E., Rollinson, E.J., Laybourn, A.J., Scott, T.E., Aiello-Lammens, M.E., Gray, S.M., Mickley, J. & Gurevitch, J. (2012) Biological invasions: a field synopsis, systematic review, and database of the literature. *Ecology and Evolution* 3, 182–196.

MacArthur, R.H. (1955) Fluctuations of animal populations and a measure of community stability. *Ecology* 36, 533–536.

MacDonald, I.A.W. & Cooper, J. (1995) Insular lessons for global biodiversity conservation with particular reference to alien invasions. In: *Islands: Biological Diversity and Ecosystem Function* (eds P.M. Vitousek, L.L. Loope & H. Adsersen), pp. 189–203. Springer, New York.

Mack, R.N., Simberloff, D., Lonsdale, W.M., Evans, H., Clout, M. & Bazzaz, F.A. (2000) Biotic invasions: causes, consequences, and control. *Ecological Applications* 10, 689–710.

Marable, M.K., Belant, J.L., Godwin, D. & Wang, G. (2012) Effects of resource dispersion and site familiarity on movements of translocated wild turkeys on fragmented landscapes. *Behavioural Processes* 91, 119–124.

Maran, T., Skumatov, D., Gomez, A., Põdra, M., Abramov, A.V. & Dinets, V. (2016) *Mustela lutreola*. *The IUCN Red List of Threatened Species* 2016, e. T14018A45199861. http://dx.doi.org/10.2305/IUCN. UK.2016-1.RLTS.T14018A45199861.en. Accessed on 2 January 2017.

Marshall, W.H. (1963) *The Ecology of Mustelids in New Zealand*. DSIR Information Series 38. New Zealand Department of Scientific and Industrial Research, Wellington.

Mason, R.A.B., Cooke, J., Moles, A.T. & Leishman, M.R. (2008) Reproductive output of invasive versus native plants. *Global Ecology and Biogeography* 17, 633–640.

Masuda, R. (2009) *Martes melampus* (Wagner, 1840). In: *The Wild Mammals of Japan* (eds S.D. Ohdachi, Y. Ishibashi, M.A. Iwasa & T. Saitoh), pp. 250–251. Shoukadoh, Kyoto.

Masuda, R. & Watanabe, S. (2009) *Mustela itatsi* Temminck, 1844. *The Wild Mammals of Japan* (eds S.D. Ohdachi, Y. Ishibashi, M.A. Iwasa & T. Saitoh), pp. 240–241. Shoukadoh, Kyoto.

Masuda, R., Lin, L., Pei, K.J., Chen, Y., Chang, S., Kaneko, Y., Yamazaki, K., Anezaki, T., Yachimori, S. & Oshida, T. (2010) Origins and founder effect on the

Japanese masked palm civet *Paguma larvata* (Viverridae, Carnivora), revealed from a comparison with its molecular phylogeography in Taiwan. *Zoological Science* 27, 499–505.

Mayol, J., Álvarez, C. & Manzano, X. (2009) Presence and control of the coati, *Nasua nasua* L., and other carnivores introduced in recent times in Mallorca. *Bolletí de la Societat d'Història Natural de les Balears* 52, 183–191.

McDonald, R.A., Abramov, A.V., Stubbe, M., Herrero, J., Maran, T., Tikhonov, A., Cavallini, P., Kranz, A., Giannatos, G., Kryštufek, B. & Reid, F. (2016) *Mustela nivalis. The IUCN Red List of Threatened Species* 2016, e.T70207409A45200499. http://dx.doi.org/10.2305/ IUCN.UK.2016-1.RLTS.T70207409A45200499.en. Accessed on 2 January 2017.

McFadden, K.W., Gompper, M.E. Valenzuela, D.G. & Morales, J.C. (2008) Evolutionary history of the Critically Endangered Cozumel dwarf carnivores inferred from mitochondrial DNA analyses. *Journal of Zoology* 276, 176–186.

McKinley, D. (1959) Historical note on the Bahama raccoon. *Journal of Mammalogy* 40, 248–249.

Medina, F.M. & Martín, A. (2010) A new invasive species in the Canary Islands: a naturalized population of ferrets *Mustela furo* in La Palma Biosphere Reserve. *Oryx* 44, 41–44.

Mel'chinov, M.S. (1958) [Attempt of introduction of Vitim sable]. *Trudy Instituta Biologii Yakutskogo Filiala Sibirskogo Otdeleniya Akademii Nauk SSSR* 4, 172–186. (In Russian).

Millar, J.S. (1977) Adaptive features of mammalian reproduction. *Evolution* 31, 370–386.

Millar, J.S. & Hickling, G.J. (1990) Fasting endurance and the evolution of mammalian body size. *Functional Ecology* 4, 5–12.

Milne, L.J. & Milne, M. (1962) *The Balance of Nature*. Macdonald, London.

Mladenoff, D.J., Haight, R.G., Sickley, T.A. & Wydeven, A.P. (1997) Causes and implications of species restoration in altered ecosystems. *BioScience* 47, 21–31.

Monakhov, V.G. (2016) *Martes zibellina. The IUCN Red List of Threatened Species* 2016, e.T41652A45213477. http://dx.doi.org/10.2305/IUCN.UK.2016-1.RLTS. T41652A45213477.en. Accessed on 2 January 2017.

Moreau, R.E. & Pakenham, R.H.W. (1941) The land vertebrates of Pemba, Zanzibar, and Mafia: a zoo-geographical study. *Proceedings of the Zoological Society of London* 110, 97–128.

Moulton, M.P., Cropper, W.P. & Avery, M.L. (2013) Is propagule size the critical factor in predicting introduction outcomes in passeriform birds? *Biological Invasions* 15, 1449–1458.

Moutou, F. (1987) *Encyclopédie des Carnivores de France. Volume 20: Les Carnivores des Îles Françaises d'Outre-Mer*. Société Française pour l'Étude et la Protection des Mammifères, Paris. (In French).

Morley, C.G. (2004a) Actions speak louder than words: a call for preventing further mongoose invasions in Fiji. In: *Proceedings of the 21st Vertebrate Pest Conference* (eds R.M. Timm & W.P. Gorenzel), pp. 37–41. University of California, Davis.

Morley, C.G. (2004b). Has the invasive mongoose *Herpestes javanicus* yet reached the island of Taveuni, Fiji? *Oryx* 38, 457–460.

Morley, C.G., McLenachan, P.A. & Lockhart, P.J. (2007) Evidence for the presence of a second species of mongoose in the Fiji Islands. *Pacific Conservation Biology* 13, 29–34.

Mudappa, D. & Jathanna, D. (2015) *Herpestes fuscus. The IUCN Red List of Threatened Species* 2015: e. T41612A45207051. http://dx.doi.org/10.2305/IUCN. UK.2015-4.RLTS.T41612A45207051.en. Accessed on 2 January 2017.

Mukhtarov, R.D. (1963) [Ecology and radiation of the raccoon in Bostandyk Mountains]. In: [*Game Animals of Uzbekistan*] (ed. G.S. Sultanov), pp. 25–35. Izdatelstvo Akademii nauk UzSSR, Tashkent. (In Russian).

Murphy, E.C. (1985) Bergmann's rule, seasonality, and geographic variation in body size of house sparrows. *Evolution* 39, 1327–1334.

Myers, K. (1986) Introduced vertebrates in Australia, with emphasis on the mammals. In: *Ecology of Biological Invasions: An Australian Perspective* (eds R. Groves & J. Burdon), pp. 120–136. Australian Academy of Science, Canberra.

Naumov, N.P. (1972) No title. In: *The Ecology of Animals* (ed. N.D. Levine; trans. F.K. Plous Jr). University of Illinois Press, Urbana.

Nellis, D.W. (1989) *Herpestes auropunctatus. Mammalian Species* 342, 1–6.

Nellis, D.W. & Everard, C.O.R. (1983) The biology of the mongoose in the Caribbean. *Studies on the Fauna of Curaçao and Other Caribbean Islands* 64, 1–162.

Nellis, D.W., Eichholz, N.F., Regan, T.W. & Feinstein, C. (1978) Mongoose in Florida. *Wildlife Society Bulletin* 6, 249–250.

Niethammer, G. (1963) *Die Einbürgerung von Säugetieren und Vögeln in Europa*. Paul Parey, Hamburg. (In German).

Novikov, G.A. (1956) [*Carnivorous Mammals of the Fauna of the USSR*]. Izdatelstvo Akademii nauk SSSR, Moscow–Leningrad. (In Russian).

Ogura, G., Sakashita, M. & Kawashima, Y. (1998) [External morphology and classification of mongoose on Okinawa Island]. *Mammalian Science (Honyurui Kagaku)* 38, 259–270. (In Japanese with English abstract).

Pakenham, R.H.W. (1984) *The Mammals of Zanzibar and Pemba Islands*. Printed privately, Harpenden.

Pankrat'ev, A.G. (1959) [Acclimatization of the raccoon in the forests of southern Primorye]. *Soobshcheniya Dal'nevostochnogo Filiala Sibiriskogo Otdeleniya Akademii Nauk SSSR* 11, 115–120. (In Russian).

Patou, M.-L., Wilting, A., Gaubert, P., Esselstyn, J.A., Cruaud, C., Jennings, A.P., Fickel, J. & Veron, G. (2010) Evolutionary history of the *Paradoxurus* palm civets – a new model for Asian biogeography. *Journal of Biogeography* 37, 2077–2097.

Pauli, J.N., Moss, W.E., Manlick, P.J., Fountain, E.D., Kirby, R., Sultaire, S.M., Perrig, P.L., Mendoza, J.E., Pokallus, J.W. & Heaton, T.H. (2015) Examining the uncertain origin and management role of martens on Prince of Wales Island, Alaska. *Conservation Biology* 29, 1257–1267.

Pavlov, M.P., Korsakova, I.B., Timofeev, V.V. & Safonov, V.G. (1973) No title. In: [*Acclimatization of Game Mammals and Birds in the USSR. Part 1*] (ed. I.D. Kiris). Volgo-Vyatskoe Knizhnoe Izdatelstvo, Kirov, Russia. (In Russian).

Peacock, D. & Abbott, I. (2010) The mongoose in Australia: failed introduction of a biological control agent. *Australian Journal of Zoology* 58, 205–227.

Pellock, S., Thompson, A., He, K.S., Mecklin, C.J. & Yang, J. (2013) Validity of Darwin's naturalization hypothesis relates to the stages of invasion. *Community Ecology* 14, 172–179.

Peterson, A.T. (2003) Predicting the geography of species' invasions via ecological niche modeling. *Quarterly Review in Ecology* 78, 419–433.

Pocock, R.I. (1939) *The Fauna of British India, including Ceylon and Burma. Mammalia*. Taylor & Francis, Ltd, London.

Pons, J.-M., Volobouev, V., Ducroz, J.-F., Tillier, A. & Reudet, D. (1999) Is the Guadeloupean racoon [sic] (*Procyon minor*) really an endemic species? New insights from molecular and chromosomal analyses. *Journal of Zoological Systematics and Evolutionary Research*, 37, 101–108.

Popov, V.A. (1964) [Results of acclimatization of American mink (*Mustela vison* Br.) in USSR]. In: *Prirodnye Resursy Volzhsko-Kamskogo Kraya (Zhivotnyi Mir)* (ed. V.A. Popov), pp. 5–15. Nauka, Moscow. (In Russian).

Procheş, Ş., Wilson, J.R.U., Richardson, D.M. & Rejmánek, M. (2008) Searching for phylogenetic pattern in biological invasions. *Global Ecology and Biogeography* 17, 5–10.

Pyšek, P. & Richardson, D. (2010) Invasive species, environmental change and management, and health. *Annual Review of Environment and Resources* 35, 25–55.

R Development Core Team (2019) *The R Project for Statistical Computing*. Vienna, Austria. http://www.R-project.org.

Reid, F., Helgen, K. & Kranz, A. (2016a) *Mustela erminea. The IUCN Red List of Threatened Species* 2016, e.T29674A45203335. http://dx.doi.org/10.2305/IUCN.UK.2016-1.RLTS.T29674A45203335.en. Accessed on 2 January 2017.

Reid, F., Schiaffini, M. & Schipper, J. (2016b) *Neovison vison. The IUCN Red List of Threatened Species* 2016, e.T41661A45214988. http://dx.doi.org/10.2305/IUCN.UK.2016-1.RLTS.T41661A45214988.en. Accessed on 2 January 2017.

Reiss, M. (1988) Scaling of home range size: body size, metabolic needs and ecology. *Trends in Ecology & Evolution* 3, 85–88.

Rejmánek, M. (1996) A theory of seed plant invasiveness: the first sketch. *Biological Conservation* 78, 171–181.

Reynolds, J.C. & Tapper, S.C. (1996) Control of mammalian predators in game management and conservation. *Mammal Review* 26, 127–156.

Ricciardi, A. & Mottiar, M. (2006) Does Darwin's naturalization hypothesis explain fish invasions? *Biological Invasions* 8, 1403–1407.

Ricciardi, A. & Rasmussen, J.B. (1998) Predicting the identity and impact of future biological invaders: a priority for aquatic resource management. *Canadian Journal of Fisheries and Aquatic Sciences* 55, 1759–1765.

Richards, S.A., Whittingham, M.J. & Stephens, P.A. (2011) Model selection and model averaging in behavioural ecology: the utility of the IT-AIC framework. *Behavioral Ecology and Sociobiology* 65, 77–89.

Roots, C. (1976) *Animal Invaders.* Universe Books, New York.

Roy, S. & Robertson, P.A. (2017) Matching the strategy to the scenario; case studies of mink *Neovison vison* management. *Mammal Study* 42, 71–80.

Sacher, G.A. & Staffeidt, E.F. (1974) Relation of gestation time to brain weight for placental mammals: implications for the theory of vertebrate growth. *The American Naturalist* 108, 593–615.

Sade, D.S. & Hildrech, R.W. (1965). Notes on the green monkey (*Cercopithecus aethiops sabaeus*) on St. Kitts, West Indies. *Caribbean Journal of Science* 5, 67–81.

Salgado, I. (2018) Is the raccoon (*Procyon lotor*) out of control in Europe? *Biodiversity and Conservation* 27, 2243–2256.

Samusenko, E.G. (1962) [Acclimatization and re-acclimatization of fur-game mammals in Bielorussia]. *Voprosy Ekologii* 6, 127–128. (In Russian).

Savenkov, V.V. (1987) No title. In: *Transactions of the 18th Congress of the International Union of Game Biologists*, p. 174.

Schreiber, A., Wirth, R., Riffel, M. & Van Rompaey, H. 1989. *Weasels, Civets, Mongooses, and their Relatives. An Action Plan for the Conservation of Mustelids and Viverrids.* IUCN/SSC Mustelid and Viverrid Specialist Group, IUCN, Gland.

Seaman, G.A. & Randall, J.E. (1962). The mongoose as a predator in the Virgin Islands. *Journal of Mammalogy* 43, 544–546.

Seebens, H., Blackburn, T.M., Dyer, E.E., Genovesi, P., Hulme, P.E. *et al.* (2018) Global rise in emerging alien species results from increased accessibility of new source pools. *Proceedings of the National Academy of Sciences of the United States of America* 115, E2264–E2273.

Sheffield, S.R. & King, C.M. (1994) *Mustela nivalis. Mammalian Species* 454, 1–10.

Sherman, H.B. (1954) Raccoons of the Bahama Islands. *Journal of Mammalogy* 35, 126.

Shiraishi, S. (1982). Rat control by weasels. *Saisyu to Shiiku* 44, 414–419.

Simberloff, D. (2009) The role of propagule pressure in biological invasions. *Annual Review of Ecology Evolution and Systematics* 40, 81–102.

Simberloff, D. & Rejmánek, M. (2010) *Encyclopedia of Biological Invasions.* University of California Press, Berkeley.

Simberloff, D., Dayan, T., Jones, C. & Ogura, G. (2000) Character displacement and release in the small Indian mongoose, *Herpestes javanicus. Ecology* 81, 2086–2099.

Sinitsyn, A.A. (1987) No title. In: *Transactions of the 18th Congress of the International Union of Game Biologists*, p. 185.

Skumatov, D., Abramov, A.V., Herrero, J., Kitchener, A., Maran, T., Kranz, A., Sándor, A., Saveljev, A., Savouré-Soubelet, A., Guinot-Ghestem, M., Zuberogoitia, I., Birks, J.D.S., Weber, A., Melisch, R. & Ruette, S. (2016) *Mustela putorius. The IUCN Red List of Threatened Species* 2016, e.T41658A45214384. http://dx.doi.org/10.2305/IUCN.UK.2016-1.RLTS.T41658A45214384.en. Accessed on 2 January 2017.

Smal, C.M. (1988) The American mink *Mustela vison* in Ireland. *Mammal Review* 18, 201–208.

Sol, D. (2000) Are islands more susceptible to be invaded than continents? Birds say no. *Ecography* 23, 687–692.

Solow, A., Roy, S., Bell, C., Milborrow, J. & Roberts, D. (2013) On inference about the introduction time of an introduced species with an application to the pine marten on Mull. *Biological Conservation* 159, 4–6.

Soubeyran, Y. (2008) *Espèces Exotiques Envahissantes dans les Collectivités Françaises d'Outre-mer (État des Lieux et Recommandations).* Collection Planète Nature. Comité français de l'UICN, Paris. (In French).

Spanish Ministry of Environment (2006) Martes foina *(Erxleben, 1777) de Ibiza.* http://www.magrama.gob.es/es/biodiversidad/temas/conservacion-de-especies-amenazadas/gardunha_tcm7-20933.pdf. Accessed on 19 November 2013.

Tate, G.H.H. (1947) *Mammals of Eastern Asia.* The Macmillan Company, New York.

Thompson, G.M. (1922) *The Naturalization of Plants and Animals in New Zealand.* Cambridge University Press, Cambridge.

Thompson, H.V. (1962) Wild mink in Britain. *New Scientist* 13, 130–132.

Thuiller, W., Richardson, D.M., Pysek, P., Midgley, G.F., Hughes, G.O. & Rouguet M. (2005) Niche-based modelling as a tool for predicting the risk of alien plant invasions at a global scale. *Global Change Biology* 11, 2234–2250.

Thuiller, W., Gallien, L., Boulangeat, I., De Bello, F., Münkemüller, T., Roquet, C. & Lavergne, S. (2010) Resolving Darwin's naturalization conundrum: a quest for evidence. *Diversity and Distributions* 16, 461–475.

Thulin, C.G., Simberloff, D., Barun, A., Mccracken, G., Pascal M,Islam MA (2006) Genetic divergence in the small Indian mongoose (*Herpestes auropunctatus*), a widely distributed invasive species. *Molecular Ecology* 15, 3947–3956.

Timm, R., Cuarón, A.D., Reid, F., Helgen, K. & González-Maya, J.F. (2016) *Procyon lotor. The IUCN Red List of Threatened Species* 2016, e. T41686A45216638. http://dx.doi.org/10.2305/IUCN. UK.2016-1.RLTS.T41686A45216638.en. Accessed on 2 January 2017.

Timmins, R.J., Duckworth, J.W., Chutipong, W., Ghimirey, Y., Willcox, D.H.A., Rahman, H., Long, B. & Choudhury, A. (2016) *Viverra zibetha. The IUCN Red List of Threatened Species* 2016, e.T41709A45220429. http://dx.doi.org/0.2305/IUCN.UK.2016-1.RLTS. T41709A45220429.en. Accessed on 2 January 2017.

Tinker, S.W. (1938) *Animals of Hawaii: A Natural History of the Amphibians, Reptiles and Mammals Living in the Hawaiian Islands.* Nippu Jiji, Honolulu.

Tomich, P.Q. (1969) Mammals of Hawaii. *Bishop Museum Special Publication* 57, 1–238.

Torri, H. (2009) *Paguma larvata* (Smith, 1827). In: *The Wild Mammals of Japan* (eds S.D. Ohdachi, Y. Ishibashi, M.A. Iwasa & T. Saitoh), pp. 267–268. Shoukadoh, Kyoto.

Treves, A. & Karanth, U.K. (2003) Human–carnivore conflict and perspectives on carnivore management worldwide. *Conservation Biology* 17, 1491–1499.

Tumanov, I.L. (2009) *Rare Carnivorous Mammals of Russia (Small and Middle-Sized Species).* Branko, Saint Petersburg.

Turbet, C.R. (1941) *Transactions and Proceedings of the Fiji Society of Science and Industry for the years 1938–1940*, pp. 7–12.

Tvrtković, N. & Kryštufek, B. (1990) Small Indian mongoose, *Herpestes auropunctatus*, on the Adriatic islands of Yugoslavia. *Bonner zoologische Beiträge* 41, 3–8.

Udagawa, T. (1952) No title. *Transactions of the Mammalogical Society of Japan* 3, 4.

Urich, F.W. (1931) The mongoose in Trinidad. *Tropical Agriculture (Trinidad)* 8, 95–97.

USFWS [United States Fish and Wildlife Service] (2005) *Draft Revised Recovery Plan for Hawaiian Waterbirds.* Second Draft of Second Revision. US Fish and Wildlife Service, Portland.

Valenzuela, A. & Alcover, J.A. (2013) Radiocarbon evidence for a prehistoric deliberate translocation: the weasel (*Mustela nivalis*) of Mallorca. *Biological Invasions* 15, 717–722.

Valenzuela, A. & Alcover, J.A. (2015) The chronology of the introduction of two species of *Martes* (Carnivora, Mustelidae) on the Western Mediterranean Islands: first direct radiocarbon evidence. *Biological Invasions* 17, 3093–3100.

Van Staaden, M.J. (1994) *Suricata suricata. Mammalian Species* 483, 1–8.

Van Wilgen, N.J. & Richardson, D.M. (2011) Is phylogenetic relatedness to native species important for the establishment of reptiles introduced to California and Florida? *Diversity and Distributions* 17, 172–181.

Varona, L.S. (1974) *Catalogo de los Mamiferos Viventes y Extinguidos de los Antillas.* Academia de Ciencias De Cuba, Havana. (In Spanish).

Vasil'kov, V. (1966) [The raccoon in Polesye]. *Okhota i Okhotnichie Khozyaistvo* 4, 22–23. (In Russian).

Vazquez, D.P. (2006) Exploring the relationship between niche breadth and invasion success. *Conceptual Ecology and Invasion Biology: Reciprocal Approaches to Nature* 1, 307–322.

Veron, G. & Jennings, A.J. (2017) Javan mongoose or small Indian mongoose – who is where? *Mammalian Biology* 87, 62–70.

Veron, G., Patou, M.-L., Pothet, G., Simberloff, D. & Jennings, A.P. (2007) Systematic status and biogeography of the Javan and small Indian mongooses (Herpestidae, Carnivora). *Zoologica Scripta* 36, 1–10.

Veron, G., Patou, M.-L., Simberloff, D., McLenachan, P.A., & Morley, C.G. (2010) The Indian brown mongoose, yet another invader in Fiji. *Biological Invasions* 12, 1947–1951.

Veron, G., Willsch, M., Dacosta, V., Patou, M.-L., Seymour, A., Bonillo, C., Couloux, A., Wong, S.T., Jennings, A.P., Fickel, J. & Wilting, A. (2014) The distribution of the Malay civet *Viverra tangalunga* (Carnivora: Viverridae) across Southeast Asia: natural or human-mediated dispersal? *Zoological Journal of the Linnean Society* 170, 917–932.

Veron, G., Patou, M.L. & Jennings, A.P. (Chapter 3, this volume) Systematics and evolution of the mongooses (Herpestidae, Carnivora). In: *Small Carnivores: Evolution, Ecology, Behaviour, and Conservation* (eds E. Do Linh San, J.J. Sato, J.L. Belant & M.J. Somers), Wiley–Blackwell, Oxford.

Vitousek, P., D'Antonio, L., Loope, L.L., Rejmánek, M. & Westbrooks, R. (1997) Introduced species: a significant component of human-caused global change. *New Zealand Journal of Ecology* 21, 1–16.

Wade-Smith, J. & Verts, B.J. (1982) *Mephitis mephitis. Mammalian Species* 173, 1–7.

Watari, Y., Takatsuki, S. & Miyashita, T. (2008) Effects of exotic mongoose (*Herpestes javanicus*) on the native fauna of Amami-Oshima Island, southern Japan, estimated by distribution patterns along the historical gradient of mongoose invasion. *Biological Invasions* 10, 7–17.

Watari, Y., Nagata, J. & Funakoshi, K. (2011) New detection of a 30-year-old population of introduced mongoose *Herpestes auropunctatus* on Kyushu Island, Japan. *Biological Invasions* 13, 269–276.

Williamson, M. (1996) *Biological Invasions.* Chapman & Hall, London.

Williamson, M.H., Brown, K.C., Holdgate, M.W., Kornberg, H., Southwood, R. & Mollison, D. (1986) The analysis and modelling of British invasions [and discussion]. *Philosophical Transactions of the Royal Society of London B, Biological Sciences* 314, 505–522.

Wilson, E.O. (1965) The challenge from related species. In: *The Genetics of Colonizing Species* (eds H.G. Baker & G.L. Stebbin), pp. 7–24. Academic Press, New York.

Wodzicki, K.A. (1950) *Introduced Mammals of New Zealand. An Ecological and Economic Survey.* Bulletin 98. Department of Scientific & Industrial Research, Wellington, New Zealand.

Won, C. & Smith, K.G. (1999) History and current status of mammals of the Korean Peninsula. *Mammal Review* 29, 3–33.

Wostenberg, D.J., Hopken, M.W., Shiels, A.B. & Piaggio, A.J. (2019) Using DNA to identify the source of invasive mongooses, *Herpestes auropunctatus* (Carnivora: Herpestidae) captured on Kauaʻi, Hawaiian Islands. *Pacific Science* 73, 215–223.

WWF [World Wildlife Fund] (2013) Ecoregions. http://worldwildlife.org/biomes. Accessed on 10 November 2013.

Yamada, F. & Sugimura, K. (2004) Negative impact of an invasive small Indian mongoose *Herpestes javanicus* on native wildlife species and evaluation of a control project in Amami-Ohshima and Okinawa Islands, Japan. *Global Environmental Research* 8, 117–124.

Yanushevich, A.I. (1963) [R*esults and Prospects of Acclimatization of Fur-Game Mammals in Kyrgyzstan*]. *Acclimatization of animals in USSR.* Izdatelstvo Akademii nauk KazSSR, Alma-Ata. (In Russian)

Yiming, L., Zhengjun, W. & Duncan, R. (2006) Why islands are easier to invade: human influences on bullfrog invasion in the Zhoushan archipelago and neighboring mainland China. *Oecologia* 148, 129–136.

Youngman, P.M. (1990) *Mustela lutreola. Mammalian Species* 362, 1–3.

Zeveloff, S.I. & Boyce, M.S. (1988) Body size patterns in North American mammal faunas. In: *Evolution of Life Histories of Mammals* (ed. M.S. Boyce), pp. 123–146. Yale University Press, New Haven.

Appendix 20.1 List of 253 identified and documented small carnivore (*sensu* IUCN SSC Small Carnivore Specialist Group, 2013) introductions worldwide, sorted alphabetically by family (*N* = 5), species (*N* = 24) and country (*N* = 66), and year of the first introduction. *n* = number of individuals released; Success = success (1) or failure (0) of introduction; Invaded biome and Native biome = Pa (Paleartic), Na (Nearctic), Ne (Neotropic), Af (Afrotropic), As (Australasia), In (Indo–Malayan), or Oc (Oceanic); Island = island introduction (Y) or mainland (N); ? = no information available.

Family	Species	Event no.	Introduction location	Year	n	Success	Invaded biome	Native biome	Island	Source
Herpestidae	*Urva auropunctata*	1.	Antigua and Barbuda (Antigua)[a]	1870s	?	1	Ne	In, Pa	Y	Hinton & Dunn (1967), Hoagland *et al.* (1989)
		2.	Bahamas[b,c]	?	?	1	Ne	In, Pa	Y	Hoagland *et al.* (1989)
		3.	Barbados	1882	20	1	Ne	In, Pa	Y	Hinton & Dunn (1967), Hoagland *et al.* (1989)
		4.	Bosnia and Herzegovina (Mostar)[b]	?	?	1	Pa	In, Pa	N	Tvrtković & Kryštufek (1990), Kryštufek & Tvrtković (1992)
		5.	Colombia	1951	80	1	Ne	In, Pa	N	Roots (1976)
		6.	Comoros (Grande Comore)[b]	?	?	1	Af	In, Pa	Y	Haltenorth & Diller (1994), Simberloff *et al.* (2000), Louette *et al.* (2004)
		7.	Costa Rica[b]	?	?	1	Ne	In, Pa	N	IUCN SSC ISSG (2018)
		8.	Croatia (Mljet)	1910	11	1	Pa	In, Pa	Y	Tvrtković & Kryštufek (1990)
		9.	Croatia (Korčula)	1910	?	1	Pa	In, Pa	Y	Tvrtković & Kryštufek (1990)
		10.	Croatia (Pelješac Peninsula)	1921	?	1	Pa	In, Pa	N	Tvrtković & Kryštufek (1990)
		11.	Croatia (Hvar)	1970	?	1	Pa	In, Pa	Y	Tvrtković & Kryštufek (1990)
		12.	Croatia (Brač)[b]	1926	?	0	Pa	In, Pa	Y	Tvrtković & Kryštufek (1990)
		13.	Croatia (Čiovo)[b]	By 1991	?	1	Pa	In, Pa	Y	Kryštufek & Tvrtković (1992)
		14.	Croatia (Kobrava)[b]	?	?	1	Pa	In, Pa	Y	Barun *et al.* (2011)
		15.	Croatia (Škrda)[b]	?	?	1	Pa	In, Pa	Y	Barun *et al.* (2011)
		16.	Cuba[d]	1882	?	1	Ne	In, Pa	Y	Espeut (1882), Borroto-Páez (2009)
		17.	Dominica	1880s	10	0	Ne	In, Pa	Y	Milne & Milne (1962), Hall (1981)
		18.	Dominican Republic/Haiti (Hispaniola)	1870s	?	1	Ne	In, Pa	Y	Long (2003)
		19.	Fed. of St Kitts and Nevis (Nevis)	1870s	?	1	Ne	In, Pa	Y	Hinton & Dunn (1967), Long (2003)
		20.	Fed. of St Kitts and Nevis (St Kitts)	1884	?	1	Ne	In, Pa	Y	Sade & Hildrech (1965), Hoagland *et al.* (1989)

No.	Location							References
21.	Fiji[e]	1870	?	0	Oc	In, Pa	Y	Horst et al. (2001)
22.	Fiji[e]	1883	2	1	Oc	In, Pa	Y	Turbet (1941) in Long (2003), Gorman (1975)
23.	France (La Désirade, Guadeloupe)	1870s[f]	?	1	Ne	In, Pa	Y	Long (2003)
24.	France (Grande-Terre and Basse-Terre, Guadeloupe)	1880–1885	?	1	Ne	In, Pa	Y	Varona (1974), Hall (1981), Lorvelec et al. (2021)
25.	France (Marie-Galante, Guadeloupe)	1870s	?	1	Ne	In, Pa	Y	Hinton & Dunn (1967), Long (2003)
26.	France (Martinique)	1889	?	1	Ne	In, Pa	Y	Horst et al. (2001), Lorvelec et al. (2021)
27.	France (Fajou, Guadeloupe)[b,g]	1930	?	1	Ne	In, Pa	Y	Lorvelec et al. (2004), Thulin et al. (2006)
28.	New Caledonia[b]	2010	2?	?	Oc	In, Pa	Y	Barun et al. (2011)
29.	France/Netherlands (Saint-Martin/Sint Maarten)	1885–1889	20	1	Ne	In, Pa	Y	Husson (1960), Horst et al. (2001), Nellis & Everard (1983)
30.	French Guiana	1900	?	1	Ne	In, Pa	N	Hoagland et al. (1989), Soubeyran (2008)
31.	Grenada (Grenada)	1876–1879	14	1	Ne	In, Pa	Y	Hinton & Dunn (1967), Hoagland et al. (1989)
32.	Grenada (Carriacou)[b]	By 1900	?	1	Ne	In, Pa	Y	Hoagland et al. (1989)
33.	Guyana	1882	?	1	Ne	In, Pa	N	de Vos et al. (1956)
34.	Honduras[b]	?	?	0	Ne	In, Pa	N	CABI (n.d.)
35.	Indonesia (Ambon, Maluku)	?	?	1	As	In, Pa	Y	de Vos et al. (1956), Haltenorth & Diller (1994)
36.	Italy	1952	?	0	Pa	In, Pa	N	Gaubert (2016)
37.	Jamaica (Jamaica)	1872	9	1	Ne	In, Pa	Y	Espeut (1882)
38.	Jamaica (Great Goat Island)[b,h]	1920–1925	?	1	Ne	In, Pa	Y	Hoagland et al. (1989)
39.	Japan (Okinawa)	1910	13–17	1	In	In, Pa	Y	Ogura et al. (1998), Yamada & Sugimura (2004)
40.	Japan (Amami-Oshima)	1979	30	1	In	In, Pa	Y	Abe (1991), Yamada & Sugimura (2004), Watari et al. (2008)
41.	Japan (Kyushu)[b,i]	<1979	?	1	In	In, Pa	Y	Watari et al. (2011)

(Continued)

Appendix 20.1 (Continued)

Family	Species	Event no.	Introduction location	Year	n	Success	Invaded biome	Native biome	Island	Source
		42.	Macedonia (Golem Grad)[b]	?	?	?	Pa	In, Pa	Y	Tvrtković & Kryštufek (1990)
		43.	Mauritius	1900	19	1	Af	In, Pa	Y	Lever (1985), Cheke (1987)
		44.	Panama[b]	?	?	1	Ne	In, Pa	N	Hoagland et al. (1989)
		45.	St Lucia[k]	1870s	?	1	Ne	In, Pa	Y	Hinton & Dunn (1967), Varona (1974), Hall (1981)
		46.	St Vincent and the Grenadines	1870s	?	1	Ne	In, Pa	Y	Hinton & Dunn (1967), Varona (1974), Hall (1981)
		47.	Samoa (Upolu)[b]	By 2010	?	?	Oc	In, Pa	Y	M. Bonin & J. Atherton in Barun et al. (2011)
		48.	Suriname	1900	?	1	Ne	In, Pa	N	Hinton & Dunn (1967), Husson (1978), Lever (1985)
		49.	Tanzania (Mafia Island)	~1940	?	1	Af	In, Pa	Y	Moreau & Pakenham (1941), de Vos et al. (1956)
		50.	Tonga[b]	2016	6	?	Oc	In, Pa	Y	Cranwell (2016)
		51.	Trinidad and Tobago (Trinidad?)	1870	?	0	Ne	In, Pa	Y	Husson (1960), Hoagland et al. (1989), Hays & Conant (2007)
		52.	Trinidad and Tobago (Trinidad)	By 1882	5	1	Ne	In, Pa	Y	Urich (1931), Hinton & Dunn (1967)
		53.	UK (Bermuda)[b]	?	?	1	Ne	In, Pa	Y	IUCN SSC ISSG (2018)
		54.	UK (Tortola, British Virgin Islands)[l]	1870s	?	1	Ne	In, Pa	Y	Hinton & Dunn (1967), Coblentz & Coblentz (1985), Long (2003)
		55.	USA (Puerto Rico, Puerto Rico)[m]	1877	20	1	Ne	In, Pa	Y	Long (2003)
		56.	USA (St John, U.S. Virgin Islands)[n]	1880s	?	1	Ne	In, Pa	Y	Long (2003)
		57.	USA (Hawaii, Hawaii)	1883	72	1	Oc	In, Pa	Y	Tinker (1938)
		58.	USA (Maui, Hawaii)	1883	?	1	Oc	In, Pa	Y	Tinker (1938)
		59.	USA (Molokai, Hawaii)	1883	?	1	Oc	In, Pa	Y	Tinker (1938)
		60.	USA (Oahu, Hawaii)[o]	1883	?	1	Oc	In, Pa	Y	Tinker (1938)
		61.	USA (St Croix, U.S. Virgin Islands)	1882–1884	4–8	1	Ne	In, Pa	Y	Seaman & Randall (1962), Hinton & Dunn (1967), Coblentz & Coblentz (1985)

	Location	Year	No.	Est.	Reg1	Reg2	Y/N	Reference
62.	USA (St Thomas, U.S. Virgin Islands)^p	By 1899	?	1	Ne	In, Pa	Y	Hoagland et al. (1989), Horst et al. (2001)
63.	USA (Vieques, Puerto Rico)^b	By 1899	?	1	Ne	In, Pa	Y	Hoagland et al. (1989)
64.	USA (Buck Island, U.S. Virgin Islands)^g	1910 or 1952	4	1	Ne	In, Pa	Y	Seaman & Randall (1962), Nellis & Everard (1983)
65.	USA (Dodge Island, Florida)^b	1977	3?	0	Ne	In, Pa	Y	Nellis et al. (1978)
66.	Venezuela	1926	100	0	Ne	In, Pa	N	Tvrtković & Kryštufek (1990)
Urva edwardsii^q 1.	Australia	1883	60	0	As	In, Pa	N	Peacock & Abbott (2010)
2.	Australia	1883	42	0	As	In, Pa	N	Peacock & Abbott (2010)
3.	Australia	1884	52	0	As	In, Pa	N	Peacock & Abbott (2010)
4.	Australia	1884	12	0	As	In, Pa	N	Peacock & Abbott (2010)
5.	Australia	1884	100	0	As	In, Pa	N	Peacock & Abbott (2010)
6.	Australia	1884	60	0	As	In, Pa	N	Peacock & Abbott (2010)
7.	Australia	1884	60	0	As	In, Pa	N	Peacock & Abbott (2010)
8.	Australia	1887	700	0	As	In, Pa	N	Peacock & Abbott (2010)
9.	Australia	1883	40	0	As	In, Pa	N	Peacock & Abbott (2010)
10.	Australia	1884	200	0	As	In, Pa	N	Peacock & Abbott (2010)
11.	Italy	1952	?	0	Pa	In, Pa	N	Gaubert & Zenatello (2009)
12.	Malaysia	?	?	0	In	In, Pa	N	Tate (1947)
13.	New Zealand	1870	14	0	As	In, Pa	N	Thompson (1922)
Urva fusca 1.	Fiji (Viti Levu)	1970s	2	1	Oc	In	Y	Morley et al. (2007), Veron et al. (2010)
Suricata suricatta 1.	Australia	1876	?	0	As	Af	N	Peacock & Abbott (2010)
Mephitidae *Mephitis mephitis* 1.	Canada (Prince Edward Island)	~1940	?	1	Na	Na	Y	Banfield (1977)
2.	Kyrgyzstan	1937	29	0	Pa	Na	N	Pavlov et al. (1973)
3.	Russian Federation	~1930	?	0	Pa	Na	N	Yanushevich (1963)
4.	Russian Federation	1933	26	0	Pa	Na	N	de Vos et al. (1956)
5.	Russian Federation (Pet'ov Island)	1936	3	0	Pa	Na	Y	Bromlei (1959)
6.	Russian Federation	1939	58	0	Pa	Na	N	Pavlov et al. (1973)

(Continued)

Appendix 20.1 (Continued)

Family	Species	Event no.	Introduction location	Year	n	Success	Invaded biome	Native biome	Island	Source
Mustelidae	*Martes foina*	7.	Ukraine	1937	29	0	Pa	Na	N	Pavlov et al. (1973)
		1.	Spain (Ibiza)	~1930	?	0	Pa	Pa	Y	Long (2003), Spanish Ministry of Environment (2006)
		2.	USA	1950–1960	?	1	Na	Pa	N	Long (1995)
	Martes martes	1.	Spain (Mallorca)	~700	?	1	Pa	Pa	Y	Clevenger (1993), Valenzuela & Alcover (2015)
		2.	Spain (Menorca)	?	?	1	Pa	Pa	Y	Clevenger (1993)
		3.	Russian Federation	1962	?	0	Pa	Pa	N	Kirisa (1972–1974)
		4.	UK (Isle of Mull)	~1980	?	1	Pa	Pa	Y	Solow et al. (2013)
	Martes melampus	1.	Japan (Hokkaido)	1940s	?	1	Pa	Pa	Y	Masuda (2009)
		2.	Japan (Sado Island)	?	?	1	Pa	Pa	Y	Masuda (2009)
	Mustela erminea	1.	Denmark (Strynø Kalv)	1980	?	1	Pa	Pa, Na	Y	Corbet & Harris (1991)
		2.	Netherlands (Terschelling Island)	1931	9	1	Pa	Pa, Na	Y	King & Moors (1979)
		3.	New Zealand	1880s	~1000	1	As	Pa, Na	Y	King & Moors (1979), King (2017a)
		4.	UK (Shetland Islands)	1680	?	1	Pa	Pa, Na	Y	Lever (1985)
		5.	UK (Whalsay Island)	~1800	?	0	Pa	Pa, Na	Y	Fitter (1959)
	Mustela itatsi	1.	Japan (Hokkaido)	1880	?	1	Pa	Pa	Y	Inukai (1934)
		2.	Japan (Miyake-jima)	1986	?	1	Pa	Pa	Y	Hamao & Higuchi (2013)
	Mustela lutreola	1.	France (Île du chat)	1956	3	0	Pa	Pa	Y	Lesel & Derenne (1975)
		2.	Russian Federation (Kuril Islands)	1983	?	1	Pa	Pa	Y	Schreiber et al. (1989)
	Mustela nivalis	1.	Australia	1885	?	0	As	Pa, Na	N	Hinton & Dunn (1967)
		2.	Netherlands (Terschelling Island)	1931	103	0	Pa	Pa, Na	Y	de Vos et al. (1956)
		3.	New Zealand	1890	~2400	1	As	Pa, Na	Y	Wodzicki (1950), King (2017a)
		4.	São Tomé and Príncipe[b]	?	?	1	Af	Pa, Na	Y	Dutton (1994)

Species	No.	Location	Year			Origin			Reference
	5.	Spain (Mallorca)	200 B.C.	?	1	Pa	Pa, Na	Y	Valenzuela & Alcover (2013)
Mustela putorius	1.	Australia	1885	?	0	As	Pa	N	de Vos *et al.* (1956)
	2.	Ireland (Great Saltee Island)	1949	2	0	Pa	Pa	Y	Buckley *et al.* (2007)
	3.	Jamaica	?	?	0	Na	Pa	Y	Milne & Milne (1962)
	4.	Japan (Tohoku)	1930	?	0	Pa	Pa	Y	Kaburaki (1940)
	5.	New Zealand	1880s	~300–2000	1	As	Pa	Y	Wodzicki (1950), King (2017c)
	6.	New Zealand	1960	?	0	As	Pa	Y	Flux & Fullagar (1992)
	7.	Russian Federation	1940	?	1	Pa	Pa	N	Lavrov (1946)
	8.	Spain (Canary Islands)	~1500	?	1	Pa	Pa	Y	Medina & Martin (2010)
	9.	UK (Rathlin Island)	~1980	?	1	Pa	Pa	Y	Buckley *et al.* (2007)
Mustela sibirica	1.	Japan	1930	?	1	Pa	Pa, In	Y	Inukai (1949) *in* Kuroda (1955)
	2.	Japan (Okujirijima)	1948	?	1	Pa	Pa, In	Y	Tokuda (1951) *in* Kuroda (1955)
	3.	Japan (Ishigaki-Jima)	1966	1600	1	Pa	Pa, In	Y	Long (2003)
	4.	Japan (Ryukyu)	1965	2000	0	Pa	Pa, In	Y	Hayashi (1981)
	5.	Kyrgyzstan	1941	30	0	Pa	Pa, In	N	Lavrov (1941)
	6.	Russian Federation (Sakhalin Island)	1937	?	0	Pa	Pa, In	Y	Inukai (1949) *in* Kuroda (1955)
Neovison vison	1.	Argentina	1932	?	0	Ne	Na	N	Jaksic *et al.* (2002)
	2.	Argentina	1946–1968	?	1	Ne	Na	N	Jaksic *et al.* (2002)
	3.	Argentina	1956	?	1	Ne	Na	N	Jaksic *et al.* (2002)
	4.	Argentina	1960	?	1	Ne	Na	N	Jaksic *et al.* (2002)
	5.	Azerbaijan	1938	46	1	Pa	Na	N	Long (2003)
	6.	Belarus	1955	725	1	Pa	Na	N	Samusenko (1962)
	7.	Belgium	?	?	0	Pa	Na	N	Bonesi & Palazon (2007)
	8.	Chile	1935	?	0	Ne	Na	N	Jaksic *et al.* (2002)
	9.	Chile	1940	?	0	Ne	Na	N	de Vos *et al.* (1956)
	10.	Chile	1967	?	1	Ne	Na	N	Jaksic *et al.* (2002)
	11.	Czech Republic	1960	?	1	Pa	Na	N	Bonesi & Palazon (2007)

(Continued)

Appendix 20.1 (Continued)

Family	Species	Event no.	Introduction location	Year	n	Success	Invaded biome	Native biome	Island	Source
		12.	Denmark	1925	?	1	Pa	Na	N	de Vos et al. (1956)
		13.	Finland	1925	?	1	Pa	Na	N	Thompson (1962)
		14.	France	?	?	1	Pa	Na	N	Bonesi & Palazon (2007)
		15.	Iceland	1930	?	1	Pa	Na	Y	Thompson (1962)
		16.	Ireland	1961	30	1	Pa	Na	Y	Smal (1988)
		17.	Japan	1920–1930	?	1	Pa	Na	Y	Long (2003)
		18.	Kazakhstan	1952	156	1	Pa	Na	N	Long (2003)
		19.	Kyrgyzstan	1956	46	1	Pa	Na	N	Beishebaev (1963) in Long (2003)
		20.	Kyrgyzstan	1956–1967	336	1	Pa	Na	N	Long (2003)
		21.	Netherlands	?	?	1	Pa	Na	N	Litjens (1980)
		22.	Norway	1930	?	1	Pa	Na	N	de Vos et al. (1956)
		23.	Poland	1953	?	1	Pa	Na	N	Brzezinski & Marzec (2003)
		24.	Russian Federation	?	?	0	Pa	Na	N	Izmailov (1968) in Long (2003)
		25.	Russian Federation (Urup Island)	?	?	1	Pa	Na	Y	Lavov (1962) in Long (2003), Benkovsky (1971)
		26.	Russian Federation	~1930	10000	1	Pa	Na	N	Popov (1964)
		27.	Russian Federation	1933	3000	1	Pa	Na	N	Pavlov et al. (1973)
		28.	Russian Federation	1933–1961	11000	1	Pa	Na	N	Berger (1962)
		29.	Russian Federation	1939	70	1	Pa	Na	N	Long (2003)
		30.	Russian Federation	1939	69	1	Pa	Na	N	Long (2003)
		31.	Russian Federation	1934–1964	570	0	Pa	Na	N	Long (2003)
		32.	Russian Federation	1934–1970	653	1	Pa	Na	N	Sinitsyn (1987) in Long (2003)
		33.	Russian Federation	1934–1970	328	1	Pa	Na	N	Long (2003)

No.	Country	Year	Count					Reference
34.	Russian Federation	1935	83	1	Pa	Na	N	Long (2003)
35.	Russian Federation	1935–1967	1245	1	Pa	Na	N	Long (2003)
36.	Russian Federation	1935–1968	1984	1	Pa	Na	N	Sinitsyn (1987) *in* Long (2003)
37.	Russian Federation	1936–1942	367	1	Pa	Na	N	Long (2003)
38.	Russian Federation	1936–1959	1159	1	Pa	Na	N	Long (2003)
39.	Russian Federation	1939	63	1	Pa	Na	N	Long (2003)
40.	Russian Federation	1940	2679	1	Pa	Na	N	Pavlov *et al.* (1973)
41.	Russian Federation	1948	60	1	Pa	Na	N	Long (2003)
42.	Russian Federation	1948–1956	638	1	Pa	Na	N	Sinitsyn (1987) *in* Long (2003)
43.	Russian Federation	1948–1964	711	1	Pa	Na	N	Sinitsyn (1987) *in* Long (2003)
44.	Russian Federation	1950–1969	361	1	Pa	Na	N	Sinitsyn (1987) *in* Long (2003)
45.	Russian Federation	1951	99	1	Pa	Na	N	Long (2003)
46.	Russian Federation	1952	56	1	Pa	Na	N	Long (2003)
47.	Russian Federation (Sakhalin Island)[b]	1956–1970	564	1	Pa	Na	Y	Benkovsky (1971)
48.	Russian Federation	1957	44	1	Pa	Na	N	Long (2003)
49.	Russian Federation	1959	100	1	Pa	Na	N	Long (2003)
50.	Russian Federation	1960	?	1	Pa	Na	N	Savenkov (1987) *in* Long (2003)
51.	Russian Federation	1963	686	1	Pa	Na	N	Long (2003)
52.	Russian Federation	1963	478	1	Pa	Na	N	Sinitsyn (1987) *in* Long (2003)
53.	Spain	1982	?	1	Pa	Na	N	Lever (1985)
54.	Sweden	1930	43	1	Pa	Na	N	de Vos (1956)
55.	Sweden	1932	20	1	Pa	Na	N	Gerell (1967a,b) *in* Long (2003)
56.	Sweden	1934	20	1	Pa	Na	N	Gerell (1967a,b) *in* Long (2003)
57.	UK (Northern Ireland)	1962	?	1	Pa	Na	Y	Thompson (1962)

(Continued)

Appendix 20.1 (Continued)

Family	Species	Event no.	Introduction location	Year	n	Success	Invaded biome	Native biome	Island	Source
		58.	UK (England, Wales, Scotland)[b]	From 1929	?	1	Pa	Na	Y	Long (2003)
		59.	UK (Isle of Lewis, Scotland)[b,r]	From 1929	?	1	Pa	Na	Y	Cuthbert (1973)
		60.	UK (Isle of Mull, Scotland)[b]	By 2006	?	1	Pa	Na	Y	Roy & Robertson (2017)
Procyonidae	*Nasua nasua*	1.	Chile (Robinson Crusoe Island)	1935	2–8	0	Ne	Ne	Y	de Vos et al. (1956)
		2.	Spain (Mallorca)	~1990	~8	0	Pa	Ne	Y	Mayol et al. (2009)
	Procyon lotor[s]	1.	Azerbaijan	1936–1986	515	1	Pa	Na, Ne	N	Pavlov et al. (1973)
		2.	Azerbaijan	1941	21	1	Pa	Na, Ne	N	Aliev (1955) *in* Long (2003)
		3.	Azerbaijan	1950–1957	202	1	Pa	Na, Ne	N	Pavlov et al. (1973)
		4.	Bahamas (New Providence)	<1784	?	1	Ne	Na, Ne	Y	McKinley (1959)
		5.	Bahamas (Grand Bahama)	1932–1933	2	1	Ne	Na, Ne	Y	Sherman (1954)
		6.	Bahamas (Abaco)[b]	~1990s	?	1	Ne	Na, Ne	Y	Helgen et al. (2008)
		7.	Barbados	~1665	?	1	Ne	Na, Ne	Y	Helgen & Wilson (2002)
		8.	Belarus	1936–1986	128	1	Pa	Na, Ne	N	Pavlov et al. (1973)
		9.	Belarus	1954–1958	128	1	Pa	Na, Ne	N	Samusenko (1962) *in* Long (2003)
		10.	Canada (Cox Island)	1937	?	1	Na	Na, Ne	Y	Banfield (1977)
		11.	Canada (Graham Island)	1937	?	1	Na	Na, Ne	Y	Banfield (1977)
		12.	Canada (Prince Edward Island)	1937	?	1	Na	Na, Ne	Y	Banfield (1977)
		13.	France (Grande-Terre and Basse-Terre, Guadeloupe)[b]	~1830	?	1	Ne	Na, Ne	Y	Lazell (1981), Helgen & Wilson (2002)
		14.	France (Martinique)[b]	~1954	?	1	Ne	Na, Ne	Y	Bon Saint Côme & Tanasi (1994)
		15.	France/Netherlands (Saint Martin/Sint Maarten)[b]	~1957	?	1	Ne	Na, Ne	Y	Husson (1960)

#	Location	Year						Reference
16.	France (La Désirade, Guadeloupe)[b]	?	?	1	Ne	Na, Ne	Y	Lorvelec et al. (2007)
17.	France (Marie-Galante, Guadeloupe)[b]	?	?	1	Ne	Na, Ne	Y	Moutou (1987), Bénito-Espinal (1990)
18.	Germany	1927[f]	4	1	Pa	Na, Ne	N	Lever (1985)
19.	Germany	1929[f]	5	1	Pa	Na, Ne	N	Corbet (1966)
20.	Germany	1945[f]	?	1	Pa	Na, Ne	N	Niethammer (1963)
21.	Japan (Inuyama)	1962	12	1	Pa	Na, Ne	Y	Ando & Kajiura (1985), Agetsuma (2004)
22.	Japan (Kani)	1982	40	1	Pa	Na, Ne	Y	Ando & Kajiura (1985), Agetsuma (2004)
23.	Japan (Eniwa)[b]	1979	10	1	Pa	Na, Ne	Y	Ikeda et al. (2004)
24.	Japan (Kamakura)[b]	1988	?	1	Pa	Na, Ne	Y	Ikeda et al. (2004)
25.	Kyrgyzstan	1936	22	1	Pa	Na, Ne	N	Aliev & Sanderson (1966)
26.	Kyrgyzstan	1937	33	1	Pa	Na, Ne	N	Pavlov et al. (1973)
27.	New Zealand	1905	2	0	As	Na, Ne	Y	Wodzicki (1950)
28.	Russian Federation	1936–1986	490	1	Pa	Na, Ne	N	Pavlov et al. (1973)
29.	Russian Federation (Pet'ov Island)	1940	4	0	Pa	Na, Ne	Y	Novikov (1956)
30.	Russian Federation	1950	23	1	Pa	Na, Ne	N	Pavlov et al. (1973)
31.	Russian Federation	1965	30	1	Pa	Na, Ne	N	Pavlov et al. (1973)
32.	Russian Federation	1953	16	1	Pa	Na, Ne	N	Pavlov et al. (1973)
33.	Russian Federation	1954	55	1	Pa	Na, Ne	N	Pankrat'ev (1959) in Long (2003)
34.	Russian Federation	1954	100	1	Pa	Na, Ne	N	Pavlov et al. (1973)
35.	Russian Federation	1956	486	1	Pa	Na, Ne	N	Pavlov et al. (1973)
36.	Russian Federation	1951	28	1	Pa	Na, Ne	N	Pavlov et al. (1973)
37.	Russian Federation	1954	52	1	Pa	Na, Ne	N	Vasil'kov (1966) in Long (2003)
38.	Russian Federation	1955	73	1	Pa	Na, Ne	N	Pankrat'ev (1959) in Long (2003)
39.	Spain	~2000	?	1	Pa	Na, Ne	N	Garcia et al. (2012)
40.	USA (Long Island, Alaska)	1935	?	1	Na	Na, Ne	Y	Burris (1965)
41.	USA (Singer, Alaska)	1941	8	1	Na	Na, Ne	Y	Burris (1965)

(Continued)

Appendix 20.1 (Continued)

Family	Species	Event no.	Introduction location	Year	n	Success	Invaded biome	Native biome	Island	Source
		42.	USA (Japonski, Alaska)	1950	?	1	Na	Na, Ne	Y	Burris (1965)
		43.	Uzbekistan	1953	43	1	Pa	Na, Ne	N	Mukhtarov (1963) *in* Long (2003)
Viverridae	*Civettictis civetta*	1.	São Tomé and Principe[b]	?	?	1	Af	Af	Y	Dutton (1994)
	Genetta genetta	1.	Spain	?	?	1	Pa	Af, Pa	N	Gaubert et al. (2011), Delibes et al. (2019)
		2.	Spain	?	?	1	Pa	Af, Pa	N	Gaubert et al. (2011), Delibes et al. (2019)
		3.	Spain (Cabrera)	?	?	1	Pa	Af, Pa	Y	Gaubert et al. (2009, 2016)
		4.	Spain (Ibiza)	?	?	1	Pa	Af, Pa	Y	Gaubert et al. (2009, 2016)
		5.	Spain (Mallorca)	?	?	1	Pa	Af, Pa	Y	Gaubert et al. (2009, 2016)
	Paguma larvata	1.	Japan (Honshu)	~1930	?	1	Pa	In, Pa	Y	Udagawa (1952) *in* Long (2003), Kuroda (1955), Torri (2009)
	Paradoxurus hermaphroditus	1.	Indonesia (Aru Islands)	?	?	1	As	In	Y	Lever (1985)
		2.	Indonesia (Lesser Sunda Islands)	?	?	1	As	In	Y	Schreiber et al. (1989), Patou et al. (2010)
		3.	Indonesia (Sulawesi)	?	?	1	Aa	In	Y	Schreiber et al. (1989), Patou et al. (2010)
	Viverra tangalunga	1.	Indonesia (Buru)	~1950	?	1	As	In	Y	Laurie & Hill (1954), Flannery (1995)
		2.	Indonesia (Bacan)	?	?	1	As	In	Y	Laurie & Hill (1954), Flannery (1995)
		3.	Indonesia (Halmahera)	?	?	1	As	In	Y	Laurie & Hill (1954), Flannery (1995)
		4.	Indonesia (Sulawesi)	?	?	1	As	In	Y	de Vos et al. (1956)
	Viverra zibetha	1.	India (Andaman Islands)	?	?	1	In	In	Y	Lever (1985)
	Viverricula indica	1.	Comoros (Grande Comore)	?	?	1	Af	In	Y	Gaubert et al. (2017)
		2.	France (Mayotte)	?	?	1	Af	In	Y	Louette (1999), Caceres & Decalf (2015), Gaubert et al. (2017)
		3.	Indonesia (Sumbawa)	?	?	1	As	In	Y	Laurie & Hill (1954)

4.	Madagascar	?	?	1	Af	In	Y	Haltenorth & Diller (1994), Goodman (2012), Gaubert et al. (2017)
5.	Tanzania (Mafia)	?	?	1	Af	In	Y	Kock & Stanley (2009)
6.	Tanzania (Pemba)	?	?	1	Af	In	Y	Pakenham (1984)
7.	Tanzania (Zanzibar)	?	?	1	Af	In	Y	Pakenham (1984)
8.	Yemen (Socotra)	~1600	?	1	Af	In	Y	Pocock (1939), Gaubert et al. (2017)

[a] *Urva auropunctata* was also introduced to the islets of Codrington and Green, but has since then been eradicated (Barun et al., 2011).

[b] These introductions were identified at a late stage and thus not included in the multivariate analyses.

[c] It is unclear to which Bahamian island(s) *Urva auropunctata* was introduced.

[d] *Urva auropunctata* is also present on Cayo Romano and Cayo Sabinal (Barun et al., 2011). These cays are in fact linked to Cuba through narrow land bridges and causeways, so colonization likely took place through natural dispersal.

[e] *Urva auropunctata* was initially introduced to the two main islands (Viti Levu and Vanua Levu), but is now also present on 11 smaller islands: Beqa, Drudrua, Kioa, Macuata-i-wai, Malake, Mavuva, Nananu-i-ra, Nananu-i-cake, Nasoata, Rabi and Yanuca (Morley, 2004a; Barun et al., 2011). Mongooses may have reached the smaller islands through different scenarios (Morley, 2004a,b).

[f] But see Hedges et al. (2016) for alternative scenarios of introduction, including a single, very recent event.

[g] *Urva auropunctata* has now been eradicated on this island.

[h] *Urva auropunctata* is also present on Little Goat Island (Barun et al., 2011).

[i] Only a recent find despite a relatively long-term presence.

[j] Ongoing eradication of *Urva auropunctata* on this island.

[k] *Urva auropunctata* was also introduced to the nearby islet of Praslin where it has now been eradicated. It is unclear whether the introduction to Praslin took place at the same time or possibly well after establishment on St Lucia.

[l] *Urva auropunctata* is also present on Beef Island (now linked to Tortola by a bridge) and Jost van Dyke (Barun et al., 2011). It is unclear whether it has been introduced to these islands at the same time or after the introduction to Tortola.

[m] *Urva auropunctata* was also present on the islet of Piñeros, but has since then been eradicated (Barun et al., 2011). It is unclear whether it was introduced to Piñeros at the same time or possibly only after its establishment on Puerto Rico island.

[n] *Urva auropunctata* is also present on Lovango cay (Nellis & Everard, 1983) and has now been eradicated on the islet of Leduck (Barun et al., 2011). It is unclear whether it was introduced to Lovango and Leduck concurrently or after the introduction to St John.

[o] *Urva auropunctata* is also present on the islets of Mokuelo and Ford, near Oahu (Tomich, 1969). In addition, according to Barun et al. (2011) citing USFWS (2005), the species was spotted in the 1970s on Kauai, but not since. This island was often believed to be mongoose free (Case & Bolger, 1991). Two mongooses genetically identical to individuals from Oahu were caught in 2012 (Wostenberg et al., 2019), although it seems that the island does not host mongooses at the moment (cf. Duffy et al., 2015).

[p] *Urva auropunctata* was also present on the small Water Island, where it has now been eradicated (Barun et al., 2011). It is unclear whether it was introduced to Water Island at the same time or after the introduction to St Thomas.

[q] Events No. 1–12 were documented as Indian grey mongoose, *Urva edwardsii*, releases; however, species identity is unclear; thus these data were not included in analyses.

[r] Having escaped from the fur farms mink spread throughout the 2800 km² island archipelago through the southern tip of South Uist within 40 years ((Roy & Robertson, 2017).

[s] There is uncertainty as to whether *Procyon lotor* was introduced unsuccessfully to Jamaica (Helgen & Wilson, 2003). Helgen & Wilson (2005) suggested that the raccoon found on Tres Marías Islands (Mexico) is a subspecies of *P. lotor*, and that human-mediated introduction must have taken place not too long ago.

[t] But see Fischer et al. (2015) for conflicting evidence on the dates and number of introductions.

21

Global Review of the Effects of Small Carnivores on Threatened Species

Michael V. Cove[1],* and Allan F. O'Connell[2]

[1] *Department of Applied Ecology and NC Cooperative Fish and Wildlife Research Unit, North Carolina State University, Raleigh, NC, USA*
[2] *USGS Patuxent Wildlife Research Center, Laurel, MD, USA*

SUMMARY

The absences of large carnivores from many ecosystems, human-induced landscape changes, and resource supplementation have been theorized to increase the abundance of small carnivore species around the world. Overabundant and/or unconstrained small carnivores can have significant effects on specific prey species that, in some cases, can cascade through entire ecosystems. Here, we review the effects of small carnivores on threatened species. We focus on four well-studied families (Procyonidae, Mephitidae, Mustelidae, and Herpestidae) and emphasize that this is a global conservation issue with consequences for biodiversity. We review and compare the impacts that small carnivores can have on a variety of prey taxa including small mammals, nesting avian and reptilian species, and rare invertebrates. We differentiate between native and exotic small carnivores because this is often an important distinction in terms of the impact severity and range of effects. In addition to direct lethal effects (i.e. predation), small carnivores can also impact threatened species as disease vectors and through competition or overexploitation, which can disrupt communities via ecological release or extinction. Furthermore, we explore other case studies in which small carnivores have had positive effects on threatened species and discuss studies that reveal other taxa responsible for exerting stronger negative effects on threatened prey. We offer some concluding remarks about global small carnivore conservation and emphasize the need for decision-analytic approaches and robust analyses that can improve our assessment of how populations of threatened species can be affected. To date, indirect effects are especially difficult to measure in the field and many studies have provided only anecdotal or correlative results, signalling a need for improving our scientific methodologies and management approaches.

Keywords

Herpestidae – invasive species – mesopredator release – Mephitidae – Mustelidae – predation – Procyonidae

Introduction

Nearly all members of the Order Carnivora (> 250 species worldwide) are predators and, through predation and interspecific competition, can exert a profound influence on biological communities (Treves & Karanth, 2003). Over the past few decades, there has been increasing interest in the complexity of predator interactions, the mechanisms involved and the resulting effects on biodiversity (Ritchie & Johnson, 2009; Prugh & Sivy, 2020). All too often, however, large carnivores (*sensu* predators) have captured the majority of attention from the scientific community due to their charismatic status, importance as keystone species in maintaining community structure (Morrison *et al.*, 2007), and societal fascination with large-bodied wildlife (Smith *et al.*, 2012 and references therein). Indeed, the importance of large

---

* Corresponding author.

Small Carnivores: Evolution, Ecology, Behaviour, and Conservation, First Edition. Edited by Emmanuel Do Linh San, Jun J. Sato, Jerrold L. Belant, and Michael J. Somers.
© 2022 John Wiley & Sons Ltd. Published 2022 by John Wiley & Sons Ltd.

carnivores as top predators cannot be overstated. It has been long known that their extinction can have far-reaching effects, prompting secondary extinctions, and disruption of established predator–prey relationships as suggested by the theory of trophic cascades (Terborgh & Estes, 2010; Hoeks *et al.*, 2020). Under these circumstances, small carnivores, typically relegated to subordinate roles within the structure and function of an ecosystem, can fill the role of apex or top predator (e.g. Mesopredator Release Hypothesis, MRH: Soulé *et al.*, 1988), especially in altered or island ecosystems. Although the mechanisms of competition and predation are relatively simple, their effects can be complex and small carnivores can generate their own sets of cascading effects throughout a system (Roemer *et al.*, 2009; Terborgh & Estes, 2010). Interestingly, these cascade-type effects initiated by small carnivores can be even more persistent than those induced by large (or apex) predators, since many small carnivores can exist at high densities, disperse long distances, and are adapted to a variety of environmental conditions (Palomares *et al.*, 1995). When system balance is disrupted, one of the most significant potential consequences is species endangerment. Although the ecological literature is replete with individual case histories of how small carnivores can impact threatened species, there has been relatively little in the way of a synthetic review of how small carnivores, as a guild, impact this group. There is also the possibility that the predator populations themselves can be affected once the system balance is upset.

In this chapter, we review the impacts that small carnivores can have on threatened species. We have focused our review and considered the differences between native and exotic small carnivores. The latter are also commonly referred to as introduced, alien, invasive, nuisance, adventive, or non-indigenous carnivores. Although technically there are differences among these terms, they are often used interchangeably. For uniformity in this chapter, we will use the term 'exotic' when referring to species living outside of their native range. Of course, the impacts from exotic carnivores have been well documented in ecology and spawned several theories and much debate. We review some of the key ecological theories (e.g. MRH) upon which these impacts have been based,

focusing on a few of the well-known, unique ecological settings where impacts are known to occur, concluding with some recommendations on methods and study design and areas for further research. Although the interactions can vary across time, space, and species, impacts are often due to the overabundance of the small carnivore species in question and/or their plasticity in adapting to changing conditions, where they either prey upon or out-compete the target threatened species. Furthermore, exotic species have long been regarded as a significant menace to threatened species (Vitousek *et al.*, 1997; Wilcove *et al.*, 1998; Mack *et al.*, 2000), and when small carnivores function in that role, they offer some of the best examples of trophic effects in the ecological world. Context is essential to this perspective, however, and it is important to point out that the effects of some small carnivores on threatened species are largely different where they function as native species as opposed to an environment where they have been introduced. In either case, conserving these threatened populations requires detailed information on the factors that affect survival (Conroy & Carroll, 2009).

The issue of small carnivore ecology is truly global in scope simply because small carnivores are cosmopolitan in their distribution. Accounts come from all over the map including North America (see accounts in this chapter), Europe (e.g. Santulli *et al.*, 2014), New Zealand (e.g. McLennan *et al.*, 1996; Dilks *et al.*, 2003; King & Powell, 2011) and a number of island systems where these effects are often intensified. Unfortunately, many small carnivore species are endemics of developing countries such as those in Asia and Africa, where their status and distribution have never been thoroughly assessed. Ironically, these very areas harbour the highest biodiversity, given the large tracts of rainforest and other important biomes they encompass, but these have now come under intense development pressures. In these areas, even the basic inventory information on small carnivore richness is still lacking. Remoteness, limited resources, varying cultural preferences (i.e. medicinal and fetish products), and the fact that many animal products are not sold making identification difficult, have limited our ability to gather such information (Doughty *et al.*, 2015). With the rapid development of several non-invasive sampling techniques, such as camera traps and hair snares (Long *et al.*, 2008), the basic inventory information about many of these enigmatic species is finally becoming available (e.g. Mathai *et al.*, 2010; Bahaa-el-din *et al.*, 2013; Greengrass, 2013; Kalle *et al.*, 2013; Gray *et al.*, 2014; Dhendup & Dorji, 2018; Hunowu *et al.*, 2020).

Although exotic predators are considered more dangerous than those that are native with respect to the decline and extinction of prey species (Salo *et al.*, 2007), the impact of native predators on prey can also be significant under certain conditions (e.g. MRH [Crooks & Soulé, 1999] or subsidized predators [Gompper & Vanak, 2008]). Furthermore, prey populations can vary in their response to predators, and so we focus on four well-studied families (Procyonidae, Mephitidae, Mustelidae, and Herpestidae), highlighting several case histories to demonstrate this variability. A common thread is that the sphere of human influence is itself pervasive, prompting an ever-increasing number of animal interactions that can adversely impact a variety of species, often with unforeseen consequences. From the most direct to the most subtle and complex interactions, natural systems that are out of balance inevitably trace back to the increasingly large human footprint that continues to impact the modern world (Estes *et al.*, 2011). Small carnivores and the effects that they can have on threatened species are no exception. To fully illustrate these relationships, we present a conceptual model (Figure 21.1) of the generalized trophic web showing the effects of small carnivores, loss of top predators, and other human-induced impacts on threatened species.

Mesopredator Release Hypothesis

Species interactions have spawned a number of interesting theories (e.g. apparent competition, intraguild predation, and competitive exclusion), but none have generated more interest over the last two decades than the MRH (e.g. Soulé *et al.*, 1988; Prugh *et al.*, 2009; Jachowski *et al.*, 2020). The theory is grounded in a series of occurrences where the absence of apex predators within an ecosystem allows smaller predators to increase their numbers, resulting in a concomitant reduction in a prey species and disruption of ecosystem function (for a full review, see Prugh

et al., 2009). Although hard, causal evidence is often lacking, the theory is frequently embedded in a variety of other associations (e.g. resource availability, niche partitioning), and such interactions have been purported to occur in both terrestrial and marine systems (Ritchie & Johnson, 2009). Mesopredator release often leads to negative effects on prey and is commonly reported in the context of trophic cascades, but is essentially an intraguild interaction among predators (Brashares *et al.*, 2010).

Small carnivores are often portrayed as the main culprits in the MRH equation (Ritchie & Johnson, 2009), with many of the species classified as generalist omnivores that are well adapted to co-existing with humans and existing on subsidized resources (Prange & Gehrt, 2004; Cove *et al.*, 2012a). However, given the rarity and endangered status of many of the carnivores within this group, generalizations are unproven. For example, of the 165 small carnivore species assessments done for *The IUCN Red List of Threatened Species*, over 25% ($n = 46$) are themselves listed as threatened to some degree (Schipper *et al.*, 2008).

Litvaitis & Villafuerte (1996) pointed out that top predators have been missing from much of their historic range in the eastern United States (US) for centuries, and many small carnivores such as northern raccoons, *Procyon lotor*, did not exhibit responses in abundance until more recent habitat modifications (e.g. bottom-up release). In their review of mesopredator release, Prugh *et al.* (2009) reported that 10 of the 17 North American small carnivores had experienced range increases compared to historic distributions, but that most of the increases were only marginal. However, small carnivores can also be ecologically 'released', when they are introduced either intentionally or accidentally (e.g. early attempts at biocontrol, escaped pets: Ikeda *et al.*, 2004; Hays & Conant, 2007), expanding their niche when predation or competition is relaxed, thus confirming the difference between realized and fundamental niche (Hutchinson, 1965). The prevalence (and intensity) of the MRH has also been linked to the maintenance of species diversity and system productivity (Brashares *et al.*, 2010). Small carnivores can benefit indirectly from agriculture and habitat modification by using the edge habitats created by land clearing to penetrate forests and capitalize on birds, insects, and small

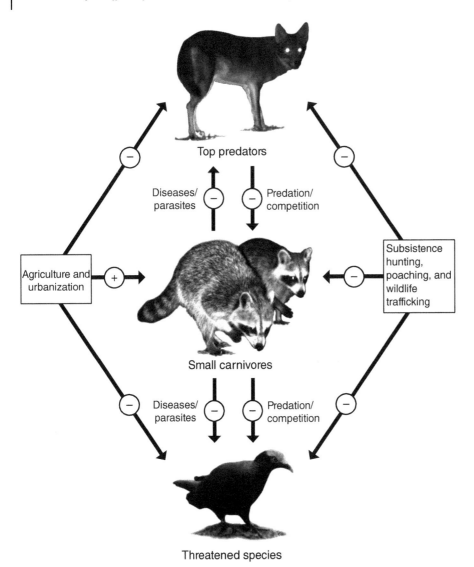

Figure 21.1 Conceptual model of a generalized trophic web with top predators (e.g. canids and felids) exerting strong effects on small carnivores (but also other members of the mesopredator guild) and small carnivores effecting threatened species (e.g. variety of taxa including some rare and specialized small carnivores themselves). Other effects include disease and parasite transmission from small carnivores to top predators and to threatened species. Additionally, the side boxes show the environmental influences and changes induced by humans and their relationships to the trophic web that are generally accepted thus far. Invasive species on islands would typically fit in the mesopredator level, but the web would likely be missing a top predator.

mammals that depredate crops and inhabit edge habitats (Dijak & Thomson, 2000; Cove *et al.*, 2012b, 2014). We suspect that regardless of circumstances, as abundances and distributional changes occur, the ecological release of small carnivores can play a large role in their effects on threatened species.

Procyonidae

The Procyonidae is a largely omnivorous taxa that are commonly cited to play major roles in mesopredator release and the decline of threatened species. The most recognizable procyonid and the most commonly cited species to affect other species and ecosystems is the northern raccoon. Raccoons are now ubiquitous across North America where they have been implicated as a major predator contributing to the decline of many threatened species (Garrott *et al.*, 1993; Goodrich & Buskirk, 1995), particularly on beach-nesting sea turtles and shorebirds (see section titled **Beach-nesting species**). The highly adaptable raccoon is now considered an important exotic species across portions of Europe, Russia, and Japan, where it is considered a threat to biodiversity (Lotze & Anderson, 1979; Ikeda *et al.*, 2004; García *et al.*, 2012). Many studies have focused on understanding the environmental drivers of raccoon occurrence and abundance because of their potential threat to migratory songbirds and other susceptible taxa, but few studies have quantified their direct effects on threatened species (Donovan *et al.*, 1997; Dijak & Thomson, 2000; Cove *et al.*, 2012b). Eagan *et al.* (2011) quantified the effects of reducing raccoon abundance, with results revealing that white-footed mice, *Peromyscus leucopus*, increased in density as a direct response to the removal. We discuss other experimental removals further in later sections.

Raccoons are expected to occur throughout most of Germany within the next 50 years, an expansion that is correlated with agriculture and habitat fragmentation. These predicted expansions into northeast Germany could pose a substantial threat to relict bog ecosystems with the critically endangered European pond turtle, *Emys orbicularis* (Fischer *et al.*, 2016). Similarly, raccoon expansion in Spain is predicted to have detrimental impacts on the endangered Spanish terrapin, *Mauremys leprosa*, waterfowl and various small mammals (García *et al.*, 2012). In Japan, raccoons have become naturalized throughout most of the country from at least three separate escape/release events since 1962 (Ikeda *et al.*, 2004). In this island setting, they prey on the endangered Japanese crayfish, *Cambaroides japonicus*, Ezo salamanders, *Hynobius retardatus*, and compete with Japanese raccoon dogs, *Nyctereutes procyonoides viverrinus* (Ikeda *et al.*, 2004). As the raccoon distribution in Japan expands, authorities fear that they may also prey on the endangered Japanese giant salamander, *Andrias japonicus*, although no studies have yet quantified the

exact role that raccoons play in the declines of threatened species across Japan.

High raccoon density is most associated with increasing predation on threatened species, but raccoons can also have strong negative effects even when occurring at low densities. For example, Hartman *et al.* (1997) observed that predation rates of ancient murrelets, *Synthliboramphus antiquus*, declined by 80% following the removal of three raccoons introduced to the Queen Charlotte Islands, Canada. Similarly, the annual mortality rates of the endangered Sandy Cay rock iguana, *Cyclura rileyi cristata*, increased from 35 to 67% between 1996 and 1997 during a period when a single raccoon was introduced to the island. The iguana population decline ended abruptly and the population has grown exponentially since the raccoon was removed (Hayes *et al.*, 2004, 2016).

While there is relatively extensive literature regarding northern raccoons and their effects on threatened species, there is yet limited information about other procyonids and their roles in trophic webs. It has been observed that white-nosed coatis, *Nasua narica*, respond to habitat changes similarly to raccoons, utilizing external resources and fragmented edge habitats and even increase in number as a result (Cove *et al.*, 2014; Pardo *et al.*, 2016). Other species of the Procyonidae likely respond similarly and their potential ecological release warrants further examination of their roles in threatened species conservation.

Mephitidae

The Mephitidae family includes 12 species of skunks and stink badgers. The most common species is the striped skunk, *Mephitis mephitis*, which is a dietary opportunist and occurs throughout North America (Wade-Smith & Verts, 1982). Other skunk species, including hog-nosed skunks, *Conepatus* spp., and spotted skunks, *Spilogale* spp., are also generalist omnivores, but the majority of their diets are composed of invertebrates, followed by small vertebrates and carrion (Kinlaw, 1995; Donadio *et al.*, 2004).

Few studies have effectively determined factors that influence the occurrence or abundance of striped skunks (but see Nichols *et al.*, 2008), or other skunk species, but aside from the *Spilogale* spp., representatives of this group are locally abundant (Dijak & Thompson, 2000). As common and potential nest predators, skunks are often believed to have strong effects on ground-nesting birds. However, Vickery *et al.* (1992) suggested that striped skunk predation of grassland bird nests in southern Maine, USA, was mostly incidental and not targeted. Striped skunks in this system forage mostly on invertebrates and only depredated nests opportunistically, which was further supported by the observation that avian remains made up less than 1.5% of the skunks' summer diet. In addition, several studies of shorebird nest success have shown that skunks are often opportunists and are not consistent in their predation of plover nests, *Charadrius* spp. (Loegering & Fraser, 1995; Neuman *et al.*, 2004), over large areas. Nevertheless, skunks can exert dramatic local effects and we suspect that their importance is largely site- and scale-dependent.

Mustelidae

Mustelids represent the largest family within the Carnivora and several species are of particular concern to threatened species, particularly where they have been introduced or somehow released inadvertently into the wild. The stoat, *Mustela erminea*, is ranked among the top 100 worst invasively destructive species (Lowe *et al.*, 2000). Stoats are voracious small predators and are most notably responsible for declines of threatened birds in mainland New Zealand and on several outlying islands (King & Murphy, 2005). Birds such as the kakapo, *Strigops habroptilus*, mohua, *Mohoua ochrocephala*, kokako, *Callaeas cinerea wilsoni*, and kiwis, *Apteryx* spp., are only a sample of the species affected by stoats. Basse *et al.* (1999) determined that recruitment among northern brown kiwis, *Apteryx mantelli*, was < 5% due to high predation of juveniles by stoats, despite low levels of predation on adult birds. Similarly, stoats were responsible for the majority of predation events of kaka, *Nestor meridionalis*, juveniles and adults on nests in an 11-year study of the species. The authors also noted that kaka survives in relatively high abundance on other islands without stoats, even in the presence of other exotic predators (Wilson *et al.*, 1998). In response to such

dramatic declines in native fauna, the New Zealand Department of Conservation has initiated several plans to eradicate stoats from important conservation islands such as Resolution Island, which could become the largest island sanctuary in the country if eradication is successful (Clayton *et al.*, 2011).

The American mink, *Neovison vison*, is native to North America where it is widely distributed and if the habitat is suitable, populations can reach reasonably high densities (1–6 individuals/km^2) (Larivière, 1999). Secretive by nature, the American mink is a generalist, voracious predator (typical of many mustelids) and is regarded as somewhat of an enigmatic species due to its cryptic colouration and semi-aquatic habits. American mink have been introduced to Europe for commercial fur farming where they escaped and are now considered an important exotic predator (Bonesi & Palazon, 2007). Mink have been implicated in accelerating the decline of several species including waterfowl, shorebirds, voles, and the endangered European mink, *Mustela lutreola* (Maran *et al.*, 1998). The decline of the European mink has been attributed to several pervasive factors such as habitat fragmentation and pollution and its range contraction was already underway before the American mink arrived in Europe (Santulli *et al.*, 2014). However, the two species are similar in many of their habits and the American mink effectively displaces the European mink through competitive exclusion. In Spain, Santulli *et al.* (2014) examined 12 years of occupancy data to determine that the presence of American mink increased the probability of extinction of its native European congener.

The sea otter, *Enhydra lutris*, is one of the largest members of the mustelid family and is well known for its role as a keystone species structuring ecological communities (Estes & Palmisano, 1974). It exerts a wide array of direct and indirect effects on species and is inexorably linked to trophic cascade theory (Anthony *et al.*, 2008). Depending on local circumstances, sea otters function as either apex predators or mesopredators (Estes *et al.*, 1998). Although the sea otter is itself listed as a threatened species in portions of its range, it has been implicated in predation on two endangered invertebrates, the white abalone, *Haliotis sorenseni*, and the black abalone, *H. carcherodii*. Although abalones are considered an important food source for sea otters, there is no evidence that otter predation alone has resulted in the extirpation of these invertebrates (NMFS, 2008). This premise is based on the fact that sea otters co-exist with several other species of abalone (Hines & Pearse, 1982; Rogers-Bennett *et al.*, 2007). However, there is speculation that sea otters and other predators exert enough pressure on abalone to restrict them to low densities, or to areas where otters cannot reach them. This scenario has come to be known as a predator pit dynamic (Holling, 1959; Bakun, 2006) in which predation drives the prey to low levels, at which the predation rate declines.

Herpestidae

The Herpestidae also represents a relatively diverse family of Carnivora, and their impacts on threatened species have made them infamous due in large part to their intended roles as biological control agents for snakes and rodents (Hays & Conant, 2007). Such introductions have created one of the world's worst invasive species, the small Indian mongoose, *Urva auropunctata* [*sensu* Veron *et al.*, 2007; see also Veron *et al.*, Chapter 3, this volume; previously *Herpestes javanicus*] (Lowe *et al.*, 2000). This species has been succesfully (and often intentionally) introduced to > 60 islands and at least two continents (see Gantchoff *et al.*, Chapter 20, this volume). As a result, it has become the focus of numerous eradication attempts (Hays & Conant, 2007; Barun *et al.*, 2011). The most noteworthy eradication efforts have occurred in Japan where the species kills a variety of threatened taxa (mammals, birds, amphibians) and has significantly impacted the endangered Amami rabbit, *Pentalagus furnessi*. The list of threatened species impacted by this carnivore is long and varied and includes the hawksbill sea turtle, *Eretmochelys imbricata* (eggs), the Mauritian pink pigeon, *Nesoenas mayeri*, the Jamaican petrel, *Pterodroma caribbaea*, and Audubon's shearwater, *Puffinus lherminieri* (GISD, 2011).

Mongoose management and eradication plans require reliable and rapid identification of incipient populations, but apparent gaps in local understanding of biosecurity and conservation can slow the reporting of these exotic predators in novel areas. For example, Watari *et al.* (2011) identified a previously undocumented population of small Indian mongoose on Kyushu Island, Japan, that had been established for at least 30 years prior to identification by the scientific community. Although long-term success for small Indian mongoose eradication is uncertain based on historical assessments (Barun *et al.*, 2011), new modelling efforts using a Bayesian framework (Fukasawa *et al.*, 2013) may offer more insight as to the efficacy of control efforts and hope for the future.

Beach-Nesting Species

With respect to the effects of small carnivores on threatened species, no single issue has been given more attention than predation of beach-nesting birds and sea turtles (Stancyk, 1982; Loegering & Fraser, 1995; Ratnaswamy *et al.*, 1997; Ratnaswamy & Warren, 1998; Martin *et al.*, 2010). Species such as piping plovers, *Charadrius melodus*, least terns, *Sternula antillarum* (endangered in some U.S. states), and loggerhead, *Caretta caretta*, green, *Chelonia mydas*, leatherback, *Dermochelys coriacea*, and hawksbill, sea turtles nest along the Atlantic and Gulf coasts of the US and throughout the Caribbean, a habitat that has been severely reduced and modified over the past century. This reduction of habitat and external food resources from beach recreation have led to increased exposure for shorebirds, turtles, and their nests to small carnivores such as raccoons, skunks, and the small Indian mongoose. Raccoons, of course, have been extensively studied especially with respect to sea turtles (Ratnaswamy & Warren, 1998; Barton & Roth, 2008; Waldstein, 2010) including the efficacy of predator removal methods (Ratnaswamy & Warren, 1998; Engeman *et al.*, 2003, 2006). Also, an in-depth behavioural study has been conducted to examine mongoose predation on Barbados (Leighton *et al.*, 2008, 2011; Leighton, 2009) that demonstrated the effect of an ecological trap for nesting turtles (Gates & Gysel, 1978; Schlaepfer *et al.*, 2002; Robertson & Hutto, 2006). In any case, many management plans now require that predator exclosures, electric fences, or deterrents, effective in many different systems, be used to keep predators away from nests and incubating adult birds (Melvin *et al.*, 1992). However, Dinsmore *et al.* (2014)

recommended that despite the success of nest exclosures in the recovery of western snowy plovers, *Charadrius alexandrinus nivosus*, in Oregon, exclosure use be limited to reduce unnecessary adult depredation. Stocking *et al.* (2017) evaluated the effectiveness of a large-scale raccoon removal to reduce nest loss of American oystercatchers, *Haematopus palliates*, on a barrier island off the coast of North Carolina, USA. They detected limited support that the raccoon removal was associated with a response in nest survival and suggested that the removal effort (~50% of the population) might have been inadequate or that a few "problem individuals" could be responsible for the majority of nest depredations. The effectiveness of removal programmes (Côté & Sutherland, 1997) and the design of sound monitoring protocols (Nichols & Williams, 2006), among other things, require further study particularly as conditions vary by locality and species.

Beyond Predation – Other Effects of Small Carnivores

Aside from predation, there are also a number of direct and indirect effects that small carnivores can have on threatened species, such as competition and disease transmission. These relationships are not as straightforward as they may seem. For example, competition among small carnivores and threatened species is not always driven by food resources, but can also depend on the availability of shelter and microhabitat refugia. In particular, nesting and denning sites can be limiting factors for many rare and endangered species, including rare small carnivores (e.g. eastern spotted skunks, *Spilogale putorius*; Lesmeister *et al.*, 2008). Indeed, generalist species such as northern raccoons and striped skunks may out-compete rare species such as the smaller-sized eastern spotted skunk, particularly in fragmented or modified habitats with limited nest sites.

There is also a high potential for intraguild effects because nearly 25% of the extant small carnivores are themselves classified as threatened species (see Do Linh San *et al.*, Chapter 1, this volume). The threatened status of these species is often a consequence of specialized diets and/or habitat degradation and fragmentation, placing the endangered small carnivores

at risk from the exploitative competition with the generalist or exotic small carnivores. For example, the majority of Malagasy small carnivores (Eupleridae) are classified as Vulnerable or Endangered (IUCN, 2020; see also Appendix A) and much of their prey are also threatened, leading to inter- and intraspecific competition within this guild. Gerber *et al.* (2012) observed shifts in the preferred activity periods of the native ring-tailed vontsira, *Galidia elegans*, in the presence of domestic/feral dogs, *Canis familiaris*, and the exotic small Indian civet, *Viverricula indica*, suggesting strong competition. Farris *et al.* (Chapter 13, this volume) reported that the small Indian civet also spatially excludes the spotted fanaloka, *Fossa fossana*, particularly in degraded forests. In fact, multi-year, guild-wide surveys in the rainforests of Madagascar have recently demonstrated that exotic carnivores are progressively replacing native carnivores. For example, no less than four (ring-tailed vontsira, broadstriped vontsira *Galidictis fasciata*, brown-tailed vontsira, *Salanoia concolor*, and falanouc, *Eupleres goudotii*) of the six native small carnivore species present in Makira Natural Park underwent a decrease in the probability of occupancy of at least 60% over a six-year study period, while the exotic small Indian civet and feral/wild cats *Felis* sp. colonized the study site (Farris *et al.*, 2017a). Similarly, the presence of domestic dogs in areas bordering Ranomafana National Park seems to be the cause of the rapid decline of ring-tailed vontsira and fosa, *Cryptoprocta ferox*, through interspecific competition (exploitative competition, kleptoparasitism, killing/predation) or disease transmission (Farris *et al.*, 2017b, 2020).

Pandemic diseases and disease ecology are areas that have precisely received more attention in the past several years (Zinsstag *et al.*, 2011). In fact, the majority of small carnivore research in the US stems from studies aimed at understanding rabies and other zoonotic diseases. Rabies (*Lyssavirus*) and canine distemper virus (CDV; *Morbillivirus*) are two of the most infectious carnivore pathogens worldwide, which both often prove to be fatal. Small carnivores, particularly those occurring at abnormally high abundance often serve as the reservoirs and vectors for these and other various pathogens. It is plausible that as reservoirs, small carnivores might expose and infect rare and threatened large carnivores and sympatric small carnivores through direct contact or their consumption by the larger guild members (Deem *et al.*, 2000; Rasambainarivo *et al.*, 2017; see also Fournier-Chambrillon *et al.*, Chapter 11, this volume). For example, the canine distemper virus has recently been determined to be a significant infectious disease of the Critically Endangered Amur tiger, *Panthera tigris altaica* (Seimon *et al.*, 2013). Though it is probable that this disease is mainly spread via feral dogs, the overlap and susceptibility of all carnivore families makes this disease a threat to all endangered carnivores and is suggestive that small carnivores likely serve as a reservoir for CDV, among other diseases. A CDV outbreak was also suggested to cause the dramatic decline of a European mink population in Navarre, Spain (Fournier-Chambrillon *et al.*, Chapter 11, this volume). Dogs and red foxes, *Vulpes vulpes*, were implicated as the main reservoirs for the disease in that study area, but the authors noted that stone martens, *Martes foina*, and European polecats, *Mustela putorius*, also serve as disease vectors affecting the Critically Endangered mink.

In addition to CDV and rabies, raccoons can carry high parasite loads (Wright & Gompper, 2005). Raccoon roundworm, *Baylisascaris procyonis*, affects intermediate hosts including small mammals. It has long been suggested that the parasitic nematode was responsible for the extirpation of the Allegheny woodrat, *Neotoma magister*, in its northern range in the US (Balcom & Yahner, 1996). This hypothesis was further supported by observations of high mortality rates among reintroduced woodrats in areas with high raccoon-roundworm contamination (Logiudice, 2003).

Are Small Carnivores Always to Blame for Impacts on Threatened Species?

Not all small carnivore effects are negative for threatened species and often trophic interactions are difficult to elucidate without removal experiments (Salo *et al.*, 2010). In some systems, small carnivores have ascended to the role of an apex predator and exert strong effects through the community. One Florida study of loggerhead sea turtle, nest predation revealed that ghost crabs, *Ocypode quadrata*, were the most common nest predators of the endangered turtles in areas with few raccoons. Raccoons reduced ghost crab abundance and although they depredated nests as well, their role in regulating the intermediate predators had a more pronounced effect on nesting success (Barton & Roth, 2008). Courchamp *et al.* (1999) used mathematical equations to model similar mesopredator release when the superpredators and mesopredators were both exotic species and the preys were endangered seabirds. The models predicted that following superpredator removal, mesopredators would increase in abundance and exert stronger negative effects on seabird nests. Empirical evidence for such an effect was found by Rayner *et al.* (2007). This relationship may be the case in areas where exotic small carnivores (i.e. small Indian mongooses, stoats) act as superpredators over rats (mesopredators) in seabird nest habitats worldwide and is useful to consider before determining management strategies to protect threatened birds (Courchamp *et al.*, 2003).

Although many studies have revealed that predation often plays a significant role in species declines, there is now evidence that unless supplemented or invasive, small carnivores are not typically the culprits. Predation, sometimes from other members of the small carnivore guild, is a leading cause of mortality of the endangered black-footed ferret, *Mustela nigripes*. In a study of steppe polecat, *Mustela eversmanii*, and black-footed ferret survival, American badgers, *Taxidea taxus*, were implicated in 30% of polecat mortalities, but only 5.5% of ferret mortalities (Eads *et al.*, 2013). Badgers utilized areas actively used by ferrets, which still suggests that they depredate the smaller carnivores or steal their prey, but the endangered small carnivores were much more susceptible to predation by coyotes, *Canis latrans* (67.1% of mortalities). Furthermore, badgers might release the endangered ferrets from exploitative competition from the exotic polecats. In southeastern Alberta, Canada, badgers were only responsible for 3% of artificial nest predation events of the endangered greater sage grouse, *Centrocercus urophasianus*, whereas 83% of predation events were attributed to ground squirrels and avian predators (Watters *et al.*, 2002). Similarly, badgers were only identified as predators of seven sage grouse nests of 87 monitored in a northern Nevada and southern Idaho study (Coates *et al.*, 2008).

With badger depredations making up such a small proportion of nest failures and black-footed ferret mortalities, badgers may not be strong drivers of extirpation or extinction of these threatened species, at least in these particular systems. Examples of small carnivores regulating smaller predators on islands and beaches, as well as evidence that badgers were not commonly responsible for the consumption of several endangered species, further suggest that the effects of small carnivore predation on threatened species are often largely context- and site-specific.

Aside from small carnivores, the loss of large predators has led to the ecological release of various other taxa that can also exert strong effects on threatened species. In their experiment to examine the effectiveness of chemical repellents to deter stone martens, red foxes, common genets, *Genetta genetta*, and wild boars, *Sus scrofa*, Vilardell *et al.* (2008) found that all Hermann's tortoise, *Testudo hermanni*, nests were depredated within 4 days. However, the authors observed wild boars to be responsible for the majority of depredations and stone martens and genets were only responsible for 2.5% of depredations in their experiment with carnivore-specific repellent. Even when boar-specific repellents were utilized, 84.7% of nests were still depredated by the wild boars further demonstrating that small carnivores were not the main offenders. Other mesopredators (felids and canids) often have much stronger effects on threatened species than small carnivores belonging to the four main families reviewed in this chapter. Feral cats, for example, have been implicated in the extinction of > 30 island endemic species (Nogales *et al.*, 2013), while red foxes, also are suggested to have driven extirpations of many Australian native species (Burbidge & Manly, 2002; Saunders *et al.*, 2010). Furthermore, multiple studies in coastal systems of the US have revealed that red foxes are the most common mammalian predators of threatened shorebird nests (Loegering & Fraser, 1995; Neuman *et al.*, 2004).

Concluding Remarks

Fundamental to all species conservation and management is gathering reliable information on the abundance and/or density of the populations that we study.

These parameters, more than any others, are key to species survival and longevity, whether they are predator or prey. Because most carnivores are secretive and elusive, and therefore difficult to sample, biologists often find it relatively easy to collect information on these species in the form of an unadjusted count or index that they attempt to strongly correlate with abundance or density. The use of indices is attractive because estimation techniques such as mark-recapture are not always practical or cost-efficient to robustly determine population parameters. However, it is well known that animal counts vary from the proportion of the true number counted and that this detection process (detectability) varies across space and time (MacKenzie *et al.*, 2002). Ignoring detectability with respect to counts of small carnivores (or any population for that matter) and extrapolating from the index to abundance or density of the entire population places the interpretation of such results on unstable ground with respect to making inference about the target population(s) (Guillera-Arroita *et al.*, 2014). This is especially true when the count or index is further linked to an impact on threatened species in the case of localized predation. For example, Schmidt *et al.* (2010) suggested that the northern raccoon was a serious threat to the endangered Lower Keys marsh rabbit, *Sylvilagus palustris hefneri*, in the Lower Keys of Florida, USA, largely based on index data in the form of track counts at study quadrats. In this case, despite no direct evidence for raccoon predation on this endangered lagomorph, an index of abundance was used to implicate raccoons as the key component in a model aimed at recovery of the endangered species. Even if these purported relationships were true, estimators that incorporate detectability into their analysis need to be given serious consideration when sampling animal populations if capture–recapture methods are not applicable. Convenience or opportunistic sampling does not lend itself to strong inference in the face of imperfect detection (e.g. Gompper *et al.*, 2006; Stein *et al.*, 2008; Güthlin *et al.*, 2014). In the case of basic inventories (which are still necessary for many species of small carnivores in remote areas), convenience or index-type sampling can be appropriate but we recommend that the goals and objectives of a particular study or survey be clearly articulated. Addressing the detection process in biological sampling must consider the assumptions under which proposed methods operate to ensure that they are valid. We recommend that, where possible, trapping and marking of individuals for capture–recapture experiments, and use of techniques such as occupancy modelling (MacKenzie *et al.*, 2006) at large scales be utilized to accurately assess the status of small carnivore populations and quantify their effects on threatened species. Finally, the complexity of species interactions and the systems they inhabit often result in management decisions that are intractable. Over the past two decades, Structured Decision Making (SDM) (Keeney & Raiffa, 1976; Ralls & Starfield 1995; Keeney, 1996; Gregory & Long, 2009), an umbrella concept covering a broad array of decision-analytical techniques, has become increasingly popular in natural resource conservation, especially with respect to the management of threatened species (Runge, 2011). Decision analysis typically revolves around a 5-step process (problem, objectives, alternatives, consequences, and trade-offs; Hammond *et al.*, 1999) but can involve other components such as facilitation or coping with uncertainty. In the case of threatened species management, two features are commonplace: decisions are recurrent, in that they need to be made repeatedly over time, and the decision almost always is plagued by some uncertainty (Runge, 2011). Under these conditions, decision analysis takes the form of adaptive management, a special case of SDM. Although there have been relatively few instances of applying SDM to predator–prey interactions, SDM can provide credible guidance for the modelling and monitoring

of population changes for small carnivores in the context of threatened species management in an adaptive framework (Martin *et al.*, 2010), and we recommend more of this approach in the future to ensure the viability of both predator and prey.

Small carnivores can exert strong effects on threatened species and hence it is important for land managers and biologists to consider strategies that reduce their effects by mitigation (e.g. fences, exclosures, and deterrents) or removal and eradication (Garrott *et al.*, 1993; Goodrich & Buskirk, 1995), but additional predator removal experiments are necessary to fully understand the dynamics of the system, especially when the predator is native to the system (Salo *et al.*, 2010). It is clear that a number of small carnivores – mostly when introduced – can exert significant impacts on some of the world's most threatened and/or endangered species. However, the current literature does not reveal strong evidence that small carnivores are the major drivers in the decline of many species, but instead are grouped into a suite of confounding factors which are mostly consequences of human-induced land cover changes, large predator reduction, and poaching/wildlife trafficking (Figure 21.1). Trophic cascades have been well documented in nature but the role of small carnivores needs further work to fully understand top-down and bottom-up roles in the ecological theory. How effectively scientists, conservationists, governments, and society at large, will study, understand, collaborate and move forward to meet the ecological needs of these mammals will determine where and how many species and populations will survive (Karanth & Chellam, 2009).

References

Anthony, R.G., Estes, J.A., Ricca, M.A., Miles, A.K. & Forsman, E.D. (2008) Bald eagles and sea otters in the Aleutian archipelago: indirect effects of trophic cascades. *Ecology* 89, 2725–2735.

Bahaa-el-din, L., Henschel, P., Aba'a, R., Abernethy, K., Bohm, T., Bout, N., Coad, L., Head, J., Inoue, E., Lahm, S., Lee, M.E., Maisels, F., Rabanal, L., Starkey, M., Taylor, G., Vanthomme, H., Nakashima, Y. & Hunter, L. (2013) Notes on the distribution and status of

small carnivores in Gabon. *Small Carnivore Conservation* 48, 19–29.

Bakun, A. (2006) Wasp-waist populations and marine ecosystem dynamics: navigating the "predator pit" topographies. *Progress in Oceanography* 68, 271–288.

Balcom, B.J. & Yahner, R.H. (1996) Microhabitat and landscape characteristics associated with the threatened Allegheny woodrat. *Conservation Biology* 10, 515–525.

Barton, B.T. & Roth, J.D. (2008) Implications of intraguild predation for sea turtle nest protection. *Biological Conservation* 141, 2139–2145.

Barun, A., Hanson, C.C., Campbell, K.J. & Simberloff, D. (2011) A review of small Indian mongoose management and eradications on islands. In: *Island Invasives: Eradication and Management* (eds C.R. Veitch, M.N. Clout & D.R. Towns), pp. 17–25. IUCN, Gland.

Basse, B., McLennan, J.A. & Wake, G.C. (1999) Analysis of the impact of stoats, *Mustela erminea*, on northern brown kiwi, *Apteryx mantelli*, in New Zealand. *Wildlife Research* 26, 227–237.

Bonesi, L. & Palazon, S. (2007) The American mink in Europe: status, impacts, and control. *Biological Conservation* 134, 470–483.

Brashares, J.S., Epps, C.W. & Stoner, C.J. (2010) Ecological and conservation implications of mesopredator release. In: *Trophic Cascades* (eds J. Terborgh & J. Estes), pp. 221–240. Island Press, Washington, DC.

Burbidge, A.A. & Manly, B.F. (2002) Mammal extinctions on Australian islands: causes and conservation implications. *Journal of Biogeography* 29, 465–473.

Clayton, R.I., Byrom, A.E., Anderson, D.P., Edge, K.A., Gleeson, D., McMurtrie, P. & Veale, A. (2011) Density estimates and detection models inform stoat (*Mustela erminea*) eradication on Resolution Island, New Zealand. In: *Island Invasives: Eradication and Management* (eds C.R. Veitch, M.N. Clout & D.R. Towns), pp. 413–417. IUCN, Gland.

Coates, P.S., Connelly, J.W. & Delehanty, D.J. (2008) Predators of Greater Sage-Grouse nests identified by video monitoring. *Journal of Field Ornithology* 79, 421–428.

Conroy, M.J. & Carroll, J.P. (2009). *Quantitative Conservation of Vertebrates*. Wiley-Blackwell, Oxford.

Côté, I.M. & Sutherland, W.J. (1997) The effectiveness of removing predators to protect bird populations. *Conservation Biology* 11, 395–405.

Courchamp, F., Langlais, M. & Sugihara, G. (1999) Cats protecting birds: modelling the mesopredator release effect. *Journal of Animal Ecology* 68, 282–292.

Courchamp, F., Chapuis, J.-L. & Pascal, M. (2003) Mammal invaders on islands: impact, control and control impact. *Biological Reviews* 78, 347–383.

Cove, M.V., Jones, B.M., Bossert, A.J., Clever, D.R. Jr, Dunwoody, R.K., White, B.C. & Jackson, V.L. (2012a) Use of camera traps to examine the mesopredator release hypothesis in a fragmented Midwestern landscape. *American Midland Naturalist* 168, 456–465.

Cove, M.V., Niva, L.M. & Jackson, V.L. (2012b) Use of probability of detection when conducting analyses of surveys of mesopredators: a case study from the Ozark Highlands of Missouri. *The Southwestern Naturalist* 57, 257–261.

Cove, M.V., Spínola, R.M., Jackson, V.L. & Saénz, J.C. (2014) The role of fragmentation and landscape changes in the ecological release of common nest predators in the Neotropics. *PeerJ* 2, e464.

Crooks, K.R. & Soulé, M.E. (1999) Mesopredator release and avifaunal extinctions in a fragmented system. *Nature* 400, 563–566.

Deem, S.L., Spelman, L.H., Yates, R.A. & Montali, R.J. (2000) Canine distemper in terrestrial carnivores: a review. *Journal of Zoo and Wildlife Medicine* 31, 441–451.

Dhendup, T. & Dorji, R. (2018) Camera-trap records of small carnivores from eastern Cambodia, 1999–2013. *Small Carnivore Conservation* 56, 36–41.

Dijak, W.D. & Thompson III, F.R. (2000) Landscape and edge effects on the distribution of mammalian predators in Missouri. *The Journal of Wildlife Management* 64, 209–216.

Dilks, P., Willans, M., Pryde, M. & Fraser, I. (2003) Large scale stoat control to protect mohua (*Mohoua ochrocephala*) and kaka (*Nestor meridionalis*) in the Eglinton Valley, Fiordland, New Zealand. *New Zealand Journal of Ecology* 27, 1–9.

Dinsmore, S.J., Lauten, D.J., Castelein, K.A., Gaines, E.P. & Stern, M.A. (2014) Predator exclosures, predator removal, and habitat improvement increase nest success of Snowy Plovers in Oregon, USA. *The Condor* 116, 619–628.

Do Linh San, E., Sato, J.J., Belant, J.L. & Somers, M.J. (Chapter 1, this volume) The world's small carnivores: definitions, richness, distribution, conservation status, ecological roles, and research efforts. In *Small Carnivores: Evolution, Ecology, Behaviour, and Conservation* (eds E. Do Linh San, J.J. Sato, J.L. Belant & M.J. Somers. Wiley-Blackwell, Oxford.

Donadio, E., Martino, S.D., Aubone, M. & Novaro, A.J. (2004) Feeding ecology of the Andean hog-nosed skunk (*Conepatus chinga*) in areas under different land use in north-western Patagonia. *Journal of Arid Environments* 56, 709–718.

Donovan, T.M., Jones, P.W., Annand, E.M. & Thompson III, F.R. (1997) Variation in local-scale edge effects: mechanisms and landscape context. *Ecology* 78, 2064–2075.

Doughty, H.L., Karpanty, S.M. & Wilbur, H.M. (2015) Local hunting of carnivores in forested Africa: a meta-analysis. *Oryx* 49, 88–95.

Eads, D.A., Biggins, D.E., Livieri, T.M. & Millspaugh, J.J. (2013) American badgers selectively excavate burrows in areas used by black-footed ferrets: implications for predator avoidance. *Journal of Mammalogy* 94, 1364–1370.

Eagan, T.S., Beasley, J.C., Olson, Z.H. & Rhodes, O.E. Jr (2011) Impacts of generalist mesopredators on the demography of small-mammal populations in fragmented landscapes. *Canadian Journal of Zoology* 89, 724–731.

Engeman, R.M., Martin, R.E., Constantin, B., Noel, R. & Woolard, J. (2003) Monitoring predators to optimize their management for marine turtle nest protection. *Biological Conservation* 113, 171–178.

Engeman, R.M., Martin, R.E., Smith, H.T., Woolard, J., Crady, C.K., Constantin, B., Stahl, M. & Groninger, N.P. (2006) Impact on predation of sea turtle nests when predator control was removed midway through the nesting season. *Wildlife Research* 33, 187–192.

Estes, J.A. & Palmisano, J.F. (1974) Sea otters: their role in structuring nearshore communities. *Science* 185, 1058–1060.

Estes, J.A., Tinker, M.T., Williams, T.M. & Doak, D.F. (1998) Killer whale predation on sea otters linking oceanic and nearshore ecosystems. *Science* 282, 473–476.

Estes, J.A., Terborgh, J., Brashares, J.S., Power, M.E., Berger, J. *et al.* (2011) Trophic downgrading of planet Earth. *Science* 333, 301–306.

Farris, Z.J., Kelly, M.J., Karpanty, S., Murphy, A., Ratelolahy, F., Andrianjakarivelo, V. & Holmes, C. (2017a) The times they are a changin': multi-year surveys reveal exotics replace native carnivores at a Madagascar rainforest site. *Biological Conservation* 206, 320–328.

Farris, Z.J., Gerber, B.D., Valenta, K., Rafaliarison, R., Razafimahaimodison, J.C., Larney, E., Rajaonarivelo, T., Randriana, Z., Wright, P.C. & Chapman, C.A. (2017b) Threats to a rainforest carnivore community: a multi-year assessment of occupancy and co-occurrence in Madagascar. *Biological Conservation* 210, 116–124.

Farris, Z.J., Gerber, B.D., Karpanty, S., Murphy, A., Wampole, E., Ratelolahy, F. & Kelly, M.J. (2020) Exploring and interpreting spatiotemporal interactions between native and invasive carnivores across a gradient of rainforest degradation. *Biological Invasions* 22, 2033–2047.

Farris, Z.J., Gerber, B.D., Ratelolahy, F., Andrianjakarivelo, V. & Kelly, M.J. (Chapter 3, this volume) Spatio-temporal overlap between a native and an exotic carnivore in Madagascar: evidence of spatial exclusion. In: *Small Carnivores: Evolution, Ecology, Behaviour, and Conservation* (eds E. Do Linh San, J.J. Sato, J.L. Belant & M.J. Somers). Wiley–Blackwell, Oxford.

Fischer, M.L., Sullivan, M.J., Greiser, G., Guerrero-Casado, J., Heddergott, M., Hohmann, U., Keuling, O., Lang, J., Martin, I., Michler, F., Winter, A. & Klein, R. (2016) Assessing and predicting the spread of non-native raccoons in Germany using hunting bag data and dispersal weighted models. *Biological Invasions* 18, 57–71.

Fournier-Chambrillon, C., Ceña, J.-C., Urra Maya, F., van de Bildt, M., Ferreras, M.C., Giralda-Carrera, G., Kuiken, T., Buisson, L., Palomares, F. & Fournier, P. (Chapter 11, this volume) A 9-year demographic and health survey of a European mink population in Navarre (Spain): role of the canine distemper virus. In: *Small Carnivores: Evolution, Ecology, Behaviour, and Conservation* (eds E. Do Linh San, J.J. Sato, J.L. Belant & M.J. Somers). Wiley–Blackwell, Oxford.

Fukasawa, K., Hashimoto, T., Tatara, M. & Abe, S. (2013) Reconstruction and prediction of invasive mongoose population dynamics from history of introduction and management: a Bayesian state-space modelling approach. *Journal of Applied Ecology* 50, 469–478.

Gantchoff, M., Libal, N.S. & Belant, J.L. (Chapter 20, this volume) Small carnivore introductions: ecological and biological correlates of success. In: *Small Carnivores: Evolution, Ecology, Behaviour, and Conservation* (eds E. Do Linh San, J.J. Sato, J.L. Belant & M.J. Somers). Wiley–Blackwell, Oxford.

García, J.T., García, F.J., Alda, F., González, J.L., Aramburu, M.J., Cortés, Y., Prieto, B., Pliego, B., Pérez, M., Herrera, J. & García-Román, L. (2012) Recent invasion and status of the raccoon (*Procyon lotor*) in Spain. *Biological Invasions* 14, 1305–1310.

Garrott, R.A., White, P.J. & Vanderbilt White, C.A. (1993) Overabundance: an issue for conservation biologists? *Conservation Biology* 7, 946–949.

Gates, J.E. & Gysel, L.W. (1978) Avian nest dispersion and fledging success in field-forest ecotones. *Ecology* 59, 871–883.

Gerber, B.D., Karpanty, S.M. & Randrianantenaina, J. (2012) Activity patterns of carnivores in the rain forests of Madagascar: implications for species coexistence. *Journal of Mammalogy* 93, 667–676.

GISD [Global Invasive Species Database] (2011) Invasive Species Specialist Group of the IUCN Species Survival Commission. http://www.issg.org/database. Accessed on 23 February 2017.

Gompper, M.E. & Vanak, A.T. (2008) Subsidized predators, landscapes of fear and disarticulated carnivore communities. *Animal Conservation* 11, 13–14.

Gompper, M.E., Kays, R.W., Ray, J.C., Lapoint, S.D., Bogan, D.A. & Cryan, J.R. (2006) A comparison of noninvasive techniques to survey carnivore communities in northeastern North America. *Wildlife Society Bulletin* 34, 1142–1151.

Goodrich, J.M. & Buskirk, S.W. (1995) Control of abundant native vertebrates for conservation of endangered species. *Conservation Biology* 9, 1357–1364.

Gray, T.N.E., Pin, C., Phan, C., Crouthers, R., Kamler, J.F. & Prum, S. (2014) Camera-trap records of small carnivores from eastern Cambodia, 1999–2013. *Small Carnivore Conservation* 50, 20–24.

Gregory, R. & Long, G. (2009). Using structured decision making to help implement a precautionary approach to endangered species management. *Risk Analysis* 29, 518–532.

Greengrass, E.J. (2013) A survey of small carnivores in the Putu Mountains, southeast Liberia. *Small Carnivore Conservation* 48, 30–36.

Guillera-Arroita, G., Lahoz-Monfort, J.J., MacKenzie, D.I., Wintle, B.A. & McCarthy, M.A. (2014) Ignoring imperfect detection in biological surveys is dangerous: a response to "Fitting and interpreting occupancy models". *PLoS One* 9, e99571.

Güthlin, D., Storch, I. & Küchenhoff, H. (2014) Toward reliable estimates of abundance: comparing index methods to assess the abundance of a mammalian predator. *PLoS One* 9, e94537.

Hammond, J.S., Keeney, R.L. & Raiffa, H. (1999) *Smart Choices: A Practical Guide to Making Better Decisions*. Harvard Business Press, Boston.

Hartman, L.H., Gaston, A.J. & Eastman, D.S. (1997) Raccoon predation on ancient murrelets on east Limestone Island, British Columbia. *The Journal of Wildlife Management* 61, 377–388.

Hayes, W.K., Carter, R.L., Cyril, S. Jr & Thornton, B. (2004) Conservation of an endangered Bahamian rock iguana. I. Population assessments, habitat restoration, and behavioral ecology. In: *Iguanas: Biology and Conservation* (eds A.C. Alberts, R.L. Carter, W.K. Hayes & E.P. Martins), pp. 232–257. University of California Press, Berkeley.

Hayes, W.K., Escobar, R.A., Fry, S.K., Fortune, E.M., Wasilewski, J.A., Tuttle, D.M., West, K.S., Iverson, J.B., Buckner, S.D. & Carter, R.L. (2016) Conservation of the endangered Sandy Cay rock iguanas (*Cyclura rileyi cristata*): invasive species control, population response, pirates, poaching, and translocation. *Herpetological Conservation and Biology* 11, 106–120.

Hays, W.S. & Conant, S. (2007) Biology and impacts of Pacific Island invasive species. 1. A worldwide review of effects of the small Indian mongoose, *Herpestes javanicus* (Carnivora: Herpestidae). *Pacific Science* 61, 3–16.

Hines, A.H. & Pearse, J.S. (1982) Abalones, shells, and sea otters: dynamics of prey populations in central California. *Ecology* 63, 1547–1560.

Hoeks, S., Huijbregts, M.A.J., Busana, M., Harfoot, M.B.J., Svenning, J.-C. & Santini, L. (2020) Mechanistic insights into the role of large carnivores for ecosystem structure and functioning. *Ecography* 43, 1752–1763.

Holling, C.S. (1959) Some characteristics of simple types of predation and parasitism. *Canadian Entomologist* 91, 385–398.

Hunowu, I., Patandung, A., Pusparini, W., Danismend, I., Cahyana, A., Abdullah, S., Johnson, C.L., Hilser, H., Rahasia, R., Gawina, J. & Linkie, M. (2020) New insights into Sulawesi's apex predator: the Sulawesi civet *Macrogalidia musschenbroekii*. *Oryx* 54, 878–881.

Hutchinson, G.E. (1965) *The Ecological Theater and the Evolutionary Play*. Yale University Press, New Haven.

Ikeda, T., Asano, M., Matoba, Y. & Abe, G. (2004) Present status of invasive alien raccoon and its impact in Japan. *Global Environmental Research* 8, 125–131.

IUCN [International Union for the Conservation of Nature] (2020) *The IUCN Red List of Threatened Species*. Version 2020-1. http://www.iucnredlist.org. Accessed on 15 March 2020.

Jachowski, D.S., Butler, A., Eng, R.Y.Y., Gigliotti, L., Harris, S. & Williams, A. (2020) Identifying mesopredator release in multi-predator systems: a review of evidence from North America. *Mammal Review* 50, 367–381.

Kalle, R., Ramesh, T., Sankar, K. & Qureshi, Q. (2013) Observations of sympatric small carnivores in Mudumalai Tiger Reserve, Western Ghats, India. *Small Carnivore Conservation* 49, 53–59.

Karanth, K.U. & Chellam, R. (2009) Carnivore conservation at the crossroad. *Oryx* 43, 1–2.

Keeney, R.L. (1996) *Value-Focused Thinking: A Path to Creative Decision Making*. Harvard University Press, Cambridge.

Keeney, R.L. & Raiffa, H. (1976) *Decisions with Multiple Objectives: Preference and Value Tradeoffs*. Wiley, New York.

King, C.M. & Murphy, E.C. (2005) Stoat. In: *The Handbook of New Zealand Mammals*. 2nd edition (ed. C.M. King), pp. 261–287. Oxford University Press, Melbourne.

King, C.M. & Powell, R.A. (2011) Managing an invasive predator pre-adapted to a pulsed resource: a model of stoat (*Mustela erminea*) irruptions in New Zealand beech forests. *Biological Invasions* 13, 3039–3055.

Kinlaw, A. (1995) *Spilogale putorius. Mammalian Species* 551, 1–7.

Larivière, S. (1999) *Mustela vison. Mammalian Species* 608, 1–9.

Leighton, P.A. (2009) *Mongoose Predation on Sea Turtle Nests: Linking Behavioural Ecology and Conservation*. PhD thesis, McGill University.

Leighton, P.A., Horrocks, J.A., Krueger, B.H., Beggs, J.A. & Kramer, D.L. (2008) Predicting species interactions from edge responses: mongoose predation on hawksbill sea turtle nests in fragmented beach habitat. *Proceedings of the Royal Society B: Biological Sciences* 275, 2465–2472.

Leighton, P.A., Horrocks, J.A. & Kramer, D.L. (2011) Predicting nest survival in sea turtles: when and where are eggs most vulnerable to predation? *Animal Conservation* 14, 186–195.

Lesmeister, D.B., Gompper, M.E. & Millspaugh, J.J. (2008) Summer resting and den site selection by eastern spotted skunks (*Spilogale putorius*) in Arkansas. *Journal of Mammalogy* 89, 1512–1520.

Litvaitis, J.A. & Villafuerte, R. (1996) Intraguild predation, mesopredator release, and prey stability. *Conservation Biology* 10, 676–677.

Loegering, J.P. & Fraser, J.D. (1995) Factors affecting Piping Plover chick survival in different brood-rearing habitats. *The Journal of Wildlife Management*, 646–655.

Logiudice, K. (2003) Trophically transmitted parasites and the conservation of small populations: Raccoon roundworm and the imperiled Allegheny woodrat. *Conservation Biology* 17, 258–266.

Long, R.A., MaCay, P., Ray, J. & Zielinski, W. (eds) (2008) *Noninvasive Survey Methods for Carnivores*. Island Press, Washington, DC.

Lotze, J.H. & Anderson, S. (1979) *Procyon lotor. Mammalian Species* 119, 1–8.

Lowe, S., Browne, M., Boudjelas, S. & De Poorter, M. (2000) *100 of the World's Worst Invasive Alien Species: A Selection from the Global Invasive Species Database*. IUCN SSC Invasive Species Specialist Group, Gland.

Mack, R.N., Simberloff, D., Lonsdale, W.M., Evans, H., Clout, M. & Bazzaz, F.A. (2000) Biotic invasions: causes, epidemiology, global consequences, and control. *Ecological Applications* 10, 689–710.

MacKenzie, D.I., Nichols, J.D., Lachman, G.B., Droege, S., Royle, J.A. & Langtimm, C.A. (2002) Estimating site occupancy rates when detection probabilities are less than one. *Ecology* 83, 2248–2255.

MacKenzie, D.I., Nichols, J.D., Royle, J.A., Pollock, K.H., Bailey, L.L. & Hines, J.E. (2006) *Occupancy Estimation and Modeling: Inferring Patterns and Dynamics of Species Occurrence*. Academic Press, Burlington.

Maran, T., Macdonald, D.W., Kruuk, H., Sidorovich, V. & Rozhnov, V.V. (1998) The continuing decline of the European mink, *Mustela lutreola*: evidence for the intraguild aggression hypothesis. In: *Behaviour and Ecology of Riparian Mammals* (eds N. Dunstone & M. Gorman), pp. 297–324. Cambridge University Press, Cambridge.

Martin, J., O'Connell, A.F. Jr, Kendall, W.L., Runge, M.C., Simons, T.R., Waldstein, A.H., Schulte, S.A., Converse, S.J., Smith, G.W., Pinion, T., Rikard, M. & Zipkin, E.F. (2010) Optimal control of native predators. *Biological Conservation* 143, 1751–1758.

Mathai, J., Hon, J., Juat, N., Peter, A. & Gumal, M. (2010) Small carnivores in a logging concession in the upper Baram, Sarawak, Borneo. *Small Carnivore Conservation* 42, 1–9.

McLennan, J.A., Potter, M.A., Robertson, H.A., Wake, G.C., Colbourne, R., Dew, L., Joyce, L., McCann, A.J., Miles, J., Miller, P.J. & Reid, J. (1996) Role of predation in the decline of kiwi, *Apteryx* spp., in New Zealand. *New Zealand Journal of Ecology* 20, 27–35.

Melvin, S.M., MacIvor, L.H. & Griffin, C.R. (1992) Predator exclosures: a technique to reduce predation at piping plover nests. *Wildlife Society Bulletin* 143–148.

Morrison, J.C., Sechrest, W., Dinerstein, E., Wilcove, D.S. & Lamoreux, J.F. (2007). Persistence of large mammal faunas as indicators of global human impacts. *Journal of Mammalogy* 88, 1363–1380.

Neuman, K.K., Page, G.W., Stenzel, L.E., Warriner, J.C. & Warriner, J.S. (2004) Effect of mammalian predator management on Snowy Plover breeding success. *Waterbirds* 27, 257–263.

NMFS [National Marine Fisheries Service] (2008) *White Abalone Recovery Plan (Haliotis sorenseni).* National Marine Fisheries Service, Long Beach.

Nichols, J.D. & Williams, B.K. (2006) Monitoring for conservation. *Trends in Ecology & Evolution* 21, 668–673.

Nichols, J.D., Bailey, L.L., O'Connell, A.F., Talancy, N.W., Grant, E.H.C., Gilbert, A.T., Annand, E.M., Husband, T.P. & Hines, J.E. (2008) Multi-scale occupancy estimation and modelling using multiple detection methods. *Journal of Applied Ecology* 45, 1321–1329.

Nogales, M., Vidal, E., Medina, F.M., Bonnaud, E., Tershy, B.R., Campbell, K.J. & Zavaleta, E.S. (2013) Feral cats and biodiversity conservation: the urgent prioritization of island management. *BioScience* 63, 804–810.

Palomares, F., Gaona, P., Ferreras P. & Delibes, M. (1995) Positive effects on game species of top predators by controlling smaller predator populations: an example with lynx, mongooses, and rabbits. *Conservation Biology* 9, 295–305.

Pardo, L.E., Cove, M.V., Spinola, R.M., de la Cruz, J.C. & Saenz, J.C. (2016) Assessing species traits and landscape relationships of the mammalian carnivore community in a neotropical biological corridor. *Biodiversity and Conservation* 25, 739–752.

Prange, S. & Gehrt, S.D. (2004) Changes in mesopredator-community structure in response to urbanization. *Canadian Journal of Zoology* 82, 1804–1817.

Prugh, L.R. & Sivy, K.J. (2020) Enemies with benefits: integrating positive and negative interactions among terrestrial carnivores. *Ecology Letters* 23, 902–918.

Prugh, L.R., Stoner, C.J., Epps, C.W., Bean, W.T., Ripple, W.J., Laliberte A.S. & Brashares, J.S. (2009) The rise of the mesopredator. *BioScience* 59, 779–791.

Ralls, K. & Starfield, A.M. (1995) Choosing a management strategy: two structured decision making methods for evaluating the predictions of stochastic simulation models. *Conservation Biology* 9, 175–181.

Rasambainarivo, F., Farris, Z.J., Andrianalizah, H. & Parker, P.G. (2017) Interactions between carnivores in Madagascar and the risk of disease transmission. *EcoHealth* 14, 691–703.

Ratnaswamy, M.J. & Warren, R.J. (1998) Removing raccoons to protect sea turtle nests: are there implications for management? *Journal of Wildlife Management* 26, 846–850.

Ratnaswamy, M.J., Warren, R.J., Kramer, M.T. & Adam, M.D. (1997) Comparisons of lethal and non-lethal techniques to reduce raccoon depredation of sea turtle nests. *Journal of Wildlife Management* 61, 368–376.

Rayner, M.J., Hauber, M.E., Imber, M.J., Stamp, R.K. & Clout, M.N. (2007) Spatial heterogeneity of mesopredator release within an oceanic island system. *Proceedings of the National Academy of Sciences of the United States of America* 104, 20862–20865.

Ritchie, E.G. & Johnson, C.N. (2009) Predator interactions, mesopredator release, and biodiversity conservation. *Ecology Letters* 12, 982–998.

Robertson, B.A. & Hutto, R.L. (2006) A framework for understanding ecological traps and an evaluation of existing evidence. *Ecology* 87, 1075–1085.

Roemer, G., Gompper, M.E. & Van Valkenburgh, B. (2009) The ecological role of the mammalian mesocarnivore. *BioScience* 59, 165–173.

Rogers-Bennett, L., Rogers, D.W. & Schultz, S.A. (2007) Modeling growth and mortality of red abalone (*Haliotis rufescens*) in northern California. *Journal of Shellfish Research* 26, 719–727.

Runge, M.C. (2011) An introduction to adaptive management for threatened and endangered species. *Journal of Fish and Wildlife Management* 2, 220–233.

Salo, P., Korpimäki, E., Banks, P.B., Nordström, M. & Dickman, C.R. (2007) Alien predators are more dangerous than native predators to prey populations. *Proceedings of the Royal Society B: Biological Sciences* 274, 1237–1243.

Salo, P., Banks, P.B., Dickman, C.R. & Korpimäki, E. (2010) Predator manipulation experiments: impacts on populations of terrestrial vertebrate prey. *Ecological Monographs* 80, 531–546.

Santulli, G., Palazón, S., Melero, Y., Gosálbez, J. & Lambin, X. (2014) Multi-season occupancy analysis reveals large scale competitive exclusion of the critically endangered European mink by the invasive non-native American mink in Spain. *Biological Conservation* 176, 21–29.

Saunders, G.R., Gentle, M.N. & Dickman, C.R. (2010) The impacts and management of foxes *Vulpes vulpes* in Australia. *Mammal Review* 40, 181–211.

Schipper, J., Hoffmann, M., Duckworth, J.W. & Conroy, J. (2008) The 2008 IUCN red listings of the world's small carnivores. *Small Carnivore Conservation* 39, 29–34.

Schlaepfer, M.A., Runge, M.C. & Sherman, P.C. (2002) Ecological and evolutionary traps. *Trends in Ecology & Evolution* 17, 474–480.

Schmidt, P.M., McCleery, R.A., Lopez, R.R. & Silvy, N. (2010) Habitat succession, hardwood encroachment and raccoons as limiting factors for Lower Keys marsh rabbits. *Biological Conservation* 143, 2703–2710.

Seimon, T.A., Miquelle, D.G., Chang, T.Y., Newton, A.L., Korotkova, I., Ivanchuk, G., Lyubchenko, E., Tupikov, A., Slabe, E. & McAloose, D. (2013) Canine distemper virus: an emerging disease in wild endangered Amur tigers (*Panthera tigris altaica*). *mBio* 4(4), e00410–13.

Smith, R., Veríssimo, D., Isaac, N.J.B. & Jones, K.E. (2012) Identifying Cinderella species: uncovering mammals with conservation flagship appeal. *Conservation Letters* 5, 205–212.

Soulé, M.E., Bolger, D.T., Alberts, A.C., Wrights, J., Sorice, M. & Hill, S. (1988) Reconstructed dynamics of rapid extinctions of chaparral-requiring birds in urban habitat islands. *Conservation Biology* 2, 75–92.

Stancyk, S.E. (1982) Non-human predators of sea turtles and their control. In: *Biology and Conservation of Sea Turtles* (ed. K.A. Bjorndal), pp. 139–152. Smithsonian Institution Press, Washington, DC.

Stein A.B., Fuller, T.K. & Marker, L.L. (2008) Opportunistic use of camera traps to assess habitat-specific mammal and bird diversity in northcentral Namibia. *Biodiversity and Conservation* 17, 3579–3587.

Stocking J.J., Simons, T.R., Parsons, A.W. & O'Connell, A.F. (2017) Managing native predators: evidence from a partial removal of raccoons (*Procyon lotor*) on the Outer Banks of North Carolina, USA. *Waterbirds* 40, 10–18.

Terborgh, J. & Estes, J.A. (eds) (2010) *Trophic Cascades: Predators, Prey, and the Changing Dynamics of Nature*. Island Press, Washington, DC.

Treves, A. & Karanth, K.U. (2003) Human–carnivore conflict and perspectives on carnivore management worldwide. *Conservation Biology* 17, 1491–1499.

Veron, G., Patou, M.-L., Pothet, G., Simberloff, D. & Jennings, A.P. (2007) Systematic status and biogeography of the Javan and small Indian mongooses (Herpestidae, Carnivora). *Zoologica Scripta* 36, 1–10.

Veron, G., Patou, M.L. & Jennings, A.P. (Chapter 3, this volume) Systematics and evolution of the mongooses (Herpestidae, Carnivora). In: *Small Carnivores: Evolution, Ecology, Behaviour, and Conservation* (eds E. Do Linh San, J.J. Sato, J.L. Belant & M.J. Somers). Wiley–Blackwell, Oxford.

Vickery, P.D., Hunter, M.L. Jr & Wells, J. (1992) Evidence of incidental nest predation and its effects on nests of threatened grassland birds. *Oikos* 63, 281–288.

Vilardell, A., Capalleras, X., Budó, J., Molist, F. & Pons, P. (2008) Test of the efficacy of two chemical repellents in the control of Hermann's tortoise nest predation. *European Journal of Wildlife Research* 54, 745–748.

Vitousek, P.M., Dantonio, C.M., Loope, L.L., Rejmanek, M. & Westbrooks, R. (1997) Introduced species: a significant component of human-caused global change. *New Zealand Journal of Ecology* 21, 1–16.

Wade-Smith, J. & Verts, B.J. (1982) *Mephitis mephitis*. *Mammalian Species* 173, 1–7.

Waldstein, A. (2010) *Raccoon Ecology and Management on Cape Lookout National Seashore, North Carolina.* MSc thesis, North Carolina State University.

Watari, Y., Nagata, J. & Funakoshi, K. (2011) New detection of a 30-year-old population of introduced mongoose *Herpestes auropunctatus* on Kyushu Island, Japan. *Biological Invasions* 13, 269–276.

Watters, M.E., McLash, T.L., Aldridge, C.L. & Brigham, R.M. (2002) The effect of vegetation structure on predation of artificial greater sage-grouse nests. *Ecoscience* 9, 314–319.

Wilcove D.S., Rothstein, D., Dubow J., Phillips, A. & Losos, E. (1998) Quantifying threats to imperiled species in the United States. *BioScience* 48, 607–615.

Wilson, P.R., Karl, B.J., Toft, R.J., Beggs, J.R. & Taylor, R.H. (1998) The role of introduced predators and competitors in the decline of Kaka (*Nestor meridionalis*) populations in New Zealand. *Biological Conservation* 83, 175–185.

Wright, A.N. & Gompper, M.E. (2005) Altered parasite assemblages in raccoons in response to manipulated resource availability. *Oecologia* 144, 148–156.

Zinsstag, J., Schelling, E., Waltner-Toews, D. & Tanner, M. (2011) From "one medicine" to "one health" and systemic approaches to health and well-being. *Preventive Veterinary Medicine* 101, 148–156.

22

The Global Consumptive Use of Small Carnivores: Social, Cultural, Religious, Economic, and Subsistence Trends from Prehistoric to Modern Times

Tim L. Hiller[1], and Stephen M. Vantassel[2]*

[1] *Wildlife Ecology Institute, Helena, MT, USA*
[2] *Wildlife Control Consultant, LLC, Lewistown, PA, USA*

SUMMARY

We reviewed an extensive set of literature that described how the global contributions of food and fur from carnivores have been dependent on subsistence, social, cultural, economic, and religious trends that have varied in space and time. In general, humans in temperate regions used small carnivores for fur, whereas humans in tropical regions used species within this group primarily for food. Human use of carnivores was often of secondary importance to the use of large herbivores, although this depended on faunal availability, the difficulty of acquisition, and other factors. During prehistory, archaeological evidence suggests that depending on species, small carnivores were utilized not only as food and fur for garments, but also for religious purposes. The shift to transcontinental trade in fur began with an increasing European presence in North America. High demand for fur during early periods led to unsustainable harvest of several species. Some species were extirpated (e.g. sea mink, *Mustela macrodon*), whereas some benefitted substantially as human activities transformed the North American landscape (e.g. striped skunks, *Mephitis mephitis*, and northern raccoon, *Procyon lotor*). In the twenty-first-century international fur trade, the northern raccoon and the American mink, *Neovison vison*, have been among the most important wild and fur-farmed species, respectively. In rural areas of less-developed countries, viverrids (e.g. common palm civet, *Paradoxurus hermaphroditus*) and herpestids (e.g. crab-eating mongoose, *Urva [= Herpestes] urva*) often are highly used for the bushmeat trade, although the lack of, or unenforced regulations may result in unsustainable harvests for some species. Regulation of international trade in small carnivores and other wildlife has been implemented in an effort to conserve endangered species, although regulatory efficacy can vary widely. Unfortunately, many species of small carnivores remain relatively unstudied or may be considered pests by some peoples.

Keywords

Bushmeat — Carnivora — fur harvest — furbearer management — subsistence

Introduction

The consumptive use of carnivores has benefitted humans in numerous ways for thousands of years. These contributions have varied in space and time and have been dependent on subsistence needs; types and availability of resources; climatic conditions; and social, cultural, economic, and religious trends within particular associations of humans. For small carnivores, patterns of consumption in space and time often relate to their use as fur for garments or adornments, particularly in temperate regions, and use of meat for nutritional value, particularly in tropical regions (e.g. Bachrach, 1949; Schieff & Baker, 1987; Colyn *et al.*, 2004). Beyond food and clothing, consumptive use also may include the utilization of animals for

*Corresponding author.

Small Carnivores: Evolution, Ecology, Behaviour, and Conservation, First Edition. Edited by Emmanuel Do Linh San, Jun J. Sato, Jerrold L. Belant, and Michael J. Somers.

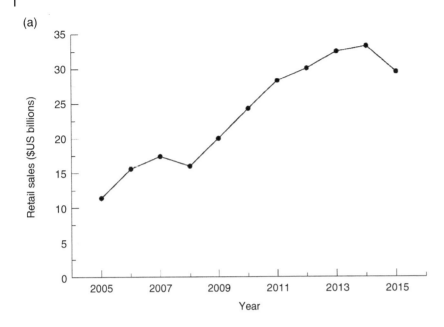

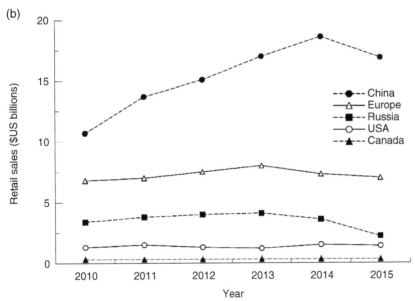

Figure 22.1 Retail sales of furs in (a) international (2005–2015); and (b) regional (2010–2015) trade. *Source:* Data based on Hansen (2017).

medicinal purposes, spiritual rituals, cultural ceremonies, or social status (e.g. McGee, 1987; Nasi *et al.*, 2008). Humans have acquired small carnivores for consumptive use through subsistence hunting and trapping, recreational harvest, commercial harvest, fur farming, and other means.

Even among modern-day humans, uses of small carnivores vary widely from fur garments worn as status symbols or alternatively as functional garments in cold climates, to a food source that helps fulfil the requirements in environments that may be lacking in availability of other sources of protein or cash income

(Loibooki *et al.*, 2002). In terms of subsistence use in less-developed countries, small carnivores may be harvested at lower rates than other wildlife species (e.g. Noss, 1998), perhaps because of lower relative abundance, greater difficulty in the capture, or lower relative benefits as a food resource based on small body size. Recreational or commercial harvest of small carnivores for fur is part of a $US 33-billion global industry (Figure 22.1a; Hansen, 2017) to meet the demands of various societies, both in terms of fashion and pragmatism. Similarly, the farming of fur, particularly the American mink, *Neovison vison*, has resulted in a substantial contribution to the fur industry, but is also highly dependent on market fluctuations and global demand. Beyond food and clothing, harvesting wildlife also has strong ties in tradition, medicine, and other cultural aspects throughout the world (Nasi *et al.*, 2008). Today, we can still readily observe a diversity of species-specific uses of small carnivores following the transition to a global network of societies and economies.

A review of how humans have used small carnivores in space and time will be as diverse of a topic as the species found within this subset of the order Carnivora. For our purposes, this subset includes species within the families Herpestidae (e.g. mongooses), Mephitidae (e.g. skunks), Mustelidae (e.g. martens, badgers, otters, mink, and weasels), Procyonidae (e.g. raccoons, coatis, ringtails, and olingos), and Viverridae (e.g. civets and genets). We also briefly mention species from the families Canidae (e.g. foxes), Didelphidae (e.g. opossums), Felidae (e.g. *Leopardus* and *Lynx*), and Nandiniidae (i.e. African palm civet, *Nandinia binotata*). Many species of small carnivores may be considered furbearers or game species. The term 'furbearer' is a classification with strong implications in management but a seemingly little foundation in ecology. Furbearer may be loosely described as a mammalian species that humans utilize primarily for their pelts (Krausman, 2013). The term 'game' has widespread use and typically refers to wildlife species that are harvested using hunting methods.

Our purpose with this chapter is not only to examine how non-domesticated species of small carnivores have been utilized by humans in a consumptive manner, but also to integrate information on other wildlife species (e.g. ungulates, mesocarnivores, and rodents)

when necessary to provide a complete understanding of complementary or competing uses. The available information on consumptive uses ranges from direct and indirect evidence collected and interpreted by archaeologists and anthropologists for uses that span thousands of years starting with prehistoric humans through time to modern-day harvest and fur-sales records describing international trade. Readers will note a disproportionate discussion related to North America and Eurasia. This is directly related not only to the relatively large amount of information available for these areas, but also to our knowledge of these regions.

The organization of so many types of information, which often described multiple consumptive uses, has been difficult but we have attempted to present this in a logical, concise, and primarily temporal manner for readers. For these reasons, we chose to describe consumptive uses based on three broad time periods: ancient, which includes pre-historical (i.e. before the invention of writing) and historical periods up to about 1000 A.D.; pre-modern, which focuses on use at the local or continental scale (about 1000 to 1599 A.D.); and the modern-day (about 1600 A.D. to present) transition to a global-trade or transcontinental economy. As with space (i.e. North America, Eurasia), there is also a disproportionate discussion by time period simply because most available information relates to what we describe as the modern-day period. We also describe how global regulation of trade in small carnivores (and other wildlife) has affected management and consumptive uses. A chapter on this topic cannot be all-inclusive and, as such, our efforts were primarily to provide a broad overview related to the known consumptive uses of small carnivores.

Ancient Evidence for the Consumptive Use of Carnivores (pre-1000 A.D.)

Information regarding the use of carnivores by humans during the pre-historical period is particularly difficult to obtain. Without the aid of written resources, our understanding of the consumptive use of carnivores is limited to inferences from archaeology and reasonable suppositions made from ethnoarchaeology. Conclusions, therefore, must be held tentatively due to the paucity and ambiguity of the evidence. The collection of direct evidence is often difficult. First, organic matter, such as bones, decays relatively rapidly, particularly in wet environments with temperature fluctuations (Behrensmeyer, 1978), or with high acidity (Klein, 1974). Second, humans may have simply eaten or otherwise used small animals at the kill site as opposed to depositing them in a garbage dump with other remains where archaeologists could find them (Wright, 1987). Similarly, the collection of evidence related to potential capture devices is difficult because such devices were made of natural materials that did not preserve well (Wadley, 2010), further complicating assessments.

Marciszak (2016) in discussing least weasel, *Mustela nivalis*, bones uncovered in a Greek cave dated to the Late Pleistocene and Mesolithic periods, summarized the findings related to mustelids by archaeologists in the Mediterranean region. He stated that *M. nivalis* was most frequently found during Neolithic and Roman period sites. Though some have suggested that the Romans introduced weasels to Egypt, archaeological evidence revealed that they were present as far back as Late Pleistocene. Interestingly, Greek myths mentioned weasel-like animals that were most likely references to *M. nivalis* (e.g. Bettini, 2013).

It is reasonable to assume that in warm or humid climates, little clothing was or is necessary for indigenous peoples as its use would seem impractical (McGee, 1987). However, in colder climates, humans would have been unable to persist without clothing and shelter, and animal skins were requisite in these harsh environments. As a group, carnivores have often been of secondary importance to large herbivores when it came to supplying the basic needs of prehistoric and modern man (Redman, 1999). This generalization seems to hold true for groups such as prehistoric peoples of North America, Eemian-period hunter-gatherers in Europe (e.g. Wright, 1987; Blasco & Peris, 2012), and elsewhere (Helms & Betts, 1987). This observation may be explained by the relative lower abundance of carnivores in comparison to herbivores, the potential for higher risk of injury when pursuing carnivores, and that the energy gained by capturing carnivores may often exceed the energetic costs expended to obtain them.

In North America, archaeological evidence suggests that prehistoric humans hunted furbearers for over 11 000 years (Wright, 1987). Despite the relative infrequency of small carnivores occurring within the faunal assemblages of archaeological sites, several species in North America were utilized to some degree. The Mummy Cave in Wyoming, United States, contained the butchered bones from the extinct noble marten, *Martes americana nobilis* (Hughes, 2009). Native American sites in south-western Pennsylvania, United States, contained bones from northern raccoons, *Procyon lotor*, North American river otters, *Lontra canadensis*, striped skunks, *Mephitis mephitis*, and other species (Gilmore, 1946). Analysis of faunal remains at Mohawk sites revealed that tribal members utilized at least 27 species of animals, including raccoons, American martens, *Martes americana*, fishers, *Pekania pennanti*, river otters, wolverines, *Gulo gulo*, and American mink (Kuhn & Funk, 2000). Reports from Europeans also confirm that Mohawks had a strong preference for meat (Kuhn & Funk, 2000). Interestingly, skunk bones were exceedingly rare or absent suggesting that the species was avoided (Gilmore, 1946; Kuhn & Funk, 2000). One of the more important species, particularly in southern agricultural areas of North America, was the raccoon. Raccoon remains were typically present within these garbage dumps, sometimes more so than North American beavers, *Castor canadensis*, and prehistoric peoples may have capitalized on the abundance of raccoons near their agricultural activities (Wright, 1987). The diverse uses of and beliefs associated with small carnivores by indigenous peoples were often group-specific and tied to climate or spiritual philosophies (McGee, 1987). Mustelids have very rarely been found at prehistoric garbage dumps, but it appears that their skulls and bones may have been used as religious medicine bundles (Wright, 1987). The river otter seems to have had religious importance to certain cultures in North America. For example, this species was associated with spirits of the dead by Athabascan-speaking people in north-western North America and this species would not be physically touched by these peoples (Wright, 1987). Similarly, to avoid consequences such as offending the animal's spirit, some North American cultures ritualistically threw certain parts or whole carcasses of semi-aquatic mammals (e.g. river otters, beavers) back into the water (Wright, 1987). One exception to the rarity of this family is the sea otter, *Enhydra lutris*. In coastal Pacific areas, this species was common and sometimes dominant within such sites (Wright, 1987).

Conversely, several peoples in India, China, and western Nepal related the red panda, *Ailurus fulgens*, to good fortune and protection and would wear furs or tails during rituals or ceremonies, whereas some people in Bhutan to this day regard an observation of a red panda as good luck and will not harm this species (Glatston & Gebauer, 2011). Although reports are somewhat conflicting, it appears as though pelts of red pandas never had commercial value despite their high quality (Glatston & Gebauer, 2011). Ancient Chinese documents have recorded the importation of Russian pelts, including sables, *Martes zibellina* (Sasaki, 1998).

In South African sites, members of the families Viverridae, Herpestidae, and Mustelidae, and other small carnivores may have been utilized during the Middle Stone Age, though this conclusion is subject to interpretation (Klein, 1974; Wadley, 2010). Questions posed have included whether these species were captured in traps or snares, and if so, were they actually non-target captures. From a religious aspect, the Hebrew people were not allowed to eat carnivores (Lockman Foundation, 1963a), nor were they allowed to use forbidden animals in sacrifices. While unlikely that Jews hunted carnivores for food, they hunted or trapped carnivores to protect livestock and for human safety (Lockman Foundation, 1963b; Negev & Gibson, 2003).

In the ninth century, Vikings living in what is known today as Norway sustained themselves in part by taxes paid in fur, which included martens and otters, from the Sami people (Hofstra & Samplonius, 1995). Researchers believe that the earliest recorded voyage made by Vikings to the White Sea of north-western coastal Russia occurred between 870 and 890, primarily to explore opportunities for fur trade (Hofstra & Samplonius, 1995). Evidence exists that at least one fur-trade expedition between the Vikings and the Medieval residents of northern Russia occurred in the early eleventh century and evidence suggests that trade continued into the twelfth century (Hotstra & Samplonius, 1995). The fur trade began to diminish in

importance as the increase in townships raised a demand for food, which was supplied by fish (Hotstra & Samplonius, 1995).

Pre-modern (About 1000–1599 A.D.)

Different Native American groups viewed furbearers with different levels of esteem. After 1500, some placed a taboo on the consumption of Virginia opossums, *Didelphis virginiana*, or skunks. However, aboriginal peoples of the United States used striped and spotted, *Spilogale* spp., skunk pelts to trim ceremonial garments, and in the south-east, wove opossum hair into textile garments (McGee, 1987). Similarly, accounts exist of Native North Americans killing skunks and opossums later during the 1700s (Brander, 1971). The trend in North America was also that wildlife species which benefitted survival through consumption of meat (e.g. ungulates, cetaceans, and pinnipeds) were the primary focus, whereas other species (e.g. American mink and American marten) did not become important until the establishment of European trade despite the previously well-established trade routes across North America (Wright, 1987).

In Europe during this period, the practice of wearing fur became so popular that those that could afford to purchase fur generally did so (Ingrams, 1924). Ingrams (1924) noted that the demand for furs and other luxury products became so great that European countries established sumptuary laws in an attempt to stem the demand. Furs of the wealthier classes included pine marten, *Martes martes*, sable, stoat, *Mustela erminea*, European otter, *Lutra lutra*, and others (Ingrams, 1924). European royalty wore fur to distinguish their status from that of commoners (Dolin, 2010). In England, King Edward III limited the wearing of furs to royalty and significant church officials. Almost 100 years later, King James I of Scotland ruled that only those with a rank of knight or above could wear fur (Dolin, 2010). Pelts from stoats, sables, and martens were only permitted for upper classes of royalty (King & Powell, 2007).

Interestingly, some Native American cultures also used fur to designate social status. Starting in the early 1500s, robes or capes made from pelts of sea otters were used to signify high social status for some indigenous cultures in North America (McGee, 1987). River otters also inspired spiritual significance during that time, with skins used as medicine bags, hats, robes, and other items, but this small carnivore was rarely eaten (McGee, 1987). American mink were similar to river otters in these respects, and some cultures prohibited their killing (McGee, 1987), but there is evidence that others were known to use mink (and raccoons) for food (e.g. Lower Chinook of the US Pacific Northwest; Ray, 1938; Lyman, 2007). Wolverines carried spiritual significance for many societies and consumption of their meat ranged from occurrence at some level to prohibition (McGee, 1987). After 1500, the skins of American badgers, *Taxidea taxus*, were commonly used for moccasins because of their durability, but consumption of meat was rare (McGee, 1987).

Native peoples clearly had a value system related to fur characteristics prior to European presence in North America (Wright, 1987). Curiously, in the Yukon after 1500, some peoples had religious beliefs that included that the use of animals only for fur was not acceptable, but rather use for both food and fur, or food only, was required (McGee, 1987), a belief that would not seem prevalent in that region today.

Modern-day Transition to International Trade (1600 A.D. to present)

By the early 1600s, the availability of fur was declining in Europe and new sources were needed to supply the demand (Dolin, 2010). Some of the demand was filled by Russia, which extended its control of the sable trade to the Pacific Ocean with its first coastal settlement in 1649 (Barger, 2008). Concurrently, the use of several types of steel traps increased in North America to capture small carnivores and other furbearers; these designs seemed to have evolved from Old World torsion traps, which were constructed of wood and powered by either plant fibres or animal sinews (Gerstell, 1985). The shift to transcontinental trade in fur began with an increasing European presence in North America. Formed in 1670, the Hudson's Bay Company in Canada was one of the most recognizable names in the

international fur trade. The Hudson's Bay Company was originally exclusively based on trade, but later encouraged its employees through financial incentives to trap small carnivores, particularly martens and wolverines (Gerstell, 1985). Other companies were major competitors in the trade of fur (and other commodities). During the late 1980s, Hudson's Bay Company sold its fur-auction businesses to what became known as North American Fur Auctions, the largest such business in North America until its withdrawal from the industry in 2019. Today, most fur is sold through major auction houses such as Kopenhagen Fur (Denmark), SAGA Furs Oyi (Finland), American Legend Cooperative (United States), Fur Harvesters Auction (Canada), and Sojuzpushnina (Russia) (International Fur Trade Federation, 2013).

The impacts of Europeans arriving in North America and establishing trade based on their values resulted in many changes to indigenous cultures. Beyond impacts related to disease and conflicts, some native peoples in North America shifted interests to trapping previously unutilized furbearers. Also, competition among trading companies eventually resulted in the elimination of the service of natives as middlemen, which had been a complex system of alliances developed primarily for political reasons through the influence of competing European countries (Ray, 1987). Events such as these eventually changed Native North American cultures such that subsistence hunting shifted to dependence on trade to acquire winter foods, and eventually certain shifts in social structures (McGee, 1987). In the south-eastern United States, intense competition among southern colonial states for the acquisition of white-tailed deer, *Odocoileus virginianus*, skins from the Cherokee for export to Britain for the leather-manufacturing industry led to a dependency of the Cherokee on, among other things, European trade goods (Dunaway, 1994). These trends are not unique to North America; however, as western colonial influences and values on African societies have also resulted in changes in traditional consumptive uses of wildlife (Sifuna, 2012), with the prohibition on consumptive use of wildlife in Kenya perhaps the most severe with regards to traditional uses.

Although sea otters were traded before the late eighteenth century, trade increased dramatically following the use of this pelt for clothing during Captain

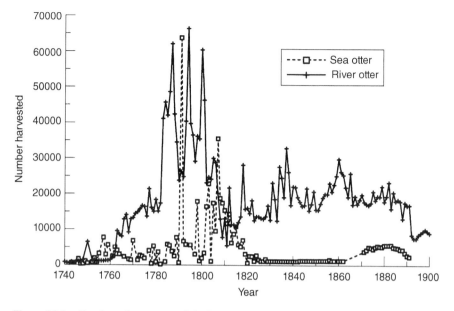

Figure 22.2   Number of sea otters, *Enhydra lutris*, and North American river otters, *Lontra canadensis*, harvested in North America from 1740 to 1900. *Source:* Data based on estimates from Novak *et al.* (1987).

James Cook's subsequent discovery of Chinese interest in these pelts during 1778 (Figure 22.2; Novak *et al.*, 1987; Barger, 2008). Britain, France, Russia, Spain, and the United States became interested in the harvest of sea otters to fulfil Chinese demand for luxuriant furs used to trim garments (Gough, 1989; Barger, 2008). Russians also traded with Eskimos in the early 1800s for furs, including sables, martens, and otters (Ray, 1975). Caywood (1967, p. 46) stated, 'At the close of the 18th century the fur trade loomed as the greatest enterprise of all times'. Substantial reductions in market demand sometimes prevented the continued population decline or potential extinction of several species. Overutilization of certain species, such as beavers for the manufacture of felt hats during 1550–1850 and sea otters from the Pacific coast of North America from about 1778 until 1820, often resulted in a transition to more plentiful fur-bearing species, such as river otters and martens (Figures 22.2 and 22.3; Novak *et al.*, 1987; Ray, 1987). Some areas, such as in the Great Lakes states, United States, experienced local declines in furbearers, including fine-furred animals (e.g. martens) due to unsustainable harvest (Gilman, 1974). In the case of the beaver, market demand transitioned from felt to silk, thereby causing a collapse in the market for this declining species

(Figure 22.4; Novak *et al.*, 1987). Clayton (1966) found that American fur exports continued to expand between 1820 and the 1930s, but not for all species. While increases in exports occurred with muskrat, *Ondatra zibethicus*, raccoon, red fox, *Vulpes vulpes*, American mink, river otter, and grey wolf, *Canis lupus*, exports declined with fisher, beaver, and marten. Populations of some of these overutilized species recovered (e.g. beavers), whereas others (e.g. sea otters) have not, probably due to multiple factors (e.g. mortalities through infectious diseases, entanglement in fishing equipment, and shark predation), despite federal protections (US Fish and Wildlife Service, 2003). One species of small carnivore, the sea mink, *Mustela macrodon*, did become extinct during the late 1800s as the direct result of unsustainable commercial harvest for the international fur trade (Turvey & Helgen, 2008). International trade in small carnivores and other species is now regulated in certain respects through the Convention on International Trade in Endangered Species of Wild Fauna and Flora (CITES), the Agreement on International Humane Trapping Standards, and other mechanisms as briefly discussed in the next section.

Many small carnivores in Europe have experienced population declines during the past several decades but others have experienced expansion due to

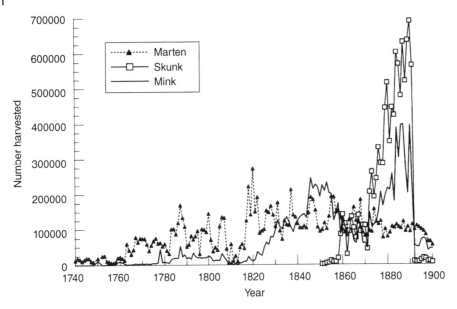

Figure 22.3 Number of American martens, *Martes americana*, striped skunks, *Mephitis mephitis*, and American mink, *Neovison vison*, harvested in North America from 1740 to 1900. *Source:* Data based on estimates from Novak *et al.* (1987).

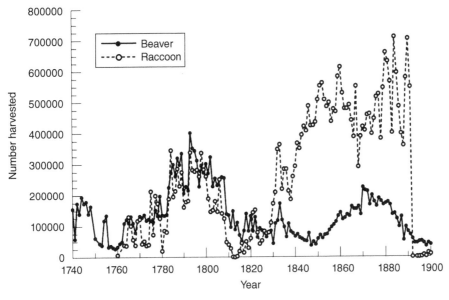

Figure 22.4 Number of North American beavers, *Castor canadensis*, and northern raccoons, *Procyon lotor*, harvested in North America from 1740 to 1900. *Source:* Data based on estimates from Novak *et al.* (1987).

conservation efforts (Proulx *et al.*, 2004). Unfortunately, many species remain relatively unstudied or may be considered pests by some peoples. In Lithuania, least weasel pelts were sold, but harvest is now prohibited, and little is known about this species (Mickevicius &

Baranauskas, 1992). Pelts of stone martens, *Martes foina*, were previously used, but some confusion existed when differentiating between this species and pine martens during surveys (Mickevicius & Baranauskas, 1992). In contrast, European polecats,

Mustela putorius, remain relatively abundant and are currently unprotected in Lithuania; anthropogenic mortalities are predominately associated with pest control efforts rather than hunting (Mickevicius & Baranauskas, 1992). Culturally, among Europeans and their North American counterparts, weasels were viewed as somewhat sinister and were often classified as vermin because they predated birds and their eggs (King & Powell, 2007). Ironically, stoats were part of the traditional garb of British justices (King & Powell, 2007). In contrast to the mixed opinions about weasels by those of European descent, Native Americans (e.g. Chugach, Cheyenne, and Lakota) held weasels in esteem (King & Powell, 2007). The Chugach tribe, in particular, believed that capturing a weasel marked one for future wealth. In Scotland, during the nineteenth century, the fur of the European badger, *Meles meles*, was used to craft sporrans, i.e. fur purses dangled from a belt and worn on the front of a kilt. The striking black-and-white scalp of the badger was even used as an adornment on sporrans worn by officers and sergeants of specific regiments (Long & Killingley, 1983). It is unclear why badger fur was used for this purpose, but it is possible that it was linked to admiration by local peoples for the strength and tenacity of this mustelid (Do Linh San, 2006).

Consumptive Use of Small Carnivores Today

As with the pre-modernization period, the consumptive use of small carnivores today is primarily related to garments, although now the discussion focuses on the transition to international trade. The human value system today with regards to fur drives this industry, both for determining which species are in demand and also the development of polarized attitudes toward the use of fur for garments, or at least the methods often employed to harvest furbearers. Species-specific market values can fluctuate dramatically over relatively short time periods (e.g. Novak *et al.*, 1987; Hiller, 2011; Beringer & Grusenmeyer, 2014), even occasionally within the same year, thereby resulting in annual fluctuations of total retail sales (Figure 22.1a,b). The current backbone of the international fur trade is generally considered to include muskrats, beavers, and northern raccoons (Figure 22.5; Novak *et al.*, 1987; Fur Institute of Canada, 2019; Statistics Canada, 2019; AFWA, 2021), and fur-farmed American mink (Obbard *et al.*, 1987; US International Trade Commission, 2004); however, the American marten and the Pacific marten, *Martes caurina* (trade name sable for both species), are typically among the most important species of small

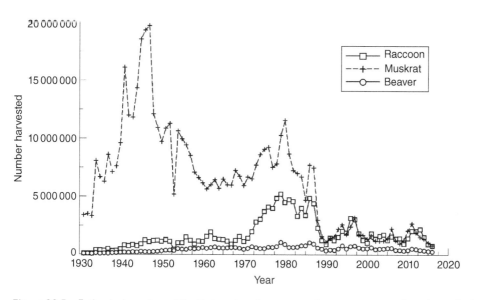

Figure 22.5 Estimated number of North American beavers, northern raccoons, and muskrats, *Ondatra zibethicus*, harvested annually in North America from 1931 to 2016. *Source:* Data based on estimates from Novak *et al.* (1987), Fur Institute of Canada (2019), Statistics Canada (2019), and AFWA (2021), and may be considered minimum harvests.

carnivores in terms of individual economic value. Northern raccoons became important in the fur trade starting around 1830 as the beaver market collapsed, and dominated the market in terms of value until the Civil War erupted in the United States (Figure 22.4; Novak *et al.*, 1987; Ray, 1987). When the Civil War concluded in 1865, both mink and skunk values and exports increased substantially (Figure 22.3; Novak *et al.*, 1987), with the primary market for the former being in Britain (Ray, 1987). Fur prices again substantially declined following World War II (Stabler & Howe, 1990).

For individual pelts, grading (and, therefore, value) is typically related to three factors: quality (e.g. pelt primness), size, and colour (Stains, 1979; Obbard, 1987). Each of these characteristics can vary geographically and seasonally, and variations in each may occur at small spatial scales within a region and even within individuals (Obbard, 1987; Worthy *et al.*, 1987). In general, fur quality is typically higher in arctic and temperate regions due to seasonal changes that increase the insulative value of their hair; the length of time that a pelt is prime (e.g. attains maximum growth) varies by species and usually increases with latitude (Bachrach, 1949; Stains, 1979). Similarly, although size may relate to individual age or sex, over a collection of individuals, size may also increase with increasing latitude, which follows Bergmann's Rule (Allee *et al.*, 1949). Although these factors may help explain why the economic importance of fur is higher in colder temperate zones in comparison to warmer tropical zones, some species of small carnivores limited to tropical regions (e.g. margay, *Leopardus wiedii*, ocelot, *Leopardus pardalis*) experienced very high levels of demand in the past because of their fur (Obbard, 1987).

The farming of mink (and Arctic foxes, *Vulpes lagopus*, and red foxes) became popular about the time of the American Civil War (1861–1865) and was in response to a market shifting from depleted species such as beavers to these small carnivores (Ray, 1987). Various pelage colours (e.g. sapphire, platinum, and blue pastel) not found naturally were produced through selective breeding for commercial purposes for both mink and foxes (see Obbard, 1987). The source of commercially produced mink came almost exclusively from American mink in North America (Shackelford, 1980).

The prevalence of fur farms has fluctuated dramatically over time, including the diversity of species, although the American mink has remained important in the commercial fur trade. In fact, 85% of all fur sold were pelts produced by fur farms based in Europe (60% of the fur-farm market; largest exporter), China (25%; largest importer), North America (10%), and Russia (5%) (International Fur Trade Federation, 2013). The top five producers (Denmark, China, Holland, United States, and Poland) of farm raised mink pelts have accounted for 75% of nearly 50 million pelts produced annually during 2005–2010 (Fur Commission USA, 2011).

Small carnivores certainly play an economic role in global trade, but they are also important to the livelihood of many people. Indeed, few options may exist for rural people in less-developed countries or remote regions to pursue activities that result in financial gain (Nasi *et al.*, 2008). Large portions of some rural populations are therefore dependent on bushmeat to meet their nutritional needs, and in some regions, small animals are an important source of food. In tropical areas, in particular, small carnivores are highly utilized in trade for food and other products. The number of small carnivores, such as viverrids and herpestids, is an important component of bushmeat in West Africa, although total biomass is relatively low (Colyn *et al.*, 2004; Bahaa-el-din *et al.*, 2013). There, small animals are either traded locally (similar to premodernization times) or eaten by the hunters themselves rather than transported for trade or other reasons (e.g. Allebone-Webb *et al.*, 2011). As a group, mammalian carnivores may also tend to be traded in local rural settings as opposed to urban markets (Macdonald *et al.*, 2011). However, as with pre-modern times, taboos exist on the consumption of certain mammalian carnivores. For example, in the Democratic Republic of the Congo, eating the meat of aquatic genet, *Genetta piscivora*, is taboo to Bambuti pygmies, except by male elders (Hart & Timm, 1978). Similarly, although some harvest and consumption of the meat of the introduced small Indian civet, *Viverricula indica*, occurs in Madagascar, it is generally of low preference and for some people, there is a cultural or traditional taboo placed on this type of consumption for this species (Randrianandrianina *et al.*, 2010).

Aside from meat, small carnivores are still harvested for different purposes throughout the African continent. Libyan striped weasels, *Ictonyx libycus*, do not seem to be used for food, but in south Tunisia, they are killed for witchcraft because they are thought to increase human male fertility, and it seems that there is some trade from Tunisia to Libya (K. de Smet *in* Cuzin, 2013). Honey badger, *Mellivora capensis*, body parts, particularly paws, skin, fat, and organs, are commonly used in traditional medicine because of the species' reputation for fearlessness and tenacity (Do Linh San *et al.*, 2016). In Gabon, African palm civet skins are used in traditional medicine, specifically to remove curses (L. Bahaa-el-din *in* Gaubert *et al.*, 2015). In some other regions, their fur is sought after to make ceremonial dresses (Malbrant & Maclatchy, 1949), wrist-bracelets, and hats, and to cover bows used for hunting (Carpaneto & Germi, 1989). In South Africa, hides and tails of the Cape genet, *Genetta tigrina*, are sometimes used to make traditional Zulu clothing items. Pieces of genet skin may also be used as stick-fight charms, or to adorn hats, whereas parts of the body of genets are used to treat ailments of eyes. Flesh consumption has also been reported by Cunningham & Zondi (1991). Similarly, these authors regarded the African striped weasel, *Poecilogale albinucha*, as one of the most-used animals in traditional medicine; in parts of its wide geographic range, the skins are commonly used by traditional healers and sangomas as a good luck charm.

In Asia, small carnivores, including data-deficient species and species of conservation concern, are typically hunted or trapped opportunistically. For example, in Laos, they are sold for food, but species often change hands many times and thus may be difficult for researchers to assess the origin of individual animals (Duckworth, 1997). A study conducted in a wildlife park in Myanmar revealed that all small carnivores (e.g. common palm civet, *Paradoxurus hermaphroditus*, crab-eating mongoose, *Urva [= Herpestes] urva*) captured as a target or non-target species were consumed locally as food or used in traditional medicine (Su, 2005). Viverrids are widely traded as food in Asia, but also for extracting civet oil for use in perfumes or for religious prayer (Robinson & Wirth, 1998; Gupta, 2004). In north-eastern India, small carnivores are hunted only for food by tribal peoples, with the only known exception being hog badgers, *Arctonyx* spp., for which the fat is also utilized for medicinal purposes (Ramakantha, 1994). Interestingly, the Kani people of south-western India do not eat the yellow-throated marten, *Martes flavigula*, because they consider the meat to be poisonous (Proulx *et al.*, 2004).

In northern Saskatchewan, Canada, the bush harvest (e.g. mammals, birds, berries, and fuelwood) by the Cree-speaking Métis accounted for about one-third of the value of their total village revenue; a portion of this income included muskrats, beavers, and Canada lynx, *Lynx canadensis*, for food and fur (Tobias & Kay, 1994). While fur may provide economic benefits to native peoples, its importance as a source of cash may not rise above 2% (Stabler & Howe, 1990).

Conservation and Management of Small Carnivores

The consumption of wildlife has some underlying conservation issues in some areas. Potential negative effects may come from illegal harvest within protected areas (e.g. national parks), unsustainable harvest of certain wildlife resources, and illegal harvest of sensitive species (Fa *et al.*, 2002; Golden, 2009; Randrianandrianina *et al.*, 2010). As noted earlier, the consumptive use of small carnivores is widespread in Africa and Asia, and there is much concern over the potential impacts on populations of these species (Robinson & Wirth, 1998) and over lack of enforcement of laws designed to protect rare species, especially those that receive little or no monitoring (e.g. Shepherd, 2008, 2012). Issues certainly are not limited to use as bushmeat or in less-developed countries. In Europe, where enforcement efficacy of trade regulations would seem high, illegal trade of protected European badgers is apparently occurring in some countries based on genetic evidence of their use in commercial shaving brushes, with manufacturers often claiming exclusive use of unprotected species such as hog badgers (Domingo-Roura *et al.*, 2006).

Sometimes, the desire to fulfil large-scale demand is so great that humans introduce non-indigenous fur-bearers into environments. As recently as the 1930s, Soviet Union introduced European mink, *Mustela lutreola*, raccoon dogs, *Nyctereutes procyonoides*, and

Kamchatka sables, *Martes zibellina kamschadalica,* into different areas of central and eastern Russia with the goal of expanding fur-harvesting opportunities (Lindemann, 1956; see also Gantchoff *et al.*, Chapter 20, this volume). Numerous other attempts to introduce non-indigenous small carnivores have occurred for this goal, including the masked palm civet, *Paguma larvata* (e.g. Japan), northern raccoon (e.g. Canada, New Zealand, several countries in Asia and Europe), and American mink (e.g. several countries in Asia, Europe, South America; see also Gantchoff *et al.*, Chapter 20, this volume, for an extensive review). In contrast, successful reintroductions of native small carnivores have also occurred, with goals of species or population conservation and restoring ecosystems (e.g. Lewis, 2006; Mowry *et al.*, 2015). Overharvest of sables resulted in a prohibition of hunting and trapping during the 1940s, but the reintroduction of approximately 20 000 sables between 1940 and 1960 occurred in an effort to restore their numbers (Monakhov, 2011). In many northern regions of the United States, fishers have been successfully reintroduced in areas where populations were extirpated decades ago because of habitat loss and unregulated harvest (e.g. Lewis *et al.*, 2012).

Three primary issues have been identified with respect to the management of furbearers: (i) increasing human populations and the resultant loss of habitat; (ii) the increasing level of public intolerance of furbearers in urban areas; and (iii) opposition from animal rights activists with regards to harvest or use of furbearers (The Wildlife Society, 2001). Small carnivores are arguably the most sensitive group with regard to these management issues. Populations of small carnivores often have low intrinsic rates of increase. This characteristic, along with long generation times and long-lived individuals, may result in species or groups of species with less resilience to harvest in comparison with species with characteristics at the other end of this spectrum (Bodmer *et al.*, 1997; Banci & Proulx, 1999). Even during the 1500s in the north-eastern United States, areas with high human densities resulted in local depletions of small carnivores and other wildlife species (McGee, 1987). Although unmeasured in many parts of the world today, this pattern could also be expected in many parts of Asia where human populations are increasing

and consequently applying more pressure to populations of small carnivores.

Harvest can play a key role in the population dynamics of small carnivores, even indirectly if harvest focuses on large animals (Nasi *et al.*, 2008). Decreasing or loss of large carnivore populations, and the potential cascade effect and mesopredator release that results (Prugh *et al.*, 2009), suggests that both large herbivores and large carnivores may play a role in small carnivore population ecology. Changes in land use and climate seem to affect small carnivore populations, both positively and negatively (e.g. Aubry *et al.*, 2007; Lawler *et al.*, 2012), and often as a synergistic mechanism. In North America, during the 1900s, the distribution of the raccoon has increased northward into the Canadian prairies, with increasing food availability caused by global warming and increasing agricultural activities appearing to be the most plausible hypothesis (Larivière, 2004). Climate change may also result in elevational shifts for generalist species, including expansion of raccoons and skunks to higher elevations (e.g. McFadden-Hiller & Hiller, 2015) and potential northward shifts in latitude by fishers (Lawler *et al.*, 2012).

Unregulated harvest in some areas of the world, often in conjunction with habitat loss, has contributed to the decline or extinction of several wildlife species (e.g. Perez, 2001). Underlying reasons may be traced to consumptive uses tied to society, culture, or religion (e.g. food, medicine, and ornamental) (Nasi *et al.*, 2008), and logging, construction of roads, and other landscape-level anthropogenic changes that serve to increase human access to wildlife. Ironically, some of these anthropogenic effects, including removal of large carnivores, may result in an increase in the abundance of certain mesocarnivore species (i.e. mesopredator release), albeit perhaps in conjunction with reductions in ecosystem function (Prugh *et al.*, 2009). Monitoring harvest has proven difficult and is either typically non-existent or has shown harvest to be unsustainable in many parts of the world, but monitoring harvest in parts of North America seems to have achieved the goal of sustainable harvests (Robinson & Wirth, 1998). In Canada and the United States, monitoring of harvest of most small carnivores is conducted by provincial and state wildlife agencies, respectively. Types of data collected can range widely by province or state and by species from basic (e.g. the number of each species harvested per year) to more complex information (e.g. collection of body parts or carcasses for individual sex and age determination, female reproductive tract assessment, and harvest effort). With sufficient data collection and analysis, informed decisions for regulatory adjustments in harvest may be made (see Hiller *et al.*, 2011; Skalski *et al.*, 2011).

After approximately a decade of debate and discussion, CITES was enforced in 1975. This international agreement, which includes about 180 countries, was implemented to prevent international trade in endangered species, but also to enact regulation in trade of other species; this responsibility was assigned for both producer and consumer countries (Wijnstekers, 2011). Species threatened with extinction, classified as Appendix I species, are traded only under very limited circumstances. Species that may not be threatened but in which trade should be controlled to avoid consumptive uses that are incompatible with survival are classified as Appendix II species. Finally, Appendix III species are those protected in at least one country and that country has requested assistance with control of trade (CITES, 2013). Some Appendix II species (e.g. river otters) have been described as 'look-alike species', where importation tags are required for species identification purposes to aid trade officials that may not be able to distinguish them from endangered species (e.g. giant otters, *Pteronura brasiliensis*). However, there has been some controversy associated with this because these so-called look-alike species are subject to the same controls as species threatened by international trade despite not being of range-wide conservation concern (e.g. AFWA, 2014).

Changes in public values concerning humanity's use of animals also impacted the methods used to capture small carnivores. Much has been written on the topic of trapping and animal welfare by a diverse range of individuals and groups. However, our focus on the debate centres on how the dispute over the consumptive use of small carnivores impacts international trade. Although animal rights and animal protectionist ideas were not new, they gained popular appeal starting in the mid-1970s following the publication of *Animal Liberation* (Singer, 1975). As the animal-protection movement grew, particularly in Europe,

officials of the European Commission in the early 1990s developed regulations that prohibited the use of foothold traps in Europe, but also the importation of pelts from certain wildlife species from countries that did not meet international trapping standards (Princen, 2004).

Countries that exported wild-captured fur took different approaches in their attempts to meet the trade requirements of the European Commission. The European Union concluded the trilateral Agreement on International Humane Trapping Standards with Canada (1998) and the Russian Federation (2008), which has the force of a treaty (European Commission, 1998; Council of the European Union, 1998; Talling & Inglis, 2009). In the United States, the largest-volume producer of wild-captured furs (US International Trade Commission, 2004), adoption of national standards was hindered by the fact that regulatory authority for furbearers resides at the state level rather than the federal level. Thus, standards or recommendations developed in the United States (i.e. Best Management Practices for Trapping, United States; AFWA, 2013; White *et al.*, 2021) have advisory force as opposed to regulatory force. Conversely, in Africa, traps are essentially unregulated and non-selective (e.g. Central African Republic; Noss, 1998) for harvesting small carnivores or certain other wildlife species for food or fur.

Although not a regulatory entity, the International Union for Conservation of Nature (IUCN) has developed *The IUCN Red List of Threatened Species* in an effort to '. . .provide information and analyses on the status, trends and threats to species in order to inform and catalyze action for biodiversity conservation' (IUCN, 2013). The Red List has a search option for users to locate information on ecology, threats, conservation actions, and other details related to small carnivores and other wildlife species.

Challenges related to consumptive uses of wildlife continue in many parts of the world, including North America, with differing human values about what constitutes animal cruelty (Vantassel, 2009). It remains to be seen how such challenges will be addressed or dismissed politically, socially, and through regulatory mechanisms, and how such decisions may impact the balance between consumptive use by humans and the conservation and management of small carnivores.

Consumptive Use of Small Carnivores in the Future

Changes in social acceptance of harvest in some countries, known or unknown levels of unsustainable harvest in others, tolerance of illegal trade, or unenforced or unenforceable laws regarding illegal trade have affected the use of certain small carnivore species in many parts of the world. The future viability of small carnivores depends to a large extent on how much value humans place on this group relative to values associated with economics, other natural resources, and other factors. As a general rule, when people have a sense of ownership and connection with a species, that species does remarkably well. One needs only to consider the success of domestic dogs, *Canis familiaris*, to recognize the truth of this point.

Unfortunately, changing public attitudes is difficult and time-consuming. The time and effort required to change attitudes may exceed the efficacy of conservation efforts for certain species in light of habitat changes, harvest practices, and other factors. Long-term success can occur only when (i) people believe that humans have a responsibility to maintain the viability of species; (ii) the culture and laws allow for creativity with and ownership of economic resources (typically satisfied by viable and open markets where the rule of law is upheld); (iii) where interested parties are willing to pay for the economic costs of preserving and maintaining species so that area residents, who are often poor, do not have to bear the full brunt of the economic costs of conservation; and (iv) assurances that when wildlife populations reach sustainable levels, consumptive users will have a way to harvest those species in a cost-effective way as part of an overall management plan. Clearly, there is a delicate balance between moral responsibilities to keep these species viable while also maintaining human interests.

Acknowledgements

We thank the editors, E. Do Linh San, J.J. Sato, J.L. Belant, and M.J. Somers, of this book for the opportunity to contribute a chapter on such a challenging topic. The thoughtful reviews of the editors, P. Jensen, G. Proulx, and B. White improved this manuscript.

References

AFWA [Association of Fish and Wildlife Agencies] (2013) Best management practices. http://fishwildlife. org/?section=best_management_practices. Accessed on 2 October 2013.

AFWA [Association of Fish and Wildlife Agencies] (2014) Case study: state and federal management of the bobcat in the context of the Convention on International Trade in Endangered Species of Wild Fauna and Flora (CITES). http://www.fishwildlife. org/files/bobcatwebsite.pdf. Accessed on 15 January 2018.

AFWA [Association of Fish and Wildlife Agencies] (2021) Furbearer management, National fur harvest database. http://www.fishwildlife.org/index. php?section=furbearer_management&activator=27. Accessed on 5 April 2021.

Allebone-Webb, S.M., Kumpel, N.F., Rist, J., Cowlishaw, G., Rowcliffe, J.M. & Milner-Gulland, E.J. (2011) Use of market data to assess bushmeat hunting sustainability in equatorial Guinea. *Conservation Biology* 25, 597–606.

Allee, W.C., Emerson, A.E., Park, O., Park, T. & Schmidt, K.P. (1949) *Principles of Animal Ecology*. W.B. Saunders, Philadelphia.

Aubry, K.B., McKelvey, K.S. & Copeland, J.P. (2007) Distribution and broadscale habitat relations of the wolverine in the contiguous United States. *Journal of Wildlife Management* 71, 2147–2158.

Bachrach, M. (1949) *Fur: A practical Treatise*. Prentice Hall, New York.

Bahaa-el-din, L., Henschel, P., Aba'a, R., Abernethy, K., Bohm, T., Bout, N., Coad, L., Head, J., Inoue, E., Lahm, S., Lee, M.E., Maisels, F., Rabanal, L., Starkey, M., Taylor, G., Vanthomme, A., Nakashima, Y. & Hunter, L. (2013) Notes on the distribution and status of small carnivores in Gabon. *Small Carnivore Conservation* 48, 19–29.

Banci, V. & Proulx, G. (1999) Resiliency of furbearers to trapping in Canada. In: *Mammal Trapping* (ed. G. Proulx), pp. 175–203. Alpha Wildlife Research & Management, Sherwood Park.

Barger, W.J. (2008) New players at the table: how Americans came to dominate early trade in the North Pacific. *Southern California Quarterly* 90, 227–257.

Behrensmeyer, A.K. (1978) Taphonomic and ecologic information from bone weathering. *Paleobiology* 4, 150–162.

Beringer, J. & Grusenmeyer, E. (2014) Furbearer program annual report. Missouri Department of Conservation, Resource Science Division. http://mdc.mo.gov/sites/default/files/resources/2012/01/fy14furbearerreport.pdf. Accessed on 9 December 2014.

Bettini, M. (2013) *Women and Weasels: Mythologies of Birth in Ancient Greece and Rome*. University of Chicago Press, Chicago.

Blasco, R. & Peris, J.F. (2012) Small and large game: human use of diverse faunal resources at Level IV of Bolomor Cave (Valencia, Spain). *Comptes Rendus Palevol* 11, 265–282.

Bodmer, R.E., Eisenberg, J.F. & Redford, K.H. (1997) Hunting and the likelihood of extinction of Amazonian mammals. *Conservation Biology* 11, 460–466.

Brander, M. (1971) *Hunting and Shooting: From Earliest Times to the Present Day*. G.P. Putnam's and Sons, New York.

Carpaneto, G.M. & Germi, F.P. (1989) The mammals in the zoological culture of the Mbuti pygmies in north-eastern Zaire. *Hystrix* 1, 1–83.

Caywood, L.R. (1967) Post-1800 sites: fur trade. *Historical Archeology* 1, 46–48.

CITES [Convention on International Trade in Endangered Species of Wild Fauna and Flora] (2013) How CITES works. http://www.cites.org/eng/disc/how.php. Accessed on 28 October 2013.

Clayton, J.L. (1966) The growth and economic significance of the American fur trade, 1790–1890. *Minnesota History* 40, 210–220.

Colyn, M., Dufour, S., Condé, P.C. & Van Rompaey, H. (2004) The importance of small carnivores in forest bushmeat hunting in the Classified Forest of Diecké, Guinea. *Small Carnivore Conservation* 31, 15–18.

Council of the European Union (1998) Council decision of 26 January 1998 concerning the conclusion of an Agreement on international humane trapping standards between the European Community, Canada and the Russian Federation and of an Agreed Minute between Canada and the European Community concerning the signing of the said Agreement (98/142/EC). http://eur-lex.europa.eu/legal-content/EN/TXT/PDF/?uri=CELEX:31998D0142&from=EN. Accessed on 28 December 2016.

Cunningham, A.B. & Zondi, A.S. (1991) *Use of Animal Parts for the Commercial Trade in Traditional Medicines*. Institute of Natural Resources, University of Natal, Pietermaritzburg.

Cuzin, F. (2013) *Poecilictis libyca* Libyan Striped Weasel. In: *The Mammals of Africa. V. Carnivores, Pangolins, Equids and Rhinoceroses* (eds J. Kingdon & M. Hoffmann), pp. 90–92. Bloomsbury, London.

Dolin, E.J. (2010) *Fur, Fortune, and Empire: The Epic History of the Fur Trade in America*. W.W. Norton & Company, New York.

Do Linh San, E. (2006) [*The Eurasian Badger: Description, Behaviour, Social Life, Protection, Observation*]. Delachaux & Niestlé, Paris. (In French).

Do Linh San, E., Begg, C., Begg, K. & Abramov, A. (2016) *Mellivora capensis*. The IUCN Red List of Threatened Species 2016, e.T41629A45210107. doi:10.2305/IUCN.UK.2016-1.RLTS.T41629A45210107.en. Accessed on 28 December 2016.

Domingo-Roura, X., Marmi, J., Ferrando, A., López-Giráldez, F., Macdonald, D.W. & Jansman, H.A.H. (2006) Badger hair in shaving brushes comes from protected Eurasian badgers. *Biological Conservation* 128, 425–430.

Duckworth, J.W. (1997) Small carnivores in Laos: a status review with notes on ecology, behaviour and conservation. *Small Carnivore Conservation* 16, 1–21.

Dunaway, W.A. (1994) The southern fur trade and the incorporation of Southern Appalachia into the world-economy, 1690–1763. *Review (Fernand Braudel Center)* 17, 215–242.

European Commission (1998) Agreement on international humane trapping standards between the European Community, Canada and the Russian Federation. http://fur.ca/wp-content/uploads/2015/09/AIHTS-Copy-of-Agreement.pdf. Accessed on 28 December 2016.

Fa, J.E., Peres, C.A. & Meeuwig, J. (2002) Bushmeat exploitation in tropical forests: an intercontinental comparison. *Conservation Biology* 16, 232–237.

Fur Commission USA (2011) World production of mink. http://old.furcommission.com/news/newsF12e.htm. Accessed on 22 October 2013.

Fur Institute of Canada (2019) https://fur.ca. Accessed on 3 August 2019.

Gantchoff, M., Libal, N.S. & Belant, J.L. (Chapter 20, this volume) Small carnivore introductions: ecological and biological correlates of success. In: *Small Carnivores: Evolution, Ecology, Behaviour, and Conservation* (eds E. Do Linh San, J.J. Sato, J.L. Belant & M.J. Somers). Wiley–Blackwell, Oxford.

Gaubert, P., Bahaa-el-din, L., Ray, J. & Do Linh San, E. (2015) *Nandinia binotata. The IUCN Red List of Threatened Species* 2015, e.T41589A45204645. doi:10.2305/IUCN.UK.2015-4.RLTS. T41589A45204645.en. Accessed on 28 December 2016.

Gerstell, R. (1985) *The Steel Trap in North America.* Stackpole Books, Harrisburg.

Gilman, R.R. (1974) The fur trade in the Upper Mississippi Valley, 1630–1850. *The Wisconsin Magazine of History* 58, 2–18.

Gilmore, R.M. (1946) Mammals in archeological colleges from southwestern Pennsylvania. *Journal of Mammalogy* 27, 227–234.

Glatston, A.R. & Gebauer, A. (2011) People and red pandas: the red panda's role in economy and culture. In: *Red Panda: Biology and Conservation of the First Panda* (ed. A.R. Glatston), pp. 11–26. Academic Press, London.

Golden, C.D. (2009) Bushmeat hunting and use in the Makira Forest, north-eastern Madagascar: a conservation and livelihoods issue. *Oryx* 43, 386–392.

Gough, B.M. (1989) India-based expeditions of trade and discovery in the North Pacific in the late Eighteenth century. *The Geographical Journal* 155, 215–223.

Gupta, B.K. (2004) Killing civets for meat and scent in India. *Small Carnivore Conservation* 31, 21.

Hansen, H.O. (2017) Global fur retail value. https://fur.ca/wp-content/uploads/2017/09/Global_Fur_Retail_July_20172.pdf. Accessed on 5 April 2021.

Hart, J.A. & Timm, R.M. 1978. Observations on the aquatic genet in Zaire. *Carnivore* 1, 130–132.

Helms, S. & Betts, A. (1987) The desert "kites" of the Badiyat Esh-Sham and North Arabia. *Paeorient* 13, 41–67.

Hiller, T.L. (2011) Oregon furbearer program report, 2010–2011. Oregon Department of Fish and Wildlife, Salem, Oregon, USA. http://www.dfw.state.or.us/resources/hunting/small_game/docs/2011_furbearer_report.pdf. Accessed on 3 December 2014.

Hiller, T.L., Etter, D.R., Belant, J.L. & Tyre, A.J. (2011) Factors affecting harvests of fishers and American martens in northern Michigan. *Journal of Wildlife Management* 75, 1399–1405.

Hofstra, T. & Samplonius, K. (1995) Viking expansion northwards: mediaeval sources. *Arctic* 48, 235–247.

Hughes, S.S. (2009) Noble marten (*Martes americana nobilis*) revisited: its adaptation and extinction. *Journal of Mammalogy* 90, 74–92.

Ingrams, F.C. (1924) Furs and the fur trade. *Journal of the Royal Society of Arts* 72, 593–605.

International Fur Trade Federation (2013) About. http://www.wearefur.com/our-trade/about-the-fur-trade. Accessed on 18 October 2013.

IUCN [International Union for Conservation of Nature] (2013) Red list overview. http://www.iucnredlist.org/about/red-list-overview#introduction. Accessed on 28 October 2013.

King, C.M. & Powell, R.A. (2007) *The Natural History of Weasels and Stoats: Ecology, Behavior, and Management.* 2nd edition. Oxford University Press, New York.

Klein, R.G. (1974) Environment and subsistence of prehistoric man in southern Cape Province, South Africa. *World Archeology* 5, 249–284.

Krausman, P.R. (2013) Defining wildlife and wildlife management. In: *Wildlife Management & Conservation: Contemporary Principles & Practice* (eds P.R. Krausman & J.W. Cain III), pp. 1–5. The Johns Hopkins University Press, Baltimore.

Kuhn, R.D. & Funk, R.E. (2000) Boning up on the Mohawk: an overview of Mohawk faunal assemblages and subsistence patterns. *Archaeology of Eastern North America* 28, 29–62.

Larivière, S. (2004) Range expansion of raccoons in the Canadian prairies: review of hypotheses. *Wildlife Society Bulletin* 32, 955–963.

Lawler, J.J., Safford, H.D. & Girvetz, E.H. (2012) Martens and fishers in a changing climate. In: *Biology and Conservation of Martens, Sables, and Fishers: A New Synthesis* (eds K.B. Aubry, W.J. Zielinski, M.G. Raphael, G. Proulx & S.W. Buskirk), pp. 371–397. Cornell University Press, Ithaca.

Lewis, J.C. (2006) Implementation plan for reintroducing fishers (*Martes pennanti*) to Olympic National Park. Washington Department of Fish and Wildlife, Olympia.

Lewis, J.C., Powell, R.A. & Zielinski, W.J. (2012) Carnivore translocations and conservation: insights from population models and field data for fishers (*Martes pennanti*). *PLoS One* 7, e32726.

Lindemann, W. (1956) Transplantation of game in Europe and Asia. *Journal of Wildlife Management* 20, 68–70.

Lockman Foundation (1963a) *New American Standard Bible*. 3rd edition. Leviticus 11:1–47. Lockman Foundation Press, La Habra.

Lockman Foundation (1963b) *New American Standard Bible*. 3rd edition. 1 Samuel 17:34. Lockman Foundation Press, La Habra.

Loibooki, M., Hofer, H., Campbell, K.L.I. & East, M.I. (2002) Bushmeat hunting by communities adjacent to the Serengeti National Park, Tanzania: the importance of livestock ownership and alternative sources of protein and income. *Environmental Conservation* 29, 391–398.

Long, C.A. & Killingley, C.A. (1983). *The Badgers of the World*. Charles C. Thomas, Springfield.

Lyman, R.L. (2007) Prehistoric mink (*Mustela vison*) trapping on the northwest coast. *Journal of Field Archaeology* 32, 91–95.

Macdonald, D.W., Johnson, P.J., Albrechtsen, L., Dutton, A., Seymour, S., Dupain, J., Hall, A. & Fa, J.E. (2011) Association of body mass with price of bushmeat in Nigeria and Cameroon. *Conservation Biology* 25, 1220–1228.

Malbrant, R. & Maclatchy, A. (1949) *Faune de l'Equateur Africain Français. Tome II. Mammifères*. Paul Lechevalier, Paris.

Marciszak, A. (2016) The common weasel *Mustela nivalis* L., 1766 from Sarakenos Cave (Greece) and the difficulty to distinguish small mustelid species in the fossil record. *Acta Zoologica Cracoviensia* 59, 25–35.

McFadden-Hiller, J.E. & Hiller, T.L. (2015) Non-invasive survey of forest carnivores in the northern Cascades of Oregon, USA. *Northwestern Naturalist* 92, 102–117.

McGee, H.F. Jr (1987) The use of furbearers by native North Americans after 1500. *Wild Furbearer Management and Conservation in North America* (eds M. Novak, J.A. Baker, M.E. Obbard & B. Malloch), pp. 13–20. Ministry of Natural Resources, Ontario.

Mickevicius, E. & Baranauskas, K. (1992) Status, abundance and distribution of mustelids in Lithuania. *Small Carnivore Conservation* 6, 11–14.

Monakhov, V.G. (2011) *Martes zibellina* (Carnivora: Mustelidae). *Mammalian Species* 43(876), 75–86.

Mowry, R.A., Schneider, T.M., Latch, E.K., Gompper, M.E., Beringer, J. & Eggert, L.S. (2015) Genetics and the successful reintroduction of the Missouri river otter. *Animal Conservation* 18, 196–206.

Nasi, R., Brown, D., Wilkie, D., Bennett, E., Tutin, C., van Tol, G. & Christopher, T. (2008) *Conservation and Use of Wildlife-based Resources: The Bushmeat Crisis*. Technical Series No. 33. Secretariat of the Convention on Biological Diversity, Montreal, and Center for International Forestry Research, Bogor.

Negev, A. & Gibson, S. (eds) (2003) *Archeological Encyclopedia of the Holy Land*. The Continuum International Publishing Group, New York.

Noss, A.J. (1998) The impacts of cable snare hunting on wildlife populations in the forests of the Central African Republic. *Conservation Biology* 12, 390–398.

Novak, M., Obbard, M.E., Jones, J.G., Newman, R., Booth, A., Satterthwaite, A.J. & Linscombe, G. (1987) Furbearer harvests in North America, 1600–1984. Ministry of Natural Resources, Ontario.

Obbard, M.E. (1987) Fur grading and pelt identification. In: *Wild Furbearer Management and Conservation in North America* (eds M. Novak, J.A. Baker, M.E. Obbard & B. Malloch), pp. 717–826. Ministry of Natural Resources, Ontario.

Obbard, M.E., Jones, J.G., Newman, R., Booth, A., Satterthwaite, A.J. & Linscombe, G. (1987) Furbearer harvest in North America. In: *Wild furbearer Management and Conservation in North America* (eds M. Novak, J.A. Baker, M.E. Obbard & B. Malloch), pp. 1007–1034. Ministry of Natural Resources, Ontario.

Perez, C.A. (2001) Synergistic effects of subsistence hunting and habitat fragmentation on Amazonian forest vertebrates. *Conservation Biology* 15, 1490–1505.

Princen, S. (2004) EC compliance with WTO law: the interplay of law and politics. *European Journal of International Law* 15, 555–574.

Proulx, G., Aubry, K., Birks, J., Buskirk, S., Fortin, C., Frost, H., Krohn, W., Mayo, L., Monakhov, V., Payer, D., Saeki, M., Santos-Reis, M., Weir, R. & Zielinski, W. (2004) World distribution and status of the genus *Martes* in 2000. In: *Martens and Fishers (Martes) in Human-Altered Environments: An International Perspective* (eds D.J. Harrison, A.K. Fuller & G. Proulx), pp. 21–76. Springer, New York.

Prugh, L.R., Stoner, C.J., Epps, C.W., Bean, W.T., Ripple, W.J., Laliberte, A.S. & Brashares, J.S. (2009) The rise of the mesopredator. *BioScience* 59, 779–791.

Ramakantha, V. (1994) Natural distribution and ecology of mustelids and viverrids in Manipur, north-eastern India. *Small Carnivore Conservation* 11, 16–18.

Randrianandrianina, F.H., Racey, P.A. & Jenkins, R.K.B. (2010) Hunting and consumption of mammals and birds by people in urban areas of western Madagascar. *Oryx* 44, 411–415.

Ray, A.J. (1987) The fur trade in North America: an overview from a historical geographical perspective. In: *Wild Furbearer Management and Conservation in North America* (eds M. Novak, J.A. Baker, M.E. Obbard & B. Malloch), pp. 21–30. Ministry of Natural Resources, Ontario.

Ray, D.J. (1975) Early maritime trade with the Eskimo of Bering Strait and the introduction of firearms. *Arctic Anthropology* 12, 1–9.

Ray, V.F. (1938) Lower Chinook ethnographic notes. *University of Washington Publications in Anthropology* 7, 29–165.

Redman, C.L. (1999) *Human Impact on Ancient Environments*. University of Arizona Press, Tucson.

Robinson, P. & Wirth, R. (1998) IUCN/SSC Wildlife Trade Programme and small carnivores. *Small Carnivore Conservation* 18, 24.

Sasaki, S. (1998) Fur animal hunting of the indigenous people in the Russian Far East: history, technology, and the economic effects. *A Step Toward Forest Conservation Strategy* 1, 495–513.

Schieff, A. & Baker, J.A. (1987) Marketing and international fur markets. In: *Wild Furbearer Management and Conservation in North America* (eds M. Novak, J.A. Baker, M.E. Obbard & B. Malloch), pp. 862–877. Ministry of Natural Resources, Ontario.

Shackelford, R.M. (1980) Domestic production of mink and foxes. University of Wisconsin, Madison. http://www.poultryscience.org/docs/pba/1952-2003/1980/1980%20Shackelford.pdf. Accessed on 22 October 2013.

Shepherd, C.R. (2008) Civets in trade in Medan, North Sumatra, Indonesia (1997–2001) with notes on legal protection. *Small Carnivore Conservation* 38, 34–36.

Shepherd, C.R. (2012) Observations of small carnivores in Jakarta wildlife markets, Indonesia, with notes on trade in Javan ferret badger *Melogale orientalis* and on the increasing demand for common palm civet *Paradoxurus hermaphroditus* for civet coffee production. *Small Carnivore Conservation* 47, 38–41.

Sifuna, N. (2012) The future of traditional customary uses of wildlife in modern Africa: a case study of Kenya and Botswana. *Advances in Anthropology* 2, 31–38.

Singer, P. (1975) *Animal Liberation: A New Ethics for our Treatment of Animals*. Avon Publishers, New York.

Skalski, J.R., Millspaugh, J.J., Clawson, M.V., Belant, J.L., Etter, D.R., Frawley, B.J. & Friedrich, P.D. (2011) Abundance trends of American martens in Michigan based on statistical population reconstruction. *Journal of Wildlife Management* 75, 1767–1773.

Stains, H.J. (1979) Primeness in North American furbearers. *Wildlife Society Bulletin* 7, 120–124.

Stabler, J.C. & Howe, E.C. (1990) Native participation in northern development: the impending crisis in the NWT. *Canadian Public Policy* 16, 262–283.

Statistics Canada (2019) Archived: number and values of pelts produced. https://www150.statcan.gc.ca/t1/tbl1/en/tv.action?pid=3210029301. Accessed on 9 April 2019.

Su, S. (2005) Small carnivores and their threats in Hlawga Wildlife Park, Myanmar. *Small Carnivore Conservation* 33, 6–13.

Talling, J.C. & Inglis, I.R. (2009) *Improvements to Trapping Standards*. Directorate-General for Environment, European Commission. http://citeseerx.ist.psu.edu/viewdoc/download?doi=10.1.1.306.7379&rep=rep1&type=pdf. Accessed on 28 December 2016.

Tobias, T.N. & Kay, J.J. (1994) The bush harvest in Pinehouse, Saskatchewan, Canada. *Arctic* 47, 207–221.

Turvey, S. & Helgen, K. (2008) *Neovison macrodon. The IUCN Red List of Threatened Species*. Version 2013.1. http://www.iucnredlist.org. Accessed on 28 October 2013.

U.S. Fish and Wildlife Service (2003) *Final Revised Recovery Plan for the Southern Sea Otter (Enhydra lutris nereis)*. U.S. Fish and Wildlife Service, Portland.

U.S. International Trade Commission (2004) *Industry and Trade Summary: Furskins*. USITC Publication 3666, Washington, DC.

Vantassel, S.M. (2009) *Dominion Over Wildlife? An Environmental-Theology of Human–Wildlife Relations*. Wipf and Stock, Eugene.

Wadley, L. (2010) Were snares and traps used in the Middle Stone Age and does it matter? A review and a case study from Sibudu, South Africa. *Journal of Human Evolution* 58, 179–192.

White, H.B., Batcheller, G.R., Boggess, E.K., Brown, C.L., Butfiloski, J.W., Decker, T.A., Erb, J.D., Fall, M.W., Hamilton, D.A., Hiller, T.L., Hubert, G.F. Jr, Lovallo, M.J., Olson, J.F. & Roberts, N.M. (2021) Best management practices for trapping furbearers in the United States. *Wildlife Monographs* 207, 3–59.

Wijnstekers, W. (2011) *The Evolution of CITES.* 9th edition. International Council for Game and Wildlife Conservation, Budapest. http://www.cites.org/eng/resources/publications.php. Accessed on 25 October 2013.

Wildlife Society (2001) *Trapping and Furbearer Management in North American Wildlife Conservation.* The Northeast Section of The Wildlife Society. http://www.fishwildlife.org/files/Trapping-and-Furbearer_Management-in-North-American-Wildllife-Conservation.pdf. Accessed on 26 October 2011.

Worthy, G.A., Rose, J. & Stormshak, F. (1987) Anatomy and physiology of fur growth: the pelage priming process. In: *Wild Furbearer Management and Conservation in North America* (eds M. Novak, J.A. Baker, M.E. Obbard & B. Malloch), pp. 827–841. Ministry of Natural Resources, Ontario.

Wright, J.V. (1987) Archaeological evidence for the use of furbearers in North America. In: *Wild Furbearer Management and Conservation in North America* (eds M. Novak, J.A. Baker, M.E. Obbard & B. Malloch), pp. 3–12. Ministry of Natural Resources, Ontario.

23

Conservation Status of the North American River Otter in the United States and Canada: Assessing Management Practices and Public Perceptions of the Species

Emily A. Bricker[1], Thomas L. Serfass[1],*, Zoe L. Hanley[2], Sadie S. Stevens[1], Kelly J. Pearce[3], and Jennifer A. Bohrman[1]

[1] *Department of Biology and Natural Resources, Frostburg State University, Frostburg, MD, USA*
[2] *Northwest Regional Office, Defenders of Wildlife, White Salmon, WA, USA*
[3] *Marine-Estuarine-Environmental Sciences Graduate Program, University of Maryland, College Park, MD, USA*

SUMMARY

The historic range of the North American river otter, *Lontra canadensis*, included much of the North American continent, from arctic Alaska and northern Canada to the southern United States (US). However, overharvest and perturbations to aquatic environments contributed to the decline and, in some cases, the extirpation of river otter populations through substantial portions of the species' former range. The last update to the conservation and management status of the river otter in the US was done by Raesly (2001), which was found to include several errors concerning state river otter reintroductions. There had not been a comprehensive study of the conservation and management status of the river otter in Canada. We conducted telephone and email surveys with furbearer biologists in the US and Canada to update the conservation and management status of river otter populations among US states and Canadian provinces and territories. The river otter has recovered in many areas and now occupies at least portions of its historic range in each jurisdiction, except Prince Edward Island, with populations reported as stable, expanding, or a combination thereof. This increase in river otter distribution and abundance was facilitated by a combination of reintroduction projects involving 22 states, improvements in aquatic habitat quality, and the natural expansion of native populations. Recovery of river otter populations has resulted in an increase of legal trapping seasons among states, increasing from 31 states in 2006 to 37 in 2013. Eleven states and one province calculated river otter population estimates, and research is needed to further develop and refine existing approaches for monitoring populations. Public complaints about river otters were reported in 43 states and 5 provinces, but were most often classified as infrequent or rare. The conservation status of river otter populations in North America has improved considerably and constitutes a conservation success story. However, the historic vulnerability of the river otter to various perturbations – particularly overharvest – is well documented, and future conservation strategies should include careful field monitoring of the species' distribution and population status to identify threats that could hinder or reverse its ongoing recovery. An Addendum is included at the end of the chapter to facilitate an update of recent changes and offer additional thoughts pertaining to the conservation of river otters.

Keywords

Conservation — furbearer — *Lontra canadensis* — overharvest — public complaints — population status — reintroduction

* Corresponding author

Small Carnivores: Evolution, Ecology, Behaviour, and Conservation, First Edition. Edited by Emmanuel Do Linh San, Jun J. Sato, Jerrold L. Belant, and Michael J. Somers.
© 2022 John Wiley & Sons Ltd. Published 2022 by John Wiley & Sons Ltd.

Introduction

Prior to European settlement, the North American river otter, *Lontra canadensis* (hereafter river otter; Figure 23.1), ranged throughout much of North America, from arctic Alaska to the southern United States (US), with populations occupying a variety of aquatic systems in each of the 48 conterminous US states and Alaska, and all Canadian provinces ($n = 10$) and territories ($n = 3$) (Hall, 1981; Melquist & Dronkert, 1987; Foster-Turley *et al.*, 1990; Melquist *et al.*, 2003; Kruuk, 2006). By the early to mid-1900s, the species had experienced substantial population declines, or complete extirpations, throughout large portions of its historic range, particularly in the interior US and southern Canada (Melquist & Dronkert, 1987; Tesky, 1993; Melquist *et al.*, 2003; Kruuk, 2006; Bricker, 2014). These declines were caused primarily through the combined detrimental effects of overharvesting by trappers, disturbances to riparian habitats (e.g. deforestation), and water pollution. Though difficult to address cause and effect circumstances of historical river otter declines with absolute certainty, the severity of the declines was potentially influenced by interactions among these perturbations and various landscape factors, most notably habitat quality and availability (i.e. the abundance, types, and productivity of aquatic habitats). For example, river otter populations seem to have been particularly vulnerable to overharvest and experienced the most severe declines in the central US, probably because the land area in

this region possessed fewer and less diverse aquatic habitats (i.e. primarily riverine systems), which likely supported fewer river otters than other regions. Disturbances to riparian habitats and water pollution associated with intensive agricultural development were probably additional important contributing factors to river otter declines in this region. Relative to the central US, coastal marshes along the gulf coast of the south-eastern US, and lacustrine and palustrine systems in the Great Lakes region of the north-central US provided a greater abundance and variety of aquatic systems. These regions may have been able to support higher densities of river otters because the aquatic systems were distributed in a manner that limited trapper access and were thus better able to sustain populations through periods of intensive harvest. Although river otter populations also declined, or experienced extirpations, in some areas of the south-east and the north-central US, viable populations still persisted in at least – often substantial – portions of these regions (Melquist *et al.*, 2003). Generally, improvements in water quality related to the federal 'Clean Water Act' in 1972 (USEPA, 2015) and better protection of riparian areas associated with federal legislation designed to minimize environmental degradation associated with agriculture (e.g. the 'Farm Bill'; see Chapter 13 *in* Bolen & Robinson, 2003) likely contributed to the recovery of river otter populations in many areas.

In 1977, the river otter was listed as an Appendix II species by CITES (Convention on International Trade in Endangered Species of Wild Fauna and Flora), a designation which requires mandatory tagging of pelts intended for export (Nilsson, 1980; CITES, 2013). The CITES II listing appears to have been a primary factor in motivating state wildlife agencies to increase conservation actions targeted at declining river otter populations. These actions depended on circumstances unique to particular states, and included initiatives such as elevating the conservation status to higher protection levels, adopting more conservative harvest strategies, and implementing reintroduction projects in states where river otters were completely or partially extirpated (Melquist & Hornocker, 1983; DiStefano, 1987; Ralls, 1990; Butler, 1991; Raesly, 2001; Serfass *et al.*, 2003). In 1976, Colorado (CO; see Table 23.1 for postal code abbreviations used hereafter for all US states) initiated the first river otter

Figure 23.1 North American river otter, *Lontra canadensis*. *Source*: Photo © Thomas L. Serfass.

Table 23.1 State abbreviations used in the text, figures, and tables based on the United States Official Postal Guide (USPS, 2015).

Abbreviation – State	Abbreviation – State	Abbreviation – State
AK – Alaska	MA – Massachusetts	OK – Oklahoma
AL – Alabama	MD – Maryland	OR – Oregon
AR – Arkansas	ME – Maine	PA – Pennsylvania
AZ – Arizona	MI – Michigan	RI – Rhode Island
CA – California	MN – Minnesota	SC – South Carolina
CO – Colorado	MO – Missouri	SD – South Dakota
CT – Connecticut	MS – Mississippi	TN – Tennessee
DE – Delaware	MT – Montana	TX – Texas
FL – Florida	NC – North Carolina	UT – Utah
GA – Georgia	ND – North Dakota	VA – Virginia
IA – Iowa	NE – Nebraska	VT – Vermont
ID – Idaho	NH – New Hampshire	WA – Washington
IL – Illinois	NJ – New Jersey	WI – Wisconsin
IN – Indiana	NM – New Mexico	WV – West Virginia
KS – Kansas	NV – Nevada	WY – Wyoming
KY – Kentucky	NY – New York	
LA – Louisiana	OH – Ohio	

reintroduction project (Tischbein, 1976), followed by reintroductions in 21 other states (Berg, 1982; Melquist & Dronkert, 1987; Raesly, 2001; Serfass *et al.*, 2003; Bricker, 2014). By 2001, 19 states completed their reintroduction project followed by PA in 2004 and NY in 2013 (Raesly, 2001; Bricker, 2014) and NM could conceivably expand reintroduction efforts (Bricker, 2014). Raesly (2001), who completed the only comprehensive review of river otter reintroductions in the US, reported that by 2001, 4121 river otters had been reintroduced, with numbers of individuals released ranging from 11 in KS to 845 in MO. Through successful reintroductions, natural colonization, or a combination thereof, river otters were reported to be occupying at least portions of their historic range within every state, with the exception of NM, at the completion of the Raesly (2001) survey.

Commercial harvesting of furbearing animals, most typically by trapping, is an established component of wildlife management in the US and Canada. At the time of Raesly's (2001) survey of US river otter reintroduction projects, 29 states had trapping seasons for populations of river otters. Among those 29 states, 22 (Raesly [2001] incorrectly recorded 21) had sustained viable populations throughout substantial portions of their state and were not involved in reintroduction projects. Six states (MD, MN, NY, NC, TN, and VA) maintained ongoing trapping seasons for native/remnant populations, while simultaneously conducting reintroduction projects in areas where river otters were extirpated. In 1996, MO, where trapping had previously been prohibited state-wide, became the first state to initiate a trapping season on populations that originated primarily from reintroductions (Raesly, 2001).

In contrast to the US, there has never been a detailed evaluation of the distribution, conservation status, and management of river otters in Canada (Belanger *et al.*, 2010). However, as recently as 2007, river otters were reported as occupying all provinces and territories except Prince Edward Island, and commercial harvests occurred in all jurisdictions where river otters were present (Melquist & Dronkert, 1987; Slough & Jung, 2007).

Our primary objective was to update the distribution, conservation status, and the management of river otters in the 48 conterminous US states and Alaska since Raesly (2001), and provide a similar evaluation for Canada. Our assessment involved all US states and Canadian provinces and territories. We address the following topics pertaining to river otters for each jurisdiction: (i) current distribution; (ii) legal conservation and management status; (iii) harvest trends; (iv) public attitudes and conflicts; and (iv) changes in distribution and management status since Raesly (2001). Raesly (2001) represents the only summary of river otter reintroduction projects, and is frequently cited in river otter-related literature. However, the information presented by Raesly (2001) does not appear to have undergone meaningful scrutiny for accuracy, which is important to ensure appropriate, future interpretation of river otter reintroduction projects. We, therefore, evaluated whether errors concerning state reintroduction projects occurred in Raesly (2001). We conclude by integrating our experience with river otter conservation and research and a literature review to identify important conservation and management needs and concerns that may be considered for the future development and implementation of conservation and management plans for the North American river otter.

Methods

During 2011, we completed a telephone survey of wildlife biologists (primarily furbearer biologists from governmental agencies), representing each US state. This survey was designed to assess the following topics pertaining to river otters: (i) distribution and conservation status, and population origins; (ii) legal status and harvest levels; (iii) population status and size, with particular emphasis on evaluating the extent of approaches used to monitor populations as a basis for decision-making; and (iv) the existence and extent of public complaints about river otters. The survey consisted of open-ended, discrete, and partially closed-ended questions where choices were provided, but the respondent could elaborate and develop an individualized response (Dillman, 1978).

From February–May 2014, we conducted a survey administered by email, which included the same questions as the previous survey to update the 2011 assessment of river otters in the US. We extended this survey to Canadian provinces and territories, reaching wildlife biologists by phone or email. In cases where we could not get updated responses from state biologists in 2014, especially concerning harvest information, we derived information from state management websites or the National Furbearer Harvest Statistics Database (AFWA, 2013). The 2014 survey was also designed to verify the accuracy of information presented in Raesly (2001) pertaining to the status and outcomes of river otter reintroduction projects in the US. Therefore, in addition to questions from the 2011 survey, biologists representing states that had been involved with river otter reintroduction projects prior to 2001 were provided with the Raesly (2001) publication, and were asked to review the document for accuracy (i.e. to identify any errors).

Population Status and Origin of Populations

Conservation activities presumably influenced the status and distribution of river otter populations, especially following the completion of reintroduction projects. To determine population status, we asked respondents to indicate whether the river otter population was believed to be declining, stable, expanding, or to use an intermediary when distinctions were not clear between categories (e.g. stable–expanding). We asked whether the source of the river otter population within the biologist's respective state, province or territory was entirely native/remnant, reintroduced, or a combination of the two (i.e. native/remnant populations persisted in some portion of the state, but extirpations in other regions necessitated reintroductions).

Legal Conservation Status and Harvest

The legal conservation status of a species determines the need for specific conservation practices, be it to determine whether population enhancement is necessary, or to justify legal harvest. To address this issue, we asked biologists to provide the most recent designated legal status of the river otter and indicate whether the designation varied among regions of the jurisdiction. Specific terminology sometimes differs among jurisdictions, but generally, the terminology was easily interpreted as various levels of either protected or harvested. It was particularly important to assess changes in legal status for the states since Raesly (2001). For states where river otters were harvested, we asked biologists to provide information on trapping season lengths, bag limits, and harvest numbers from 2006 to 2012 (2006–2007 and 2012–2013 trapping seasons, respectively). We also compared river otter harvest information in the US to that in Canada and pelt prices on the fur market from 2006 to 2012. River otters are sometimes unintentionally captured by trappers pursuing other furbearing animals. Such by-catches can occur in regions where river otters are protected or legally harvested. To gain insight into river otter mortality associated with unintentional captures, we asked biologists to provide any information available on the frequency of such captures. We also asked whether recommended procedures for avoiding accidental captures were disseminated to trappers.

Population Monitoring

Sustainable harvest of wildlife depends on careful monitoring of the population status of harvested species. To assess the extent and process by which protected and harvested river otter populations are monitored, biologists were asked whether the state or province/territory possessed an updated, region-wide estimate for the river otter population and, if so, to provide the method(s) used to derive that estimate. We were particularly interested in knowing if population estimates (or any techniques to assess population trends) measured were accompanied and validated by field studies. We assumed information pertaining to the annual number of river otters killed during trapping seasons was routinely collected by most jurisdictions and did not focus on that aspect of monitoring.

Public Complaints

As a charismatic predator, the river otter has been presented to the public in many ways, ranging from descriptions of their positive role in aquatic ecosystems, to negative depictions of their feeding habits (Hamilton, 1999; Goedeke, 2005; Serfass *et al.*, 2014). Hence, we were interested in determining the type and extent of public complaints about river otters,

especially those associated with concerns over their impact on fish populations, as well as concerns regarding depredation of fish-rearing facilities or private ponds. Therefore, we asked biologists participating in the survey to provide insight or actual data pertaining to the perceived or real public perception, management issues, and public complaints about river otters.

Comparison to Raesly (2001)

To assess changes in the population and management status of river otters from 2000 to 2014, we compared our data pertaining to population and management status to that of Raesly (2001). Given the success of reintroduction projects and improved management practices for river otters in the US described by Raesly (2001), we presumed that the distribution and size of river otter populations had the potential to expand rapidly, which correspondingly would influence the species' conservation status among states. Monitoring such changes throughout management jurisdictions (in this case, among individual states) is critical to developing a wider perspective that will facilitate the implementation of holistic approaches to species conservation.

Retrospective evaluation and interpretation of the reintroduction process, and the long-term outcomes of those projects, necessitates a complete and accurate understanding of the methods utilized during the projects and the consequent fates of reintroduced populations. For example, the genetic make-up of a reintroduced population is determined by the composition of the source population(s), and evaluating genetic changes in a reintroduced population over time is dependent on knowing the origins of individuals that served as founders of that population. Hence, to ensure that false information about reintroduction projects was not perpetuated, we asked furbearer biologists representing reintroduction states to review the Raesly (2001) survey, and report any errors pertaining to their reintroduction process.

Data Interpretation and Validation

We calculated the percentage of responses for categorical data, and means and standard deviations for continuous data (e.g. harvest levels in the US and Canada).

The outcomes are presented as descriptive statistics in the text. Complex information for patterns, trends, and comparisons of key outcomes was portrayed in figure or tabular format.

Results

Surveys were completed for the 48 conterminous states and AK in the 2011 survey, and for the 45 (92%) of those states and all provinces/territories, except Nunavut, in the 2014 survey. The Canadian territory, Nunavut, was not included in this survey as we were unable to procure contact information for a furbearer biologist in this region. Though we lack an official survey for this jurisdiction, we were able to derive from websites that the territory has a stable, trapped river otter population. Michigan, NY, WA, and WI did not respond to our 2014 survey, so we used data from the original 2011 survey supplemented with complete or partial information pertinent to the survey available from wildlife-agency websites. For example, information derived from the websites of these four states generally was adequate to assess whether river otters were the focus of any public complaints, but not in determining the frequency of complaints. River otter harvest information was not available from the Furbearer Harvest Database or from the state wildlife agency website for WA after 2009. Thus, harvest information presented for that state is limited to information provided in the 2011 survey, which includes data from 2006–2009.

Population Status and Origin of Populations

Outcomes of surveys showed that river otters occupy at least portions of every state, province, and territory within the species' historic range, except for Prince Edward Island. River otter populations within the various jurisdictions were reported as stable in 16 (32%) states and 9 (75%) provinces/territories, expanding in 20 (41%) states and 1 (8%) province/territory, or a combination thereof, depending on the region within a jurisdiction, in 13 (27%) states and 2 (17%) provinces/territories (Figure 23.2), with none of the jurisdictions reporting declining population trends. Current river otter populations originated exclusively from native/remnant populations in 27 (55%) US states (AK, AL, AR, CA, CT, DE, FL, GA, ID, LA, MA, ME, MI, MS, MT, ND, NH, NJ, NV, OR, RI, SC, TX, VT, WA, WI, WY) and all Canadian provinces/territories with the exception of Prince Edward Island; exclusively from reintroduced populations in 8 (16%) states (AZ, CO, IN, KS, NE, NM, OH, WV); or a combination of reintroduced and native/remnant populations in 14 (29%) states (IA, IL, KY, MD, MN, MO, NC, NY, OK, PA, SD, TN, UT, VA).

The river otter populations in some states listed as native may sometimes have been derived by dispersal of individuals from adjacent states. For example, the current population of river otters in ND is considered to have been founded through expansion of populations in MN (Brandt *et al.*, 2014). Likewise, reintroduced populations in some states have undoubtedly been supplemented by river otters dispersing from adjacent states.

Legal Conservation Status and Harvest

As of 2014, commercial trapping of river otters was allowed in 37 states (increasing from 31 states in 2006; Figure 23.3) and all provinces/territories where the species occurs. In the 12 states where legal harvest was prohibited, river otters were classified as a species of least concern (AZ, ND, PA), protected furbearer with a closed season (IN, NM, RI, WY), vulnerable (UT), threatened (CO, NE, SD), or a species of special concern/furbearer (CA) (Figure 23.4). Among the 37 states that harvested river otters, three states (KS, OH, WV) had exclusively reintroduced populations and 11 states (IA, IL, KY, MD, MN, MO, NY, NC, OK, TN, and VA) had a combination of reintroduced and native river otter populations. Five states (MD, MN, NY, NC, and VA) implemented reintroductions, but never legally prohibited trapping of remnant/native populations, only prohibiting harvest at reintroduction sites (Figure 23.5). In states that had a combination of reintroduced and native/remnant populations that were both protected from trapping, the opening of trapping seasons encompassed regions with populations originating from either source – the respective populations presumably had either merged or were considered independently viable when trapping was initiated.

Figure 23.2 Population trends of the North American river otter within each continental US state, and Canadian province and territory based on information derived through wildlife agencies in each of these geopolitical jurisdictions through May 2014. (*Note:* The river otter was never present in Hawaii and, at the time of the survey, it was considered extirpated from Prince Edward Island, Canada).

From 2006 to 2012, the US harvested a total of 170 894 river otters ($\bar{x} = 24\,413$; SD = 6642; range: 17 055–35 128). Canada harvested a total of 82 698 river otters ($\bar{x} = 11\,814$; SD = 1283; range: 9604–13 934) during this same period (Figure 23.6). The US harvested an average of 12 885 (SD = 5847) more river otters per year than Canada between 2006 and 2012. The state with the highest river otter harvest was MO ($\bar{x} = 2431.7$; SD = 1209.6) and the state with the lowest harvest was NV ($\bar{x} = 12.9$; SD = 11.2) (Figure 23.3). The largest river otter harvest in Canada was in Ontario ($\bar{x} = 4552.4$; SD = 1123.1) and the lowest was

in the Yukon Territory ($\bar{x} = 7.3$; SD = 2.5). States that had reintroduced river otters contributed 47% of the total US harvest from 2006 to 2012. The only three states with harvest above an average of 2000 river otters (IL, MN, MO) from 2006 to 2012 had partially reintroduced populations; IL has only had a single trapping season (2012/13) (Figure 23.3).

River otter harvests in the US and Canada decreased from 2006 to 2008, but increased between 2009 and 2012 – from 17 055 to 35 128 in the US, and from 10 378 to 12 020 in Canada (Figure 23.6). The overall average increase in harvest per jurisdiction was 806 (SD = 210)

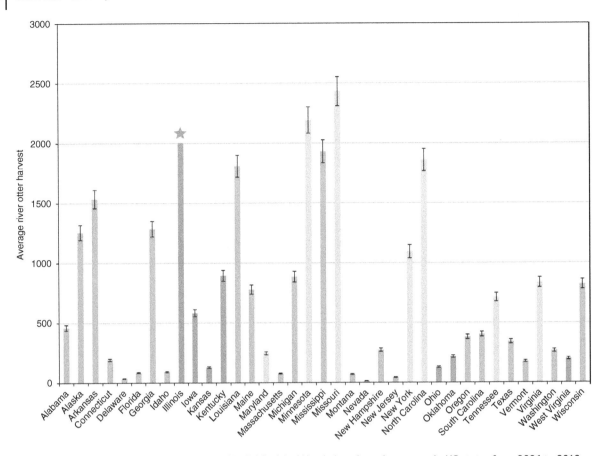

Figure 23.3 The average harvest (number of individuals) of North American river otters in US states from 2006 to 2012 based on information derived through wildlife agencies in each of those states. River otters in these states are derived from either: (i) entirely native/remnant populations, (ii) a combination of native/remnant and reintroduced populations, or (iii) entirely reintroduced populations (Raesly, 2001). States depicted in light blue are comprised of native/remnant populations and have had ongoing, long-term trapping seasons for river otters (except Idaho, which had a closed trapping season at the time of the Raesly [2001] survey). States depicted in light green or light purple are comprised either of native/remnant and reintroduced populations or entirely reintroduced populations. States depicted in light green had trapping seasons at the time of Raesly's (2001) survey of the management status of river otters in the US. River otter trapping seasons were closed in states depicted in light purple at the time of Raesly (2001). Illinois (indicated by the light purple star) initiated a single trapping season in 2012; therefore, the bar only represents the harvest for that year. Washington has not provided river otter harvest records to the public in any capacity since 2009; so, the bar represents harvest from the years 2006–2009. (*Note:* All reintroduced populations of river otters were initially protected from trapping, but remnant/native populations in some states were being trapped while reintroductions were taking place elsewhere in a state [Raesly, 2001]).

river otters in the US and 1142 (SD = 142) individuals in Canada from 2009 to 2012. In the US, this increase in harvest noticeably corresponded with increasing annual pelt prices (Figure 23.6). Four states (LA, ME, MI, NY) did not exhibit this increase, with harvests in these states dropping between 2011 and 2012. ME exhibited the most drastic decrease in harvest (48%) during this period, from 1234 river otters taken by trappers in 2011 to 646 in 2012.

Incidental captures and accidental kills were reported in 39 (80%) states and 8 (67%) provinces/territories. The nature of these incidental captures/accidental kills consisted of capture in traps intended for the North American beaver, *Castor canadensis*, automobile collisions, and retaliatory killings by pond owners who experienced river otter depredation. Of the 39 states that experienced incidental captures, 22 (56%) recommended to trappers some form of

Figure 23.4 The legal status of the North American river otter within each US state, and Canadian province and territory based on information derived through wildlife agencies in each of these geopolitical jurisdictions through May 2014. (*Notes:* The river otter was never present in Hawaii and at the time of the survey considered extirpated from Prince Edward Island (PEI), Canada; see Addendum for an update on states allowing trapping seasons for river otters and status of river otters on PEI).

prevention technique (e.g. trap or trapping-setting modifications) to minimize incidental river otter captures. Two of the eight provinces/territories (25%) that experienced incidental captures, recommended prevention techniques.

Population Monitoring

Eleven states and a single province indicated having estimates for river otter populations (Table 23.2). Of the 37 states with trapping seasons and 12 states where legal trapping is prohibited, 26 (70%) and 11 (92%), respectively, do not have population estimates for the species. New Brunswick was the only province to possess a population estimate for the river otter. Population estimates were based on basic modelling ($n = 3$), basic modelling parameterized with harvest information ($n = 6$), extrapolation of a population density from densities estimated in other states ($n = 2$), or unspecified methods ($n = 1$) (Table 23.2). None of the population estimates was based on information derived from field studies or validated through field investigations.

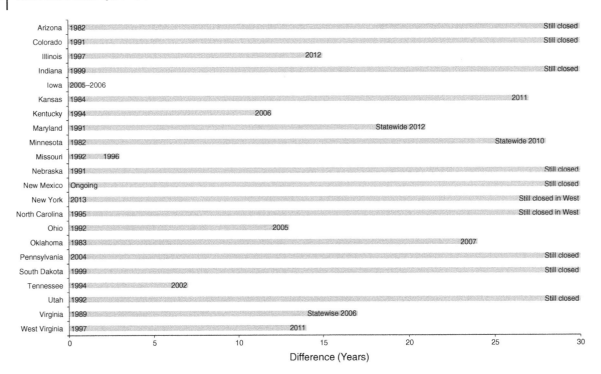

Figure 23.5 Time elapsed between the completion of reintroduction projects for the North American river otter in 22 US states and the status of trapping seasons (i.e. year trapping was instated or remains prohibited [designated as 'Still Closed'] on the reintroduced populations based on information derived from wildlife agencies in each of these states through May 2014. River otters were initially protected from trapping at all reintroduction areas. However, some reintroduction states retained remnant/native populations (in non-reintroduction areas) that alternatively have been subject to trapping pre- and post-reintroductions (i.e. MD, MN, NY, NC, and VA) or were protected prior to reintroductions (i.e. IA, IL, KY, MO, OK, and TN). A 'Statewide' designation indicates the states that initiated trapping seasons on reintroduced populations, but trapping was ongoing with native/remnant populations. A 'Still closed in West' designation pertains to states where trapping of native/remnant populations was ongoing, but reintroductions occurred in the western part of those states and those populations are still protected from trapping. (*Note:* See Addendum for an update on states allowing trapping seasons for river otters).

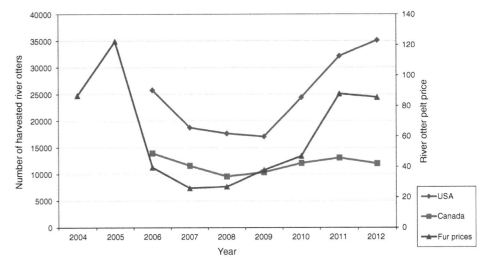

Figure 23.6 The North American river otter harvest levels and average pelt prices in the United States and Canada, and the average river otter pelt price from 2006 to 2012, based on the information derived through wildlife agencies in the 'continental US states, and Canadian provinces and territories.

Table 23.2 The US states and Canadian province whose wildlife agencies reported having calculated population estimates for the North American river otter, *Lontra canadensis*, as of May 2014. The estimate and estimation methods are provided.

State	Population estimate	Method used
Alaska	No state total estimate. Coastal populations range from 32–44 and 36–65 otters/100 km shoreline	Basic modelling parameterized with harvest information
California	1000–10 000	Basic modelling
Idaho	Over 8000	Extrapolation of population density from the 1980s
Illinois	10 865 (Projected 15 331–18 852 in 2014)	Basic modelling parameterized with harvest information
Iowa	7000–12 000	Basic modelling parameterized with harvest information
Kansas	1400	Basic modelling
Minnesota	12 348	Basic modelling
Missouri	15 000–20 000	Basic modelling parameterized with harvest information
New Brunswick	3000–4000	No method provided
New Jersey	936 in freshwater waterways; 1725 in tidal waters	Extrapolation of population density from density estimates in other states
Ohio	8000	Basic modelling parameterized with harvest information
Wisconsin	10 100	Basic modelling parameterized with harvest information

Population estimates ranged from 3000 to 4000 (New Brunswick) to 15 000–20 000 (MO) (Table 23.2).

Public Complaints and Depredation

Forty-three (88%) of the 49 surveyed states experienced depredation and/or public complaints pertaining to the river otter. Of the 12 Canadian provinces and territories surveyed, five (45%) experienced depredation and/or public complaints.

Complaints about river otters consisted of depredation of fish at hatcheries (fish farms) and private ponds, use of boats and boat houses as den sites, game fish depredation, defecation on boat docks, lobster-trap depredation, duckling depredation in storm water ponds, depredation of shorebird hens, and a single instance of direct confrontation (aggressive behaviour) of a river otter toward a human (Table 23.3). Complaints were classified as infrequent/rare (< 10/year), occasional (10–30/year), or frequent (> 30/year). Twenty-two states and five provinces/territories reported complaints as infrequent or occasional (Table 23.3). Four (80%) of the five states that experienced frequent complaints had implemented reintroductions (IA, IN, MO, OK) (Table 23.3). Depredation at fish hatcheries and farms (AL, NC, OK) and depredation at private ponds (IA, MO, OK) were the only complaints reported as frequent in multiple states (Table 23.3). Fisherman complaints of otters consuming a large amount of game fish and otters defecating on docks were reported as frequent in MO and OK, respectively (Table 23.3).

The perceived public perceptions of river otters were reported as mostly positive by furbearer biologists in 39 (80%) states and all Canadian provinces and territories, except Quebec and the Yukon Territory, which reported perceptions as being neutral. Public perceptions of otters were reported to be mostly neutral in five states (AK, IN, MD, NV, NJ), divided between positive and negative in three states (KY, MO, WA), and mostly negative in one state (AZ). No response to this question was received from MI.

Comparison to Raesly (2001)

Since 2001, 16 (33%) of 49 states updated their river otter population status to a more favourable classification, and no state increased the species' protection

Table 23.3 Frequency and nature of the public complaints and management issues reported by wildlife agencies in the US states, and Canadian provinces and territories pertaining to the North American river otter through May 2014.

Complaint or management issue	Infrequent/very rare (< 10/year)	Occasional (10–30/year)	Frequent (> 30/year)
Depredation at hatcheries/fish farms	AR, CO, FL, ID, LA, MA, MN, MS, MO, MT, RI, UT, VA, Nova Scotia, Ontario	NC	AL, OK
Private pond depredation	AL, DE, FL, IL, MT, NH, NJ, OH, OR, RI, UT, VA, WI	KS, SC, WV	IA, IN, MO, OK
Use of boat or boathouses as den sites	OR, Manitoba, Ontario, British Columbia	AK	
Fisherman complaints of game fish depredation	ID, IA, MD, RI, UT, VA	Alberta	MO
Use of dock as a latrine	AR, MN, NJ, OR, Ontario, British Columbia		OK
Depredation of lobster traps	Nova Scotia		
Depredation of ducklings in storm water ponds	Manitoba		
Depredation of shorebird hens	Nova Scotia		
Direct confrontation with otter	MT		

status (Table 23.4). Since Raesly's (2001) survey: (i) a river otter population has been established in NM through reintroduction efforts; (ii) the river otter is no longer considered endangered in CO, IN, NE and OH; (iii) OH has changed the status of the river otter from the state endangered to a trapped furbearer; and (iv) OK made the transition from a protected species to a trapped furbearer (Table 23.4).

Since Raesly (2001), reported trends in river otter populations have changed from stable to expanding in three states (MI, ND, NJ), and from expanding to stable in four states (AR, CT, SC, WA) (Table 23.5). Raesly (2001) only provided trends for states with completely native/remnant populations, so comparisons could not be made for all surveyed states in 2014 (Figure 23.2). There has been a 32% increase (from 29 to 37) in the number of states with open river otter trapping seasons since Raesly (2001). Four states with reintroduced river otter populations have opened trapping seasons since 2001 (IL, KS, OK, WV) (Tables 23.4 and 23.5).

There were several errors and discrepancies in the reporting of aspects of reintroduction projects by Raesly (2001). They include missing sources for reintroduced river otters (CO and WV), incorrect sources for reintroduced river otters (CO and IL), incorrect release periods (KS and OH), incorrect number of

Table 23.4 The US states where the population status of the North American river otter was revised from 2001 to 2014. The 2001 statuses are from Raesly (2001) and the revised statuses (through May 2014) are from information obtained through wildlife agencies in the respective states.

State	Status 2001	Status 2014
AZ	Species of Special Concern	Least Concern
CA	Not Designated	Species of Special Concern
CO	State Endangered	State Threatened
IA	State Threatened	Trapped Furbearer
IL	State Threatened	Trapped Furbearer
IN	State Endangered	Protected Furbearer
KA	Experimental Population	Trapped Furbearer
ND	Not Designated	Least Concern
NE	State Endangered	State Threatened
NM	Extirpated	Protected Furbearer
OH	State Endangered	Trapped Furbearer
OK	Protected Species	Trapped Furbearer
PA	Species at Risk	Least Concern
UT	Not Designated	Vulnerable
WV	Protected Species	Trapped Furbearer
WY	Protected Species	Data Deficient

Table 23.5 The US states where perceived population trends of the North American river otter changed from 2001 to 2014. Comparisons could only be made with states included in Raesly (2001), which focused on states with entirely native/remnant river otter populations. Changes in population trends are based on comparing those reported by Raesly (2001) to those derived from wildlife agencies through May 2014.

State	Population trend 2001	Population trend 2014
AR	Increasing	Stable
CT	Increasing	Stable
GA	Increasing	Stable–Expanding
MI	Stable	Expanding
MT	Increasing	Stable–Expanding
ND	Stable	Expanding
NJ	Stable	Expanding
SC	Increasing	Stable
WA	Increasing	Stable
WI	Increasing	Stable–Expanding
WY	Increasing	Stable–Expanding

Table 23.6 Corrections to errors identified by wildlife agencies in CO, IL, KS, OH, SD, and WV in Raesly's (2001) portrayal of reintroduction projects conducted in the US. Entries in bold are elements omitted from the Raesly (2001) survey, and those that are slashed were included in the publication and are incorrect.

State	Dates	No. of individuals	No. of sites
CO	1976–1991	109	**AK, CA**, ~~LA~~, **MI, MN**, NF, **OR, WA**, WI
IL	1994–1997	346	LA, ~~KY~~
KS[a]	1983–**1985**	**19**	ID, MN
OH	1986–**1993**	123	AR, LA
SD[b]	1998–1999	**35**	LA
WV	1984–1997	245	**LA**, MD, NC, SC, VA

[a] The wildlife biologist providing information for the wildlife agency in Kansas reported that river otter populations in the state were comprised of solely reintroduced individuals, but Raesly (2001) indicated populations as being comprised of a combination of reintroduced and native/remnant individuals.

[b] The wildlife biologist providing information for the wildlife agency in South Dakota identified additional errors in Raesly (2001), including: (i) the correct number of reintroduction sites was 2, not 1; (ii) genetic concerns were not a source criterion when choosing reintroduction sites; and (iii) historic distribution was a site-selection criterion not mentioned.

river otters released (KS, OH, and SD), incorrect number of release sites (SD), missing information concerning the choice of site criterion (SD), and incorrect source criterion (SD) (Table 23.6). KS was also misrepresented concerning the composition of river otter populations within the state (Table 23.6).

Discussion

High participation rates by wildlife biologists from the various jurisdictions, and information accessible from wildlife-agency websites, provided a relatively efficient way to assess the status and distribution of river otters in the US and Canada. Outcomes of our survey demonstrate that the conservation status of the river otter has improved substantially following the first reintroduction project in Colorado in 1976 (Tischbein, 1976) and has continued to improve since Raesly (2001).

Conservation and Population Status

Overharvesting and perturbations to aquatic habitats contributed to widespread declines and extirpations of river otter populations throughout large portions of the species' historic range in North America, with the most substantial declines occurring in the US. Implementation of more progressive conservation strategies, including the reintroduction of river otters in 22 states and improvements in the quality of aquatic and riparian habitats, have contributed to the recovery of river otter populations throughout much of the US. The success of these conservation activities is exemplified by the river otter currently occupying at least portions of its historic range in all major geopolitical jurisdictions of the continental US and Canada, with the exception of Prince Edward Island.

Legal Conservation Status and Harvest

Personnel in wildlife management agencies are in a challenging position because they are expected, based on the Public Trust Doctrine, to balance the interests of primary stakeholders (i.e. hunters, anglers, and trappers) with those of the wider citizenry they represent (Smith, 2011). River otters are given the general classification of a furbearer, a designation that implies intent to include legal trapping for fur among management activities (regardless of protection status at a particular time). However, allowing a trapping season regulated by a governmental wildlife management agency is dependent on information confirming that population levels are suitable to sustain a regulated harvest, and restrictions on trapping are expected when that criterion is not met. Thus, as river otter populations increase through natural expansion or aided by reintroductions, the reinstatement of trapping seasons should be anticipated with the North American philosophy under which furbearers are managed. Since 2006, the number of states with an open river otter trapping season increased from 31 to 37, indicating that state wildlife management agencies feel confident that populations are adequate to sustain harvest. The increasing number of states with river otter trapping seasons corresponded with a reduction in the protection status afforded the species.

Trappers in the US consistently harvested more river otters than those in Canada. The increasing value of river otter pelts in the fur market during the survey period likely explains the rise in the US harvest through the generation of greater incentive for trappers to pursue river otters (Beringer & Blair, 2013; Sasse, 2013) (Figure 23.6). However, there was not an obvious association between harvest and fur prices in Canada (Figure 23.6). Prices for river otter pelts fluctuated between 2005 and 2012, from a maximum of $112 in 2005, a minimum of about $27 in 2008, and then increasing to about $85 in 2012 (Beringer & Blair, 2013; Sasse, 2013). The increase in river otter harvest in the US may also be attributed to the initiation of new trapping seasons and the liberalizing of season lengths and bag limits in portions or the entirety of states.

Population Monitoring

Science-based harvest management relies on the premise that wildlife populations can be killed sustainably. To ensure that this criterion is met, some level of monitoring to understand changes in the size,

age and gender structure, as well as the distribution of harvested populations is necessary. The Appendix II listing of the river otter by CITES mandates the tagging of all river otter pelts intended for export outside of the US and Canada (Nilsson, 1980; CITES, 2013). Records kept as part of the pelt tagging process ensure that wildlife agencies in jurisdictions where river otter are legally trapped will document the number of individuals harvested annually.

Thirty-seven states and all Canadian provinces/territories, except New Brunswick, did not have a population estimate for the river otter. Generally, few, or vaguely defined, details were provided for how population estimates were derived. Of the 11 states and the single province with population estimates (Table 23.1), the methods used to obtain the estimates were derived from basic modelling, often parameterized with basic harvest information (i.e. age structure and gender composition determined through necropsies performed on carcasses of river otters provided by trappers), or by extrapolating population estimates from population densities previously published for river otters. For example, NJ reported using population densities estimated for river otters in ID (Melquist & Hornocker, 1983) and the coastal regions of TX (Foy, 1984) and LA (Shirley *et al.*, 1988) to derive state-wide population estimates.

Generally, there was no evidence obtained through our survey indicating that meaningful efforts have been undertaken to calibrate population estimates through field studies. Captive studies have demonstrated the potential to extract DNA from scats of river otters (Fike *et al.*, 2004); application of this approach in the field offers the possibility of determining the number of river otters occupying a particular area or set of habitat conditions. The use of genetic-based field work to determine habitat occupancy of river otter populations may serve as a basis for more meaningful extrapolation of population estimates across a variety of aquatic landscapes. Mowry *et al.* (2011) pioneered the use of genetic technology as a basis for estimating river otter densities along several riverine systems in MO. The application of this technology will be improved through a more concrete understanding of the variation in scat-marking habits among river otters inhabiting specific areas. This information can be used to calibrate population indices that account for differences in scat-marking by individuals based on gender and season (e.g. Olson *et al.*, 2005, 2009; Stevens & Serfass, 2008).

Population estimates are expensive and difficult to calculate for species that are elusive, highly dispersed, and often occur at low population densities, like the river otter (Kohn *et al.*, 1999; Piggott & Taylor, 2003). Consequently, adequate funding is needed, and often a limiting factor, for detailed monitoring of the population status of species with these characteristics. One Canadian furbearer biologist stated that the river otter (and other furbearers) is a 'second class citizen' compared to big game species, in terms of funding and interest for research. Such viewpoints appear to be tacitly confirmed by the general lack of interest of wildlife agencies in most jurisdictions to develop population estimates for the river otter at present or in the future (Gallant, 2007). Given recent concerns expressed about the initiation of river otter trapping seasons with limited supporting justification (e.g. Serfass *et al.*, 2017, pp. 563–568), we would anticipate that studies will be initiated or other rationale applied to demonstrate that river otter populations can be sustained when subjected to legal trapping for fur.

In lieu of developing population estimates validated by field studies, our literature review revealed that field-sign surveys (e.g. detecting scats at latrines, but also other signs such as tracks in the snow) have been used in many areas of North America to reliably determine the presence or absence of river otters (e.g. Reid *et al.*, 1987; Shackelford & Whitaker, 1997; Swimley *et al.*, 1998; Melquist *et al.*, 2003; Gallant *et al.*, 2008; Stevens & Serfass, 2008; Stevens *et al.*, 2011a; Just *et al.*, 2012). These types of evaluations should be incorporated with studies intended to determine population densities through extracting DNA from scats as a means of enhancing approaches used to monitor the overall size and distribution of populations.

Public Complaints

The expansion of river otter populations is likely to increase interactions between people and species. River otters have sometimes been accused by anglers to be harmful to game fish and blamed for depredating fish at private ponds and commercial and public fish hatcheries (i.e. fish-rearing facilities) (Serfass *et al.*, 2014). However, the extent of concerns by anglers, or the severity of instances of accused depredation, has seldom been evaluated (Ashcraft, 2012; Serfass *et al.*, 2014; Pearce *et al.*, 2017). Our study identified depredation at private ponds and fish hatcheries as the complaints most commonly reported as frequent, especially in central US states with reintroduced river otter populations. The most common prevention method among these states was the issuance of nuisance permits, which enabled the citizen to trap the river otter responsible for the complaint. Indiana indicated that they suggest other prevention methods (e.g. fencing for ponds) or provide information pertaining to river otter ecology to mitigate frequent public complaints. Conflict-prevention methods reported to have been employed by some state wildlife agencies to address infrequent or occasional complaints included: exclusion devices such as traditional or electric fences, increasing pond complexity (i.e. increasing refuge for fish against predation), or hiring trappers during the legal trapping season to remove the offending animal.

Though some Canadian provinces reported complaints, biologists mentioned that hatcheries and the general public tended to not complain, 'were used to [human–otter interactions]', or were more interested in employing conflict-prevention methods (e.g. fencing, creation of fish refuge in ponds) at their facilities. Complaints were much more frequent in the US, and people seemed slightly less tolerant of negative river otter interactions. Kellert (1996) noted that people tend to have more negative opinions of carnivorous species with initial and increasing, direct or perceived contact, such as when the species is expanding its range, as is the case with recovering river otter populations. Additionally, those opinions tend to decline as the duration of human–carnivore conflict increases, especially in areas without an extensive history with that carnivore, such as those that experienced extirpations. As people in Canada have always lived with river otters, their opinions may be moderated by familiarity, as indicated by fewer negative attitudes expressed concerning the species. Over time, public opinions in the US may reach a stasis as the public learns to tolerate river otters. Educational efforts that encourage conflict–prevention techniques and general river otter awareness may help ensure that the lasting attitude in the US is a positive one (Kellert *et al.*, 1996; Serfass *et al.*, 2003; Serfass *et al.*, 2014).

Despite reports of complaints in both the US and Canada, a majority of biologists perceived that the general public held a positive perception of river otters. Unfortunately, these positive perceptions have generally not been accompanied in newspaper articles focusing on depredation caused by the species, which have the potential to negatively influence the overall opinion of the public generally naïve to issues pertaining to wildlife ecology, conservation, and management (Serfass *et al.*, 2014). Although biologists generally perceived the public as holding a positive view toward river otters, this perception, or any contradictory, negative messages portrayed in some state newspapers, may not align with actual public opinions, which were not assessed by our survey. However, a survey of anglers' attitudes toward river otters at reintroduction sites in PA indicated extremely favourable attitudes towards the species (Serfass *et al.*, 2014), and demonstrates the importance of conducting social surveys to properly assess human attitudes towards wildlife.

Contrasts with Raesly (2001)

Improvements in the overall status of river otters in the US since Raesly (2001) are encouraging for the long-term conservation of the species. Population trends were only provided for states with completely native/remnant populations in the Raesly (2001) survey, limiting us from making comparisons for states with reintroduced river otter populations in our 2014 survey. The omission of states with reintroduced populations in Raesly (2001) may have been influenced by the assumption that reintroduced populations were most likely expanding or stable–expanding. The rapid and positive changes in the conservation status of the river otter reported by many wildlife agencies demonstrate that populations of some species can initially expand with proper protection, such as the case with reintroduced river otter populations. Such an outcome also underlines a lingering question associated with a potential management concern: Why did remnant/native populations not expand as rapidly as reintroduced ones? Determining factors that have served as underlying causes to limit the expansion of native/remnant river otter populations should be a research

priority, particularly in the US, where the need for reintroduction projects in some jurisdictions may have been negated if management actions had been taken to better facilitate the natural range expansion of these populations. Understanding the factors that may limit the expansion of native/remnant river otter populations will greatly influence future management decisions for river otters.

A questionnaire survey such as that used in our study is the only realistic method to obtain a general assessment of the conservation and management statuses of wide-ranging species, such as the river otter, over a large geographic area. The accuracy of these data is important because wildlife agencies are likely to use the provided information, and examples of other agencies, as a guide for decision-making regarding species management. We documented several important errors in Raesly (2001) concerning river otter reintroductions in the US, which have been unknowingly perpetuated through frequent citations in the literature (e.g. Bluett *et al.*, 2004; Bischoff, 2006; Johnson *et al.*, 2007; Brandt *et al.*, 2014).

Conservation Concerns and Research Needs

The river otter has transitioned from a species of conservation concern in many areas of North America to one that is now widely trapped, including states where the species was reintroduced. Decisions to initiate river otter harvests have been justified based on a variety of factors, including the rapid expansion of populations, even into suboptimal habitat conditions (e.g. Triska *et al.*, 2011), and portrayals of increasing river otter–human conflicts. Harvest of river otters has been demonstrated to be sustainable in many states. However, there were issues identified during our evaluation that should be considered by wildlife biologists in states that allow the trapping of river otters, particularly those where trapping was recently instated. We identified three concerns in particular regarding the management of river otters: (i) relatively few states that harvest river otters have formal, field-based monitoring protocols for meaningfully assessing either the density or distribution of river otter populations; (ii) the validity of justifications for allowing harvest to alleviate conflicts with anglers, owners of farm ponds, and managers/owners of fish-rearing facilities (Serfass *et al.*, 2014); and (iii) the lack of a large-scale, genetic assessment to determine how the use of river otters from various sources for reintroductions, and their interaction with remnant/native populations, will influence the genetic structure of river otter populations in the future.

Population Monitoring

Although desirable, we do not generally consider population estimates necessary to design sustainable harvest strategies for many species but believe some, ongoing field-based monitoring procedures for assessing population trends are desirable. Such consideration is particularly important for species existing at low densities and associated with habitat conditions often limited in the landscape, such as river otters. Increases in river otter harvest have furthered the need for implementing reliable, field-based approaches for monitoring the long-term status of river otter populations (Bricker, 2014). River otters scent-mark at latrines by depositing scats (spraints), urine, and anal gland secretions (Gorman & Trowbridge, 1989; Melquist *et al.*, 2003; Ben-David *et al.*, 2005). Latrines are visually conspicuous and easily detected in riparian areas. The prominence and the relative ease by which latrines can be located and identified has contributed to their popular use in sign surveys for river otters (Dubuc *et al.*, 1990; Newman & Griffin, 1994; Swimley *et al.*, 1998).

Latrine, and other activity-sign, surveys generally are considered reliable for detecting the presence of river otters (Shackelford & Whitaker, 1997; Swimley *et al.*, 1998; Melquist *et al.*, 2003; Gallant *et al.*, 2007, 2008; Stevens & Serfass, 2008), but can be time-consuming and costly because of the logistics associated with accessing riparian habitats, particularly in remote areas. To minimize logistical constraints associated with accessing survey sites, some states have conducted sign surveys in riverine systems at bridge-crossings (e.g. Clark *et al.*, 1987; Shacklford & Whitaker, 1997; Roberts *et al.*, 2008) as a primary method for evaluating the occurrence of river otters. Only recently, however, have studies been conducted to assess various factors pertaining to the efficacy of

surveys to detect river otter signs near bridges (e.g. Gallant *et al.*, 2008; Crimmins *et al.*, 2009; Stevens *et al.*, 2011a; Just *et al.*, 2012). At the least, states, where trapping of river otters is allowed, should incorporate approaches identified in the aforementioned studies as a basis for implementing and refining the monitoring of river otter populations. Such monitoring should especially be encouraged in states where reintroduced river otter populations are harvested. The implementation of systematic monitoring in areas now occupied by river otters that previously would have been considered suboptimal habitat should also be stressed. Current optimism about river otters being able to tolerate a wider range of aquatic habitat disturbances may be misleading and unfounded in that such disturbed areas could represent sink habitats; where populations are sustained by dispersing individuals and not through adequate levels of reproduction and survival by individuals occupying the area.

River Otter–Human Conflicts

The reintroduction of mammalian predators has often been met with controversy by citizens near reintroduction sites, primarily because of concern for depredation of livestock and pets, and that predation may adversely impact populations of game species. The reintroduction of river otters in many states has, in some cases, been negatively depicted in the media because of the species' predatory (i.e. fish-eating) habits. The successful reintroduction of river otters in MO, OH, KY, and IL was followed by strikingly similar patterns of negative media messages suggesting that river otter predation was having widespread negative impacts on commercially reared fish and game fish important to anglers. Management actions, including opening trapping seasons, subsequently were implemented in these states purportedly to alleviate the public concern and animosity portrayed in the media about river otters. Regardless of causative factors, wildlife agency personnel should be responsible for implementing initiatives aimed at correcting misconceptions about wildlife. We contend that the wildlife agency personnel are best positioned to counter unfounded public discourse concerning controversial wildlife, but this has not always been the case in the aforementioned states. Serfass *et al.* (2014) reviews issues related to negative messaging associated with human–river otter conflicts in some states, and uses a case study from the Pennsylvania River Otter Reintroduction Project to demonstrate conservation benefits that can be derived by applying social science theory, and associated methodologies, as a basis for determining and accurately depicting public attitudes towards predators and predator reintroduction projects. For example, a social survey of anglers in PA revealed that a large majority had very favourable opinions about the river otter and its reintroduction in PA (Serfass *et al.*, 2014). This type of assessment should be considered by wildlife agencies in states experiencing apparent river otter–human conflicts with the intent of assessing if the extent of perceived concerns (e.g. based on negative portrayals in the media) is reflective of overall public opinions.

Ultimately, understanding and portraying actual public opinions about the river otter (and other species) can serve to counter the propagation of unfounded negative media messaging about the species. However, Serfass *et al.* (2017, pp. 563–568) review aspects of river otter trapping in relation to the North American Model of Wildlife Conservation, providing examples of the manner in which a segment of the wildlife conservation system in the US has promoted 'conflict' associated with river otter predation on fish to justify and gain support for implementing trapping seasons on the species. Although the public has very favourable attitudes towards river otters in PA (Serfass *et al.*, 2014), the Pennsylvania Game Commission (PGC) has nonetheless promoted a river otter trapping season in part for 'damage management', at fish-rearing facilities (Hardisky, 2013). Such justification stands in contrast to an extensive study by Pearce *et al.* (2017) demonstrating that river otters seldom posed an economic threat to fish-rearing facilities in PA.

River Otter Genetics

River otters from LA have most commonly been used as a source for reintroduction projects in the US; about 64% of river otters reintroduced in the US were obtained from this state. These river otters have been adaptable, surviving and reproducing in climates and habitat conditions substantially different from LAs. Winters in LA are generally mild, but river otters from this state have been released and established populations in more northern states such as OH and PA (Raesly, 2001), where winter conditions typically include snow and ice cover. The frequent use of river otters from LA has contributed to increased distribution of the species across the continent, but may also have contributed to the homogenization of the species' genetic diversity (Brandt *et al.*, 2014). Originally, 19 subspecies of river otter were recognized in North America (Hall & Kelson, 1959), but the number of subspecies was subsequently modified to six by Hall (1981). How the use of river otters from various sources for reintroductions and their future interaction with remnant/native populations will influence the genetic structure and subspecies delineations is unknown. Serfass *et al.* (1998) and Brandt *et al.* (2014) discussed genetic implications for river otter reintroductions in North America and Mowry *et al.* (2015) reviewed the genetic integrity of a reintroduced river otter population in MO. Ultimately, a large-scale, genetic assessment should be initiated to determine the genetic composition of remnant and reintroduced river otter populations. This investigation should serve as a baseline to complement subsequent genetic monitoring for investigating the level of genetic introgression that occurs as reintroduced populations expand and interact with remnant populations.

Conclusion

About 40 years ago, river otters were a species of conservation concern in many areas of North America. The implementation of more progressive conservation strategies and the improvement in the quality of aquatic and riparian habitats have contributed to the recovery of river otter populations in many areas of the US and Canada. The species now occupies at least a portion, or most, of its historic range in North America. Many states now harvest river otters, including states where river otters were reintroduced. However, the historic vulnerability of the river otter to overharvest, and the depiction of the river otter as a nuisance in several states (Serfass *et al.*, 2014), requires vigilance to ensure that the river otter remains a story of conservation success. Future conservation strategies should include careful field monitoring of the species' distribution and population status and educational outreach programmes that accurately depict the ecological role of river otters in aquatic ecosystems to mitigate potential threats that could hinder the species' recovery.

Acknowledgements

We would like to thank the dedicated wildlife biologists from governmental agencies in each state, province, and territory who generously volunteered their time and knowledge to participate in our survey. The Department of Biology and Natural Resources at Frostburg provided administrative and facilities support during the completion of surveys and the writing of this Chapter. We are extremely appreciative of the constructive input provided by the three reviewers of our manuscript and are particularly grateful to Dr. Emmanuel Do Linh San for his guidance and support throughout the review and revision process.

Addendum: Conservation Status of the North American River Otter – An Update

Thomas L. Serfass, Emily A. Bricker, and Zoe L. Hanley

The publication date of 'Small Carnivores: Evolution, Ecology, Behaviour, and Conservation' was delayed, contributing to the need to update certain aspects pertaining to the conservation status of the river otter presented in the accompanying Chapter. This Addendum facilitates such an update, focusing primarily on examining the expansion of trapping seasons among states where river otters were protected in 2014 (when survey results reported in the Chapter were completed), but also to assess any new information relevant to the conservation of river otters. We derived information for the addendum primarily from internet searches, which served to locate and review state-wildlife-agency hunting and trapping digests (i.e. guides to hunting and trapping regulations) and media reports (primarily newspaper articles) for information on the trapping of river otters. Internet searches and literature reviews served to assess if there was any recent evidence of new reintroduction efforts, natural recolonization of river otters to Prince Edward Island (the only jurisdiction reporting an extirpated river otter population at the time of the previous survey), and the report of a little known population in Mexico. We also reviewed two recent publications pertaining to the management of river otters in the United States (Erb *et al.*, 2018; Roberts *et al.*, 2020). Certain common themes and messages pertaining to the portrayal of trapping emerged during the review of media reports, Erb *et al.* (2018), and Roberts *et al.* (2020), which we interpret with regard to river otter conservation. These interpretations derive from our (authors of this Addendum) opinions and insights, and not necessarily those of others authoring the associated Chapter.

Expansion of Trapping

At the completion of our original survey, 12 of the 49 states within the historic range of the river otter protected the species from trapping. Since that survey, 4 more states initiated trapping seasons (IN in 2015; PA in 2016; ND in 2017; and SD in 2020) with populations in AZ, CA, CO, NE, NM, RI, UT, and WY still protected from legal trapping. The river otter population in IN is derived from reintroductions and that in ND through colonization from surrounding jurisdictions (see Brandt *et al.*, 2014). The trapping season in PA was initiated on a remnant (native) population in the northeastern portion of the state that had been legally protected from trapping (and other killings) since 1952, and expanded river otter trapping to include reintroduced populations in northwestern PA during early 2021. As a prelude to the initiation of trapping seasons, some states conducted graduate-level, university-based studies of river otter populations. For example, immediately prior to initiating the trapping season PA derived a population estimate ($n = 259$; 95% asymptotic Wald-type CI = 175–603 [Forman, 2015]) for the remnant population in the northeast, apparently serving as justification for implementing the aforementioned trapping season. South Dakota removed the river otter from the state list of threatened mammals in May 2020 and concurrently announced plans to initiate a trapping season later in the year (Mercer, 2020). The management authority in SD (Department of Game, Fish, and Parks) is primarily using reported sightings and the accidental killing of river otters in traps set for other legally trapped species as justification for initiating a river otter trapping season (South Dakota Department of

Game, Fish, and Parks, 2020). From 2012 to 2019, the number of reported sightings (including accidentally killings in traps) ranged from 32 to 42, with the majority of reports occurring in three eastern counties (South Dakota Department of Game, Fish, and Parks, 2020). Nebraska has recently completed a study assessing the distribution of river otters (Bieber *et al.*, 2018), which we anticipate will serve as a basis for justifying the initiation of a trapping season. The trapping season for river otters in NC is apparently now statewide.

Reintroduction Projects

River otter reintroduction efforts are largely concluded, although NM is contemplating the expansion of reintroduction efforts to an additional drainage and, during March and April 2021, conducted a supplemental release of nine river otters live-captured in LA to genetically enrich the state's existing reintroduced population (Melissa Savage, personal communication, 2 August 2021). Roberts *et al.* (2020) recently indicated that a total of 23 reintroduction projects had taken place in the US, apparently mistaking the population that naturally recolonized in ND as originating from reintroductions in that state (see Brandt *et al.*, 2014).

Natural Recolonization and Evidence of a Population in Mexico

Recent evidence suggests that river otters have the potential to naturally recolonize the Province of Prince Edward Island (PEI). Three male otters recently were reported captured by beaver trappers on PEI – 1 during Fall 2016, and 2 in early/spring 2017 (Canadian Wildlife Health Cooperative [CWHC], 2017). Pioneering individuals presumably would have originated in New Brunswick or Nova Scotia. There also is newly reported evidence suggesting that a native population of North American river otters occurs in the Upper Rio Conchos basin of northern Mexico (Gallo-Reynoso *et al.*, 2019).

Portrayals of River Otter Conservation

Trapping is a predominant theme in portrayals of river otters in media releases and other publications directly or indirectly associated with state wildlife agencies, particularly in the US. These repetitive portrayals often serve to promote the virtues of recreational trapping in relation to the conservation of river otters through what appear as coordinated, common themes seemingly intended to ameliorate public concerns about trapping river otters and recreational trapping in general. Common messages encountered include that trapping: (i) is highly regulated; (ii) is science-based; (iii) is humane (based on rigorous trap testing); and (iv) has facilitated the recovery of otter populations. We generally do not dispute the sustainability of regulated, recreational trapping of river otters in North America or that some traps may meet current international humane standards for capturing river otters. However, we do have ethical concerns about what appears as a coordinated marketing approach to justify and promote the trapping of river otters – in effect, the success of restoration efforts appears to be serving as a basis for using the river otter as a 'flagship species' to gain support for trapping in general. This addendum is necessarily brief and therefore cannot serve as an appropriate format for expressing concerns in detail. As examples, however, Serfass *et al.* (2014) raised specific concerns pertaining to the vilification of river otters prior to initiation of trapping seasons in some states, and Serfass *et al.* (2017, pp. 563–568) reviewed approaches applied to justify and promote river otter trapping. Following are examples of promotional rhetoric that has often accompanied portrayals of river otter conservation in the US, particularly in relation to the implementation of trapping seasons.

1) ***River otter recovery and the US system of wildlife conservation.*** The recovery of river otters in the US is undeniably a conservation success story. Portrayals of the recovery espouse the virtues of the furbearer management system practised in the US as uniquely contributing to the recovery of otter populations. Seldom mentioned is the extensive recovery of the Eurasian otter, *Lutra lutra*, that occurred throughout much of Europe during virtually the same time period as the recovery of river otters in the US. Unlike the river otter, rarely were reintroduction projects implemented to restore Eurasian otter populations. Instead, the recovery generally resulted from natural expansion of remnant populations, facilitated by initiatives to

improve water quality and riparian habitat conditions, and curtailing the killing of otters by anglers. Design and implementation of these habitat improvement and social initiatives often were part of planning to facilitate the expansion of otter populations (e.g. Reuther, 1995). Recreational trapping was not an intended part of the successful recovery of otters in Europe. Europeans evaluated the expansion of otter populations through a standardized, multi-national monitoring program. Recently, river otters have been recovering in the state of CA. Reintroductions have not been part of that recovery, which instead resulted from the natural expansion of remnant populations. A private NGO (River Otter Ecology Project) independently and systematically documented the expansion of the population over time by engaging citizen scientists to assist in field-monitoring efforts (Bouley *et al.*, 2015; Carroll *et al.*, 2020). From these examples, multiple approaches can serve to recover, conserve, and effectively monitor otter populations.

2) ***Expansion and promotion of river otter trapping in the US.*** Expansion of recreational river otter trapping in the US coincides with efforts to promote recreationally trapping in general (see Serfass *et al.*, 2017, pp. 563–568 for examples). In fact, the recovery of river otters has in many ways served to promote the virtues of trapping in furbearer management by fostering an association between trapping for conservation and trapping for fur. Two statements from Organ *et al.* (2015, p. 48) serve to elucidate promotional aspects derived from such an association: 'These are the same traps [those used to capture river otter for reintroduction projects] used by the public to harvest furbearers'. and '. . ., [river otters] were captured unharmed using long-spring foothold traps (below, left) with offset jaws'. The first statement serves to infer that if used for conservation a trap would be suitable (i.e. meet humane standards) for trapping in general. The second statement serves to further connect trapping for conservation with trapping for fur, but from our experiences is factually incorrect in that river otters trapped in LA for reintroduction projects typically were caught in standard no. 11 (double long-spring) foothold traps (i.e., jaws were not offset as indicated in the statement)

(see Serfass *et al.*, 1996, 2017, pp. 557–558). Serfass *et al.* (2017, pp. 563–568) further reviewed promotional aspects and other current topics pertaining to trapping river otters for fur. The Association of Fish and Wildlife Agencies (AFWA) has established a process of 'Best Management Practices for Trapping' (BMPs) as part of an initiative to establish humane standards for trapping (AFWA, 2006, 2014), in part related to public concerns pertaining to humane issues and trapping. The expansion of river otter trapping in the US has been accompanied by various rhetoric seemingly designed to ameliorate potential public opposition to the initiation of trapping seasons (see Serfass *et al.*, 2014, 2017, pp. 563–568).

3) ***Monitoring river otter populations.*** Monitoring river otter populations in the US is primarily based on recording annual kills during trapping seasons, unintended killings during trapping seasons for other furbearing animals, and other forms of mortality such as collisions with vehicles. Evaluation of the number of river otters killed during trapping seasons is consistent among years, fulfilling Convention on International Trade in Endangered Species of Wild Fauna and Flora (CITES) requirements to tag river otter pelts intended for export (CITES, 2013). Catch-per-unit effort (i.e. number of animals trapped in relation to trapping intensity) and other indices relevant to population monitoring often are derived from river otters killed by trapping and have been applied in assessing population trends (Erb *et al.*, 2018; Roberts *et al.*, 2020), but field-based monitoring is limited. Although there is a compilation of independent studies pertinent to field monitoring (e.g. Swimley *et al.*, 1998; DePue & Ben-David, 2010; Stevens *et al.*, 2011a; Crowley *et al.*, 2012), there are no standardized, repeatedly used field-based approaches in place (within or among primary management jurisdictions) to meaningfully monitor trends and factors influencing trends of river otter populations. Roberts *et al.* (2020) mapped the distribution of river otters in the US by their presence or absence at the county level, but provided no detail on procedures used to acquire data used to populate the counties. The map depicts the presence or absence of river otters by county, with no consideration of

the availability of suitable aquatic habitats as the ultimate limiting factor accompanying the portrayal. Although useful for demonstrating overall range expansion for promoting the success of river otter recovery efforts, such depiction provides limited utility for monitoring beyond an extremely coarse scale or to evaluate specific factors contributing to population trends. Also, by not considering the availability of suitable aquatic habitats and adjacent uplands, the depiction is an extreme exaggeration of the river otter's actual occurrence on the landscape. Hubbard & Serfass (2005) demonstrated how such county-level depictions can severely overestimate the occurrence of river otters in the landscape, and describe a GIS-derived grid-based approach applicable for systematically monitoring and realistically depicting population trends over time. Europe has a well-established, standardized approach for monitoring otter populations within and among countries, which is repeatable over time and facilitates habitat evaluations to enable assessment of factors influencing the presence or absence of otters. Europeans have a long history of refining and using this approach to effectively and systematically track the continent-wide expansion of otter populations. The following sample of journal article titles serves as an example of the landscape-level monitoring approach developed and applied for monitoring otter populations in an integrated manner throughout Europe: 'Otters, their habitat and conservation in northeast Greece' (Macdonald & Mason, 1985); 'The use of latrines to survey populations of otters *Lutra lutra*' (Kruuk *et al.*, 1986); 'Habitat networking: a new chance for the otter in Europe?' (Reuther, 1995); 'Some results of the 1991 and 1999 otter (*Lutra lutra*) surveys in the river Ise catchment, Lower-Saxony, Germany' (Reuther & Roy, 2001); 'Survey and habitat evaluation for a peripheral population of the Eurasian otter in Italy' (Loy *et al.*, 2004); 'Growth in otter (*Lutra lutra*) population in the UK as shown by long-term monitoring' (Mason & Macdonald, 2004); 'Monitoring of the otter recolonization of Poland' (Romanowski, 2006). Expansion of river otter trapping in the US warrants the development of a standardized, multi-jurisdictional field-based approach for monitoring river otters comparable to that for the Eurasian otter in Europe.

4) ***Public Trust Doctrine and river otters.*** Roberts *et al.* (2020) invoke the North American Model of Wildlife Conservation's (NAM) version of the 'Public Trust Doctrine' (PTD) as guiding the management of river otters in the US. This interpretation is understandable and expected considering the consumptive use (i.e. hunting and trapping) focus of NAM and the manner in which supporters of NAM have applied the PTD to justify a consumptive-use philosophy in managing wildlife (see Treves *et al.*, 2015 for a critique of PTD as applied to carnivores). Serfass *et al.* (2018) established concerns pertaining to NAM's focus on consumptive use in relation to limiting access of non-hunting/trapping citizens to decision-making pertaining to wildlife policy. Ultimately, the intended application of the PTD is to serve as a framework for entrusting the government (serving as the trustee) to conserve wildlife as a benefit to the entire public (the beneficiaries), with wildlife agency professionals (primarily biologists) serving as managers of the trust (i.e., the wildlife resource) (Smith, 2011). Expansion of trapping predominates among recent management actions undertaken by state wildlife agencies. Trappers represent about 0.05% of the US population (Responsive Management, 2015), but are the primary beneficiaries of most management actions implemented by state wildlife agencies. In contrast to trapping, non-consumptive values of river otters (e.g. flagships to promote aquatic conservation and education [Stevens *et al.*, 2011b]) have received relatively little attention from state wildlife agencies pertaining to river otters. The river otter is well liked by the general public (Serfass *et al.*, 2014), and conservation groups outside the framework of state wildlife agencies are recognizing other values of river otters (e.g. Chesapeake Bay Otter Watch [n.d.]; Elizabeth River Project's Otter Spotter program [n.d.]; River otter Ecology Project [n.d.]) in educating the public about aquatic conservation and ecology and promote clean-water initiatives. Given the non-consumptive purposes demonstrated by these projects and the overall popularity of river otters (Serfass *et al.*, 2014), we discerned no overriding public interest or support for the recreational killing of river otters to explain motivations for expansion of trapping seasons. The recreational killing of

wildlife is a primary, underlying principle of the wildlife conservation system represented by NAM. Under such a system, a trapping season should be anticipated ensuing the recovery of a 'furbearing' animal. Nonetheless, the entirety of motives underlying the seemingly integrated expansion of trapping is difficult to discern with certainty and can only be speculative. A quote from a furbearer biologist responsible for overseeing the implementation of the new trapping season for river otters in PA may offer additional insight:

'A series of new trapping opportunities has helped keep interest up, according to Lovallo [furbearer biologist].

Many people who were involved in the trapping heydays of the 1970s and 1980s may not find today's prices that appealing, but the Game Commission has done a great job of offering new opportunities in the state, thanks to the conservation efforts of trappers', he said. **'In 2000, we were able to offer the first**

bobcat trapping season. In 2005, cable traps were legalized and offered new opportunities. In 2016, we had our first controlled fisher trapping, and just this year we were able to add some opportunities for river otters. Each time there is a new opportunity that our resources can sustain, we see an influx of new trappers, and they seem to stick around well afterward' (Zaktansky, 2019).

Efforts to expand and promote river otter trapping to enhance recruitment and retention of trappers represents a disproportionate and inequitable application of the PTD towards a small fraction of beneficiaries (i.e. trappers). Such inequity is evident by the relatively little attention afforded non-consumptive wildlife enthusiasts in relation to aspects of river otter conservation that go beyond fur trapping. Enlightened application of PTD principles in relation to river otters should seemingly also entail consideration of the values and interests of beneficiaries beyond those of fur trappers.

References

AFWA [Association of Fish and Wildlife Agencies] (2006) Best management practices for trapping in the United States. https://www.fishwildlife.org/application/files/5015/2104/8473/Introduction_comp.pdf. Accessed on 3 March 2020.

AFWA [Association of Fish and Wildlife Agencies] (2013) National furbearer harvest statistics database. In: *Furbearer Management Resources.* Association of Fish and Wildlife Agencies, Washington, DC. http://fishwildlife.org/?section=furbearer_management_resources. Accessed on 5 May 2014.

AFWA [Association of Fish and Wildlife Agencies] (2014) Best management practices for trapping river otter in the United States. https://www.fishwildlife.org/application/files/8015/2105/3073/Otter_BMP_2014_F.pdf. Accessed on 3 March 2020.

Ashcraft, K.J. (2012) *Otter Visitation at Fish-Rearing Facilities in Pennsylvania.* MSc thesis, *Frostburg State University, Frostburg.*

Belanger, M., Askin, N., Tan, L. & Wittnich, C. (2010) The history and current status of otter research within Canada based on peer-reviewed journal articles. *IUCN Otter Specialist Group Bulletin* 27, 127–133.

Ben-David, M., Blundell, G.M., Kern, J.W., Maier, J.A.K., Brown, E.D. & Jewett, S.C. (2005) Communication in river otters: creation of variable resource sheds for terrestrial communities. *Ecology* 86, 1331–1345.

Berg, W.E. (1982) Reintroduction of fisher, pine marten, and river otter. In: *Midwest Furbearer Management* (ed. G.C. Sanderson), pp. 159–173. North Central Section of the Wildlife Society, Bloomington.

Beringer, J. & Blair, J. (2013) Furbearer program annual report. Missouri Department of Conservation, Resource Science Division, Missouri.

Bieber, N.R., Wilson, S.P. & Allen, C.R. (2018) River otter distribution in Nebraska. *Wildlife Society Bulletin*, 42, 136–143.

Bischoff, R. (2006) Status of the northern river otter in Nebraska. *The Prairie Naturalist* 35, 117–120.

Bluett, R.D., Nielsen, C.K., Gottfried, R.W., Miller, C.A. & Woolf, A. (2004) Status of the river otter (*Lontra* canadensis) in Illinois, 1998–2004. *Transactions of the Illinois State Academy of Science* 97, 209–217.

Bolen, E.G. & Robinson, W.L. (2003) *Wildlife Ecology and Management*. 5th edition. Prentice Hall, Upper Saddle River.

Bouley, P., Isadore, M. & Carroll, T. (2015) Return of North American river otters, *Lontra canadensis*, to coastal habitats of the San Francisco Bar area, California. *Northwestern Naturalist* 96, 1–12.

Brandt, J.R., Brandt, A.L., Ammer, F.K., Roca, A.L. & Serfass, T.L. (2014) Impact of population expansion on genetic diversity and structure of river otters (*Lontra canadensis*) in central North America. *Journal of Heredity* 105, 39–47.

Bricker, E.A. (2014) *Conservation Status of the North American River Otter (*Lontra canadensis*) in the United States and Canada: Assessing Management Practices and Public Perceptions of the Species*. MSc thesis, Frostburg State University, Frostburg.

Butler, D.R. (1991) The reintroduction of the beaver into the south. *Southeastern Geographer* 31, 39–43.

Carroll, T., Hellwig, E. & Isadore, M. (2020) An approach for long-term monitoring of recovering populations of Nearctic river otters (*Lontra canadensis*) in the San Francisco Bay area California, *Northwestern Naturalist* 101, 77–91.

Chesapeake Bay Otter Watch (n.d.) https://www.dropbox.com/s/qtyibm3y3wu1th8/Otter%20Video% 20SERC %20w%20Edits.mp4?dl=0. Accessed on 3 March 2020.

CITES [Convention on the International Trade in Endangered Species of Wild Flora and Fauna] (2013) How CITES works. http://www.cites.org/eng/disc/how.php. Accessed on 3 March 2020.

Clark, J.D., Jenkins, J.H., Bush, P.B. & Moser, E.B. (1987) Pollution trends in river otter in Georgia. *Proceedings of the Annual Conference of the South East Association Fish and Wildlife Agencies* 35, 71–79.

Crimmins, S.M., Roberts, N.M., Hamilton, D.A. & Mynsberge, A.R. (2009) Seasonal detection rates of river otters (*Lontra canadensis*) using bridge-site and random-site surveys. *Canadian Journal of Zoology* 87, 993–999.

Crowley, S., Johnson, C.J. & Hodder, D. (2012) Spatial and behavioral scales of habitat selection and activity by river otters at latrine sites. *Journal of Mammalogy* 93, 170–182.

CWHC [Canadian Wildlife Health Cooperative] (2017) Accidental bycatches identify rare returns of river otters to PEI. The blog of the Canadian Wildlife Health Cooperative. http://blog.healthywildlife.ca/accidental-bycatches-identify-rare-returns-river-otters-pei/. Accessed on 3 March 2020.

DePue, J.E. & Ben-David, M. (2010) River otter latrine site selection in arid habitats of western Colorado, USA. *Journal of Wildlife Management* 74, 1763–1767.

Dillman, D.A. (1978) *Mail and Telephone Surveys – The Total Design Method*. Wiley, New York.

DiStefano, J.J. (1987) Wild furbearer management in the Northeastern United States. In: *Wild Furbearer Management and Conservation in North America* (eds M. Novak, M.E. Obbard & B. Mallock), pp. 1077–1089. Ministry of Natural Resources, Toronto.

Dubuc, L.J., Krohn, W.B. & Owen, R.B. Jr (1990) Predicting occurrence of river otters by habitat on Mount Desert Island, Maine. *Journal of Wildlife Management* 54, 594–599.

Elizabeth River Project (n.d.). Otter spotter program. https://elizabethriver.org/otterspotter?page=5. Accessed on 3 March 2020.

Erb, J., Roberts, N.M. & Dwyer, C. (2018) An otterly successful conservation story – return of North American river otters. *The Wildlife Professional* 12, 45–49.

Fike, J.A., Serfass, T.L., Beheler, A.S. & Rhodes, O.E. Jr (2004) Genotyping error rates associated with alternative sources of DNA for the North American river otter. *IUCN Otter Specialist Group Bulletin* 21A.

Forman, N.S. (2015) *River Otter Population Monitoring in Northeastern Pennsylvania using Non-Invasive Genetic Sampling and Spatial Capture-Recapture Models*. MSc thesis, The Pennsylvania State University, University Park, State College.

Foster-Turley, P., Macdonald, S. & Mason, C. (eds) (1990) *Otters: An Action Plan for their Conservation*. IUCN SCC Otter Specialist Group. Kelvyn Press, Broadview.

Foy, M.K. (1984) *Seasonal Movement, Home Range, and Habitat Use of River Otter in Southeastern Texas*. MSc thesis, Texas A&M University, College Station.

Gallant, D. (2007) Species-wise disparity in scientific knowledge about otters: an obstacle to optimal management and conservation actions? *IUCN Otter Specialist Group Bulletin* 24, 5–13.

Gallant, D., Vasseur, L. & Bérubé, C.H. (2007) Unveiling the limitations of scat surveys to monitor social species: a case study on river otters. *The Journal of Wildlife Management* 71, 258–265.

Gallant, D., Vasseur, L. & Bérubé, C.H. (2008) Evaluating bridge survey ability to detect river otter *Lontra canadensis* presence: a comparative study. *Wildlife Biology* 14, 61–69.

Gallo-Reynoso, J.P., Macías-Sánchez, S., Nuñez-Ramos, V.A., Loya-Jaquez, A., Barba-Acuña, I.D., Armenta-Méndez, L.d.C, Guerrero-Flores, J.J., Ponce-García, G. & Gardea-Bejar, A.A. (2019) Identity and distribution of the Nearctic otter (*Lontra canadensis*) at the Rio Conchos Basin, Chihuahua, Mexico. *Therya* 10, 243–253.

Goedeke, T.L. (2005) Devils, angels, or animals: the social construction of otters in conflict over management. In: *Mad About Wildlife: Looking at Social Conflict Over Wildlife* (eds A. Herda-Rapp & T.L. Goedeke), pp. 22–50. Brill Academic Publishers, Boston.

Gorman, M.L. & Trowbridge, B.J. (1989) The role of odor in the social lives of carnivores. In: *Carnivore Behavior, Ecology, and Evolution*. Volume 1 (ed. J.L. Gittleman), pp. 57–88. Cornell University Press, Ithaca.

Hall, E.R. (1981) *The Mammals of North America*. 2nd Edition. John Wiley and Sons, New York.

Hall, E.R. & Kelson, K.R. (1959) *The Mammals of North America*. Ronald Press Company, New York.

Hamilton, D. (1999) Controversy in times of plenty. *Missouri Conservationist* 60, 17–24.

Hardisky, T.S. (2013) River otter management plan in Pennsylvania. Pennsylvania Game Commission Bureau of Wildlife Management in Pennsylvania. http://www.pgc.pa.gov/HuntTrap/Trappingand Furbearers/Documents/PA%20Otter%20 Management% 20Plan%20Draft.pdf. Accessed on 10 February 2018.

Hubbard, B. & Serfass, T. (2005) Assessing the distribution of reintroduced populations of river otters in Pennsylvania (USA) – development of a landscape-level approach. *IUCN Otter Specialist Group Bulletin* 21, 63–69.

Johnson, S.A., Walker, H.D., Hudson, C.M., Hewitt, T.R. & Thompson, J.S. (2007) Prospects for restoring river otters in Indiana. *Proceedings of the Indiana Academy of Science* 116, 71–83.

Just, E.H., Stevens, S.S., Spinola, R.M. & Serfass, T.L. (2012) Detecting river otter latrines near bridges: does habitat and season influence survey success? *Wildlife Biology* 3, 264–271.

Kellert, S.R., Black, M., Rush, C.R. & Bath, A.J. (1996) Human culture and large carnivore conservation in North America. *Conservation Biology* 10, 977–990.

Kohn, M.H., York, E.C., Kamradt, D.A., Haught, G., Sauvajot, R.M. & Wayne, R.K. (1999) Estimating population size by genotyping feces. *Proceedings of the Royal Society B: Biological Sciences* 266, 657–663.

Kruuk, H. (2006) Otters: *Ecology, Behaviour and Conservation*. Oxford University Press, New York.

Kruuk, H., Conroy, J.W.H., Glimmerveen, U. & Ouwerkerk, E.J. (1986) The use of latrines to survey populations of otters *Lutra lutra*. *Biological Conservation* 35, 187–194.

Loy, A., Bucci, L., Carranza, M.L., De Castro, G., Di Marzio, P. & Reggiani, G (2004) Survey and habitat evaluation for a peripheral population of the Eurasian otter in Italy. *IUCN Otter Specialist Group Bulletin* 21A.

Macdonald, S.M. & Mason, C.F. (1985) Otters, their habitat and conservation in northeast Greece. *Biological Conservation* 31, 191–210.

Mason, C.F. & Macdonald, S.M. (2004) Growth in otter (*Lutra lutra*) populations in the UK as shown by long-term monitoring. *AMBIO: A Journal of the Human Environment* 33, 148–152.

Melquist, W.E. & Dronkert, A.E. (1987) River otter. In: *Wild Furbearer Management and conservation in North America* (eds M. Novak, M.E. Obbard & B. Mallock), pp. 625–64. Ministry of Natural Resources, Ontario.

Melquist, W.E. & Hornocker, M.G. (1983) Ecology of river otters in west central Idaho. *Wildlife Monographs* 83, 3–60.

Melquist, W.E., Pelechla, P.J. & Toweill, V. (2003) River Otter (*Lontra canadensis*). In: *Wild Mammals of North America: Biology, Management, and Conservation* (eds G.A. Feldhamer, B.C. Thompson & J.A. Chapman), pp. 708–734. The Johns Hopkins University Press, Baltimore.

Mercer, B. (2020) River otter taken off South Dakota List of Threatened Species, offered for limited trapping. Capital News Bureau. https://www.keloland.com/ news/capitol-news-bureau/river-otter-taken-off-south-dakota-list-of-threatened-species-offered-for-limited-trapping/. Accessed on 29 May 2020.

Mowry, R.A., Gompper, M.E., Beringer, J. & Eggert, L.S. (2011) River otter population size estimation using noninvasive latrine surveys. *Journal of Wildlife Management* 75, 1625–1636.

Mowry, R.A., Schneider, T.M., Latch, E.K., Gompper, M.E., Beringer, J. & Eggert, L.S. (2015) Genetics and the successful reintroduction of the Missouri river otter. *Animal Conservation* 18, 196–206.

Newman, D.G. & Griffin, C.R. (1994) Wetland use by river otters in Massachusetts. *Journal of Wildlife Management* 58, 18–23.

Nilsson, G. (1980) *River Otter Research Workshop.* Florida State Museum, Gainesville.

Olson, Z.II., Stevens, S.S. & Serfass, T.L. (2005) Do juvenile Nearctic river otters (*Lontra canadensis*) contribute to fall scent marking? *The Canadian Field-Naturalist* 119, 457–459.

Olson, Z.H., Serfass, T.L. & Rhodes, O.E. Jr (2009) Seasonal variation in latrine site visitation and scent marking by Nearctic river otters (*Lontra canadensis*). *IUCN Otter Specialist Group Bulletin* 25, 109–119.

Organ, J.F., Decker, T., Langlois, S. & Mirick, P.G. (2015) Trapping and furbearer management in North American Wildlife Conservation. https://www.dec.ny.gov/docs/wildlife_pdf/trapfurmgmt.pdf. Accessed on 3 March 2020.

Pearce, K.J., Serfass, T.L., Ashcraft, S.A. & Stevens, S.S. (2017) Applying social and ecological approaches to evaluate factors influencing river otter (*Lontra canadensis*) visitation to fish-rearing facilities in Pennsylvania. *European Journal of wildlife Research* 63, 33.

Piggott, M.P. & Taylor, A.C. (2003) Extensive evaluation of faecal preservation and DNA extraction methods in Australian native and introduced species. *Australian Journal of Zoology* 51, 341–355.

Raesly, E.J. (2001) Progress and status of river otter reintroduction projects in the United States. *The Wildlife Society Bulletin* 29, 856–892.

Ralls, K. (1990) Reintroductions. In: *Otters: An Action Plan for their Conservation* (eds P. Foster-Turley, S. Macdonald & C. Mason), pp. 20–21. IUCN SCC Otter Specialist Group. Kelvyn Press, Broadview.

Reid, D.G., Bayer, M.D., Code, T.E. & McLean, B. (1987) A possible method for estimating river otter, *Lutra canadensis*, populations using snow tracks. *Canadian Field-Naturalist* 101, 576–580.

Responsive Management (2015) Trap use, furbearers trapped, and trapper characteristics in the United States in 2015. Responsive Management National Office, Harrisonburg, Virginia, USA.

Reuther, C. (1995) Habitat networking: a new chance for the otter in Europe? *Hystrix* 7, 229–238.

Reuther, C. & Roy, A. (2001) Some results of the 1991 and 1999 otter (*Lutra lutra*) surveys in the river Ise catchment, Lower-Saxony, Germany. *IUCN Otter Specialist Group Bulletin* 18, 28–40.

River Otter Ecology Project. (n.d.). https://riverotterecology.org/. Accessed 1 March 2020.

Roberts, N.M., Crimmins, S.M., Hamilton, D.A. & Gallagher, E. (2008) An evaluation of bridge-sign surveys to monitor river otter (*Lontra canadensis*) populations. *American Midland Naturalist* 160, 358–363.

Roberts, N.M., Lovallo, M.J. & Crimmins, S.M. (2020) River otter status, management, and distribution in the Unites States: evidence of large scale population increase and range expansion. *Journal of Fish and Wildlife Management* 11, 279–286.

Romanowski, J. (2006) Monitoring of the otter recolonization of Poland. *Hystrix* 17, 37–46.

Sasse, B. (2013) 2012–2013 Furbearing animal report. Arkansas Game and Fish Commission. Wildlife Management Division, Mayflower.

Serfass, T.L., Brooks, R.P., Swimley, T.J., Rymon, L.M. & Hayden, A.H. (1996) Considerations for capturing, handling, and translocating river otters. *Wildlife Society Bulletin* 24, 25–31.

Serfass, T.L., Novak, J.M., Johns, P.E. & Brooks, R.P. (1998). Genetic variation among river otter populations in North America: considerations for reintroduction projects. *Journal of Mammalogy* 79, 736–746.

Serfass, T.L., Brooks, R.P., Rymon, L.M. & Rhodes, O.E. Jr (2003) River otters in Pennsylvania, USA: lessons for predator re-introduction. In: *Proceedings of the European Otter Conference: Returning the Otter in Europe – Where and How?* (eds J.W.H. Conroy, G.M. Yoxon, A.C. Gutleb & J. Ruiz-Olmo). Journal of the International Otter Survival Fund, Isle of Skye.

Serfass, T.L., Bohrman, J.A., Stevens, S.S. & Bruskotter, J.T. (2014) Otters and anglers can share the stream! The role of social science in dissuading negative messaging about reintroduced predators. *Human Dimensions of Wildlife* 19, 532–544.

Serfass, T.L., Wright, L., Pearce, K.J. & Duplaix, N. (2017) Animal welfare issues pertaining to the trapping of otters for conservation, research and fur.

In: *Marine Mammal Welfare* (ed. A. Butterworth), pp. 543–571. Springer-Verlag, Heidelberg.

Serfass, T.L., Brooks, R.P. & Bruskotter, J.T. (2018) North American Model of Wildlife Conservation: empowerment and exclusivity hinder advances in Wildlife Conservation. *Canadian Wildlife Biology and Management* 7, 101–118.

Shackelford, J. & Whitaker, J. (1997) Relative abundance of the northern river otter, *Lutra canadensis*, in three drainage basins of southeastern Oklahoma. *Proceedings of the Oklahoma Academy of Science* 77, 93–98.

Shirley, M.G., Linscombe, R.G., Kinler, N.W., Knaus, R.M. & Wright, V.L. (1988) Population estimates of river otters in a Louisiana coastal marshland. *Journal of Wildlife Management* 52, 512–515.

Slough, B.G. & Jung, T.J. (2007) Diversity and distribution of the terrestrial mammals of the Yukon Territory: a review. *The Canadian Field-Naturalist* 121, 119–127.

Smith, C.A. (2011) The role of state wildlife professionals under the public trust doctrine. *Journal of Wildlife Management* 75, 1539–1543.

South Dakota Department of Game, Fish and Parks (2020) South Dakota river otter management plan. South Dakota Department of Game, Fish and Parks, Wildlife Division Report Number 2020-02, Pierre.

Stevens, S.S. & Serfass, T.L. (2008) Visitation patterns and behavior of Nearctic river otters (*Lontra canadensis*) at latrines. *Northeastern Naturalist* 15, 1–12.

Stevens, S.S., Just, E.H., Cordes, R.C., Brooks, R.P. & Serfass, T.L. (2011a) The influence of habitat quality on the detection of river otter (*Lontra canadensis*) latrines near bridges. *American Midland Naturalist* 166, 435–445.

Stevens, S.S., Organ J.F. & Serfass, T.L. (2011b) Otters as flagships: social and cultural considerations. *IUCN Otter Specialist Group Bulletin* 28, 150–161.

Swimley, T.J., Serfass, T.L., Brooks, R.P. & Tzilkowski, W.M. (1998) Predicting river otter latrine sites in Pennsylvania. *Wildlife Society Bulletin* 26, 836–845.

Tesky, J.L. (1993) *Lutra [Lontra] canadensis*. In: *Fire Effects Information System*. USDA Forest Service Rocky Mountain Research Station, Missoula. http://www.fs.fed.us/database/feis/animals/mammal/luca/all.html. Accessed on 5 May 2014.

Tischbein, G. (1976) More river otter transplanted to Colorado's west slope. *Wildlife News – Colorado Division of Wildlife* 1, 2.

Treves, A., Chapron, G., López-Bao, J.V., Shoemaker, C., Goeckner, A.R. & Bruskotter, J.T. (2015) Predators and the public trust. *Biological Reviews of the Cambridge Philosophical Society* 92, 248–270.

Triska, M.D., Loughry, S.C. & Serfass, T.L. (2011) River otters use agricultural field along the Turtle River in eastern North Dakota as crossover and latrine area. *The Prairie Naturalist* 43, 52–55.

USEPA [United States Environmental Protection Agency] (2015) Summary of the Clean Water Act. http://fishwildlife.org/?section=furbearer_management_resources. Accessed on 25 February 2015.

USPS [United States Postal Service] (2015) Appendix B: Two-letter State and Possession abbreviations. http://pe.usps.com/text/pub28/28apb.htm. Accessed on 2 March 2015.

Zaktansky, J. (2019) Outdoors: trapping is still popular in some parts of Pennsylvania. The Herald. https://www.sharonherald.com/sports/outdoors-trapping-is-still-popular-in-some-parts-of-pennsylvania/article_5a5e0096-f052-11e9-86df-d3976a40ea2f.html. Accessed on 3 March 2020.

Part VI

Appendices: The World's Small Carnivores

Appendix A

Species List and Selected Attributes

List of the 232 small carnivore species ($<21.5\,\text{kg}$) that are currently recognized by the IUCN (2021). For additional species recognized by other sources, see Appendix B. CR = Critically Endangered, EN = Endangered, V = Vulnerable, NT = Near Threatened, LC = Least Concern, DD = Data Deficient. *Compiler:* E. Do Linh San.

Small Carnivores: Evolution, Ecology, Behaviour, and Conservation, First Edition. Edited by Emmanuel Do Linh San, Jun J. Sato, Jerrold L. Belant, and Michael J. Somers.

Ailuridae

Family and scientific name[a]	Authority[b]	English name(s)[c]	Average body mass in kg[d]	Distribution[e]	Number of countries[f]	IUCN Red List categorization 2011–2019[g]	Population trend[g]	Chapter(s) in present book[h]
Ailurus fulgens	F.G. Cuvier, 1825	**Red panda**[i] [lesser panda; fire fox; golden/bear dog]	4.5	Asia	5	EN	Decreasing	(1), **2**, 22
Canidae								
Atelocynus microtis	(Sclater, 1883)	**Short-eared dog** [short-eared fox]	9.5	South America	5	NT	Decreasing	[1]
Canis [= Lupulella] adustus	Sundevall, 1847	**Side-striped jackal**	8.95	Africa	37	LC	Stable	[1]
Canis aureus	Linnaeus, 1758	Eurasian **golden jackal** [Asiatic/common/Indian jackal]	11	Asia & Europe	46+1	LC	Increasing	(1), 7, 16, (18)
Canis latrans	Say, 1823	**Coyote** [brush wolf; prairie wolf]	12.9	Central & North America	10	LC	Increasing	(1), 6, 7, 10, 17, (18), 21
Canis lupaster	Hemprich & Ehrenberg, 1833	**African (golden) wolf** [golden wolf]	11	Africa	24	LC	Decreasing	(1), 17
Canis [= Lupulella] mesomelas	Schreber, 1775	**Black-backed jackal** [silver-backed jackal]	8.5	Africa	17	LC	Stable	(1), 12, **17**, (18)
Canis simensis	Rüppell, 1840	**Ethiopian wolf** [Simien jackal/fox; Abyssinian wolf/red fox]	15.25	Africa	1	EN	Decreasing	(1), 7, 17
Cerdocyon thous	(Linnaeus, 1766)	**Crab-eating fox** [crab-eating zorro; savannah fox]	6.5	South America	12	LC	Stable	[1], (18)
Cuon alpinus	(Pallas, 1811)	**Dhole** [Asiatic wild dog]	15.5	Asia	20	EN	Decreasing	(1), 7, **16**
Lycalopex [= Pseudalopex] culpaeus	Molina, 1782	**Culpeo** fox [culpaeo; Andean fox]	8.6	South America	6	LC	Stable	[1], (18)
Lycalopex [= Pseudalopex] fulvipes	Martin, 1837	**Darwin's fox**	2.9	South America	1	EN	Decreasing	(1)
Lycalopex [= Pseudalopex] griseus	Gray, 1837	**South American grey fox** [chilla; Argentine/Southern grey fox]	3.75	South America	3	LC	Stable	[1], (18)

Species	Author, year	Common name		Distribution		IUCN	Trend	References
Lycalopex [= Pseudalopex] gymnocercus	G. Fischer, 1814	Pampas fox [Azara's fox]	5.5	South America	5	LC	Stable	[1], 9, (18)
Lycalopex [= Pseudalopex] sechurae	Thomas, 1900	Sechuran fox [Sechura/Peruvian (desert) fox]	3.4	South America	2	NT	Unknown	[1], (18)
Lycalopex [= Pseudalopex] vetulus	Lund, 1842	Hoary fox [hoary zorro; small-toothed dog]	3.25	South America	1	NT	Decreasing	[1]
Nyctereutes procyonoides	(Gray, 1834)	Raccoon dog [tanuki]	7.7	Asia & Europe	7 (29)	LC	Stable	[1], (5) 7, (18), 22
Otocyon megalotis	(Desmarest, 1822)	Bat-eared fox	4.4	Africa	12	LC	Stable	[1], 5, 12, 17, (18)
Speothos venaticus	(Lund, 1842)	Bush dog [savannah/vinegar dog]	6.5	South America	11+1	NT	Decreasing	[1], (5)
Urocyon cinereoargenteus	(Schreber, 1775)	Northern grey fox [tree fox]	4.5	North, Central & South America	12	LC	Stable	[1], [5] 7, (18)
Urocyon littoralis	(Baird, 1857)	Island grey fox [Channel Islands fox]	1.9	North America	1	NT	Increasing	[1]
Vulpes bengalensis	(Shaw, 1800)	Bengal fox [Indian fox]	2.8	Asia	4	LC	Decreasing	[1]
Vulpes cana	Blanford, 1877	Blanford's fox [King/Royal/Afghan fox]	1.2	Asia & Africa	12	LC	Stable	[1]
Vulpes chama	(A. Smith, 1833)	Cape fox	2.65	Africa	4+2	LC	Stable	[1], 17, (18)
Vulpes corsac	(Linnaeus, 1768)	Corsac fox [corsac; steppe fox]	2.4	Asia	11	LC	Unknown	[1], (18)
Vulpes ferrilata	Hodgson, 1842	Tibetan fox [sand fox]	4.35	Asia	3	LC	Unknown	[1]
Vulpes [= Alopex] lagopus	Linnaeus, 1758	Arctic fox [polar/white/blue fox]	4.9	North America, Europe & Asia	9	LC	Stable	[1], 4, (18), 22
Vulpes macrotis	Merriam, 1888	Kit fox [desert fox]	2.15	North America	2	LC	Decreasing	[1], 7
Vulpes pallida	(Cretzschmar, 1826)	Pale fox [pallid fox; African sand fox]	2.55	Africa	14+1 (1)	LC	Unknown	[1]
Vulpes rueppellii	(Schinz, 1825)	Rüppell's fox [(Rüppell's sand fox]	1.7	Africa & Asia	25	LC	Stable	[1]
Vulpes velox	(Say, 1823)	Swift fox	2.28	North America	1	LC	Stable	[1], 7, 17

(Continued)

(Continued)

Family and scientific name[a]	Authority[b]	English name(s)[c]	Average body mass in kg[d]	Distribution[e]	Number of countries[f]	IUCN Red List categorization 2011–2019[g]	Population trend[g]	Chapter(s) in present book[h]
Vulpes vulpes	(Linnaeus, 1758)	**Red fox** [cross/silver/common fox]	8.7	Asia, Europe, North America, Africa & Oceania	85+1 (2)	LC	Stable	(1), 4, (5), 7, 10, 11, **14**, (18), 21, 22
Vulpes zerda	(Zimmermann, 1780)	**Fennec (fox)**	1.35	Africa	11	LC	Stable	[1]
Eupleridae								
Cryptoprocta ferox	Bennett, 1833	**Fossa** [fosa]	7.05	Madagascar	1	VU	Decreasing	(1), 13, 21
Eupleres goudotii	Doyère, 1835	**(Eastern) falanouc**	1.85	Madagascar	1	VU	Decreasing	(1), 13, 21
Eupleres major	Lavauden, 1929	**Western/Giant falanouc**	3.7	Madagascar	1	EN	Decreasing	(1), 1, 3
Fossa fossana	(P.L.S. Müller, 1776)	**(Spotted) fanaloka** [Malagasy (striped) civet]	1.6	Madagascar	1	VU	Decreasing	(1), **13**, 21
Galidia elegans	I. Geoffroy Saint-Hilaire, 1837	**Ring-tailed vontsira** [ring-tailed mongoose]	0.78	Madagascar	1	LC	Decreasing	[1], 3, 13, 21
Galidictis fasciata	(Gmelin, 1788)	**Broad-striped vontsira** [broad-striped Malagasy mongoose]	0.63	Madagascar	1	VU	Decreasing	(1), 3, 13, 21
Galidictis grandidieri	Wozencraft, 1986	**Grandidier's vontsira** [Grandidier's mongoose; Giant-striped mongoose]	1.35	Madagascar	1	EN	Decreasing	(1), 3
Mungotictis decemlineata	(A. Grandidier, 1867)	**Bokiboky** [Narrow-striped boky/mongoose]	0.6	Madagascar	1	EN	Decreasing	(1), 3
Salanoia concolor	(I. Geoffroy Saint-Hilaire, 1837)	**Brown-tailed vontsira** [brown-tailed mongoose; salano]	0.78	Madagascar	1	VU	Decreasing	(1), 3, 13, 21
Felidae								
Caracal [= Profelis] aurata	(Temminck, 1827)	**African golden cat**	10.65	Africa	14+9	VU	Decreasing	[1]
Caracal caracal	(Schreber, 1776)	**Caracal**	16.1	Africa & Asia	60+1	LC	Unknown	[1], 12, 17

			Mass	Distribution	No.	IUCN	Trend	References
Catopuma [= Pardofelis] badia	(Gray, 1874)	**(Borneo) bay cat**	1.95	Asia	2+1	EN	Decreasing	(1), [5]
Catopuma [= Pardofelis] temminckii	(Vigors & Horsfield, 1827)	**Asiatic golden cat**	12.15	Asia	12	NT	Decreasing	[1], **16**
Felis bieti	Milne-Edwards, 1892	**Chinese mountain cat** [Chinese desert/steppe cat]	7.75	Asia	1	VU	Decreasing	[1]
Felis chaus	Schreber, 1777	**Jungle cat** [swamp/reed cat]	7.4	Asia & Africa	29+1	LC	Decreasing	[1]
Felis margarita	Loche, 1858	**Sand cat**	2.38	Africa & Asia	19+5	LC	Unknown	[1]
Felis nigripes	Burchell, 1824	**Black-footed cat** [small-spotted cat]	1.73	Africa	3+4	VU	Decreasing	[1], 17
Felis silvestris	Schreber, 1777	**(European) wild cat**	4.85	Africa, Asia & Europe	101+2	LC	Decreasing	(1)[k], **4**, (5), 7[k], 12[l], **17**[l]
Herpailurus [= Puma] yagouaroundi	(É. Geoffroy Saint-Hilaire, 1803)	**Jaguarundi** [eyra (cat)]	5.55	South, Central & North America	19+2	LC	Decreasing	[1], [5], 6, 16
Leopardus colocola[m]	(Molina, 1782)	Central Chilean[n] **Pampas cat** [colocolo]	2.7	South America	8+1	NT	Decreasing	(1)
Leopardus geoffroyi	(d'Orbigny & Gervais, 1844)	**Geoffroy's cat**	5.2	South America	6	LC	Stable	[1], 9
Leopardus guigna	(Molina, 1782)	**Guiña** [guina; kodkod]	2.15	South America	2	VU	Decreasing	[1]
Leopardus guttulus	(Hensel, 1872)	**Southern tiger cat** [southern tigrina]	2.5	South America	3	VU	Decreasing	(1)
Leopardus jacobita	(Cornalia, 1865)	**Andean mountain cat**	4	South America	4	EN	Decreasing	(1)
Leopardus pardalis	(Linnaeus, 1758)	**Ocelot**	12.6	South, Central & North America	22	LC	Decreasing	[1], 7, 12, 16, 22
Leopardus tigrinus	(Schreber, 1775)	**Northern tiger cat** [northern tigrina; oncilla; little-spotted cat]	2.5	South & Central America	11+1	VU	Decreasing	(1), [5]
Leopardus wiedii	(Schinz, 1821)	**Margay**	3.6	South, Central & North America	20+1	NT	Decreasing	[1], [5], 16, 22

(Continued)

(Continued)

Family and scientific name[a]	Authority[b]	English name(s)[c]	Average body mass in kg[d]	Distribution[e]	Number of countries[f]	IUCN Red List categorization 2011–2019[g]	Population trend[g]	Chapter(s) in present book[h]
Leptailurus serval	(Schreber, 1776)	**Serval**	12	Africa	40+3	LC	Stable	[1], 12
Lynx canadensis	Kerr, 1792	**Canada/Canadian lynx**	11.15	North America	2	LC	Stable	[1], 5, 22
Lynx lynx	(Linnaeus, 1758)	**Eurasian/European lynx**	21	Asia & Europe	45	LC	Stable	(1), 4
Lynx pardinus	(Temminck, 1827)	**Iberian lynx** [Spanish/Pardel lynx]	12.3	Europe	2	EN	Increasing	(1), 16
Lynx rufus	(Schreber, 1777)	**Bobcat** [bay/red lynx]	10.95	North America	3	LC	Stable	[1], 5, 7, 10, (18)
Neofelis diardi	(G. Cuvier, 1823)	**Sunda/Diardi's Clouded leopard**	17.5	Asia	3	VU	Decreasing	(1), 16
Neofelis nebulosa	(Griffith, 1821)	**Indochinese/Mainland clouded leopard**	17.5	Asia	11	VU	Decreasing	(1), **16**
Otocolobus manul	(Pallas, 1776)	**Pallas's cat** [manul; steppe cat]	3.9	Asia	12+6	LC	Decreasing	[1]
Pardofelis marmorata	(Martin, 1837)	**Marbled cat**	3.75	Asia	13	NT	Decreasing	[1], [5], 16
Prionailurus bengalensis	(Kerr, 1792)	**(Indochinese/Mainland) leopard cat**	3.83	Asia	23+1	LC	Stable	(1), [5], **16, 19**
Prionailurus planiceps	(Vigors & Horsfield, 1827)	**Flat-headed cat**	1.85	Asia	3+1	EN	Decreasing	(1), [5]
Prionailurus rubiginosus	(I. Geoffroy Saint-Hilaire, 1831)	**Rusty-spotted cat**	1.3	Asia	3	NT	Decreasing	[1]
Prionailurus viverrinus	(Bennett, 1833)	**Fishing cat**	10.55	Asia	11	VU	Decreasing	(1)
Herpestidae								
Atilax paludinosus	(G. [Baron] Cuvier, 1829)	**Marsh/Water mongoose**	3.25	Africa	38	LC	Decreasing	(1), 3, [5], 12, **15**, 19
Bdeogale crassicauda	Peters, 1852	**Bushy-tailed mongoose**	1.7	Africa	7 (+1)	LC	Unknown	(1), **3**
Bdeogale jacksoni	(Thomas, 1894)	**Jackson's mongoose**	2.5	Africa	3	NT	Decreasing	[1], **3**

Bdeogale nigripes	Pucheran, 1855	**Black-legged mongoose** [black-footed mongoose]	3.4	Africa	7	LC	Decreasing	[1], **3, 15**
Bdeogale omnivora	Heller, 1913	**Sokoke bushy-tailed mongoose** [Sokoke dog mongoose]	1.7	Africa	2	VU	Decreasing	(1), 3
Crossarchus alexandri	Thomas & Wroughton, 1907	**Alexander's cusimanse** [Alexander's kusimanse]	1.5	Africa	4	LC	Decreasing	[1], 3, 7
Crossarchus ansorgei	Thomas, 1910	**Ansorge's/Angolan cusimanse** [Angolan mongoose]	1.05	Africa	2	LC	Decreasing	[1], 3
Crossarchus obscurus	F.G. Cuvier, 1825	**Common cusimanse** [common kusimanse; long-nosed/West African cusimanse]	0.73	Africa	5	LC	Unknown	[1], 3
Crossarchus platycephalus	Goldman, 1984	**Flat-headed cusimanse** [flat-headed kusimanse]	1	Africa	7	LC	Unknown	[1], 3, 15
Cynictis penicillata	(G. [Baron] Cuvier, 1829)	**Yellow mongoose**	0.67	Africa	5+1	LC	Stable	[1], **3**, 7, 17
Dologale dybowskii	(Pousargues, 1893)	**Pousargues's mongoose** [savannah mongoose]	0.35	Africa	4	DD	Unknown	(1), 3
Galerella [= Herpestes] flavescens	(Bocage, 1889)	**Kaokoveld/Angolan slender mongoose** [black (slender) mongoose; large red mongoose]	0.73	Africa	2	LC	Unknown	(1), 3
Galerella [= Herpestes] ochracea	(J.E. Gray, 1848)	**Somali(an) slender mongoose**	0.53	Africa	3	LC	Unknown	[1], 3
Galerella [= Herpestes] pulverulenta	(Wagner, 1839)	**Cape/small grey mongoose**	0.87	Africa	3	LC	Stable	[1], 3, 7, 17, 19
Galerella [= Herpestes] sanguinea	(Rüppell, 1835)	**(Common) slender mongoose** [black-tipped mongoose]	0.58	Africa	39	LC	Stable	[1], 3, [5], 7, 12, 15, 17
Helogale hirtula	Thomas, 1904	**Somali(an) dwarf mongoose** [Ethiopian/desert dwarf mongoose]	0.29	Africa	3	LC	Unknown	[1], 3

(Continued)

(Continued)

Family and scientific name[a]	Authority[b]	English name(s)[c]	Average body mass in kg[d]	Distribution[e]	Number of countries[f]	IUCN Red List categorization 2011–2019[g]	Population trend[g]	Chapter(s) in present book[h]
Helogale parvula	(Sundevall, 1847)	**(Common) dwarf mongoose**	0.28	Africa	15	LC	Stable	[1], 3, [5], 7, 17
Herpestes ichneumon	(Linnaeus, 1758)	**Egyptian mongoose** [large grey mongoose; ichneumon]	3.15	Africa, Europe & Asia	51+2	LC	Stable	[1], 3, [5], **15**, (18), 19
Ichneumia albicauda	(G. [Baron] Cuvier, 1829)	**White-tailed mongoose**	3.5	Africa & Asia	38	LC	Stable	[1], 3, [5], 12, (18), 19
Liberiictis kuhni	Hayman, 1958	**Liberian mongoose**	2.3	Africa	3	VU	Decreasing	[1], 3
Mungos gambianus	(Ogilby, 1835)	**Gambian mongoose**	1.6	Africa	10	LC	Stable	[1], 3
Mungos mungo	(Gmelin, 1788)	**Banded mongoose** [striped mongoose]	1.4	Africa	32	LC	Stable	(1), 3, [5], **7**, 12, 17
Paracynictis selousi	(de Winton, 1896)	**Selous's mongoose**	1.8	Africa	8	LC	Unknown	[1], 3, [5]
Rhynchogale melleri	(Gray, 1865)	**Meller's mongoose**	2.35	Africa	8	LC	Unknown	[1], **3**, [5]
Suricata suricatta	(Schreber, 1776)	**Meerkat/Suricate** [slender-tailed meerkat; grey meerkat]	0.8	Africa	4	LC	Stable	(1), 3, **7**, 17
Urva [= Herpestes] auropunctata	(Hodgson, 1836)	**Small Indian mongoose**	0.48	Asia, Africa, Europe, Central & South America, Oceania	12 (29+8)	LC	Unknown	(1), **3**, 19, **20**
Urva [= Herpestes] brachyura	(Gray, 1836)	**Short-tailed mongoose**	2.5	Asia	3	NT	Decreasing	(1), 3, [5], **19**
Urva [= Herpestes] edwardsii	(É. Geoffroy Saint-Hilaire, 1818)	**Indian grey mongoose** [common grey mongoose]	1.45	Asia	13 (+1)	LC	Stable	[1], 3, **20**
Urva [= Herpestes] fusca	(Waterhouse, 1838)	**(Indian) brown mongoose**	1.9	Asia & Oceania	2	LC	Stable	(1), 3, 19, **20**
Urva [= Herpestes] javanica	(É. Geoffroy Saint-Hilaire, 1818)	**Javan mongoose** [small Asian mongoose]	0.73	Asia	10	LC	Unknown	[1], **3**, 16, **19**

Scientific name	Authority	Common name		Region		IUCN	Trend	References
Urva [= Herpestes] semitorquata	(Gray, 1846)	**Collared mongoose**	3	Asia	2+2	NT	Decreasing	(1), 3, [5], **19**
Urva [= Herpestes] smithii	(Gray, 1837)	**Ruddy mongoose**	2.23	Asia	3	LC	Unknown	[1], 3
Urva [= Herpestes] urva	(Hodgson, 1836)	**Crab-eating mongoose**	3.5	Asia	13	LC	Decreasing	(1), 3, **19**, 22
Urva [= Herpestes] vitticollis	(Bennett, 1835)	**Striped-necked mongoose**	2.55	Asia	2	LC	Stable	[1], 3
Xelogale [= Herpestes] naso	Allen, 1919	**Long-nosed**/snouted **mongoose**	3.2	Africa	10	LC	Decreasing	[1], **3**, **15**, 19
Hyaenidae								
Proteles cristatus[o]	(Sparrman, 1783)	**Aardwolf**	10.85	Africa	17+2	LC	Stable	[1], 7, 12, 17
Mephitidae								
Conepatus chinga	(Molina, 1782)	**Molina's hog-nosed skunk** [Andean hog-nosed skunk]	2	South America	7	LC	Decreasing	(1), 6, **9**
Conepatus humboldtii	Gray, 1837	**Humboldt's hog-nosed skunk** [Patagonian skunk]	1.5	South America	2	LC	Stable	(1)
Conepatus leuconotus	(Lichtenstein, 1832)	**North**[p] **American hog-nosed skunk** [common/white-backed hog-nosed skunk]	3.25	North & Central America	6	LC	Decreasing	(1), **6**
Conepatus semistriatus	(Boddaert, 1785)	**Amazonian hog-nosed skunk**	2.45	South, Central & North America	11	LC	Unknown	[1], [5], 6
Mephitis macroura	Lichtenstein, 1832	**Hooded skunk**	1.7	North & Central America	6	LC	Increasing	[1]
Mephitis mephitis	(Schreber, 1776)	**Striped skunk**	3.05	North America	3	LC	Stable	[1], [5], 6, **20**, 22
Mydaus javanensis	(Desmarest, 1820)	**Sunda stink badger** [Malayan/Indonesian stink badger; teledu]	2.5	Asia	2+1	LC	Stable	[1], 6
Mydaus marchei	(Huet, 1887)	**Palawan stink badger** [Philippine stink badger skunk badger; pantot; teledu]	1.68	Asia	1	LC	Stable	[1]

(Continued)

(Continued)

Family and scientific name[a]	Authority[b]	English name(s)[c]	Average body mass in kg[d]	Distribution[e]	Number of countries[f]	IUCN Red List categorization 2011–2019[g]	Population trend[g]	Chapter(s) in present book[h]
Spilogale angustifrons	Howell, 1902	**Southern spotted skunk** [Central American spotted skunk]	0.39	Central & North America	7	LC	Stable	[1]
Spilogale gracilis	Merriam, 1890	**Western spotted skunk**	0.63	North & Central America	2	LC	Decreasing	[1], 6, 22
Spilogale putorius	(Linnaeus, 1758)	**Eastern spotted skunk**	0.55	North America	3	VU	Decreasing	[1], 22
Spilogale pygmaea	Thomas, 1898	**Pygmy (spotted) skunk**	0.18	North America	1	VU	Decreasing	[1]
Mustelidae								
Aonyx capensis	(Schinz, 1821)	**African[q]/Cape clawless otter** [African small-clawed otter]	15.5	Africa	32	NT	Decreasing	(1), 7, 17
Aonyx [= Amblonyx] cinereus	(Illiger, 1815)	**Asian/Oriental small-clawed otter**	3.1	Asia & Europe	16 (1)	VU	Decreasing	[1], [5]
Aonyx congicus	Lönnberg, 1910	**Congo clawless otter** [Congo small-clawed otter; swamp otter]	19.5	Africa	9+2	NT	Decreasing	(1), 15
Arctonyx albogularis	(Blyth, 1853)	**Northern/Chinese hog badger**	7.5	Asia	3+1	LC	Decreasing	(1), 16, 22
Arctonyx collaris	F.G. Cuvier, 1825	**(Greater)' hog badger**	11	Asia	7+2	VU	Decreasing	(1), 7, 22
Arctonyx hoevenii	(Hubrecht, 1891)	**Sumatran hog badger**	6	Asia	1	LC	Stable	(1)
Eira barbara	(Linnaeus, 1758)	**Tayra** [grey-headed tayra; eira]	4.85	South, Central & North America	20	LC	Decreasing	[1], (5), (18)
Galictis cuja	(Molina, 1782)	**Lesser grison**	1.75	South America	6	LC	Unknown	[1], 2
Galictis vittata	(Schreber, 1776)	**Greater grison**	2.7	South, Central & North America	16+1	LC	Stable	[1], 2, [5]
Gulo gulo	(Linnaeus, 1758)	**Wolverine** [glutton; skunk bear]	12.4	North America, Asia & Europe	9	LC	Decreasing	[1], **4**, 22

Species	Authority	Common name		Distribution		IUCN	Trend	References
Hydrictis [= Lutra] maculicollis	(Lichtenstein, 1835)	**Spotted-necked otter**	4.75	Africa	32+4	NT	Decreasing	[1]
Ictonyx libycus	(Hemprich & Ehrenberg, 1833)	**Libyan (striped) weasel** [Saharan/North African striped weasel; Saharan striped polecat]	0.4	Africa	15	LC	Unknown	[1], 6, 22
Ictonyx striatus	(Perry, 1810)	**Zorilla** [striped polecat]	0.95	Africa	39	LC	Stable	[1], [5], 12, 17
Lontra canadensis	(Schreber, 1777)	**North American (river) otter** [Nearctic/Canadian otter]	9.45	North America	2+1	LC	Stable	(1), [5] 7, 22, **23**
Lontra felina	(Molina, 1782)	**Marine otter** [chungungo]	4.5	South America	3	EN	Decreasing	(1)
Lontra longicaudis	(Olfers, 1818)	**Neotropical** river otter	8.5	South, Central & North America	20	NT	Decreasing	(1), 7, (18)
Lontra provocax	(Thomas, 1908)	**Southern**/Patagonian river otter [huillín]	11.25	South America	2	EN	Decreasing	(1)
Lutra lutra	(Linnaeus, 1758)	**Eurasian/European otter** [common otter]	11.5	Asia, Europe & Africa	79	NT	Decreasing	(1), **4, 7,** 14, 22
Lutra sumatrana	(Gray, 1865)	**Hairy-nosed otter**	6.5	Asia	5+2	EN	Decreasing	(1), [5]
Lutrogale perspicillata	(I. Geoffroy Saint-Hilaire, 1826)	Indian **smooth-coated otter** [smooth otter]	9	Asia	15	VU	Decreasing	[1], 2
Lyncodon patagonicus	(de Blainville, 1842)	**Patagonian weasel** [huroncito]	0.23	South America	2	LC	Unknown	[1], 2
Martes americana	(Turton, 1806)	**American marten**	0.8	North America	2	LC	Decreasing	[1], [5], (18), 22
Martes flavigula	(Boddaert, 1785)	**Himalayan yellow-throated marten** [kharza]	2.15	Asia	19	LC	Decreasing	[1], [5], **16**, (18), 22
Martes foina	(Erxleben, 1777)	**Stone/Beech marten**	1.7	Asia, Europe & North America	55 (1)	LC	Stable	(1), **4**, [5], **10**, 11, **14**, (18), **20**, 22
Martes gwatkinsii	Horsfield, 1851	**Nilgiri marten**	2	Asia	1	VU	Stable	[1]
Martes martes	(Linnaeus, 1758)	Eurasian/European **pine marten**	1.3	Europe & Asia	43	LC	Stable	(1), **4**, [5], 7, 10, 11, **14**, (18), **20**, 22

(Continued)

(Continued)

Family and scientific name[a]	Authority[b]	English name(s)[c]	Average body mass in kg[d]	Distribution[e]	Number of countries[f]	IUCN Red List categorization 2011–2019[g]	Population trend[g]	Chapter(s) in present book[h]
Martes melampus	(Wagner, 1840)	**Japanese marten** [yellow marten; Tsushima Island marten]	1.2	Asia	1 (+1)	LC	Stable	[1], (18), 20
Martes zibellina	(Linnaeus, 1758)	Japanese **sable**	1.25	Asia	6	LC	Increasing	[1], 20, 22
Meles anakuma	Temminck, 1844	**Japanese badger** [anakuma; anaguma]	11.5	Asia	1	LC	Decreasing	(1), 7
Meles leucurus	(Hodgson, 1847)	**(Northeast)**[s] **Asian badger**	6.25	Asia	8	LC	Unknown	(1)
Meles meles	(Linnaeus, 1758)	**European badger** [Eurasian[t] badger]	10.25	Europe & Asia	50+2	LC	Stable	(1), **4**, **7**, **8**, 10, 14, (18), 22
Mellivora capensis	(Schreber, 1776)	**Honey badger** [ratel]	10.35	Africa & Asia	62+2	LC	Decreasing	(1), 2, 6, 7, 12, 15, **17**, 22
Melogale cucphuongensis	Nadler, Streicher, Stefen, Schwierz & Roos, 2011	**Vietnam/Cuc Phuong ferret badger**	?	Asia	2	DD	Unknown	(1)
Melogale everetti	(Thomas, 1895)	**Bornean ferret badger** [Kinabalu/Everett's ferret badger]	1.5	Asia	1	EN	Decreasing	(1)
Melogale moschata	(Gray, 1831)	**Small-toothed/Chinese ferret badger**	1.2	Asia	6+1	LC	Stable	[1], (18)
Melogale orientalis	(Horsfield, 1821)	**Javan ferret badger**	1.5	Asia	1	LC	Unknown	[1]
Melogale personata	I. Geoffroy Saint-Hilaire, 1831	**Large-toothed/Burmese ferret badger**	2.25	Asia	9	LC	Unknown	[1], 16
Mustela africana	Desmarest, 1818	**Amazon weasel** [tropical weasel]	0.2	South America	4+1	LC	Unknown	[1], [5]
Mustela altaica	Pallas, 1811	**Altai** mountain **weasel** [Alpine weasel]	0.24	Asia	10+2	NT	Decreasing	[1]

Species	Author	Common name		Distribution		IUCN	Population trend	References
Mustela erminea	Linnaeus, 1758	**Stoat** [ermine; short-tailed weasel]	0.22	North America, Europe, Asia & Oceania	54 (1)	LC	Stable	(1), **4**, [5], **20**, 22
Mustela eversmanii	Lesson, 1827	**Steppe polecat** [steppe weasel]	0.8	Asia & Europe	20+1	LC	Decreasing	[1], **4**, [5], 7, 21
Mustela felipei	Izor & de la Torre, 1978	**Colombian weasel** [Felipe's weasel]	0.14	South America	2	VU	Decreasing	[1]
Mustela frenata	Lichtenstein, 1831	**Long-tailed weasel**	0.27	North, Central & South America	16	LC	Stable	[1], [5]
Mustela itatsi	Temminck, 1844	**Japanese weasel**	0.59	Asia	1	NT	Decreasing	(1), (18), **20**
Mustela kathiah	Hodgson, 1835	**Yellow-bellied weasel** [mountain weasel]	0.23	Asia	9	LC	Stable	(1)
Mustela lutreola	(Linnaeus, 1761)	**European mink**	0.75	Europe	6	CR	Decreasing	(1), **4, 11**, 13, **20**, 22
Mustela lutreolina	Robinson & Thomas, 1917	**Indonesian mountain weasel**	0.32	Asia	1	LC	Stable	[1]
Mustela nigripes	(Audubon & Bachman, 1851)	**Black-footed ferret**	0.93	North America	1	EN	Decreasing	(1), 21
Mustela nivalis	Linnaeus, 1766	**Least weasel** [common weasel]	0.16	Asia, Europe, Africa & Oceania	61 (4)	LC	Stable	(1), 2, **4**, (5), 5, **20**, 22
Mustela nudipes	Desmarest, 1822	**Malay weasel**	1	Asia	4	LC	Decreasing	[1], [5]
Mustela putorius	Linnaeus, 1758	**Western/European polecat** [common polecat ferret]	1.05	Europe, Asia, Africa & Oceania	40 (3)	LC	Decreasing	(1), **4**, [5], 7, 11, 14, **20**, 22
Mustela russelliana	Thomas, 1911	**Sichuan weasel**		Asia	1	DD	Unknown	(1)
Mustela sibirica	Pallas, 1773	**Siberian weasel** [Siberian polecat; Himalayan wease ; kolinsky]	0.59	Asia	11+2 (1)	LC	Stable	[1], 7, 16, **20**
Mustela strigidorsa	Gray, 1853	**Stripe-backed weasel** [back-striped weasel]	1.35	Asia	7	LC	Stable	[1], 16
Mustela subpalmata	Hemprich & Ehrenberg, 1833	**Egyptian weasel**	0.16	Africa	1	LC	Increasing	(1)

(Continued)

(Continued)

Family and scientific name[a]	Authority[b]	English name(s)[c]	Average body mass in kg[d]	Distribution[e]	Number of countries[f]	IUCN Red List categorization 2011–2019[g]	Population trend[g]	Chapter(s) in present book[h]
Mustela tonkinensis	Björkegren, 1941	**Tonkin weasel**		Asia	1	DD	Unknown	(1)
Neovison [= Mustela] vison	(Schreber, 1777)	**American mink**	1.38	North & South America, Europe, & Asia	2 (28+6)	LC	Stable	[1], [5], 11, 13, **20**, 22
Pekania [= Martes] pennanti	(Erxleben, 1777)	**Fisher**	3.4	North America	2	LC	Unknown	[1], [5], 10, 22
Poecilogale albinucha	(Gray, 1864)	**(African) striped weasel** [white-napped weasel]	0.3	Africa	17	LC	Unknown	[1], 6, 7, 22
Taxidea taxus	(Schreber, 1777)	**North American badger**	8	North America	3	LC	Decreasing	(1), 17, 21, 22
Vormela peregusna	(Güldenstädt, 1770)	**European marbled polecat**	0.51	Asia & Europe	25	VU	Decreasing	[1], 4
Nandiniidae								
Nandinia binotata	(Gray, 1830)	**African/two-spotted palm civet** [nandinia]	2.1	Africa	28+1	LC	Unknown	[1], **7**, **15**, (18), 22
Prionodontidae								
Prionodon linsang	(Hardwicke, 1821)	**Banded linsang**	0.7	Asia	5	LC	Decreasing	[1], [5]
Prionodon pardicolor	Hodgson, 1841	**Spotted linsang**	0.88	Asia	9	LC	Decreasing	[1]
Procyonidae								
Bassaricyon alleni	Thomas, 1880	**Eastern lowland olingo** [Allen's olingo]	1.25	South America	7	LC	Decreasing	(1)
Bassaricyon gabbii	J.A. Allen, 1876	**Northern olingo** [bushy-tailed olingo]	1.3	Central America	3+2	LC	Decreasing	(1), [5]
Bassaricyon medius	Thomas, 1909	**Western lowland olingo** [Panamanian/middle olingo]	1.1	South & Central America	3	LC	Decreasing	[1]
Bassaricyon neblina	Helgen, Pinto, Kays, Helgen, Tsuchiya, Quinn, Wilson & Maldonado, 2013	**Olinguito** [Andean olingo]	0.93	South America	2	NT	Decreasing	(1)

Species	Authority	Common name		Region		IUCN	Population trend	References
Bassariscus astutus	(Lichtenstein, 1830)	**Ringtail** (ring-tailed cat)	0.95	North America	2	LC	Unknown	[1], 7, (18)
Bassariscus sumichrasti	(Saussure, 1860)	Central American **cacomistle**	0.95	Central & North America	8	LC	Unknown	[1], [5]
Nasua narica	(Linnaeus, 1766)	**White-nosed coati** [coatimundi]	4.55	North, Central & South America	10	LC	Decreasing	[1], (5)
Nasua nasua	(Linnaeus, 1766)	**South American coati** [brown-nosed coati; ring-tailed coati]	4.6	South America	12+1	LC	Decreasing	[1], (5), (18), **20**
Nasuella meridensis	(Thomas, 1901)	**Eastern mountain coati**		South America	1	EN	Decreasing	(1)
Nasuella olivacea	(Gray, 1865)	**Western mountain coati**	1.25	South America	2+1	NT	Decreasing	(1)
Potos flavus	(Schreber, 1774)	**Kinkajou**	3	South America	17	LC	Decreasing	(1), (5), (18)
Procyon cancrivorus	(G. [Baron] Cuvier, 1798)	**Crab-eating raccoon**	5.4	South America	15	LC	Decreasing	(1), [5]
Procyon lotor	(Linnaeus, 1758)	**(Northern) raccoon** [common raccoon]	6.35	North, Central & South America, Europe & Asia	10 (12)	LC	Increasing	(1), (5), 7, (18), **20**, 22
Procyon pygmaeus	Merriam, 1901	**Pygmy raccoon** [Cozumel raccoon]	3.5	North America	1	CR	Decreasing	(1)
Viverridae								
Arctictis binturong	(Raffles, 1821)	**Binturong** [bearcat]	14.5	Asia	13+1	VU	Decreasing	(1), **16**, (18), 19
Arctogalidia trivirgata	(Gray, 1832)	**Small-toothed palm civet**	2.25	Asia	10+2	LC	Decreasing	[1], [5], (18), 19
Chrotogale owstoni	Thomas, 1912	**Owston's** palm **civet**	3.35	Asia	2+2	EN	Decreasing	(1), 19
Civettictis civetta	(Schreber, 1776)	**African civet**	13.5	Africa	37 (1)	LC	Unknown	(1), 7, **12**, **15**, (18)
Cynogale bennettii	Gray, 1837	Sunda **otter civet**	4	Asia	3+1	EN	Decreasing	(1), [5], 19
Diplogale hosei	(Thomas, 1892)	**Hose's** palm **civet**	1.3	Asia	3	VU	Decreasing	[1], **19**
Genetta abyssinica	(Rüppell, 1836)	**Ethiopian/Abyssinian genet**	1.65	Africa	5	DD	Unknown	(1)

(Continued)

(Continued)

Family and scientific name[a]	Authority[b]	English name(s)[c]	Average body mass in kg[d]	Distribution[e]	Number of countries[f]	IUCN Red List categorization 2011–2019[g]	Population trend[g]	Chapter(s) in present book[h]
Genetta angolensis	Bocage, 1882	**Miombo/Angolan genet**	2	Africa	6	LC	Unknown	[1]
Genetta bourloni	Gaubert, 2003	**Bourlon's genet**	1.75	Africa	4	VU	Decreasing	[1]
Genetta cristata	Hayman in Sanborn, 1940	**Crested** servaline **genet**	2.5	Africa	4	VU	Decreasing	[1]
Genetta genetta	(Linnaeus, 1758)	**Common/small-spotted genet**	1.85	Africa, Europe & Asia	39 (5+4)	LC	Stable	(1), [5], 7, 12, 17, (18), **20**, 21
Genetta johnstoni	Pocock, 1908	**Johnston's genet**	2.4	Africa	6	NT	Decreasing	[1], [5], 7
Genetta maculata	(Gray, 1830)	**Rusty-spotted genet** [(Central African) large-spotted genet]	2.25	Africa	31	LC	Unknown	[1], 12, **15**, (18)
Genetta pardina	I. Geoffroy Saint-Hilaire, 1832	**Pardine genet** [West African large-spotted genet]	3.1	Africa	11	LC	Unknown	[1]
Genetta piscivora	(J.A. Allen, 1919)	**Aquatic genet** [fishing genet]	1.5	Africa	1	NT	Decreasing	(1), 22
Genetta poensis	Waterhouse, 1838	**King genet**	2.25	Africa	5	DD	Unknown	(1)
Genetta servalina	Pucheran, 1855	**Servaline genet**	2.65	Africa	11	LC	Unknown	[1], **15**
Genetta thierryi	Matschie, 1902	**Hausa genet** [houssa/ Thierry's genet]	1.4	Africa	13	LC	Unknown	[1]
Genetta tigrina	(Schreber, 1776)	**Cape genet** [South African large-spotted genet]	1.75	Africa	2	LC	Stable	[1], 7, 22
Genetta victoriae	Thomas, 1901	**Giant forest genet**	3	Africa	2+1	LC	Unknown	[1]
Hemigalus derbyanus	(Gray, 1837)	**Banded** palm **civet**	2	Asia	5	NT	Decreasing	[1], [5], **19**
Macrogalidia musschenbroekii	(Schlegel, 1877)	**Sulawesi (palm) civet** [Giant/Celebes palm civet]	5	Asia	1	VU	Decreasing	[1], 19
Paguma larvata	(C.E.H. Smith, 1827)	**Masked palm civet**	4	Asia	15 (1)	LC	Decreasing	[1], [5], 16, (18), 19, **20**, 22

Species	Authority	English name	Body mass	Continent	Countries	IUCN	Trend	References
Paradoxurus hermaphroditus	(Pallas, 1777)	**Common palm civet** [Asian/Indian palm civet; musang; toddy cat]	3.5	Asia	16 (2+1)	LC	Decreasing	(1), [5], 7, (18), **19, 20**, 22
Paradoxurus jerdoni	Blanford, 1885	**Brown palm civet** [Jerdon's palm civet]	3.15	Asia	1	LC	Stable	[1]
Paradoxurus zeylonensis	(Pallas, 1778)	**Golden palm civet**	3.6	Asia	1	LC	Unknown	(1)
Poiana leightoni	Pocock, 1908	**West African/Leighton's oyan/linsang**	0.6	Africa	2	VU	Decreasing	[1]
Poiana richardsonii	(Thomson, 1842)	**Central African oyan** [African/Richardson's linsang]	0.48	Africa	6	LC	Unknown	[1], **15**
Viverra civettina	Blyth, 1862	**Malabar** large-spotted **civet**	7.3	Asia	(+1)	CR	Decreasing	(1)
Viverra megaspila	Blyth, 1862	**Large-spotted civet**	8.5	Asia	5+2 (+1)	EN	Decreasing	(1), **19**
Viverra tangalunga	Gray, 1832	**Malay(an) civet** [Oriental civet]	5	Asia	2+1 (2)	LC	Stable	(1), [5], **19, 20**
Viverra zibetha	Linnaeus, 1758	**Large Indian civet**	8.5	Asia	11 (+1)	LC	Decreasing	[1], 12, (18), **19, 20**
Viverricula indica	(É. Geoffroy Saint-Hilaire, 1803)	**Small Indian civet** [lesser oriental civet; rasse]	3	Asia & Africa	15 (5)	LC	Stable	(1), [5], **13**, (18), **19, 20**, 22

[a] When relevant, alternative genus names are given in square brackets. Although it would be identical for several pairs of alternative taxonomic treatments, the gender of the species name provided agrees with the first-written genus.

[b] The authority refers to the taxonomic treatment based on the first-written genus.

[c] The most common English names are indicated in bold. Terms written in light font are rarely employed. Alternative, but uncommon English names are provided within square brackets. Terms given in parentheses, being in bold or light font, are facultatively used (but see below for some recommendations).

[d] The average body mass has been calculated by averaging minimum and maximum values provided in Wilson & Mittermeier (2009) and Hunter & Barret (2011, 2018). Therefore, values listed here may not be representative of local populations where most individuals tend to have a body mass that is nearer to the range-wide minimum or maximum values, respectively. In addition, because the quantity (and therefore representativity) of data available for some species was not optimal, some of the values provided here are only so for approximate body mass ranking and comparative purposes.

[e] Continents are listed in decreasing order of importance as per surface area occupied by the corresponding species. North America is defined here as Canada, the USA and Mexico, while Central America includes the Caribbean Islands. Ocean a comprises Australasia, New Zealand, Polynesia, Melanesia and Micronesia.

[f] These values correspond to the number of native countries; however, note that the maintenance or presence of some species in some native countries may have been facilitated by reintroductions, restocking and even introductions in non-native areas. The number of additional non-native countries in which the concerned species have been successfully introduced is given in parentheses; this number also includes countries where the origin of the species is uncertain, as such anomalous cases may suggest that introductions occurred. Numerals preceded by the + sign correspond to the number of countries (native and/or non-native) in which the species' presence is uncertain. Information about native countries and introductions were derived from IUCN (2021), updated with information from recent literature. Actual values may slightly differ depending on recent and unreported country-wide extirpations or first records, notably enabled by camera-trapping. A few mistakes may also have gone undetected during the species assessments. For a review of both successful and failed introductions in selected small carnivore families, see Gantchoff *et al.* (Chapter 20, this volume).

[g] Although the conservation status and population trends of the species listed here have mostly been assessed in 2014–2016, there are some exceptions, with a few assessments taking place between 2011 and 2013, and others in 2018 and 2019. Please refer to https://www.iucn.org for more details.

[h] Numbers in bold indicate that the species is one of the main focus of the related chapter. Numbers in light font indicate that the species is mentioned in the text, in a table and/or in a figure. The presence of parentheses indicates that the species is *also* included in meta-analyses, whereas the presence of square brackets indicates that the species is *only* included in meta-analyses.

[i] If the existence of two species of red pandas is validated by the scientific community (see Appendix B), *Ailurus fulgens* should then be referred to as 'Himalayan' red panda.

[j] The populations of wild cats occurring in Africa and Asia have been suggested to belong to a separate species, the African wild cat, *Felis lybica* (see Appendix B). To avoid any confusion, the use of the term 'European' is now recommended when referring to *F. silvestris*.

[k] This chapter deals with both the European and the African wild cat, *Felis lybica*.

[l] These two chapters only deal with the African wild cat, *Felis lybica*.

[m] This species was previously referred to as *Leopardus colocolo*. However, *colocola* was the name initially used and likely intended for the pampas cat (see Kitchener et al., 2017, p. 51, for details).

[n] Owing to the likely validity of four additional species of Pampas cat (see Nascimento *et al.*, 2021), the use of the terms 'Central Chilean' is recommended when referring to *L. colocola*.

[o] Contrarily to what is indicated by IUCN (2021), some recent compendiums (Wozencraft, 2005; Wilson & Mittermeier, 2009; Hunter & Barrett, 2011, 2018), some online encyclopedias, and numerous animal diversity and taxonomy websites, the correct scientific name of the aardwolf is not *Proteles cristata* but *Proteles cristatus*. Indeed, according to Article 31.2 of the International Commission for Zoological Nomenclature, a species name that is a Latin adjective 'must agree in gender with the generic name with which it is at any time combined' (Werdelin *et al.*, 2021). Thus, because *Proteles* is masculine, the species name must be masculine too, i.e. *cristatus* (while *cristata* is feminine).

[p] Since they are found across the Americas, all *Conepatus* species are 'American' skunks. We recommend the use of the English name 'North American skunk' when referring to *C. leuconotus*, as it is the only representative of this genus that is almost entirely restricted to North America.

[q] Due to the recognition of Congo clawless otter, *Aonyx congicus*, as a separate species, the term 'Cape' should be preferred to 'African' when referring to *A. capensis*.

[r] Due to the recent split of *Arctonyx* into three species, the use of the term 'Greater' is recommended when referring to *A. collaris*.

[s] A fourth species has recently been recognized in the *Meles* genus, namely the Southwest Asian badger, *M. canescens*. To avoid any confusion, the use of the term 'Northeast' is therefore recommended when referring to *M. leucurus*.

[t] Due to the split of *Meles* into four species and the mostly European distribution of *M. meles*, the term 'Eurasian' should be avoided when referring to this species.

References

Gantchoff, M., Libal, N.S. & Belant, J.L. (Chapter 20, this volume) Small carnivore introductions: ecological and biological correlates of success. In: *Small Carnivores: Evolution, Ecology, Behaviour, and Conservation* (eds E. Do Linh San, J.J. Sato, J.L. Belant & M.J. Somers), Wiley–Blackwell, Oxford.

Hunter, L. & Barrett, P. (2011) *A Field Guide to the Carnivores of the World*. 1st edition. New Holland Publishers, London.

Hunter, L. & Barrett, P. (2018) *A Field Guide to the Carnivores of the World*. 2nd edition. Bloomsbury, London.

IUCN [International Union for Conservation of Nature] (2021) *The IUCN Red List of Threatened Species. Version 2021-1*. https://www.iucnredlist.org. Accessed on 24 August 2021.

Kitchener, A.C., Breitenmoser-Würsten, C., Eizirik, E., Gentry, A., Werdelin, L. *et al.* (2017) A revised taxonomy of the Felidae. The final report of the Cat Classification Task force of the IUCN/SSC Cat Specialist Group. *Cat News Special Issue* 11, 1–80.

Nadler, T., Streicher, U., Stefen, C., Schwierz, E. & Roos, C. (2011) A new species of ferret-badger, genus *Melogale*, from Vietnam. *Der Zoologische Garten N. F.* 80, 271–286.

Nascimento, F.O., Feng, J. & Feijó, A. (2021) Taxonomic revision of the pampas cat *Leopardus colocola* complex (Carnivora: Felidae): an integrative approach. *Zoological Journal of the Linnean Society* 191, 575–611.

Werdelin, L., Kitchener, A.C., Abramov, A. Veron, G. & Do Linh San, E. (2021) The scientific name of the aardwolf is *Proteles cristatus. African Journal of Wildlife Research* 51, 149–152.

Wilson, D.E. & Mittermeier, R.A. (eds) (2009) *Handbook of the Mammals of the World. Volume 1. Carnivores*. Lynx, Barcelona.

Wozencraft, W.C. (1986) A new species of striped mongoose from Madagascar. *Journal of Mammalogy* 67, 561–571.

Wozencraft, W.C. (2005) Order Carnivora. In: *Mammal Species of the World: A taxonomic and Geographic Reference*. 3rd edition (eds D.E. Wilson & D.M. Reeder), pp. 532–628. The Johns Hopkins University Press, Baltimore.

Appendix B

Contentious Taxonomic Cases

A detailed (but non-exhaustive) list of the main small carnivoran taxa ($n = 72$) that have been – to date – the subject of discussions as to whether they should be attributed species or subspecies level[a]. Y = yes, N = no. *Compiler: E. Do Linh San.*

Family and scientific name	Authority	Common name(s)	Distribution	Recognized as a species by IUCN (2021)	Brief information about the state of knowledge and selected references
Ailuridae					
Ailurus styani	Thomas, 1902	Chinese/Styan's red panda	China, Myanmar	N	Regarded as a separate species from (Himalayan) red panda, *A. fulgens*, based on differences in cranial and fur-colouration characters (Groves, 2011) and genomic evidence (Hu *et al.*, 2020).
Canidae					
Canis dingo	Meyer, 1793	Dingo	Asia, Oceania	N	Previously regarded as a separate species or a subspecies of the domestic dog (i.e. *C. familiaris dingo*) or of the grey wolf (i.e. *C. lupus dingo*). Molecular studies have shown that the *dingo* clade originated from an ancient breed of East Asian domestic dogs (Savolainen *et al.*, 2004; Oskarsson *et al.*, 2012; Greig *et al.*, 2018). Some authors argue that this taxon should be regarded as a species in its own right (Crowther *et al.*, 2014; Smith *et al.*, 2019), while others provide compelling evidence that dingos are free-ranging dogs (Fan *et al.*, 2016; Jackson *et al.*, 2017, 2019, 2021).

Small Carnivores: Evolution, Ecology, Behaviour, and Conservation, First Edition. Edited by Emmanuel Do Linh San, Jun J. Sato, Jerrold L. Belant, and Michael J. Somers.
© 2022 John Wiley & Sons Ltd. Published 2022 by John Wiley & Sons Ltd.

(Continued)

Family and scientific name	Authority	Common name(s)	Distribution	Recognized as a species by IUCN (2021)	Brief information about the state of knowledge and selected references
Canis hallstromi	Troughton, 1957	New Guinea singing/ highland dog	Papua (Indonesia), Papua New Guinea	N	Regarded as a separate species (Koler-Matznick *et al.,* 2003), a subspecies of the dingo (*C. dingo hallstromi*; Koler-Matznick & Stinner, 2011) or a subspecies of the grey wolf corresponding to the dingo (*C. lupus dingo*). However, since these canids share the same origin as the Australian dingo, they should also be regarded as free-roaming dogs, *C. familiaris* (Alvares *et al.*, 2019; and see references above).
Canis lupaster	Hemprich & Ehrenberg, 1833	African golden wolf	Africa	Y	Previously considered equivalent to the golden jackal, *C. aureus*. Several studies have progressively shown that it is a separate species (Rueness *et al.*, 2011; Gaubert *et al.*, 2012; Koepfli *et al.*, 2015; Viranta *et al.*, 2017; Atickem *et al.*, 2018; Gopalakrishnan *et al.*, 2018). The prior scientific name, *C. anthus* (Cuvier, 1820), is often given to this newly recognized species, but is regarded as dubious (*nomen dubium*) by Viranta *et al.* (2017).
Canis [= Lupulella] schmidti	Noack, 1897	'Eastern' black-backed jackal	Eastern Africa	N	Atickem *et al.* (2018) suggested that *C. mesomelas schmidti* may need to be elevated to species level based on disjunct distributions and substantial genetic differences (short fragment of the Cyt*b* gene) with the nominate subspecies *C. m. mesomelas* ('Southern' black-backed jackal). In addition, Dinets (2015) and Viranta *et al.* (2017) indicated that African jackals should be placed in the genus *Lupulella* Hilzheimer, 1906.
Canis oriens	Way & Lynn, 2016	Eastern/ Northeastern coyote	Canada, USA	N	What has also been referred to as 'coywolf' is a hybrid of coyote, *C. latrans*, grey wolf, *C. lupus*, eastern wolf, *C. lycaon* (or *C. lupus lycaon*), and domestic dog, *C. familiaris* (Wilson *et al.*, 2012; Wheeldon *et al.*, 2013; Monzón *et al.*, 2014). The correct taxonomy to be used for these morphologically and genetically distinct canids is debated (Way & Lynn, 2016; Wheeldon & Patterson, 2017). *Canis oriens* is a *nomen nudum* and hence unavailable anyway.

(Continued)

Family and scientific name	Authority	Common name(s)	Distribution	Recognized as a species by IUCN (2021)	Brief information about the state of knowledge and selected references
Nyctereutes viverrinus	(Desmarest, 1820)	Japanese raccoon dog	Japan	N	The chromosomal, molecular and morphological differences between Japanese and mainland populations suggest that Japanese raccoon dogs should be classified as a distinct species (Ward *et al.*, 1987; Wada & Imai, 1991; Kim *et al.*, 2013, 2015).
Vulpes fulva	(Desmarest, 1820)	North American red fox	Canada, USA	N	Differences in colouration, morphology, behaviour, ecology and genetic characters, as well as restricted hybridization with non-native red foxes (*V. vulpes* from European origin) strongly suggest that native North American red foxes are a separate species (Kamler & Ballard, 2002; Aubry *et al.*, 2009; Sacks *et al.*, 2011; Statham *et al.*, 2014).
Eupleridae					
Eupleres major	Lavauden, 1929	Western falanouc	Madagascar	Y	Only known from a few museum specimens and recent camera-trapping records (Evans *et al.*, 2013; Merson *et al.*, 2018). Goodman & Helgen (2010) suggested to elevate *E. goudotii major* to species level, but molecular studies of Veron & Goodman (2018) have shown that the studied samples do not differ from that of *E. goudotii* (Eastern falanouc).
Galidictis grandidieri	Wozencraft, 1986	Grandidier's vontsira	Madagascar	Y	Regarded as a separate species by Wozencraft (1986, 2005), but molecular studies by Veron *et al.* (2017) have now shown that this taxon should be considered as a subspecies of broad-striped vontsira, namely *G. fasciata grandidieri*.
Salanoia durrelli	Durbin, Funk, Hawkins, Hills, Jenkins, Moncrieff & Ralainasolo, 2010	Durrell's vontsira	Madagascar	N	Described by Durbin *et al.* (2010), but studies by Veron *et al.* (2017) have shown that it does not differ genetically from *S. concolor* (brown-tailed vontsira).

(Continued)

(Continued)

Family and scientific name	Authority	Common name(s)	Distribution	Recognized as a species by IUCN (2021)	Brief information about the state of knowledge and selected references
Felidae					
Felis bieti	Milne-Edwards, 1892	Chinese mountain cat	China	Y	Regarded as a subspecies of European wild cat, *F. silvestris*, by Driscoll *et al.* (2007) and Yu *et al.* (2021). However, morphological, molecular and biogeographical data strongly support the view that *F. bieti* is a separate species (Kitchener & Rees, 2009; Kitchener *et al.*, 2017).
Felis lybica	Forster, 1780	African wild cat	Africa, Asia	N	Tentatively treated by Kitchener *et al.* (2017) as a separate species from *F. silvestris* based on an interpretation of results from previous molecular studies (Driscoll *et al.*, 2007). Yu *et al.* (2021) have confirmed that divergence from *F. silvestris* and *F. bieti* occurred ~1.5 mya based on whole genomes, but these authors conservatively regarded the African wild cat as a subspecies.
Leopardus braccatus	(Cope, 1889)	Pantanal cat (Brazilian Pampas cat)	Argentina, Bolivia, Brazil, Paraguay	N	*L. braccatus* and *L. pajeros* were proposed splits from Pampas cat, *L. colocola, sensu lato* based on morphological characteristics (Garcia-Perea, 1994), but follow-up genetic analyses did not seem to fully support separation to species level (Johnson *et al.*, 1999b; Napolitano *et al.*, 2008; Cossíos *et al.*, 2009; Ruiz-García *et al.*, 2013; da Silva Santos *et al.*, 2018). However, based on multiple lines of evidence derived from morphology, molecular phylogeny, biogeography and climatic niche datasets, Nascimento *et al.* (2021) advocated for the validity of both species, and recognized *L. garleppi* and *L. munoai* as additional species. However, divergence between these taxa is ~0.5 mya or less.
Leopardus garleppi	(Matschie, 1912)	Garlepp's Pampas cat (Northern Pampas cat)	Argentina, Chile, Ecuador	N	
Leopardus munoai	(Ximénez, 1961)	Muñoa's Pampas cat (Uruguayan pampas cat)	Argentina, Brazil, Uruguay	N	
Leopardus pajeros	(Desmarest, 1816)	Southern Pampas cat	Argentina, Chile	N	
Leopardus emiliae	(Thomas, 1914)	Eastern tiger cat	Brazil	N	Regarded as a separate species from northern tiger cat, *L. tigrinus*, by Nascimento & Feijó (2017) based on differences in cranial and fur-colouration characters. More research is required.
Leopardus guttulus	(Hensel, 1872)	Southern tiger cat	Argentina, Brazil, Paraguay	Y	Regarded as a separate species from *L. tigrinus* based on morphological and genetic differences (Trigo *et al.*, 2013; Nascimento & Feijó, 2017).

(Continued)

Family and scientific name	Authority	Common name(s)	Distribution	Recognized as a species by IUCN (2021)	Brief information about the state of knowledge and selected references
Neofelis diardi	(G. Cuvier, 1823)	Sunda clouded leopard	Borneo, Sumatra	Y	Now regarded as a separate species from the Indochinese clouded leopard, *N. nebulosa*, based on morphological and molecular studies (Buckley-Beason *et al.*, 2006; Kitchener *et al.*, 2006; Wilting *et al.*, 2007; Christiansen, 2008).
Prionailurus iriomotensis	(Imaizumi, 1967)	Iriomote cat	Iriomote Island (Japan)	N	Originally described as a distinct species based on morphology (Imaizumi, 1967). Now considered a subspecies of the Indochinese leopard cat, *P. bengalensis*, based on genetic analysis (Masuda & Yoshida, 1995; Johnson *et al.*, 1999a; Patel *et al.*, 2017).
Prionailurus javanensis	(Desmarest, 1816)	Sunda leopard cat	Sumatra, Java, Bali, Borneo, four Philippine islands, possibly the Malay Peninsula south of the Kra Isthmus	N	Now regarded as a separate species from *P. bengalensis* based on molecular, morphological and biogeographical grounds (Luo *et al.*, 2014; Kitchener *et al.*, 2017; Patel *et al.*, 2017).
Herpestidae					
Bdeogale omnivora	Heller, 1913	Sokoke bushy-tailed mongoose	Kenya, Tanzania	Y	Regarded as a subspecies of bushy-tailed mongoose, *B. crassicauda*, by Kingdon (1977) and Wozencraft (2005), but treated as a separate species by the same or other authors (Engel & Van Rompaey, 1995; Kingdon, 1997; Göller, 2005; Taylor, 2013) based on differences in fur colouration and body size.
Galerella nigrata	(Thomas, 1928)	Black mongoose	Namibia	N	Considered as the same species as Angolan slender mongoose, *G. flavescens*, by Crawford-Cabral (1996), but regarded as a separate species by others (Rathbun & Cowley, 2008; Rapson *et al.*, 2012).
Urva [= Herpestes] hosei	(Jentink, 1903)	Hose's mongoose	Sarawak (Borneo, Malaysia)	N	This putative species is known from a single specimen (Jentink, 1903) of dubious origin. Both morphological and molecular investigations suggest that it belongs to the short-tailed mongoose, *Urva brachyura* (Patou *et al.*, 2009; Veron *et al.*, 2015a); *U. brachyura* from Borneo (subspecies *U. b. rajah*) may correspond to a separate, unnamed species.

(Continued)

(Continued)

Family and scientific name	Authority	Common name(s)	Distribution	Recognized as a species by IUCN (2021)	Brief information about the state of knowledge and selected references
Urva [= Herpestes] palustris	(Ghose, 1965)	Bengal/marsh mongoose	India	N	Described as a separate species by Ghose (1965), but regarded as a subspecies of *Urva auropunctata* by Veron & Jennings (2017). Mallick (2011) summarizes the morphological and behavioural peculiarities of this possibly threatened taxon. Genetic data are needed to clarify its taxonomic status.
Urva [= Herpestes] parva	(Jentink, 1895)	Calamian mongoose	Calamian Islands (Philippines)	N	Name given by Jentink (1895) to a single specimen. Based on morphological grounds, Veron *et al.* (2015a) suggest that it belongs to the collared mongoose and should be named *Urva semitorquata parva*.
Urva [= Mungos] palawana	(Allen, 1910)	Palawan mongoose	Palawan Islands (Philippines)	N	Species described by Allen (1910) and named *Mungos palawanus*. Regarded as a synonym of *Herpestes brachyurus* (now *Urva brachyura*) by Corbet & Hill (1992). Molecular and morphological studies by Veron *et al.* (2015a) however suggest that it rather belongs to the collared mongoose, possibly a subspecies that should be named *Urva semitorquata parva*.
Hyaenidae					
Proteles septentrionalis	(Rothschild, 1902)	Eastern aardwolf	East Africa	N	Allio *et al.* (2021) proposed to elevate *P. cristatus septentrionalis* to species level based on vicariant distributions and substantial genetic differences with the nominate subspecies *P. c. cristatus* (Southern aardwolf). Validation through a deeper investigation of morphological and behavioural differences between both taxa is needed.
Mephitidae					
Conepatus humboldtii	Gray, 1837	Humboldt's hog-nosed skunk	Argentina, Chile	Y	Considered by Schiaffini *et al.* (2013) to be the same species as Molina's hog-nosed skunk, *C. chinga*, based on mitochondrial DNA, craniodental geometric morphometrics, and pelage colouration pattern.
Conepatus mesoleucus	(Lichtenstein, 1832)	Western hog-nosed skunk	North & Central America	N	Found to represent the same species as North American hog-nosed skunk, *C. leuconotus*, by Dragoo *et al.* (2003) based on cranial morphology and mitochondrial DNA.

(Continued)

Family and scientific name	Authority	Common name(s)	Distribution	Recognized as a species by IUCN (2021)	Brief information about the state of knowledge and selected references
Spilogale interrupta	(Rafinesque, 1820)	Plains spotted skunk	USA, S Canada (?)	N	First described as a separate species by Rafinesque (1820) under the name *Mephitis interrupta*. Later, long considered as a subspecies of Eastern spotted skunk, *S. putorius* (Hall & Kelson, 1959; Van Gelder, 1959; Hall, 1981; Wozencraft, 2005; Dragoo, 2009). Recently regarded as a species in its own right based on multilocus nuclear and whole mitogenome data analyses (McDonough *et al.*, 2022).
Spilogale leucoparia	Merriam, 1890	Desert spotted skunk	USA, N Mexico	N	Described as a separate species by Merriam(1890) and regarded as such by Howell (1906). Later considered as a subspecies of *S. putorius* (Van Gelder, 1959; Hall, 1981) or Western spotted skunk, *S. gracilis* (Wozencraft, 2005; Dragoo, 2009). Recently regarded as a species in its own right based on multilocus nuclear and whole mitogenome data analyses (McDonough *et al.*, 2022).
Spilogale yucatanensis	Burt, 1938	Yucatán spotted skunk	Mexico (Yucatán Peninsula), N Belize (?), N Guatemala (?)	N	Described as a subspecies of Southern spotted skunk, *S. angustifrons*, by Burt (1938) and long considered as such (Hall & Kelson, 1959; Wozencraft, 2005; Dragoo,2009) or as a subspecies of *S. putorius* (Van Gelder, 1959; Hall, 1981). Recently regarded as a species in its own right based on multilocus nuclear and whole mitogenome data analyses (McDonough *et al.*, 2022).
Mustelidae					
Aonyx congicus	Lönnberg, 1910	Congo clawless otter	Equatorial Africa	Y	Regarded as conspecific with Cape Clawless Otter, *A. capensis* (Davis, 1978; Larivière & Jennings, 2009), or as a separate species based on tooth size, skin and restricted genetic differences (Rosevear, 1974; Van Zyll de Jong, 1987; Jacques *et al.*, 2009).
Arctonyx albogularis	(Blyth, 1853)	Northern hog badger	China, India, Mongolia	Y	These two species were split from greater hog badger, *A. collaris,* based on craniometric analyses, qualitative craniodental features, external comparisons, and geographical and ecological considerations (Helgen *et al.*, 2008a). Genetic data are needed to confirm this new classification.
Arctonyx hoevenii	(Hubrecht, 1891)	Sumatran hog badger	Sumatra (Indonesia)	Y	

(Continued)

(Continued)

Family and scientific name	Authority	Common name(s)	Distribution	Recognized as a species by IUCN (2021)	Brief information about the state of knowledge and selected references
Lutra nippon	Imaizumi & Yoshiyuki, 1989	Japanese otter	Japan	N	Described as a separate species from Eurasian otter, *L. lutra*, based on skull differences by Imaizumi & Yoshiyuki (1989). Molecular studies provided evidence of genetic differentiation that was considered at the boundary between subspecific and specific levels (Suzuki *et al.*, 1996; Waku *et al.*, 2016; Park *et al.*, 2019). This putative species has not been observed in the wild since 1979 and is now believed to be extinct in Japan (Ando, 2008).
Martes caurina	(Merriam, 1890)	Pacific marten	Canada, USA	N	Included with American marten, *M. americana*, by Wright (1953) on the grounds of inferred hybridization between both forms, but recently reconsidered as a separate species based on genetic and morphometric differences (Carr & Hicks, 1997; Stone & Cook, 2002; Dawson & Cook, 2012; Colella *et al.*, 2018).
Meles anakuma	Temminck, 1844	Japanese badger	Japan	Y	These three species were split from European badger, *M. meles*, grounded on paleontological and/or morphological and genetic evidence (Kawamura *et al.*, 1989; Kurose *et al.*, 2001; Abramov 2002, 2003; Abramov & Puzachenko, 2006, 2013; Marmi *et al.*, 2006; Del Cerro *et al.*, 2010; Tashima *et al.* 2010, 2011).
Meles canescens	Blanford, 1875	Southwest Asian badger	Europe, Asia	N	
Meles leucurus	(Hodgson, 1847)	Asian badger	Asia	Y	
Melogale cucphuongensis	Nadler, Streicher, Stefen, Schwierz & Roos, 2011	Vietnam/Cuc Phuong ferret badger	Vietnam, China	Y	New species discovered recently by Nadler *et al.* (2011). The studied specimen exhibited morphological and genetic differentiation from known sympatric *Melogale* species. Genetically confirmed specimens found in few localities of Vietnam and SE China (Li *et al.*, 2019; Rozhnov *et al.*, 2019). Further specimens and research are required.
Mustela aistoodonnivalis	Wu & Kao, 1991	*Not yet attributed*	China	N	Putative new species (Wu & Kao, 1991; Wang, 2003; Groves, 2007) or equivalent to Sichuan weasel, *M. russelliana* (Abramov & Duckworth, 2016). Probably more related to yellow-bellied weasel, *M. kathiah*, than least weasel, *M. nivalis*.

(Continued)

Family and scientific name	Authority	Common name(s)	Distribution	Recognized as a species by IUCN (2021)	Brief information about the state of knowledge and selected references
Mustela itatsi	Temminck, 1844	Japanese weasel	Japan	Y	Long regarded as a subspecies of Siberian weasel, *M. sibirica*, but morphometric and genetic studies strongly support its elevation to species level (Abramov, 2000; Kurose *et al.*, 2000; Sato *et al.*, 2003; Suzuki *et al.*, 2011; Masuda *et al.*, 2012).
Mustela russelliana	Thomas, 1911	Sichuan weasel	China	Y	Described as a species but then relegated to the status of a subspecies of *M. nivalis* (e.g. Abramov & Baryshnikov, 2000; Wang, 2003). Later reconsidered as a species by Groves (2007) under a phylogenetic species concept, based on measurements. Probably more related to *M. kathiah* than *M. nivalis*.
Mustela subhemachalana	Hodgson, 1837	Himalayan weasel	Afghanistan, Bhutan (?), China (Tibet), India, Nepal, Pakistan (?)	N	Described as a separate species, but long time regarded as a subspecies of Siberian weasel, M. sibirica. Recent multivariate analyses revealed that specimens from the western Himalayas (Kashmir, Nepal and Sikkim) are morphologically distinct from all other populations of M. sibirica and can be treated as a distinct species (Abramov *et al.*, 2018). Molecular studies are required to confirm these results.
Mustela subpalmata	Hemprich & Ehrenberg, 1833	Egyptian weasel	Egypt	Y	Recognized as a separate species based on morphometric differences (van Zyll de Jong, 1992; Reig, 1997; Abramov & Baryshnikov, 2000), but recent molecular analyses suggest that the studied sample does not genetically differ from *M. nivalis* (Rodrigues *et al.*, 2016).
Mustela tonkinensis	Björkegren, 1942	Tonkin weasel	Vietnam	Y	Known by a single specimen initially described as a new species (Björkegren, 1942) and then treated as a subspecies of *M. nivalis* (e.g. Abramov, 2006). Reconsidered as a species by Groves (2007) on the basis of skull differences and under a phylogenetic species concept. Additional specimens and genetic data are needed to clarify its taxonomic status.

(Continued)

(Continued)

Family and scientific name	Authority	Common name(s)	Distribution	Recognized as a species by IUCN (2021)	Brief information about the state of knowledge and selected references
Procyonidae					
Bassaricyon beddardi	Pocock, 1921	Beddard's olingo	Brazil, Guyana, Venezuela	N	Not considered as a species anymore based on morphometric and genetic studies (Helgen *et al.*, 2013); synonym to *B. alleni* (eastern lowland olingo).
Bassaricyon lasius	Harris, 1932	Harris's olingo	Costa Rica	N	Not considered as separate species anymore based on morphometric and genetic studies (Helgen *et al.*, 2013); both taxonomic treatments are synonym to *B. gabbii* (northern olingo).
Bassaricyon pauli	Enders, 1936	Chiriqui olingo	Panama	N	
Bassaricyon neblina	Helgen, Pinto, Kays, Helgen, Quinn, Wilson & Maldonado, 2013	Olinguito	Colombia, Ecuador	Y	Populations previously believed to belong to *B. alleni*, but recently identified as corresponding to a separate species based on morphometric and genetic studies (Helgen *et al.*, 2013).
Nasua nelsoni	Merriam, 1901	Dwarf coati	Cozumel Island (Mexico)	N	Considered an endemic species from Cozumel and regarded as Critically Endangered by some authors (Cuaron *et al.*, 2004, 2009; McFadden *et al.*, 2008, 2010). However, genetic studies failed to confirm the distinctiveness of the Cozumel coati from the Yucatan populations of white-nosed coatis, *N. narica* (McFadden *et al.*, 2008; Nigenda-Morales *et al.*, 2019).
Nasuella meridensis	(Thomas, 1901)	Eastern mountain coati	Venezuela	Y	Species split from western mountain coati, *N. olivacea*, based on morphological and molecular comparisons (Helgen *et al.*, 2009). *Nasuella* possibly should be subsumed under *Nasua*.
Procyon insularis	Merriam, 1898	Tres Marías raccoon	Tres Marías Islands (Mexico)	N	These populations of the northern raccoon, *P. lotor*, established following introductions by humans and are now regarded as insular forms (Pons *et al.*, 1999; Helgen & Wilson, 2003, 2005; Helgen *et al.*, 2008b).
Procyon maynardi	Bangs, 1898	Bahamian raccoon	Bahamas	N	
Procyon minor	Miller, 1911	Guadeloupean raccoon	Lesser Antilles	N	

(Continued)

Family and scientific name	Authority	Common name(s)	Distribution	Recognized as a species by IUCN (2021)	Brief information about the state of knowledge and selected references
Procyon pygmaeus	Merriam, 1901	Pygmy raccoon	Cozumel Island (Mexico)	Y	Treated as a separate species from *P. lotor* based on morphometric and genetic differences (MacFadden *et al.*, 2008; McFadden & Meiri, 2013). However, a reanalysis of the genetic data supports a conspecific status with *P. lotor* (Louppe *et al.*, 2020).
Viverridae					
Arctogalidia stigmatica	(Temminck, 1853)	Bornean small-toothed palm civet	Borneo (Brunei, Indonesia, Malaysia)	N	Veron *et al.* (2015b) found that *Arctogalidia trivirgata* samples from Borneo formed a separate, divergent clade. Although molecular differentiation was close to the species divergence seen in other carnivorans, the authors advocated the need for further genetic studies (with more samples and nuclear markers) to clarify its status.
Arctogalidia trilineata	(Wagner, 1841)	Javan small-toothed palm civet	Java (Indonesia)	N	Eaton *et al.* (2010) suggested that the Javan form of *A. trivirgata* – with its distinctive coat colouration – is a full species endemic to the island. Molecular divergence observed to date only warrants subspecific level (Veron *et al.* 2015b), but further data would be needed to confirm this assessment.
Cynogale lowei	Pocock, 1933	Lowe's otter civet	Vietnam	N	Roberton *et al.* (2017) demonstrated that this species does not exist as the single specimen used to describe *C. lowei* was, in fact, a juvenile Eurasian otter, *Lutra lutra*.
Genetta felina	(Thunberg, 1811)	Feline genet	Southern Africa	N	Considered either a separate species (Gaubert *et al.*, 2005; Jennings & Veron, 2009) or a subspecies of common genet, *G. genetta* (Delibes & Gaubert, 2013).
Genetta 'letabae'	Thomas & Schwann, 1906	*Not yet attributed*	Southern Africa	N	Both taxa were proposed to correspond to separate species or a complex of species (Gaubert *et al.*, 2005); exact scientific names still need to be established.
Genetta 'schoutedeni'	(Crawford-Cabral, 1970)	*Not yet attributed*	Equatorial Africa	N	

(*Continued*)

(Continued)

Family and scientific name	Authority	Common name(s)	Distribution	Recognized as a species by IUCN (2021)	Brief information about the state of knowledge and selected references
Paradoxurus aureus	F. Cuvier, 1822	Golden wet-zone palm civet	Sri Lanka	N	Groves *et al.* (2009) proposed that there are four species of 'golden palm civets' in Sri Lanka – mostly based on coat-colour variation – and that the name *Paradoxurus zeylonensis* does not apply to any of those species. Molecular analyses by Veron *et al.* (2015c) have shown that these phenetic variations are not related to any genetic differentiation and that *P. zeylonensis* is a valid name for this unique species.
Paradoxurus montanus	Kelaart, 1852	Sri Lankan brown palm civet	Sri Lanka	N	
Paradoxurus stenocephalus	Groves, Rajapaksha & Manemandra-Arachchi, 2009	Golden dry-zone palm civet	Sri Lanka	N	
Paradoxurus sp. nov.?	Groves, Rajapaksha & Manemandra-Arachchi, 2009	*Not yet attributed*	Sri Lanka	N	
Paradoxurus lignicolor	Miller, 1903	Mentawai palm civet	Mentawai Islands (Indonesia)	N	Described as a separate species by Corbet & Hill (1992), then included in *Paradoxurus hermaphroditus* (Patou *et al.*, 2010), and now regarded as a subspecies of *P. philippinensis* (Veron *et al.*, 2015c).
Paradoxurus musangus	(Raffles, 1821)	Sumatran palm civet	Southeast Asia	N	Recently proposed species splits from common palm civet, *P. hermaphroditus*, based on morphological data and molecular studies by Veron *et al.* (2015c).
Paradoxurus philippinensis	Jourdan, 1837	Philippine palm civet	Brunei, Indonesia, Malaysia, Philippines	N	
Viverra civettina	Blyth, 1862	Malabar civet	India	Y	Nandini & Mudappa (2010) argued that it is possibly the large-spotted civet, *V. megaspila*, maybe transported to the Western Ghats (southern India). The type and other historical museum specimens were all from zoos with unknown origin. The species was rediscovered in the late 1980s, with a few skins obtained in the Western Ghats; no further records were obtained since the early 1990s.

[a] Note that this list does not include the domestic/feral cat, domestic/feral dog, and domestic/feral ferret, which are either regarded as subspecies of the European/African wild cat, *Felis silvestris/lybica,* grey wolf, *Canis lupus,* and European polecat, *Mustela putorius,* respectively, or species in their own right, i.e. *Felis catus, Canis familiaris* and *Mustela furo,* respectively. Following Gentry *et al.* (2004), we have chosen the latter option in this book.

References

Abramov, A.V. (2000) The taxonomic status of the Japanese Weasel, *Mustela itatsi* (Carnivora, Mustelidae). *Zoologicheskii Zhurnal* 79, 80–88. (In Russian with English abstract).

Abramov, A.V. (2002) Variation of the baculum structure of the Palaearctic badger (Carnivora, Mustelidae, *Meles*). *Russian Journal of Theriology* 1, 57–60.

Abramov, A.V. (2003) The head colour pattern of the Eurasian badgers (Mustelidae, *Meles*). *Small Carnivore Conservation* 29, 5–7.

Abramov, A.V. (2006) Taxonomic remarks on two poorly known Southeast Asian weasels (Mustelidae, Mustela). *Small Carnivore Conservation* 34&35, 22–24.

Abramov, A.V. & Baryshnikov, G.F. (2000) Geographic variation and intraspecific taxonomy of weasel *Mustela nivalis* (Carnivora, Mustelidae). *Zoosystematica Rossica* 8, 365–402.

Abramov, A.V. & Duckworth, J.W. (2016) *Mustela russelliana. The IUCN Red List of Threatened Species* 2016, e.T70207817A70207828. http://dx.doi.org/10.2305/IUCN.UK.2016-1.RLTS.T70207817A70207828.en. Accessed on 22 March 2019.

Abramov, A.V. & Puzachenko, A.Yu. (2006) Geographical variability of skull and taxonomy of Eurasian badgers (Mustelidae, *Meles*). *Zoologicheskii Zhurnal* 85, 641–655.

Abramov, A.V. & Puzachenko, A.Yu. (2013) The taxonomic status of badgers (Mammalia, Mustelidae) from southwest Asia based on cranial morphometrics, with the rediscription of *Meles canescens*. *Zootaxa* 3681, 44–58.

Abramov, A.V., Puzachenko, A.Yu. & Masuda, R. (2018) Cranial variation in the Siberian weasel *Mustela sibirica* (Carnivora, Mustelidae) and its possible taxonomic implications. *Zoological Studies* 57, e14.

Allen, J.A. (1910) Mammals from Palawan Island, Philippine Islands. *Bulletin of the American Museum of Natural History* 28, 13–17.

Allio, R., Tilak, M.K., Scornavacca, C., Avenant, N.L., Kitchener, A.C., Corre, E., Nabholz, B. & Delsuc, F. (2021) High-quality carnivore genomes from roadkill samples enable species delimitation in aardwolf and bat-eared fox. *eLife* 10, e63167.

Alvares, F., Bogdanowicz, W., Campbell, L.A.D., Godinho, R., Hatlauf, J., Jhala, Y.V., Kitchener, A.,

Koepfli, K., Krofel, M., Senn, H., Sillero-Zubiri, C., Viranta, S. & Werhahn, G. (2019) *Old World Canis* spp. *with Taxonomic Ambiguity: Workshop Conclusions and Recommendations*. CIBIO, Vairão, Portugal, May 2019. http://www.canids.org/Old_world_canis_taxonomy_workshop.pdf. Accessed on 4 April 2020.

Ando, M. (2008) [*The Japanese Otter: Lessons From Its Extinction*]. University of Tokyo Press, Tokyo. (In Japanese).

Atickem, A., Stenseth, N.C., Drouilly, M., Bock, S., Roos, C., & Zinner, D. (2018) Deep divergence among mitochondrial lineages in African jackals. *Zoologica Scripta* 47, 1–8.

Aubry, K.B., Statham, M.J., Sacks, B.N., Perrines, J.D. & Wisely, S.M. (2009) Phylogeography of the North American red fox: vicariance in Pleistocene forest refugia. *Molecular Ecology* 18, 2668–2686.

Björkegren, B. (1942) On a new weasel from northern Tonkin. *Arkiv för Zoologi* 33B 15, 1–4.

Buckley-Beason, V.A., Johnson, W.E., Nash, W.G., Stanyon, R., Menninger, J.C. *et al.* (2006) Molecular evidence for species-level distinctions in clouded leopards. *Current Biology* 16, 2371–2376.

Burt, W.H. (1938) A new spotted skunk (*Spilogale*) from Yucatan. *Occasional Papers of the Museum of Zoology, University of Michigan* 384, 1–3.

Carr, S.M. & Hicks, S.A. (1997) Are there two species of pine marten in North America? Genetic and evolutionary relationship within *Martes*. In: *Martes: Taxonomy, Ecology, Techniques, and Management. Proceedings of the Second International Martes Symposium* (eds G. Proulx, H.N. Bryant & P.M. Woodard), pp. 15–28. Provincial Museum of Alberta, Edmonton.

Christiansen, P. (2008) Species distinction and evolutionary differences in the clouded leopard (*Neofelis nebulosa*) and Diard's clouded leopard (*Neofelis diardi*). *Journal of Mammalogy* 89, 1435–1446.

Colella, J.P., Johnson, E.J. & Cook, J.A. (2018) Reconciling molecules and morphology in North American *Martes. Journal of Mammalogy* 99, 1323–1335.

Corbet, G. & Hill, J. (1992) *The Mammals of the Indomalayan Region*. Oxford University Press, Oxford.

Cossíos, D., Lucherini, M., Ruiz-García, M. & Angers, B. (2009) Influence of ancient glacial periods on the Andean fauna: the case of the pampas cat (*Leopardus colocolo*). *BMC Evolutionary Biology* 9, 68.

Crawford-Cabral, J. (1996) The species of *Galerella* (Mammalia: Carnivora: Herpestinae) occurring in the southwestern corner of Angola. *Garcia de Orta, Serie de Zoologia, Lisboa* 21, 7–17.

Crowther, M.S., Fillios, M., Colman, N. & Letnic, M. (2014) An updated description of the Australian dingo (*Canis dingo* Meyer, 1793). *Journal of Zoology*, 293, 192–203.

Cuarón, A.D., Martínez-Morales, M.A., Mcfadden, K.W., Gompper, M.E. & Valenzuela, D. (2004) The status of dwarf carnivores on Cozumel Island, Mexico. *Biodiversity and Conservation* 13, 317–331.

Cuarón, A.D., Valenzuela-Galván, D., García-Vasco, D., Copa, M.E., Bautista, S. *et al.* (2009) Conservation of the endemic dwarf carnivores of Cozumel Island, Mexico. *Small Carnivore Conservation* 41, 15–21.

Davis, J.A. (1978) A classification of the otters. In: *Otters: Proceedings of the First Working Meeting of the Otter Specialist Group* (ed. N. Duplaix), pp. 14–33. IUCN, Morges.

Dawson, N.G. & Cook, J.A. (2012) Behind the genes: diversification of North American martens (*Martes americana* and *M. caurina*). In: *Biology and Conservation of Martens, Sables and Fishers: A New Synthesis* (eds K.B. Aubry, W.J. Zielinski, M.G. Raphael, G. Proulx & S.W. Buskirk), pp. 23–38. Cornell University Press, Ithaca.

da Silva Santos, A., Trigo, T.C., de Oliveira, T.G., Silveira, L. & Eizirik, E. (2018) Phylogeographic analyses of the pampas cat (*Leopardus colocola*; Carnivora, Felidae) reveal a complex demographic history. *Genetics and Molecular Biology* 41, 273–287.

Del Cerro, I., Ferrando, A., Marmi, J., Chashchin, P., Taberlet, P. & Bosch, M. (2010) Nuclear and mitochondrial phylogenies provide evidence for four species of Eurasian badgers (Carnivora). *Zoologica Scripta* 39, 415–425.

Delibes, M. & Gaubert, P. (2013) *Genetta genetta* Common genet (small-spotted genet). In: *Mammals of Africa. Volume V. Carnivores, Pangolins, Equids and Rhinoceroses* (eds J. Kingdon & M. Hoffmann), pp. 224–229. Bloomsbury, London.

Dinets, V. (2015) The *Canis* tangle: a systematics overview and taxonomic recommendations. *Vavilovskii Zhurnal Genetiki i Selektsii – Vavilov Journal of Genetics and Breeding* 19, 286–291.

Dragoo, J.W. (2009) Family Mephitidae (Skunks). In: *Handbook of the Mammals of the World. Volume 1. Carnivores* (eds D.E. Wilson & R.A. Mittermeier), pp. 532–562. Lynx, Barcelona.

Dragoo, J.W., Honeycutt, R.L. & Schmidly, D.J. (2003) Taxonomic status of white-backed hog-nosed skunks, genus *Conepatus* (Carnivora: Mephitidae). *Journal of Mammalogy* 84, 159–176.

Driscoll, C.A., Menotti-Raymond, M., Roca, A.L. Hupe, K., Johnson, W.E., Geffen, E., Harley, E.H., Delibes, M., Pontier, D., Kitchener, A.C., Yamaguchi, N., O'Brien, S.J. & Macdonald, D.W. (2007) The near Eastern origin of cat domestication. *Science* 317, 519–523.

Durbin, J., Funk, S.M., Hawkins, F., Hills, D.M., Jenkins, P.D., Moncrieff, C.B. & Ralainasolo, F.B. (2010) Investigations into the status of a new taxon of *Salanoia* (Mammalia: Carnivora: Eupleridae) from the marshes of Lac Alaotra, Madagascar. *Systematics and Biodiversity* 8, 341–355.

Eaton, J.A., Wüst, R., Wirth, R., Shepherd, C.R., Semiadi, G., Hall, J. & Duckworth, J.W. (2010) Recent records of the Javan small-toothed palm civet *Arctogalidia* (*trivirgata*) *trilineata*. *Small Carnivore Conservation* 43, 16–22.

Engel, T. & Van Rompaey, H. (1995) New records of the rare Sokoke bushy-tailed mongoose, *Bdeogale crassicauda omnivora* in the coastal Shimba Hills National Reserve and at Diani Beach, Kenya. *Small Carnivore Conservation* 12, 12–13.

Evans, B., Rakotondraparany, F., Cole, L., Graham, S., Long, P. & Gandola, R. (2013) The carnivores of Mariarano forest, Madagascar: first insights. *Small Carnivore Conservation* 49, 15–19.

Fan, Z., Silva, P., Gronau, I., Wang, S., Armero, A.S. *et al.* (2016) Worldwide patterns of genomic variation and admixture in gray wolves. *Genome Research* 26, 163–173.

Garcia-Perea, R. (1994) The pampas cat group (Genus *Lynchailurus* Severtzov, 1858) (Carnivora: Felidae), a systematic and biogeographic review. *American Museum Novitates* 3096, 1–35.

Gaubert, P., Taylor, P.J. & Veron, G. (2005) Integrative taxonomy and phylogenetic systematics of the genets (Carnivora, Viverridae, *Genetta*): a new classification of the most speciose carnivoran genus in Africa. In: *African Biodiversity: Molecules, Organisms, Ecosystems [Proceedings of the 5th International Symposium in Tropical Biology, Museum Koenig, Bonn]* (eds B.A. Huber, B.J. Sinclair & K.-H. Lampe), pp. 371–383. Springer Verlag, Berlin.

Gaubert, P., Bloch, C., Benyacoub, S., Abdelhamid, A., Pagani, P., Djagoun, C.A.M.S., Couloux, A. & Dufour, S. (2012) Reviving the African wolf *Canis lupus lupaster* in North and West Africa: a mitochondrial lineage ranging more than 6,000 km wide. *PLoS One* 7, e42740.

Gentry, A., Clutton-Brock, J. & Groves, C.P. 2004. The naming of wild animal species and their domestic derivatives. *Journal of Archaeological Science* 31, 645–651.

Ghose, R.K. (1965) A new species of mongoose (Mammalia: Carnivora: Viverridae) from West Bengal, India. *Proceedings of the Zoological Society of Calcutta* 18, 173–178.

Göller, O.Z. (2005) Potential sighting of the Sokoke dog mongoose *Bdeogale omnivora* in the East Usambara Mountains, Tanzania. *Journal of East African Natural History* 94, 235–238.

Goodman, S.M. & Helgen, K.M. 2010. Species limits and distribution of the Malagasy carnivoran genus *Eupleres* (Family Eupleridae). *Mammalia* 74, 177–185.

Gopalakrishnan, S., Sinding, M.S., Ramos-Madrigal, J., Niemann, J., Samaniego Castruita, J.A. *et al.* 2018. Interspecific gene flow shaped the evolution of genus *Canis*. *Current Biology* 28, 3441–3449.

Greig, K., Gosling, A., Collins, C.J., Boocock, J., McDonald, K. *et al.* (2018) Complex history of dog (*Canis familiaris*) origins and translocations in the Pacific revealed by ancient mitogenomes. *Scientific Reports* 8, 9130.

Groves, C. (2007) On some weasels *Mustela* from eastern Asia. *Small Carnivore Conservation* 37, 21–25.

Groves, C. (2011) The taxonomy and phylogeny of the genus *Ailurus*. In: *Red Panda: Biology and Conservation of the First Panda* (ed. A.R. Glatston), pp. 101–124. Academic Press, London.

Groves, C.P., Rajapaksha, C. & Manemandra-Arachchi, K. (2009) The taxonomy of the endemic golden palm civet of Sri Lanka. *Zoological Journal of the Linnean Society* 155, 238–251.

Hall, E.R. (1981) *The Mammals of North America.* John Wiley and Sons, New York.

Hall, E.R. & Kelson, K.R. (1959) *The Mammals of North America.* The Ronald Press Company, New York.

Hassanin, A., Veron, G., Ropiquet, A., Jansen van Vuuren, B., Lécu, A., Goodman, S.M., Haider, J. & Nguyen, T.T. (2021) Evolutionary history of Carnivora (Mammalia, Laurasiatheria) inferred from mitochondrial genomes. *PLoS One* 16, e0240770.

Helgen, K.M. & Wilson, D.E. (2003) Taxonomic status and conservation relevance of the raccoons (*Procyon* spp.) of the West Indies. *Journal of Zoology* 259, 69–76.

Helgen, K.M. & Wilson, D.E. (2005) A systematic and zoogeographic overview of the raccoons of Mexico and Central America. In: *Contribuciones Mastozoologicas: en Homenaje a Bernardo Villa* (eds V. Sanchez-Cordero & R.A. Medellin), pp. 221–236. Instituto de Biología e Instituto de Ecología, UNAM, Mexico.

Helgen, K.M., Lim, N.T.-L. & Helgen, L.E. (2008a) The hog-badger is not an edentate: systematics and evolution of the genus *Arctonyx* (Mammalia: Mustelidae). *Zoological Journal of the Linnean Society* 154, 353–385.

Helgen, K.M., Maldonado, J.E., Wilson, D.E. & Buckner, S.D. (2008b) Molecular confirmation of the origin and invasive status of West Indian raccoons. *Journal of Mammalogy* 89, 282–291.

Helgen, K.M., Kays, R., Helgen, L.E., Tsuchiya-Jerep, M.T.N., Pinto, C.M., Koepfli, K.-P., Eizirik, E. & Maldonado, J.E. (2009) Taxonomic boundaries and geographic distributions revealed by an integrative systematic overview of the mountain coatis, *Nasuella* (Carnivora: Procyonidae). *Small Carnivore Conservation* 41, 65–74.

Helgen, K.M., Pinto, C.M., Kays, R., Helgen, L.E., Tsuchiya, M.T.N., Quinn, A., Wilson, D.E. & Maldonado, J.E. (2013) Taxonomic revision of the olingos (*Bassaricyon*), with description of a new species, the Olinguito. *ZooKeys* 324, 1–83.

Howell, A.H. (1906) Revision of the skunks of the genus *Spilogale*. *North American Fauna* 26, 1–55.

Hu, Y., Thapa, A., Fan, H., Ma, T., Wu, Q., Ma, S., Zhang, D., Wang, B., Li, M., Yan, L. & Wei, F. (2020) Genomic evidence for two phylogenetic species and long-term population bottlenecks in red pandas. *Science Advances* 6, eaax5751.

Imaizumi, Y. (1967) A new genus and species of cat from Iriomote, Ryukyu Islands. *Journal of the Mammalogical Society of Japan* 3, 74.

Imaizumi, Y. & Yoshiyuki, M. (1989) Taxonomic status of the Japanese otter (Carnivora, Mustelidae), with a description of a new species. *Bulletin of the National Science Museum. Series A, Zoology* 15, 177–188.

IUCN [International Union for Conservation of Nature] (2021) *The IUCN Red List of Threatened Species. Version 2021-1*. https://www.iucnredlist.org. Accessed on 24 August 2021.

Jackson, S.M., Groves, C.P., Fleming, P.J., Aplin, K.P., Eldridge, M.D., Gonzalez, A. & Helgen, K.M. (2017) The wayward dog: is the Australian native dog or dingo a distinct species? *Zootaxa* 4317, 201–224.

Jackson, S.M., Groves, C.P., Fleming, P.J., Aplin, K.P., Eldridge, M.D., Gonzalez, A. & Helgen, K.M. (2019) The dogma of dingoes – Taxonomic status of the dingo: a reply to Smith *et al. Zootaxa* 4564, 198–212.

Jackson, S.M., Fleming, P.J., Eldridge, M.D., Archer, M., Ingleby, S., Johnson, R.N. & Helgen, K.M. (2021) Taxonomy of the dingo: it's an ancient dog. *Australian Zoologist* 41, 347–357.

Jacques, H., Veron, G., Alary, F. & Aulagnier, S. (2009) The Congo clawless otter (*Aonyx congicus*) (*Mustelidae: Lutrinae*): a review of its systematics, distribution and conservation status. *African Zoology* 44, 159–170.

Jennings, A.P. & Veron, G. (2009) Family Viverridae (Civets, genets and oyans). In: *Handbook of the Mammals of the World. Volume 1. Carnivores* (eds D.E. Wilson & R.A. Mittermeier), pp. 174–232. Lynx, Barcelona.

Jentink, F.A. (1895) On two mammals from the Calamianes Islands. *Notes from the Leyden Museum* 17, 41–48.

Jentink, F.A. (1903). A new bornean *Herpestes*. *Notes from Leyden Museum* 23, 223–228.

Johnson, W.E., Shinyashiki, F.S., Menotti Raymond M., Driscoll, C., Leh, C., Sunquist, M., Johnston, L., Bush, M., Wildt, D., Yuhki, N. & O'Brien, S.J. (1999a) Molecular genetic characterization of two insular Asian cat species, Bornean bay cat and Iriomote cat. In: *Evolutionary Theory and Process: Modern perspectives. Papers in Honour of Eviatar Nevo.* (ed. S.P. Wasser), pp. 223. Kluwer Academic Publisher, Dordrecht.

Johnson, W.E., Pecon Slattery, J., Eizirik, E., Kim, J.-H., Menotti Raymond M., Bonacic, C., Cambre, R., Crawshaw, P., Nunes, A., Seuanez, H.N., Martins Moreira, A., Seymour, K.L., Simon, F., Swanson, W. & O'Brien, S.J. (1999b) Disparate phylogeographic patterns of molecular genetic variation in four closely related South American small cat species. *Molecular Ecology* 8, 79–94.

Kamler, J.F. & Ballard, W.B. (2002) A review of native and nonnative red foxes in North America. *Wildlife Society Bulletin* 30, 370–379.

Kawamura, Y., Kamei, T. & Taruno, H. (1989) Middle and Late Pleistocene mammalian faunas in Japan. *Quaternary Research* 28, 317–326.

Kim, S.-I., Park, S.-K., Lee, H., Oshida, T., Kimura, J., Kim, Y.-J., Nguyen, S.T., Sashika, M. & Min, M.-S. (2013) Phylogeography of Korean raccoon dogs: implications of peripheral isolation of a forest mammal in East Asia. *Journal of Zoology* 290, 225–235.

Kim, S.-I., Oshida, T., Lee, H., Min, M.-S. & Kimura, J. (2015) Evolutionary and biogeographical implications of variation in skull morphology of raccoon dogs (*Nyctereutes procyonoides*, Mammalia: Carnivora). *Biological Journal of the Linnean Society* 116, 856–872.

Kingdon, J. (1977) *East African Mammals. An Atlas of Evolution in Africa. Volume IIIA. Carnivores.* The University of Chicago Press, Chicago.

Kingdon, J. (1997) *The Kingdon Field Guide to African Mammals.* Academic Press, London.

Kitchener, A.C. & Rees, E.E. (2009) Modelling the dynamic biogeography of the wildcat: implications for taxonomy and conservation. *Journal of Zoology* 279, 144–155.

Kitchener, A.C., Beaumont, M.A. & Richardson, D. (2006) Geographical variation in the clouded leopard, *Neofelis nebulosa*, reveals two species instead of one. *Current Biology* 16, 2377–2383.

Kitchener, A.C., Breitenmoser-Würsten, C., Eizirik, E., Gentry, A., Werdelin, L. *et al.* (2017) A revised taxonomy of the Felidae. The final report of the Cat Classification Task force of the IUCN/SSC Cat Specialist Group. *Cat News Special Issue* 11, 1–80.

Koepfli, K.P., Pollinger, J., Godinho, R., Robinson, J., Lea, A. *et al.* (2015) Genome-wide evidence reveals that African and Eurasian golden jackals are distinct species. *Current Biology* 25, 2158–2165.

Koler-Matznick, J. & Stinner, M. (2011) First report of captive New Guinea dingo (*Canis dingo hallstromi*) den-digging and parental behavior. *Zoo Biology* 30, 445–450.

Koler-Matznick, J., Brisbin, I.L. Jr, Feinstein, M. & Bulmer, S. (2003) An expanded description of the New Guinea singing dog (*Canis hallstromi* Troughton, 1957). *Journal of Zoology* 261, 109–110.

Kurose, N., Abramov, A.V. & Masuda, R. (2000) Intrageneric diversity of the cytochrome *b* gene and phylogeny of Eurasian species of the genus *Mustela* (Mustelidae, Carnivora). *Zoological Science* 17, 673–679.

Kurose, N., Kaneko, Y., Abramov, A.V., Siriaroonrat, B. & Masuda, R. (2001) Low genetic diversity in Japanese populations of the Eurasian Badger *Meles meles* (Mustelidae, Carnivora) revealed by mitochondrial cytochrome *b* gene sequences. *Zoological Science* 18, 1145–1152.

Larivière, S. & Jennings, A.P. (2009) Family Mustelidae (Weasels and relatives). In: *Handbook of the Mammals of the World. Volume 1. Carnivores* (eds D.E. Wilson & R.A. Mittermeier), pp. 564–656. Lynx, Barcelona.

Li,S., Yu, G.-H., Liu, S. & Jin, C.-S. (2019) First record of the ferret-badger *Melogale cucphuongensis* Nadler et al., 2011 (Carnivora: Mustelidae), with description of a new subspecies, in southeastern China. *Zoological Research* 40, 575–579.

Louppe, V., Baron, J., Pons, J.-M. & Veron, G. (2020) New insights on the geographical origins of the Caribbean raccoons. *Journal of Zoological Systematics and Evolutionary Research* 10, 7461.

Luo, S.-J., Zhang, Y., Johnson, W.E., Miao, L., Martelli, P., Antunes, A., Smith, J.L.D. & O'Brien, S.J. (2014) Sympatric Asian felid phylogeography reveals a major Indochinese–Sundaic divergence. *Molecular Ecology* 23, 2072–2092.

Mallick, J.K. (2011) New records and conservation status review of the endemic Bengal Mongoose *Herpestes palustris* Ghose, 1965 in southern West Bengal, India. *Small Carnivore Conservation* 45, 31–48.

Marmi, J., Lopez-Giraldez, F., Macdonald, D.W., Calafell, F., Zholnerovskaya, E. & Domingo-Roura, X. (2006) Mitochondrial DNA reveals a strong phylogeographic structure in the badger across Eurasia. *Molecular Ecology* 15, 1007–1020.

Masuda, R. & Yoshida, M.C. (1995) Two Japanese wildcats, the Tsushima cat and the Iriomote cat, show the same mitochondrial DNA lineage as the leopard cat *Felis bengalensis*. *Zoological Science* 12, 655–659.

Masuda, R., Kurose, N., Watanabe, S., Abramov, A.V., Han, S.H., Lin, L.K. & Oshida, T. (2012) Molecular phylogeography of the Japanese Weasel, *Mustela itatsi* (Carnivora: Mustelidae), endemic to the Japanese islands, revealed by mitochondrial DNA analysis. *Biological Journal of the Linnean Society* 107, 307–321.

McDonough, M.M., Ferguson, A.W., Dowler, R.C., Gompper, M. & Maldonado, J. (2022) Phylogenomic systematics of the spotted skunks (Carnivora, Mephitidae, *Spilogale*): additional species diversity and Pleistocene climate change as a major driver of diversification. *Molecular Phylogenetics and Evolution* 167, 107266.

McFadden, K.W. & Meiri, S. (2013) Dwarfism in insular carnivores: a case study of the pygmy raccoon. *Journal of Zoology* 289, 213–221.

McFadden, K.W., Gompper, M.E. Valenzuela, D.G. & Morales, J.C. (2008) Evolutionary history of the Critically Endangered Cozumel dwarf carnivores inferred from mitochondrial DNA analyses. *Journal of Zoology* 276, 176–186.

McFadden, K.W., García-Vasco, D., Cuarón, A.D., Valenzuela-Galván, D., Medellín, R.A. & Gompper, M.E. (2010) Vulnerable island carnivores: the endangered endemic dwarf procyonids from Cozumel Island. *Biodiversity and Conservation* 19, 491–502.

Merson, S.D., Macdonald, D.W. & Dollar, L.J. (2018) Novel photographic and morphometric records of the Western falanouc *Eupleres major* in Ankarafantsika National Park, Madagascar. *Small Carnivore Conservation* 56, 60–67.

Monzón, J., Kays, R. & Dykhuizen, D. (2014) Assessment of coyote–wolf–dog admixture using ancestryinformative diagnostic SNPs. *Molecular Ecology* 23, 182–197.

Nadler, T., Streicher, U., Stefen, C., Schwierz, E. & Roos, C. (2011) A new species of ferret-badger, genus *Melogale*, from Vietnam. *Der Zoologische Garten N. F.* 80, 271–286.

Nandini, R. & Mudappa, D. (2010) Mystery or myth: a review of history and conservation status of the Malabar Civet *Viverra civettina* Blyth, 1862. *Small Carnivore Conservation* 43, 47–59.

Napolitano, C., Bennett, M., Johnson, W.E., O'Brien, S.J., Marquet, P.A., Barria, I., Poulin, E. & Iriarte, A. (2008) Ecological and biogeographical inferences on two sympatric and enigmatic Andean cat species using genetic identification of faecal samples. *Molecular Ecology* 17, 678–690.

Nascimento, F.O. & Feijó, A. (2017) Taxonomic revision of the tigrina *Leopardus tigrinus* (Schreber, 1775) species group (Carnivora, Felidae). *Papéis Avulsos de Zoologia* 57, 231–264.

Nascimento, F.O., Feng, J. & Feijó, A. (2021) Taxonomic revision of the pampas cat *Leopardus colocola* complex (Carnivora: Felidae): an integrative approach. *Zoological Journal of the Linnean Society* 191, 575–611.

Nigenda-Morales, S.F., Gompper, M.E., Valenzuela-Galván, D., Lay, A.R., Kapheim, K.M., Hass, C., Booth-Binczik, S.D., Binczik, G.A., Hirsch, B.T., McColgin, M. & Koprowski, J.L. (2019) Phylogeographic and diversification patterns of the white-nosed coati (*Nasua narica*): evidence for south-to-north colonization of North America. *Molecular Phylogenetics and Evolution* 131, 149–163.

Oskarsson, M.C.R., Klütsch, C.F.C., Boonyaprakob, U., Wilton, A., Tanabe, Y. & Savolainen, P. (2012) Mitochondrial DNA data indicate an introduction through mainland Southeast Asia for Australian dingoes and Polynesian domestic dogs. *Proceedings of the Royal Society of London, Series B Biological Sciences* 279 (1730), 967–974.

Park, H.-C., Kurihara, N., Kim, K.S., Min, M.-S., Han, S., Lee, H. & Kimura, J. (2019) What is the taxonomic status of East Asian otter species based on molecular evidence?: focus on the position of the Japanese otter holotype specimen from museum. *Animal Cells and Systems* 23, 228–234.

Patel, R.P., Wutke, S., Lenz, D., Mukherjee, S., Ramakrishnan, U., Veron, G., Fickel, J., Wilting, A. & Förster, D.W. (2017) Genetic structure and phylogeography of the leopard cat (*Prionailurus bengalensis*) inferred from mitochondrial genomes. *Journal of Heredity* 108, 349–360.

Patou, M.-L., McLenachan, P.A., Morley, C.G., Couloux, A., Jennings, A.P. & Veron, G. (2009) Molecular phylogeny of the Herpestidae (Mammalia, Carnivora) with a special emphasis on the Asian *Herpestes*. *Molecular Phylogenetics and Evolution* 53, 69–80.

Patou, M.-L., Wilting, A., Gaubert, P., Esselstyn, J.A., Cruaud, C., Jennings, A.P., Fickel, J. & Veron, G. (2010) Evolutionary history of the *Paradoxurus* palm civets – a new model for Asian biogeography. *Journal of Biogeography* 37, 2077–2097.

Pons, J.-M., Volobouev, V., Ducroz, J.-F., Tillier, A. & Reudet, D. (1999) Is the Guadeloupean racoon [sic] (*Procyon minor*) really an endemic species? New insights from molecular and chromosomal analyses. *Journal of Zoological Systematics and Evolutionary Research* 37, 101–108.

Rapson, S.A., Goldizen, A.W. & Seddon, J. (2012) Species boundaries and hybridization between the black mongoose (*Galerella nigrata*) and the slender mongoose (*Galerella sanguinea*). *Molecular Phylogenetics and Evolution* 65, 831–839.

Rathbun, G.B. & Cowley, T.E. (2008) Behavioural ecology of the black mongoose (*Galerella nigrata*) in Namibia. *Mammalian Biology* 73, 444–450.

Reig, S. (1997) Biogeographic and evolutionary implications of size variation in North American least weasels (*Mustela nivalis*). *Canadian Journal of Zoology* 75, 2036–2049.

Roberton, S.I., Gilbert, M.T.P., Campos, P.F., Salleh, F.M., Tridico, S. & Hills, D. (2017) Lowe's otter civet *Cynogale lowei* does not exist. *Small Carnivore Conservation* 54, 42–58.

Rodrigues, M., Bos, A.R., Hoath, R., Schembri, P.J., Lymberakis, P., Cento, M., Ghawar, W., Ozkurt, S.O., Santos-Reis, M., Merilä, J. & Fernandes, C. (2016)

Taxonomic status and origin of the Egyptian weasel (*Mustela subpalmata*) inferred from mitochondrial DNA. *Genetica* 144, 191–202.

Rozhnov, V.V., Korablev, M.P. & Abramov, A.V. (2019) Systematics and distribution of ferret badgers *Melogale* (Mammalia, Mustelidae) in Vietnam: first genetic data. *Doklady Biological Sciences* 485, 47–51.

Rueness, E.K., Asmyhr, M.G., Sillero-Zubiri, C., Macdonald, D.W., Bekele, A., Atickem, A. & Stenseth, N.C. (2011) The cryptic African wolf: *Canis aureus lupaster* is not a golden jackal and is not endemic to Egypt. *PLoS One* 6, e16385.

Ruiz-García, M., Cossíos, D., Lucherini, M., Yañez, J., Pinedo-Castro, M. & Angers, B. (2013) Population genetics and spatial structure in two Andean cats (the pampas cat, *Leopardus pajeros*, and the Andean mountain cat, *L. jacobita*) by means of nuclear and mitochondrial markers and some notes on biometrical markers. In: *Molecular Population Genetics, Evolutionary Biology and Biological Conservation on Neotropical Carnivores* (eds M. Ruiz-García & J.M. Shostell), pp. 187–244. Nova, New York.

Sacks, B.N., Moore, M., Statham, M.J. & Wittmer H.U. (2011) A restricted hybrid zone between native and introduced red fox *Vulpes vulpes* populations suggests reproductive barriers and competitive exclusion. *Molecular Ecology* 20, 326–341.

Savolainen, P., Leitner, T., Wilton, A.N., Matisoo-Smith, E. & Lundeberg, J. (2004) A detailed picture of the origin of the Australian dingo, obtained from the study of Mitochondrial DNA. *Proceedings of the National Academy of Sciences of the United States of America* 101, 12387–12390.

Sato, J.J., Hosoda, T., Wolsan, M., Tsuchiya, K., Yamamoto, Y. & Suzuki, H. (2003) Phylogenetic relationships and divergence time among mustelids (Mammalia; Carnivora) based on nucleotide sequences of the nuclear interphotoreceptor retinoid binding protein and mitochondrial cytochrome *b* genes. *Zoological Science* 20, 243–264.

Schiaffini, M.I., Gabrielli, M., Prevosti, F.J., Cardoso, Y.P., Castillo, D., Bo, R., Casanave, E. & Lizarralde, M. (2013) Taxonomic status of southern South American *Conepatus* (Carnivora: Mephitidae). *Zoological Journal of the Linnean Society* 167, 327–344.

Smith, B., Cairns, K.M., Adams, J.W., Newsome, T.M., Fillios, M. *et al.* (2019) Taxonomic status of the Australian dingo: the case for *Canis dingo* Meyer, 1793. *Zootaxa* 4564, 173–197.

Statham, M.J., Murdoch, J., Janecka, J., Aubry, K.B., Edwards, C.J., Soulsbury, C.D., Berry, O., Wang, Z., Harrison, D., Pearch, M., Tomsett, L., Chupasko, J. & Sacks, B.N. (2014) Range-wide multilocus phylogeography of the red fox reveals ancient continental divergence, minimal genomic exchange and distinct demographic histories. *Molecular Ecology* 23, 4813–4830.

Stone, K.D. & Cook, J.A. (2002) Molecular evolution of Holarctic martens (genus *Martes*, Mammalia: Carnivora: Mustelidae). *Molecular Phylogenetics and Evolution* 24, 169–179.

Suzuki, T., Yuasa, H. & Machida, Y. (1996) Phylogenetic position of the Japanese river otter *Lutra nippon* inferred from the nucleotide sequence of 224 bp of the mitochondrial cytochrome b gene. *Zoological Science* 13, 621–626.

Suzuki, S., Abe, M. & Motokawa, M. (2011) Allometric comparison of skulls from two closely related weasels, *Mustela itatsi* and *M. sibirica*. *Zoological Science* 28, 676–688.

Tashima, S., Kaneko, Y., Anezaki, T., Baba, M., Yachimori, S. & Masuda, R. (2010) Genetic diversity among the Japanese badger (*Meles anakuma*) populations, revealed by microsatellite analysis. *Mammal Study* 35, 221–226.

Tashima, S., Kaneko, Y., Anezaki, T., Baba, M., Yachimori, S., Abramov, A.V., Saveljev, A.P. & Masuda, R. (2011) Phylogeographic sympatry and isolation of the Eurasian badgers (*Meles*, Mustelidae, Carnivora): implication for an alternative analysis using maternally as well as paternally inherited genes. *Zoological Science* 28, 293–303.

Taylor, M.E. (2013) *Bdeogale omnivora* Sokoke bushy-tailed mongoose. In: *Mammals of Africa. Volume V. Carnivores, Pangolins, Equids and Rhinoceroses* (eds J. Kingdon & M. Hoffmann), pp. 328–330. Bloomsbury, London.

Trigo, T.C., Schneider, A., de Oliveira, T.G., Lehugeur, L.M., Silveira, L., Freitas, T.R. & Eizirik, E. (2013) Molecular data reveal complex hybridization and a cryptic species of Neotropical wild cat. *Current Biology* 23, 2528–2533.

Van Gelder, R.G. (1959) A taxonomic revision of the spotted skunks (genus *Spilogale*). *Bulletin of the American Museum of Natural History* 117, 229–392.

van Zyll de Jong, C.G. (1987) A phylogenetic study of the *Lutrinae* (*Carnivora; Mustelidae*) using morphological data. *Canadian Journal of Zoology* 65, 2536–2544.

van Zyll de Jong, C.G. (1992) A morphometric analysis of cranial variation in Holarctic weasels (*Mustela nivalis*). *Zeitschrift für Säugetierkunde* 57, 77–93.

Veron, G. & Goodman, S.M. (2018) One or two species of the rare Malagasy carnivoran *Eupleres* (Eupleridae)? New insights from molecular data. *Mammalia* 82, 107–112.

Veron, G. & Jennings, A.J. (2017) Javan mongoose or small Indian mongoose – who is where? *Mammalian Biology* 87, 62–70.

Veron, G., Patou, M.-L., Debruyne, R., Couloux, A., Fernandez, D.A.P., Wong, S.T., Fuchs, J. & Jennings, A.P. (2015a) Systematics of the Southeast Asian mongooses (Herpestidae, Carnivora) – Solving the mystery of the elusive collared mongoose and Palawan mongoose. *Zoological Journal of the Linnean Society* 173, 236–248.

Veron, G., Patou, M.-L. & Jennings, A.P. (2015b) Molecular systematics of the small-toothed palm civet (*Arctogalidia trivirgata*) reveals a strong divergence of Bornean populations. *Mammalian Biology* 80, 347–354.

Veron, G., Patou, M.-L., Tóth, M., Goonatilake, M. & Jennings, A.P. (2015c) How many species of *Paradoxurus* civets are there? New insights from India and Sri Lanka. *Journal of Zoological Systematics and Evolutionary Research* 53, 161–174.

Veron, G., Dupré, D., Jennings, A.P., Gardner, C.J., Hassanin, A. & Goodman, S.M. (2017) New insights into the systematics of Malagasy mongoose-like carnivorans (Carnivora, Eupleridae, Galidiinae) based on mitochondrial and nuclear DNA sequences. *Journal of Zoological Systematics and Evolutionary Research* 55, 250–264.

Veron, M., Patou, M.L. & Jennings, A.P. (Chapter 3, this volume) Systematics and evolution of the mongooses (Herpestidae, Carnivora). In: *Small Carnivores: Evolution, Ecology, Behaviour, and Conservation* (eds E. Do Linh San, J.J. Sato, J.L. Belant & M.J. Somers), Wiley-Blackwell, Oxford.

Viranta, S., Atickem, A., Werdelin, L. & Stenseth, N.C. (2017) Rediscovering a forgotten canid species. *BMC Zoology* 2, 6.

Wada, M.Y. & Imai, H.T. (1991) On the Robertsonian polymorphism found in the Japanese raccoon dog (*Nyctereutes procyonoides viverrinus*). *Japanese Journal of Genetics* 66, 1–11.

Waku, D., Segawa, T., Yonezawa, T., Akiyoshi, A., Ishige, T., Ueda, M., Ogawa, H., Sasaki, H., Ando, M., Kohno, N. & Sasaki, T. (2016) Evaluating the phylogenetic status of the extinct Japanese otter on the basis of mitochondrial genome analysis. *PLoS One* 11, e0149341.

Wang, Y.X. (2003) *A Complete Checklist of Mammal Species and Subspecies in China (A Taxonomic and Geographic Reference)*. China Forestry Publishing House, Beijing.

Way, J.G. & Lynn, W.S. (2016) Northeastern coyote/coywolf taxonomy and admixture: a meta-analysis. *Canid Biology & Conservation* 19, 1–7.

Wheeldon, T.J. & Patterson, B.R. (2017) Comment on "northeastern coyote/coywolf" taxonomy and admixture. *Canid Biology & Conservation* 20, 14–15.

Wheeldon, T.J., Rutledge, L.Y., Patterson, B.R., White, B.N. & Wilson, P.J. (2013) Y-chromosome evidence supports asymmetric dog introgression into eastern coyotes. *Ecology and Evolution* 3, 3005–3020.

Wilson, P.J., Rutledge, L.Y., Wheeldon, T.J., Patterson, B.R. & White, B.N. (2012) Y-chromosome evidence supports widespread signatures of three-species *Canis* hybridization in eastern North America. *Ecology and Evolution* 2, 2325–2332.

Wilting, A., Buckley-Beason, V.A., Feldhaar, H., Gadau, J., O'Brien, S.J. & Linsenmair, K.E. (2007) Clouded leopard phylogeny revisited: support for species recognition and population division between Borneo and Sumatra. *Frontiers in Zoology* 4, 15.

Wozencraft, W.C. (1986) A new species of striped mongoose from Madagascar. *Journal of Mammalogy* 67, 561–571.

Wozencraft, W.C. (2005) Order Carnivora. In: *Mammal Species of the World: A Taxonomic and Geographic Reference*. 3rd edition (eds D.E. Wilson & D.M. Reeder), pp. 532–628. The Johns Hopkins University Press, Baltimore.

Wright, P.L. (1953) Intergradation between *Martes americana* and *Martes caurina* in Western Montana. *Journal of Mammalogy* 34, 74–86.

Wu, J. & Kao, Y. (1991). [A new species of mammals in China: lacked-teeth pygmy weasel (*Mustela aistoodonnivalis* sp. nov.)]. *Journal of Northwest University* 21, 87–92. (In Chinese with English abstract).

Yu, H., Xing, Y.-T., Meng, H., He, B., Li, W.-J., Qi, X.-Z., Zhao, J.-Y., Zhuang, Y., Xu, X., Yamaguchi, N., Driscoll, C.A., O'Brien, S.J. & Luo, S.-J. (2021) Genomic evidence for the Chinese mountain cat as a wildcat conspecific (*Felis silvestris bieti*) and its introgression to domestic cats. *Science Advances* 7, eabg0221.

Index

The index does not cover the chapter abstracts, as well as the short introductions to each chapter presented in Chapter 1 (pages 21–26). Page numbers in italics refer to Figures and those in bold to Tables, including their captions and footnotes. However, when included in a broad page range focusing on a key topic, figures and tables are not repeated separately. All currently recognized small carnivore species are listed in Appendix A (pp. 539–557) and therefore the corresponding page numbers have been omitted from the index. Similarly, all contentious taxa (species vs subspecies levels) can be found in Appendix B (pp. 559–579) and are not repeated here unless they were mentioned in the text. Lastly, note that Appendix 18.1 lists a wide range of national parks, nature reserves and plant species not mentioned in the text, and therefore not listed in the index.

Small Carnivores: Evolution, Ecology, Behaviour, and Conservation, First Edition. Edited by Emmanuel Do Linh San, Jun J. Sato, Jerrold L. Belant, and Michael J. Somers.
© 2022 John Wiley & Sons Ltd. Published 2022 by John Wiley & Sons Ltd.